Frontiers in Number Theory, Physics, and Geometry I

Pierre Cartier   Bernard Julia
Pierre Moussa   Pierre Vanhove   (Eds.)

Frontiers in Number Theory, Physics, and Geometry I

On Random Matrices, Zeta Functions, and Dynamical Systems

Springer

Pierre Cartier
I.H.E.S.
35 route de Chartres
F-91440 Bures-sur-Yvette
France
e-mail: *cartier@ihes.fr*

Pierre Moussa
Service de Physique Théorique
CEA/Saclay
F-91191 Gif-sur-Yvette
France
e-mail: *moussa@spht.saclay.cea.fr*

Bernard Julia
LPTENS
24 rue Lhomond
75005 Paris
France
Bernard.Julia@lpt.ens.fr

Pierre Vanhove
Service de Physique Théorique
CEA/Saclay
F-91191 Gif-sur-Yvette
France
e-mail: *pierre.vanhove@cern.ch*

Cover photos:
G. Pólya (courtesy of G.L. Alexanderson); Eugene P. Wigner (courtesy of M. Wigner).

Corrected Second Printing 2006

Mathematics Subject Classification (2000): 11A55, 11K50, 11M41, 15A52, 37C27, 37C30, 58B34, 81Q50, 81R60

ISBN-13 978-3-642-06219-3 e-ISBN-13 978-3-540-31347-2

Springer is a part of Springer Science+Business Media
springer.com

Softcover reprint of the hardcover 1st edition 2006

Cover design: *Erich Kirchner*, Heidelberg

Printed on acid-free paper

Preface

The present book contains fifteen contributions on various topics related to Number Theory, Physics and Geometry. It presents, together with a forthcoming second volume, most of the courses and seminars delivered at the meeting entitled "Frontiers in Number Theory, Physics and Geometry", which took place at the Centre de Physique des Houches in the french Alps March 9-21, 2003.

The relation between mathematics and physics has a long history. Let us mention only ordinary differential equations and mechanics, partial differential equations in solid and fluid mechanics or electrodynamics, group theory is essential in crystallography, elasticity or quantum mechanics...

The role of number theory and of more abstract parts of mathematics such as topological, differential and algebraic geometry in physics has become prominent more recently. Diverse instances of this trend appear in the works of such scientists as V. Arnold, M. Atiyah, M. Berry, F. Dyson, L. Faddeev, D. Hejhal, C. Itzykson, V. Kac, Y. Manin, J. Moser, W. Nahm, A. Polyakov, D. Ruelle, A. Selberg, C. Siegel, S. Smale, E. Witten and many others.

In 1989 a first meeting took place at the Centre de Physique des Houches. The triggering idea was due at that time to the late Claude Itzykson (1938-1995). The meeting gathered physicists and mathematicians, and was the occasion of long and passionate discussions.

The seminars were published in a book entitled "Number Theory and Physics", J.-M. Luck, P. Moussa, and M. Waldschmidt editors, Springer Proceedings in Physics, Vol. 47, 1990. The lectures were published as a second book entitled "From Number Theory to Physics", with C. Itzykson joining the editorial team, Springer (2nd edition 1995).

Ten years later the evolution of the interface between theoretical physics and mathematics prompted M. Waldschmidt, P. Cartier and B. Julia to renew the experience. However the emphasis was somewhat shifted to include in particular selected chapters at the interface of physics and geometry, random matrices or various zeta- and L- functions. Once the project of the new meeting entitled "Frontiers in Number Theory, Physics and Geometry" received support from the European Union the High level scientific conference was organized in Les Houches.

The Scientific Committee for the meeting “Frontiers in Number Theory, Physics and Geometry”, was composed of the following scientists: Frits Beukers, Jean-Benoît Bost, Pierre Cartier, Predrag Cvitanovic, Michel Duflo, Giovanni Gallavotti, Patricio Leboeuf, Werner Nahm, Ivan Todorov, Claire Voisin, Michel Waldschmidt, Jean-Christophe Yoccoz, and Jean-Bernard Zuber. The Organizing Committee included:

Bernard Julia (LPTENS, Paris scientific coordinator),
Pierre Moussa (SPhT CEA-Saclay), and
Pierre Vanhove (CERN and SPhT CEA-Saclay).

During two weeks, five lectures or seminars were given every day to about seventy-five participants. The topics belonged to three main domains:

1. Dynamical Systems, Number theory, and Random matrices, with lectures by E. Bogomolny on Quantum and arithmetical chaos, J. Conrey on L-functions and random matrix theory, J.-C. Yoccoz on Interval exchange maps, and A. Zorich on Flat surfaces;
2. Polylogarithms and Perturbative Physics,
with lectures by P. Cartier on Polylogarithms and motivic aspects, W. Nahm on Physics and dilogarithms, and D. Zagier on Polylogarithms;
3. Symmetries and Non-pertubative Physics, with lectures by
A. Connes on Galoisian symmetries, zeta function and renormalization,
R. Dijkgraaf on String duality and automorphic forms,
P. Di Vecchia on Gauge theory and D-branes,
E. Frenkel on Vertex algebras, algebraic curves and Langlands program,
G. Moore on String theory and number theory,
C. Soulé on Arithmetic groups.

In addition seminars were given by participants many of whom could have given full sets of lectures had time been available. They were: Z. Bern, A. Bondal, P. Candelas, J. Conway, P. Cvitanovic, H. Gangl, G. Gentile, D. Kreimer, J. Lagarias, M. Marcolli, J. Marklof, S. Marmi, J. McKay, B. Pioline, M. Pollicott, H. Then, E. Vasserot, A. Vershik, D. Voiculescu, A. Voros, S. Weinzierl, K. Wendland, A. Zabrodin.

We have chosen to reorganize the written contributions in two parts according to their subject. These naturally lead to two different volumes. The present volume is the first one, let us now briefly describe its contents.

This volume is itself composed of three parts including each lectures and seminars covering one theme. In the first part, we present the contributions on the theme “Random matrices : from Physics to Number Theory”. It begins with lectures by E. Bogomolny, which review three selected topics of quantum chaos, namely trace formulas with or without chaos, the two-point spectral correlation function of Riemann zeta function zeroes, and the two-point spectral correlation functions of the Laplace-Beltrami operator for modular domains leading to arithmetic chaos. The lectures can serve as a non-formal introduction to mathematical methods of quantum chaos. A general introduction to arithmetic groups will appear in the second volume. There are then lectures by J. Conrey who examines relations between random-matrix theory and families of arithmetic L-functions (mostly in characteristics zero), that is Dirichlet series satisfying functional equations similar to those obeyed by the Riemann zeta-function. The relevant L-functions are those associated with cusp-forms. The moments of L-functions are related to correlation functions of eigenvalues of random matrices.

Then follow a number of seminar presentations: by J. Marklof on some energy level statistics in relation with almost modular functions; by H. Then on arithmetic quantum chaos in a particular three-dimensional hyperbolic domain, in relation to Maass waveforms. Next P. Wiegmann and A. Zabrodin study the large N expansion for normal and complex matrix ensembles. D. Voiculescu reviews symmetries of free probability models. Finally A. Vershik presents some random (resp. universal) graphs and metric spaces.

In the second part "Zeta functions: a transverse tool", the theme is zeta-functions and their applications.

First the lectures by A. Connes were written up in collaboration with M. Marcolli and have been divided into two parts. The lectures contain the most up-to date research work by the authors, including a lot of original material as well as the basic material in this exciting subject.

The second one will appear in the second volume as it relates to renormalization of quantum field theories. In their first chapter they introduce the noncommutative space of commensurability classes of Q-lattices and the arithmetic properties of KMS states in the corresponding quantum statistical mechanical system. In the 1-dimensional case this space gives the spectral realization of zeroes of zeta-functions. They give a description of the multiple phase transitions and arithmetic spontaneous symmetry breaking in the case of Q-lattices of dimension two. The system at zero temperature settles onto a classical Shimura variety, which parametrizes the pure phases of the system. The noncommutative space has an arithmetic structure provided by a rational subalgebra closely related to the modular Hecke algebra. The action of the symmetry group involves the formalism of superselection sectors and the full noncommutative system at positive temperature. It acts on values of the ground states at the rational elements via the Galois group of the modular field.

Then we report seminars given by A. Voros on zeta functions built on Riemann zeroes; by J. Lagarias on Hilbert spaces of entire functions and Dirichlet L-functions; and by M. Pollicott on Dynamical zeta functions and closed orbits for geodesic and hyperbolic flows.

In the third part "Dynamical systems: interval exchanges, flat surfaces and small divisors", are gathered all the other contributions on dynamical systems. The lectures by A. Zorich provide an extensive self-contained introduction to the geometry of Flat surfaces which allows a description of flows on compact Riemann surfaces of arbitrary genus. The course by J.-C. Yoccoz analyzes Interval exchange maps such as the first return maps of these flows. Ergodic properties of maps are connected with ergodic properties of flows. This leads to a generalization to surfaces of higher genus of the irrational flows on the two dimensional torus. The adaptation of a continued fraction like algorithm to this situation is a prerequisite to extension of small divisors techniques to higher genus cases.

Finally we conclude this volume with seminars given by G. Gentile on Brjuno numbers and dynamical systems and by S. Marmi on Real and Complex Brjuno functions. In both talks either perturbation of irrational rotations or twist maps are considered, with fine details on arithmetic conditions (Brjuno condition and Brjuno numbers) for stability of trajectories under perturbations of parameters, and on the size of stability domains in the parametric space (Brjuno functions).

The following institutions are most gratefully acknowledged for their generous financial support to the meeting:

Département Sciences Physiques et Mathématiques and the Service de Formation permanente of the Centre National de la Recherche Scientifique; École Normale Supérieure de Paris; Département des Sciences de la matière du Commissariat à l'Énergie Atomique; Institut des Hautes Etudes Scientifiques; National Science Foundation; Ministère de la Recherche et de la Technologie and Ministère des Affaires Étrangères; The International association of mathematical physics and most especially the Commission of the European Communities.

Three European excellence networks helped also in various ways. Let us start with the most closely involved "Mathematical aspects of Quantum chaos", but the other two were "Superstrings" and "Quantum structure of spacetime and the geometric nature of fundamental interactions".

On the practical side we thank CERN Theory division for allowing us to use their computers for the webpage and registration process. We are also grateful to Marcelle Martin, Thierry Paul and the staff of les Houches for their patient help. We had the privilege to have two distinguished participants: Cécile de Witt-Morette (founder of the Les Houches School) and the late Bryce de Witt whose communicative and critical enthusiasm were greatly appreciated.

Paris
July 2005

Pierre Cartier
Bernard Julia
Pierre Moussa
Pierre Vanhove

Note from the Publisher

The first printing of this book had to be withdrawn and replaced by this corrected 2nd printing, when unanticipated numbering errors were discovered in several articles. These errors were traced back to a simple small LaTeX command in one article, that had – mistakenly – been assumed "harmless".

Heidelberg
May 2006

List of Contributors

List of Authors: (following the order of appearance of the contributions)

- E. Bogomolny, *Laboratoire de Physique Théorique et Modèles Statistiques Université de Paris XI, Bât. 100, 91405 Orsay Cedex, France*
- J. Brian Conrey, *American Institute of Mathematics, Palo Alto, CA, USA*
- Jens Marklof, *School of Mathematics, University of Bristol, Bristol BS8 1TW, U.K.*
- H. Then, *Abteilung Theoretische Physik, Universität Ulm, Albert-Einstein-Allee 11, 89069 Ulm, Germany*
- P. Wiegmann, *James Frank Institute and Enrico Fermi Institute of the University of Chicago, 5640 S.Ellis Avenue, Chicago, IL 60637, USA Landau Institute for Theoretical Physics, Moscow, Russia*
  A. Zabrodin, *Institute of Biochemical Physics, Kosygina str. 4, 119991 Moscow, Russia and ITEP, Bol. Cheremushkinskaya str. 25, 117259 Moscow, Russia*
- D. Voiculescu, *Department of Mathematics University of California at Berkeley Berkeley, CA 94720-3840, USA*
- A.M. Vershik, *St.Petersburg Mathematical Institute of Russian Academy of Science Fontanka 27 St.Petersburg, 191011, Russia*
- A. Connes, *Collège de France, 3, rue Ulm, F-75005 Paris, France I.H.E.S. 35 route de Chartres F-91440 Bures-sur-Yvette, France*
  M. Marcolli, *Max–Planck Institut für Mathematik, Vivatsgasse 7, D-53111 Bonn, Germany*
- A. Voros, *CEA, Service de Physique Théorique de Saclay (CNRS URA 2306) F-91191 Gif-sur-Yvette Cedex, France*
- J.C. Lagarias, *Department of Mathematics, University of Michigan, Ann Arbor,MI 48109-1109 USA*
- M. Pollicott, *Department of Mathematics, Manchester University, Oxford Road, Manchester M13 9PL UK*
- J.-C. Yoccoz, *Collège de France, 3 Rue d'Ulm, F-75005 Paris, France*

- A. Zorich, *IRMAR, Université de Rennes 1, Campus de Beaulieu, 35042 Rennes, France*
- G. Gentile, *Dipartimento di Matematica, Università di Roma Tre, I-00146 Roma, Italy*
- S.Marmi, *Scuola Normale Superiore, Piazza dei Cavalieri 7, I-56126 Pisa, Italy*
  P. Moussa, *Service de Physique Théorique, CEA/Saclay, F-91191 Gif-sur-Yvette, France*
  J.-C. Yoccoz, *Collège de France, 3 Rue d'Ulm, F-75005 Paris, France*

Editors:

- Bernard Julia, *LPTENS, 24 rue Lhomond 75005 Paris, France,* `Bernard.Julia@lpt.ens.fr`
- Pierre Cartier, *I.H.E.S. 35 route de Chartres F-91440 Bures-sur-Yvette,* `France, cartier@ihes.fr`
- Pierre Moussa, *Service de Physique Théorique, CEA/Saclay, F-91191 Gif-sur-Yvette,* `France, moussa@spht.saclay.cea.fr`
- Pierre Vanhove, *Service de Physique Théorique, CEA/Saclay, F-91191 Gif-sur-Yvette,* `France, Pierre.vanhove@cern.ch`

Contents

Part I

Random Matrices: from Physics to Number Theory

Quantum and Arithmetical Chaos

Eugene Bogomolny

Laboratoire de Physique Théorique et Modèles Statistiques
Université de Paris XI, Bât. 100, 91405 Orsay Cedex, France
eugene.bogomolny@lptms.u-psud.fr

Summary. The lectures are centered around three selected topics of quantum chaos: the Selberg trace formula, the two-point spectral correlation function of Riemann zeta function zeros, and the Laplace–Beltrami operator for the modular group. The lectures cover a wide range of quantum chaos applications and can serve as a non-formal introduction to mathematical methods of quantum chaos.

Introduction

Quantum chaos is a nickname for the investigation of quantum systems which do not permit exact solutions. The absence of explicit formulas means that underlying problems are so complicated that they cannot be expressed in terms of known ($\simeq$ simple) functions. The class of non-soluble systems is very large and practically any model (except a small set of completely integrable systems) belongs to it. An extreme case of quantum non-soluble problems appears naturally when one considers the quantization of classically chaotic systems which explains the word 'chaos' in the title.

As, by definition, for complex systems exact solutions are not possible, new analytical approaches were developed within quantum chaos. First, one may find relations between different non-integrable models, hoping that for certain questions a problem will be more tractable than another. Second, one considers, instead of exact quantities, the calculation of their smoothed values. In many cases such coarse graining appears naturally in experimental settings and, usually, it is more easy to treat. Third, one tries to understand statistical properties of quantum quantities by organizing them in suitable ensembles. An advantage of such an approach is that many different models may statistically be indistinguishable which leads to the notion of statistical universality.

The ideas and methods of quantum chaos are not restricted only to quantum models. They can equally well be applied to any problem whose analytical solution either is not possible or is very complicated. One of the most spectacular examples of such interrelations is the application of quantum chaos to number theory, in particular, to the zeros of the Riemann zeta function. Though a hypothetical quantum-like system whose eigenvalues coincide with the imaginary part of Riemann zeta function zeros has not (yet!) been found, the Riemann zeta function is, in many aspects, similar to dynamical zeta functions and the investigation of such relations already mutually enriched both quantum chaos and number theory (see e.g. the calculation by Keating and Snaith of moments of the Riemann zeta function using random matrix theory [43]).

The topics of these lectures were chosen specially to emphasize the interplay between physics and mathematics which is typical in quantum chaos.

In Chapter I different types of trace formulas are discussed. The main attention is given to the derivation of the Selberg trace formula which relates the spectral density of automorphic Laplacian on hyperbolic surfaces generated by discrete groups with classical periodic orbits for the free motion on these surfaces. This question is rarely discussed in the physics literature but is of general interest because it is the only case where the trace formula is exact and not only a leading semiclassical contribution as for general dynamical systems. Short derivations of trace formulas for dynamical systems and for the Riemann zeta function zeros are also presented in this chapter.

According to the well-known conjecture [17] statistical properties of eigenvalues of energies of quantum chaotic systems are described by standard random matrix ensembles depending only on system symmetries. In Chapter II we discuss analytical methods of confirmation of this conjecture. The largest part of this Chapter is devoted to a heuristic derivation of the 'exact' two-point correlation function for the Riemann zeros. The derivation is based on the Hardy–Littlewood conjecture about the distribution of prime pairs which is also reviewed. The resulting formula agrees very well with numerical calculations of Odlyzko.

In Chapter III a special class of dynamical systems is considered, namely, hyperbolic surfaces generated by arithmetic groups. Though from the viewpoint of classical mechanics these models are the best known examples of classical chaos, their spectral statistics are close to the Poisson statistics typical for integrable models. The reason for this unexpected behavior is found to be related with exponential degeneracies of periodic orbit lengths characteristic for arithmetical systems. The case of the modular group is considered in details and the exact expression for the two-point correlation function for this problem is derived.

To be accessible for physics students the lectures are written in a nonformal manner. In many cases analogies are used instead of theorems and complicated mathematical notions are illustrated by simple examples.

I. Trace Formulas

Different types of trace formulas are the cornerstone of quantum chaos. Trace formulas relate quantum properties of a system with their classical counterparts. In the simplest and widely used case the trace formula expresses the quantum density of states through a sum over periodic orbits and each term in this sum can be calculated from pure classical mechanics.

In general, dynamical trace formulas represent only the leading term of the semiclassical expansion in powers of $\hbar$. The computation of other terms is possible though quite tedious [1]. The noticeable exception is the free motion on constant negative curvature surfaces generated by discrete groups where the trace formula (called the Selberg trace formula) is exact. The derivation of this formula is the main goal of this Section.

For clarity, in Sect. 1 the simplest case of the rectangular billiard is briefly considered and the trace formula for this system is derived. The derivation is presented in a manner which permits to generalize it to the Selberg case of constant negative curvature surfaces generated by discrete groups which is considered in details in Sect. 2. In Sects. 3 and 4 the derivations of the trace formula for, respectively, classically integrable and chaotic systems are presented. In Sect. 5 it is demonstrated that the density of Riemann zeta function zeros can be written as a sort of trace formula where the role of periodic orbits is played by prime numbers. Section 6 is a summary of this chapter.

1 Plane Rectangular Billiard

To clarify the derivation of trace formulas let us consider in details a very simple example, namely, the computation of the energy spectrum for the plane rectangular billiard with periodic boundary conditions.

This problem consists of solving the equation

$$(\Delta + E_{\mathbf{n}})\Psi_{\mathbf{n}}(x,y) = 0 \tag{1}$$

where $\Delta = \partial^2/\partial x^2 + \partial^2/\partial y^2$ is the usual two-dimensional Laplacian with periodic boundary conditions

$$\Psi_{\mathbf{n}}(x+a,y) = \Psi_{\mathbf{n}}(x,y+b) = \Psi_{\mathbf{n}}(x,y) \tag{2}$$

where a and b are sizes of the rectangle.

The plane wave

$$\Psi_{\mathbf{n}}(x,y) = e^{\mathrm{i}k_1 x + \mathrm{i}k_2 y}$$

is an admissible solution of (1). Boundary conditions (2) determine the allowed values of the momentum $\mathbf{k}$

$$k_1 = \frac{2\pi}{a} n_1 , \quad k_2 = \frac{2\pi}{b} n_2 ,$$

with $n_1, n_2 = 0, \pm 1, \pm 2, \ldots$, and, consequently, energy eigenvalues are

$$E_{n_1 n_2} = \left(\frac{2\pi}{a} n_1 \right)^2 + \left(\frac{2\pi}{b} n_2 \right)^2 . \tag{3}$$

The first step of construction of trace formulas is to consider instead of individual eigenvalues their density defined as the sum over all eigenvalues which explains the word 'trace'

$$d(E) \equiv \sum_{n_1, n_2 = -\infty}^{+\infty} \delta(E - E_{n_1 n_2}) . \tag{4}$$

To transforms this and similar expressions into a convenient form one often uses the Poisson summation formula

$$\sum_{n=-\infty}^{+\infty} f(n) = \sum_{m=-\infty}^{+\infty} \int_{-\infty}^{+\infty} \mathrm{e}^{2\pi i m n} f(n) \mathrm{d}n . \tag{5}$$

An informal proof of this identity can, for example, be done as follows.

First

$$\sum_{n=-\infty}^{+\infty} f(n) = \int_{-\infty}^{+\infty} f(x) g(x) \mathrm{d}x$$

where $g(x)$ is the periodic δ-function

$$g(x) = \sum_{n=-\infty}^{+\infty} \delta(x - n) .$$

As any periodic function with period 1, $g(x)$ can be expanded into the Fourier series

$$g(x) = \sum_{m=-\infty}^{+\infty} \mathrm{e}^{2\pi i m x} c_m .$$

Coefficients c_m are obtained by the integration of $g(x)$ over one period

$$c_m = \int_{-1/2}^{+1/2} g(y) \mathrm{e}^{-2\pi i m y} \mathrm{d}y = 1$$

which gives (5).

By applying the Poisson summation formula (5) to the density of states (4) one gets

$$d(E) = \sum_{m_1,m_2=-\infty}^{+\infty} \int\!\!\int \mathrm{e}^{2\pi \mathrm{i}(m_1 n_1 + m_2 n_2)} \times$$
$$\times\, \delta\left(E - \left(\frac{2\pi}{a} n_1\right)^2 - \left(\frac{2\pi}{b} n_2\right)^2\right) \mathrm{d}n_1 \mathrm{d}n_2 \,.$$

Perform the following substitutions: $E = k^2$, $n_1 = ar\cos\varphi/2\pi$, and $n_2 = br\sin\varphi/2\pi$. Then $\mathrm{d}n_1\mathrm{d}n_2 = abr\mathrm{d}r\mathrm{d}\varphi/(2\pi)^2$ and

$$\begin{aligned} d(E) &= \frac{\mu(D)}{(2\pi)^2} \sum_{m_1,m_2=-\infty}^{+\infty} \int\!\!\int \mathrm{e}^{\mathrm{i}(m_1 a\cos\varphi + m_2 b\sin\varphi)r}\, \delta(k^2 - r^2) r\mathrm{d}r\mathrm{d}\varphi \\ &= \frac{\mu(D)}{2(2\pi)^2} \sum_{m_1,m_2=-\infty}^{+\infty} \int_0^{2\pi} \mathrm{e}^{\mathrm{i}k\sqrt{(m_1a)^2+(m_2b)^2}\cos\varphi} \mathrm{d}\varphi \\ &= \frac{\mu(D)}{4\pi} \sum_{m_1,m_2=-\infty}^{+\infty} J_0(kL_p)\,, \end{aligned}$$

where $\mu(D) = ab$ is the area of the rectangle,

$$J_0(x) = \frac{1}{2\pi} \int_0^{2\pi} \mathrm{e}^{\mathrm{i}x\cos\varphi} \mathrm{d}\varphi$$

is the Bessel function of order zero (see e.g. [32], Vol. 2, Sect. 7), and

$$L_p = \sqrt{(m_1 a)^2 + (m_2 b)^2}$$

is (as it is easy to check) the length of a periodic orbit in the rectangle with periodic boundary conditions.

Separating the term with $m_1 = m_2 = 0$ one concludes that the eigenvalue density of the rectangle with periodic boundary conditions can be written as the sum of two terms

$$d(E) = \bar{d}(E) + d^{(osc)}(E)\,,$$

where

$$\bar{d}(E) = \frac{\mu(D)}{4\pi} \tag{6}$$

is the smooth part of the density and

$$d^{(osc)}(E) = \frac{\mu(D)}{4\pi} \sum_{\text{p.o.}} J_0(kL_p)\,, \tag{7}$$

is the oscillating part equal to a sum over all periodic orbits in the rectangle.

As

$$J_0(z) \stackrel{z\to\infty}{\longrightarrow} \sqrt{\frac{2}{\pi z}} \cos\left(z - \frac{\pi}{4}\right)$$

the oscillating part of the level density in the semiclassical limit $k \to \infty$ takes the form

$$d^{(osc)}(E) = \frac{\mu(D)}{\sqrt{8\pi k}} \sum_{\text{p.o.}} \frac{1}{\sqrt{L_p}} \cos\left(kL_p - \frac{\pi}{4}\right) . \tag{8}$$

Let us repeat the main steps which lead to this trace formula. One starts with an explicit formula (like (3)) which expresses eigenvalues as a function of integers. Using the Poisson summation formula (5) the density of states (4) is transformed into a sum over periodic orbits. In Sect. 3 it will be demonstrated that exactly this method can be applied for any integrable system in the semiclassical limit where eigenvalues can be approximated by the WKB formulas.

More Refined Approach

The above method of deriving the trace formula for the rectangular billiard can be applied only if one knows an explicit expression for eigenvalues. For chaotic systems this is not possible and another method has to be used.

Assume that one has to solve the equation

$$(E_{\mathbf{n}} - \hat{H})\Psi_{\mathbf{n}}(\mathbf{x}) = 0$$

for a certain problem with a Hamiltonian $\hat{H}$. Under quite general conditions eigenfunctions $\Psi_{\mathbf{n}}(\mathbf{x})$ can be chosen orthogonal

$$\int \Psi_{\mathbf{n}}(\mathbf{x})\Psi^*_{\mathbf{m}}(\mathbf{x})\mathrm{d}\mathbf{x} = \delta_{nm}$$

and they form a complete system of functions

$$\sum_{\mathbf{n}} \Psi_{\mathbf{n}}(\mathbf{x})\Psi^*_{\mathbf{n}}(\mathbf{y}) = \delta(\mathbf{x} - \mathbf{y}) .$$

The Green function of the problem, by definition, obeys the equation

$$(E - \hat{H})G_E(\mathbf{x}, \mathbf{y}) = \delta(\mathbf{x} - \mathbf{y})$$

and the same boundary conditions as the original eigenfunctions. Its explicit form can formally be written through exact eigenfunctions and eigenvalues as follows

$$G_E(\mathbf{x}, \mathbf{y}) = \sum_{\mathbf{n}} \frac{\Psi_{\mathbf{n}}(\mathbf{x})\Psi^*_{\mathbf{n}}(\mathbf{y})}{E - E_n + \mathrm{i}\epsilon} . \tag{9}$$

The $+\mathrm{i}\epsilon$ prescription determines the so-called retarded Green function.

Example

To get used to Green functions let us consider in details the calculation of the Green function for the free motion in f-dimensional Euclidean space. This Green function obeys the free equation

$$(E+\hbar^2\Delta)G_E^{(0)}(\mathbf{x},\mathbf{y})=\delta(\mathbf{x}-\mathbf{y}) \ . \tag{10}$$

Let us look for the solution of the above equation in the form $G_E^{(0)}(\mathbf{x},\mathbf{y})=G(r)$ where $r=|\mathbf{x}-\mathbf{y}|$ is the distance between two points.

Simple calculations shows that for $r\neq 0$ $G(r)$ obeys the equation

$$\frac{\mathrm{d}^2G}{\mathrm{d}r^2}+\frac{f-1}{r}\frac{\mathrm{d}G}{\mathrm{d}r}+\frac{k^2}{\hbar^2}G=0$$

where $E=k^2$.

After the substitution

$$G(r)=r^{1-f/2}g\left(\frac{k}{\hbar}r\right)$$

one gets for $g(z)$ the Bessel equation (see e.g. [32], Vol. 2, Sect. 7)

$$\frac{\mathrm{d}^2g}{\mathrm{d}z^2}+\frac{1}{z}\frac{\mathrm{d}g}{\mathrm{d}z}+\left(1-\frac{\nu^2}{z^2}\right)g=0 \tag{11}$$

with $\nu=|f/2-1|$.

There are many solutions of this equation. The above $+\mathrm{i}\epsilon$ prescription means that when $k\to k+i\epsilon$ with a positive ϵ the Green function has to decrease at large distances. It is easy to see that $G(r)$ is proportional to $\mathrm{e}^{\pm ikr/\hbar}$ at large r. The $+i\epsilon$ prescription selects a solution which behaves at infinity like $\mathrm{e}^{+ikr/\hbar}$ with positive k. The required solution of (11) is the first Hankel function (see [32], Vol. 2, Sect. 7)

$$g(z)=C_fH_\nu^{(1)}(z) \tag{12}$$

where C_f is a constant and $H_\nu^{(1)}(z)$ has the following asymptotics for large and small z

$$H_\nu^{(1)}(z)\stackrel{z\to\infty}{\longrightarrow}\sqrt{\frac{2}{\pi z}}\mathrm{e}^{\mathrm{i}(z-\pi\nu/2-\pi/4)}$$

and

$$H_\nu^{(1)}(z)\stackrel{z\to 0}{\longrightarrow}\begin{cases}-\mathrm{i}2^\nu\Gamma(\nu)z^{-\nu}/\pi\ , & \nu\neq 2\\ 2\mathrm{i}\ln z/\pi\ , & \nu=2\end{cases} .$$

The overall factor in (12) has to be computed from the requirement that the Green function will give the correct δ-function contribution in the right hand side of (10). This term can appear only in the result of differentiation of the Green function at small r where it has the following behaviour

$$G(r) \stackrel{r\to 0}{\longrightarrow} G_0(r) = A_f r^{2-f}$$

with

$$A_f = C_f \frac{2^\nu \hbar^\nu \Gamma(\nu)}{\mathrm{i}\pi k^\nu} .$$

One should have

$$\hbar^2 \Delta G_0(r) = \delta(\mathbf{r}) . \tag{13}$$

Multiplying this equality by a suitable test function $f(r)$ quickly decreasing at infinity one has

$$\hbar^2 \int f(r) \Delta G_0(r) \mathrm{d}\mathbf{r} = f(0) .$$

Integrating by parts one obtains

$$\hbar^2 \int \frac{\partial}{\partial x_\mu} f(r) \frac{\partial}{\partial x_\mu} G_0(r) \mathrm{d}\mathbf{r} = -f(0) .$$

As both functions $f(r)$ and $G_0(r)$ depend only on the modulus of $\mathbf{r}$ one finally finds

$$\hbar^2 \int_0^\infty \frac{\mathrm{d}f(r)}{\mathrm{d}r} \frac{\mathrm{d}G_0(r)}{\mathrm{d}r} r^{f-1} \mathrm{d}r S_{f-1} = -f(0)$$

where S_{f-1} is the volume of the $(f-1)$-dimensional sphere $x_1^2 + \ldots + x_f^2 = 1$. Using (13) one concludes that in order to give the δ-function term A_f has to obey

$$\hbar^2 A_f (f-2) S_{f-1} = -1 .$$

One of the simplest method of calculation of S_{f-1} is the following identity

$$\int_{-\infty}^\infty \mathrm{e}^{-x_1^2} \mathrm{d}x_1 \int_{-\infty}^\infty \mathrm{e}^{-x_2^2} \mathrm{d}x_2 \ldots \int_{-\infty}^\infty \mathrm{e}^{-x_f^2} \mathrm{d}x_f = \pi^{f/2} .$$

By changing Cartesian coordinates in the left hand side to hyper-spherical ones we obtain

$$\int_0^\infty \mathrm{e}^{-r^2} r^{f-1} \mathrm{d}r S_{f-1} = \pi^{f/2}$$

which gives

$$S_{f-1} = \frac{2\pi^{f/2}}{\Gamma(f/2)}$$

where $\Gamma(x)$ is the usual gamma-function (see e.g. [32], Vol. 1, Sect. 1).

Combining together all terms and using the relation $x\Gamma(x) = \Gamma(x+1)$ one gets the explicit expression for the free Green function in f dimensions

$$G_E^{(0)}(\mathbf{x}, \mathbf{y}) = \frac{k^\nu}{4i\hbar^2 (2\pi\hbar r)^\nu} H_\nu^{(1)} \left(\frac{k}{\hbar} |\mathbf{x} - \mathbf{y}| \right) \tag{14}$$

where $\nu = |f/2 - 1|$. In particular, in the two-dimensional Euclidean space

$$G_E^{(0)}(\mathbf{x},\mathbf{y}) = \frac{1}{4i\hbar^2} H_0^{(1)}\left(\frac{k}{\hbar}|\mathbf{x}-\mathbf{y}|\right) . \tag{15}$$

Another method of calculation of the free Green function is based on (9) which for the free motion is equivalent to the Fourier expansion

$$G_E^{(0)}(\mathbf{x},\mathbf{y}) = \int \frac{\mathrm{d}\mathbf{p}}{(2\pi\hbar)^f} \frac{\mathrm{e}^{\mathrm{i}\mathbf{p}(\mathbf{x}-\mathbf{y})/\hbar}}{E-p^2+\mathrm{i}\epsilon} . \tag{16}$$

Performing angular integration one obtains the same formulas as above.

The knowledge of the Green function permits to calculate practically all quantum mechanical quantities. In particular, using

$$\mathrm{Im}\,\frac{1}{x+\mathrm{i}\varepsilon} \stackrel{\varepsilon\to 0}{\longrightarrow} -\pi\delta(x)$$

one gets that the eigenvalue density is expressed through the exact Green function as follows

$$d(E) = -\frac{1}{\pi}\mathrm{Im}\int_D G_E(\mathbf{x},\mathbf{x})\mathrm{d}\mathbf{x} . \tag{17}$$

This general expression is the starting point of all trace formulas.

For the above model of the rectangle with periodic boundary conditions the exact Green function has to obey

$$\left(\frac{\partial^2}{\partial x^2} + \frac{\partial^2}{\partial y^2} + E\right)G_E(x,y;x',y') = \delta(x-x')\delta(y-y') \tag{18}$$

and the periodic boundary conditions

$$G_E(x+na, y+mb; x', y') = G_E(x,y;x',y') \tag{19}$$

for all integer m and n.

The fact important for us later is that the rectangular billiard with periodic boundary conditions can be considered as the result of the factorization of the whole plane (x,y) with respect to the group of integer translations

$$x \to x + na, \quad y \to y + mb \tag{20}$$

with integer m and n.

The factorization of the plan (x,y) with respect to these transformations means two things. First, any two points connected by a group transformation is considered as one point. Hence (19) fulfilled. Second, inside the rectangle there is no points which are connected by these transformations. In mathematical language the rectangle with sizes (a,b) is the fundamental domain of the group (20).

Correspondingly, the exact Green function for the rectangular billiard with periodic boundary conditions equals the sum of the free Green function over all elements of the group of integer translations (20)

$$G_E(x,y;x',y') = \sum_{n,m=-\infty}^{\infty} G_E^{(0)}(x+na, y+mb; x', y')\ .$$

Here $G_E^{(0)}(\mathbf{x},\mathbf{x}')$ is the Green function corresponding to the free motion without periodic boundary conditions. To prove formally that it is really the exact Green function one has to note that (i) it obeys (18) because each term in the sum obeys it, (ii) it obeys boundary conditions (19) by construction (provided the sum converges), and (iii) inside the initial rectangle only identity term can produce a δ-function contribution required in (18) because all other terms will give δ-functions outside the rectangle.

The next steps are straightforward. The free Green function for the two-dimensional Euclidean plane has the form (15). From (17) it follows that the eigenvalue density for the rectangular billiard is

$$\begin{aligned} d(E) &= -\frac{1}{\pi}\mathrm{Im}\int_D G_E(\mathbf{x},\mathbf{x})\mathrm{d}\mathbf{x} = \\ &= \frac{1}{4\pi}\sum_{mn}\int_D \mathrm{Im}\ H_0^{(1)}\left(k\sqrt{(ma)^2+(nb)^2}\right)\mathrm{d}\mathbf{x} = \\ &= \frac{\mu(D)}{4\pi} + \frac{\mu(D)}{4\pi}\sum\nolimits'_{\mathrm{p.o.}} J_0(kL_p) \end{aligned} \qquad (21)$$

which coincides exactly with (6) and (7) obtained directly from the knowledge of the eigenvalues.

The principal drawback of all trace formulas is that the sum over periodic orbits does not converge. Even the sum of the squares diverges. The simplest way to treat this problem is to multiply both sides of (21) by a suitable test function $h(E)$ and integrate them over E. In this manner one obtains

$$\sum_n h(E_n) = \frac{\mu(D)}{4\pi}\int_0^{\infty} h(E)dE + \frac{\mu(D)}{4\pi}\sum_{\mathrm{p.o.}}\int_0^{\infty} h(E)J_0(\sqrt{E}L_p)\mathrm{d}E\ .$$

When the Fourier harmonics of $h(E)$ decrease quickly the sum over periodic orbits converges and this expression constitutes a mathematically well defined trace formula. Nevertheless for approximate calculations of eigenvalues of energies one can still use 'naive' trace formulas by introducing a cut-off on periodic orbit sum. For example, in Fig. 1 the result of numerical application of the above trace formula is presented. In performing this calculation one uses the asymptotic form of the oscillating part of the density of state (8) with only 250 first periodic orbits. Though additional oscillations are clearly seen, one can read off this figure the positions of first energy levels for the problem considered. In the literature many different methods of resummation of trace formulas were discussed (see e.g. [19] and references therein).

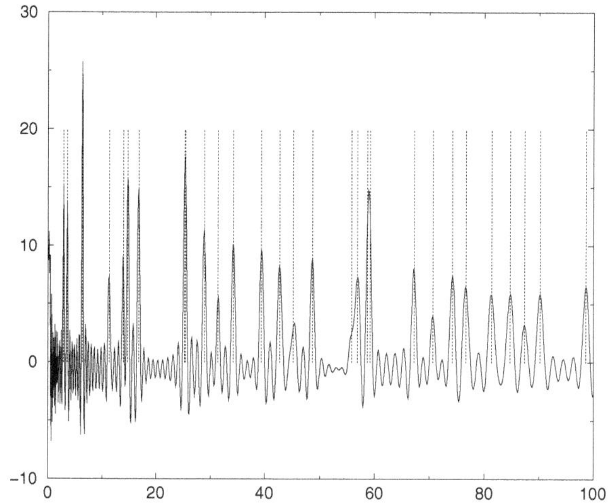

Fig. 1. The trace formula for the rectangular billiard with periodic boundary conditions calculated by taking into account 250 different periodic orbits. Dotted lines indicate the position of exact energy levels

2 Billiards on Constant Negative Curvature Surfaces

The crucial point in the second method of derivation of the trace formula for the rectangular billiard with periodic boundary conditions was a representation of the exact Green function as a sum of a free Green function over all images of the initial point. This method of images can be applied for any problem which corresponds to a factorization of a space over the action of a discrete group. In the Euclidean plane (i.e. the space of zero curvature) there exist only a few discrete groups. Much more different discrete groups are possible in the constant negative curvature (hyperbolic) space. Correspondingly, one can derive the trace formula (called the Selberg trace formula) for all hyperbolic surfaces generated by discrete groups.

The exposition of this Section follows closely [20]. In Sect. 2.1 hyperbolic geometry is non-formally discussed. The important fact is that on hyperbolic plane there exist an infinite number of discrete groups (see e.g. [42]). Their properties are mentioned in Sect. 2.2. In Sect. 2.3 the classical mechanics on hyperbolic surfaces is considered and in Sect. 2.4 the notion of quantum problems on such surfaces is introduced. The construction of the Selberg trace formula for hyperbolic surfaces generated by discrete groups consists of two steps. The first is the explicit calculation of the free hyperbolic Green function performed in Sect. 2.5. The second step includes the summation over all group transformations. In Sect. 2.6 it is demonstrated that the identity group element gives the mean density of states. Other group elements contribute to the oscillating part of the level density and correspond to classical periodic orbits for the motion on systems considered. The relation between group elements and periodic orbits is not unique. All conjugated matrices correspond to one periodic orbit. The summation over classes of conjugated elements is done in Sect. 2.7. Performing necessary integrations in Sect. 2.8 one gets the famous Selberg trace formula. Using this formula in Sect. 2.9 we compute the asymptotic density of periodic orbits for discrete groups. In Sect. 2.10 the construction of the Selberg zeta function is presented. The importance of this function follows from the fact that its non-trivial zeros coincide with eigenvalues of the Laplace–Beltrami operator automorphic with respect to a discrete group (see Sect. 2.11). Though the Selberg zeta function is defined formally only in a part of the complex plan, it obeys a functional equation (Sect. 2.12) which permits the analytical continuation to the whole complex plane.

2.1 Hyperbolic Geometry

The standard representation of the constant negative curvature space is the Poincaré upper half plane model (x, y) with $y > 0$ (see e.g. [7] and [42]) with the following metric form

$$\mathrm{d}s^2 = \frac{1}{y^2}(\mathrm{d}x^2 + \mathrm{d}y^2) .$$

The geodesic in this space (= the straight line) connecting two points is the arc of circle perpendicular to the abscissa axis which passes through these points (see Fig. 2). The distance $d(\mathbf{x}, \mathbf{y})$ between two points $\mathbf{x} = (x_1, y_1)$ and $\mathbf{y} = (x_2, y_2)$ is defined as the length of the geodesic connecting these points. Explicitly

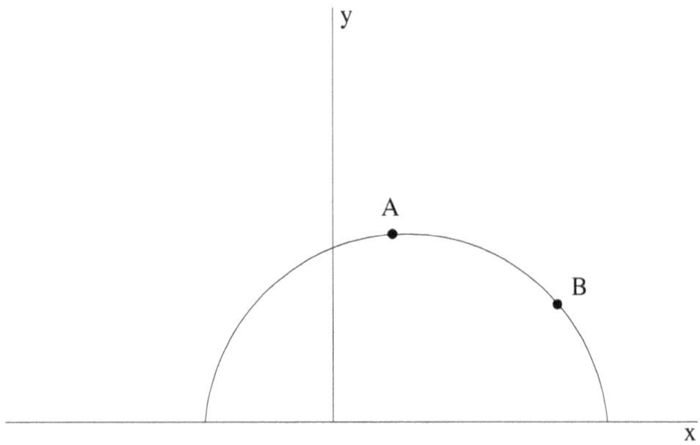

Fig. 2. The Poincaré model of constant negative curvature space. Solid line indicates the geodesic passing through points A and B

$$\cosh d(\mathbf{x},\mathbf{y}) = 1 + \frac{(x_1-x_2)^2+(y_1-y_2)^2}{2y_1y_2} = 1 + \frac{|z_1-z_2|^2}{2\mathrm{Im}\ z_1 \mathrm{Im}\ z_2} \tag{22}$$

where in the last equation one combined coordinates (x, y) into a complex number $z = x + iy$.

In the Euclidean plane the distance between two points remains invariant under 3-parameter group of rotations and translations. For constant negative curvature space the distance (22) is invariant under fractional transformations

$$z \to z' = g(z) \equiv \frac{az+b}{cz+d} \tag{23}$$

with real parameters a, b, c, d. This invariance follows from the following relations

$$z_1' - z_2' = \frac{az_1+b}{cz_1+d} - \frac{az_2+b}{cz_2+d} = (ad-bc)\frac{z_1-z_2}{(cz_1+d)(cz_2+d)} ,$$

and

$$y' = \frac{1}{2i}(z' - z'^*) = (ad-bc)\frac{y}{|cz+d|^2} .$$

Substituting these expressions to (22) one concludes that the distance between two transformed points z_1', z_2' is the same as between initial points z_1, z_2.

As fractional transformations are not changed under the multiplication of all elements a, b, c, d by a real factor, one can normalize them by the requirement

$$ad - bc = 1 .$$

In this case the distance preserving transformations are described by 2×2 matrices with real elements and unit determinant

$$g = \begin{pmatrix} a & b \\ c & d \end{pmatrix}, \quad \text{and} \quad \det g \equiv ad - bc = 1 .$$

It is easy to check that the result of two successive fractional transformations (23) corresponds to the usual multiplication of the corresponding matrices.

The collection of all such matrices forms a group called the projective special linear group over reals and it is denoted by PSL(2,$\mathbb{R}$). 'Linear' in the name means that it is a matrix group, 'special' indicates that the determinant equals 1, and 'projective' here has to remind that fractional transformations (23) are not changed when all elements are multiplied by ± 1 which is equivalent that two matrices $\pm\mathbf{1}$ corresponds to the identity element of the group.

The free classical motion on the constant negative curvature surface is defined as the motion along geodesics (i.e. circles perpendicular to the abscissa axis). The measure invariant under fractional transformations is the following differential form

$$\mathrm{d}\mu = \frac{\mathrm{d}x\mathrm{d}y}{y^2} . \tag{24}$$

This measure is invariant in the sense that if two regions, D and D', are related by a transformation (23), $D' = g(D)$, the measures of these two regions are equal, $\mu(D') = \mu(D)$.

The operator invariant with respect to distance preserving transformations (23) is called the Laplace–Beltrami operator and it has the following form

$$\Delta_{LB} = y^2 \left(\frac{\partial^2}{\partial x^2} + \frac{\partial^2}{\partial y^2} \right) . \tag{25}$$

Its invariance means that

$$\Delta_{LB} f(g(z)) = \Delta_{LB} f(z)$$

for any fractional transformation $g(z)$.

Practically all notions used for the Euclidean space can be translated to the constant negative curvature case (see e.g. [7]).

2.2 Discrete Groups

A rectangle (a torus) considered in Sect. 1 was the result of the factorization of the free motion on the plane by a discrete group of translations (20). Exactly in the same way one can construct a finite constant negative surface by factorizing the upper half plane by the action of a discrete group $\in PSL(2,\mathbb{R})$.

A group is discrete if (roughly speaking) there is a finite vicinity of every point of our space such that the results of all the group transformations (except the identity) lie outside this vicinity. The images of a point cannot approach each other too close.

Example

The group of transformation of the unit circle into itself. The group consists of all transformations of the following type

$$z \to g(n)z, \quad \text{and} \quad g(n) = \exp(2\pi i \alpha n) ,$$

where α is a constant and n is an integer. If α is a rational number $\alpha = M/N$, $g(n)$ can take only a finite number of values $(g(n))^N = 1$ and the corresponding group is discrete. But if α is an irrational number, the images of any point cover the whole circle uniformly and the group is not discrete.

Modular Group

Mathematical fact: in the upper half plane there exists an infinite number of discrete groups (see e.g. [42]). As an example let us consider the group of 2×2 integer matrices with unit determinant

$$\begin{pmatrix} m & n \\ k & l \end{pmatrix}, \quad m, n, k, l \text{ are integers and } ml - nk = 1 .$$

This is evidently a group. It is called the modular group PSL(2,$\mathbb{Z}$) ($\mathbb{Z}$ means integers) and it is one of the most investigated groups in mathematics.

This group is generated by the translation $T: \ z \to z+1$ and the inversion $S: \ z \to -1/z$ (see e.g. [42]) which are represented by the following matrices

$$T: \begin{pmatrix} 1 & 1 \\ 0 & 1 \end{pmatrix}, \quad S: \begin{pmatrix} 0 & 1 \\ -1 & 0 \end{pmatrix}.$$

These matrices obey defining relations

$$S^2 = -1 \ , \ (ST)^3 = 1$$

and are generators in the sense that any modular group matrix can be represented as a product of a certain sequence of matrices corresponding to S and T.

Fundamental Region

Similarly to the statement that the rectangular billiard is a fundamental domain of integer translations, one can construct a fundamental domain for any discrete group.

By definition the fundamental domain of a group is defined as a region on the upper half plane such that (i) for all points outside the fundamental domain there exists a group transformation that puts it to fundamental domain and (ii) no two points inside the fundamental domain are connected by group transformations.

The fundamental domain for the modular group is presented in Fig. 3. In general, the fundamental region of a discrete group has a shape of a polygon built from segments of geodesics. Group generators identify corresponding sides of the polygon.

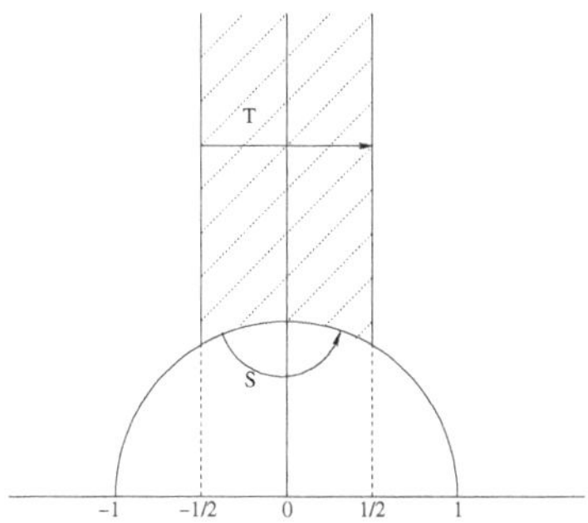

Fig. 3. Fundamental domain of the modular group. The indicated parts are identified by the corresponding generators

2.3 Classical Mechanics

Assume that we have a discrete group G with corresponding matrices $M \in G \in \mathrm{PSL}(2,\mathbb{R})$

$$M = \begin{pmatrix} a & b \\ c & d \end{pmatrix} .$$

The factorization over action of the group means that points z and z' where

$$z' = \frac{az+b}{cz+d} \tag{26}$$

are identified i.e. they are considered as one point. The classical motion on the resulting surface is the motion (with unit velocity) on geodesics (semi-circles perpendicular to the real axis) inside the fundamental domain but when a trajectory hits a boundary it reappears from the opposite side as prescribed by boundary identifications.

For each hyperbolic matrix $M \in G$ with $|\mathrm{Tr}\, M| > 2$ one can associate a periodic orbit defined as a geodesics which remains invariant under the corresponding transformation. The equation of such invariant geodesics has the form

$$c(x^2+y^2)+(d-a)x-b=0 . \tag{27}$$

This equation is the only function which has the following property. If $z = x+iy$ belongs to this curve then

$$z' = \frac{az+b}{cz+d}$$

also belongs to it.

The length of the periodic orbit is the distance along these geodesics between a point and its image. Let z' as above be the result of transformation (26) then the distance between z and z' is

$$\cosh l_p = 1 + \frac{|z-z'|^2}{2yy'} .$$

But $y' = y/|cz+d|^2$ and

$$z - \frac{az+b}{cz+d} = \frac{c(x+\mathrm{i}y)^2-(d-a)(x+\mathrm{i}y)-b}{cz+d} = y\frac{-2cy+\mathrm{i}(d-a+2cx)}{cz+d} .$$

Here we have used the fact that point z belongs to the periodic orbit (i.e. its coordinates obey (27)). Therefore

$$\cosh l_p = 1+\frac{1}{2}|-2cy+\mathrm{i}(d-a+2acx)|^2 =$$
$$= 1+\frac{1}{2}[4bc+(d-a)^2] = \frac{1}{2}(a+d)^2-1 .$$

Notice that the length of periodic orbit does not depend on an initial point and is a function only of the trace of the corresponding matrix.

Finally one gets

$$2\cosh\frac{l_p}{2} = |\mathrm{Tr}\ M|\ . \tag{28}$$

Periodic orbits are defined only for hyperbolic matrices with $|\mathrm{Tr}\ M| > 2$. For discrete groups only a finite number of elliptic matrices with $|\mathrm{Tr}\ M| < 2$ can exist (see [42]).

To each hyperbolic group matrix one can associate only one periodic orbit but each periodic orbit corresponds to infinitely many group matrices. This is due to the fact that z and $g(z)$ for any group transformation have to be considered as one point. Therefore all matrices of the form

$$SMS^{-1}$$

for all $S \in G$ give one periodic orbit. These matrices form a class of conjugated matrices and periodic orbits of the classical motion are in one-to-one correspondence with classes of conjugated matrices.

2.4 Quantum Problem

The natural 'quantum' problem on hyperbolic plane consists in considering the same equation as in (1) but with the substitution of the invariant Laplace–Beltrami operator (25) instead of the usual Laplace operator

$$\left(y^2\left(\frac{\partial^2}{\partial x^2} + \frac{\partial^2}{\partial y^2}\right) + E_n\right)\Psi_n(x,y) = 0$$

for the class of functions invariant (= automorphic) with respect to a given discrete group G

$$\Psi_n(x',y') = \Psi_n(x,y)$$

where $z' = x' + iy'$ is connected with $z = x + iy$ by group transformations

$$z' = \frac{az+b}{cz+d}\ .$$

It is easy to check that the Laplace–Beltrami operator (25) is self-adjoint with respect to the invariant measure (24), i.e.

$$\int \Psi^*(\Delta\Psi)\mathrm{d}\mu = \int(\Delta\Psi^*)\Psi\mathrm{d}\mu$$

and all eigenvalues E_n are real and $E_n \geq 0$.

2.5 Construction of the Green Function

As in the case of plain rectangular billiards the construction of the Green function requires two main steps.

- The computation of the exact Green function for the free motion on the whole upper half plane.
- The summation of the free Green function over all images of the initial point under group transformations.

The free hyperbolic Green function obeys the equation

$$(\Delta_{LB} + E)G_E^{(0)}(\mathbf{x}, \mathbf{x}') = \delta(\mathbf{x} - \mathbf{x}')$$

and should depend only on the (hyperbolic) distance between points $\mathbf{x}, \mathbf{x}'$

$$u = \cosh d(\mathbf{x}, \mathbf{x}') = 1 + \frac{(x - x')^2 + (y - y')^2}{2yy'} \ .$$

After simple calculations one gets that $G(y)$ with $y \neq 0$ obeys the equation for the Legendre functions (see e.g. [32], Vol.1, Sect. 3)

$$(1 - u^2)\frac{\mathrm{d}^2 G}{\mathrm{d}u^2} - 2u\frac{\mathrm{d}G}{\mathrm{d}u} + l(l+1)G = 0$$

where

$$E = \frac{1}{4} + k^2 = -l(l+1)$$

and

$$l = -\frac{1}{2} - \mathrm{i}k \ .$$

As for the plane case the required solution of the above equation should grow as e^{ikd} when $d \to \infty$ and should behave like $\ln d/2\pi$ when $d \to 0$. From [32], Vol.1, Sect. 3 it follows that

$$G_E^{(0)}(\mathbf{x}, \mathbf{x}') = -\frac{1}{2\pi} Q_{-\frac{1}{2} - \mathrm{i}k}(\cosh d(\mathbf{x}, \mathbf{x}')) \ .$$

Here $Q_{-\frac{1}{2}-ik}(\cosh d)$ is the Legendre function of the second kind with the integral representation [32], Vol. 1 (3.7.4)

$$Q_{-\frac{1}{2}-\mathrm{i}k}(\cosh d) = \frac{1}{\sqrt{2}} \int_d^\infty \frac{\mathrm{e}^{\mathrm{i}kr}\,\mathrm{d}r}{\sqrt{\cosh r - \cosh d}}$$

and the following asymptotics

$$Q_{-\frac{1}{2}-\mathrm{i}k}(\cosh d) \stackrel{d \to 0}{\longrightarrow} -\log d$$

and

$$Q_{-\frac{1}{2}-\mathrm{i}k}(\cosh d) \stackrel{d\to\infty}{\longrightarrow} \sqrt{\frac{\pi}{2k\sinh d}}\mathrm{e}^{\mathrm{i}(kd-\pi/4)} .$$

The automorphic Green function is the sum over all images of one of the points

$$G_E(\mathbf{x},\mathbf{x}') = \sum_g G_E^{(0)}(\mathbf{x}, g(\mathbf{x}'))$$

where the summation is performed over all group transformations.

2.6 Density of State

Using the standard formula (17)

$$d(E) = -\frac{1}{\pi}\int_D \mathrm{Im}\ G_E(\mathbf{x},\mathbf{x})\mathrm{d}\mu$$

one gets the expression for the density of states as the sum over all group elements

$$d(E) = \frac{1}{2\sqrt{2}\pi^2}\sum_g \int_D \frac{\mathrm{d}x\mathrm{d}y}{y^2}\left(\int_{d(z,g(z))}^{\infty} \frac{\sin kr\mathrm{d}r}{\sqrt{\cosh r - \cosh d(z,g(z))}}\right) .$$

Mean Density of States

The mean density of states corresponds to the identity element of our group. In this case $g(z) = z$ and $d(z, g(z)) = 0$. Therefore

$$\begin{aligned}\bar{d}(E) &= \frac{1}{2\sqrt{2}\pi^2}\int_D \frac{\mathrm{d}x\mathrm{d}y}{y^2}\int_0^{\infty}\frac{\sin kr}{\sqrt{\cosh r - 1}}\mathrm{d}r \\ &= \frac{\mu(D)}{(2\pi)^2}\int_0^{\infty}\frac{\sin kr}{\sinh(r/2)}\mathrm{d}r\end{aligned}$$

where

$$\mu(D) = \int_D \frac{\mathrm{d}x\mathrm{d}y}{y^2}$$

is the (hyperbolic) area of the fundamental domain.

The last integral is

$$\int_0^{\infty}\frac{\sin kr}{\sinh(r/2)}\mathrm{d}r = \pi\tanh\pi k$$

and the mean density of states takes the form

$$\bar{d}(E) = \frac{\mu(D)}{4\pi}\tanh\pi k .$$

When $k\to\infty$ it tends to $\mu(D)/4\pi$ as for the plane case.

2.7 Conjugated Classes

The most tedious step is the computation of the contribution from non-trivial fractional transformations.

Let us divide all group matrices into classes of conjugated elements. It means that all matrices having the form

$$g' = SgS^{-1}$$

where S belong to the group are considered as forming one class.

Two classes either have no common elements or coincide. This statement is a consequence of the fact that if

$$S_1 g_1 S_1^{-1} = S_2 g_2 S_2^{-1}$$

then $g_2 = S_3 g_1 S_3^{-1}$ where $S_3 = S_1^{-1} S_2$. Therefore g_2 belongs to the same class as g_1 and group matrices are split into classes of mutually non-conjugated elements.

The summation over group elements can be rewritten as the double sum over classes of conjugated elements and the elements in each class. Let g be a representative of a class. Then the summation over elements in this class is

$$\sum_S \int_D f(z, SgS^{-1}(z)) \mathrm{d}\mu$$

and the summation is performed over all group matrices S provided there is no double counting in the sum. The latter means that matrices S should be such that they do not contain matrices for which

$$S_1 g S_1^{-1} = S_2 g S_2^{-1}$$

or the matrix $S_3 = S_1^{-1} S_2$ commutes with matrix g

$$S_3 g = g S_3 \ .$$

Denote the set of matrices commuting with g by S_g. They form a subgroup of the initial group G as their products also commute with g. To ensure the unique decomposition of group matrices into non-overlapping classes of conjugated elements the summation should be performed over matrices S such that no two of them can be represented as

$$S_2 = s S_1$$

and s belongs to S_g. This is equivalent to the statement that we sum over all matrices but the matrices sS are considered as one matrix. It means that we factorize the group over S_g and consider the group G/S_g.

As the distance is invariant under simultaneous transformations of both coordinates

$$d(z, z') = d(S(z), S(z'))$$

one has

$$d(z, g(z)) = d(S(z), Sg(z)) = d(y, SgS^{-1}(y))$$

where $y = S(z)$.

These relations give

$$\int_D f(d(y, SgS^{-1}(y)))\mathrm{d}\mu = \int_{S^{-1}(D)} f(z, g(z))\mathrm{d}\mu$$

and the last integral is taken over the image of the fundamental domain under the transformation S^{-1}. Therefore

$$\sum_S \int_D f(d(y, SgS^{-1}(y)))\mathrm{d}\mu = \sum_S \int_{S^{-1}(D)} f(d(z, g(z)))\mathrm{d}\mu \,.$$

For different S images $S^{-1}(D)$ are different and do not overlap. The integrand does not depend on S and

$$\sum_S \int_D f(d(y, SgS^{-1}(y)))\mathrm{d}\mu = \int_{D_g} f(d(z, g(z)))\mathrm{d}\mu$$

where

$$D_g = \sum_S S^{-1}(D) \,.$$

The sum of all images $S^{-1}(D)$ will cover the whole upper half plane but we have to sum not over all S but only over S factorized by the action the group of matrices commuting with a fixed matrix g. Therefore the sum will be a smaller region.

Any matrix g can be written as a power of a primitive element

$$g = g_0^n$$

and it is (almost) evident that matrices commuting with g are precisely the group of matrices generated by g_0. This is a cyclic Abelian group consisting of all (positive, negative, and zero) powers of g_0

$$S_g = g_0^m \,, \quad m = 0, \pm 1, \pm 2, \ldots$$

and as a discrete group it has a fundamental domain FD_g.

Therefore

$$\sum_{S \in G/S_g} \int_D f(d(y, SgS^{-1}(y)))\mathrm{d}\mu = \int_{FD_g} f(d(z, g(z)))\mathrm{d}\mu \,.$$

In the left hand side the integration is taken over the fundamental domain of the whole group G and the summation is done over matrices from G factorized by the subgroup S_g of matrices which commutes with a fixed matrix g. In the right hand side there is no summation but the integration is performed over the (large) fundamental domain of the subgroup S_g.

2.8 Selberg Trace Formula

We have demonstrated that the density of states of the hyperbolic Laplace–Beltrami operator automorphic over a discrete group can be represented as

$$d(E) = \bar{d}(E) + \sum_g d_g(E)$$

where

$$d_g(E) = \frac{1}{2\sqrt{2}\pi^2}\int_{FD_g} \mathrm{d}\mu \int_{d(z,g(z))}^{\infty} \frac{\sin kr}{\sqrt{\cosh r - \cosh d(z,g(z))}}\mathrm{d}r$$

and the summation is performed over classes of conjugated matrices.

Let us consider the case of hyperbolic matrices $g = g_0^m$ (i.e. $|\mathrm{Tr}\ g_0| > 2$). By a suitable matrix B such matrix can be transform to the diagonal form

$$Bg_0B^{-1} = \begin{pmatrix} \lambda_0 & 0 \\ 0 & \lambda_0^{-1} \end{pmatrix} .$$

For hyperbolic matrices λ_0 is real and $|\lambda_0| > 1$. By the same transformation the matrix g will be transformed to

$$BgB^{-1} = \begin{pmatrix} \lambda & 0 \\ 0 & \lambda^{-1} \end{pmatrix}$$

and $\lambda = \lambda_0^m$.

Assume that g is in the diagonal form. Then $g(z) = \lambda^2 z$ and

$$\cosh d(z, g(z)) = 1 + \frac{(\lambda^2 - 1)^2(x^2 + y^2)}{2\lambda^2 y^2} .$$

Because λ_0 is real the transformation $z' = \lambda_0^2 z$ gives $y' = \lambda_0^2 y$ and the fundamental domain of $S_g = \lambda_0^{2m} z$ has the form of a horizontal strip $1 < y < \lambda_0^2$ indicated in Fig. 4. Now

$$d_g(E) = \int_{-\infty}^{\infty} \mathrm{d}x \int_1^{\lambda_0^2} F\left(\frac{(\lambda^2 - 1)^2(x^2 + y^2)}{\lambda^2 y^2}\right) \frac{\mathrm{d}y}{y^2} .$$

Introducing a new variable $\xi = xy$ one gets

$$\begin{aligned} d_g(E) &= \int_1^{\lambda_0^2} \frac{\mathrm{d}y}{y} \int_{-\infty}^{\infty} F\left((1+\xi^2)\frac{(\lambda^2-1)^2}{\lambda^2}\right) \mathrm{d}\xi = \\ &= \ln \lambda_0^2 \int_{-\infty}^{\infty} F\left((1+\xi^2)\frac{(\lambda^2-1)^2}{\lambda^2}\right) \mathrm{d}\xi . \end{aligned}$$

After the substitution

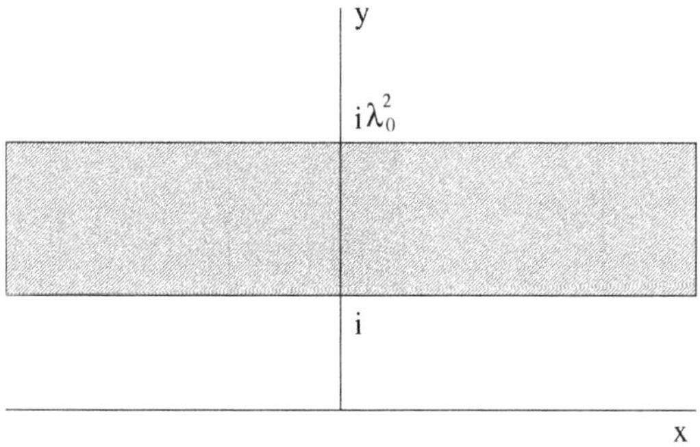

Fig. 4. Fundamental domain of multiplication group

$$u = (1+\xi^2)\frac{(\lambda^2-1)^2}{\lambda^2}$$

one obtains

$$d_g(E) = \frac{\ln\lambda_0^2}{\sqrt{u_0}}\int_{u_0}^{\infty}\frac{F(u)}{\sqrt{u-u_0}}\mathrm{d}u$$

where

$$u_0 = \frac{(\lambda^2-1)^2}{\lambda^2} = \lambda^2 + \frac{1}{\lambda^2} - 2\,.$$

The variable u is connected with the distance by $\cosh d = 1 + u/2$ and the function $F(2(\cosh d - 1))$ has the form

$$F(2(\cosh d-1)) = \frac{1}{2\sqrt{2}\pi^2}\int_d^{\infty}\frac{\sin kr}{\sqrt{\cosh r - \cosh d}}\mathrm{d}r\,.$$

Introduce a variable τ connected with r as u is connected with d

$$\cosh\tau = 1 + \frac{r}{2}\,,\quad \frac{\mathrm{d}r}{\mathrm{d}\tau} = \frac{1}{\sqrt{\tau^2+4\tau}}\,.$$

It gives

$$F(u) = \frac{1}{2\pi^2}\int_u^{\infty}\frac{\sin kr(\tau)}{\sqrt{(\tau-u)(\tau^2+4\tau)}}\mathrm{d}\tau$$

and

$$d_g(E) = \frac{\ln\lambda_0^2}{2\pi^2\sqrt{u_0}}f(u_0)$$

where

$$f(w) = \int_w^{\infty}\frac{\mathrm{d}u}{\sqrt{u-w}}\int_u^{\infty}\frac{\sin kr(\tau)}{\sqrt{(\tau-u)(\tau^2+4\tau)}}\mathrm{d}\tau\,.$$

Changing the order of integration one obtains

$$f(w) = \int_w^\infty \frac{\sin kr(\tau)}{\sqrt{\tau^2+4\tau}} \mathrm{d}\tau \int_w^\tau \frac{\mathrm{d}u}{\sqrt{(u-w)(\tau-u)}} \,.$$

The last integral is a half of the residue at infinity

$$\int_w^\tau \frac{\mathrm{d}u}{\sqrt{(u-w)(\tau-u)}} = \pi$$

and

$$f(w) = \pi \int_w^\infty \frac{\sin kr(\tau)}{\sqrt{\tau^2+4\tau}} \mathrm{d}\tau = \pi \int_{l_p}^\infty \sin(kr)\mathrm{d}r = \frac{\pi}{k}\cos kl_p \,.$$

Here l_p is the minimal value of r corresponding to u_0

$$\cosh l_p = 1 + \frac{u_0}{2} = 1 + \frac{1}{2}(\lambda^2 + \frac{1}{\lambda^2} - 2) = \frac{1}{2}(\lambda + \frac{1}{\lambda})^2 - 1$$

or

$$2\cosh l_p = \lambda + \frac{1}{\lambda} \equiv \text{ Tr } g$$

i.e. l_p is the length of periodic orbit associated with the matrix g.

Therefore

$$d_g(E) = \frac{\ln \lambda_0^2}{2\pi k\sqrt{\lambda+\lambda^{-1}-2}} \cos kl_p = \frac{l_p^{(0)}}{4\pi k \sinh l_p/2} \cos kl_p$$

where $l_p^{(0)}$ is the length of the primitive periodic orbit associated with g_0.

Combining all terms together one finds that the eigenvalues density of the Laplace–Beltrami operator automorphic with respect to a discrete group with only hyperbolic matrices has the form

$$d(E) = \frac{\mu(D)}{4\pi} \tanh \pi k + \sum_{\text{p.p.o.}} \frac{l_p}{4\pi k} \sum_{n=1}^\infty \frac{\cos(knl_p)}{\sinh(nl_p/2)} \,.$$

The oscillating part of the density is given by the double sum. The first summation is done over all primitive periodic orbits (p.p.o.) and the second sum is performed over all repetitions of these orbits. Here k is the momentum related with the energy by $E = k^2 + 1/4$.

To obtain mathematically sound formula and to avoid problems with convergence it is common to multiply both parts of the above equality by a test function $h(k)$ and to integrate over $\mathrm{d}E = 2k\mathrm{d}k$. To assume the convergence the test function $h(r)$ should have the following properties

- The function $h(r)$ is a function analytical in the region $|\text{Im } r| \le 1/2 + \delta$ with certain $\delta > 0$.
- $h(-r) = h(r)$.
- $|h(r)| \le A(1+|r|)^{-2-\delta}$.

The left hand side of the above equation is

$$\int d(E)h(k)\mathrm{d}E = \sum_n \delta(E-E_n)h(k)\mathrm{d}E = \sum_n h(k_n)\,.$$

In the right hand side one obtains

$$\int h(k)\frac{\cos kl}{2\pi k}k\mathrm{d}k = \frac{1}{2\pi}\int_{-\infty}^{\infty} h(k)\mathrm{e}^{-\mathrm{i}kl}\mathrm{d}k\,.$$

The final formula takes the form

$$\sum_n h(k_n) = \frac{\mu(D)}{2\pi}\int_{-\infty}^{\infty} kh(k)\tanh(\pi k)\mathrm{d}k + \\ + \sum_{\text{p.p.o.}} l_p \sum_{n=1}^{\infty} \frac{1}{2\sinh(nl_p/2)} g(nl_p) \tag{29}$$

where k_n is related with eigenvalue E_n as follows

$$E_n = k_n^2 + \frac{1}{4}$$

and $g(l)$ is the Fourier transform of $h(k)$

$$g(l) = \frac{1}{2\pi}\int_{-\infty}^{\infty} h(k)\mathrm{e}^{-\mathrm{i}kl}\mathrm{d}k\,.$$

This is the famous Selberg trace formula. It connects eigenvalues of the Laplace–Beltrami operator for functions automorphic with respect to a discrete group having only hyperbolic elements with classical periodic orbits.

2.9 Density of Periodic Orbits

To find the density of periodic orbits for a discrete group let us choose the test function $h(r)$ in (29) as

$$h(r) = \mathrm{e}^{-(r^2+1/4)T} \equiv \mathrm{e}^{-ET}$$

with a parameter $T > 0$. Its Fourier transforms is

$$g(u) = \frac{1}{2\pi}\int_{-\infty}^{\infty} h(k)\mathrm{e}^{-\mathrm{i}ku}\mathrm{d}k = \frac{\mathrm{e}^{-T/4}}{2\sqrt{\pi T}}\mathrm{e}^{-u^2/4T}\,.$$

In the left hand side of the Selberg trace formula one obtains

$$\sum_n \mathrm{e}^{-E_nT} = 1 + \sum_{E_n>0} \mathrm{e}^{-E_nT}$$

where we take into account that for any discrete group there is one zero eigenvalue corresponding to a constant eigenfunction. Therefore when $T \to \infty$ the above sum tends to one

$$\sum_n \mathrm{e}^{-E_n T} \stackrel{T\to\infty}{\longrightarrow} 1\,.$$

One can easily check that in the right hand side of (29) the contribution of the smooth part of the density goes to zero at large T and the contribution of periodic orbits is important only for primitive periodic orbits with $n = 1$. The latter is

$$\frac{\mathrm{e}^{-T/4}}{2\sqrt{\pi T}} \sum_{\mathrm{p}} l_p \mathrm{e}^{-l_p^2/4T - l_p/2} = \frac{\mathrm{e}^{-T/4}}{2\sqrt{\pi T}} \int_0^\infty l\mathrm{e}^{-l^2/4T - l/2} \rho(l)\mathrm{d}l$$

where $\rho(l)$ is the density of periodic orbits. Hence the Selberg trace formula states that

$$\lim_{T\to\infty} \frac{\mathrm{e}^{-T/4}}{2\sqrt{\pi T}} \int_0^\infty l\mathrm{e}^{-l^2/4T - l/2} \rho(l)\mathrm{d}l = 1\,.$$

Assume that $\rho(l) = b\mathrm{e}^{al}/l$ with certain constants a and b. Then from the above limit it follows that $a = b = 1$ which demonstrates that the density of periodic orbits for a discrete group increases exponentially with the length

$$\rho(l) = \frac{\mathrm{e}^l}{l}\,.$$

2.10 Selberg Zeta Function

Among many applications of the Selberg trace formula let us consider the construction of the Selberg zeta function.

Choose as test function $h(k)$ the function

$$h(k) = \frac{1}{k^2 + \alpha^2} - \frac{1}{k^2 + \beta^2}\,.$$

Its Fourier transform is

$$g(l) = \frac{1}{2\alpha}\mathrm{e}^{-\alpha|l|} - \frac{1}{2\beta}\mathrm{e}^{-\beta|l|}\,.$$

The Selberg trace formula gives

$$\sum_n \left(\frac{1}{k_n^2 + \alpha^2} - \frac{1}{k_n^2 + \beta^2} \right) =$$
$$= \frac{\mu(D)}{2\pi} \int_{-\infty}^{\infty} k \tanh \pi k \left(\frac{1}{k^2 + \alpha^2} - \frac{1}{k^2 + \beta^2} \right) \mathrm{d}k +$$
$$+ \sum_{\mathrm{p.p.o.}} \sum_{n=1}^{\infty} \frac{l_p}{2\sinh n l_p/2} \left(\frac{\mathrm{e}^{-\alpha l_p}}{2\alpha} - \frac{\mathrm{e}^{-\beta l_p}}{2\beta} \right)\,.$$

The Selberg zeta function is defined as the following formal product

$$Z(s) = \prod_{\text{p.p.o}} \prod_{m=0}^{\infty} (1 - \mathrm{e}^{-l_p(s+m)}) \,. \tag{30}$$

One has

$$\begin{aligned} \frac{1}{Z}\frac{dZ}{ds} &= \sum_{\text{p.p.o.}} \sum_{m=0}^{\infty} \frac{l_p \mathrm{e}^{-l_p(s+m)}}{1-\mathrm{e}^{-l_p(s+m)}} = \sum_{\text{p.p.o.}} l_p \sum_{n=1}^{\infty}\sum_{m=0}^{\infty} \mathrm{e}^{-l_p(s+m)n} = \\ &= \sum_{\text{p.p.o.}} l_p \sum_{n=1}^{\infty} \frac{\mathrm{e}^{-l_p n s}}{1-\mathrm{e}^{-l_p n}} = \sum_{\text{p.p.o.}} l_p \sum_{n=1}^{\infty} \frac{1}{2\sinh n l_p/2} \mathrm{e}^{-l_p n(s-1/2)} \,. \end{aligned}$$

Choose $\alpha = s - 1/2$ and $\beta = s' - 1/2$ then

$$\begin{aligned} &\sum_n \left(\frac{1}{k_n^2 + (s-1/2)^2} - \frac{1}{k_n^2 + (s'-1/2)^2} \right) = \\ &= \frac{\mu(D)}{4\pi} \int_{-\infty}^{\infty} k \tanh \pi k \left(\frac{1}{k^2 + (s-1/2)^2} - \frac{1}{k^2 + (s'-1/2)^2} \right) \mathrm{d}k + \\ &+ \frac{1}{2s-1}\frac{Z(s)'}{Z(s)} - \frac{1}{2s'-1}\frac{Z(s')'}{Z(s')} \,. \end{aligned}$$

The integral in the right hand side can be computed by the residues

$$\int_{-\infty}^{\infty} k \tanh \pi k \left(\frac{1}{k^2 + (s-1/2)^2} - \frac{1}{k^2 + (s-1/2)^2} \right) \mathrm{d}k = f(s) - f(s')$$

where $f(s)$ is the sum over residues from one pole $k = i(s-1/2)$ and from poles $k_n = i(n+1/2)$ of $\tanh \pi k$

$$\begin{aligned} f(s) &= 2\pi \mathrm{i} \left[\frac{1}{2} \tanh[\mathrm{i}\pi(s-1/2)] + \frac{\mathrm{i}}{\pi} \sum_{n=0}^{\infty} \frac{n+1/2}{(s-1/2)^2 - (n+1/2)^2} \right] = \\ &= \pi \cot \pi s - \sum_{n=1}^{\infty} \frac{1}{s-n} + \sum_{n=1}^{\infty} \frac{1}{s+n} \,. \end{aligned}$$

But

$$\pi \cot \pi s = \sum_{n=1}^{\infty} \frac{1}{s-n} + \sum_{n=1}^{\infty} \frac{1}{s+n} \,,$$

therefore

$$f(s) = 2 \sum_{n=1}^{\infty} \frac{1}{s+n} \,.$$

Using these relations one gets the identity valid for all values of s and s'

$$\frac{1}{2s-1}\frac{Z'(s)}{Z(s)} = \frac{1}{2s'-1}\frac{Z'(s')}{Z(s')} - \frac{\mu(D)}{2\pi}\sum_{n=0}^{\infty}\left(\frac{1}{s+n} - \frac{1}{s'+n}\right) + \\ + \sum_n \left(\frac{1}{k_n^2+(s-1/2)^2} - \frac{1}{k_n^2+(s'-1/2)^2}\right) . \tag{31}$$

The right hand side of this identity has poles at $s = 1/2 + \mathrm{i}k_n$ and $s = -n$. The same poles have to be present in the left hand side. If

$$\frac{Z'(s)}{Z(s)} \to \frac{\nu_k}{s - s_k}$$

then

$$Z(s) \to (s - s_k)^{\nu_k} \quad \text{when } s \to s_k \; .$$

When $\nu_k > 0$ (resp. $\nu_k < 0$) point s_k is a zero (resp. a pole) of the Selberg zeta function $Z(s)$.

2.11 Zeros of the Selberg Zeta Function

Combining all poles one concludes that the Selberg zeta function for a group with only hyperbolic elements have two different sets of zero. The first consists of non-trivial zeros

$$s = 1/2 \pm \mathrm{i}k_n ,$$

coming from eigenvalues of the Laplace–Beltrami operator for automorphic functions. The second set includes a zero from $E = 0$ eigenvalue and zeros from the smooth term. These zeros are called trivial zeros and they are located at points

$$s = -m \;\; (m = 1, 2, \ldots)$$

with multiplicity $\nu_m = (2m+1)\mu(D)/2\pi$, at point $s = 0$ with multiplicity $\nu_0 = \mu(D)/2\pi$ and a single zero at $s = 1$. These multiplicities are integers because the area of a compact fundamental domain $\mu(D) = 4\pi(g-1)$ where g is the genus of the surface.

The structure of these zeros is presented schematically at Fig. 5.

2.12 Functional Equation

The infinite product defining the Selberg zeta function (30) converges only when Re $s > 1/2$. Nevertheless the Selberg zeta function can be analytically continued to the whole complex plane s with the aid of (31).

Put $s' = 1 - s$ in (31). The sum over eigenvalues cancels and $f(s) - f(1 - s) = 2\pi \cot \pi s$. Therefore

$$\frac{1}{2s-1}\left(\frac{Z'(s)}{Z(s)} + \frac{Z'(1-s)}{Z(1-s)}\right) = -\frac{\mu(D)}{2}\cot \pi s$$

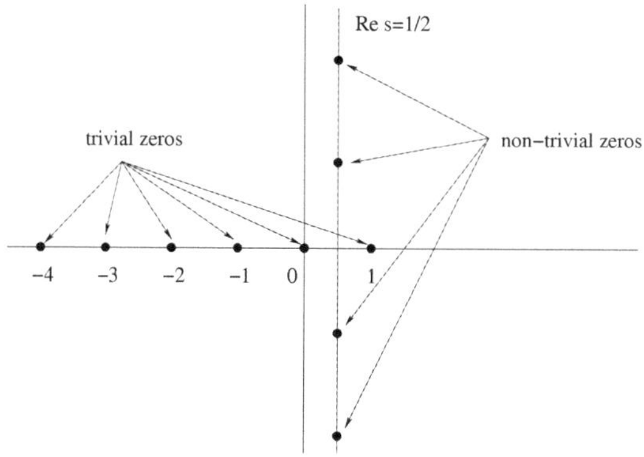

Fig. 5. Zeros of the Selberg zeta function

which is equivalent to the following relation (called functional equation)

$$Z(s) = \varphi(s) Z(1-s) \tag{32}$$

where

$$\frac{\varphi'(s)}{\varphi(s)} = -\mu(D)(s - \frac{1}{2}) \cot \pi s$$

and $\varphi(1/2) = 1$.

Explicitly

$$\varphi(s) = \exp\left(\mu(D) \int_0^{s-1/2} u \tan \pi u du \right) .$$

Therefore if one knows the Selberg zeta function when Re $s > 1$ (32) gives its continuation to the mirror region Re $s < 0$.

3 Trace Formulas for Integrable Dynamical Systems

A f-dimensional system is called integrable if its classical Hamiltonian can be written as a function of action variables only

$$H(\mathbf{I}) = H(I_1, \ldots, I_f) .$$

In this representation the classical equations of motion take especially simple form

$$\dot{\mathbf{I}} = -\frac{\partial H}{\partial \boldsymbol{\varphi}} = 0 , \quad \dot{\boldsymbol{\varphi}} = \frac{\partial H}{\partial \mathbf{I}} = \boldsymbol{\omega} .$$

The semiclassical quantization consists of fixing the values of the action variables

$$I_j = \hbar(n_j + \frac{\mu_j}{4})$$

where n_j are integers and μ_j are called the Maslov indices.

In this approximation eigenvalues of energy of the system are a function of these integers

$$E(\mathbf{n}) = H\left(\hbar(n_1 + \frac{\mu_1}{4}), \ldots, \hbar(n_f + \frac{\mu_f}{4})\right) .$$

The eigenvalue density is the sum over all integers n_j

$$d(E) = \sum_{\mathbf{n}} \delta(E - H(\hbar(\mathbf{n} + \frac{1}{4}\boldsymbol{\mu})) .$$

Using the Poisson summation formula (5) one transforms this expression as follows

$$\begin{aligned} d(E) &= \sum_{\mathbf{N}} \int e^{2\pi i \mathbf{N}\mathbf{n}} \delta(E - H(\hbar(\mathbf{n} + \frac{1}{4}\boldsymbol{\mu}))\mathrm{d}\mathbf{n} = \\ &= \frac{1}{\hbar^f} \sum_{\mathbf{N}} \mathrm{e}^{-\mathrm{i}\pi \mathbf{N}\boldsymbol{\mu}/2} \int \mathrm{e}^{2\pi i \mathbf{N}\mathbf{I}/\hbar} \int \delta(E - H(\mathbf{I}))\mathrm{d}\mathbf{I} \end{aligned} \tag{33}$$

where the summation is taken over f integers N_j.

3.1 Smooth Part of the Density

The term with $\mathbf{N} = 0$ in (33) corresponds to the smooth part of the density

$$\bar{d}(E) = \frac{1}{\hbar^f} \int \delta(E - H(\mathbf{I}))\mathrm{d}\mathbf{I} .$$

As $\mathrm{d}\mathbf{I}\mathrm{d}\boldsymbol{\varphi}$ is the canonical invariant, $\mathrm{d}\mathbf{I}\mathrm{d}\boldsymbol{\varphi} = \mathrm{d}\mathbf{p}\mathrm{d}\mathbf{q}$ where $\mathbf{p}$ and $\mathbf{q}$ are the momenta and coordinates and, because $\int \mathrm{d}\boldsymbol{\varphi} = (2\pi)^f$, the formula for the smooth part of the level density can be rewritten in the Thomas-Fermi form

$$\bar{d}(E) = \int \delta(E - H(\mathbf{p}, \mathbf{q})) \frac{\mathrm{d}\mathbf{p}\mathrm{d}\mathbf{q}}{(2\pi\hbar)^f} . \tag{34}$$

The usual interpretation of this formula is that each quantum state occupies $(2\pi\hbar)^f$ volume on the constant energy surface. For general systems (34) represents the leading term of the expansion of the smooth part of the level density when $\hbar \to 0$. Other terms can be found e.g. in [5]. See also [14] for the resummation of such series for certain models.

3.2 Oscillating Part of the Density

In the semiclassical approximation $\hbar \to 0$ terms with $\mathbf{N} \neq 0$ in (33) can be calculated by the saddle point method. Our derivation differs slightly from the one given in [9]. First it is convenient to represent δ-function as follows

$$\delta(x) = \frac{1}{2\pi\hbar} \int_{-\infty}^{\infty} \mathrm{e}^{\mathrm{i}\alpha x/\hbar} \mathrm{d}\alpha \,.$$

Then

$$d^{(osc)}(E) = \frac{1}{2\pi\hbar^{f+1}} \sum_{\mathbf{N}} \mathrm{e}^{-\mathrm{i}\pi\mathbf{N}\boldsymbol{\mu}/2} \int_{-\infty}^{\infty} \mathrm{d}\alpha \int \mathrm{e}^{\mathrm{i}S(\mathbf{I},\alpha)/\hbar} \mathrm{d}\mathbf{I}$$

where the effective action, $S(\mathbf{I}, \alpha)$, is

$$S(\mathbf{I}, \alpha) = 2\pi\mathbf{N}\mathbf{I} + \alpha(E - H(\mathbf{I})) \,.$$

The integration over $\mathbf{I}$ and α can be performed by the saddle point method. The saddle point values, $\mathbf{I}_{sp}$ and α_{sp}, are determined from equations

$$\frac{\partial S}{\partial \alpha} = E - H(\mathbf{I}_{sp}) = 0 \,, \quad \frac{\partial S}{\partial \mathbf{I}} = 2\pi\mathbf{N} - \alpha_{sp}\boldsymbol{\omega}_{sp} = 0 \,.$$

The first equation shows that in the leading approximation $\mathbf{I}_{sp}$ belongs to the constant energy surface and the second equation selects special values of $\mathbf{I}_{sp}$ for which frequencies ω_j are commensurable

$$\boldsymbol{\omega}_{sp} = \frac{2\pi}{\alpha_{sp}} \mathbf{N} \,.$$

Together the saddle point conditions demonstrate that in the limit $\hbar \to 0$ the dominant contribution to the term with fixed integer vector $\mathbf{N}$ comes from the classical periodic orbit with period

$$T_p = \alpha_{sp}$$

and the saddle point action coincides with the classical action along this trajectory

$$S_{sp} = 2\pi\mathbf{N}\mathbf{I}_{sp} \,.$$

To compute remaining integrals it is necessary to expand the full action up to quadratic terms on deviations from the saddle point values. One has

$$S(\mathbf{I}_{sp} + \delta\mathbf{I}, \alpha_{sp} + \delta\alpha) = S_{sp} + \frac{T_p}{2}(\delta I_i H_{ij} \delta I_j) - \delta\alpha(\omega_j \delta I_j)$$

where the summation over repeating indexes is assumed. H_{ij} is the matrix of the second derivatives of the Hamiltonian computed at the saddle point

$$H_{ij} \equiv \frac{\partial^2 H}{\partial I_i \partial I_j} \Big|_{\mathbf{I}=\mathbf{I}_{sp}} \,.$$

The following steps are straightforward

$$\int \mathrm{d}\delta\mathbf{I}\mathrm{d}\delta\alpha \exp\left(\frac{\mathrm{i}}{\hbar}S(\mathbf{I},\alpha)\right) =$$

$$= \mathrm{e}^{\mathrm{i}S_{sp}/\hbar}\int \mathrm{d}\delta\alpha \int \mathrm{d}\delta\mathbf{I} \exp\left(\frac{\mathrm{i}}{2\hbar}T_p(\delta I_i H_{ij}\delta I_j) - \frac{\delta\alpha}{\hbar}(\omega_j\delta I_j)\right) =$$

$$= \left(\frac{2\pi\hbar}{T_p}\right)^{f/2} \frac{\mathrm{e}^{\mathrm{i}S_{sp}/\hbar}}{\sqrt{|\det H_{ij}|}}\int \delta\alpha \exp\left(-\frac{\mathrm{i}}{2\hbar T_p}(\delta\alpha)^2(\omega_i H^{-1}_{ij}\omega_j) + \frac{\mathrm{i}}{4}\pi\beta'\right) =$$

$$= \left(\frac{2\pi\hbar}{T_p}\right)^{f/2} \frac{\sqrt{2\pi\hbar T_p}}{\sqrt{|\det H_{ij}|(\omega_k H^{-1}_{kl}\omega_l)}} \exp\left(\frac{\mathrm{i}}{\hbar}S_{sp} + \frac{\mathrm{i}}{4}\pi\beta\right) =$$

$$= \frac{(2\pi)^{(f-1)/2}\hbar^{(f+1)/2}}{T_p^{(f-3)/2}|(N_i Q_{ij} N_j)|^{1/2}} \exp\left(\frac{\mathrm{i}}{\hbar}S_{sp} + \frac{\mathrm{i}}{4}\pi\beta\right)$$

where $Q_{ij} = H^{-1}_{ij}\det H$ called the co-matrix of H_{ij} is the determinant obtained from H_{ij} by omitting the i-th row and the j-th column. The phase β is the signature of H_{ij} minus the sign of $(\omega H^{-1}\omega)$.

The final expression for the oscillating part of the level density of an integrable system with a Hamiltonian $H(\mathbf{I})$ is

$$d^{(osc)}(E) = \sum_{\mathbf{N}} P_{\mathbf{N}} \exp\left(\mathrm{i}\frac{S_p}{\hbar} - \mathrm{i}\frac{\pi}{4}\mathbf{N}\boldsymbol{\mu} + \mathrm{i}\frac{\pi}{4}\beta\right)$$

where $S_p = 2\pi\mathbf{N}\mathbf{I}$ is the action over a classical periodic orbit with fixed winding numbers and

$$P_{\mathbf{N}} = \left(\frac{2\pi}{\hbar T_p}\right)^{(f-3)/2} \frac{1}{\hbar^2|(N_i Q_{ij} N_j)|^{1/2}}\ .$$

The summation over integer vectors $\mathbf{N}$ is equivalent to the summation over all classical periodic orbit families of the system.

4 Trace Formula for Chaotic Systems

We will not discuss further chaotic systems in full generality (see e.g. [28]). For our purposes it is sufficient to consider those systems for which all periodic orbits are isolated and unstable. To compute the eigenvalue density for such a chaotic system one has to start with general expression (17)

$$d(E) = -\frac{1}{\pi}\int \mathrm{Im}\ G_E(\mathbf{x},\mathbf{x})\mathrm{d}\mathbf{x}$$

which relates the quantum density with the Green function of the system, $G_E(\mathbf{x},\mathbf{y})$, obeying the Schroedinger equation with a δ-function term in the right hand side

$$(E - \hat{H})G_E(\mathbf{x}, \mathbf{y}) = \delta(\mathbf{x} - \mathbf{y}) \ .$$

For concreteness let us consider the usual case

$$\hat{H} = -\hbar^2 \Delta + V(\mathbf{x}) \ .$$

The exact Green function can be computed exactly only in very limited cases. For generic systems the best which can be achieved is the calculation of the Green function in the semiclassical limit $\hbar \to 0$.

4.1 Semiclassical Green Function

Let us try to obey the Schroedinger equation in the following form (see [33])

$$G_E(\mathbf{x}, \mathbf{y}) = A(\mathbf{x}, \mathbf{y}) \mathrm{e}^{\mathrm{i}S(\mathbf{x},\mathbf{y})/\hbar} \tag{35}$$

where the prefactor $A(\mathbf{x}, \mathbf{y})$ can be expanded into a power series of $\hbar$.

Separating the real and imaginary parts of the Schroedinger equation one gets two equations

$$\big(E - (\nabla S)^2 - V(\mathbf{x})\big) + \hbar^2 \Delta A = 0$$

and

$$2\nabla S \nabla A + \Delta S A = 0 \ .$$

In the leading order in $\hbar$ the first equation reduces to the Hamilton-Jacobi equation for the classical action $S(\mathbf{x}, \mathbf{y})$

$$E = (\nabla S)^2 + V(\mathbf{x}) \ .$$

It is well known that the solution of this equation can be obtained in the following way.

Find the solution of the usual classical equations of motion

$$\ddot{\mathbf{x}} = -\frac{\partial V}{\partial \mathbf{x}}$$

with energy E which starts at a fixed point $\mathbf{y}$ and ends at a point $\mathbf{x}$. Then

$$S(\mathbf{x}, \mathbf{y}) = \int_{\mathbf{y}}^{\mathbf{x}} \mathbf{p} d\mathbf{x}$$

where $\mathbf{p}$ is the momentum and the integral is taken over this trajectory.

Instead of proving this fact we illustrate it on an example of the free motion. The free motion equations $\ddot{\mathbf{x}} = 0$ have a general solution

$$\mathbf{x} = \mathbf{k}t + \mathbf{y}$$

with a fixed vector $\mathbf{k}$. One has

$$\mathbf{k} = \frac{\mathbf{x} - \mathbf{y}}{t}$$

and the conservation of energy $|\mathbf{k}|^2 = E$ determines the time of motion

$$t = \frac{|\mathbf{x} - \mathbf{y}|}{\sqrt{E}} .$$

Therefore

$$S(\mathbf{x}, \mathbf{y}) = \sqrt{E}|\mathbf{x} - \mathbf{y}|$$

which, evidently, is the solution of the free Hamilton–Jacobi equation.

The next order equation

$$2\nabla S \nabla A + \Delta S A = 0$$

is equivalent to the conservation of current. Indeed, for the semiclassical wave function (35)

$$\mathbf{J} = \frac{1}{2\mathrm{i}}(\Psi^* \nabla \Psi - \Psi \nabla \Psi^*) = A^2 \nabla S$$

and

$$\nabla \mathbf{J} = A(2\nabla A \nabla S + A \Delta S) = 0 .$$

The solution of the above transport equation has the form

$$A(\mathbf{x}, \mathbf{y}) = \frac{\pi}{(2\pi\hbar)^{(f+1)/2}} \left| \frac{1}{k_i k_f} \det \left(-\frac{\partial^2 S}{\partial t_{i\perp} \partial t_{f\perp}} \right) \right|^{1/2}$$

where $t_{i\perp}$ and $t_{f\perp}$ are coordinates perpendicular to the trajectory in the initial, $\mathbf{y}$, and final, $\mathbf{x}$, points respectively and k_i, k_f are the initial and final momenta. The derivation of this formula can be found e.g. in [33]. The overall prefactor in this formula can be fixed by comparing with the asymptotics of the free Green function (14) at large distances.

The final formula for the semiclassical Green function takes the form

$$G_E(\mathbf{x}, \mathbf{y}) = \sum_{\substack{\text{classical} \\ \text{trajectories}}} \frac{\pi}{(2\pi\hbar)^{(f+1)/2}} \left| \frac{1}{k_i k_f} \det \left(-\frac{\partial^2 S}{\partial t_{i\perp} \partial t_{f\perp}} \right) \right|^{1/2} \times$$

$$\times \exp \left(\frac{\mathrm{i}}{\hbar} S_{cl}(\mathbf{x}, \mathbf{y}) - \frac{\mathrm{i}}{4} \pi \mu \right)$$

where the sum is taken over all classical trajectories with energy E which connect points $\mathbf{y}$ and $\mathbf{x}$. μ is the Maslov index which, roughly speaking, counts the number of points along the trajectory where semiclassical approximation cannot be applied.

4.2 Gutzwiller Trace Formula

The knowledge of the Green function permits the calculation of the density of eigenstates by the usual formula (17)

$$d(E) = -\frac{1}{\pi}\int \text{Im } G_E(\mathbf{x},\mathbf{x})\text{d}\mathbf{x}\,.$$

The Green function $G_E(\mathbf{x},\mathbf{y})$ at points $\mathbf{x}$ and $\mathbf{y}$ very close to each other has two different contributions (see Fig. 6). The first comes from very short

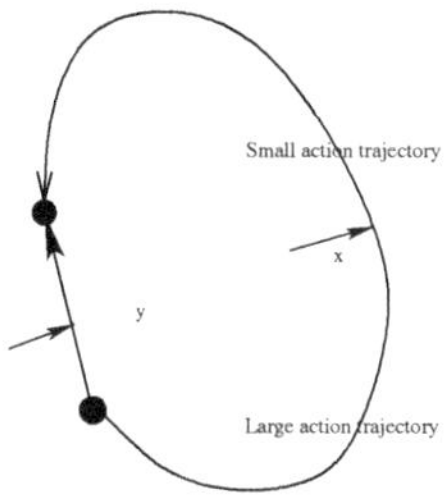

Fig. 6. Small and large action contributions to the Green function for nearby points

trajectories where semiclassical approximation cannot, in general, be applied. The second is related with long trajectories. The first contribution can be computed by using the Thomas–Fermi (local) approximation for the Green function. In this approximation one uses the local formula (cf. (16))

$$G_E(\mathbf{x},\mathbf{y}) \stackrel{\mathbf{y}\to\mathbf{x}}{\longrightarrow} \int \frac{\text{d}\mathbf{p}}{(2\pi\hbar)^f}\frac{\text{e}^{\text{i}\mathbf{p}(\mathbf{x}-\mathbf{y})/\hbar}}{(E-H(\mathbf{p},\mathbf{x})+i\epsilon)}\,.$$

Therefore

$$\text{Im } G_E(\mathbf{x},\mathbf{x}) = -\pi\int\frac{\text{d}\mathbf{p}}{(2\pi\hbar)^f}\delta\left(E-H(\mathbf{p},\mathbf{x})\right)$$

and the smooth part of the level density in the leading approximation equals the phase-space volume of the constant energy surface divided by $(2\pi\hbar)^f$

$$\bar{d}(E) = \int\frac{\text{d}\mathbf{p}\text{d}\mathbf{x}}{(2\pi\hbar)^f}\delta\left(E-H(\mathbf{p},\mathbf{x})\right)\,.$$

The contribution from long classical trajectories with finite actions corresponds to the oscillating part of the density and can be calculated using the semiclassical approximation of the Green function (35).

One has

$$d^{(osc)}(E) = -\frac{1}{\pi}\text{Im}\sum_{\substack{\text{classical}\\ \text{trajectories}}}\int A(\mathbf{x},\mathbf{x})\text{e}^{\text{i}S(\mathbf{x},\mathbf{x})/\hbar}\text{d}\mathbf{x}\,.$$

When $\hbar \to 0$ the integration can be performed in the saddle point approximation. The saddles are solutions of the equation

$$\left[\frac{\partial S(\mathbf{x},\mathbf{y})}{\partial \mathbf{x}} + \frac{\partial S(\mathbf{x},\mathbf{y})}{\partial \mathbf{y}}\right]_{\mathbf{y}=\mathbf{x}} = 0 \ .$$

But

$$\frac{\partial S(\mathbf{x},\mathbf{y})}{\partial \mathbf{x}} = \mathbf{k}_f \ , \quad \frac{\partial S(\mathbf{x},\mathbf{y})}{\partial \mathbf{y}} = -\mathbf{k}_i$$

where $\mathbf{k}_f$ and $\mathbf{k}_i$ are the momenta in the final and initial points respectively.

Hence the saddle point equations select special classical orbits which start and end in the same point with the same momentum. It means that the saddles are classical periodic orbits of the system and

$$S_{sp} = S_p \ .$$

To calculate the integral around one particular periodic orbit it is convenient to split the integration over the whole space to one integration along the orbit and $(f-1)$ integrations in directions perpendicular to the orbit. For simplicity we consider the two-dimensional case.

The change of the action when a point is at the distance y from the periodic orbit is

$$\delta S = \frac{1}{2} y^2 \frac{\partial^2 S(y,y)}{\partial y^2}|_{y=0}$$

where $S(y,y)$ is the classical action for a classical orbit in a vicinity of the periodic orbit (see Fig. 7). To compute such derivatives it is useful to use

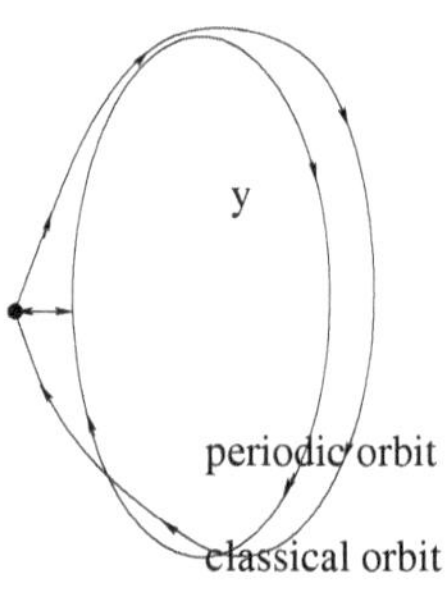

Fig. 7. A periodic orbit and a closed classical orbit in its vicinity

the monodromy matrix, m_{ij}, which relates initial and final coordinates and momenta in a vicinity of periodic orbit in the linear approximation

$$\begin{pmatrix} \delta y_f \\ \delta p_f \end{pmatrix} = \begin{pmatrix} m_{11} & m_{12} \\ m_{21} & m_{22} \end{pmatrix} \begin{pmatrix} \delta y_i \\ \delta p_i \end{pmatrix} \ .$$

As the classical motion preserves the canonical invariant dpdq it follows that $\det M = 1$.

One has

$$\delta y_f = m_{11}\delta y_i + m_{12}\delta p_i \,,$$
$$\delta p_f = m_{21}\delta y_i + m_{22}\delta p_i \,.$$

But

$$p_i = -\frac{\partial S}{\partial y_i}\,, \quad p_f = \frac{\partial S}{\partial y_f}\,.$$

Therefore

$$\delta p_i = -\frac{\partial^2 S}{\partial y_i^2}\delta y_i - \frac{\partial^2 S}{\partial y_i \partial y_f}\delta y_f\,, \quad \delta p_f = \frac{\partial^2 S}{\partial y_i \partial y_f}\delta y_i + \frac{\partial^2 S}{\partial y_f^2}\delta y_f\,.$$

From comparison of these two expression one obtains the expressions of the second derivatives of the action through monodromy matrix elements

$$\frac{\partial^2 S}{\partial y_i \partial y_f} = -\frac{1}{m_{12}}\,, \quad \frac{\partial^2 S}{\partial y_i^2} = \frac{m_{11}}{m_{12}}\,, \quad \frac{\partial^2 S}{\partial y_f^2} = \frac{m_{22}}{m_{12}}\,.$$

Substituting these expressions to the contribution to the trace formula from one periodic orbit one gets (in two dimensions)

$$d_p^{(osc)}(E) = \frac{1}{\mathrm{i}(2\pi \mathrm{i}\hbar)^{3/2}} \int |m_{12}|^{-1/2} \exp(\frac{\mathrm{i}}{\hbar}S_p + \mathrm{i}\frac{m_{11}+m_{22}-2}{2\hbar\, m_{12}}y^2)\mathrm{d}y\frac{\mathrm{d}x}{k(x)}$$

where x and y are respectively coordinates parallel and perpendicular to the trajectory.

Computing the resulting integrals one obtains

$$d_p^{(osc)}(E) = \frac{T_p}{\pi\hbar}\frac{\mathrm{e}^{\mathrm{i}S_p/\hbar - \mathrm{i}\pi\mu_p/2}}{\sqrt{|m_{11}+m_{22}-2|}}$$

where $T_p = \int \mathrm{d}x/k(x)$ is the geometrical period of the trajectory.

Finally the Gutzwiller trace formula takes the form (valid in arbitrary dimensions)

$$d^{(osc)}(E) = \sum_{\substack{\text{primitive}\\ \text{periodic}\\ \text{orbits}}} \frac{T_p}{\pi\hbar}\sum_{n=1}^{\infty}\frac{1}{|\det(M_p^n - 1)|^{1/2}}\cos\left[n(\frac{S_p}{\hbar} - \frac{\pi}{2}\mu_p)\right]\,.$$

In the derivation of this formula we assumed that all periodic orbits are unstable and M_p is the monodromy matrix for a primitive periodic orbit.

5 Riemann Zeta Function

The trace-like formulas exist not only for dynamical systems but also for the Riemann zeta function (and other number-theoretical zeta functions as well).

The Riemann zeta function is a function of complex variable s defined as follows

$$\zeta(s) = \sum_{n=1}^{\infty} \frac{1}{n^s} = \prod_p (1 - p^{-s})^{-1} \tag{36}$$

where the product is taken over prime numbers. The second equality (called the Euler product) is a consequence of the unique factorization of integers into a product of prime numbers.

This function converges only when $\mathrm{Re} s > 1$ but can analytically be continued in the whole complex s-plane.

5.1 Functional Equation

The possibility of this continuation is connected with the fact that the Riemann zeta function satisfies the important functional equation

$$\zeta(s) = \varphi(s)\zeta(1-s) \tag{37}$$

where

$$\varphi(s) = 2^s \pi^{s-1} \sin\left(\frac{\pi s}{2}\right) \Gamma(1-s)\ . \tag{38}$$

We present one of numerous method of proving this relation (see e.g. [55]).

When $\mathrm{Re}\ s > 0$ one has the equality

$$\int_0^{\infty} x^{s/2-1} \mathrm{e}^{-\pi n^2 x} \mathrm{d}x = \frac{\Gamma(s/2)}{n^s \pi^{s/2}}$$

where $\Gamma(x)$ is the Gamma function (see e.g. [32], Vol. 1, Sect. 1). Therefore if $\mathrm{Re}\ s > 1$

$$\frac{\Gamma(s/2)\zeta(s)}{\pi^{s/2}} = \int_0^{\infty} x^{s/2-1} \Psi(x) \mathrm{d}x$$

where $\Psi(x)$ is given by the following series

$$\Psi(x) = \sum_{n=1}^{\infty} \mathrm{e}^{-\pi n^2 x}\ .$$

Using the Poisson summation formula (5) one obtains

$$\sum_{n=-\infty}^{\infty} \mathrm{e}^{-\pi n^2 x} = \frac{1}{\sqrt{x}} \sum_{n=-\infty}^{\infty} \mathrm{e}^{-\pi n^2 / x}$$

which leads to the identity

$$2\Psi(x)+1=\frac{1}{\sqrt{x}}\left(2\Psi(\frac{1}{x})+1\right) .$$

Hence

$$\begin{aligned}\xi(s)\equiv\pi^{-s/2}\Gamma(\frac{1}{2}s)\zeta(s)&=\int_0^1 x^{s/2}\Psi(x)\mathrm{d}x+\int_1^\infty x^{s/2}\Psi(x)\mathrm{d}x=\\ &=\int_0^1 x^{s/2}\left(\frac{1}{\sqrt{x}}\Psi(\frac{1}{x})+\frac{1}{2\sqrt{x}}-\frac{1}{2}\right)\mathrm{d}x+\int_1^\infty x^{s/2}\Psi(x)\mathrm{d}x=\\ &=\frac{1}{s-1}-\frac{1}{s}+\int_0^1 x^{s/2-3/2}\Psi(\frac{1}{x})\mathrm{d}x+\int_1^\infty x^{s/2}\Psi(x)\mathrm{d}x=\\ &=\frac{1}{s(s-1)}+\int_1^\infty\left(x^{-s/2-1/2}+x^{s/2-1}\right)\Psi(x)\mathrm{d}x .\end{aligned}$$

The last integral is convergent for all values of s and gives the analytical continuation of the Riemann zeta function to the whole complex s-plane, the only singularity being the pole at $s=1$ with unit residue

$$\zeta(s)\stackrel{s\to 1}{\longrightarrow}\frac{1}{s-1} .$$

(The pole at $s=0$ is canceled by the pole of $\Gamma(s/2)$ giving $\zeta(0)=-1/2$.)

One of important consequences of the above formula of analytical continuation is that it does not change under the substitution $s\to 1-s$. Therefore for all values of s

$$\xi(s)=\xi(1-s)$$

or

$$\zeta(s)=\varphi(s)\zeta(1-s)$$

where

$$\varphi(s)=\pi^{s-1/2}\frac{\Gamma(1/2-s/2)}{\Gamma(s/2)} \tag{39}$$

By standard formulas (see e.g. [32], Vol. 1, 1.2.5, 1.2.15)

$$\Gamma(x)\Gamma(1-x)=\frac{\pi}{\sin\pi x} ,\quad \Gamma(2x)=2^{2x-1}\pi^{-1/2}\Gamma(x)\Gamma(x+\frac{1}{2})$$

the last expression can be transformed to (38) which proves the functional equation (37).

From the functional equation (37) is follows that $\zeta(s)$ has 'trivial' zeros at negative even integers (except zero) $s=-2,-4,\ldots$ which appear from $\sin(\pi s/2)$ in (38). All other non-trivial zeros, $\zeta(s_n)=0$, are situated in the so-called critical strip $0<$ Re $s<1$. If one denotes these zeros as $s_n=1/2+\mathrm{i}\gamma_n$ then functional equation together with the fact that $\zeta(s)^*=\zeta(s^*)$ state that in general there exit 4 sets of zeros: γ_n , $-\gamma_n$, γ_n^* , $-\gamma_n^*$.

According to the famous *Riemann conjecture* (see e.g. [55]) all nontrivial zeros of $\zeta(s)$ lie at the symmetry line Re $s=1/2$ or γ_n are all real quantities.

Numerical calculations confirms this conjecture for exceptionally large number of zeros (see e.g. [47] and the web site of Odlyzko [48]) but a mathematical proof is still absent.

5.2 Trace Formula for the Riemann Zeros

Let us fix a test function $h(r)$ exactly as it was done for the Selberg trace formula in Sect. 2.8 i.e.

- $h(r)$ is a function analytical in the region $|\mathrm{Im}\ r| \leq 1/2 + \delta$,
- $h(-r) = h(r)$,
- $|h(r)| \leq A(1 + |r|)^{-2-\delta}$.

Denote as in that Section

$$g(u) = \frac{1}{2\pi} \int_{-\infty}^{+\infty} h(r) \mathrm{e}^{-\mathrm{i}ru} \mathrm{d}r$$

and define

$$H(s) = \int_{-\infty}^{+\infty} g(u) \mathrm{e}^{(s-1/2)u} \mathrm{d}u \ .$$

Now let us compute the integral

$$\frac{1}{2\pi \mathrm{i}} \oint \mathrm{d}s H(s) \frac{\zeta'(s)}{\zeta(s)}$$

where the contour of integration is taken over the rectangle $-\eta \leq \ \mathrm{Re}\ s \leq 1+\eta$ and $-T \leq \ \mathrm{Im}\ s \leq T$ with $0 < \eta < \delta$ and $T \to +\infty$. Inside this rectangle there are poles of $\zeta'(s)/\zeta(s)$ coming from non-trivial zeros of the Riemann zeta function, $s_n = 1/2 + \mathrm{i}\gamma_n$, and the one from the pole of $\zeta(s)$ at $s = 1$. The total contribution from these poles is

$$\sum_n h(\gamma_n) - h(-\frac{\mathrm{i}}{2}) \ .$$

One can check that the limit $T \to \infty$ exists and, consequently, one has the identity

$$\sum_n h(\gamma_n) - h(-\frac{\mathrm{i}}{2}) = \frac{1}{2\pi \mathrm{i}} \int_{1+\eta-\mathrm{i}\infty}^{1+\eta+\mathrm{i}\infty} \mathrm{d}s H(s) \frac{\zeta'(s)}{\zeta(s)} - \frac{1}{2\pi \mathrm{i}} \int_{-\eta-\mathrm{i}\infty}^{-\eta+\mathrm{i}\infty} \mathrm{d}s H(s) \frac{\zeta'(s)}{\zeta(s)} \ .$$

Let us substitute in the second integral the functional equation (37) with $\varphi(s)$ from (39). One has

$$\frac{\zeta'(s)}{\zeta(s)} = \ln \pi - \frac{\zeta'(1-s)}{\zeta(1-s)} - \frac{1}{2} \left[\frac{\Gamma'}{\Gamma} \left(\frac{s}{2} \right) + \frac{\Gamma'}{\Gamma} \left(\frac{1-s}{2} \right) \right] \ .$$

Now all integrals converge and one can move the integration contour till $s = 1/2 + ir$ with real r. In this manner one obtains

$$\frac{1}{4\pi \mathrm{i}} \int_{-\eta-\mathrm{i}\infty}^{-\eta+\mathrm{i}\infty} \mathrm{d}s H(s) \left[\frac{\Gamma'}{\Gamma}\left(\frac{s}{2}\right) + \frac{\Gamma'}{\Gamma}\left(\frac{1-s}{2}\right) \right] =$$

$$= h(\frac{\mathrm{i}}{2}) + \frac{1}{2\pi} \int_{-\infty}^{+\infty} h(r) \frac{\Gamma'}{\Gamma}\left(\frac{1}{4} + \frac{\mathrm{i}}{2} r\right) \mathrm{d}r \,.$$

The first term in the right hand side of this equality is due to the appearance of the pole of $\Gamma(s/2)$ at $s = 0$ when the integration contour shifted till $s = 1/2 + ir$. Also we have used that $h(-r) = h(r)$.

For terms with the Riemann zeta function one can use the expansion which follows from (36)

$$\frac{\zeta'(s)}{\zeta(s)} = -\sum_p \ln p \sum_{n=1}^{\infty} p^{-ns} \,.$$

Shifting the integration contour as above (i.e. till $s = 1/2 + ir$), using that $g(-u) = g(u)$, and combining all terms together one gets the following Weil explicit formula for the Riemann zeros

$$\sum_{\substack{\text{non-trivial} \\ \text{zeros}}} h(\gamma_n) = \frac{1}{2\pi} \int_{-\infty}^{\infty} h(r) \frac{\Gamma'}{\Gamma}\left(\frac{1}{4} + \frac{\mathrm{i}}{2} r\right) \mathrm{d}r + h(\frac{\mathrm{i}}{2}) + h(-\frac{\mathrm{i}}{2}) -$$

$$- g(0) \ln \pi - 2 \sum_{\text{primes}} \ln p \sum_{n=1}^{\infty} \frac{1}{p^{n/2}} g(n \ln p) \,.$$

Here γ_n are related with non-trivial zeros of the Riemann zeta function, s_n, as follows

$$s_n = \frac{1}{2} + \mathrm{i}\gamma_n \,.$$

This formula is an analog of usual trace formulas as it relates zeros of the Riemann zeta function defined in a quite complicated manner with prime numbers which are a common notion.

The similarity with dynamical trace formulas is more striking if one assumes the validity of the Riemann conjecture which states that γ_n are real quantities (which in a certain sense can be considered as energy levels of a quantum system). In 'semiclassical' limit $r \to \infty$ using the Stirling formula (see e.g. [32], Vol. 1, 1.9.4)

$$\ln \Gamma(z) \stackrel{|z| \to \infty}{\longrightarrow} (z - \frac{1}{2}) \ln z - z + \frac{1}{2} \ln 2\pi$$

one obtains that the density of Riemann zeros

$$d(E) = \sum_n \delta(E - \gamma_n)$$

can be expressed by the following 'physical' trace formula valid at large E

$$d(E) = \bar{d}(E) + d^{(osc)}(E)$$

where

$$\bar{d}(E) = \frac{1}{2\pi} \ln \frac{E}{2\pi} + \text{ corrections },$$

and

$$d^{(osc)}(E) = -\frac{1}{\pi} \sum_p \sum_{n=1}^{\infty} \frac{\ln p}{p^{n/2}} \cos(En \ln p)$$

where the summation is performed over all prime numbers.

5.3 Chaotic Systems and the Riemann Zeta Function

By comparing the above equations with the trace formulas of chaotic systems one observes (see e.g. [38], [12], [13]) a remarkable correspondence between different quantities in these trace formulas

- periodic orbits of chaotic systems $\leftrightarrow$ primes,
- periodic orbit period $T_p \quad \leftrightarrow \quad \ln p$,
- convergence properties of both formulas are also quite similar.

The number of periodic orbits with period less than T for chaotic systems is asymptotically

$$N(T_p < T) = \frac{\mathrm{e}^{hT}}{hT},$$

where the constant h is called the topological entropy.

The number of prime numbers less than x is given by the prime number theorem (see e.g. [55])

$$N(p < x) = \frac{x}{\ln x} .$$

As $\ln p \equiv T_p$ this expression has the form similar to number of periodic orbits of chaotic systems with $h = 1$

$$N(T_p < T) = \frac{\mathrm{e}^{T}}{T} .$$

Due to these similarities number-theoretical zeta functions play the role of simple (but by far non-trivial) models of quantum chaos.

Notice that the overall signs of the oscillating part of trace formulas for the Riemann zeta function and dynamical systems are different. According to Connes [30] it may be interpreted as Riemann zeros belong not to a spectrum of a certain self-adjoint operator but to an 'absorption' spectrum. Roughly speaking it means the following. Let us assume that the spectrum of a 'Riemann Hamiltonian' is continuous and it covers the whole axis. But exactly when eigenvalues equal Riemann zeros corresponding eigenfunctions of this

Hamiltonian vanish. Therefore these eigenvalues do not belong to the spectrum and Riemann zeros correspond to such missing points similarly to black lines (forming absorption spectra) which are visible when light passes through an absorption media. In Connes' approach the 'Riemann Hamiltonian' may be very simple (see also [15]) but the functional space where it has to be defined is extremely intricate.

6 Summary

Trace formulas can be constructed for all 'reasonable' systems. They express the quantum density of states (and other quantities as well) as a sum over classical periodic orbits. All quantities which enter trace formulas can be computed within pure classical mechanics.

Trace formulas consist of two terms

$$d(E) = \bar{d}(E) + d^{(osc)}(E) \,.$$

The smooth part of the density, $\bar{d}(E)$, for all dynamical systems is given by the Thomas–Fermi formula (plus corrections if necessary)

$$\bar{d}(E) = \int \frac{\mathrm{d}\mathbf{p}\mathrm{d}\mathbf{x}}{(2\pi\hbar)^f}\delta\left(E - H(\mathbf{p},\mathbf{x})\right) \,.$$

For integrable systems the oscillating part of the density, $d^{(osc)}(E)$, is

$$d^{(osc)}(E) = \sum_{\mathbf{N}} \left(\frac{2\pi}{\hbar T_p}\right)^{(f-3)/2} \frac{1}{\hbar^2\sqrt{|(N_i Q_{ij} N_j)|}} \exp\left(\mathrm{i}\frac{S_p}{\hbar} - \mathrm{i}\frac{\pi}{4}\mathbf{N}\boldsymbol{\mu} + \mathrm{i}\frac{\pi}{4}\beta\right)$$

where $S_p = 2\pi\mathbf{N}\mathbf{I}$ is the action over a classical periodic orbit with fixed winding numbers $\mathbf{N}$ and Q_{ij} is the co-matrix of the matrix of the second derivatives of the Hamiltonian.

For chaotic systems $d^{(osc)}(E)$ is represented as a sum over all classical periodic orbits

$$d^{(osc)}(E) = \sum_{\text{p.p.o.}} \frac{T_p}{\pi\hbar} \sum_{n=1}^{\infty} \frac{1}{|\det(M_p^n - 1)|^{1/2}} \cos\left(n\frac{S_p}{\hbar} - n\frac{\pi}{2}\mu_p\right)$$

where S_p is the classical action along a primitive periodic trajectory and M_p is its monodromy matrix.

Usually trace formulas represent the dominant contribution when $\hbar \to 0$. They are exact only in very special cases as for constant negative curvature surfaces generated by discrete groups where they coincide with the Selberg trace formula. For a group with only hyperbolic elements

$$\bar{d}(E) = \frac{\mu(D)}{4\pi}\tanh \pi k$$

where $\mu(D)$ is the area of the fundamental domain of the group and

$$d^{(osc)}(E) = \sum_{\text{p.p.o.}} \frac{l_p}{4\pi k} \sum_{n=1}^{\infty} \frac{\cos(knl_p)}{\sinh(nl_p/2)}$$

where l_p are lengths of periodic orbits.

The formulas similar to trace formulas exist also for number-theoretical zeta functions (assuming the generalized Riemann conjecture). In particular, for the Riemann zeta function

$$\bar{d}(E) = \frac{1}{2\pi} \ln \frac{E}{2\pi}$$

and

$$d^{(osc)}(E) = -\frac{1}{\pi} \sum_{\text{prime}} \sum_{n=1}^{\infty} \frac{\ln p}{p^{n/2}} \cos(En \ln p) \ .$$

The principal difficulty of all trace formulas is the divergence of the sums over periodic orbits. To obtain a mathematically meaningful formula one considers instead of the singular density of states its smoothed version defined as a sum over all eigenvalues of a suitable chosen smooth test-function. When its Fourier harmonics decrease quickly the resulting formula represent a well defined object.

Suggestions for Further Readings

- A very detailed account of trace formulas derived by multiple scattering method can be found in a series of papers by Balian and Bloch [8].
- A concise mathematical review of hyperbolic geometry is given in [42].
- Explicit forms of the Selberg trace formula for general discrete groups with elliptic and parabolic elements are presented in two volumes of Hejhal's monumental work [39] which contains practically all known information about the Selberg trace formula.
- In [38] one can find a mathematical discussion about different relations between number-theoretical zeta functions and dynamical systems.

II. Statistical Distribution of Quantum Eigenvalues

Wigner and Dyson in the fifties had proposed to describe complicated (and mostly unknown) Hamiltonian of heavy nuclei by a member of an ensemble of random matrices and they argued that the type of this ensemble depends only on the symmetry of the Hamiltonian. For systems without time-reversal invariance the relevant ensemble is the Gaussian Unitary Ensemble (GUE), for systems invariant with respect to time-reversal the ensemble is the Gaussian Orthogonal Ensemble (GOE) and for systems with time-reversal invariance but with half-integer spin energy levels have to be described according to the Gaussian Symplectic Ensemble (GSE) of random matrices.

For these classical ensembles all correlation functions which determine statistical properties of eigenvalues E_n can be written explicitly (see e.g. [46], [16]). The simplest of them is the one-point correlation function or the mean level density, $\bar{d}(E)$, which is the probability density of finding a level in the interval $(E,\ E+\mathrm{d}E)$. When $\bar{d}(E)$ is known one can construct a new sequence of levels, e_n, called unfolded spectrum as follows

$$e_n = \int^{E_n} \bar{d}(E)\mathrm{d}E\ .$$

This artificially constructed sequence has automatically unit local mean density which signifies that the mean level density (provided it is a smooth function of E) plays a minor role in describing statistical properties of a spectrum at small intervals.

The two-point correlation function, $R_2(\epsilon)$, is the probability density of finding two levels separated by a distance in the interval $(\epsilon,\ \epsilon+\mathrm{d}\epsilon)$. The characteristic properties of the above ensembles is the phenomenon of level repulsion which manifest itself in the vanishing of the two-point correlation function at small values of argument

$$R_2(\epsilon) \xrightarrow{\epsilon\to 0} \epsilon^{\beta}$$

where the parameter $\beta = 1, 2$, and 4 for, respectively, GOE, GUE, and GSE. This behaviour is in contrast with the case of the Poisson statistics of independent random variables where

$$R_2(\epsilon) \xrightarrow{\epsilon\to 0} \bar{d}(E) \neq 0\ .$$

For later use we present the explicit form of the two-point correlation function for GUE with mean density $\bar{d}$

$$\tilde{R}_2(\epsilon) = \bar{d}^2 + \bar{d}\delta(\epsilon) + \bar{R}_2(\epsilon) + R_2^{(osc)}(\epsilon) \tag{40}$$

where the smooth part of the connected two-point correlation function is given by

$$\bar{R}_2(\epsilon) = -\frac{1}{2\pi^2\epsilon^2} \tag{41}$$

and its oscillating part is

$$R_2^{(osc)}(\epsilon) = \frac{\mathrm{e}^{2\pi i \bar{d}\epsilon} + \mathrm{e}^{-2\pi i \bar{d}\epsilon}}{4\pi^2\epsilon^2} \ . \tag{42}$$

The term $\bar{d}\delta(\epsilon)$ in (40) corresponds to taking into account two identical levels and it is universal for all systems without spectral degeneracy. It is a matter of convenience to include it to $R_2(\epsilon)$ or not. When one adopts the definition (45) the appearance of such terms is inevitable.

Another useful quantity is the two-point correlation form factor defined as the Fourier transform of the two-point correlation function (unfolded to the unit density)

$$K(t) = \int_{-\infty}^{\infty} R_2(x)\mathrm{e}^{2\pi i t x} dx \ . \tag{43}$$

For convenience one introduces a factor 2π in the definition of time.

In Fig. 8 the two-point correlation form factors for usual random matrix ensembles are presented. Their explicit formulas can be found in [46], [16]. For these classical ensembles small-t behaviour of the form factors is

$$K(t) \xrightarrow{t\to 0} \frac{2}{\beta} t \tag{44}$$

with the same β as above.

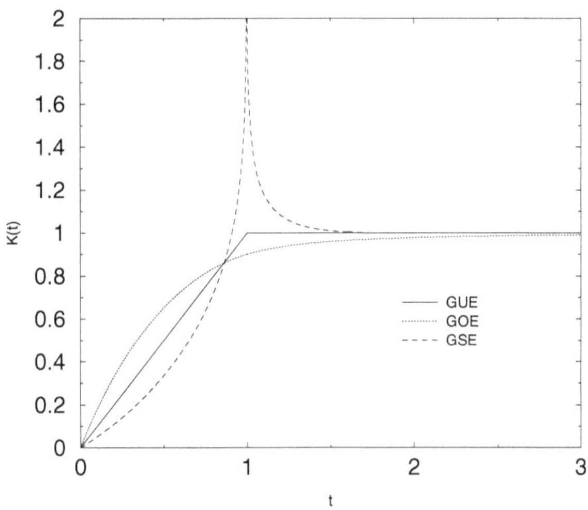

Fig. 8. Two point correlation form factor of classical random matrix ensembles

The nearest-neighbor distribution, $p(s)$, is defined as the probability density of finding two levels separated by distance s but, contrary to the two-point correlation function, no levels inside this interval are allowed. For classical ensembles the nearest-neighbor distributions can be expressed through solutions of certain integral equations and numerically they are close to the Wigner surmise (see e.g. [16])

$$p(s) = as^{\beta}e^{-bs^2}$$

where β is the same as above and constants a and b are determined from normalization conditions

$$\int_0^{\infty} p(s)\mathrm{d}s = \int_0^{\infty} sp(s)\mathrm{d}s = 1\ .$$

These functions are presented at Fig. 9 together with the Poisson prediction for this quantity $p(s) = \mathrm{e}^{-s}$.

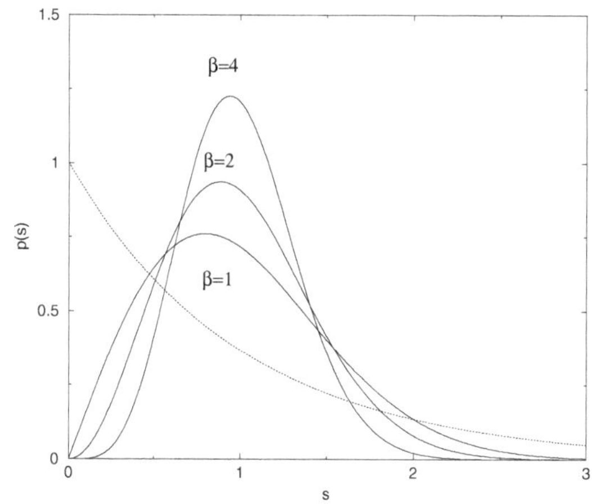

Fig. 9. Nearest-neighbor distribution for the standard random matrix ensembles. Dotted line – the Poisson prediction

Though random matrix ensembles were first introduced to describe spectral statistics of heavy nuclei later it was understood that the same conjectures can be applied also for simple dynamical systems and today's standard accepted conjectures are the following

- The energy levels of classically integrable systems on the scale of the mean level density behave as independent random variables and their distribution is close to the Poisson distribution [10].
- The energy levels of classically chaotic systems are not independent but on the scale of the mean level density they are distributed as eigenvalues of random matrix ensembles depending only on symmetry properties of the system considered [17].

- For systems without time-reversal invariance the distribution of energy levels should be close to the distribution of the Gaussian Unitary Ensemble (GUE) characterized by quadratic level repulsion.
- For systems with time-reversal invariance the corresponding distribution should be close to that of the Gaussian Orthogonal Ensemble (GOE) with linear level repulsion.
- For systems with time-reversal invariance but with half-integer spin energy levels should be described according to the Gaussian Symplectic Ensemble (GSE) of random matrices with quartic level repulsion.

These conjectures are well confirmed by numerical calculations.

The purpose of this Chapter is to investigate methods which permit to obtain spectral statistics analytically. For a large part of the Section we follow [25]. In Sect. 1 a formal expression is obtained which relates correlation functions with products of trace formulas. In Sect. 1.1 the simplest approximation to compute such products is discussed. It is called the diagonal approximation and it consists of taking into account only terms with exactly the same actions. Unfortunately, for chaotic systems this approximation can be used, strictly speaking, only for very small time estimated in Sect. 1.2. To understand the behaviour of the correlation functions for longer time more complicated methods of calculation of non-diagonal terms have to be developed. In Sect. 2 this goal is achieved for the Riemann zeta function. To obtain the information about correlations of prime pairs we use the Hardy–Littlewood conjecture which is reviewed in Sect. 2.1. The explicit form of the two-point correlation function for the Riemann zeros is obtained in Sect. 2.2. In Sect. 3 it is demonstrated that the obtained expression very well agrees with numerical calculations of spectral statistics for Riemann zeros.

1 Correlation Functions

Formally n-point correlation functions of energy levels are defined as the probability density of having n energy levels at given positions. Because the density of states, $d(E)$, is the probability density of finding one level at point E, correlation functions are connected to the density of states as follows

$$R_n(\epsilon_1, \epsilon_2, \ldots, \epsilon_n) = \langle d(E+\epsilon_1) d(E+\epsilon_2) \ldots d(E+\epsilon_n) \rangle \ . \tag{45}$$

The brackets $\langle \ldots \rangle$ denote a smoothing over an appropriate energy window

$$\langle f(E) \rangle = \int f(E') \sigma(E-E') \mathrm{d}E'$$

with a certain function $\sigma(E)$. Such smoothing means that one considers eigenvalues of quantum dynamical systems at different intervals of energy as forming a statistical ensemble.

The function $\sigma(E)$ is assumed to fulfill the normalization condition

$$\int \sigma(E)dE = 1$$

and to be centered around zero with a width ΔE obeying inequalities

$$\Delta E_q \ll \Delta E \ll \Delta E_{cl} \ll E\,. \qquad (46)$$

Here ΔE_q has to be of the order of the mean level spacing, $\Delta E_q \approx 1/\bar{d}$, and ΔE_{cl} denotes the energy scale at which classical dynamics changes noticeably. A schematic picture of $\sigma(E)$ is represented at Fig. 10.

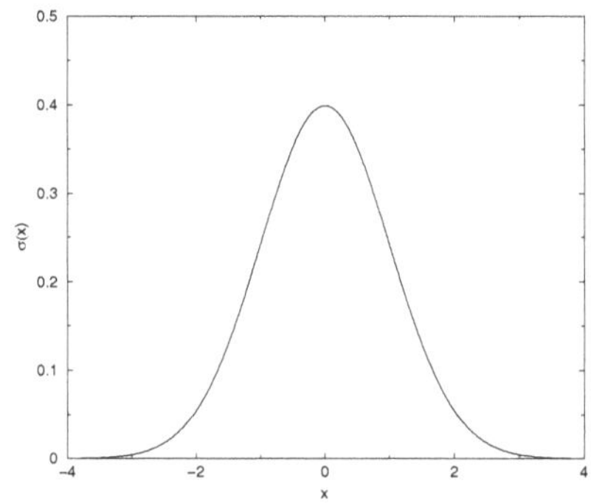

Fig. 10. Schematic form of smoothing function

The trace formula for the density of states of chaotic systems was discussed in Chapter I and it has the form

$$d(E) = \bar{d}(E) + \sum_{p,n} A_{p,n} \mathrm{e}^{\mathrm{i}nS_p(E)/\hbar} + \text{c.c.}$$

where the summation is performed over all primitive periodic orbits and its repetitions, and

$$A_{p,n} = \frac{T_p}{2\pi\hbar|\det(M_p^n - 1)|^{1/2}} \mathrm{e}^{-\pi\mathrm{i}n\mu_p/2}\,. \qquad (47)$$

Substituting this expression in the formula for the two-point correlation function one gets

$$R_2(\epsilon_1, \epsilon_2) = \bar{d}^2 + \\ + \sum_{p_i, n_i} A_{p_1,n_1} A^*_{p_2,n_2} \left\langle \exp \frac{\mathrm{i}}{\hbar}(n_1 S_{p_1}(E+\epsilon_1) - n_2 S_{p_2}(E+\epsilon_2)) \right\rangle + \text{c.c.}$$

and the terms with the sum of actions are assumed to be washed out by the smoothing procedure.

Expanding the actions and taking into account that $\partial S(E)/\partial E = T(E)$ where $T(E)$ is the classical period of motion one finds

$$R_2^{(c)}(\epsilon_1,\epsilon_2) = \sum_{p_i,n_i} A_{p_1,n_1}A^*_{p_2,n_2}\left\langle \exp\frac{\mathrm{i}}{\hbar}(n_1 S_{p_1}(E) - n_2 S_{p_2}(E))\right\rangle$$
$$\times \exp\frac{\mathrm{i}}{\hbar}(n_1 T_{p_1}(E)\epsilon_1 - n_2 T_{p_2}(E)\epsilon_2) + \mathrm{c.c.}\,.$$

Here $R_2^{(c)}(\epsilon_1,\epsilon_2)$ is the connected part of the two-point correlation function $R_2(\epsilon_1,\epsilon_2) = \bar{d}^2 + R_2^{(c)}(\epsilon_1,\epsilon_2)$.

The most difficult part is the computation of the mean value of terms with the difference of actions

$$\left\langle \exp\frac{\mathrm{i}}{\hbar}(n_1 S_{p_1}(E) - n_2 S_{p_2}(E))\right\rangle\,.$$

1.1 Diagonal Approximation

Berry [11] proposed to estimate such sums in an approximation (called the diagonal approximation) by taking into account only terms with *exactly* the same actions having in mind that terms with different values of actions will be small after the smoothing.

Let g be the mean multiplicity of periodic orbit actions. Then the connected part of the two-point correlation function in the diagonal approximation is

$$R_2^{(diag)}(\epsilon) = g\sum_{p,\;n\geq 1}|A_{p,n}|^2 \mathrm{e}^{\mathrm{i}nT_p(E)\epsilon/\hbar} + \mathrm{c.c.}\,. \tag{48}$$

Here $\epsilon = \epsilon_1 - \epsilon_2$ and the sum is taken over all primitive periodic orbits.

From (48) it follows that the two-point correlation form factor

$$K(t) = \int_{-\infty}^{+\infty} R_2(\epsilon)\mathrm{e}^{2\pi\mathrm{i}t\epsilon}\mathrm{d}\epsilon.$$

in the diagonal approximation equals the following sum over classical periodic orbits

$$K^{(diag)}(t) = 2\pi g\sum_{p,n}|A_{p,n}|^2\delta\left(2\pi t - \frac{nT_p(E)}{\hbar}\right) + \mathrm{c.c.}\,. \tag{49}$$

According to the Hannay-Ozorio de Almeida sum rule [34] sums over periodic orbits of a chaotic systems can be calculated by using the local density of periodic orbits related with the monodromy matrix, M_p, as follows

$$\mathrm{d}\rho_p = \frac{\mathrm{d}T_p}{T_p}|\det(M_p - 1)|\,.$$

Using (47) one gets

$$K^{(diag)}(t) = \frac{g}{2\pi\hbar}\int T_p\delta(2\pi t - \frac{T_p}{\hbar})dT_p = gt$$

where g is the mean multiplicity of periodic orbits (i.e. the mean proportion of periodic orbits with exactly the same action). For generic systems without time-reversal invariance there is no reasons for equality of actions for different periodic orbits and $g = 1$ but for systems with time-reversal invariance each orbit can be traversed in two directions therefore in general for such systems $g = 2$. Comparing these expressions one concludes that the diagonal approximation reproduces the correct small-t behavior of form-factors of classical ensembles (cf. (44)).

Unfortunately, $K^{(diag)}(t)$ grows with increasing t but the exact form-factor for systems without spectral degeneracy should tend to $\bar{d}$ for large t. This is a consequence of the following arguments. According to (45)

$$R_2(\epsilon) = \left\langle \sum_{m,n}\delta(E-E_n)\delta(E+\epsilon-E_m)\right\rangle =$$

$$= \left\langle \sum_{m,n}\delta(E-E_n)\delta(\epsilon-E_m+E_n)\right\rangle .$$

If there is no levels with exactly the same energy the second δ-function in the right hand side of this equation tends to $\delta(\epsilon)$ when $\epsilon \to 0$ and the first one gives $\bar{d}$. Therefore

$$R_2(\epsilon) \to \bar{d}\delta(\epsilon) , \quad \text{when } \epsilon \to 0$$

which is equivalent to the following asymptotics of the form factor

$$K(t) \to \bar{d} , \quad \text{when } t \to \infty .$$

This evident contradiction clearly indicates that the diagonal approximation for chaotic systems cannot be correct for all values of t and more complicated tools are needed to obtain the full form factor.

1.2 Criterion of Applicability of Diagonal Approximation

One can give a (pessimistic) estimate till what time the diagonal approximation can be valid by the following method. The main ingredient of the diagonal approximation is the assumption that after smoothing all off-diagonal terms give negligible contribution. This condition is almost the same as the condition of the absence of quantum interference. But it is known that the quantum interference is not important for times smaller than the Ehrenfest time which is of the order of

$$t_E \approx \frac{1}{\lambda_0}\ln(1/\hbar),$$

where λ_0 is a (classical) constant of the order of the Lyapunov exponent defined in such a way that the mean splitting of two nearby trajectories at time t grows as $\exp(\lambda_0 t)$. For billiards $(ka)^{-1}$, where a is of the order of system size, plays the role of $\hbar$ and $\lambda_0 = k\lambda$ where k is the momentum and λ determines the deviation of two trajectories with length $L = kt$. The constant λ which we also called the Lyapunov exponent is independent on k for billiards and

$$t_E \approx \frac{1}{\lambda k} \ln(ka) \;.$$

In the semiclassical limit $k \to \infty$ the Ehrenfest time and, consequently, the time during which one can use the diagonal approximation tends to zero as $\ln k/k$.

More careful argumentation can be done as follows. The off-diagonal terms can be neglected if

$$\left| \left\langle \exp \frac{\mathrm{i}}{\hbar}(S_{p_1}(E) - S_{p_2}(E)) \right\rangle \right| \ll 1 \;.$$

But this quantity is small provided the difference of periods of two orbits $\Delta T = T_{p1} - T_{p_2}$ times the energy window ΔE used in the definition of smoothing procedure is large

$$\frac{1}{\hbar}(T_{p1} - T_{p_2})\Delta E \gg 1 \;. \tag{50}$$

For billiards $T_p = L_p/k$ and this condition means that one has to consider all periodic orbits such that their difference of lengths is

$$L_{p1} - L_{p_2} \gg \frac{\hbar k}{\Delta E} \;.$$

But the number of periodic orbits with length L for chaotic systems grows exponentially

$$N(L_p < L) = \frac{e^{\lambda L}}{\lambda L}$$

where λ is a constant of the order of the Lyapunov exponent. Therefore in the interval $L, L + \delta l$ there is $e^{\lambda L} \delta l / L$ orbits and the mean difference of lengths between orbits with lengths less than L is of the order of

$$\Delta L = L \exp(-\lambda L) \;.$$

To fulfill the above condition one has to restrict the maximum length of periodic orbits, L_m, by

$$L_m \exp(-\lambda L_m) \approx \frac{k\hbar}{\Delta E} \;.$$

In the limit of large L_m with logarithmic accuracy this relation gives

$$L_m \approx \frac{1}{\lambda} \ln \frac{\Delta E}{k\hbar\lambda} \tag{51}$$

which corresponds to the maximal time till the diagonal approximation can be applied

$$t_m = \frac{L_m}{k} \sim \frac{1}{\lambda k} \ln \frac{\Delta E}{\lambda k} \, .$$

As $\Delta E \ll E = k^2$, $t_m < t_E$.

Another important time scale for bounded quantum systems is called the Heisenberg time, t_H. It is the time during which one can see the discreteness of the spectrum

$$t_H = 2\pi \bar{d} \, .$$

As for billiards $\bar{d}$ is a constant

$$t_E \ll t_H \, .$$

For the Riemann zeta function the situation is better because (i) in this case 'momentum' plays the role of 'energy' (the 'action' $E \ln p$ is linear in E and not proportional to $\sqrt{E}$ as for dynamical systems) and (ii) the density of states for the Riemann zeta function is $(\ln(E/2\pi))/(2\pi)$.

The analog of (50) in this case is

$$(\ln p_1 - \ln p_2) \Delta E \gg 1 \, .$$

It means that to apply the diagonal approximation prime numbers have to be such that the difference between any two of them obeys

$$\frac{\delta p}{p} \Delta E \gg 1.$$

The difference between primes near p is of the order of $\ln p$. Hence from the above inequalities it follows that diagonal approximation can be used till time $t_m = \ln p_m$ where p_m is such that

$$\frac{\ln p_m}{p_m} \geq \frac{1}{\Delta E} \, .$$

Or with logarithmic precision $p_m \leq \Delta E$. As $\Delta E \leq E$ (see (46)), $p_m \sim E$ and the maximum time

$$t_m \sim \ln E = 2\pi \bar{d}(E)$$

i.e. the diagonal approximation for the Riemann zeta function is valid till the Heisenberg time which agrees with the Montgomery theorem [45].

This type of estimates clearly indicates that the diagonal approximation for chaotic dynamical systems can not, strictly speaking, be used to obtain an information about the form-factor for large value of t. Only the short-time behaviour of correlation functions can be calculated by this method. (Notice that for GUE systems the diagonal approximation gives the expected answer till the Heisenberg time but it just signifies that one has to find special reasons why all other terms cancel.)

2 Beyond the Diagonal Approximation

The simplest and the most natural way of semi-classical computation of the two-point correlation functions is to find a method of calculating off-diagonal terms. We shall discuss here this type of computation on the example of the Riemann zeta function where much more information than for dynamical systems is available (for the latter see [21] and [25]).

The trace formula for the Riemann zeta function may be rewritten in the form

$$d^{(osc)}(E) = -\frac{1}{\pi}\sum_{n=1}^{\infty}\frac{1}{\sqrt{n}}\Lambda(n)\cos(E\ln n)$$

where

$$\Lambda(n) = \begin{cases} \ln p, & \text{if } n = p^k \\ 0, & \text{otherwise} \end{cases} .$$

The connected two-point correlation function of the Riemann zeros, $R_2^{(c)} = R_2 - \bar{d}^2$, is

$$R_2^{(c)}(\epsilon_1,\epsilon_2) = \frac{1}{4\pi^2}\sum_{n_1,n_2}\frac{\Lambda(n_1)\Lambda(n_2)}{\sqrt{n_1 n_2}}\left\langle e^{i(E+\epsilon_1)\ln n_1 - i(E+\epsilon_2)\ln n_2}\right\rangle + \text{c.c.} .$$

The diagonal approximation corresponds to taking into account terms with $n_1 = n_2$

$$R_2^{(diag)}(\epsilon_1,\epsilon_2) = \frac{1}{4\pi^2}\sum_{n}\frac{\Lambda^2(n)}{n}e^{i(\epsilon_1-\epsilon_2)\ln n} + \text{c.c.} =$$

$$= \frac{1}{4\pi^2}\sum_{p,m}\frac{\ln^2 p}{p^m}e^{i(\epsilon_1-\epsilon_2)m\ln p} + \text{c.c.} .$$

This expression may be transformed as follows (cf. [2])

$$R_2^{(diag)}(\epsilon) = -\frac{1}{4\pi^2}\frac{\partial^2}{\partial\epsilon^2}\ln\Delta(\epsilon)$$

where

$$\Delta(\epsilon) = |\zeta(1+i\epsilon)|^2\Phi^{(diag)}(\epsilon) ,$$

and function $\Phi^{(diag)}(\epsilon)$ is given by a convergent sum over prime numbers

$$\Phi^{(diag)}(\epsilon) = \exp\left(2\sum_{p}\sum_{m=1}^{\infty}\frac{1-m}{m^2 p^m}\cos(m\epsilon\ln p)\right) .$$

In the limit $\epsilon \to 0$, $\zeta(1+i\epsilon) \to (i\epsilon)^{-1}$ and $\Phi^{(diag)}(\epsilon) \to \text{const}$. Therefore in this limit

$$R_2^{(diag)}(\epsilon) \to -\frac{1}{2\pi^2\epsilon^2}$$

which agrees with the smooth part of the GUE result (41).

The off-diagonal contribution takes the form

$$R_2^{(off)}(\epsilon_1,\epsilon_2) = \sum_{n_1\neq n_2} \frac{\Lambda(n_1)\Lambda(n_2)}{4\pi^2\sqrt{n_1 n_2}} \left\langle \mathrm{e}^{\mathrm{i}E\ln(n_1/n_2)+\mathrm{i}(\epsilon_1\ln n_1-\epsilon_2\ln n_2)} \right\rangle + \text{c.c.}\,.$$

The term $\exp(\mathrm{i}E\ln(n_1/n_2))$ oscillates quickly if n_1 is not close to n_2. Denoting

$$n_1 = n_2 + r$$

and expanding all smooth functions on r one gets

$$R_2^{(off)}(\epsilon) = \frac{1}{4\pi^2}\sum_{n,r} \frac{\Lambda(n)\Lambda(n+r)}{n} \left\langle \mathrm{e}^{\mathrm{i}Er/n+\mathrm{i}\epsilon\ln n} \right\rangle + \text{c.c.}$$

where $\epsilon = \epsilon_1 - \epsilon_2$.

The main problem is clearly seen here. The function

$$F(n,r) = \Lambda(n)\Lambda(n+r)$$

is quite a wild function as it is nonzero only when both n and $n+r$ are powers of prime numbers. As we have assumed that $r \ll n$, the dominant contribution to the two-point correlation function will come from the mean value of this function over all n, i.e. one has to substitute into $R_2^{(off)}(\epsilon)$ instead of $F(n,r)$ its mean value

$$\alpha(r) = \lim_{N\to\infty}\frac{1}{N}\sum_{n=1}^{N}\Lambda(n)\Lambda(n+r)\,.$$

2.1 The Hardy–Littlewood Conjecture

Fortunately the explicit expression for this function comes from the famous Hardy–Littlewood conjecture. There are two different methods which permit to 'find' this conjecture. We start with the original Hardy-Littlewood derivation [35].

First, let us recall two known facts. The number of prime numbers less that a given number $N(p<x)$ is asymptotically (see e.g. [55])

$$N(p<x) = \frac{x}{\ln x}\,.$$

Conveniently it can also be expressed in the following form

$$\lim_{N\to\infty}\frac{1}{N}\sum_{n=1}^{N}\Lambda(n) = 1\,.$$

The number of prime number $N_{q,r}(p<x)$ in arithmetic progression of the form $mq+r$ with $(r,q)=1$ and $r<q$ is given by the following asymptotic formula (see e.g. [31])

$$N_{q,r}(p<x)=\frac{x}{\varphi(q)\ln x}$$

where $\varphi(n)$ is the Euler function which counts integers less than n and co-prime with n

$$\varphi(n)=n\prod_{p|n}\left(1-\frac{1}{p}\right).$$

As above, this relation can be rewritten in the equivalent form

$$\lim_{N\to\infty}\frac{1}{N}\sum_{m=1}^{N}\Lambda(mq+r)=\frac{1}{\varphi(q)}. \tag{52}$$

In the Hardy-Littlewood method [35] one introduces the function

$$f(x)=\sum_{n=1}^{\infty}\Lambda(n)x^n$$

which converges for all complex x such that $|x|<1$.

In the circle method of Hardy and Littlewood [35] one considers the behaviour of this function close to the unit circle when the phase of x is near a rational number $2\pi p/q$ with co-prime integers p and q. One gets

$$f(\mathrm{e}^{-u}\mathrm{e}^{2\pi \mathrm{i}p/q+\mathrm{i}\delta})=\sum_{n=1}^{\infty}\Lambda(n)\mathrm{e}^{-nu}\mathrm{e}^{2\pi \mathrm{i}np/q+\mathrm{i}n\delta}$$

with $u,\delta\to 0$.

In the exponent there is a quickly changing function $2\pi np/q$. It is quite natural to consider n from the arithmetic progression

$$n=mq+r$$

with fixed q and $r<q$. In this case

$$f(\mathrm{e}^{-u}\mathrm{e}^{2\pi \mathrm{i}p/q+\mathrm{i}\delta})=\sum_{m,\ r}\Lambda(mq+r)\mathrm{e}^{-(mq+r)(u-\mathrm{i}\delta)}\mathrm{e}^{2\pi \mathrm{i}rp/q}.$$

Substituting instead of $\Lambda(mq+r)$ its mean value (52) one gets

$$f(\mathrm{e}^{-u}\mathrm{e}^{\mathrm{i}2\pi p/q+\mathrm{i}\delta})\approx\frac{1}{\varphi(q)}\sum_{(r,\ q)=1}\mathrm{e}^{2\pi \mathrm{i}rp/q}\int_0^{\infty}\mathrm{e}^{-n(u-\mathrm{i}\delta)}\mathrm{d}n=\frac{\mu(q)}{\varphi(q)(u-\mathrm{i}\delta)}.$$

In the last step we use that fact that [36]

$$\sum_{(r,\ q)=1} \mathrm{e}^{2\pi \mathrm{i} r/q} = \mu(q)$$

where $\mu(q)$ is the Möbius function defined through the factorization of q on prime factors

$$\mu(q) = \begin{cases} 1 & \text{if } q=1 \\ (-1)^k & \text{if } q = p_1 \dots p_k \\ 0 & \text{if } q \text{ is divisible on } p^2 \end{cases} .$$

The final expression means that function $f(x)$ has a pole singularity at the unit circle at every rational point.

The knowledge of $f(x)$ permits formally to compute the mean value of the product of two Λ-functions.

Let

$$J_r(R) = \frac{1}{2\pi} \int_0^{2\pi} f(R\mathrm{e}^{\mathrm{i}\varphi}) f(R\mathrm{e}^{-\mathrm{i}\varphi}) \mathrm{e}^{-\mathrm{i}r\varphi} \mathrm{d}\varphi = R^r \sum_m \Lambda(m+r)\Lambda(m) R^{2m} .$$

As the function $f(x)$ has a pole singularity at the unit circle at every rational point one can try to approximate this integral by the sum over singularities

$$\begin{aligned} J_r(\mathrm{e}^{-u}) &= \frac{1}{2\pi} \int_0^{2\pi} f(R\mathrm{e}^{\mathrm{i}\varphi}) f(R\mathrm{e}^{-\mathrm{i}\varphi}) \mathrm{e}^{-\mathrm{i}r\varphi} \mathrm{d}\varphi = \\ &= \frac{1}{2\pi} \sum_{(p,\ q)=1} \int f(\mathrm{e}^{-u+\mathrm{i}2\pi p/q+\mathrm{i}\delta}) f(\mathrm{e}^{-u-2\pi \mathrm{i} p/q-\mathrm{i}\delta}) \mathrm{e}^{-\mathrm{i}r(2\pi p/q+\mathrm{i}\delta)} \mathrm{d}\delta = \\ &= \frac{1}{2\pi} \sum_{(p,\ q)=1} \mathrm{e}^{2\pi \mathrm{i} r p/q} \left(\frac{\mu(q)}{\varphi(q)} \right)^2 \int \frac{\mathrm{d}\delta}{u^2+\delta^2} = \\ &= \frac{1}{2u} \sum_{(p,\ q)=1} \mathrm{e}^{2\pi \mathrm{i} p r/q} \left(\frac{\mu(q)}{\varphi(q)} \right)^2 . \end{aligned}$$

Therefore

$$\sum_{n=1}^{\infty} \Lambda(n)\Lambda(n+r) \mathrm{e}^{-2nu} \xrightarrow{u \to 0} \frac{1}{2u} \sum_{(p,\ q)=1} \mathrm{e}^{2\pi \mathrm{i} r p/q} \left(\frac{\mu(q)}{\varphi(q)} \right)^2$$

from which it follows that

$$\lim_{N\to\infty} \frac{1}{N} \sum_{n=1}^{N} \Lambda(n)\Lambda(n+r) = \alpha(r)$$

where

$$\alpha(r) = \sum_{q=1}^{\infty} \left(\frac{\mu(q)}{\varphi(q)} \right)^2 \sum_{(p,\ q)=1} e^{2\pi \mathrm{i} r p/q} . \tag{53}$$

Using properties of such singular series one can prove [35] that for even r $\alpha(r) = 0$ and for odd r it can be represented as the following product over prime numbers

$$\alpha(r) = C_2 \prod_{p|r} \frac{p-1}{p-2} \tag{54}$$

where the product is taken over all prime divisors of r bigger than 2 and C_2 is the so-called twin prime constant

$$C_2 = 2 \prod_{p>2} (1 - \frac{1}{(p-1)^2}) \approx 1.32032\ldots . \tag{55}$$

Instead of demonstrating the formal equivalence of (53) and (54) we present another heuristic 'derivation' based on the probabilistic interpretation of prime numbers which gives directly (54) and (55).

The argumentation consists on the following steps.

- Probability that a given number is divisible by a prime p is

$$\lim_{N\to\infty} \frac{1}{N} [\text{number of integers divisible by } p \le N] = \frac{1}{p} .$$

  In general to find such probabilities it is necessary to consider only the residues modulo p and find how many of them obey the requirement.
- Probability that a given number is not divisible by a prime p is

$$1 - \frac{1}{p} .$$

- Probability that a number is not divisible by primes $p_1, p_2, \ldots p_k$ is

$$\prod_{j=1}^{k} (1 - \frac{1}{p_j}) . \tag{56}$$

The above formula is correct for any finite collection of primes but for computations with infinite number of primes it may be wrong.

For example, when used naively it gives that

- probability that a number x is a prime is

$$\prod_{p<\sqrt{x}} (1 - \frac{1}{p}) .$$

This prime number 'theorem' is false because from it it follows that the number of primes less than x is [55]

$$\Pi(x) = x \prod_{p<\sqrt{x}} (1 - \frac{1}{p}) \stackrel{x\to\infty}{\longrightarrow} \frac{x}{\ln x} 2\mathrm{e}^{-\gamma}$$

which differs from the true prime number theorem by a factor $2e^{-\gamma} \approx 1.123$ where γ is the Euler constant. The origin of this discrepancy is related with the approximation frequently used above: $[x/p] = x/p$ where $[x]$ is the integer part of x. Instead of (56) one should have $\prod_p(1 - [x/p]/x)$. For a finite number of primes and $x \to \infty$ it tends to (56). But when the number of primes considered increases with x errors are accumulated giving a constant factor.

Nevertheless one could try to use probabilistic arguments by forming artificially convergent quantities. One has

- Probability that x and $x + r$ are primes is

$$\lim_{N\to\infty} \frac{1}{N} [\text{ number of integers } x < N \text{ such that } x \text{ and } x+r \text{ are primes }] .$$

Let us consider a prime p. Two cases are possible. Either $p\,|r$ or $p\not|r$. In the first case the probability that both number x and $x + r$ are not divisible by p is the same as the probability that only number x is not divisible by p which is

$$\prod_{p|r} \left(1 - \frac{1}{p}\right) .$$

When $p\not|r$ one has to remove two numbers from the set of residues as $x = 0, 1, .., p-1 \pmod p$ and $x+r = 0, 1, .., p-1 \pmod p$. Therefore the probability that both numbers x and $x + r$ are not divisible by a prime p is

$$\prod_{p\not|r} \left(1 - \frac{2}{p}\right) .$$

Finally

- Probability that both x and $x + r$ are primes $= \prod_{p|r} \frac{p-1}{p} \prod_{p\not|r} \frac{p-2}{p}$.

To find a convergent expression we divide both sides by the probability that numbers x and $x + r$ are independently prime numbers computed also in the probabilistic approximation. The latter quantity is

$$[\text{ Probability that } x \text{ is prime and } x \le N\ ] = \prod_p \frac{p-1}{p} .$$

Therefore

$$\frac{[\text{ Probability that both } x \text{ and } x+r \text{ are primes with } x, x+r \le N\ ]}{[\text{ Probability that } x \text{ is prime }]^2} \approx$$

$$\approx \prod_{p|r} \frac{p-1}{p} \prod_{p\not|r} \frac{p-2}{p} \prod_p (\frac{p}{p-1})^2 = 2 \prod_{p>2} \left(1 - \frac{1}{(p-1)^2}\right) \prod_{p|r} \frac{p-1}{p-2} .$$

As the denominator in the above expression is $1/\ln^2 N$ it follows that the probability that both x and $x + r$ are primes with $x \le N$, and $x + r \le N$ is asymptotically

$$\frac{\alpha(r)}{\ln^2 N}$$

with the same function $\alpha(r)$ as in (54).

We stress that the Hardy–Littlewood conjecture is still not proved. Even the existence of infinite number of twin primes (primes separated by 2) is not yet proved while the Hardy–Littlewood conjecture states that their density is $C_2/\ln^2 N$.

2.2 Two-Point Correlation Function of Riemann Zeros

Taking the above expression of the Hardy–Littlewood conjecture as granted we get

$$R_2^{(off)}(\epsilon) = \frac{1}{4\pi^2}\sum_{n\geq 1}\frac{1}{n}\mathrm{e}^{\mathrm{i}\epsilon \ln n}\sum_r \alpha(r)\mathrm{e}^{\mathrm{i}Er/n} + \text{c.c.}\,.$$

After substitution the formula for $\alpha(r)$ and performing the sum over all r one obtains

$$R_2^{(off)}(\epsilon) = \frac{1}{4\pi^2}\sum_n \frac{1}{n}\mathrm{e}^{\mathrm{i}\epsilon \ln n}\sum_{(p,q)=1}\left(\frac{\mu(q)}{\varphi(q)}\right)^2 \delta\left(\frac{p}{q}-\frac{E}{2\pi n}\right) + \text{c.c.}$$

where the summation is taken over all pairs of mutually co-prime positive integers p and q (without the restriction $p < q$).

Changing the summation over n to the integration permits to transform this expression to contributions of values of n where

$$\frac{p}{q} - \frac{E}{2\pi n} = 0\,.$$

In this approximation

$$R_2^{(off)}(\epsilon) = \frac{1}{4\pi^2}\mathrm{e}^{\mathrm{i}\epsilon \ln E/2\pi}\sum_{(p,q)=1}\left(\frac{\mu(q)}{\varphi(q)}\right)^2\left(\frac{q}{p}\right)^{1+\mathrm{i}\epsilon} + \text{c.c.}\,.$$

Using the formula (which is a mathematical expression of the inclusion–exclusion principle)

$$\sum_{(p,q)=1} f(p) = \sum_{k=1}^{\infty}\sum_{\delta | q} f(k\delta)\mu(\delta)$$

and taking into account that $2\pi\bar{d} = \ln(E/2\pi)$ one obtains

$$R_2^{(off)}(\epsilon) = \frac{1}{4\pi^2}|\zeta(1+\mathrm{i}\epsilon)|^2\mathrm{e}^{2\pi\mathrm{i}\bar{d}\epsilon}\Phi^{(off)}(\epsilon) + \text{c.c.} \tag{57}$$

where function $\Phi^{(off)}(\epsilon)$ is given by a convergent product over primes

$$\Phi^{(off)}(\epsilon) = \prod_p \left(1 - \frac{(1-p^{\mathrm{i}\epsilon})^2}{(p-1)^2}\right)$$

and $\Phi^{(off)}(0) = 1$.

In the limit of small ϵ

$$R_2^{(off)}(\epsilon) = \frac{1}{(2\pi\epsilon)^2}\left(\mathrm{e}^{2\pi\mathrm{i}\bar{d}\epsilon} + \mathrm{e}^{-2\pi\mathrm{i}\bar{d}\epsilon}\right)$$

which exactly corresponds to the GUE results for the oscillating part of the two-point correlation function (42).

The above calculations demonstrate how one can compute the two-point correlation function through the knowledge of correlation function of periodic orbit pairs. For the Riemann case one can prove under the same conjectures that all n-point correlation functions of Riemann zeros tend to corresponding GUE results [22].

3 Summary

Trace formulas can formally be used to calculate spectral correlation functions for dynamical systems. In particular, the two-point correlation function is the product of two densities of states

$$R_2(\epsilon) \equiv \langle d(E+\epsilon)d(E)\rangle \ .$$

The diagonal approximation consists of taking into account in such products only terms with exactly the same action. For chaotic systems this approximation is valid only for very small time. In particular, it permits to obtain the short-time behaviour of correlation form factors which agrees with predictions of standard random matrix ensembles.

The main difficulty in such approach to spectral statistics is the necessity to compute contributions from non-diagonal terms which requires the knowledge of correlation functions of periodic orbits with nearby actions.

For the Riemann zeta function zeros it can be done using the Hardy–Littlewood conjecture which claims that the number of prime pairs p and $p+r$ such that $p<N$ for large N is asymptotically

$$\alpha(r)\frac{N}{\ln^2 N}$$

where $\alpha(r)$ (with even r) is given by the product over all odd prime divisors of r

$$\alpha(r) = C_2 \prod_{p|r} \frac{p-1}{p-2}$$

and

$$C_2 = 2\prod_{p>2}\left(1-\frac{1}{(p-1)^2}\right) .$$

Using this formula one gets that the two-point correlation function of Riemann zeros is

$$R_2(\epsilon) = \bar{d}^2(E) + R_2^{(diag)}(\epsilon) + R_2^{(off)}(\epsilon)$$

where the diagonal part

$$R_2^{(diag)}(\epsilon) = -\frac{1}{4\pi^2}\frac{\partial^2}{\partial\epsilon^2}\ln\left[|\zeta(1+\mathrm{i}\epsilon)|^2\Phi^{(diag)}(\epsilon)\right]$$

and non-diagonal part

$$R_2^{(off)}(\epsilon) = \frac{1}{4\pi^2}|\zeta(1+\mathrm{i}\epsilon)|^2\mathrm{e}^{2\pi\mathrm{i}\bar{d}\epsilon}\Phi^{(off)}(\epsilon) + \text{c.c.} .$$

The functions $\Phi^{(diag)}(\epsilon)$ and $\Phi^{(off)}(\epsilon)$ are given by convergent products over all primes

$$\Phi^{(diag)}(\epsilon) = \exp\left(2\sum_p\sum_{m=1}^{\infty}\frac{1-m}{m^2p^m}\cos(m\epsilon\ln p)\right)$$

and

$$\Phi^{(off)}(\epsilon) = \prod_p\left(1-\frac{(1-p^{\mathrm{i}\epsilon})^2}{(p-1)^2}\right) .$$

In [25] a few other methods were developed to 'obtain' the two-point correlation function for Riemann zeros. These methods were based on different ideas and certain of them can be generalized for dynamical systems. Though neither of the methods can be considered as a strict mathematical proof, all lead to the same expression (57).

It is also of interest to check numerically the above formulas. When numerical calculations are performed one considers usually correlation functions for the unfolded spectrum. For the two-point correlation function this procedure corresponds to the following transformation

$$R_2^{(\text{unfolded})}(\epsilon) = \frac{1}{\bar{d}^2(E)}R_2\left(\frac{\epsilon}{\bar{d}(E)}\right) .$$

At Fig. 11 we present the two-point correlation function for $2\cdot 10^8$ zeros near the 10^{23}-th zero computed numerically by Odlyzko [49] together with the GUE prediction for this quantity

$$R_2^{\text{GUE}}(\epsilon) = 1-\left(\frac{\sin\pi\epsilon}{\pi\epsilon}\right)^2 .$$

At Figs. 12-15 we present the difference between the two-point correlation function computed numerically and the GUE prediction. At Fig. 16 we present

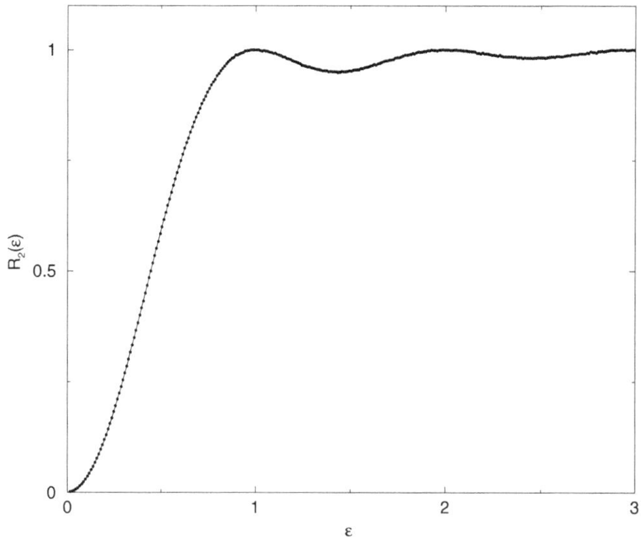

Fig. 11. Two point correlation function of the Riemann zeros near the 10^{23}-th zero (dots) and the GUE prediction (solid line)

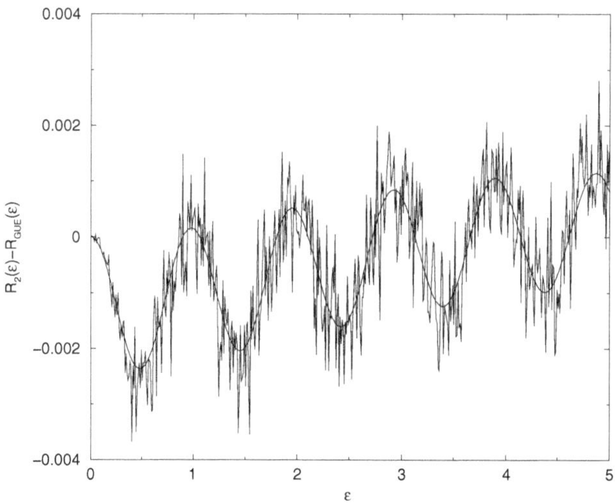

Fig. 12. The difference between the two point correlation function of the Riemann zeros and the GUE prediction in the interval $0 < \epsilon < 5$. The solid line is the difference between the 'exact' correlation function and the GUE prediction in this interval

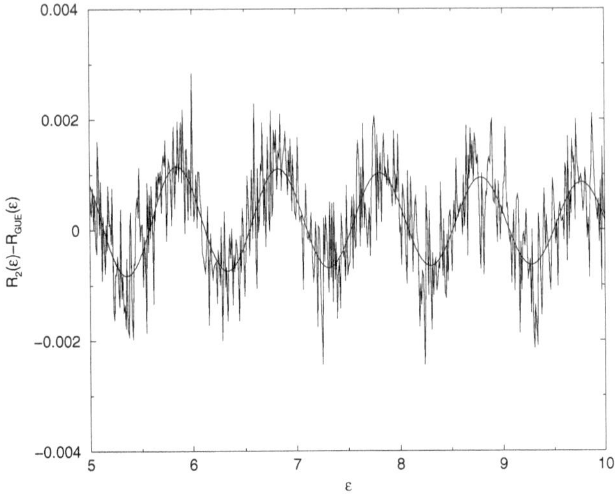

Fig. 13. The same as at Fig. 12 but in the interval $5 < \epsilon < 10$

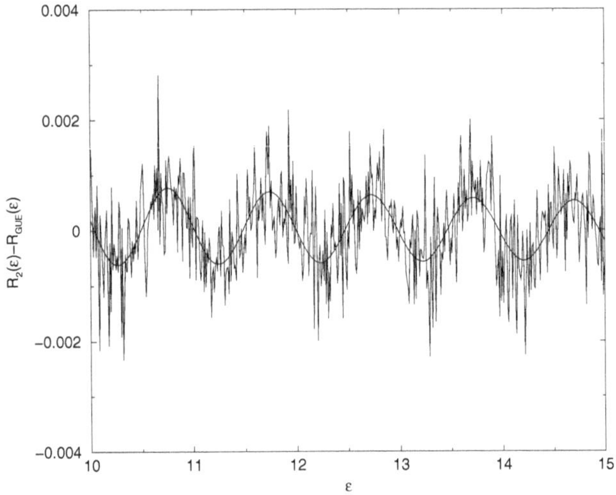

Fig. 14. The same as at Fig. 12 but in the interval $10 < \epsilon < 15$

the difference between numerically computed two-point correlation function and the 'exact' function and at Fig. 17 the histogram of differences is given. Notice that these differences are structure less and the histogram corresponds practically exactly to statistical errors inherent in the calculation of the two-point correlation functions which signifies that the obtained formula agrees very well with the numerics.

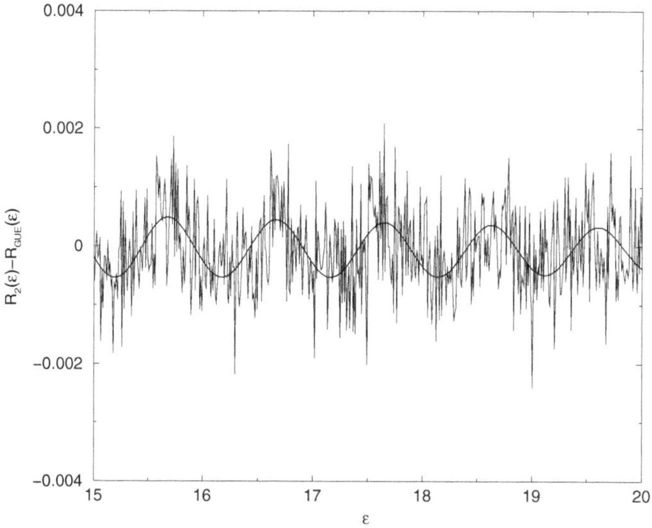

Fig. 15. The same as at Fig. 12 but in the interval $15 < \epsilon < 20$

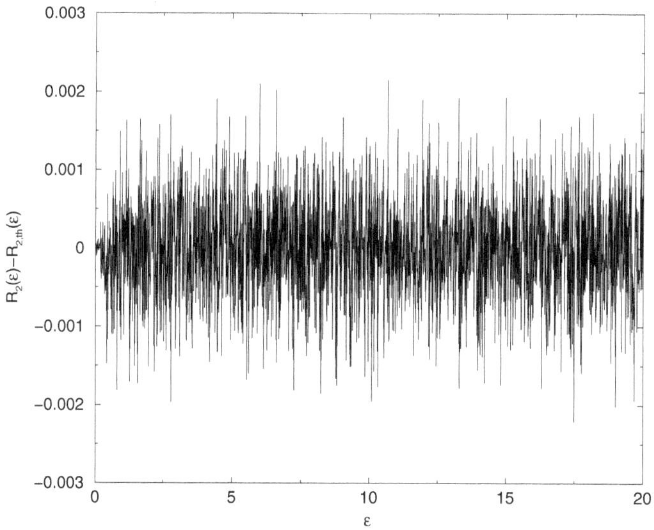

Fig. 16. The difference between numerically computed two-point correlation function and the 'exact' function

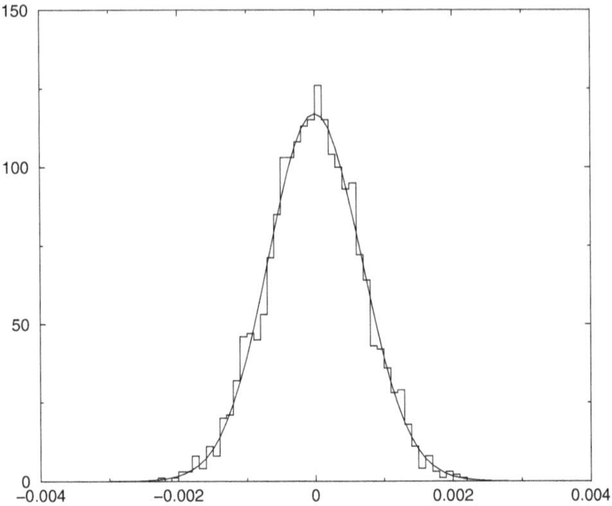

Fig. 17. The histogram of the deviations of the numerically computed two-point correlation function of Riemann zeros and the 'exact' formula. Solid line is the Gaussian fit to the histogram

III. Arithmetic Systems

As was discussed above it is well accepted that spectral statistics of classically chaotic systems in the universal limit coincides with spectral statistics of the usual random matrix ensembles. But it is also known (see e.g. [7], [29] and references therein) that the motion on constant negative curvature surfaces generated by discrete groups (considered in Chapter I is the best example of classical chaos. Consequently, models on constant negative curvature seem to be ideal tools to check the conjecture on spectral fluctuations of classically chaotic systems. Their classical motion is extremely chaotic and time-reversal invariant and *a priori* assumption was that all of them should have energy levels distributions close to predictions of the Gaussian orthogonal ensemble (GOE) of random matrices.

Nevertheless when the first large scale numerical calculations were performed [3], [52] they clearly indicated that for certain hyperbolic models the spectral statistics were quite close to Poisson statistics typical for integrable systems.

As an example we present in Fig. 18 the nearest-neighbor distribution for the hyperbolic triangle with angles $(\pi/2, \pi/3, 0)$ corresponding to the well-known modular triangle with Dirichlet boundary conditions. The agreement with Poisson prediction is striking although classical motion for this system is perfectly chaotic.

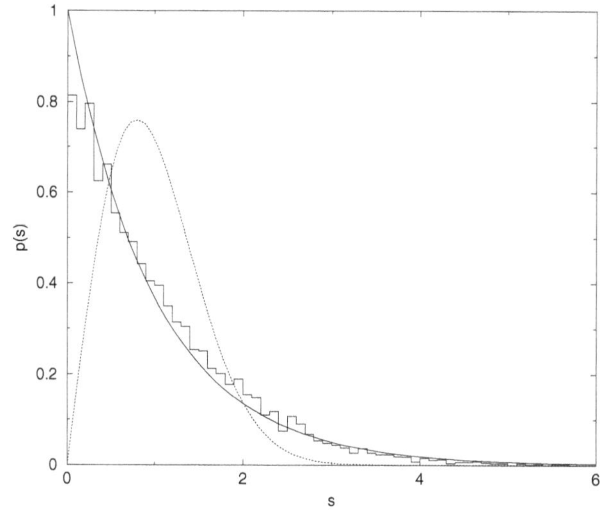

Fig. 18. The nearest neighbor distribution for 10000 first levels of the triangle $(\pi/2, \pi/3, 0)$ (the modular triangle). Solid line – the Poisson distribution. Dotted line – the GOE distribution

The purpose of this Chapter is to show that this strange behaviour is the consequence of exponentially large exact degeneracy of periodic orbit lengths in systems considered [18]. In all hyperbolic surfaces where the Poisson-like statistics was observed there is on average $\exp(l/2)$ classical periodic orbits with exactly the same length l. It will be demonstrated that this is the characteristic property of models generated by the so-called arithmetic groups. In these lectures we shall consider only discrete subgroups of $SL(2,R)$ whereas in the second volume C. Soulé will present a more general definition of arithmeticity. As classical mechanics is not sensitive to lengths of periodic orbits all these models remain completely chaotic. But the cumulative effect of interference of degenerate periodic orbits changes drastically the quantum mechanical properties.

This Chapter is based on [24]. In Sect. 1 simple calculations prove exponential degeneracy of periodic orbit lengths for the modular group. The main peculiarity of the modular group matrices is that their traces are integers. Therefore if one considers all matrices with $|\mathrm{Tr}\ M| < X$ the number of different traces increases at most linearly with X. In Sect. 2 it is shown that this property is typical for all arithmetic groups. An informal mini-review of such groups is given in this Section and it is demonstrated that for all these groups exponentially many periodic orbits have exactly the same length. From the results of [53] it follows that there is exactly 85 triangles generated by discrete arithmetic groups. All triangular models were the Poisson-like spectral statistics was numerically observed are in this list. In Sect. 3 it is shown that in the diagonal approximation the two-point correlation form factor of arithmetic systems jumps very quickly to the Poisson value thus confirming unusual nature of arithmetic systems. In Sect. 4 the exact two-point correlation function for the modular domain is calculated. The correlations of multiplicities are obtained by a generalization of the Hardy–Littlewood method discussed in Sect. 2.1. The resulting formula proves that in the universal limit the two-point correlation function of eigenvalues of the Laplace–Beltrami operator automorphic with respect to the modular group tends to the Poisson prediction. Arithmetic groups have many other interesting properties. In particular, for all arithmetic groups it is possible to construct an infinite number of mutually commuting operators which commute also with the Laplace–Beltrami operator. Properties of these operators called the Hecke operators are discussed in Sect. 5. The Jacquet–Langlands correspondence between different arithmetic groups is mentioned in Sect. 6. In Sect. 7 non-arithmetic models are briefly discussed.

1 Modular Group

The modular group is the group of all 2×2 matrices

$$M = \begin{pmatrix} m & n \\ k & l \end{pmatrix}$$

with integer m, n, k, l and the unit determinant $ml - nk = 1$.

The periodic orbits correspond in a unique way to the conjugacy classes of hyperbolic elements of the group (see Sect. 2.3). The length of periodic orbit l_p is related with the trace of a representative matrix of the conjugacy class M as follows

$$2\cosh\frac{l_p}{2} = |\mathrm{Tr}\ M|\ .$$

As all elements of modular group matrices are integers, the trace is also an integer

$$|\mathrm{Tr} M| = n\ . \tag{58}$$

Here the arithmetical nature of the group clearly comes into the play. This simple property is very important. It signifies that for the modular group there is just a quite restrictive set of all possible traces and, consequently, of periodic orbit lengths. For modular group the number of possible different lengths is the number of different integers less than $2\cosh L/2$ (see (58)), hence

$$N_{\text{dif. lengths}} = 2\cosh\frac{L}{2} \stackrel{L\to\infty}{\longrightarrow} \mathrm{e}^{L/2}\ .$$

On the other hand, for any discrete group the number of periodic orbits of length less than L grows asymptotically as

$$N(l_p < L) = \frac{\mathrm{e}^L}{L}\ .$$

Let $g(l)$ be the multiplicity of periodic orbits with length l. One has obvious relations valid for large L

$$\sum_{l<L} g(l) = \frac{e^L}{L}, \quad \sum_{l<L} 1 = \mathrm{e}^{L/2}$$

where the summation extends over different lengths of periodic orbits counted without taking multiplicity into account.

Let us define the *mean* multiplicity $\langle g(l) \rangle$ as the following ratio

$$\langle g(l) \rangle = \frac{\text{Number of periodic orbits with } l < l_p < l + \Delta l}{\text{Number of different lengths with } l < l_p < l + \Delta l} . \tag{59}$$

Asymptotically for large L the previous formulas gives

$$\langle g(l) \rangle = 2\frac{\mathrm{e}^{l/2}}{l}$$

which demonstrates that periodic orbit lengths for the modular group are exponentially degenerated.

2 Arithmetic Groups

The crucial feature which led to the exponential degeneracy of periodic orbit lengths for the modular group was the fact that traces of modular group matrices were integers which was a direct consequence of the arithmetic nature of modular group. But 2×2 matrix groups with integer elements are exhausted by the modular group and its subgroups.

Nevertheless, one can construct a quite large class of discrete groups with strong arithmetic properties by considering groups which are not equal to 2×2 integer matrices but which permit a *representation* by $n \times n$ integer matrices ($n > 2$).

The existence of such representation means that for each 2×2 group matrix, g, one can associate a $n\times n$ matrix with integer entries, $M(g)$, in such a way that the matrix associated to the product of two group matrices equals the product of two matrices associated to the corresponding factors

$$M(ab) = M(a) \times M(b)$$

for all a and b from the group considered and $M(\mathbf{1}) = \mathbf{1}$.

To define general arithmetic groups we need a few definitions.

- A subset of a group Γ is called a subgroup if it forms itself a group.
- A subgroup g of a group Γ is called a subgroup of finite index $(k+1)$ if Γ can be represented as a finite union

$$\Gamma = g + g\gamma_1 + \ldots + g\gamma_k$$

with $\gamma_k \in \Gamma$.

- Two groups are called commensurable if they have a common subgroup which is of finite index in both of them.

Groups which have a representation by integer matrices and all groups commensurable with them are called arithmetic groups. This Section is devoted to the investigation of their properties.

In Sect. 2.1 a non-formal review of algebraic fields is given and in Sect. 2.2 the construction of quaternion algebras over algebraic fields is shortly discussed. It appears that all arithmetic groups can be obtained from quaternion algebras with division and in Sect. 2.3 the necessary and sufficient conditions that a given group will be an arithmetic group is presented. Using these conditions in Sect. 2.4 it is proved that periodic orbit lengths for all arithmetic groups have the same exponential degeneracy (up to a constant factor) as for the modular group.

2.1 Algebraic Fields

Everybody is familiar with usual rational numbers

$$u = \frac{p}{q}$$

with integer p and q. Their important properties are that (i) the sum and the product of any two rational numbers also have the same form and (ii) all elements except 0 have an inverse (i.e. the division is always possible). From mathematical viewpoint rational numbers form a field called $\mathbb{Q}$.

Algebraic fields of finite degree, $\mathbb{F}$, are a generalization of this reference field obtaining by adding to the set of rational numbers a root α of an irreducible polynomial

$$\sum_{k=0}^{n} c_k \alpha^k = 0 \tag{60}$$

with integer coefficients c_k. This field is denoted $\mathbb{F} = \mathbb{Q}(\alpha)$.

Each element $u \in \mathbb{Q}(\alpha)$ can be represented by the sum

$$u = \sum_{i=0}^{n-1} b_i \alpha^i$$

where the b_i are usual rationals. The summation and the multiplication of these elements are done as with usual numbers except that all powers of α larger than $n-1$ have to be reduced using the defining equation (60).

Integers of the field $\mathbb{Q}(\alpha)$ are its elements which obey a polynomial equation with integer coefficients with an additional condition that the highest power coefficient equals one (such polynomials are called monic polynomials).

In general, algebraic integers, ω, of a field of degree n are freely generated by n linearly independent elements of the field β_k with integer coefficients (in mathematical language it means that they form a free $\mathbb{Z}$-module of rank n). Explicitly

$$\omega = \sum_{k=0}^{n-1} m_k \beta_k \tag{61}$$

where all m_k are usual integers. In simple cases $\beta_k = \alpha^k$ and

$$\omega = \sum_{k=0}^{n-1} m_k \alpha^k$$

with integer or semi-integer coefficients m_k. Algebraic integers like usual integers form a ring (not a field) because the division is not always possible.

The polynomial equation defining the field (60) has n different roots α_i, $i = 0, 1, ..., n-1$ with $\alpha_0 = \alpha$. Any relation between elements of the field remains unchanged under the transformations

$$\phi_i : \ \alpha \to \alpha_i$$

where one substitutes in all expressions instead of one root α another root α_i. These transformations are called isomorphisms or embeddings of this field into $\mathbb{C}$ and they are the only transformations respecting the laws of the field.

Example

Add to the field of rational numbers $\mathbb{Q}$ one root of the quadratic equation

$$x^2 = d$$

where d is a square-free integer. Elements of this field $\mathbb{Q}(\sqrt{d})$ can be written as

$$u = p + q\sqrt{d}$$

with p and q rationals. Let

$$\omega = a + b\sqrt{d}$$

be integers of this field. To find values of a and b one notes that ω obeys the quadratic equation

$$\omega^2 - 2a\omega + a^2 - db^2 = 0 \ .$$

To describe algebraic integers the coefficients of this equation: $2a$ and $a^2 - db^2$, have to be usual integers. Depending on d two types of solutions are possible.

- If $d \equiv 2$ or $d \equiv 3 \pmod 4$ then a and b have to be integers and

$$\omega = m + n\sqrt{d}$$

with integers m and n.

- If $d \equiv 1 \pmod 4$ then both a and b can be demi-integers and

$$\omega = \frac{m}{2} + \frac{n}{2}\sqrt{d}$$

with integers $m \equiv n \pmod 2$.
To avoid the last restriction this expression can be rewritten in the form (61)

$$\omega = m + n\frac{1+\sqrt{d}}{2} \tag{62}$$

with arbitrary integers m and n.

As this field is defined by an equation of the second degree it has two isomorphisms

$$\phi_0 : p + q\sqrt{d} \longrightarrow p + q\sqrt{d} \, ,$$
$$\phi_1 : p + q\sqrt{d} \longrightarrow p - q\sqrt{d} \, .$$

Because the product of two algebraic integers is also an algebraic integer from (61) it follows that all algebraic integers permit a representation by matrices with integer elements in such a way that the matrix representing a product of two integers equals the product of matrices representing each factor.

For example, for the above considered case of $\mathbb{Q}(\sqrt{d})$ with $d \not\equiv 1 \pmod 4$ one can associate with an integer of this field, $\omega = m + n\sqrt{d}$, a 2×2 matrix

$$M(\omega) = \begin{pmatrix} m & n \\ dn & m \end{pmatrix} . \tag{63}$$

It is easy to check that this is the true representation of field integers because $M(\omega_1\omega_2) = M(\omega_1)M(\omega_2)$ and $M(1) = \mathbf{1}$.

When $d \equiv 1 \pmod 4$ the integers have the form (62) and one can check that the matrix representation can be chosen as follows

$$M(\omega) = \begin{pmatrix} m & n \\ \frac{d-1}{4}n & m+n \end{pmatrix} .$$

2.2 Quaternion Algebras

Algebras are more general objects than fields. A (vector) algebra of finite dimension d is defined as formal sum

$$\gamma = x_1\mathbf{i}_2 + x_2\mathbf{i}_2 + \ldots + x_d\mathbf{i}_d \, .$$

Here x_j belong to a basis field $\mathbb{F}$ and $\mathbf{i}_j$ are formal objects (vectors) with a prescribed multiplication table

$$\mathbf{i}_j\mathbf{i}_k = \sum_{p=1}^{d} C_{jk}^{p}\mathbf{i}_p$$

where C_{jk}^{p} are from the basis field. The sum and the product of any two elements of an algebra belong to it. General algebras should be neither commutative, nor associative.

An algebra is called a normed algebra if there exists a function, $N(\gamma)$, which associates to any element of the algebra a number from the basis field such that the norm of the product equals the product of the norms of both factors

$$N(\gamma_1\gamma_2) = N(\gamma_1)N(\gamma_2)\ .$$

An algebra is called a division algebra if the division is always possible (except a zero element).

Finite dimensional normed division algebras over real numbers are exhausted by the following three possibilities (the Frobenius theorem [44]).

- Commutative and associative division algebras are isomorphed either to the usual field of real numbers $\mathbb{R}$ or to the field of complex numbers $\mathbb{C}$. In the latter case the algebra is given by

  $$\gamma = x_1 + x_2\mathrm{i}$$

  with $\mathrm{i}^2 = -1$. The norm in this case is

  $$N(\gamma) = x_1^2 + x_2^2\ .$$

- Non-commutative but associative division algebras are isomorphed to the quaternion algebra

  $$\gamma = x_1 + x_2\mathbf{i} + x_3\mathbf{j} + x_4\mathbf{k} \tag{64}$$

  where

  $$\mathbf{i}^2 = \mathbf{j}^2 = \mathbf{k}^2 = -1\ ,\ \mathbf{k} = \mathbf{ij} = -\mathbf{ji}\ . \tag{65}$$

  The norm of the quaternion algebra is

  $$N(\gamma) = x_1^2 + x_2^2 + x_3^2 + x_4^2\ .$$

- Non-associative normed division algebras are isomorphed to the octonion algebra

  $$\gamma = \sum_{k=1}^{8} x_k \mathbf{i}_k$$

  with a complicated multiplication table and the norm given by the sum of 8 squares

  $$N(\gamma) - \sum_{k=1}^{8} x_k^2\ .$$

Similarly, for an algebraic field $\mathbb{F}$ of finite degree there exist quaternion normed algebras defined similarly to Eqs. (64) and (65) [56]. These algebras are labeled by two elements $a, b \in \mathbb{F}$ and it is a four-dimensional non-commutative algebra with basis $(\mathbf{1}, \mathbf{i}, \mathbf{j}, \mathbf{k})$ as in (64) with the following multiplication table

$$\mathbf{i}^2 = a\,,\ \ \mathbf{j}^2 = b\,,\ \ \mathbf{k} = \mathbf{ij} = -\mathbf{ji}\,. \tag{66}$$

Such algebra is denoted by $\left(\frac{a,b}{\mathbb{F}}\right)$ and its norm is

$$N(\gamma) = x_1^2 - ax_2^2 - bx_3^2 + abx_4^2\,. \tag{67}$$

The matrix representation of the quaternion algebra (66) is obtained by the isomorphism

$$\mathbf{i} \to \begin{pmatrix} \sqrt{a} & 0 \\ 0 & -\sqrt{a} \end{pmatrix}\,,\ \mathbf{j} \to \begin{pmatrix} 0 & 1 \\ b & 0 \end{pmatrix}\,,\ \mathbf{k} = \mathbf{ij} \to \begin{pmatrix} 0 & \sqrt{a} \\ -b\sqrt{a} & 0 \end{pmatrix}\,.$$

Explicitly

$$\gamma = \begin{pmatrix} x_1 + x_2\sqrt{a} & x_3 + x_4\sqrt{a} \\ b(x_3 - x_4\sqrt{a}) & x_1 - x_2\sqrt{a} \end{pmatrix} \tag{68}$$

with $x_1, x_2, x_3, x_4 \in \mathbb{F}$. As it is a representation of the quaternion algebra the product of these matrices also has the same form. In this representation the norm of the algebra (67) equals the determinant of the matrix (68)

$$N(\gamma) = \det \gamma\,.$$

From an algebraic field $\mathbb{F}$ of finite degree one can build also another simple set of matrices called M(2,$\mathbb{F}$) given by 2×2 matrices with entries from $\mathbb{F}$

$$\begin{pmatrix} x_1 & x_2 \\ x_3 & x_4 \end{pmatrix}\,. \tag{69}$$

Are the two sets (68) and (69) different or are they isomorphic? For example, if $a = u^2$ and $u \in \mathbb{F}$ the set (68) is, evidently, within M(2,$\mathbb{F}$).

Let us show that if $\sqrt{a} \notin \mathbb{F}$ and if there exist certain elements $q_1, q_2, q_3, q_4 \in \mathbb{F}$ such that the determinant of the matrix (68) is zero

$$\det(\gamma) = q_1^2 - q_2^2 a - b(q_3^2 - q_4^2 a) = 0 \tag{70}$$

then matrices (68) are isomorphic to M(2,$\mathbb{F}$) [42].

Indeed, from the above expression it follows that in this case b has the form

$$b = (q_1^2 - q_2^2 a)(q_3^2 - q_4^2 a)^{-1} = (u_1 + u_2\sqrt{a})(u_1 - u_2\sqrt{a})$$

where

$$u_1 + u_2\sqrt{a} = (q_1 + q_2\sqrt{a})(q_3 + q_4\sqrt{a})^{-1}\,.$$

As all fractions of field elements belong to the field $\mathbb{F}$, u_1 and u_2 are also elements of $\mathbb{F}$. Now one can check that

$$\begin{pmatrix} x_1 + x_2\sqrt{a} & x_3 + x_4\sqrt{a} \\ b(x_3 - x_4\sqrt{a}) & x_1 - x_2\sqrt{a} \end{pmatrix} = S^{-1}MS$$

where

$$M = \begin{pmatrix} x_1 + x_3 u_1 + x_4 u_2 a & x_2 - x_3 u_2 - x_4 u_1 \\ a(x_2 + x_3 u_2 + x_4 u_1) & x_1 - x_3 u_1 - x_4 u_2 a \end{pmatrix}$$

and S is a fixed (independent of x_i) matrix

$$S = \begin{pmatrix} u_1 + u_2\sqrt{a} & 1 \\ \sqrt{a}(u_1 + u_2\sqrt{a}) & -\sqrt{a} \end{pmatrix} .$$

The importance of such representation lies in the fact that the matrix M contains only elements of our basis field $\mathbb{F}$ and does not contain $\sqrt{a}$. Therefore, it belongs to M(2,$\mathbb{F}$) and the set of matrices γ (68) is the conjugation of matrices from M(2,$\mathbb{F}$) by a fixed matrix S. In other words, when the equation $\det(\gamma) = 0$ has a solution $\gamma \in \mathbb{F}$ expression (68) is just a complicated way of writing matrices from M(2,$\mathbb{F}$). These considerations demonstrate that in order to construct a group different from M(2,$\mathbb{F}$) it is necessary to require that there exist *no* elements from the basis field such that the determinant (70) equals zero. Or, equivalently, any matrix (68) should have an inverse element. In the language of quaternion algebra this property corresponds to the division algebra for which any element has an inverse.

As for real fields this condition is quite restrictive and an explicit answer can be obtained only in simple cases. Let us consider for example the case when $\mathbb{F}$ is the field of usual rational numbers $\mathbb{F} = \mathbb{Q}$. The following theorem [42] gives a series of division algebras over $\mathbb{Q}$.

Let b be a prime number and a be an integer such that the equation

$$x^2 \equiv a \pmod{b}$$

has no integer solution. Then the pair (a, b) defines a division algebra over $\mathbb{Q}$ or equivalently the equation (70)

$$x_1^2 - x_2^2 a - b(x_3^2 - x_4^2 a) = 0 \tag{71}$$

has only zero rational solutions.

To prove the theorem note that due to homogeneity of this equation it is sufficient to consider integer solutions x_1, x_2, x_3, x_4 without common factors. From (71) it follows that

$$x_1^2 \equiv x_2^2 a \pmod{b} .$$

Consider first the case when b does not divide x_2, $b \nmid x_2$. As b is assumed to be a prime, $x_2^{-1} \pmod{b}$ exists and $(x_1/x_2)^2 \equiv a \pmod{b}$ which contradicts our assumption. Hence $b \,|\, x_2$ but then $b \,|\, x_1$ and

$$x_3^2 \equiv x_4^2 a \pmod{b} .$$

The same arguments give that $b \,|\, x_4$ and $b \,|\, x_3$ which contradicts the assumption about the absence of common factors of x_i. Therefore there is no rational solution of (71) and the quaternion algebra defined by a and b is a division algebra.

Quaternion algebras with division are analogs of algebraic fields. How one can define integers of a quaternion algebra?

We have seen above that algebraic integers of a field of degree n form a free $\mathbb{Z}$-module of rank n i.e. they can be represented as a sum of n elements of the field with integer coefficients (see (61). Similarly one can define 'integers' of a quaternion algebra over such field as a free $\mathbb{Z}$-module of rank $4n$ (which generates the whole algebra). For technical reasons they are called not integers but 'the order' in the algebra. The word 'integers' in algebras is reserved for elements for which the trace and the determinant of matrix (68) are integers of the basis field. Different orders exist and the one which is not contained in any other order is called the maximal order.

The simplest case appears when a and b are integers of the basis field $\mathbb{F}$. Then matrices of the form

$$\begin{pmatrix} x_1 + x_2\sqrt{a} & x_3 + x_4\sqrt{a} \\ b(x_3 - x_4\sqrt{a}) & x_1 - x_2\sqrt{a} \end{pmatrix}$$

where all x_k are integers of $\mathbb{F}$ form an order of the algebra $\left(\frac{a,b}{\mathbb{F}}\right)$ (but not necessarily the maximal order).

Matrices of the order in a division quaternion algebra with unit determinant form a group. Each matrix of this group belongs to the order and, therefore, is defined by $4n$ integers. The product of two group matrices have the same form and corresponds to a certain transformation of integers defining both matrices. It means that these groups can be represented by $4n \times 4n$ matrices with integer elements.

All such groups, all their subgroups, and all groups commensurable with them are *discrete arithmetic groups* with finite fundamental domain [42].

Example

As $x^2 \pmod 5$ takes only values $0, 1, 4$ the equation

$$x^2 \equiv 3 \pmod 5$$

has no integer solution. Hence the pair (3,5) defines a division algebra over $\mathbb{Q}$.

A simple order of this algebra has the form

$$\begin{pmatrix} m + n\sqrt{3} & k + l\sqrt{3} \\ 5(k - l\sqrt{3}) & m - n\sqrt{3} \end{pmatrix} \tag{72}$$

with integer m, n, k, l. When one considers these matrices with the unit determinant

$$m^2 - 3n^2 - 5k^2 + 15l^2 = 1$$

they form a discrete arithmetic group Γ_1 with a finite fundamental area.

The order (72) is not the maximal order. The latter can be chosen e.g. as follows

$$\begin{pmatrix} \frac{1}{2}(m+n\sqrt{3}) & \frac{1}{2}(k+l\sqrt{3}) \\ \frac{5}{2}(k-l\sqrt{3}) & \frac{1}{2}(m-n\sqrt{3}) \end{pmatrix} \tag{73}$$

with integer m, n, k, l such that $m \equiv k \pmod 2$ and $n \equiv l \pmod 2$. Matrices (73) with the unit determinant

$$m^2 - 3n^2 - 5k^2 + 15l^2 = 4$$

constitute another discrete arithmetic group Γ_2 whose fundamental domain is smaller than for the group (72) as Γ_1 is a subgroup of Γ_2.

Using the representation (63) one concludes that to each 2×2 matrix of the group Γ_1 one can associate the 4×4 matrix with integer elements

$$M(\gamma) = \begin{pmatrix} m & n & k & l \\ 3n & m & 3l & k \\ 5k & -5l & m & -n \\ -15k & 5k & -3n & m \end{pmatrix} .$$

It is straightforward to check that (i) $M(\gamma_1\gamma_2) = M(\gamma_1)M(\gamma_2)$ for all γ_1 , $\gamma_2 \in \Gamma_1$, (ii) $M(\mathbf{1}) = \mathbf{1}$, and (iii) $\det(M) = (\det(\gamma))^2 = 1$. Together these expressions mean that this group is an arithmetic group.

2.3 Criterion of Arithmeticity

For general fields the situation is more complicated. To explain the general criterion of arithmetic groups let us first stress a difference between usual integers and algebraic integers.

The usual integers correspond to a discrete set of points. But for general algebraic integers this is not the case. For example, in the field $\mathbb{Q}(\sqrt{2})$ integers have the form $n + m\sqrt{2}$ with integer n and m. But it is evident that one can construct sequences of these algebraic integers converging to zero, e.g. $(\sqrt{2}-1)^k = M_k - N_k\sqrt{2} \to 0$ when $k \to \infty$. Therefore the set of algebraic integers is not discrete as it has finite accumulation points.

How can one deal with such problem? The main point is that these small terms become large under the transformation

$$\sqrt{2} \to -\sqrt{2} \tag{74}$$

Let consider in the above example not all algebraic integers $n+m\sqrt{2}$ but only those which after transformation (74) remain bounded

$$|n - m\sqrt{2}| < \text{constant} .$$

It is clear that now arbitrary small integers are excluded and one gets a discrete set of points.

For more general fields the transformation (74) is generalized to all non-trivial isomorphisms of the field. To remove arbitrary small elements one has to require that for all isomorphisms of the field (except the identity), ϕ_i, transformed values of integers (61) are restricted

$$\left| \sum_{k=0}^{n-1} m_k \phi_i(\beta_k) \right| < \text{constant} . \tag{75}$$

In order to be sure that all small numbers are removed it is necessary that all roots of defining equation (60) are real. Otherwise, changing a root to its complex conjugate may not change modulus of integers.

These considerations make reasonable that in order to construct a *discrete* subset of algebraic integers (without finite accumulation points) it is necessary that (i) the field be a totally real field (i.e. all roots of defining equation (60) are real) and (ii) for all non-trivial isomorphisms of the field transformed integers remain bounded as in (75).

A precise criterion of arithmeticity obtained by Takeuchi [53] is quite similar (see also [6] and [24] for particular examples).

Takeuchi proved that a group Γ is an arithmetic group if and only if the traces of group matrices have the following properties

- All $\mathrm{Tr}(\gamma)$ are integers of a totally real algebraic field of finite degree.
- For any non-trivial isomorphism ϕ of this field that changes some $|\mathrm{Tr}(\gamma)|$ for certain γ, the value of the transformed trace satisfies $|\phi(\mathrm{Tr}(\gamma))| \le 2$.

There are two types of arithmetic groups. Non-compact groups, built from $SL(2,\mathbb{Z})$, and compact ones built from quaternion algebra different from $M(2,\mathbb{Q})$.

The above criterion is quite effective, in particular, it permits to find all possible arithmetic groups with triangular fundamental domains [53]. There are 85 triangular hyperbolic surfaces generated by discrete arithmetic groups. All of them are given in Table 1.

2.4 Multiplicities of Periodic Orbits for General Arithmetic Groups

The geometrical length of the periodic orbit, l, is connected with the trace of class of conjugate matrices by (28). When $l \to \infty$

$$\exp \frac{l}{2} = |\mathrm{Tr}(\gamma)| .$$

Let us prove that for an arithmetic group the number of possible values of group matrix traces obeys

$$N(|\mathrm{Tr}(\gamma)| \le R) \stackrel{R\to\infty}{\longrightarrow} C \cdot R$$

Table 1. The list of arithmetic triangles from [53]. (n, m, p) in the first column corresponds to the three angles $(\pi/n, \pi/m, \pi/p)$. The second column indicates the algebraic field from which is built the corresponding arithmetic group.

(m,n,p)	$\mathbb{F}$
(2,3,∞) (2,4,∞) (2,6,∞) (2,∞,∞) (3,3,∞) (3,∞,∞) (4,4,∞) (6,6,∞) (∞,∞,∞)	$\mathbb{Q}$
(2,4,6) (2,6,6) (3,4,4) (3,6,6)	$\mathbb{Q}$
(2,3,8) (2,4,8) (2,6,8) (2,8,8) (3,3,4) (3,8,8) (4,4,4) (4,6,6) (4,8,8)	$\mathbb{Q}(\sqrt{2})$
(2,3,12) (2,6,12) (3,3,6) (3,4,12) (3,12,12) (6,6,6)	$\mathbb{Q}(\sqrt{3})$
(2,4,12) (2,12,12) (4,4,6) (6,12,12)	$\mathbb{Q}(\sqrt{3})$
(2,4,5) (2,4,10) (2,5,5) (2,10,10) (4,4,5) (5,10,10)	$\mathbb{Q}(\sqrt{5})$
(2,5,6) (3,5,5)	$\mathbb{Q}(\sqrt{5})$
(2,3,10) (2,5,10) (3,3,5) (5,5,5)	$\mathbb{Q}(\sqrt{5})$
(3,4,6)	$\mathbb{Q}(\sqrt{6})$
(2,3,7) (2,3,14) (2,4,7) (2,7,7) (2,7,14) (3,3,7) (7,7,7)	$\mathbb{Q}(\cos\pi/7)$
(2,3,9) (2,3,18) (2,9,18) (3,3,9) (3,6,18) (9,9,9)	$\mathbb{Q}(\cos\pi/9)$
(2,4,18) (2,18,18) (4,4,9) (9,18,18)	$\mathbb{Q}(\cos\pi/9)$
(2,3,16) (2,8,16) (3,3,8) (4,16,16) (8,8,8)	$\mathbb{Q}(\cos\pi/8)$
(2,5,20) (5,5,10)	$\mathbb{Q}(\cos\pi/10)$
(2,3,24) (2,12,24) (3,3,12) (3,8,24) (6,24,24) (12,12,12)	$\mathbb{Q}(\cos\pi/12)$
(2,5,30) (5,5,15)	$\mathbb{Q}(\cos\pi/15)$
(2,3,30) (2,15,30) (3,3,15) (3,10,30) (15,15,15)	$\mathbb{Q}(\cos\pi/15)$
(2,5,8) (4,5,5)	$\mathbb{Q}(\sqrt{2},\sqrt{5})$
(2,3,11)	$\mathbb{Q}(\cos\pi/11)$

with a constant C depending on the group. The traces of matrices of arithmetic groups are dispatched as usual integers among real numbers.

Let Γ be an arithmetic group. The set of traces $\{\mathrm{Tr}(\gamma),\ \gamma \in \Gamma\}$ are integers of an algebraic field $\mathbb{F}$

$$t_0 = \sum_{i=0}^{n-1} m_i \beta_i$$

where m_i are integers and β_i are linearly independent elements of the field. Consider the simplest case $\beta_i = \alpha^i$ then

$$\mathrm{Tr}(\gamma) \equiv t_0 = \sum_{i=0}^{n-1} m_i \alpha^i \ .$$

For a field of degree n there exist $n-1$ non-trivial isomorphisms $\phi_k : \alpha \to \alpha_k$ where α_k is a root of the defining polynomial different from α.

Suppose that all such transformations change $|\mathrm{Tr}(\gamma)|$. According to the criterion of Takeuchi all transformed traces satisfy

$$|t_k| \leq 2$$

where

$$t_k \equiv \phi_k(\mathrm{Tr}(\gamma)) = \sum_{i=0}^{n-1} m_i \alpha_k^i \ .$$

Consider these equations as transformations from variables t_i to new variables m_i [27]. The volume elements in these two representations are related as

$$\mathrm{d}t_0 \mathrm{d}t_1 ... \mathrm{d}t_{n-1} = |\mathcal{J}| \mathrm{d}m_0 \mathrm{d}m_1 ... \mathrm{d}m_{n-1}$$

where

$$\mathcal{J} = \det\left(\frac{\partial t_j}{\partial m_k}\right)$$

is called the discriminant of the field and in our case (when $\beta_i = \alpha^i$)

$$\mathcal{J} = \det(\alpha_k^j)|_{k,j=0,\dots,n-1} = \prod_{i \neq j}(\alpha_i - \alpha_j) \ .$$

As m_i are integers the volume of the smallest cell is one, and the total number of possible integer solutions is asymptotically

$$N(|\mathrm{Tr}(\gamma)| \leq R) = N(|t_0| \leq R, |t_j| \leq 2) \simeq C \cdot R$$

where $C = 2^n/\mathcal{J}$.

For any surface of finite area generated by a discrete group the total number of periodic orbits with length less than a given value is asymptotically the following

$$N_{tot}(l < L) \stackrel{L \to \infty}{\longrightarrow} \frac{\mathrm{e}^L}{L} \ .$$

The number of periodic orbits with *different* lengths is the same as the number of group matrix traces

$$N_{\text{diff. lengths}}(l < L) \sim C \cdot \mathrm{e}^{L/2}$$

Let $g(l)$ be the multiplicity of periodic orbits with length l. Then

$$\sum_{l<L} g(l) = \frac{\mathrm{e}^L}{L} \text{ and } \sum_{l<L} 1 = C\mathrm{e}^{L/2}$$

where the summation is done over different lengths.

Finally the *mean* multiplicity of arithmetic systems defined as in (59) has the following asymptotics

$$\langle g \rangle = \frac{(\sum_{l<L} g(l))'}{(\sum_{l<L} 1)'} \sim \frac{2\mathrm{e}^{L/2}}{CL} \, . \tag{76}$$

Thus we demonstrate that the arithmetic nature of arithmetic groups leads to exponential multiplicities of periodic orbit lengths.

For generic systems one usually does not expect any degeneracy of periodic orbit lengths except the ones which follow from exact symmetries of the model. For example, systems with time-reversal invariance, in general, should have the mean multiplicity equal to 2, which corresponds to the same geometrical periodic orbits spanned in two directions. Therefore, arithmetic systems are *very exceptional* in this respect as they display exponentially large multiplicities of periodic orbit lengths. Notice, nevertheless, that according to Horowitz-Randol theorem [41], [51] this degeneracy is unbounded for any surface generated by a discrete group. However degeneracies of this theorem are much smaller than exponential.

3 Diagonal Approximation for Arithmetic Systems

The large multiplicities of periodic orbit lengths in arithmetical systems seem to have no importance in classical mechanics. These systems are as chaotic as any other models of free motion on constant negative curvature surfaces with finite area. Nevertheless, the quantum spectra of these systems are anomalous: is it connected to these degeneracies? In this Section we estimate the quantum two-point correlation form factor for arithmetic systems in the diagonal approximation as was done in Sect. 1.1 for generic chaotic systems.

Assume that there exist $g(l)$ periodic orbits with the same length l. Exactly as it was done in Sect. 1.1 one gets the following expression for the diagonal approximation of the two-point correlation function

$$R_2^{(diag)}(\epsilon) = \sum_{p,n} |A_{p,n}(l_p)|^2 g(l_p) \mathrm{e}^{\mathrm{i}nT_p\epsilon} + \text{c.c.} \tag{77}$$

where the summation is done over all periodic orbits. The only difference with (48) is that in Sect. 1.1 it was assumed that g is a constant but here the multiplicity $g - g(l)$.

Define the two-point correlation form factor as the Fourier transform of $R_2(\epsilon)$

$$K(t) = \int_{-\infty}^{+\infty} R_2(\epsilon) \mathrm{e}^{\mathrm{i}t\epsilon} \, .$$

This definition differs from the previous one by the absence of the factor 2π in the exponent. For later purposes it is more convenient.

Equation (77) leads to the following expression for the two-point correlation form factor in the diagonal approximation

$$K^{(diag)}(t) = 2\pi \sum_{p,n} |A_{p,n}(l_p)|^2 g(l_p) \delta \left(t - \frac{nl_p}{2k} \right) . \tag{78}$$

From (76) it follows that the mean multiplicity of periodic orbit lengths for arithmetic systems is asymptotically

$$\langle g(l_p) \rangle = \frac{2\mathrm{e}^{l_p/2}}{Cl_p}$$

with a model dependent constant C (for the modular group $C = 1$).

For any models generated by discrete groups the summation over all periodic orbits is asymptotically equals the integration with the following measure

$$\sum_{l_p} \to \int \frac{dl}{l} \mathrm{e}^l .$$

Taking into account that when $l \to \infty$ the term with $n = 1$ dominates and (see (47))

$$A_{p,1}(l) \stackrel{l \to \infty}{\longrightarrow} \frac{l\mathrm{e}^{-l/2}}{4\pi k}$$

one obtains that in the diagonal approximation

$$K^{(diag)}(t) \sim \frac{\mathrm{e}^{kt}}{2\pi kC} . \tag{79}$$

It means that the correlation form factor $K(t)$ for arithmetic systems grows much faster than was usually assumed and that for time of order of the Ehrenfest time it becomes of the order of 1.

The simplest approximation to the full form factor is the following

$$K(t) = \begin{cases} K^{diag}(t) & \text{for } t < t^* \\ \bar{d} & \text{for } t > t^* \end{cases}$$

where t^* is defined by the requirement that $K^{diag}(t^*) = \bar{d}$

$$t^* \sim \frac{1}{k} \ln(2\pi kC\bar{d}) .$$

For the true Poisson statistics $K(t)$ always equals $\bar{d}$. For usual integrable systems $K(t)$ increases to this value during the time of the order of shortest periodic orbit periods, $t^* \sim 1/k$. For arithmetic systems $K(t)$ jumps to the universal saturation value in a time of order of the Ehrenfest time which has an additional logarithm of the momentum.

Therefore, spectral statistics of arithmetic systems is much closer to the Poisson prediction typical for integrable systems than to any of standard random matrix ensembles conjectured for generic ergodic systems.

4 Exact Two-Point Correlation Function for the Modular Group

The diagonal approximation gives quite crude estimate of the form factor. For the modular group it is possible to compute explicitly the two-point correlation function [23]. The calculations are based on a generalization of the Hardy–Littlewood method and depend strongly on the number-theoretical properties of the multiplicities of the periodic orbits of the modular group. In Sect. 4.1 using the Selberg trace formula the two-point correlation form factor is expressed through the two-point correlation function of multiplicities of periodic orbit lengths for the modular group. In Sect. 4.2 the latter is calculated by a certain generalization of the Hardy–Littlewood method. Quite tedious explicit formulas are given in Sect. 4.3 and the final expression for the two-point correlation form factor is presented in Sect. 4.4.

4.1 Basic Identities

The modular group has been considered in Sect. 1. It is the group of all 2×2 matrices with integer elements and unit determinant. The periodic orbits of the modular group correspond in a unique way to the conjugacy classes of hyperbolic elements of the modular group. The length of periodic orbit l_p is related with the trace of a representative matrix of the conjugacy class as follows

$$|\mathrm{Tr}M| = 2\cosh l_p/2 \, .$$

As all elements of modular group matrices are integers the trace is also an integer

$$|\mathrm{Tr}M| = n \, .$$

In Sect. 1 it was demonstrated that the *mean* multiplicity of periodic orbit length for the modular group is

$$\langle g(l)\rangle = 2\frac{\mathrm{e}^{l/2}}{l} \, .$$

Denote by n the trace of a given conjugacy class and by $g(n)$ the number of distinct conjugacy classes corresponding to trace n. As n goes as $e^{L/2}$ when $n \to \infty$ one concludes that

$$\langle g(n)\rangle \stackrel{n\to\infty}{\longrightarrow} \frac{n}{\ln n} \, . \tag{80}$$

According to the Selberg trace formula the density of eigenvalues for the modular surface $d(E) = \bar{d}(E) + d^{(osc)}(E)$ where the oscillating part of the density is represented by the following formal sum

$$d^{(osc)}(E) = \frac{2}{\pi k}\sum_n g(n)\frac{\ln n}{n}\cos(2k\ln n) \, .$$

From (80) it follows that mean value of $g(n)\ln n/n$ is one. Therefore we define

$$\alpha(n) = g(n)\frac{\ln n}{n} ,$$

so

$$d^{(osc)}(E) = \frac{1}{\pi k}\sum_n \alpha(n)\cos(2k\ln n)$$

and $\langle \alpha(n)\rangle = 1$.

As it was done in Sect. 1 one gets

$$R_2(\epsilon_1,\epsilon_2) = \bar{d}^2 + R_2^c(\epsilon_1,\epsilon_2)$$

where

$$R_2^c(\epsilon_1,\epsilon_2) = \frac{1}{(2\pi k)^2}\sum_{n_1,n_2}\alpha(n_1)\alpha(n_2)\Big\langle \mathrm{e}^{2\mathrm{i}(k_1\ln n_1+k_2\ln n_2)} + \\ + \mathrm{e}^{2\mathrm{i}(k_1\ln n_1-k_2\ln n_2)}\Big\rangle + \text{c.c.}$$

and

$$k_i \approx \sqrt{E+\epsilon_i} \stackrel{E\to\infty}{\longrightarrow} k + \epsilon_i/2k .$$

As was discussed in Sect. 1 due to the energy average the first term will be washed out and the second one gives contributions only when

$$n_2 = n_1 + r \quad \text{with } r \ll n_1 .$$

Finally $R_2^c(\epsilon_1,\epsilon_2) = \overline{R}_2(\epsilon)$ where $\epsilon = \epsilon_1 - \epsilon_2$ and

$$\overline{R}_2(\epsilon) = \frac{1}{4\pi^2k^2}\sum_n\sum_r \alpha(n)\alpha(n+r)\exp\left(-2\mathrm{i}\frac{kr}{n} + \mathrm{i}\epsilon\frac{\ln n}{k}\right) + \text{c.c.} .$$

Let assume that the following mean value exists

$$\gamma(r) = \lim_{N\to\infty}\frac{1}{N}\sum_{n=1}^{N}\alpha(n)\alpha(n+r) .$$

The dominant contribution to the two-point correlation function corresponds to replace the product $\alpha(n)\alpha(n+r)$ by its mean value $\gamma(r)$

$$\overline{R}_2(\epsilon) \approx \frac{1}{4\pi^2k^2}\int_{n0}^{\infty} dn \sum_{r=-\infty}^{\infty}\gamma(r)e^{-2ikr/n}\exp\left(\mathrm{i}\epsilon\frac{\ln n}{k}\right) + \text{ c.c.} \tag{81}$$

where we have used a continuum approximation for n starting formally from a certain fixed $n_0 \gg 1$, since only large values of n make a significant contribution.

Define a (real) function $f(x)$ as follows

$$f(x) = \sum_{r=-\infty}^{\infty} \gamma(r) e^{-irx} \,. \tag{82}$$

This function has the meaning of the Fourier transform of the two-point correlation function for multiplicities of the modular group.

After changing variable $n \to \mathrm{e}^{uk}$ in (81) one gets that the two-point correlation function for the modular group is expressed through $f(x)$ as follows

$$\overline{R}_2(\epsilon) \approx \frac{1}{2\pi^2 k} \int_0^{\infty} \mathrm{e}^{ku} f(2ke^{-ku}) \cos \epsilon u \; du \tag{83}$$

and the two-point correlation form factor is

$$K(t) = \int_{-\infty}^{\infty} \overline{R}_2(\epsilon) \mathrm{e}^{\mathrm{i}\epsilon t} \mathrm{d}\epsilon = \frac{1}{2\pi k} \mathrm{e}^{kt} f(2ke^{-kt}) \,. \tag{84}$$

Therefore all non-trivial information is contained in functions $\gamma(r)$ or $f(x)$.

The simplest diagonal approximation is to assume that the $\alpha(n)$ are essentially uncorrelated, that is, $\gamma(r)$ is zero for $r \neq 0$. This gives for $f(x)$ a constant value which leads to an exponential growth of $K(t)$ as in (79). But from general considerations $K(t)$ obtained from a discrete spectrum has to saturate to a constant value for $t \to \infty$, consequently, the diagonal approximation cannot be correct for large t.

4.2 Two-Point Correlation Function of Multiplicities

The purpose of this Section is to calculate the two-point correlation function of modular group multiplicities, $\gamma(r)$, whose Fourier harmonics according to (84) determines the two-point correlation form factor.

The calculation will be done by a generalization of the Hardy-Littlewood method for prime pairs discussed in Sect. 2.1. As for primes one has to perform the three following steps.

The First Step

Define the mean value of $\alpha(n)$ when n runs over integers of the form $mq + r$ for fixed q and $r < q$ in the following way

$$\alpha(q;r) - \lim_{N\to\infty} \frac{1}{N} \sum_{m=0}^{N-1} \alpha(mq + r) \,.$$

Since $\langle \alpha(n) \rangle = 1$

$$\sum_{r=0}^{q-1} \alpha(q;r) = q \,.$$

Let M_q be the set of 2×2 matrices with entries being integers modulo q and having determinant equals one modulo q. These matrices form a group under multiplication modulo q which is sometimes called the modular group.

Define also $M_{q,r}$ to be the set of elements of M_q with trace equal to r modulo q. One can prove [23] that

$$\alpha(q;r) = \frac{q|M_{q,r}|}{|M_q|}$$

where $|M|$ is the number of elements of a set M.

The intuitive meaning of this result is the following: $g(n)$ is the number of conjugacy classes of modular matrices of trace n. To each modular matrix, one can associate an element of M_q in a unique way simply by taking the entries of the matrix modulo q. If n is equal to r modulo q, then all these matrices will belong to $M_{q,r}$. If we therefore assume that the matrices of the modular group cover the set M_q in some sense uniformly, the result follows. More careful treatment has been performed in [23].

Example

Let us consider $q = 2$. Integers modulo 2 are 0 and 1. The group M_2 consists of the following matrices

$$\begin{pmatrix} 1 & 0 \\ 0 & 1 \end{pmatrix}, \begin{pmatrix} 1 & 1 \\ 0 & 1 \end{pmatrix}, \begin{pmatrix} 1 & 0 \\ 1 & 1 \end{pmatrix}, \begin{pmatrix} 0 & 1 \\ 1 & 0 \end{pmatrix}, \begin{pmatrix} 0 & 1 \\ 1 & 1 \end{pmatrix}, \begin{pmatrix} 1 & 1 \\ 1 & 0 \end{pmatrix}.$$

The dimension of the group M_2, i.e. the total number of matrices, $|M_2| = 6$. Among these matrices there are four matrices with zero trace (mod 2), i.e. $|M_{2,0}| = 4$, and two matrices with trace equals 1 (mod 2), $|M_{2,1}| = 2$. Therefore

$$\alpha(2;0) = \frac{2 \cdot 4}{6} = \frac{4}{3}, \quad \alpha(2;1) = \frac{2 \cdot 2}{6} = \frac{2}{3}.$$

The Second Step

Define as in the Hardy-Littlewood method the following function

$$\Phi(z) = \sum_{n=0}^{\infty} \alpha(n) z^n.$$

Since $\langle \alpha(n) \rangle = 1$ the convergence radius of this series is equal to one.

The importance of this function follows from the integral

$$J_r(e^{-u}) = \mathrm{e}^{ru} \int_0^{2\pi} \frac{\mathrm{d}\phi}{2\pi} \Phi^* \left(\mathrm{e}^{-u+\mathrm{i}\phi}\right) \Phi \left(\mathrm{e}^{-u-\mathrm{i}\phi}\right) \mathrm{e}^{-\mathrm{i}r\phi} = \sum_{n=1}^{\infty} \alpha(n)\alpha(n+r)\mathrm{e}^{-2nu}$$

whose right-hand side by a Tauberian theorem is connected with the two-point correlation function of multiplicities, $\gamma(r)$.

The essence of the Hardy–Littlewood approach is the investigation of the function $\Phi(z)$ when $z = \exp(-u + \mathrm{i}\epsilon + 2\pi \mathrm{i}p/q)$ with $u \to 0$ and $\epsilon \to 0$, where p and q are co-prime integers. The main step is then to write n in the form $mq + r$ with r lying between 0 and $q - 1$ and prove that in the expression for $\Phi(z)$ the dominant contribution as u and ϵ go to zero will be given by the *mean* value of $\alpha(mq + r)$, that is, one may substitute it by $\alpha(q; r)$.

Accepting this, one has that as $u \to 0$ and $\epsilon \to 0$

$$\begin{aligned}\Phi\left(\exp\left(-u + 2\pi \mathrm{i}p/q + \mathrm{i}\epsilon\right)\right) &= \sum_{r=0}^{q-1}\sum_{m=0}^{\infty} \alpha(mq + r)\mathrm{e}^{-(u-\mathrm{i}\epsilon)(mq+r)}\mathrm{e}^{2\pi \mathrm{i}rp/q} = \\ &= \sum_{r=0}^{q-1} \alpha(q; r)\mathrm{e}^{2\pi \mathrm{i}pr/q}\frac{1}{q}\int_0^{\infty} \mathrm{d}n\, \mathrm{e}^{-(u-\mathrm{i}\epsilon)n} = \\ &= \frac{\beta(p, q)}{u - \mathrm{i}\epsilon}\end{aligned}$$

where

$$\beta(p, q) = q^{-1}\sum_{r=0}^{q-1} \alpha(q; r) \exp\left(2\pi \mathrm{i}\frac{p}{q}r\right) .$$

Hence $\Phi(z)$ has a pole singularity at all rational points on the unit circle.

The Third Step

Divide the unit circle in intervals $I_{p,q}$ centered around $\exp(2\pi \mathrm{i}p/q)$, where p and q are co-prime integers with $p < q$. If one neglects all terms in each interval except the pole term and extends the integration over ϵ to the whole line, one gets

$$\begin{aligned}J_r(\mathrm{e}^{-u}) &= \mathrm{e}^{ru} \sum_{(p,q)=1} \int_{-\infty}^{\infty} \frac{\mathrm{d}\epsilon}{2\pi}\frac{|\beta(p, q)|^2}{u^2 + \epsilon^2}\mathrm{e}^{\mathrm{i}r(2\pi p/q+\epsilon)} = \\ &= \frac{1}{2u} \sum_{(p,q)=1} |\beta(p, q)|^2 \exp\left(2\pi \mathrm{i}\frac{p}{q}r\right) .\end{aligned}$$

Finally one obtains that

$$\gamma(r) = \sum_{(p,q)=1} |\beta(p, q)|^2 \exp\left(2\pi \mathrm{i}\frac{p}{q}r\right) .$$

The sum is performed over all q, and p co-prime to q with $0 < p < q$.

This is the two-point correlation function of *multiplicities* of the periodic orbits for the modular group . All other quantities of interest can be obtained from it. In particular, the function $f(x)$ (82) is given by

$$f(x) = 2\pi \sum_{(p,q)=1} |\beta(p, q)|^2\delta\left(x - 2\pi\frac{p}{q}\right)$$

where the summation is done over all p and q co-prime, without the restriction $p < q$.

According to (83) and (84) the knowledge of $f(x)$ determines immediately the two-point correction function the form factor of modular domain eigenvalues.

4.3 Explicit Formulas

Let us define the so-called Kloosterman sums

$$S(n,m;c) = \sum_{(d,c)=1} \exp\left(\frac{2\pi \mathrm{i}}{c}(nd + md^{-1})\right)$$

where the summation is taken over all $d < c$ co-prime with c and d^{-1} is an integer modulo c which obeys $d^{-1}d = 1 \pmod{c}$.

One can show (see [23]) that $\beta(p,q)$ can be expressed through these sums in the following way

$$\beta(p,q) = \frac{1}{q^2 \prod_{\omega|q}(1-\omega^{-2})} S(p,p;q)$$

where ω are the prime divisors of q.

The function $\gamma(r)$ can be written as

$$\gamma(r) = \sum_{n=1}^{\infty} A_r(n)$$

where $A_r(q)$ is given by

$$A_r(q) = \sum_{p:(p,q)=1} |\beta(p,q)|^2 \exp\left(2\pi \mathrm{i} r \frac{p}{q}\right) \; .$$

One can prove that $A_r(q)$ is multiplicative function of q, i.e. $A_r(n_1 n_2) = A_r(n_1)A_r(n_2)$ provided $(n_1 n_2) = 1$, therefore one needs to know only its values on powers of primes and $\gamma(r)$ can be rewritten as the infinite product over all prime numbers

$$\gamma(r) = \prod_p (1 + \sum_{k=1}^{\infty} A_r(p^k))$$

To present a closed expression for $A_r(q)$ let us introduce the standard definition of the Legendre symbol

$$\left(\frac{a}{q}\right) = \begin{cases} 1\,, & \text{if } a \equiv x^2 \pmod{q} \text{ has a solution } a \not\equiv 0 \pmod{q} \\ 0\,, & \text{if } a \equiv 0 \pmod{q} \\ -1\,, & \text{otherwise} \end{cases} \; .$$

The meaning of this symbol is perhaps best understood by saying that the number of *distinct* solutions of the equation $x^2 \equiv a \pmod{q}$ is $1 + (a/q)$.

A fairly tedious evaluation of $A_r(q)$ (see [23] for details) gives the following formulas.

Let $q = p^n$ where p is an odd prime. Then for $n = 1$ one has

$$A_r(p) = \frac{1}{(p^2-1)^2}\left[p\sum_{x=0}^{p-1}\left(\frac{(x^2-4)((x+r)^2-4)}{p}\right) - 1\right] .$$

For $n \geq 2$ we have, letting t be an arbitrary non-zero number modulo p,

$$A_r(p^n) = \frac{1}{p^{2n}(1-p^{-2})}\begin{cases} 2(1-1/p)\ ,\ r \equiv 0 & \pmod{p^n} \\ -2/p\ ,\ r \equiv tp^{n-1} & \pmod{p^n} \\ \epsilon(n,p)(1-1/p)\ ,\ r \equiv \pm 4 & \pmod{p^n} \\ -\epsilon(n,p)/p\ ,\ r \equiv \pm 4 + tp^{n-1} & \pmod{p^n} \end{cases}$$

where $\epsilon(n,p)$ takes the value -1 if n is odd and p is of the form $4k+3$ and is equal to 1 in all other cases. For $p = 2$, we list down individual cases for low powers and eventually state a general rule

$$A_r(2) = \frac{1}{9}\begin{cases} 1\ ,\ r \equiv 0 & \pmod{2} \\ -1\ ,\ r \equiv 1 & \pmod{2} \end{cases} ,$$

$$A_r(4) = \frac{1}{18}\begin{cases} 1\ ,\ r \equiv 0 & \pmod{4} \\ -1\ ,\ r \equiv 2 & \pmod{4} \end{cases} ,$$

$$A_r(8) = 0 ,$$

$$A_r(16) = \frac{1}{9\cdot 16}\begin{cases} 1\ ,\ r \equiv 0 & \pmod{16} \\ -1\ ,\ r \equiv 8 & \pmod{16} \end{cases} ,$$

$$A_r(32) = 0 ,$$

and finally, for the general case $n \geq 6$

$$A_r(2^n) = \frac{1}{9\cdot 2^{2n-4}}\begin{cases} 2\ ,\ r \equiv 0 & \pmod{2^n} \\ -2\ ,\ r \equiv 2^{n-1} & \pmod{2^n} \\ 1\ ,\ r \equiv \pm(4+2^{n-2}) & \pmod{2^n} \\ -1\ ,\ r \equiv \pm(4+2^{n-2}+2^{n-1}) & \pmod{2^n} \end{cases} .$$

All terms not explicitly shown equal zero. In [50] these formulas were proved by a different method.

4.4 Two-Point Form Factor

These formulas give the explicit expression for the two–point correlation form factor

$$K(t) = \frac{1}{2\pi^2 k}\sum_{(p,q)=1}\left|\frac{q}{p}\beta(p,q)\right|^2 \delta(t - t_{p,q}) .$$

where

$$t_{p,q} = \frac{1}{k} \ln \frac{kq}{\pi p} \, .$$

In the limit $k \to \infty$ and t fixed, the dominant contribution comes from terms with $p/q \ll 1$. Smoothing over such values one can show (see [23]) that in this limit $K(t)$ tends to the constant Poisson value

$$K(t) = \frac{A}{4\pi}$$

where $A = \pi/3$ is the area of the fundamental region of the modular group. For small t (of the order of the Ehrenfest time $\ln k/k$) $K(t)$ has number–theoretical oscillations due to cumulative contributions of degenerate periodic orbits. For very small values of t (of the order of $1/k$) the two-point form factor has δ peaks connected with short periodic orbits.

Though the modular group is by no means a generic system, it is the first ergodic dynamical system for which it was possible to compute explicitly the distribution of the energy levels.

5 Hecke Operators

Arithmetic groups have many interesting properties. In particular, for all arithmetic groups it is possible to construct an infinite number of mutually commuting operators which commute also with the Laplace–Beltrami operator. These operators are of pure arithmetic origin and are called the *Hecke operators* [37], [54].

In a certain sense these operators permit to 'understand' why arithmetic systems have the Poisson statistics typical only for integrable systems. The point is that integrable systems are systems with sufficiently large number of independent commuting operators and Hecke operators may be viewed as a manifestation of a kind of arithmetic integrability of arithmetic systems which does the Poisson statistics for these models natural [27]. Unfortunately, precise relations along this line seem to be impossible.

Let us consider informally the construction of Hecke operators for the modular group. Choose two matrices A and B from the modular group with the same trace. As they have the same trace and determinant, they have the same eigenvalues and there exists a matrix γ such that

$$\gamma A \gamma^{-1} = B \quad \text{or} \quad \gamma A = B\gamma \quad \text{and} \quad \det(\gamma) \neq 0 \, . \tag{85}$$

If A and B are not conjugated in the modular group, $\gamma \notin \mathrm{PSL}(2, \mathbb{Z})$. But matrix γ can be chosen as a matrix with integer elements but with the determinant $\neq 1$.

Example

Consider the following simple matrices

$$A = \begin{pmatrix} 3 & 1 \\ 2 & 1 \end{pmatrix} \,, \quad B = \begin{pmatrix} 3 & 2 \\ 1 & 1 \end{pmatrix} \,.$$

General form of matrices γ which obey (85) is

$$\gamma = \begin{pmatrix} 2\alpha + 2\beta & \alpha \\ \alpha & \beta \end{pmatrix}$$

with arbitrary α and β.

It is clear that there exists no $\gamma \in \mathrm{PSL}(2, \mathbb{Z})$ but choosing different integer values of α and β one can construct an infinite number of integer matrices with determinant $\neq 1$ which obeys (85). For example,

$$\gamma = \begin{pmatrix} 0 & 1 \\ 1 & 1 \end{pmatrix} \ (\det = -1) \,, \ \gamma = \begin{pmatrix} 4 & 1 \\ 1 & 1 \end{pmatrix} \ (\det = 3) \ldots \,.$$

These considerations demonstrate that when dealing with the modular group it is quite natural to consider matrices with integer elements but with the determinant different from one

$$M_p = \left\{ \begin{pmatrix} a & b \\ c & d \end{pmatrix} \ a, b, c, d \text{ integers}, ad - bc = p \right\} \,.$$

Matrices M_p with $p \neq 1$ do not form a group because their product has not the same form.

A matrix $m_p \in M_p$ can uniquely be represented in the form

$$m_p = \mu \alpha_p \tag{86}$$

where $\mu \in \mathrm{PSL}(2, \mathbb{Z})$ and α_p is one of matrices from the following finite set

$$\alpha_p = \left\{ \begin{pmatrix} a & b \\ 0 & d \end{pmatrix} \ a, b, d \text{ integers} \ , \ ad = p \,, \ d > 0 \,, \ 0 \leq b \leq d - 1 \right\} \,. \tag{87}$$

Instead of proving this fact let us transform a simple matrix

$$m_3 = \begin{pmatrix} 4 & 1 \\ 1 & 1 \end{pmatrix}$$

to the form (86). General proof (see e.g. [54]) follows the same steps. First, it is necessary to find a matrix

$$\mu' = \begin{pmatrix} \mu_1 & \mu_2 \\ \mu_3 & \mu_4 \end{pmatrix}$$

such that (i) $\det \mu' = 1$ and (ii)

$$\begin{pmatrix} \mu_1 & \mu_2 \\ \mu_3 & \mu_4 \end{pmatrix} \begin{pmatrix} 4 & 1 \\ 1 & 1 \end{pmatrix} = \begin{pmatrix} a & b \\ 0 & d \end{pmatrix} .$$

The condition of the zero of low-left element gives the equation $4\mu_3 + \mu_4 = 0$ and because μ_4 and μ_3 are coprime they can be chosen as follows: $\mu_4 = 4$ and $\mu_3 = -1$. Unit determinant condition gives $\mu_1 = k$ and $\mu_2 = 1 - 4k$ with an arbitrary integer k. Finally, $b = 1 - 3k$ and the smallest positive b modulo 3 corresponds to $k = 0$. Hence, our matrix m_3 has the following representation

$$\begin{pmatrix} 4 & 1 \\ 1 & 1 \end{pmatrix} = \begin{pmatrix} 4 & -1 \\ 1 & 0 \end{pmatrix} \begin{pmatrix} 1 & 1 \\ 0 & 3 \end{pmatrix} .$$

An important property of the set M_p is that when one multiplies a matrix from this set by a matrix from the modular group the resulting matrix also belongs to M_p

$$M_p g = M_p \quad \text{for all } g \in \mathrm{PSL}(2, \mathbb{Z}) .$$

Let $\Psi(z)$ be an *automorphic function* of the modular group, i.e.

$$\Psi(gz) = \Psi(z) , \quad \text{for all } g \in \ \mathrm{PSL}(2, \mathbb{Z}) .$$

Then it is easy to see that the function

$$\Psi'(z) \equiv (T_p\Psi)(z) = \frac{1}{\sqrt{p}} \sum_{a,b,d} \Psi\left(\frac{az+b}{d}\right)$$

where the summation is performed over all $ad = p, d > 0, 0 \le b \le d-1$ will also be an automorphic function for the modular group. This is a consequence of the fact that in the right-hand side of this expression there is effectively the summation over all matrices from M_p. As M_p does not change after multiplication by a modular group matrix $\Psi'(z)$ is an automorphic function for the modular group. $(T_p\Psi)(z)$ is a kind of symmetrization of $\Psi(z)$ over images of z by all elements of M_p and the operators T_p are called Hecke operators.

These operators form a *commutative algebra* with the following product (see e.g. [54])

$$T_n T_m = \sum_{d|(n,m)} T_{nm/d^2} \tag{88}$$

where the summation is done over all divisors of the greatest common divisor of m and n. The most important case corresponds to Hecke operators with prime indices because all the others can be obtained from (88).

When p is a prime number

$$(T_p\Psi)(z) = \frac{1}{\sqrt{p}} \left[\Psi(pz) + \sum_{0 \le j < p} \Psi\left(\frac{z+j}{p}\right) \right] .$$

Since Hecke operators involve only fractional transformations all of them commute with the Laplace–Beltrami operator. Consequently, if $\Psi(x,y)$ is an eigenfunction of the Laplace–Beltrami operator, then $(T_p\Psi)(x,y)$ will also be an eigenfunction with the same eigenvalue. If there is no spectral degeneracy (which strongly suggested by numerics) every eigenfunction of the Laplace–Beltrami operator is in the same time an eigenfunction of all Hecke operators

$$(T_p\Psi_n)(x,y) = \lambda_p(n)\Psi_n(x,y) \,. \tag{89}$$

It is known (see e.g. [54]) that eigenfunctions of the Laplace–Beltrami operator for the modular group have the following Fourier expansion

$$\Psi_n(x,y) = y^{1/2}\sum_{p=-\infty}^{\infty} c_p(n)K_{s_n-1/2}(2\pi p y)e^{2\pi i p x}$$

where the eigenvalue of the Laplace–Beltrami operator $E_n = s_n(s_n-1)$ and $K_\nu(x)$ is the Hankel function.

One has $z = x+iy$ and $(az+b)/d = (ax+b)/d + iay/d$, therefore

$$(T_m\Psi_n)(x,y) = \frac{1}{\sqrt{m}}\sum_{a,b,d}\left(\frac{ay}{d}\right)^{1/2}\sum_p c_p(n)K_{s_n-1/2}\left(2\pi p\frac{ay}{d}\right)e^{2\pi i p(ax+b)/d}$$

where the first summation is performed over all a,b,d as in (87).

The summation over b gives zero if d does not divide p. Otherwise

$$(T_m\Psi_n)(x,y) = y^{1/2}\sum_{d|p,d|m} c_p(n)K_{s_n-1/2}(2\pi y p m/d^2)e^{2\pi i p m x/d^2} \,.$$

Let $k = m/d$ and $u = pm/d^2$. Then $p = mu/k^2$ and

$$(T_m\Psi_n)(x,y) = y^{1/2}\sum_u\sum_{k|(m,u)} c_{mu/k^2}(n)K_{s_n-1/2}(2uy)e^{2\pi i u x} \,.$$

If $T_m\Psi_n = \lambda_m(n)\Psi_n$ then by comparing the first Fourier coefficient one gets

$$c_m(n) = \lambda_m(n)c_1 \,.$$

Assuming $c_1 \neq 0$ and using a convenient normalization $c_1 = 1$ one concludes that *eigenvalues of the Hecke operators coincide with the Fourier coefficients.*

We note also that similarly to the construction of the Selberg trace formula one can build the trace formulas for Hecke operators (see e.g. [24] and references therein). Such trace formula schematically has the form (cf. (29))

$$\sum_n \lambda_p(n)h(k_n) = \frac{1}{\sqrt{p}}\sum_{\text{hyperbolic}}\frac{l_p}{2\sinh(L_p/2)}g(L_p)$$
$$+ \text{ smooth, parabolic, and elliptic terms} \,.$$

Here $h(k)$ is a test function like in Sect. 2.8 and $g(l)$ is its Fourier transform. In the left-hand side the summation is performed over all eigenvalues $E_n =$

$k_n^2 + 1/4$ of the Laplace–Beltrami operator and $\lambda_p(n)$ is the eigenvalue of the Hecke operator T_p (89) applied to the eigenfunction of the Laplace–Beltrami operator with eigenvalue E_n. In the right-hand side the summation is done over all 'hyperbolic' matrices from M_p with $\mathrm{Tr}\ m_p \neq p+1$. L_p is the 'length' associated with matrix m_p

$$2\cosh(L_p/2) = |\mathrm{Tr}\ m_p|/\sqrt{p}$$

and l_p is the minimal length of modular group matrices commuting with m_p.

6 Jacquet–Langlands Correspondence

Another curious fact about arithmetic groups is the Jacquet–Langlands correspondence (see [40]) which claims that for an arithmetic group derived from a quaternion group over $\mathbb{Q}$ (with a finite fundamental domain) one can find a subgroup of the modular group (with infinite fundamental domain) in such a way that amongst all automorphic eigenvalues of the Laplace–Beltrami operator for this modular subgroup one can find all eigenvalues of the compact arithmetic group.

The simplest arithmetic group Γ derived from quaternion algebra over $\mathbb{Q}$ with division is (see Sect. 2.2)

$$\Gamma = \begin{pmatrix} k_1 + k_2\sqrt{a} & k_3 + k_4\sqrt{a} \\ b(k_3 - k_4\sqrt{a}) & k_1 - k_2\sqrt{a} \end{pmatrix}$$

where b is a prime number, a is an integer such that the equation $x^2 \equiv a \pmod{b}$ has no integer solution (e.g. $a = 3$, $b = 5$), and integers k_i are such that

$$\det(\gamma) = k_1^2 - ak_2^2 - bk_3^2 + abk_4^2 = 1 \,.$$

Denote $z = x + \mathrm{i}y$, $\tau = u + \mathrm{i}v$ $(y, v > 0)$ and define for *all* n_j

$$\alpha = n_1 + n_2\sqrt{a}\,,\ \beta = n_3 + n_4\sqrt{a}\,,$$

$$\gamma = b(n_3 - n_4\sqrt{a})\,,\ \delta = n_1 - n_2\sqrt{a}\,.$$

Fix an arbitrary z_0 and compute the following kernel

$$\Phi(\tau, z) = \sum_{n_j=-\infty}^{+\infty} \exp K(\tau, z)$$

where

$$K(\tau, z) = -\pi \mathrm{Im}\tau \frac{|\alpha z_0 + \beta - z(\gamma \bar{z}_0 + \delta)|^2}{\mathrm{Im}\ z \mathrm{Im} z_0} + 2\pi\mathrm{i}\bar{\tau}(\alpha\delta - \beta\gamma)\,.$$

Here $\bar{z}$ is the complex conjugate of z.

Let $\psi_n(z)$ be an eigenfunction of the Laplace–Beltrami operator automorphic with respect to the quaternion group Γ. It means that

- $(\Delta_{L-B} + E_n)\,\psi_n(z) = 0$,
- $\psi_n(Mz) = \psi_n(z)$ for all $M \in \Gamma$.

Then the function

$$\Psi(\tau) = \mathrm{Im}\tau \int_{\mathcal{D}} \Phi(\tau, z)\psi_n(z)\frac{dxdy}{y^2}$$

where the integral is taken over the fundamental domain of the group Γ is an eigenfunction the Laplace–Beltrami operator with the same eigenvalue E_n but automorphic with respect to the *congruence* subgroup of the modular group $\Gamma_0(4ab)$ where

$$\Gamma_0(N) = \begin{pmatrix} m & n \\ k & l \end{pmatrix} \in \mathrm{SL}(2, \mathbb{Z})$$

with an additional condition that

$$k \equiv 0 \;(\bmod N)\;.$$

Direct (but tedious) proof of this statement can be found in [40].

7 Non-Arithmetic Triangles

In the precedent Section we have seen that arithmetic systems have the Poisson spectral statistics. But what about non-arithmetic models?

Let us consider, as example, the so-called *Hecke* triangles which are hyperbolic triangles with angles $(0, \pi/2, \pi/n)$. All of them tesselate the upper half-plane and are fundamental domains of the discrete groups generated by reflections across their sides. The modular billiard is one of them corresponding to $n = 3$. Similar to it they all have an infinite cusp.

According to Table 1 the Hecke triangles are arithmetic only for $n = 3, 4, 6, \infty$. All these arithmetic triangles have an exponential degeneracies of periodic orbit lengths which leads to the Poisson-like statistics of energy levels.

The simplest non-arithmetic Hecke triangle is the one with $n = 5$. At Fig. 19 we present the results of numerical calculations of the nearest-neighbor distribution for 6000 first energy levels for this triangle with the Dirichlet boundary conditions. For others Hecke triangles one gets similar pictures. It is clearly seen that numerics agrees very well with the predictions of the Gaussian Orthogonal ensembles of random matrices as it should be for generic chaotic models.

But what are the multiplicities of periodic orbit lengths for non-arithmetic Hecke triangles? As these model are not-arithmetic, one would expect that their length multiplicities should be equal to two as for generic time-reversal invariant systems. Nevertheless numerical calculations (see [24] for details) demonstrated that this is not always the case. At Fig. 20 we present the numerically computed mean length multiplicities for the Hecke triangles with

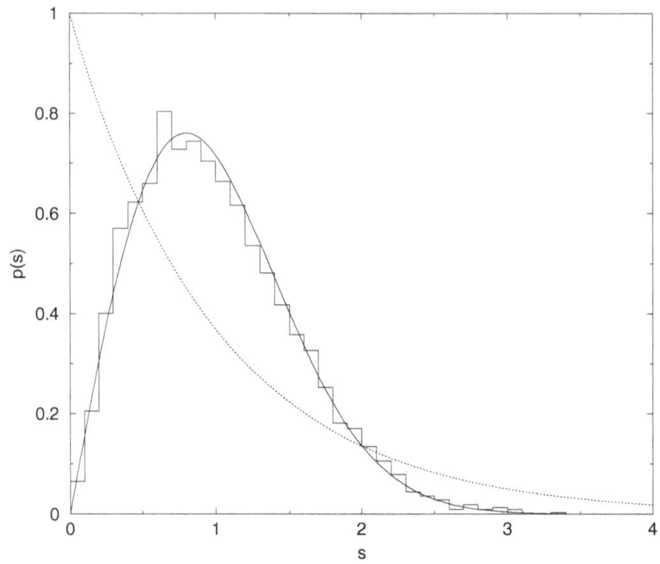

Fig. 19. The nearest-neighbor distribution of 6000 energy levels for the non-arithmetic Hecke triangular billiard with $n = 5$. The solid line – the GOE prediction. Dotted line – the Poisson result

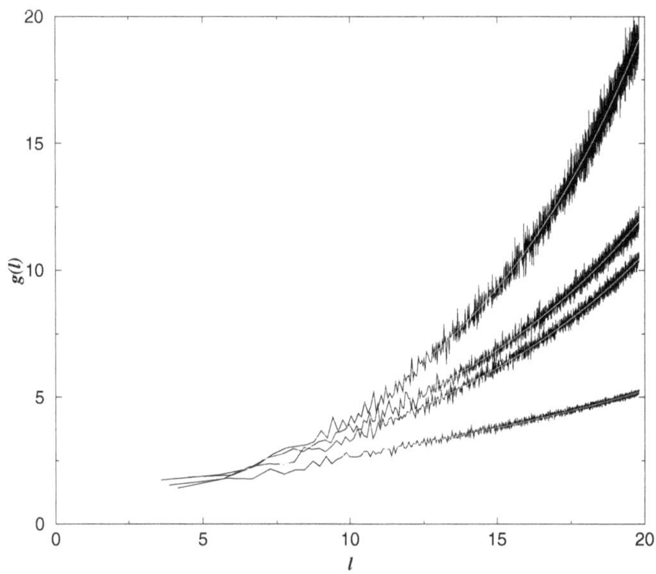

Fig. 20. Mean length multiplicities of periodic orbits for the Hecke triangles with (from top to bottom) $n = 12$, $n = 5$, $n = 8$, and $n = 10$. White lines are numerical fits (90)

$n = 5,\ 8,\ 10,\ 12$ for lengths $l < 20$. White lines indicate a two-parameter fit to these data in the form $\bar{g}(l) = a_n e^{b_n l}$

$$\begin{aligned} n = 5 \ &: \ \bar{g}(l) \approx 1.235 e^{.114 l} \ , \\ n = 8 \ &: \ \bar{g}(l) \approx 1.095 e^{.114 l} \ , \\ n = 10 \ &: \ \bar{g}(l) \approx 1.143 e^{.065 l} \ , \\ n = 12 \ &: \ \bar{g}(l) \approx .986 e^{.150 l} \ . \end{aligned} \tag{90}$$

These expressions fit numerical data in the given interval of lengths quite well and indicate that for, at least, certain Hecke triangles mean length multiplicity increases exponentially. We stress that (90) are only the best least-square numerical fits and no attempts were made to determine the accuracy of coefficients.

The discussion of the origin of such unexpected multiplicities for non-arithmetic triangles is beyond the scope of these lectures (on this subject see [26]). However it is of interest to understand why exponentially large multiplicities of periodic orbit lengths do not contradict the observed GOE behaviour of spectral statistics (cf. Fig. 19).

Assume that a system has an exponentially large number of periodic orbits with the same length l increasing as

$$g(l) \sim \frac{\mathrm{e}^{\lambda l}}{l}$$

with $\lambda \leq 1/2$.

Let us repeat the arguments of Sect. 1.2 for this case with exact degeneracicies. In Sect. 1.2 it was demonstrated that periodic orbits with different lengths can be treated in the diagonal approximation if

$$l_{p_1} - l_{p_2} \gg \frac{k}{\Delta E} \tag{91}$$

where k is the momentum and ΔE is the width of the energy average inherent in the definition of correlation functions of dynamical systems.

As the density of orbits with *different* lengths is

$$\rho_{\text{diff. lengths}} \approx \frac{\mathrm{e}^{l}}{g(l) l} \sim \mathrm{e}^{(1-\lambda) l}$$

it follows that the inequality (91) is valid till maximal length

$$l_m \sim \frac{1}{1-\lambda} \ln \frac{\Delta E}{k} \sim \frac{1}{1-\lambda} \ln k \ . \tag{92}$$

Notice that due to assumed large multiplicity l_m is different from (51).

From (78) it follows that the two-point correlation form factor in the diagonal approximation up to numerical factor is

$$K(t) \sim \frac{k}{l}|A(l)|^2 g(l)\mathrm{e}^l$$

where $t = l/2k$ and $A(l) \sim l\mathrm{e}^{l/2}/k$. Combining all terms together one obtains that during the maximal time of applicability of the diagonal approximation $t_m = l_m/2k$ with l_m from (92) the form factor increases till

$$K(t_m) \sim \frac{\mathrm{e}^{\lambda l_m}}{k} \sim k^{(2\lambda-1)/(1-\lambda)} \ .$$

Hence, if $\lambda = 1/2$ as for arithmetic systems the two-point correlation form factor during the time of validity of the diagonal approximation increases till a constant value of the order of 1. But if $\lambda < 1/2$ the form factor for the time of validity can reach only a value of the order of $k^{-\nu}$ with $\nu = (1-2\lambda)/(1-\lambda) > 0$. As $k \to \infty$ this value tends to zero and no apparent contradiction with standard random matrix ensembles can be derived within the diagonal approximation.

8 Summary

Arithmetic groups are a special sub-class of discrete groups characterized by the existence of a representation by matrices with integer elements. A readable mathematical review of such groups is given in [42]. There are two types of arithmetic groups. The first includes groups commensurable with the modular group and having non-compact fundamental domains with infinite cusps. The second type of compact arithmetic groups combines groups commensurable with groups derived from quaternion algebras with division. These groups have finite fundamental domains.

From classical viewpoint the free motion on surfaces generated by arithmetic groups is as chaotic as for any hyperbolic surfaces. But quantum mechanics on these arithmetic surfaces is very special. In particular, spectral statistics of the Laplace–Beltrami operator automorphic with respect to arithmetic group is described by the Poisson statistics typical for integrable systems and not by the random matrix statistics typical for chaotic models.

The origin of this peculiarity can be traced to the existence in arithmetic systems of a very large number of periodic orbits with exactly the same length. For all arithmetic groups the mean multiplicity of periodic orbits with length l behaves like $\mathrm{e}^{l/2}/l$. This has to be compared with the total density of periodic orbits which for all discrete groups is e^l/l. It is the cumulative effect of the interference of many periodic orbits with the same length which changes drastically the spectral statistics.

In the diagonal approximation the two-point correlation form factor $K(t)$ for arithmetic systems at small t increases exponentially like e^{kt}/k and during the Ehrenfest time (which is the limit of applicability of the diagonal approximation) reaches a constant value.

More detailed information can be obtained for the modular group where it is possible to compute the two-point correlation form factor analytically. The final answer is

$$K(t) = \frac{1}{2\pi^2 k} \sum_{(p,q)=1} \left| \frac{q}{p} \beta(p,q) \right|^2 \delta(t - t_{p,q})$$

where

$$t_{p,q} = \frac{1}{k} \ln \frac{kq}{\pi p}$$

and $\beta(p,q)$ is a number-theoretical function given in Sect. 4.3.

This formula means that the two-point form factor for the modular group is a sum over δ-functions at special points $t_{p,q}$ situated in a vicinity of the Ehrenfest time. The set of δ-functions is dense but the largest peaks correspond to the smallest ratios p/q. Nevertheless, small peaks with $p/q \ll 1$ are much more numerous and integrally they dominate. In the limit t fixed and $k \to \infty$ $K(t) \to \bar{d}$ thus confirming the Poisson nature of the spectral statistics of the modular group.

Arithmetic groups have many interesting properties. Hecke operators and the Jacquet–Langlands correspondence are the most remarkable.

Acknowledgement

It is a pleasure to thank Charles Schmit for his aid in the preparation of these lectures and A.M. Odlyzko for presenting numerical data of the two-point correlation function of Riemann zeros. Laboratoire de Physique Théorique et Modèles Statistique is Unité Mixte de Recherche de l'Université Paris XI et du CNRS (UMR 8626).

References

1. D. Alonso and P. Gaspard, $\hbar$-Expansion for the Periodic Orbit Quantization of Chaotic Systems, Chaos 3 (1993) 601-612; Erratum *ibid* 4 (1994) 105.
2. A.V. Andreev and B.L. Altshuler, Spectral Statistics beyond Random Matrix Theory, Phys. Rev. Lett. 75 (1995) 902-905.
3. R. Aurich, M. Sieber, and F. Steiner, Quantum Chaos on the Hadamard-Gutziller problem, Phys. Rev. Lett. 61 (1988) 483.
4. A. Aurich and F. Steiner, On the Periodic Orbits of a Strongly Chaotic System, Physica D 32 (1988) 451.
5. H.P. Baltes and E.R. Hilf, Spectra of Finite Systems, Wissenschaftverlag, Mannheim, 1976.
6. R. Aurich, E. Bogomolny, and F. Steiner, Periodic Orbits on the Regular Octagon, Physica D 48 (1991) 91-101.

7. N.L. Balasz and A. Voros, Chaos on the Pseudosphere, Phys. Rep. 143 (1986) 109.
8. R. Balian and C. Bloch, Distribution of eigenfrequencies for the wave equation in a finite domain: I Three-dimensional problems with smooth boundary surface, Ann. Phys. 60 (1970) 401; II Electromagnetic Field. Riemannian spaces, Ann. Phys. 64 (1971) 271; III Eigenfrequency density fluctuations, Ann. Phys. 69 (1972) 76; *ibid* Asymptotic evaluation of the Green's function for large quantum numbers, Ann. Phys. 63 (1971) 592; *ibid* Solutions of the Schrödinger equation in terms of classical paths, Ann. Phys. 85 (1974) 514.
9. M.V. Berry and M. Tabor, Closed Orbits and the Regular Bound Spectrum, Proc. R. Soc. Lond. A 349 (1976) 101-123.
10. M.V. Berry and M. Tabor, Level Clustering in the Regular Spectrum, Proc. R. Soc. London A 356 (1977) 375-394.
11. M.V. Berry, Semiclassical Theory of Spectrum Rigidity, Proc. R. Soc. London A 400 (1985) 229-251.
12. M.V. Berry, Semiclassical formula for the number variance of the Riemann zeros, Nonlinearity, 1 (1988) 399-407.
13. M.V. Berry, Some Quantum-to-Classical Asymptotics, in [28] (1989) 251-303.
14. M.V. Berry and C.J. Howls, High Orders of the Weyl Expansion for Quantum Billiards, Resurgence of Periodic Orbits, and the Stokes Phenomenon, Proc. R. Soc. Lond A 447 (1994) 527-555.
15. M.V. Berry and J.P. Keating, $H = xp$ and the Riemann zeros, in 'Supersymmetry and trace formulas', eds. I.V. Lerner and J.P. Keating, Plenum, New York (1999) 355-367.
16. O. Bohigas, Random Matrix Theories and Chaotic Dynamics, in [28] (1989) 87-199.
17. O. Bohigas, M.-J. Giannoni, and C. Schmit, Characteristic of Chaotic Quantum Spectra and Universality of Level Fluctuations Law, Phys. Rev. Lett. 52 (1984) 1: Spectral Properties of the Laplacian and Random Matrix Theory, J. Physique Lett. 45 (1984) L-1015.
18. E. Bogomolny, B. Georgeot, M.J. Giannoni, and C. Schmit, Chaotic Billiards Generated by Arithmetic Groups, Phys. Rev. Lett. 69 (1992) 1477-1480.
19. E.Bogomolny and C. Schmit, Semiclassical Computation of High-Excited Energy Levels, Nonlinearity 6 (1993) 523-547.
20. E. Bogomolny, Introduction to models on constant negative curvature surfaces, in Quantum Dynamics of Simple Systems, The Forty Fourth Scottish Universities Summer School in Physics, Stirling, 1994, Eds. G-L Oppo, S. M. Barnett, E. Riis, and M. Wilkinson.
21. E. Bogomolny and J.P. Keating, Gutzwiller's Trace Formula and Spectral Statistics: Beyond the Diagonal Approximation, Phys. Rev. Lett. 77 (1996) 1472-1475.
22. E. Bogomolny and J.P. Keating, Random Matrix Theory and the Riemann Zeros I: Nonlinearity 8 (1995) 1115-1131; Random Matrix Theory and the Riemann Zeros II: n-point correlations, Nonlinearity 9 (1996) 911-935.
23. E. Bogomolny, F. Leyvraz, and C. Schmit, Distribution of Eigenvalues for the Modular Group, Commun. Math. Phys. 176 (1996) 577-617.
24. E. Bogomolny, B. Georgeot, M.-J. Giannoni, and C. Schmit, Arithmetic Chaos, Phys. Rep. 291 (1997) 219-324.
25. E. Bogomolny, Spectral Statistics and Periodic Orbits, in Proceedings of the International School of Physics 'Enrico Fermi', Varenna (1999), New Directions

in Quantum Chaos, Eds. G. Casati, I. Guarneri, and U. Smilansky, IOS Press, Amsterdam, Oxford, Tokyo, Washington, 2000, 333-369.
26. E. Bogomolny and C. Schmit, Multiplicities of Periodic Orbits Lengths for Non-Arithmetic Models, J. Phys. A: Math. Gen. 37 (2004) 4501-4526.
27. J. Bolte, G. Steil, and F. Steiner, Arithmetic Chaos and Violations of Universality in Energy Level Statistics, Phys. Rev. Lett. 69 (1992) 2188.
28. Chaos and Quantum Physics, Proceedings of the Les Houches Summer School (1989) Eds. M.-J. Giannoni, A. Voros, J. Zinn-Justin. North Holland, Amsterdam, London, New York, Tokyo, 1991.
29. Y. Colin de Verdière, Hyperbolic Geometry in Two-Dimensions and Trace Formulas, in [28] (1989) 305-330.
30. A. Connes, Trace Formula in Non-Commutative Geometry and the Zeros of the Riemann Zeta Function, (1997) arXiv: math.NT/9811068.
31. H. Davenport, Multiplicative Number Theory, revised by H. Montgomery, Springer-Verlag, New York, Heidelberg, Berlin, 1980.
32. A. Erdélyi et al. Higher Transcendental Functions Vol. 1 (Bateman Manuscript Project). McGraw–Hill, New York, 1953.
33. M.C. Gutzwiller, Chaos in Classical and Quantum Mechanics, Springer, New York, 1990.
34. J.H. Hannay and A.M. Ozorio de Almeida, J. Phys. A 17 (1984) 3429-3440.
35. G.H. Hardy and J.E. Littlewood, Some Problems of 'Partitio Numerorum'; III: On the expression of a Number as a Sum of Primes, Acta Mathematica, 44 (1923) 1-70.
36. G.H. Hardy and E.M. Wright, An Introduction to the Theory of Numbers, Clarenton Press, Oxford, 1979.
37. E. Hecke, Lectures on Dirichlet Series, Modular Functions and Quadratic Forms, Vandenhoeck and Ruprecht, Gottingen, 1983.
38. D. Hejhal, The Selberg Trace Formula and the Riemann Zeta Function, Duke Math. J. 43 (1976) 441-482.
39. D. Hejhal, The Selberg Trace Formula for PSL(2,R), Vol. 1, Lectures Notes in Mathematics 548 (1979); Vol. 2, *ibid* 1001 (1983).
40. D. Hejhal, A classical approach to a well known spectral correspondence on quaternion groups, in Number Theory, D.V. Chudnovsky, G.V. Chudnovsky, H. Cohn, M.B. Nathanson Eds., Lectures Notes in Mathematics 1135 (1985) 127.
41. R.D. Horowitz, Characters of Free Groups Represented in the Two-Dimensional Special Linear Group, Comm. Pure Appl. Math. 25 (1972) 635.
42. S. Katok, Fuchsian Groups, University of Chicago Press, Chicago and London, 1992.
43. J.P. Keating and N.C. Snaith, Random Matrix Theory and $\zeta(1/2+it)$, Commun. Math. Phys. 214 (2000) 57-89.
44. A.G. Kurosh, Lectures in General Algebra, Pergamon Press, Oxford, London, Edinburgh, New York, Paris, Frankfurt, 1965.
45. H.L. Montgomery, The pair correlation of zeros of zeta-function, Proc. Symp. Pure Math. (1973) 181-193.
46. M.L. Mehta, Random Matrices and the Statistical Theory of Energy levels, Academic Press, New York, 1967.
47. A.M. Odlyzko, On the Distribution of Spacing Between Zeros of Zeta Function, Math. of Comp. 48 (1987) 273.
48. The web site of A.M. Odlyzko: www.dtc.umn.edu/∼odlyzko/.

49. A.M. Odlyzko, private communication (2003).
50. Manfred Peter, The Correlation Between Multiplicities of Closed Geodesics on the Modular Surface (2001) arXiv: math.NT/0104234.
51. B. Randol, The Length Spectrum of Riemann Surface is Always of Unbounded Multiplicity, Proc. Amer. Math. Soc. 78 (1980) 455.
52. C. Schmit, Quantum and Classical Properties of Some Billiards on the Hyperbolic Plane, in [28] (1989) 331-369.
53. K. Takeuchi, On some Discrete Subgroups of SL(2,R), J. Fa. Sci. Univ. Tokyo Sect. 1A 16 (1969) 97-100; A Characterization of Arithmetic Fuchsian Groups, J. Math. Soc. Japan 27 (1975) 600-612; Arithmetic Triangle Groups, J. Math. Soc. Japan 29 (1977) 91-106; Commensurability Classes of Arithmetic Triangles Groups, J. Fac. Sci. Univ. Tokyo Sect. 1A 24 (1977) 201-212.
54. A. Terras, Harmonic Analysis on Symmetric Spaces and Applications, Springer, Berlin, 1979.
55. E. C. Titchmarsh, The Theory of the Riemann Zeta–Function. Oxford, Clarenton Press, 1951.
56. M.F. Vignéras, Arithmétique des algèbres de quaternions, Lectures Note in Mathematics, 800 (1980).

Notes on L-functions and Random Matrix Theory

J. Brian Conrey

American Institute of Mathematics, Palo Alto, CA conrey@aimath.org

1 Introduction

The connection between L-functions and random matrix theory began with the 1972 discovery by Montgomery and Dyson that the zeros of the Riemann zeta-function seem to be distributed on the 1/2-line like the eigenvalues of large random hermitian matrices (the GUE ensemble studied by physicists). Since then there has been a lot of development. In the 1990s Hejhal generalized Montgomery's work to triple correlation and Rudnick and Sarnak to n-correlation for arbitrary n. In the last five years though, the subject has really taken off due to several developments. One was the conjectures of Keating-Snaith and Conrey-Farmer on moments of zeta- and L-functions and another was the development of the notion of symmetry type of families of L-functions by Katz-Sarnak. Added to these was the work of Conrey-Ghosh, Conrey-Gonek, Duke-Friedlander-Iwaniec, Kowalski-Michel-Vanderkam, Jutila, Motohashi, Ivic, Soundararajan, Rubinstein, and others on moments of families of L-functions which provide the evidence for the random matrix conjectures.

We now have at our disposal very accurate models for the behavior of L-functions (i.e. the distribution of values including zeros) and can use these to make arithmetical predictions. Three examples of conjectures on L-functions that have been motivated by random matrix theory will illustrate some of this development.

Conjecture 1 [CFKRS1]: *Let $\zeta(s)$ denote the Riemann zeta-function. Then, for any $\epsilon > 0$,*

$$\frac{1}{T}\int_0^T |\zeta(1/2+it)|^6 \, dt = P_3(\log \frac{T}{2\pi}) + O_\epsilon(T^{\epsilon-1/2})$$

where

$$\begin{aligned} P_3(x) = {} & 0.00000570852\,x^9 + 0.00040502\,x^8 + 0.011072\,x^7 + 0.148400\,x^6 \\ & + 1.04592\,x^5 + 3.98438\,x^4 + 8.607319\,x^3 + 10.274330\,x^2 \\ & + 6.593913\,x + 0.916515. \end{aligned}$$

Conjecture 2 [CKRS]: *Let $E : y^2 = x^3 + Ax + B$ be an elliptic curve, where A and B are integers. For a squarefree integer d, let E_d be the twisted elliptic curve which has equation $d^2y = x^3 + Ax + B$. Let r_d denote the rank of the* *elliptic curve E_d; this is just the number of independent points with rational coordinates required to generate the group of all rational points on the curve. Now restrict attention to those d^* for which the rank is at least 2. Then we conjecture (a) that there exist constants C_E and ν_E which depend only on E such that*

$$\sum_{d^* \leq x} 1 \sim C_E x^{3/4} (\log x)^{\nu_E}$$

and (b) *that for any prime p, the limit as $x \to \infty$ of the ratio of $d^* \leq x$ with d^* being a square modulo p to the $d^* \leq x$ which are non-square modulo p is equal to*

$$\sqrt{\frac{p+1-a_p}{p+1+a_p}}$$

where $a_p = p - N_p$ and N_p is the number of solutions of $y^2 \equiv x^3 + Ax + B$ modulo p.

Conjecture 3 [U]: *There exists a number $c > 0$, a sequence of $N \to \infty$ with an elliptic curve E_N of conductor N such that the rank of E_N is at least $c\frac{\log N}{\log \log N}$.*

The first two of these conjectures have been numerically tested extensively with excellent agreement.

One of the purposes of these notes will be to explain the relationship of conjectures like these to random matrix theory. Another purpose is to gather in one place a collection of useful formulas and examples for researchers working in these fields. We begin with the random matrix side of the story and finish with the L-function side; there are a few brief comments about finite field zeta-functions in the middle. Much of the material is taken from [C], [CFKRS], and [CKRS].

2 Random Matrix Theory

2.1 The Classical Groups

In these notes we will be interested in the classical compact matrix groups with their Haar measures. In the case of the unitary group, this ensemble coincides with what is called CUE in the physics literature. However, the orthogonal and symplectic groups we are interested in are different from COE and CSE in the physics literature. Below we define the groups and give their Haar measures and indicate some of the statistics we are interested in and how to compute them.

- The **unitary group** $\mathrm{U}(N)$ is the group of $N \times N$ matrices U with entries in $\mathbb{C}$ for which $UU^* = I$ where U^* denotes the conjugate transpose of U, i.e. if $U = (u_{i,j})$, then $U^* = (\overline{u_{j,i}})$.

- The **special orthogonal group** SO(N). This is the subgroup of U(N) consisting of real matrices with determinant 1. SO($2N$) leads to the symmetry type O$^+$ and SO($2N$+1) leads to the symmetry type O$^-$. In these notes we will deal only with SO($2N$) and O$^+$; we will abuse notation somewhat and refer to this as O.
- The **symplectic group** USp($2N$) is the subgroup of U($2N$) of matrices U for which $UZU^t = Z$ where U^t denotes the transpose of U and

$$Z = \begin{pmatrix} 0 & I_N \\ -I_N & 0 \end{pmatrix}$$

Since all of our matrices are unitary, their eigenvalues all have modulus 1. The $N \times N$ matrix $A \in \mathrm{U}(N)$ has N eigenvalues on the unit circle, $e^{i\theta_1}, \dots e^{i\theta_N}$ with the eigenangles satisfying $0 \le \theta_1 \le \theta_2 \cdots \le \theta_N < 2\pi$. Thus, the average spacing between the eigenangles is $2\pi/N$. In the cases of Sp($2N$) and SO($2N$) the eigenvalues come in pairs $e^{\pm i\theta_j}$ with $0 \le \theta_1 \le \cdots \le \theta_N < \pi$ and the average spacing between the eigenangles is π/N. If we want to scale the eigenangles to have average spacing 1, then we consider $\tilde{\theta_j} = N\theta_j/\pi$ or $N\theta_j/(2\pi)$, whichever is appropriate.

2.2 The Weyl Integration Formula

In order to do analysis on these groups we need to know how to work with the measures. Weyl's formula provides a convenient way to reduce the integrations to ordinary multiple integrals.

The conjugacy classes of $N \times N$ unitary matrices can be parametrized by their N eigenvalues on the unit circle. Any configuration of N points on the unit circle corresponds to a conjugacy class in U(N). If $f(A) = f(\theta) = f(\theta_1, \dots \theta_N)$ is a symmetric function of N variables, then Weyl's formula for the Haar measure gives

$$\int_{U(N)} f(A) dA_{\mathrm{U}(N)} = \frac{1}{(2\pi)^N N!} \int_{[0,2\pi]^N} f(\theta) \prod_{1 \le j < k \le N} |e^{i\theta_j} - e^{i\theta_k}|^2 \, d\theta_1 \dots d\theta_N$$

where $dA_{\mathrm{U}(N)}$ is the Haar measure. Similarly, on Sp($2N$) and SO($2N$) we have respectively

$$\int_{\mathrm{Sp}(N)} f(A) dA_{\mathrm{Sp}(2N)} = \frac{2^{N^2}}{\pi^N N!} \int_{[0,\pi]^N} f(\theta) \prod_{j<k} (\cos\theta_j - \cos\theta_k)^2 \prod_{j=1}^N \sin^2\theta_j \prod_{j=1}^N d\theta_j \,;$$

$$\int_{SO(2N)} f(A) dA_{\mathrm{SO}(2N)} = \frac{2^{(N-1)^2}}{\pi^N N!} \int_{[0,\pi]^N} f(\theta) \prod_{j<k} (\cos\theta_j - \cos\theta_k)^2 \prod_{j=1}^N d\theta_j \,.$$

For example, if you wanted to compute the average of the square of the absolute value of the trace of 3×3 unitary matrices you would compute the integral

$$\frac{1}{6(2\pi)^3}\int_0^{2\pi}\int_0^{2\pi}\int_0^{2\pi}|e^{i\theta_1}+e^{i\theta_2}+e^{i\theta_3}|^2$$
$$|e^{i\theta_1}-e^{i\theta_2}|^2|e^{i\theta_1}-e^{i\theta_3}|^2|e^{i\theta_2}-e^{i\theta_3}|^2\;d\theta_1\;d\theta_2\;d\theta_3;$$

which equals 1.

Alternate Formulation of the Haar Measure

The Weyl formulas above are especially useful for computing moments of characteristic polynomials, as we shall soon see. For computing "local statistics" such as the neighbor spacing, especially in the large N limit, it is useful to have the Weyl formulas for the Haar measures written in an alternate form.

Let

$$L_{\mathrm{U}(N)}(\theta_j,\theta_k)=\exp\left(\tfrac{-i(N-1)(\theta_j-\theta_k)}{2}\right)\sum_{n=0}^{N-1}e^{in(\theta_j-\theta_k)}=\frac{\sin\frac{N(\theta_j-\theta_k)}{2}}{\sin\frac{\theta_j-\theta_k}{2}}$$

$$\begin{aligned}L_{\mathrm{Sp}(2N)}(\theta_j,\theta_k)&=2\sum_{n=1}^{N}\sin n\theta_j\sin n\theta_k\\&=\frac{\sin(N+\frac12)(\theta_j-\theta_k)}{2\sin\frac{\theta_j-\theta_k}{2}}-\frac{\sin(N+\frac12)(\theta_j+\theta_k)}{2\sin\frac{\theta_j+\theta_k}{2}}\end{aligned}$$

and

$$\begin{aligned}L_{\mathrm{SO}(2N)}(\theta_j,\theta_k)&=1+2\sum_{n=1}^{N-1}\cos n\theta_j\cos n\theta_k\\&=\frac{\sin(N-\frac12)(\theta_k-\theta_k)}{2\sin\frac{\theta_j-\theta_k}{2}}+\frac{\sin(N+\frac12)(\theta_j+\theta_k)}{2\sin\frac{\theta_j+\theta_k}{2}}.\end{aligned}$$

Then

$$dA_{\mathrm{U}(N)}=\frac{1}{(2\pi)^N N!}\det_{N\times N}(L_{\mathrm{U}(N)}(\theta_j,\theta_k))\prod_{j=1}^{N}d\theta_j$$

$$dA_{\mathrm{Sp}(2N)}=\frac{1}{\pi^N N!}\det_{N\times N}(L_{\mathrm{Sp}(2N)}(\theta_j,\theta_k))\prod_{j=1}^{N}d\theta_j$$

$$dA_{\mathrm{SO}(2N)}=\frac{1}{\pi^N N!}\det_{N\times N}(L_{\mathrm{SO}(2N)}(\theta_j,\theta_k))\prod_{j=1}^{N}d\theta_j$$

A key thing to note is that since

$$\lim_{N\to\infty}N\sin\tfrac{x}{N}=\sin x$$

it follows that

$$\lim_{N\to\infty} \frac{1}{N^n} \det_{n\times n} L_{\mathrm{U}(N)}\left(\frac{2\pi x_j}{N}, \frac{2\pi x_k}{N}\right) = \det_{n\times n}\left(\frac{\sin\pi(x_j-x_k)}{\pi(x_j-x_k)}\right).$$

Similarly,

$$\lim_{N\to\infty} \frac{1}{N^n} \det_{n\times n} L_{\mathrm{Sp}(2N)}\left(\frac{\pi x_j}{N}, \frac{\pi x_k}{N}\right) = \det_{n\times n}\left(\frac{\sin\pi(x_j-x_k)}{\pi(x_j-x_k)} - \frac{\sin\pi(x_j+x_k)}{\pi(x_j+x_k)}\right)$$

and

$$\lim_{N\to\infty} \frac{1}{N^n} \det_{n\times n} L_{\mathrm{SO}(2N)}\left(\frac{\pi x_j}{N}, \frac{\pi x_k}{N}\right) = \det_{n\times n}\left(\frac{\sin\pi(x_j-x_k)}{\pi(x_j-x_k)} + \frac{\sin\pi(x_j+x_k)}{\pi(x_j+x_k)}\right).$$

2.3 Four Statistics

In these notes we are particularly interested in theorems and conjectures about moments and values of characteristic polynomials of random matrices and L-functions (these represent a “global” perspective) and we are also interested in things like spacing of consecutive eigenvalues and zeros (these represent a “local” perspective). In this section we describe some local statistics of interest. Note that in the physics literature “n-level density” and “n-correlation” for GUE, GOE, and GSE are not distinguished because the associated measures are rotationally invariant. Here, however, the distinction is important.

There is a simple combinatorial relation (described in Sect. 2.6) between neighbor-spacing statistics and correlation statistics.

For the matrix groups $U(N)$, $SO(2N)$, and $Sp(2N)$ we are interested in the statistics “nearest neighbor” and “n-level correlation” are the same for all three groups whereas “n-level density” and “jth eigenvalue” are different for all three groups. In the physics literature these statistics are different for all three matrix ensembles GUE, GOE, and GSE. Now we describe these statistics.

Suppose we have a sequence $\mathcal{T} = \{T_N\}_{N=1}^{\infty}$ of N-tuples of numbers $T_N = (t_{N,1}, t_{N,2}, \dots, t_{N,N})$ where $t_{N,1} \le t_{N,2} \le \dots \le t_{N,N}$ and such that $t_{N,N} - t_{N,1} \sim N$ as $N \to \infty$, so that the average spacing is 1.

- The n**-level density** of $\mathcal{T}$ is $W(x_1, \dots x_n)$ means that

$$\lim_{N\to\infty} \sum_{\substack{(i_1,\dots i_n), i_j\le N\\ i_j\ne i_k}} f(t_{N,i_1}, \dots t_{N,i_n})$$
$$= \int_{\mathbb{R}^n} f(x_1,\dots,x_n) W(x_1,\dots,x_n)\, dx_1,\dots,dx_n\,.$$

  for any compactly supported smooth symmetric function $f : \mathbb{R}^n \to \mathbb{C}$.
- The j**-th lowest zero density** is $\nu_j(x)$ means that for any compactly supported smooth function $f : \mathbb{R} \to \mathbb{C}$

$$\lim_{N\to\infty} \frac{1}{N} \sum_{n\le N} f(t_{n,j}) = \int_0^\infty f(x)\nu_j(x)\, dx\,.$$

- The **consecutive spacing density** or **nearest neighbor density** is $\mu(x)$ means that for any compactly supported smooth function $f : \mathbb{R}^n \to \mathbb{C}$ we have

$$\lim_{N\to\infty} \frac{1}{N} \sum_{i\leq N-1} f(t_{N,i+1} - t_{N,i}) = \int_0^\infty f(x)\mu(x)\, dx\,.$$

Wigner conjectured that $\mu(x)$ is equal to $\frac{\pi}{2}xe^{-\frac{\pi}{4}x^2}$. This conjecture, which is very accurate for small x, is known as Wigner's surmise.

- The n**-correlation density** is $V(x_1,\ldots,x_n)$ means that for any smooth symmetric function f that depends only on the differences of the variables (i.e. $f(x_1+u,\ldots,x_n+u) = f(x_1,\ldots,x_n)$ for all u), and is rapidly decaying on the hyperplane $P_n : \{(x_1,\ldots,x_n : \sum x_i = 0\}$, we have, as $N\to\infty$,

$$\lim_{N\to\infty} \frac{1}{N} \sum_{\substack{t_{i_1},\ldots t_{i_n}\in T_N \\ i_j\neq i_k}} f(t_{i_1},\ldots t_{i_n})$$
$$= \int_{P_n} f(x_1,\ldots,x_n)V(x_1,\ldots,x_n)\, dx_1 \ldots\, dx_{n-1}$$

as $N\to\infty$. The spacing and n-correlation densities are universal, i.e. the same for each of O, Sp, and U, whereas the n-level and j-th lowest zero densities depend on the symmetry type.

2.4 Formulas for the Density Functions

In this section we compile the formulas for the density functions of the statistics described in the last section. In the next two sections we describe some of the computations leading to these formulae.

The 1-level density functions are

$$W(\mathrm{O})(x) = 1 + \frac{\sin 2\pi x}{2\pi x}\,,$$
$$W(\mathrm{Sp})(x) = 1 - \frac{\sin 2\pi x}{2\pi x}\,,$$
$$W(\mathrm{U})(x) = 1\,.$$

- The n–**level density** is

$$W_\epsilon(x_1,\ldots,x_n) = \det_{n\times n}\left(\frac{\sin\pi(x_i - x_j)}{\pi(x_i - x_j)} - \epsilon\frac{\sin\pi(x_i + x_j)}{\pi(x_i + x_j)}\right)$$

where $\epsilon = 0$ for U; $\epsilon = 1$ for Sp; $\epsilon = -1$ for O^+.

- The **lowest zero density** is $\nu_1(x)$ where

$$\nu_1(x) = -\frac{d}{dx}\prod_{j=0}^{\infty}(1-\lambda_j(x)) \qquad \mathrm{U}\,;$$

$$\nu_1(x) = -\frac{d}{dx}\prod_{j=0}^{\infty}(1-\lambda_{2j+1}(2x)) \qquad \mathrm{Sp};$$

$$\nu_1(x) = -\frac{d}{dx}\prod_{j=0}^{\infty}(1-\lambda_{2j}(2x)) \qquad \mathrm{O},$$

where $1 \ge \lambda_0(x) \ge \lambda_1(x) \cdots \ge 0$ are the eigenvalues of the integral equation

$$\int_{-x/2}^{x/2} \frac{\sin \pi(t-u)}{\pi(t-u)} f(u)\, du = \lambda(x) f(t)$$

- The **consecutive spacing density** is

$$\mu(x) = \frac{d^2}{dx^2}\prod_{j=0}^{\infty}(1-\lambda_j(x)).$$

- The n-**correlation density** is $V(x_1,\dots,x_n) = W_0(x_1,\dots,x_n)$,

2.5 Gaudin's Lemma

The critical device for calculating the density functions for the "local" statistics is due to Gaudin and is described in this section.

Associated to each $N \times N$ unitary matrix A are its N eigenvalues $\exp(i\theta_j)$ where $0 \le \theta_1 \le \cdots \le \theta_N \le 2\pi$. We let $X(A)$ denote this sequence of θ. We integrate a function $F(A)$ over $\mathrm{U}(N)$ by parametrizing the group by the θ_i and using Weyl's formula to convert the integral into an N-fold integral over the θ_i.

Often one wants to integrate with respect to Haar measure over a group G a function $\tilde{f}(A) = \tilde{f}(\theta_1,\dots,\theta_N)$ of N variables that is "lifted" from a function f of n variables:

$$\tilde{f}(\theta_1,\dots,\theta_N) = \sum_{1\le i_1<i_2<\cdots<i_n\le N} f(\theta_{i_1},\dots,\theta_{i_n})$$

where the sum is over all possible n-tuples $(i_1,\dots,i_n)$ of distinct integers between 1 and N. Gaudin's lemma gives a simplification of this computation from an N-fold integral to an n-fold integral. Define the measure $dA_{N,\mathrm{U}(n)}$ on $\mathrm{U}(n)$ by

$$dA_{N,\mathrm{U}(n)} = \frac{1}{n!(2\pi)^n}\det_{n\times n}(L_{\mathrm{U}(N)}(\theta_j,\theta_k))\prod_{j=1}^{n} d\theta_j,$$

the measure $dA_{2N,\mathrm{Sp}(2n)}$ on $\mathrm{Sp}(2n)$ by

$$dA_{2N,\mathrm{Sp}(2n)} = \frac{1}{n!\pi^n}\det_{n\times n}(L_{\mathrm{Sp}(2N)}(\theta_j,\theta_k))\prod_{j=1}^{n} d\theta_j,$$

and the measure $dA_{2N,\mathrm{SO}(2n)}$ on $\mathrm{SO}(2n)$ by

$$dA_{2N,\mathrm{SO}(2n)} = \frac{1}{n!\pi^n} \det_{n\times n}\left(L_{\mathrm{SO}(2N)}(\theta_j,\theta_k)\right) \prod_{j=1}^{n} d\theta_j,$$

Then Gaudin's Lemma asserts that

$$\int_{\mathrm{U}(N)} \tilde{f}(A) dA_{\mathrm{U}(N)} = \int_{\mathrm{U}(n)} f(A) dA_{N,\mathrm{U}(n)}$$

and

$$\int_{\mathrm{Sp}(2N)} \tilde{f}(A) dA_{\mathrm{Sp}(2N)} = \int_{\mathrm{Sp}(2n)} f(A) dA_{2N,\mathrm{Sp}(2n)}$$

and

$$\int_{\mathrm{SO}(2N)} \tilde{f}(A) dA_{\mathrm{SO}(2N)} = \int_{\mathrm{SO}(2n)} f(A) dA_{2N,\mathrm{SO}(2n)}$$

One way to view Gaudin's Lemma is just as a collection of integration formulas; for example

$$\binom{N}{n} \int_{[0,2\pi]^{N-n}} \det_{N\times N}\left(S_N(\theta_j-\theta_k)\right) \frac{d\theta_{n+1}\dots d\theta_N}{(2\pi)^N N!} = \det_{n\times n}\left(S_N(\theta_j-\theta_k)\right)\frac{1}{(2\pi)^n n!}$$

where $S_N(x) = (\sin Nx/2)/(\sin x/2)$.

We illustrate the utility of Gaudin's formula by computing the n-level density function for $\mathrm{U}(N)$. Let $f(x_1,\dots,x_n)$ be a suitable test function. To compute the n-level density we need to evaluate

$$\lim_{N\to\infty} \int_{U(N)} \sum_{\substack{(i_1,\dots,i_n)\\ i_j\neq i_k}} f\left(\tfrac{N\theta_{i_1}}{2\pi},\dots,\tfrac{N\theta_{i_n}}{2\pi}\right) \prod_{j<k} |e^{i\theta_k} - e^{i\theta_j}|^2 \, d\theta_1 \dots \, d\theta_N$$

where we have rescaled the eigenangles of the matrix $A \in U(N)$ so that they have mean spacing 1. The sum over tuples $(i_1,\dots,i_n)$ with the i_j distinct is $n!$ times the sum over ordered tuples $1 \le i_1 < i_2 < \dots < i_N \le N$. Thus by Gaudin's lemma and after using the new expression for the Haar measure and changing variables $\theta_j \to 2\pi x_j/N$, the above is equal to

$$\lim_{N\to\infty} \frac{1}{N^n} \int_{[0,N]^n} f(x_1,\dots,x_n) \det_{n\times n} L_{\mathrm{U}(N)}(2\pi x_j/N, 2\pi x_k/N) \prod_{j=1}^{n} dx_j.$$

Thus, our integral is

$$= \int_{\mathbb{R}^n} f(x_1,\dots,x_n) \det_{n\times n}\left(\frac{\sin\pi(x_j-x_k)}{\pi(x_j-x_k)}\right) \, dx_1 \dots \, dx_n,$$

so that $W_{U,n}(x_1,\dots,x_n) - \det_{n\times n}\left(\frac{\sin\pi(x_j-x_k)}{\pi(x_j-x_k)}\right)$. Similar calculations lead to the density functions for $\mathrm{Sp}(2N)$ and $\mathrm{SO}(2N)$.

As a second example we sketch the computation of the pair–correlation statistic for each group. Let $G_\epsilon(N)$ stand for $U(N)$ if $\epsilon=0$, $Sp(2N)$ if $\epsilon=1$, and $SO(2N)$ if $\epsilon=-1$. Let $f(x,y)=f(y,x)=g(x-y)$ where g is even, smooth, and compactly supported (or of rapid decay). To compute the pair-correlation statistic for the G_ϵ we need to evaluate

$$\lim_{N\to\infty}\frac{1}{N}\int_{G(N)}\sum_{\substack{i_1,i_2\in\{1,\dots,N\}\\ i_1\neq i_2}} f\left(\tfrac{N\theta_{i_1}}{(2-|\epsilon|)\pi},\tfrac{N\theta_{i_2}}{(2-|\epsilon|)\pi}\right)dA_{G(N)}.$$

By Gaudin's lemma and a change of variables, this is

$$=\lim_{N\to\infty}\frac{1}{N^3}\int_0^N\int_0^N g(x_1-x_2)\det_{2\times 2}\left(L_{G(N)}\left(\tfrac{(2-|\epsilon|)\pi x_j}{N},\tfrac{(2-|\epsilon|)\pi x_k}{N}\right)\right)dx_1\,dx_2.$$

For large N,

$$\frac{1}{N^2}\det_{2\times 2}\left(L_{G(N)}\left(\tfrac{(2-|\epsilon|)\pi x_j}{N},\tfrac{(2-|\epsilon|)\pi x_k}{N}\right)\right)\approx$$

$$\det\begin{pmatrix}1-\epsilon\frac{\sin 2\pi x_1}{2\pi x_1} & \frac{\sin\pi(x_1-x_2)}{\pi(x_1-x_2)}-\epsilon\frac{\sin\pi(x_1+x_2)}{\pi(x_1+x_2)}\\ \frac{\sin\pi(x_1-x_2)}{\pi(x_1-x_2)}-\epsilon\frac{\sin\pi(x_1+x_2)}{\pi(x_1+x_2)} & 1-\epsilon\frac{\sin 2\pi x_2}{2\pi x_2}\end{pmatrix}.$$

Now let $u=x_1-x_2$ and $v=x_2$ in the double integral. Then the double integral is asymptotically

$$\frac{1}{N}\int_{-N}^{N}g(u)\int_{\min\{0,-u\}}^{\max\{N,N-u\}}\det\begin{pmatrix}1-\epsilon\frac{\sin 2\pi(u+v)}{2\pi(u+v)} & \frac{\sin\pi u}{\pi u}-\epsilon\frac{\sin\pi(u+2v)}{\pi(u+2v)}\\ \frac{\sin\pi u}{\pi u}-\epsilon\frac{\sin\pi(u+2v)}{\pi(u+2v)} & 1-\epsilon\frac{\sin 2\pi(u+v)}{2\pi(u+v)}\end{pmatrix}dv\,du.$$

The integrals with respect to v of the terms with a $\sin(u+v)$ or $\sin(u+2v)$ will be bounded. The other terms are constant with respect to v. Since the length of the integration in the v-integral is $N-|u|$, our expression is

$$\sim\int_{-N}^{N}g(u)\left(1-\frac{|u|}{N}\right)\det\begin{pmatrix}1 & \frac{\sin\pi u}{\pi u}\\ \frac{\sin\pi u}{\pi u} & 1\end{pmatrix}du$$

$$\to\int_{-\infty}^{\infty}g(u)\det\begin{pmatrix}1 & \frac{\sin\pi u}{\pi u}\\ \frac{\sin\pi u}{\pi u} & 1\end{pmatrix}du.$$

Finally, the last expression is

$$=\frac{1}{2}\int_{x_1+x_2=0}f(x_1,x_2)\det\begin{pmatrix}1 & \frac{\sin\pi(x_1-x_2)}{\pi(x_1-x_2)}\\ \frac{\sin\pi(x_1-x_2)}{\pi(x_1-x_2)} & 1\end{pmatrix}dx_1.$$

Similarly, in computing the n-correlation statistic we are led to consider

$$\frac{1}{N}\int\cdots\int_{[0,N]^n}f(x_1,\dots,x_n)\det_{n\times n}\left(\frac{\sin\pi(x_j-x_k)}{\pi(x_j-x_k)}-\epsilon\frac{\sin\pi(x_j+x_k)}{\pi(x_j+x_k)}\right)dx_1\dots dx_n$$

as $N\to\infty$.

2.6 Some Notation From Katz-Sarnak and a Combinatorial Identity

The purpose of this section is to explain some notation from Katz-Sarnak and to give a combinatorial lemma which shows how to pass between neighbor spacing statistics and correlation statistics. Let $a \geq 0$ be an integer and $s \geq 0$ real. Let $G(N)$ denote one of the groups $U(N)$, $SO(2N)$ or $Sp(2N)$ and for $A \in G(N)$ let $X(A)$ be the sequence $X(1) \leq X(2) \leq \ldots X(N)$ be the sequence of eigenangles in increasing order (from 0 to 2π for $U(N)$ and from 0 to π for $SO(2N)$ or $Sp(2N)$. Define

$$\mathrm{Sep}(a)(s) := \#\{(i,j) : 1 \leq i < j \leq N, j - i = a + 1, X(j) - X(i) = s\}$$

and

$$\mathrm{Clump}(a)(s) := \{1 \leq i_0 < i_1 < \cdots < i_{a+1} \leq N : X(i_{a+1}) - X(i_0) = s\}.$$

Clearly Sep is related to neighbor spacing statistics and Clump is related to correlation statistics. Their relationship is given by

$$\mathrm{Clump}(a)(s) = \sum_{b \geq a} \binom{b}{a} \mathrm{Sep}(b)(s).$$

This leads to the identities between generating functions:

$$\sum_{a=0}^{\infty} \mathrm{Clump}(a)(s) T^a = \sum_{b=0}^{\infty} \mathrm{Sep}(b)(s)(T+1)^b$$

and

$$\sum_{b=0}^{\infty} \mathrm{Sep}(b)(s) T^b = \sum_{a=0}^{\infty} \mathrm{Clump}(a)(s)(T-1)^a$$

Katz and Sarnak further define

$$\mathrm{Sep}(a, f) := \mathrm{Sep}(a, f, N, X) = \int f(s) d\mathrm{Sep}(a, s)$$

for a one variable integrable function f. From this definition, it is an easy matter to scale the sequence X so as to have mean spacing 1 (replace $X(A)$ by $NX(A)/(2\pi)$ when $G(N) = U(N)$, for example); then integrate over $G(N)$ with the Haar measure, divide by N and let $N \to \infty$ to obtain the ath neighbor spacing statistic. Similarly with Clump and the a-correlation function. Katz and Sarnak also define these statistics for vectors a and multi-variable integrable functions f.

2.7 First Eigenvalue and Neighbor Spacings

We sketch some of the ideas needed to compute the density functions for these statistics. (This is taken from Mehta's book.) Initially, we work with the unitary group, for which the measure is rotationally invariant. Let

$$B_N(\alpha) = \int_{[\alpha, 2\pi-\alpha]^N} dA_{\mathrm{U}(N)}$$

so that $B_N(\alpha)$ is the measure of the set of unitary $N \times N$ matrices with all eigenangles in $[\alpha, 2\pi - \alpha]$. Now we use the fact that $dA_{\mathrm{U}(N)}$ is essentially the square of a Vandermonde determinant:

$$\begin{aligned}(2\pi)^N N! dA_{\mathrm{U}(N)} &= \prod_{1\le j<k\le N} |e^{i\theta_k} - e^{i\theta_j}|^2 = \prod_{1\le j<k\le N} (e^{i\theta_k} - e^{i\theta_j})(e^{-i\theta_k} - e^{-i\theta_j}) \\ &= \left|\det\left(e^{i(j-1)\theta_k}\right)\right|^2 .\end{aligned}$$

Now we use Gram's identity in the form: for an interval $\mathcal{S}$ and functions ϕ,

$$\frac{1}{N!}\int_{\mathcal{S}^N} |\det_{N\times N}(\phi_j(x_k))|^2 \, dx_1 \ldots dx_N = \det_{N\times N}\left(\int_{\mathcal{S}} \phi_j(x)\overline{\phi_k(x)}\, dx\right).$$

(By the way, Gram's formula with $S = [0, 2\pi]$ and $\phi_j(x) = e^{i(j-1)x}$ gives a quick proof that the Haar measure for the unitary group has total mass 1.) Thus,

$$\begin{aligned}B_N(\alpha) &= \frac{1}{(2\pi)^N}\det_{N\times N}\left(\int_\alpha^{2\pi-\alpha} e^{i(k-j)y}\, dy\right) \\ &= \det_{N\times N}\left(\delta_{jk} - \frac{1}{2\pi}\int_{-\alpha}^{\alpha} e^{i(k-j)y}\, dy\right)\end{aligned}$$

This determinant is just the characteristic polynomial, evaluated at 1, of the matrix $A = (a_{jk})$ with entries $a_{jk} = \frac{1}{2\pi}\int_{-\alpha}^{\alpha} e^{i(k-j)y}\, dy$ and so is equal to $\prod_{j=1}^N (1 - \lambda_{j,N}(\alpha))$ where the $\lambda_{j,N}(\alpha)$ are the eigenvalues of A. It is easy to check that the eigenvalues of A are the same as the eigenvalues of the integral operator

$$(K\psi)(x) = \frac{1}{2\pi}\int_{-\alpha}^{\alpha}\sum_{j=0}^{N-1} e^{ij(x-y)}\psi(y)\, dy$$

which acts on the N-dimensional vector space of trigonometric polynomials of degree $N-1$; an eigenvector $\mathbf{v} = (v_0, \ldots v_{N-1})$ of A with eigenvalue λ corresponds to an eigenfunction $\psi(x) = \sum_{j=0}^{N-1} v_j e^{ijx}$ with the same eigenvalue.

The kernel function

$$\sum_{j=0}^{N-1} e^{ij(x-y)} = L_{\mathrm{U}(N)}(x, y)$$

satisfies

$$\lim_{N\to\infty} L_{\mathrm{U}(N)}(2\pi x/N, 2\pi y/N) = \frac{\sin\pi(x-y)}{\pi(x-y)}.$$

Thus, we are led to consider the eigenvalues of the integral equation with the kernel function $\frac{\sin\pi(x-y)}{\pi(x-y)}$. Then it can be deduced (see [KS] for the details of the limiting process) that

$$B(\alpha) := \lim_{N\to\infty} B_N(N\alpha/(2\pi)) = \prod_{n=1}^{\infty}(1-\lambda_n(\alpha))$$

where the $\lambda_n(\alpha)$ are the eigenvalues of the integral operator with kernel $\frac{\sin\pi(x-y)}{\pi(x-y)}$ on the interval $[-\alpha,\alpha]$. This formula gives a rapidly converging infinite product for the desired density function. See Mehta [Meh] for tables, graphs and more information about these functions. The function $B(\alpha)$ is the probability that no rescaled eigenvalues are smaller than α in absolute value. Then it is a simple matter to conclude that the probability of the smallest eigenvalue being α is $-B'(\alpha)$ and the probability density function for the nearest neighbor spacing is $B''(\alpha)$.

This argument works for the unitary group because of the rotational invariance of the measure. However, we have already pointed out that the nearest neighbor spacing can be determined combinatorially from the correlation functions and that the correlation functions are the same for all three of our $G(N)$. Therefore, this calculation gives the nearest neighbor spacing statistic for all three groups.

For the lowest eigenvalue these calculations can be carried out for the other groups using the kernels $\frac{\sin\pi(x-y)}{\pi(x-y)} - \epsilon\frac{\sin\pi(x+y)}{\pi(x+y)}$.

2.8 The Selberg Integral

For computing global statistics such as moments of characteristic polynomials we require Selberg's integral, which is a generalization of the Euler's beta-integral. There are many versions of Selberg's integral; three of them follow (see [Meh]).

$$\int_0^1 \cdots \int_0^1 |\Delta(x)|^{2\gamma} \prod_{j=1}^{n} x_j^{\alpha-1}(1-x_j)^{\beta-1}\, dx_1 \ldots dx_n$$
$$= \prod_{j=0}^{n-1} \frac{\Gamma(1+\gamma+\gamma j)\Gamma(\alpha+j\gamma)\Gamma(\beta+j\gamma)}{\Gamma(1+\gamma)\Gamma(\alpha+\beta+(n+j-1)\gamma)}$$

for $\Re\alpha > 0, \Re\beta > 0, \Re\gamma > -\min(\frac{1}{n}, \frac{\Re\alpha}{n-1}, \frac{\Re\beta}{n-1})$. Here

$$\Delta(x) = \prod_{1\le j<\ell\le n}(x_j - x_\ell).$$

Alternatively,

$$\int_{-1}^{1}\cdots\int_{-1}^{1}\prod_{1\le i<j\le N}|x_i-x_j|^{2\gamma}\prod_{j=1}^{n}(1-x_j)^{\alpha-1}(1+x_j)^{\beta-1}dx_j$$
$$=2^{\gamma n(n-1)+n(\alpha+\beta-1)}\prod_{j=0}^{n-1}\frac{\Gamma(1+\gamma+j\gamma)\Gamma(\alpha+j\gamma)\Gamma(\beta+j\gamma)}{\Gamma(1+\gamma)\Gamma(\alpha+\beta+\gamma(n+j-1))}.$$

Another version has

$$\int_{0}^{\infty}\cdots\int_{0}^{\infty}|\Delta(x)|^{2\gamma}\prod_{j=1}^{n}x_j^{\alpha-1}(1+x)^{-\alpha-\beta-2\gamma(n-1)}\prod_{j=1}^{n}dx_j$$
$$=\prod_{j=0}^{n-1}\frac{\Gamma(1+\gamma+\gamma j)\Gamma(\alpha+j\gamma)\Gamma(\beta+j\gamma)}{\Gamma(1+\gamma)\Gamma(\alpha+\beta+(n+j-1)\gamma)}.$$

2.9 Characteristic Polynomials of Random Matrices

In this section we describe the properties of characteristic polynomials of unitary matrices, expressing things in a way that highlights the connection with number theory. You may find it helpful to refer to section 4 for the basic properties of L-functions.

Let

$$\Lambda(s)=\Lambda_A(s)=\det(I-As)$$

denote the characteristic polynomial of an $N\times N$ matrix A. We want to compare characteristic polynomials with L-functions, so here we outline some of the relevant properties of these characteristic polynomials.

If we expand $\Lambda(s)$, we obtain

$$\Lambda(s)=\sum_{n=0}^{N}a_n s^n,$$

which is analogous to the Dirichlet series representation for L-functions.

- Analytic continuation: Since $\Lambda(s)$ is a polynomial, it is an entire function.
- Functional equation: Since A is unitary, we have

  $$\Lambda(s)=(-1)^N s^N\det A\det(I-A^\dagger s^{-1}),$$

  and so, writing

  $$\det A=e^{i\phi}$$

  (where unitarity implies that $\phi\in R$), we have

  $$\Lambda(s)=(-1)^N s^N e^{i\phi}\overline{\Lambda}(\tfrac{1}{s}).$$

This plays the same role for $\Lambda(s)$ as the functional equation for L-functions: it represents a symmetry with respect to the unit circle ($s = re^{i\alpha} \to \frac{1}{s} = \frac{1}{r}e^{-i\alpha}$).
As just indicated, the critical line corresponds to the unit circle, and furthermore, the half-plane to the right of the critical line corresponds to the interior of the unit circle.

- Critical values: The critical point for $\Lambda(s)$ is $s = 1 = e^{i\cdot 0}$, and $\Lambda(1)$ is the critical value.
- Location of zeros: Since A is unitary, its eigenvalues all have modulus 1, so the zeros of $\Lambda(s)$ lie on the unit circle.
- Average spacing of zeros: Since the $N \times N$ matrix A has N eigenvalues on the unit circle, the average spacing between zeros of $\Lambda_A(s)$ is $2\pi/N$. When modeling a family of L-functions, we choose N as a function of the conductor of $L(s)$ so that the L-function and the characteristic polynomial have the same average spacing between their zeros.
- Approximate functional equation
$$\sum_{n=0}^{N} a_n s^n = (-1)^N e^{i\phi} \sum_{n=0}^{N} \overline{a_n} s^{N-m}$$
and so
$$a_n = (-1)^N e^{i\phi} \overline{a_{N-n}}.$$
Hence, when N is odd, we have
$$\Lambda(s) = \sum_{n=0}^{\frac{N-1}{2}} a_n s^n + (-1)^N e^{i\phi} s^N \sum_{n=0}^{\frac{N-1}{2}} \overline{a_n} s^{-n},$$
which corresponds to the approximate functional equation for L-functions. When N is even, there is an additional term: $a_{\frac{N}{2}} s^{\frac{N}{2}}$.

The above discussion applies to any unitary matrix. We also consider matrices which, in addition to being unitary, are also either symplectic or orthogonal. We use these three ensembles of matrices to model families of L-functions. While the notion of "family of L-functions" has not yet been made precise, we give several natural examples in Sect. 4. Associated to each family is a "symmetry type" which identifies the matrix ensemble which will be used to model the family. This correspondence is most easily seen in terms of the sign of the functional equation, which is analogous to the determinant of the matrix. If A is unitary symplectic, then $\det A = 1$ (i.e. $\phi = 0$), and if A is orthogonal, then $\det A = \pm 1$. Correspondingly, the functional equations for L-functions with unitary symmetry involve a (generally complex) phase factor, whereas for L-functions with symplectic symmetry this phase factor is unity, and in the case of orthogonal symmetry it is either $+1$ or -1.

2.10 Moments of Characteristic Polynomials

Keating and Snaith [KSn1] and [KSn2] calculated exact formulas for the moments of characteristic polynomials of our ensembles for the purpose of comparing with moments of families of L-functions. These are calculated from Weyl's formulas together with appropriate uses of Selberg's integral formula.

$$\begin{aligned} M_{\mathrm{U}(N)}(\lambda) &= \int_{\mathrm{U}(N)} |\det(I - Ae^{it}))|^{2\lambda}\, dA \\ &= \prod_{j=1}^{N} \frac{\Gamma(j)\Gamma(j+2\lambda)}{\Gamma(j+\lambda)^2}, \end{aligned}$$

$$\begin{aligned} M_{\mathrm{Sp}(2N)}(\lambda) &= \int_{\mathrm{Sp}(2N)} |\det(I - A)|^{\lambda}\, dA \\ &= 2^{2N\lambda} \prod_{j=1}^{N} \frac{\Gamma(1+N+j)\Gamma(1/2+\lambda+j)}{\Gamma(1/2+j)\Gamma(1+\lambda+N+j)}, \end{aligned}$$

$$\begin{aligned} M_{\mathrm{SO}(2N)}(\lambda) &= \int_{\mathrm{SO}(2N)} |\det(I - A)|^{\lambda}\, dA \\ &= 2^{N\lambda} \prod_{j=1}^{N} \frac{\Gamma(N+j-1)\Gamma(\lambda+j-1/2)}{\Gamma(j-1/2)\Gamma(\lambda+j+N-1)}. \end{aligned}$$

In the first formula, t is any real number and the right side is independent of t. The right side of this first formula can be re-written in the case λ is an integer k as a polynomial in N:

$$g(k,N) := \frac{(N+1)(N+2)^2 \dots (N+k)^k (N+k+1)^{k-1} \dots (N+2k-1)}{1 \cdot 2^2 \cdot 3^3 \dots k^k (k+1)^{k-1} \dots (2k-1)}.$$

For fixed k, this is asymptotic to

$$\frac{N^{k^2}}{1 \cdot 2^2 \cdot 3^3 \dots k^k (k+1)^{k-1} \dots (2k-1)} = \frac{g_k}{k^2!} N^{k^2}$$

as $N \to \infty$. Keating and Snaith identified the number g_k as the critical constant that was missing from number theorist's attempts to formulate a conjecture for the $2k$th moment of the Riemann zeta-function. We will see that the role of N is played by $\log \frac{T}{2\pi}$ in the case of the zeta-function. Similar remarks apply to the moments for the other groups and their relation to moments of families of L-functions.

2.11 Lower Order Terms and Permutation Sums

The polynomial $g(k, N)$ above can be expanded and expressions for the lower order terms in N explicitly displayed; however, it transpires that this is not a good way to approach lower order terms for moments of L-functions. For the more detailed analysis of lower order terms in our moment formulas for L-functions we need a different method than Selberg's integral. Basically, it is necessary to consider shifted moments and to look at a kind of additive reformulation of $g(k, N)$.

The formulas we arrive at are slightly complicated but can be simplified through some integral formulas which we relate here. Here is an example of one of our formulas. (Recall that $\Lambda_A(s) = \det(I - As)$.)

$$\begin{aligned}\int_{U(N)} &\Lambda_A(e^{-\alpha})\Lambda_A(e^{-\beta})\Lambda_{A^*}(e^{-\gamma})\Lambda_{A^*}(e^{-\delta})\, dA_N \\ &= Z(\alpha,\beta,\gamma,\delta) + e^{N(\alpha+\gamma)}Z(-\gamma,\beta,-\alpha,\delta) + e^{N(\alpha+\delta)}Z(-\delta,\beta,\gamma,-\alpha) \\ &\quad + e^{N(\beta+\gamma)}Z(\alpha,-\gamma,-\beta,\delta) + e^{N(\beta+\delta)}Z(\alpha,-\delta,\gamma,-\beta) \\ &\quad + e^{N(\alpha+\beta+\gamma+\delta)}Z(-\gamma,-\delta,-\alpha,-\beta)\end{aligned}$$

where

$$Z(\alpha,\beta,\gamma,\delta) = z(1+\alpha+\gamma)z(1+\alpha+\delta)z(1+\beta+\gamma)z(1+\gamma+\delta)$$

with $z(x) = 1/(1-e^x)$.

We will see that an analogous formula (with error term) holds for the shifted fourth moment of the zeta-function, but with $z(x)$ replaced by $\zeta(1+x)$ and with N replaced by $\log\frac{t}{2\pi}$ and with the inclusion of an arithmetic factor in each term.

Notice that the main term here consists of six terms. The sum of these six terms can be expressed as the residue of a four-fold contour integral as described below.

Suppose $F(a; b) = F(a_1, \ldots, a_k; b_1, \ldots, b_k)$ is a function of $2k$ variables, which is symmetric with respect to the first k variables and also symmetric with respect to the second set of k variables. Suppose also that F is regular near $(0, \ldots, 0)$. Suppose further that $f(s)$ has a simple pole of residue 1 at $s = 0$ but is otherwise analytic in a neighborhood about $s = 0$. Let

$$G(a_1, \ldots, a_k; b_1, \ldots b_k) = F(a_1, \ldots; \ldots, b_k)\prod_{i=1}^{k}\prod_{j=1}^{k} f(a_i - b_j).$$

If for all $1 \le i, j \le k$, $\alpha_i - \alpha_{j+k}$ is contained in the region of analyticity of $f(s)$ then

$$\begin{aligned}\sum_{\sigma\in\Xi} G(\alpha_{\sigma(1)}, \ldots, \alpha_{\sigma(k)}; \alpha_{\sigma(k+1)} \ldots \alpha_{\sigma(2k)}) &= \frac{(-1)^k}{k!^2}\frac{1}{(2\pi i)^{2k}} \times \\ \oint\cdots\oint \frac{G(z_1, \ldots, z_k; z_{k+1}, \ldots, z_{2k})\Delta(z_1, \ldots, z_{2k})^2}{\prod_{i=1}^{2k}\prod_{j=1}^{2k}(z_i - \alpha_j)}&\, dz_1 \ldots dz_{2k},\end{aligned}$$

where one integrates about small circles enclosing the α_j's, and where Ξ is the set of $\binom{2k}{k}$ permutations $\sigma \in S_{2k}$ such that $\sigma(1) < \cdots < \sigma(k)$ and $\sigma(k+1) < \cdots < \sigma(2k)$.

The above applies to the Unitary case. The next formula is useful in the Symplectic and Orthogonal cases.

Suppose F is a symmetric function of k variables, regular near $(0, \ldots, 0)$, and $f(s)$ has a simple pole of residue 1 at $s = 0$ and is otherwise analytic in a neighborhood of $s = 0$, and let

$$G(a_1, \ldots, a_k) = F(a_1, \ldots, a_k) \prod_{1 \le i \le j \le k} f(a_i + a_j)$$

or

$$G(a_1, \ldots, a_k) = F(a_1, \ldots, a_k) \prod_{1 \le i < j \le k} f(a_i + a_j).$$

If $\alpha_i + \alpha_j$ are contained in the region of analyticity of $f(s)$ then

$$\sum_{\epsilon_j \in \{-1,1\}} G(\epsilon_1 \alpha_1, \ldots, \epsilon_k \alpha_k) = \frac{(-1)^{k(k-1)/2}}{(2\pi i)^k} \frac{2^k}{k!}$$

$$\oint \cdots \oint G(z_1, \ldots, z_k) \frac{\Delta(z_1^2, \ldots, z_k^2)^2 \prod_{j=1}^k z_j}{\prod_{i=1}^k \prod_{j=1}^k (z_i - \alpha_j)(z_i + \alpha_j)} dz_1 \cdots dz_k,$$

and

$$\sum_{\epsilon_j \in \{-1,1\}} (\prod_{j=1}^k \epsilon_j) G(\epsilon_1 \alpha_1, \ldots, \epsilon_k \alpha_k) = \frac{(-1)^{k(k-1)/2}}{(2\pi i)^k} \frac{2^k}{k!}$$

$$\oint \cdots \oint G(z_1, \ldots, z_k) \frac{\Delta(z_1^2, \ldots, z_k^2)^2 \prod_{j=1}^k \alpha_j}{\prod_{i=1}^k \prod_{j=1}^k (z_i - \alpha_j)(z_i + \alpha_j)} dz_1 \cdots dz_k,$$

where the path of integration encloses the $\pm\alpha_j$'s.

3 Zeta and L-Functions Over Finite Fields

In transition to our discussion of L-functions we mention briefly the finite field analogues. These zeta-and L-functions are polynomials with all roots on a circle and so in fact are characteristic polynomials of matrices (orthogonal and symplectic). Deligne's equidistribution theorem allowed Katz and Sarnak to calculate local statistics for these zeros, which turn out to be the same as the statistics for the matrix groups, after scaling and taking large N limits.

Sarnak and Katz developed their theory of symmetry types of families of L-functions by studying zeta and L-functions over finite fields. In this context they were able to rigorously prove that the zeros of the families obeyed the statistics of the random matrix models; the relevant random matrix group depending on the "geometric monodromy" of the family of L-functions. We give a brief overview of some of their results.

To begin with consider the case of curves over finite fields. Let $\mathbb{F}_q$ be the finite field with q elements. Consider a homogeneous form $F(X,Y,Z)$ in three variables of degree d over $\mathbb{F}_q$. This corresponds to a curve of genus $g = d(d-1)/2$ provided that F and its partial derivatives have no zeros in common. The zeta-function for F is

$$Z(F/\mathbb{F}_q, T) = \exp\left(\sum_{n=1}^{\infty} \frac{N_n}{n} T^n\right)$$

where N_n is the number of solutions of $F(X,Y,Z) = 0$ in $\mathbb{F}_q^3$ (where two solutions (X,Y,Z) and (X',Y',Z') are the same if $(X',Y',Z') = \lambda(X,Y,Z)$ for some λ. Then, Z is a rational function of T of the form $P(T)/(1-T)(1-qT)$. Moreover,

$$P(T) = \prod_{j=1}^{2g} (1 - \alpha_j T)$$

where $|\alpha_j| = \sqrt{q}$ – this is the Riemann Hypothesis for Z. Writing $\alpha_j = \sqrt{q}e^{i\phi_j}$ with $0 \le \phi_1 \le \phi_2 \le \cdots \le \phi_{2g} < 2\pi$, we can study the distribution of the angles ϕ_j.

If we suitably normalize the angles and average with respect to all curves of genus g, then when we take the limit as $q \to \infty$ and then as $g \to \infty$ the resulting nearest neighbor distribution is the same as for the unitary group.

The lowest lying zero statistic for this collection is identical to that for Sp.

Another example of symmetry type Sp arises from curves of the form $y^2 = f(x)$ where $f(x)$ is monic and square-free.

An example where the orthogonal symmetry type arises is for $Dy^2 = x(x-1)(x-t)$ – quadratic twists of an elliptic curve. Here D is a fundamental discriminant and t varies

The critical component to their results is Deligne's equidistribution theorem.

See the book by Katz and Sarnak for more details.

4 L-functions

The definition of L-function which we give below is a slight modification of what has come to be called the "Selberg class" Let $s = \sigma + it$ with σ and t real. An *L-function* is a Dirichlet series

$$L(s) = \sum_{n=1}^{\infty} \frac{\lambda_n}{n^s},$$

with $\lambda_n \ll_\epsilon n^\epsilon$ for every $\epsilon > 0$, which has three additional properties.

- Analytic continuation: $L(s)$ continues to a meromorphic function of finite order, with at most finitely many poles, and all poles are located on the $\sigma = 1$ line.
- Functional equation: There is a number ε with $|\varepsilon| = 1$, and a function $\gamma_L(s)$ of the form

$$\gamma_L(s) = P(s)Q^s \prod_{j=1}^{k} \Gamma(w_j s + \mu_j)$$

where P is a polynomial whose only zeros in $\sigma > 0$ are at the poles of $L(s)$, $Q > 0$, $w_j > 0$, and $\Re\mu_j \geq 0$, such that

$$\xi_L(s) := \gamma_L(s)L(s)$$

is entire, and

$$\xi_L(s) = \varepsilon\overline{\xi_L}(1-s),$$

where $\overline{\xi_L}(s) = \overline{\xi_L(\overline{s})}$.
It is sometimes convenient to write the functional equation in asymmetric form:

$$L(s) = \varepsilon X_L(s)\overline{L}(1-s),$$

where $X_L(s) = \dfrac{\overline{\gamma_L}(1-s)}{\gamma_L(s)}$.
- Euler product: For $\sigma > 1$ we have

$$L(s) = \prod_p L_p(1/p^s),$$

where the product is over the primes p, and

$$L_p(1/p^s) = \sum_{k=0}^{\infty} \frac{\lambda_{p^k}}{p^{ks}} = \exp\left(\sum_{k=1}^{\infty} \frac{b_{p^k}}{p^{ks}}\right),$$

where $b_n \ll n^\theta$ with $\theta < \frac{1}{2}$.
Note that $L(s) \equiv 1$ is the only constant L-function, the set of L-functions is closed under products, and if $L(s)$ is an L-function then so is $L(s+iy)$ for any real y. An L-function is called *primitive* if it cannot be written as a nontrivial product of L-functions, and it can be shown, assuming Selberg's orthonormality conjectures, that any L-function has a unique representation as a product of primitive L-functions. It is believed that L-functions only arise from arithmetic objects, such as characters, automorphic forms, and automorphic representations. Very little is known about L-functions beyond those cases which have been shown to be arithmetic.
There are several interesting consequences of the above properties, and many conjectures which have been established in few (or no) cases. We highlight some of those properties which have random matrix analogues as described in Sect. 2.8.

- Location of zeros: Since $\xi_L(s)$ is entire, $L(s)$ must vanish at the poles of the Γ-functions in the γ_L factor. These are known as the *trivial zeros* of the L-function. By the functional equation and the Euler product, the only other possible zeros of $L(s)$ lie in the *critical strip* $0 \le \sigma \le 1$. By the argument principle, the number of nontrivial zeros with $0 < t < T$ is asymptotically $(W/\pi)T \log T$, where $W = \sum w_j$. The *Riemann Hypothesis* for $L(s)$ asserts that the nontrivial zeros of $L(s)$ lie on the *critical line* $\sigma = \frac{1}{2}$. The much weaker (but still deep) assertion that $L(s) \neq 0$ on $\sigma = 1$ has been proven for arithmetic L-functions.
- Average spacing of zeros: By the zero counting result described above, the average gap between zeros of $L(s)$ with imaginary part T is $\pi/W \log T$. In the notation of (1.1.2), the average spacing between the first few zeros of $L(s)$ is $\log Q + W + \sum |\Im(\mu_j)|$.
- Critical values: The value $L(\frac{1}{2})$ is called the *critical value* of the L-function. The significance of $s = \frac{1}{2}$ is that it is the symmetry point of the functional equation. The mean values we study in this paper are averages of (powers of) critical values of L-functions, where the average is taken over a "family" of L-functions.
  Note. If the set $\{\mu_j\}$ is stable under complex conjugation and the λ_n are real, then ε is commonly called the *sign of the functional equation*. If the sign is -1 then $L(s)$ has an odd order zero at $s = \frac{1}{2}$; more generally, if the sign is not 1 then $L(\frac{1}{2}) = 0$. When $L(\frac{1}{2})$ vanishes, it is common to use the term 'critical value' for the first nonzero derivative $L^{(j)}(\frac{1}{2})$, but in this paper we use 'critical value' to mean 'value at the critical point.'
- Approximate functional equation and analytic conductor: A standard tool for studying analytic properties of L-functions is an approximate functional equation for $L(s)$:

$$L(s) = \sum_{1 \le n \le x} \frac{\lambda_n}{n^s} + \varepsilon X_L(s) \sum_{1 \le n \le y} \frac{\overline{\lambda_n}}{n^{1-s}} + error \quad term$$

  where $xy = (t/2\pi)^{2W}$. The name comes from the fact that the right side looks like $L(s)$ if x is large, and like $\varepsilon X_L(s)\overline{L}(1-s)$ if x is small, which suggests the asymmetric form of the functional equation. The quantity $(t/2\pi)^{2W}$ is called the analytic conductor for this L-function. In general, there is a conductor associated with a family of L-functions; this conductor varies "continuously" with the family. Even though the family is often a discrete family, this is a useful concept; for example, the analytic conductor is the basic parameter which goes to ∞ in our discussion of moments.

Below we give some specific examples. In each example we try to give an up-to-date account of what is known about these L-functions, especially with regard to their moments. In every situation our knowledge of moments is consistent with the conjectures made in [CFKRS] where we give recipes for how to determine all of the main terms for averages of L-functions over families.

An amazing fact emerges: when dealing with L-functions, each family has its own basic harmonic detector which on the surface appears to be very different from the detectors from other families, yet in the end somehow functions the same. When trying to prove asymptotic formulas for moments in a family of L-functions, at each step to a higher moment a new aspect or feature of the harmonic detector comes into view and assumes center stage yet somehow conspires to contribute appropriately to the same simple formula that is analogous to the RMT formulas. Moreover, in studying two different families but with the same symmetry type, the end formulas, apart from arithmetic constants, are always the same. We refer the reader to [CFKRS] for a detailed description of the recipes for these conjectures.

4.1 The Riemann Zeta-Function

The Riemann zeta-function is given by

$$\zeta(s) := 1 + \frac{1}{2^s} + \frac{1}{3^s} + \cdots = \sum_{n=1}^{\infty} \frac{1}{n^s}.$$

The series converges in the half-plane where the real part of s is larger than 1. Riemann proved that $\zeta(s)$ has an analytic continuation to the whole plane apart from a simple pole at $s = 1$. Moreover, he proved that $\zeta(s)$ satisfies a *functional equation* which in its symmetric form is given by

$$\xi(s) := \tfrac{1}{2}s(s-1)\pi^{-\frac{s}{2}}\Gamma\left(\frac{s}{2}\right)\zeta(s) = \xi(1-s)$$

where $\Gamma(s)$ is the usual Gamma-function. The zeta-function had been studied previously by Euler and others, but only as a function of a real variable. In particular, Euler proved that

$$\begin{aligned}\zeta(s) &= \left(1 + \frac{1}{2^s} + \frac{1}{4^s} + \frac{1}{8^s} + \dots\right)\left(1 + \frac{1}{3^s} + \frac{1}{9^s} + \dots\right)\left(1 + \frac{1}{5^s} + \dots\right)\dots \\ &= \prod_p \left(1 - \frac{1}{p^s}\right)^{-1}\end{aligned}$$

where the infinite product (called the *Euler product*) is over all the prime numbers. The product converges when the real part of s is greater than 1 and is an analytic version of the fundamental theorem of arithmetic, which states that every integer can be factored into primes in a unique way. The Euler product implies that there are no zeros of $\zeta(s)$ with real part greater than 1; the functional equation implies that there are no zeros with real parts less than 0, apart from the *trivial zeros* at $s = -2, -4, -6, \dots$. Thus, all of the complex zeros are in the *critical strip* $0 \le \Re s \le 1$. The functional equation shows that the complex zeros are symmetric with respect to the line $\Re s = \frac{1}{2}$.

Riemann calculated the first few complex zeros $\frac{1}{2}+i14.134\ldots$, $\frac{1}{2}+i21.022\ldots$ and proved that the number $N(T)$ of zeros with imaginary parts between 0 and T is

$$N(T) = \frac{T}{2\pi}\log\frac{T}{2\pi e} + \frac{7}{8} + S(T) + O(1/T)$$

where $S(T) = \frac{1}{\pi}\arg\zeta(1/2+iT)$ is computed by continuous variation starting from $\arg\zeta(2) = 0$ and proceeding along straight lines, first up to $2+iT$ and then to $1/2+iT$. Riemann also proved that $S(T) = O(\log T)$. Note for future reference that at a height T the average gap between zero heights is $\sim 2\pi/\log T$. Riemann suggested that the number $N_0(T)$ of zeros of $\zeta(1/2+it)$ with $0 < t \le T$ seemed to be about

$$\frac{T}{2\pi}\log\frac{T}{2\pi e};$$

and then made his conjecture that all of the zeros of $\zeta(s)$ in fact lie on the 1/2-line; this is the Riemann Hypothesis.

Weil's Explicit Formula

André Weil proved the following formula which is a generalization of Riemann's formula mentioned above and which specifically illustrates the dependence between primes and zeros. Let h be an even function which is holomorphic in the strip $|\Im t| \le 1/2+\delta$ and satisfying $h(t) = O((1+|t|)^{-2-\delta})$ for some $\delta > 0$, and let

$$g(u) = \frac{1}{2\pi}\int_{-\infty}^{\infty} h(r)e^{-iur}\, dr.$$

Then we have the following duality between primes and zeros:

$$\sum_\gamma h(\gamma) = 2h(\tfrac{i}{2}) - g(0)\log\pi + \frac{1}{2\pi}\int_{-\infty}^{\infty} h(r)\frac{\Gamma'}{\Gamma}(\tfrac{1}{4}+\tfrac{1}{2}ir)\, dr - 2\sum_{n=1}^{\infty}\frac{\Lambda(n)}{\sqrt{n}}g(\log n).$$

In this formula, a zero is written as $\rho = 1/2+i\gamma$ where $\gamma \in \mathbb{C}$; of course RH is the assertion that all of the γ are real. Also, von Mangoldt's function $\Lambda(n)$ is equal to $\log p$ if n is a power of the prime p and is 0 if n is not a power of a prime ($\Lambda(1) = 0$.) Using this duality Weil gave a criterion for RH: *RH if and only if*

$$\sum_\gamma h(\gamma) > 0$$

for every (admissible) function h of the form $h(r) = h_0(r)\overline{h_0(\overline{r})}$.

Orthogonality

Deriving statistics for families of L-functions are built around orthogonality relations. We can consider the Riemann zeta-function $\zeta(s)$ (or any L-function by itself) as being a "family" in the parameter t where $s = \sigma + it$. Then, the basic orthogonality relation is given by

$$\int_0^T m^{it} n^{-it}\, dt = \begin{cases} T \text{ if } m = n \\ \frac{(m/n)^{iT} - 1}{i \log(m/n)} \text{ if } m \neq n \end{cases}$$

The mean-value theorem of Montgomery and Vaughan asserts that

$$\int_0^T \left| \sum_{n=1}^N \lambda_n n^{it} \right|^2 dt = \sum_{n=1}^N (T + O(n)|\lambda_n|^2.$$

Approximate Functional Equation

Hardy and Littlewood proved what they called an approximate functional equation for $\zeta(s)$:

$$\zeta(s) = \sum_{n \leq \sqrt{\frac{|t|}{2\pi}}} \frac{1}{n^s} + \chi(s) \sum_{n \leq \sqrt{\frac{|t|}{2\pi}}} \frac{1}{n^{1-s}} + O(|t|^{1/4-\sigma})$$

where $\chi(s) = \chi(1-s)^{-1}$ is the factor from the asymmetric form of the functional equation:

$$\chi(1-s) = 2(2\pi)^{-s}\Gamma(s)\cos \pi s/2.$$

This formula is useful for proving moment theorems. The analytic conductor is $t/(2\pi)$.

General Remarks on Approximate Functional Equations

In general, an approximate functional equation is obtained as follows. Suppose that we have an L-function $L(s) = \sum_{n=1}^\infty \lambda_n n^{-s}$ where the series converges absolutely for $\sigma > 1$ which is entire and satisfies the functional equation $L(s) = X(s)L(1-s)$. Let $G(z)$ be a suitable function which satisfies $G(0) = 1$ and is analytic in $-1-\delta < \Re z < 1+\delta$ for some $\delta > 0$. Consider

$$I(s, Y) = \frac{1}{2\pi i} \int_{(1)} L(s+z)G(z)Y^z \frac{dz}{z}$$

where we are thinking of s having $\Re s \approx 1/2$. On the one hand, we integrate term-by-term (note that $1 + \Re s > 1$) and have

$$I(s, x) = \sum_{n=1}^\infty \frac{\lambda_n}{n^s} \tilde{G}(n/Y)$$

where

$$\tilde{G}(r) = \frac{1}{2\pi i}\int_{(1)} G(z) r^{-z}\frac{dz}{z}.$$

The "suitability" of G is supposed to ensure convergence here. On the other hand, we move the path of integration to $\sigma = -1$, crossing the pole at $s = 0$ with residue $L(s)$; then we change variables $z \to -z$ and use the functional equation to obtain

$$I(s, X) = L(s) + \frac{1}{2\pi i}\int_{(1)} L(1-s+z)X(s-z)G(-z)Y^{-z}\frac{dz}{z}.$$

Let $H_s(z) = X(s-z)G(-z)/X(s)$ and assume further that $H(s,z)$ is analytic in the strip $-1-\delta < \Re z < 1+\delta$. (Note that this does not really entail an additional assumption because $X(z) = A^z \prod \Gamma(w_i(1-z)+\mu_i)/\Gamma(w_i z + \mu_i)$ for some $w_i > 0$ and complex μ_i with $\Re\mu_i \geq 0$. Therefore, the leftmost poles of $X(z)$ will have real parts ≥ 1.) Integrating term-by-term again we finally obtain

$$L(s) = \sum_{n=1}^{\infty} \frac{\lambda_n}{n^s}\tilde{G}(n/Y) + X(s)\sum_{n=1}^{\infty}\frac{\lambda_n}{n^{1-s}}\tilde{H}_s(nY).$$

The suitability of G is such that the transforms $\tilde{G}(y)$ and $\tilde{H}_s(y)$ should decay as $y \to \infty$ so that the series converge. Generally speaking these transforms should be about 1 for small y and about 0 for larger y. In this way, we obtain our "approximate" functional equations.

Note that this technique works fine for products

$$L_1(s_1)L_2(s_2)\dots L_k(s_k)$$

too.

Moment Theorems

The second moment of $|\zeta(1/2+it)|$ was proven by Hardy and Littlewood and refined by Ingham. It states that

$$\frac{1}{T}\int_0^T |\zeta(1/2+it)|^2 \, dt = \log\frac{T}{2\pi} + 2\gamma - 1 + O(T^{1/2}).$$

The exponent on the error term has been improved to slightly less than 1/3.

If we consider the shifted moment we get a more general statement: Suppose that $|\alpha|, |\beta| \ll 1/\log T$. Then

$$\begin{aligned}\int_1^T &\zeta(1/2+it+\alpha)\zeta(1/2-it-\beta)\, dt \\ &= \int_1^T \left(\zeta(1+\alpha-\beta) + \left(\frac{t}{2\pi}\right)^{\beta-\alpha}\zeta(1+\beta-\alpha)\right) dt \\ &\quad + O(T^{1/2})\end{aligned}$$

The asymptotics of the fourth power moment was first achieved by Ingham:

$$\frac{1}{T}\int_0^T |\zeta(1/2+it)|^4 \, dt \sim \frac{1}{2\pi^2}\log^4 T.$$

Subsequent works of Atkinson and Heath-Brown revealed that

$$\frac{1}{T}\int_0^T |\zeta(1/2+it)|^4 \, dt = \int_0^T P_2\left(\log\frac{t}{2\pi}\right) \, dt + O(T^{7/8+\epsilon})$$

where P_2 is a polynomial of degree 4. This polynomial was computed explicitly by Conrey:

$$\begin{aligned}
P_2(x) &= \frac{1}{2\pi^2}x^4 + \frac{8}{\pi^4}\left(\gamma\pi^2 - 3\zeta'(2)\right)x^3 \\
&+ \frac{6}{\pi^6}\left(-48\gamma\zeta'(2)\pi^2 - 12\zeta''(2)\pi^2 + 7\gamma^2\pi^4 + 144\zeta'(2)^2 - 2\,\gamma_1\pi^4\right)x^2 \\
&+ \frac{12}{\pi^8}\Bigg(6\gamma^3\pi^6 - 84\gamma^2\zeta'(2)\pi^4 + 24\gamma_1\zeta'(2)\pi^4 - 1728\zeta'(2)^3 + 576\gamma\zeta'(2)^2\pi^2 \\
&+ 288\zeta'(2)\zeta''(2)\pi^2 - 8\zeta'''(2)\pi^4 - 10\gamma_1\gamma\pi^6 - \gamma_2\pi^6 - 48\gamma\zeta''(2)\pi^4\Bigg)x \\
&+ \frac{4}{\pi^{10}}\Bigg(-12\zeta''''(2)\pi^6 + 36\gamma_2\zeta'(2)\pi^6 + 9\gamma^4\pi^8 + 21\gamma_1^2\pi^8 + 432\zeta''(2)^2\pi^4 \\
&+ 3456\gamma\zeta'(2)\zeta''(2)\pi^4 + 3024\gamma^2\zeta'(2)^2\pi^4 - 36\gamma^2\gamma_1\pi^8 - 252\gamma^2\zeta''(2)\pi^6 \\
&+ 3\gamma\gamma_2\pi^8 + 72\gamma_1\zeta''(2)\pi^6 + 360\gamma_1\gamma\zeta'(2)\pi^6 - 216\gamma^3\zeta'(2)\pi^6 \\
&- 864\gamma_1\zeta'(2)^2\pi^4 + 5\gamma_3\pi^8 + 576\zeta'(2)\zeta'''(2)\pi^4 - 20736\gamma\zeta'(2)^3\pi^2 \\
&- 15552\zeta''(2)\zeta'(2)^2\pi^2 - 96\gamma\zeta'''(2)\pi^6 + 62208\zeta'(2)^4\Bigg)
\end{aligned}$$

The point of displaying this formula is for comparison with how simple the formula becomes when recast in terms of shifts below. We note that numerically,

$$P_2(x) = 0.05066\,x^4 + 0.69886\,x^3 + 2.42596\,x^2 + 3.22790\,x + 1.312424$$

Following work of Motohashi [M2] it was discovered how to put the shifted mean value of $\zeta(s)$ into a nice symmetric form. Let $s = 1/2 + it$ and let $\alpha, \beta, \gamma, \delta \ll 1/\log T$. Then

$$\begin{aligned}
&\int_0^T \zeta(s+\alpha)\zeta(s+\beta)\zeta(1-s+\gamma)\zeta(1-s+\delta)\,dt \\
&\quad = \int_0^T \big(Z(\alpha,\beta,\gamma,\delta) + \tau^{-\alpha-\gamma}Z(-\gamma,\beta,-\alpha,\delta) + \tau^{-\alpha-\delta}Z(-\delta,\beta,\gamma,-\alpha) \\
&\qquad + \tau^{-\beta-\gamma}Z(\alpha,-\gamma,-\beta,\delta) + \tau^{-\beta-\delta}Z(\alpha,-\delta,\gamma,-\beta) \\
&\qquad + \tau^{-\alpha-\beta-\gamma-\delta}Z(-\gamma,-\delta,-\alpha,-\beta)\big)\,dt + O(T^{7/8})
\end{aligned}$$

where $\tau = \sqrt{\frac{t}{2\pi}}$ and

$$Z(\alpha,\beta,\gamma,\delta) = \frac{\zeta(1+\alpha+\gamma)\zeta(1+\alpha+\delta)\zeta(1+\beta+\gamma)\zeta(1+\gamma+\delta)}{\zeta(2+\alpha+\beta+\gamma+\delta)}.$$

This formula should be compared with the analogous formula for the shifted fourth moment of unitary characteristic polynomials presented in Sect. 2.11.

Conrey and Ghosh [CGh] conjectured that

$$\int_0^T |\zeta(1/2+it)|^6 \, dt \sim 42 \prod_p \left(1-\frac{1}{p}\right)^4 \left(1+\frac{4}{p}+\frac{1}{p^2}\right) \frac{\log^9 T}{9!}$$

and in general that

$$\int_0^T |\zeta(1/2+it)|^{2k} \, dt \sim g_k a_k \frac{\log^{k^2} T}{k^2!}$$

where

$$a_k = \prod_p \left(1-\frac{1}{p}\right)^{(k-1)^2} \sum_{j=0}^{k-1} \binom{k-1}{j}^2 p^{-j}$$

for some integer g_k.

Conrey and Gonek [CGo] conjectured that

$$\int_0^T |\zeta(1/2+it)|^8 \, dt \sim 24024 \prod_p \left(1-\frac{1}{p}\right)^9 \left(1+\frac{9}{p}+\frac{9}{p^2}+\frac{1}{p^3}\right) \frac{\log^{16} T}{16!}$$

Keating and Snaith made the key connection with random matrix theory and conjectured that

$$g_k = k^2! \prod_{j=0}^{k-1} \frac{j!}{(j+k)!}.$$

Conrey, Farmer, Keating, Rubinstein, and Snaith [CFKRS1] have found a way to express all of the lower order terms as well. Their conjecture is

$$\int_0^T |\zeta(1/2+it)|^{2k} \, dt = \int_0^T P_k\left(\log \tfrac{t}{2\pi}\right) \, dt + O(T^{\frac{1}{2}+\epsilon})$$

as $T \to \infty$, where P_k is the polynomial of degree k^2 given by the $2k$-fold residue

$$P_k(x) = \frac{(-1)^k}{k!^2} \frac{1}{(2\pi i)^{2k}} \oint \cdots \oint \frac{G(z_1,\ldots,z_{2k}) \Delta(z_1,\ldots,z_{2k})^2}{\prod_{j=1}^{2k} z_j^{2k}} e^{\frac{x}{2}\sum_{j=1}^{k} z_j - z_{j+k}} \, dz_1 \ldots dz_{2k},$$

where one integrates over small circles about $z_i = 0$, with

$$G(\alpha_1,\ldots,\alpha_{2k}) = A_k(\alpha_1,\ldots,\alpha_{2k})\prod_{i=1}^{k}\prod_{j=1}^{k}\zeta(1+\alpha_i-\alpha_{j+k}),$$

and A_k is the Euler product which is absolutely convergent for $\sum_{j=1}^k |\alpha_j| + |\beta_j| < 1/2$, defined by $A_k(\alpha) =$

$$\prod_p \prod_{i=1}^{k}\prod_{j=1}^{k}\left(1-\frac{1}{p^{1+\alpha_i-\alpha_{j+k}}}\right)\int_0^1 \prod_{j=1}^{k}\left(1-\frac{e(\theta)}{p^{1/2+\alpha_j}}\right)^{-1}\left(1-\frac{e(-\theta)}{p^{1/2-\alpha_{j+k}}}\right)^{-1} d\theta.$$

More generally, with $s = 1/2 + it$,

$$\begin{aligned}\int_0^T \zeta(s+\alpha_1)\ldots\zeta(s+\alpha_k)\zeta(1-s+\alpha_{k+1})\ldots\zeta(1-s+\alpha_{2k})\ dt \\ = \int_0^T P_k\left(\alpha, \log\tfrac{t}{2\pi}\right)\ dt + O(T^{\frac{1}{2}+\epsilon}),\end{aligned}$$

where $P_k(\alpha, x) =$

$$\frac{(-1)^k}{k!^2}\frac{1}{(2\pi i)^{2k}}\oint\cdots\oint \frac{G(z_1,\ldots,z_{2k})\Delta(z_1,\ldots,z_{2k})^2}{\prod_{j=1}^{2k}\prod_{i=1}^{2k}(z_j-\alpha_i)} e^{\frac{x}{2}\sum_{j=1}^k z_j - z_{j+k}}\, dz_1\ldots dz_{2k},$$

with the path of integration being small circles surrounding the poles α_i. For example,

$$\begin{aligned}P_3(x) &= 0.00000570852\,x^9 + 0.00040502\,x^8 + 0.011072\,x^7 + 0.148400\,x^6 \\ &\quad + 1.04592\,x^5 + 3.98438\,x^4 + 8.607319\,x^3 + 10.274330\,x^2 \\ &\quad + 6.593913\,x + .916515.\end{aligned}$$

When $|\zeta(1/2+it)|^6$ is integrated numerically from 0 to 2,350,000 the ratio between the actual value and the conjectured value is 1.00017.

Pair Correlation of Zeros

In 1972, Hugh Montgomery was investigating the spacings between zeros of the zeta-function in an attempt to solve the class number problem. Montgomery's theorem is that, under the assumption of the Riemann Hypothesis, if $|\alpha| < 1$, then

$$F(\alpha, T) = \frac{1}{N(T)}\int_{-\infty}^{\infty}\left|\sum_{\gamma\le T} T^{i\alpha\gamma} w(t-\gamma)\right|^2 dt = T^{-2\alpha}\log T(1+o(1)) + |\alpha| + o(1)$$

as $T \to \infty$. Here w is a suitable weight function (Montgomery used $w(x) = \sqrt{2/\pi}4/(4+x^2)$.) Montgomery conjectured, based on the above and on conjectures for the distribution of twin primes and other prime pairs that $F(\alpha, T) = 1 + o(1)$ for $|\alpha| \geq 1$. From this conjecture, he deduced that

$$\sum_{\frac{2\pi\alpha}{\log T} < \gamma - \gamma' \leq \frac{2\pi\beta}{\log T}} 1 \sim N(T) \int_\alpha^\beta \left(1 - \left(\frac{\sin \pi u}{\pi u}\right)^2\right) du.$$

The sum on the left counts the number of pairs $0 < \gamma, \gamma' < T$ of ordinates of zeros with normalized spacing between positive numbers $0 < \alpha < \beta$. The integral on the right is the pair-correlation function from random matrix theory. It was this fortuitous discovery, made in a conversation between Montgomery and Freeman Dyson at tea-time at the Institute for Advanced Study, that set in motion the circle of ideas involving L-functions and RMT.

Higher Correlations

In 1996 Rudnick and Sarnak [RS] made some interesting progress on the GUE conjecture. To explain their result, number the ordinates of the zeros of $\zeta(s)$: $0 < \gamma_1 \leq \gamma_2 \leq \ldots$. Introduce a scaling $\tilde{\gamma} = \gamma \frac{\log \gamma}{2\pi}$ so that the $\tilde{\gamma}$ have asymptotic mean spacing 1. Then Rudnick and Sarnak proved that

$$\lim_{T\to\infty} \frac{1}{T} \sum_{\substack{\gamma_{j_1}, \ldots, \gamma_{j_n} \leq T \\ j_m \neq j_n}} f(\tilde{\gamma}_{j_1}, \ldots, \tilde{\gamma}_{j_n}) = \int_{P_n} W_{U,n}(\mathbf{x}) f(\mathbf{x})\, d\mathbf{x}$$

where $W_{U,n}(\mathbf{x}) = W_{U,n}(x_1, \ldots, x_n)$ is the n-correlation function for the Gaussian Unitary Ensemble and where f is any function satisfying (1) $f(\mathbf{x} + t(1, \ldots, 1)) = f(\mathbf{x})$ for $t \in \mathbb{R}$; (2) f is smooth and symmetric in the variables and decays rapidly as $x \to \infty$ in the hyperplane $P_n := \{(x_1, \ldots, x_n) : \sum_{j=1}^n x_j = 0\}$; (3) the Fourier transform $\hat{f}(\mathbf{u})$ of f is supported in $\sum_{j=1}^n |u_j| < 2$. The condition (1) assures that f is a function of the differences of the γ_j. This result agrees with RMT.

4.2 Dirichlet L-functions

In order to see the full analogy between L-functions and Random Matrix Theory, it is necessary to consider a variety of families of L-functions with different symmetry types. The simplest L-function after the ζ-function is the Dirichlet L-function for the non-trivial character of conductor 3:

$$L(s, \chi_{-3}) = 1 - \frac{1}{2^s} + \frac{1}{4^s} - \frac{1}{5^s} + \frac{1}{7^s} - \frac{1}{8^s} + - \ldots.$$

This can be written as an Euler product

$$L(s,\chi_{-3}) = \prod_{p\equiv 1 \bmod 3}(1-p^{-s})^{-1}\prod_{p\equiv 2 \bmod 3}(1+p^{-s})^{-1},$$

satisfies the functional equation

$$\xi(s,\chi_{-3}) := \left(\tfrac{\pi}{3}\right)^{-\frac{s}{2}}\Gamma(\tfrac{s+1}{2})L(s,\chi_3) = \xi(1-s,\chi_{-3}),$$

and is expected to have all of its non-trivial zeros on the 1/2-line. A similar construction works for any primitive Dirichlet character. In general, a Dirichlet character is a completely multiplicative periodic function $\chi : \mathbb{N} \to \mathbb{C}$; i.e. $\chi(mn) = \chi(m)\chi(n)$ for all m, n and $\chi(m+q) = \chi(m)$ for some integer q. It is the *primitive* characters which lead to the arithmetic L-functions. We now describe how to construct the primitive characters. For each $q \ge 1$ there are precisely

$$\psi(q) = \sum_{d|q}\mu(d)\phi(q/d)$$

primitive characters to the modulus q. If q has the factorization $q = p_1^{e_1}\dots p_r^{e_r}$, then any primitive character χ mod q has a unique representation as a product $\chi = \chi_1 \cdots \chi_r$ where χ_j is a primitive character modulo $p_j^{e_j}$. We now describe how to construct the primitive characters modulo p^e. If p is odd, then the number of integers less than or equal to p^e and relatively prime to p^e is given by $\phi(p^e) = p^e - p^{e-1}$. These reduced residues modulo p^e form a multiplicative group which is cyclic; let g be a generator of this group (i.e. a *primitive root* of p^e.) We can specify any character χ modulo p^e by saying what the value of $\chi(g)$ is (clearly this value must be a $\phi(p^e)$ root of unity). The primitive characters are those for which $\chi(g) = \exp(2\pi i a/\phi(p^e))$ where $(a, \phi(p^e)) = 1$. For $p = 2$, the reduced residues modulo 2^e do not form a cyclic group unless $e = 1$ or 2. If $e \ge 3$ then the reduced residues are given by $\pm 5^j$ with $j = 0, 1, \dots 2^{e-2}$. The primitive characters χ modulo 2^e are determined by the value of $\chi(5) = \exp(2\pi i a/2^{e-2})$ with $1 \le a \le 2^{e-2}$ odd and by the value of $\chi(-1) = \pm 1$. This describes all primitive characters. For each primitive character χ mod q the *Gauss sum* is given by

$$\tau(\chi) = \sum_{n=1}^{q}\chi(n)e(n/q).$$

It satisfies $|\tau(\chi)| = \sqrt{q}$; we write $\tau(\chi) = \epsilon_\chi\sqrt{q}$. The Dirichlet L-function is given by

$$L(s,\chi) = \sum_{n=1}^{\infty}\frac{\chi(n)}{n^s} = \prod_p\left(1-\frac{\chi(p)}{p^s}\right)^{-1}$$

for $\sigma > 1$. Odd characters are those for which $\chi(-1) = -1$; even characters have $\chi(-1) = 1$. The functional equation for an even character is

$$\xi(s,\chi) := (\pi/\sqrt{q})^{-s/2}\Gamma(s/2)L(s,\chi) = \epsilon_q\xi(1-s,\overline{\chi}).$$

For an odd character, the functional equation is

$$\xi(s,\chi) := (\pi/\sqrt{q})^{-s/2}\Gamma((s+1)/2)L(s,\chi) = \epsilon_q \xi(1-s,\overline{\chi}).$$

All Characters

We have described the primitive characters above. Imprimitive characters arise in two ways. First, the principal character χ_0 modulo p defined by $\chi_0(n) = 0$ if $p \mid n$ and$= +1$ if $p \nmid n$ is an imprimitive character. Second, a primitive character modulo p^e regarded as a character modulo p^f where $f > e$ is an imprimitive character. Finally, the product of a primitive character with an imprimitive character is an imprimitive character. Any character χ (primitive or imprimitive) which satisfies $\chi(m+q) = \chi(m)$ is called a character modulo q. There are $\phi(q)$ characters modulo q. The analytic conductor is $qt/(2\pi)$; in the context of averaging over the characters, the t part is suppressed at that the conductor is $q/(2\pi)$.

Orthogonality Relations

The basic orthogonality relation is then expressed by: if $(mn,q) = 1$, then

$$\sum_{\chi \bmod q} \chi(m)\overline{\chi(n)} = \begin{cases} \phi(q) \text{ if } m = n \bmod q \\ 0 \text{ if } m \neq n \bmod q \end{cases}$$

For primitive characters, this takes the shape: if $(mn,q) = 1$,then

$$\sum_{\chi \bmod q}^{\star} \chi(m)\overline{\chi(n)} = \sum_{d|(q,m-n)} \phi(d)\mu(q/d).$$

The Polya-Vinogradov inequality asserts that

$$|\sum_{n=1}^{N} \chi(n)| \ll q^{1/2}\log q$$

for any non-principal character χ mod q.

The large sieve inequality asserts that

$$\sum_{q\le Q} \frac{q}{\phi(q)} \sum_{\chi \bmod q}^{\star} \left|\sum_{n=1}^{N} a_n\chi(n)\right|^2 \le (Q^2+N)\sum_{n=1}^{N} |a_n|^2.$$

Pair Correlation of Zeros

Ali Ozluk, in his thesis [Oz] gave a generalization to Dirichlet L-functions of Montgomery's Pair Correlation results. Let

$$F_K(\alpha,Q)=\frac{1}{N_K(Q)}\sum_{q\le Q}\frac{1}{\phi(q)}\sum_{\chi \bmod q}\left|\sum_{\gamma}K(\tfrac{1}{2}+i\gamma)Q^{i\alpha\gamma}\right|^2,$$

where the inner sum is over the imaginary parts γ of zeros of $L(s,\chi)$. Here $K(s)$, is a suitable weight function and $N_K(Q)=\sum_\gamma K(1/2+i\gamma)$ is the normalization factor. The main result of the paper is an asymptotic formula for $F_K(\alpha,Q)$ in the interval $|\alpha|<2$. For $|\alpha|\le 1$, the result obtained is an analogous to Montgomery's result. For $1<|\alpha|<2$ Ozluk shows that $F_k(\alpha,Q)=1+O(1/\log Q)$, which supports Montgomery's conjecture.

Moments

For the second moment Heath-Brown [H-B2] showed that

$$\sum_{\chi \bmod p}^{\star}|L(1/2,\chi)|^2=(p-1)\left(\log\frac{p}{8\pi}+\gamma\right)+2\zeta(1/2)^2p^{1/2}+O(1)$$

and for the fourth moment, he proved ([H-B3] that

$$\sum_{\chi \bmod p}^{\star}|L(1/2,\chi)|^4=\frac{p-1}{2\pi^2}\log^4 p+O(\log^3 p)$$

in analogy with the second and fourth moments of the Riemann zeta-function. Here the * indicates that the sum is over primitive characters modulo the prime p. An analogous formula can be proven in the case of the second moment for primitive characters modulo a composite number q; however, the analogue for the fourth moment for all large moduli q has not yet been proven.

A second moment for shifted L-functions averaged over primitive characters has been proven by Conrey (unpublished); this formula agrees with the analogous formula for the shifted second moment of unitary characteristic polynomials:

$$\frac{2}{\psi(q)}\sum_{\substack{\chi \bmod q\\ \chi(-1)=1}}^{\star}L(1/2+\alpha,\chi)L(1/2+\beta,\overline{\chi})$$
$$=\zeta(1+\alpha+\beta)\prod_{p|q}(1-\frac{1}{p^{1+\alpha+\beta}})+$$
$$X^+(q,\alpha,\beta)\zeta(1-\alpha-\beta)\prod_{p|q}(1-\frac{1}{p^{1-\alpha-\beta}})+O(q^{\epsilon-1/2}).$$

Here

$$X^+(q,\alpha,\beta)=\frac{4}{q}\left(\frac{q}{2\pi}\right)^{1-\alpha-\beta}\Gamma(\tfrac{1}{2}-\alpha)\Gamma(\tfrac{1}{2}-\beta)\cos\tfrac{\pi}{2}(\tfrac{1}{2}-\alpha)\cos\tfrac{\pi}{2}(\tfrac{1}{2}-\beta)$$

is the factor from the functional equation, as expected.

4.3 Real Primitive Characters

A special role is played by the real or quadratic Dirichlet characters. These we denote by χ_d where d is a fundamental discriminant: d can be positive or negative, is either odd, square-free, and congruent to 1 modulo 4, or is 4 times a square-free integer congruent to 2 or 3 modulo 4. Thus, the sequence of positive fundamental discriminants begins $d = 1, 5, 8, 12, 13, 17, 21, 24, 28, \ldots$ and the sequence of negative fundamental discriminants begins d=-3, -4, -7, -8, -11, -15, -19,-20, -23, -24, ... The character χ_d only takes on the values $+1, 0, -1$; it is primitive with the modulus $|d|$. If $d > 0$, then χ_d is an even character and if $d < 0$ it is an odd character. The character χ_d is the character associated with the quadratic field $Q(\sqrt{d})$. In particular, the prime p *splits* or factors in this field if $\chi_d(p) = +1$; it remains prime if $\chi_d(p) = -1$; and it *ramifies* or is a square if $\chi_d(p) = 0$. The real characters χ_d can be decomposed into a product of characters $\chi_{-4}, \chi_8, \chi_{-8}$, and $\chi_{\pm p}$ for odd primes p where $\chi_{\pm p}(n) = \left(\frac{n}{p}\right)$ is the Legendre symbol ($= 1$ if n is a non-zero square modulo p, and $= -1$ if n is a non-zero non-square modulo p, $= 0$ if $p \mid n$). The character $\chi_{-4}(n)$ is 0 for even n, is $+1$ for n congruent to 1 modulo 4, and is -1 for n congruent to 3 modulo 4. The character $\chi_8(n)$ is 0 for even n, is $+1$ for n congruent to ± 1 modulo 8, and is -1 for n congruent to ± 3 modulo 8. Finally, $\chi_{-8}(n)$ is 0 for even n, is $+1$ for n congruent to 1 or 3 modulo 8, and is -1 for n congruent to 5 or 7 modulo 8.

Orthogonality

First of all, the number $N_q^+(x)$ of fundamental discriminants d with $0 < d \le x$ and $(d, q) = 1$ satisfies $N_q^+(x) \sim \frac{3}{\pi^2}\frac{\phi(q)}{q}x$ and similarly the number $N_q^-(x)$ of negative fundamental discriminants d with $0 < -d < x$ and $(d, q) = 1$ satisfies $N_q^-(x) \sim \frac{3}{\pi^2}\frac{\phi(q)}{q}x$.

By the Polya-Vinogradov inequality,

$$\sum_{0<d\le x} \chi_d(n) = \begin{cases} N_n^+(x) \text{ if } n \text{ is a square} \\ O(n^{1/2+\epsilon}) \text{ if } n \text{ is not a square} \end{cases}$$

Heath-Brown [H-B4] proved a very useful large-sieve type inequality:

$$\sum_{|d|\le Q} \left| \sum_{n=1}^{N} a_n \chi_d(n) \right|^2 \ll_\epsilon (QN)^\epsilon (Q+N) \sum_{n_1 n_2 = \Box} |a_{n_1} a_{n_2}|.$$

Soundararajan [So] gave a Poisson summation formula for smooth (real) character sums:

$$\sum_d \chi_d(n) F(d/X) = \frac{X}{n} \sum_{k=-\infty}^{\infty} \hat{F}(kX/n) \tau_k(n)$$

where

$$\tau_k(n) = \sum_{a=1}^{n} \left(\frac{a}{n}\right) e(ak/n)$$

is a Gauss sum.

Approximate Functional Equations

For integers $j \geq 1$ put $\omega_j(0) = 1$ and for $\xi > 0$ define $\omega_j(\xi)$ by

$$\omega_j(\xi) = \frac{1}{2\pi i} \int_{(c)} \left(\frac{\Gamma(\frac{s}{2} + \frac{1}{4})}{\Gamma(\frac{1}{4})} \right)^j \xi^{-s} \frac{ds}{s}$$

where c is any positive real number. As usual, $d_j(n)$ will denote the j-th divisor function; that is the coefficient of n^{-s} in the Dirichlet series expansion of $\zeta(s)^j$. For integers $j \geq 1$, we define

$$A_j(d) = \sum_{n=1}^{\infty} \frac{d_j(n)\chi_d(n)}{\sqrt{n}} \omega_j\left(n \left(\frac{\pi}{|d|} \right)^{\frac{j}{2}} \right).$$

Then for all integers $j \geq 1$,

$$L(\tfrac{1}{2}, \chi_d)^j = 2A_j(d).$$

The analytic conductor is $|d|/(2\pi)$.

Moments

Jutila [Jut] proved that

$$\sum_{0<d<X} L(1/2, \chi_d) = \frac{P(1)}{4\zeta(2)} X \left(\log \frac{X}{\pi} + \frac{\Gamma'}{\Gamma}(1/4) + 4\gamma - 1 + 4\frac{P'(1)}{P(1)} \right)$$
$$+ O(X^{3/4+\epsilon})$$

where $P(s) = \prod_p \left(1 - \frac{1}{(p+1)p^s}\right)$ and for the same sum but with negative discriminants $-d < X$, the term $\frac{\Gamma'}{\Gamma}(1/4)$ is replaced by $\frac{\Gamma'}{\Gamma}(3/4)$. In the same paper, he obtains that

$$\sum_{0<d<X} L(1/2, \chi_d)^2 = \frac{c}{\zeta(2)} X \log^3 X + O(X(\log X)^{5/2+\epsilon})$$

and the same for the sum over negative discriminants where

$$c = \frac{1}{48} \prod_p (1 - (4p^2 - 3p + 1)/(p^4 + p^3)).$$

Soundararajan [So] obtains a full degree three polynomial for the mean-square, a full degree six polynomial for the cubic moment, and a conjecture for the fourth moment. These results and conjecture can be summarized, using $D^* = \sum_{|d|\le D} 1$, as:

$$\frac{1}{D^*}\sum_{|d|\le D} L(\tfrac12,\chi_d) \sim a_1 \log(D^{\frac12});$$

$$\frac{1}{D^*}\sum_{|d|\le D} L^2(\tfrac12,\chi_d) \sim 2\frac{a_2 \log^3(D^{\frac12})}{3!};$$

$$\frac{1}{D^*}\sum_{|d|\le D} L^3(\tfrac12,\chi_d) \sim 16\frac{a_3 \log^6(D^{\frac12})}{6!};$$

$$\frac{1}{D^*}\sum_{|d|\le D} L^4(\tfrac12,\chi_d) \sim 768\frac{a_4 \log^{10}(D^{\frac12})}{10!}.$$

The general conjecture stated by Keating and Snaith using random matrix theory is:

$$\frac{1}{D^*}\sum_{|d|\le D} L^k(\tfrac12,\chi_d) \sim \prod_{\ell=1}^{k}\frac{\ell!}{2\ell!} a_k \log^{k(k+1)/2}(D)$$

where

$$a_k = \prod_p \frac{\left(1-\frac1p\right)^{\frac{k(k+1)}{2}}}{\left(1+\frac1p\right)}\left(\frac{\left(1-\frac{1}{\sqrt p}\right)^{-k}+\left(1+\frac{1}{\sqrt p}\right)^{-k}}{2}+\frac1p\right).$$

The conjectures all agree.

More generally, we conjecture that

$$\sum_{0<-d<D} L(1/2,\chi_d)^k = \sum_{0<-d<D} Q_k(\log \tfrac{|d|}{2\pi}) + O(D^{1/2+\epsilon})$$

as $D\to\infty$, where Q_k is the polynomial of degree $k(k+1)/2$ given by the k-fold residue

$$Q_k(x) = \frac{(-1)^{k(k-1)/2}2^k}{k!}\frac{1}{(2\pi i)^k}\oint\cdots\oint \frac{G_-(z_1,\dots,z_k)\Delta(z_1^2,\dots,z_k^2)^2}{\prod_{j=1}^{k} z_j^{2k-1}} e^{\frac{x}{2}\sum_{j=1}^k z_j}\, dz_1\dots dz_k,$$

where

$$G_-(\alpha_1,\dots,\alpha_k) = A_k(\alpha_1,\dots,\alpha_k)\prod_{j=1}^{k}\left(\frac{\Gamma(\frac34+\frac{\alpha_j}{2})2^{\alpha_j}}{\Gamma(\frac34-\frac{\alpha_j}{2})}\right)^{\frac12}\prod_{1\le i\le j\le k}\zeta(1+\alpha_i+\alpha_j),$$

and A_k is the Euler product which is absolutely convergent for $\sum_{j=1}^k |\alpha_j| < 1/2$, defined by

$$A_k(\alpha_1,\ldots,\alpha_k) = \prod_p \prod_{1\le i\le j\le k}\left(1-\frac{1}{p^{1+\alpha_i+\alpha_j}}\right)$$
$$\times\left(\frac{1}{2}\left(\prod_{j=1}^k\left(1-\frac{1}{p^{\frac{1}{2}+\alpha_j}}\right)^{-1}+\prod_{j=1}^k\left(1+\frac{1}{p^{\frac{1}{2}+\alpha_j}}\right)^{-1}\right)+\frac{1}{p}\right)\left(1+\frac{1}{p}\right)^{-1}.$$

Still more generally, we conjecture that

$$\sum_{0<-d<D}\xi(1/2+\alpha_1,\chi_d)\ldots\xi(1/2+\alpha_k,\chi_d) = \sum_{0<-d<D}^{\star} Q_k(\alpha,\log\tfrac{|d|}{2\pi})+O(D^{1/2+\epsilon}),$$

where

$$Q_k(\alpha,x) = \frac{(-1)^{k(k-1)/2}2^k}{k!}\frac{1}{(2\pi i)^k}$$
$$\times\oint\cdots\oint\frac{G_-(z_1,\ldots,z_k)\Delta(z_1^2,\ldots,z_k^2)^2\prod_{j=1}^k z_j}{\prod_{\ell=1}^k\prod_{j=1}^k(z_j-\alpha_\ell)(z_j+\alpha_\ell)}e^{\frac{x}{2}\sum_{j=1}^k z_j}\,dz_1\ldots dz_k,$$

where the path of integration encloses the $\pm\alpha$'s. There is a similar conjecture for the analogous sum over positive fundamental discriminants. For this conjecture G_- is replaced by G_+, where

$$G_+(\alpha_1,\ldots,\alpha_k) = A_k(\alpha_1,\ldots,\alpha_k)\prod_{j=1}^k\left(\frac{\Gamma(\frac{1}{4}+\frac{\alpha_j}{2})2^{\alpha_j}}{\Gamma(\frac{1}{4}-\frac{\alpha_j}{2})}\right)^{\frac{1}{2}}\prod_{1\le i\le j\le k}\zeta(1+\alpha_i+\alpha_j),$$

and A_k is as before.

4.4 Modular L-functions

Ramanujan's tau-function, defined implicitly by

$$x\prod_{n=1}^\infty(1-x^n)^{24} = \sum_{n=1}^\infty \tau(n)x^n,$$

also yields an L-function. The associated Fourier series

$$\Delta(z) := \sum_{n=1}^\infty \tau(n)\exp(2\pi i n z)$$

satisfies

$$\Delta\left(\frac{az+b}{cz+d}\right) = (cz+d)^{12}\Delta(z)$$

for all integers a, b, c, d with $ad-bc=1$. A function satisfying these equations is called a *modular form* of weight 12. The associated L-function

$$L_\Delta(s) := \sum_{n=1}^{\infty} \frac{\tau(n)/n^{11/2}}{n^s} = \prod_p \left(1 - \frac{\tau(p)/p^{11/2}}{p^s} + \frac{1}{p^{2s}}\right)^{-1}$$

satisfies the functional equation

$$\xi_\Delta := (2\pi)^{-s}\Gamma(s+11/2)L_\Delta(s) = \xi_\Delta(1-s)$$

and it is expected that all of its complex zeros are on the 1/2-line. In general a cusp form of weight k for the full modular group is a holomorphic function f on the upper half-plane which satisfies

$$f\left(\frac{az+b}{cz+d}\right) = (cz+d)^k f(z)$$

for all integers a, b, c, d with $ad-bc=1$ and also has the property that $\lim_{y\to\infty} f(iy) = 0$. Cusp forms for the whole modular group exist only for even integers $k=12$ and $k \geq 16$. The cusp forms of a given weight k of this form make a complex vector space S_k of dimension $[k/12]$ if $k \neq 2 \bmod 12$ and of dimension $[k/12]-1$ if $k = 2 \bmod 12$. Each such vector space has a special basis H_k of Hecke eigenforms which consist of functions $f(z) = \sum_{n=1}^{\infty} \lambda_f(n)e(nz)$ for which

$$\lambda_f(m)\lambda_f(n) = \sum_{d|(m,n)} d^{k-1}\lambda_f(mn/d^2).$$

The Fourier coefficients $\lambda_f(n)$ are real algebraic integers of degree equal to the dimension of the vector space $= \#H_k$. Thus, when $k = 12, 16, 18, 20, 22, 26$ the spaces are one dimensional and the coefficients are ordinary integers. We can express these explicitly in terms of the Eisenstein series

$$E_4(z) = 1 + 240\sum_{n=1}^{\infty} \sigma_3(n)e(nz)$$

and

$$E_6(z) = 1 - 504\sum_{n=1}^{\infty} \sigma_5(n)e(nz)$$

where $\sigma_r(n)$ is the sum of the rth powers of the positive divisors of n:

$$\sigma_r(n) = \sum_{d|n} d^r.$$

Then, $\Delta(z)E_4(z)$ gives the unique Hecke form of weight 16; $\Delta(z)E_6(z)$ gives the unique Hecke form of weight 18; $\Delta(z)E_4(z)^2$ is the Hecke form of weight 20; $\Delta(z)E_4(z)E_6(z)$ is the Hecke form of weight 22; and $\Delta(z)E_4(z)^2E_6(z)$ is the Hecke form of weight 26. The two Hecke forms of weight 24 are given by

$$\Delta(z)E_4(z)^3 + x\Delta(z)^2$$

where $x = -156 \pm 12\sqrt{144169}$. The L-function associated with a Hecke form f of weight k is given by

$$L_f(s) = \sum_{n=1}^{\infty} \lambda_f(n)/n^{(k-1)/2}n^s = \prod_p \left(1 - \frac{\lambda_f(p)/p^{(k-1)/2}}{p^s} + \frac{1}{p^{2s}}\right)^{-1}.$$

By Deligne's theorem $\lambda_f(p)/p^{(k-1)/2} = 2\cos\theta_f(p)$ for a real $\theta_f(p)$. It is conjectured (Sato-Tate) that for each f the $\{\theta_f(p) : p \text{ prime}\}$ is uniformly distributed on $[0, \pi)$ with respect to the measure $\frac{2}{\pi}\sin^2\theta \, d\theta$. We write $\cos\theta_f(p) = \alpha_f(p) + \overline{\alpha_f(p)}$ where $\alpha_f(p) = e^{i\theta_f(p)}$; then

$$L_f(s) = \prod_p \left(1 - \frac{\alpha_f(p)}{p^s}\right)^{-1} \left(1 - \frac{\overline{\alpha_f(p)}}{p^s}\right)^{-1}.$$

The functional equation satisfied by $L_f(s)$ is

$$\xi_f(s) = (2\pi)^{-s}\Gamma(s + (k-1)/2)L_f(s) = (-1)^{k/2}\xi_f(1-s).$$

Orthogonality Relations

The Petersson inner product on the space S_k is defined by

$$\langle f, g\rangle = \iint_{\mathcal{D}} f(z)\overline{g(z)}y^k \frac{dxdy}{y^2}.$$

Here the integration is over the *fundamental domain*

$$\mathcal{D} := \{(x, y) : -1/2 \le x \le 1/2, y \ge \sqrt{1-x^2}\}.$$

Let $\mathcal{F}$ be an orthogonal basis of S_k with respect to this inner product. The Petersson formula tells us that

$$\frac{\Gamma(k-1)}{(4\pi\sqrt{mn})^{k-1}} \sum_{f\in\mathcal{F}} \frac{\lambda_f(m)\overline{\lambda_f(n)}}{\langle f, f\rangle} = \delta_{m,n} + 2\pi i^{-k} \sum_{c=1}^{\infty} \frac{S(m,n,c)}{c} J_{k-1}\left(\frac{4\pi\sqrt{mn}}{c}\right)$$

where J_{k-1} is the Bessel function of index $k-1$ and $S(m, n, c)$ is the Kloosterman sum

$$S(m, n, c) = \sum_{(x,c)=1} e((mx + n\overline{x})/c)$$

where the sum is over a set of reduced residue classes modulo cand where $\overline{x}$ satisfies $x\overline{x} = 1 \bmod c$. By a theorem of Weil, $|S(m, n, c)| \le (m, n, c)^{1/2}d(c)\sqrt{c}$ where $d(c)$ is the number of positive divisors of c.

Higher Level Modular Forms

An example of a higher level modular form is the modular form $\sum_{n=1}^{\infty}\lambda_n e(nz)$ associated to an elliptic curve $E : y^2 = x^3 + Ax + B$ where A, B are integers. The associated L-function, called the Hasse-Weil L-function, is

$$L_E(s) = \sum_{n=1}^{\infty}\frac{\lambda(n)/n^{1/2}}{n^s} = \prod_{p\nmid q}\left(1-\frac{\lambda(p)/p^{1/2}}{p^s}+\frac{1}{p^{2s}}\right)^{-1}\prod_{p|q}\left(1-\frac{\lambda(p)/p^{1/2}}{p^s}\right)^{-1}$$

where q is the conductor of the curve. The coefficients λ_n are constructed easily from λ_p for prime p; in turn the λ_p are given by $\lambda_p = p - N_p$ where N_p is the number of solutions of E when considered modulo p. The work of Wiles and others proved that these L-functions are associated to modular forms of weight 2. This modularity implies the functional equation

$$\xi_E(s) := (2\pi/\sqrt{q})^{-s}\Gamma(s+1/2)L_E(s) = \xi_E(1-s).$$

It is believed that all of the complex zeros of $L_E(s)$ are on the 1/2-line. A similar construction should work for other sets of polynomial equations but so far this has not been proven.

Level q Cusp Forms with no Multiplier System

We let $\Gamma_0(q)$ denote the group of matrices $\begin{pmatrix} a & b \\ c & d \end{pmatrix}$ with integers a, b, c, d satisfying $ad - bc = 1$ and $q \mid c$. This group is called the *Hecke congruence group of level* q. A function f holomorphic on the upper half plane satisfying

$$f\left(\frac{az+b}{cz+d}\right) = (cz+d)^k f(z)$$

for all matrices in $\Gamma_0(q)$ and $\lim_{y\to 0} f(a/q+iy) = 0$ for all rational numbers a/q is called a cusp form for $\Gamma_0(q)$; the space of these is a finite dimensional vector space $S_k(q)$. The space S_k above is the same as $S_k(1)$. Again, these spaces are empty unless k is an even integer. If k is an even integer, then

$$\dim S_k(q) = \frac{(k-1)}{12}\nu(q) + \left(\left[\frac{k}{4}\right]-\frac{k-1}{4}\right)\nu_2(q) + \left(\left[\frac{k}{3}\right]-\frac{k-1}{3}\right)\nu_3(q) - \frac{\nu_\infty(q)}{2}$$

where $\nu(q)$ is the index of the subgroup $\Gamma_0(q)$ in the full modular group $\Gamma_0(1)$:

$$\nu(q) = q\prod_{p|q}\left(1+\frac{1}{p}\right);$$

$\nu_\infty(q)$ is the number of *cusps* of $\Gamma_0(q)$:

$$\nu_\infty(q) = \sum_{d|q} \phi((d, q/d));$$

$\nu_2(q)$ is the number of inequivalent *elliptic points* of order 2:

$$\nu_2(q) = \begin{cases} 0 \text{ if } 4 \mid q \\ \prod_{p|q}(1 + \chi_{-4}(p)) \text{ otherwise} \end{cases}$$

and $\nu_3(q)$ is the number of inequivalent *elliptic points* of order 3:

$$\nu_3(q) = \begin{cases} 0 \text{ if } 9 \mid q \\ \prod_{p|q}(1 + \chi_{-3}(p)) \text{ otherwise} \end{cases}.$$

It is clear from this formula that the dimension of $S_k(q)$ grows approximately linearly with q and k. For the spaces $S_k(q)$ the issue of primitive forms and imprimitive forms arise, much as the situation with characters. In fact, one should think of the Fourier coefficients of cusp forms as being a generalization of characters. They are not periodic, but they act as harmonic detectors, much as characters do, through their orthogonality relations (below). Imprimitive cusp forms arise in two ways. Firstly , if $f(z) \in S_k(q)$, then $f(z) \in S_k(dq)$ for any integer $d > 1$. Secondly, if $f(z) \in S_k(q)$, then $f(dz) \in S_k(\Gamma_0(dq))$ for any $d > 1$. The dimension of the subspace of primitive forms is given by

$$\dim S_k^{\text{new}}(q) = \sum_{d|q} \mu_2(d) \dim S_k(q/d)$$

where $\mu_2(n)$ is the multiplicative function defined for prime powers by $\mu_2(p^e) = -2$ if $e = 1$, $= 1$ if $e = 2$, and $= 0$ if $e > 2$. The subspace of newforms has a Hecke basis $H_k(q)$ consisting of primitive forms, or newforms, or Hecke forms. These can be identified as those f which have a Fourier series

$$f(z) = \sum_{n=1}^{\infty} \lambda_f(n) e(nz)$$

where the $\lambda_f(n)$ have the property that the associated L-function has an Euler product

$$\begin{aligned} L_f(s) &= \sum_{n=1}^{\infty} \frac{\lambda_f(n)/n^{(k-1)/2}}{n^s} \\ &= \prod_{p \nmid q} \left(1 - \frac{\lambda_f(p)/p^{(k-1)/2}}{p^s} + \frac{1}{p^{2s}}\right)^{-1} \prod_{p|q} \left(1 - \frac{\lambda_f(p)/p^{(k-1)/2}}{p^s}\right)^{-1}. \end{aligned}$$

We can express this as

$$L_f(s) = \prod_p \left(1 - \frac{\alpha_f(p)}{p^s}\right)^{-1} \left(1 - \frac{\alpha'_f(p)}{p^s}\right)^{-1}$$

where if $p \nmid q$ then $\alpha'_f(p) = \overline{\alpha_f(p)}$ whereas if $p \mid q$ then $\alpha'_f(p) = 0$. The functional equation of the L-function is

$$\xi_f(s) := (2\pi/q)^{-s}\Gamma(s + (k-1)/2)L_f(s) = \pm\xi_f(1-s).$$

Now the $\pm$ depends on more than the weight k.

Examples of Cuspforms with Level > 1

We give a small table of the dimensions of the spaces $S_k(q)$ of cuspforms and of newforms $S_k^{\text{new}}(q)$ and give some explicit constructions of some of the elements of $H_k(q)$.

Table 1. Dimensions of spaces $S_k(q)$ of cusp forms

$q\backslash k$	2	4	6	8	10	12	14	16	18	20	22	24	26	28	30
1	0	0	0	0	0	1	0	1	1	1	1	2	1	2	2
2	0	0	0	1	1	2	2	3	3	4	4	5	5	6	6
3	0	0	1	1	2	3	3	4	5	5	6	7	7	8	9
4	0	0	1	2	3	4	5	6	7	8	9	10	11	12	13
5	0	1	1	3	3	5	5	7	7	9	9	11	11	13	13
6	0	1	3	5	7	9	11	13	15	17	19	21	23	25	27
7	0	1	3	3	5	7	7	9	11	11	13	15	15	17	19
8	0	1	3	5	7	9	11	13	15	17	19	21	23	25	27
9	0	1	3	5	7	9	11	13	15	17	19	21	23	25	27
10	0	3	5	9	11	15	17	21	23	27	29	33	35	39	41
11	1	2	4	6	8	10	12	14	16	18	20	22	24	26	28

Notice that all of the cusp forms of level 1 are primitive. Each such form $f(z)$ leads to $d(q)$ old forms $f(dz)$ of level q. Some of the newforms of low level and weight can be expressed as eta-products. Thus, the unique newform of level 2 and weight 8 is given explicitly by

$$e(z)\prod_{n=1}^{\infty}(1 - e(nz))^8(1 - e(2nz))^8;$$

the unique newform of weight 6 and level 3 is

$$e(z)\prod_{n=1}^{\infty}(1 - e(nz))^6(1 - e(3nz))^6;$$

the unique newform of weight 6 and level 4 is

$$e(z)\prod_{n=1}^{\infty}(1 - e(2nz))^{12};$$

Table 2. Dimensions of spaces $S_k^{\text{new}}(q)$ of primitive forms

$q\backslash k$	2	4	6	8	10	12	14	16	18	20	22	24	26	28	30
1	0	0	0	0	0	1	0	1	1	1	1	2	1	2	2
2	0	0	0	1	1	0	2	1	1	2	2	1	3	2	2
3	0	0	1	1	2	1	3	2	3	3	4	3	5	4	5
4	0	0	1	0	1	1	1	1	2	1	2	2	2	2	3
5	0	1	1	3	3	3	5	5	5	7	7	7	9	9	9
6	0	1	1	1	1	3	1	3	3	3	3	5	3	5	5
7	0	1	3	3	5	5	7	7	9	9	11	11	13	13	15
8	0	1	1	2	2	3	3	4	4	5	5	6	6	7	7
9	0	1	1	3	3	4	5	6	6	8	8	9	10	11	11
10	0	1	3	1	3	5	3	5	7	8	7	9	7	9	11
11	1	2	4	6	8	8	12	12	14	16	18	18	22	22	24

the unique newform of weight 4 and level 5 is

$$e(z)\prod_{n=1}^{\infty}(1-e(nz))^4(1-e(5nz))^4;$$

The first cusp form of weight 2 is for level 11. This newform corresponds to the elliptic curve of conductor 11. It can be expressed as a product

$$e(z)\prod_{n=1}^{\infty}(1-e(nz))^2(1-e(11nz))^2.$$

Orthogonality

The space $S_k(q)$ can be made into a Hilbert space by introducing the Petersson inner product, as we did earlier for the case $q=1$, except that the integral is over the fundamental domain $\mathcal{D}(q)$ of $\Gamma_0(q)$ which can be represented as a union of $\nu(q)$ copies of the fundamental domain $\mathcal{D}$. Let $\mathcal{F}_k(q)$ be an orthogonal basis of $S_k(q)$ with respect to this inner product. The Petersson formula tells us that

$$\frac{\Gamma(k-1)}{(4\pi\sqrt{mn})^{k-1}}\sum_{f\in\mathcal{F}_k(q)}\frac{\lambda_f(m)\overline{\lambda_f(n)}}{\langle f,f\rangle}$$
$$=\delta_{m,n}+2\pi i^{-k}\sum_{c\equiv 0 \bmod q}\frac{S(m,n,c)}{c}J_{k-1}\left(\frac{4\pi\sqrt{mn}}{c}\right).$$

Let

$$\psi_f(n)=\left(\frac{\Gamma(k-1)}{(4\pi n)^{k-1}}\right)^{1/2}\lambda_f(n)$$

and let

$$\mathcal{L}_f(b) = \sum_{n=1}^{N} b_n \psi_f(n).$$

Then for any complex numbers b_n

$$\sum_{f \in \mathcal{F}_k(q)} |\mathcal{L}_f(b)|^2 = \left(1 + O\left(\frac{N}{q}\right)\right) \sum_{n=1}^{N} |b_n|^2.$$

A variant of this involves summing over k for a fixed q. Suppose that g is smooth and supported in $[K, 2K]$. Then

$$2 \sum_{k \text{ even}} g(k-1) \sum_{f \in \mathcal{F}_k(q)} |\mathcal{L}_f(b)|^2 = (\hat{g}(0) + \eta(K, N)) \sum_{n=1}^{N} |b_n|^2$$

where $\hat{g}(0) = \int g(y)\, dy$ and

$$\eta(K, N) \ll \left(\frac{N}{qK^3} + \left(\frac{N}{qK^2}\right)^j\right) N \log 2N$$

for any $j \geq 0$ and the implied constant depends only on j. These results are taken from [Iwa2], Chapter 5.

1-Level Density or Low Lying Zeros

The average spacing for all the zeros of all the $L(f, s)$ with $f \in H_k(q)$ up to a fixed height t_0 is asymptotic to $2\pi / \log(k^2 q)$. Let ϕ be a test function which is even and rapidly decaying. Iwaniec, Luo, and Sarnak [ILS] proved that if the support of $\hat{\phi}$ is contained in $(-2, 2)$, then

$$\lim_{KQ \to \infty} \sum_{k=2}^{K} \sum_{q \leq Q} \frac{1}{|H_k(q)|} \sum_{f \in H_k(q), L_f(1/2 + i\gamma_f) = 0} \phi\left(\frac{\gamma_f \log k^2 Q}{2\pi}\right)$$
$$= \int_{-\infty}^{\infty} \phi(x) W(\mathrm{O})(x)\, dx\,.$$

Similar results hold if the summation is restricted to even forms or to odd forms in which case the integral on the right side has $W(\mathrm{O}^{\pm})(x)$.

It should be pointed out that the Fourier transforms of the density functions $W(\mathrm{O})(x)$, $W(\mathrm{O}^{+})(x)$, and $W(\mathrm{O}^{-})(x)$ all agree in the diagonal range; so it is only when one goes beyond the diagonal that the distinguishing features of these three symmetry types becomes apparent.

Approximate Functional Equation

Using the functional equation we can represent the central values $L_f(\frac{1}{2})$ by partial sums of lengths about $O(kq)$. To this end we choose a function $G(s)$ which is holomorphic in $|\text{Re } s| \le A$ such that

$$G(s) = G(-s)$$
$$\Gamma(\tfrac{k}{2})G(0) = 1$$
$$\Gamma(s + \tfrac{k}{2})G(s) \ll (|s| + 1)^{-2A}$$

for some $A \ge 1$. Consider the integral

$$I = \tfrac{1}{2\pi i} \int_{(1)} \xi_f(s + \tfrac{1}{2}, \chi)G(s)s^{-1}\, ds.$$

Moving the integration to the line Re $s = -1$ we derive

$$\xi_f(\tfrac{1}{2})G(0) = 2I.$$

On the other hand, integrating termwise we derive

$$I = \sum_1^{\infty} \lambda_f(n)\chi(n) \left(\frac{q}{2\pi n}\right)^{\frac{1}{2}} V\left(\frac{n}{q}\right)$$

where $V(y)$ is the inverse Mellin transform of $(2\pi)^{-s}\Gamma(s + \frac{k}{2})G(s)s^{-1}$,

$$V(y) = \frac{1}{2\pi i} \int_{(1)} \Gamma(s + \tfrac{k}{2})G(s)(2\pi y)^{-s}s^{-1}\, ds.$$

Hence we get: For any Hecke form $f \in H_k(q)$ we have

$$L_f(\tfrac{1}{2}) = 2\sum_1^{\infty} \lambda_f(n)\chi(n)n^{-\frac{1}{2}}V(n/q).$$

Observe that $V(y)$ satisfies the following bounds

$$V(y) = 1 + O(y^A),$$
$$V(y) \ll (1 + y)^{-A},$$
$$V^{(\ell)}(y) \ll y^A(1 + y)^{-2A},$$

for $0 < \ell < A$. One can choose $G(s)$ depending on k so that

$$V(y) \ll k(1 + y/k)^{-A},$$

therefore the series dies rapidly as soon as n exceeds kq. If one is not concerned with the dependence of implied constants on the parameter k then one has a simple choice $G(s) = \Gamma(k/2)^{-1}$ getting the incomplete gamma function

$$V(y) = \frac{1}{\Gamma(\frac{k}{2})} \int_{2\pi y}^{\infty} e^{-x}x^{\frac{k}{2}-1}\, dx.$$

Moments

In this section let q be a prime number. Let $\sum^{\mathrm{h}}$ denote the harmonic average over $f \in H_2(q)$, that is we weight the term associated with f in the sum by $1/\langle f, f\rangle$. This leads to simpler results. Then, work of Duke, and of Duke, Friedlander and Iwaniec, and of Kowalski, Michel and Vanderkam, we know

$$\sum_{f\in H_2(q)}^{\mathrm{h}} L(1/2,f) \sim a_1$$

$$\sum_{f\in H_2(q)}^{\mathrm{h}} L^2(1/2,f) \sim 2a_2 \log q^{\frac{1}{2}}$$

$$\sum_{f\in H_2(q)}^{\mathrm{h}} L^3(1/2,f) \sim 8a_3 \frac{\log^3 q^{\frac{1}{2}}}{3!}$$

$$\sum_{f\in H_2(q)}^{\mathrm{h}} L^4(1/2,f) \sim 128a_4 \frac{\log^6 q^{\frac{1}{2}}}{6!}$$

where $a_k = A(0,\ldots 0)$ with

$$A_k(\alpha_1,\ldots,\alpha_k) = \prod_p \prod_{1\le i<j\le k}\left(1-\frac{1}{p^{1+\alpha_i+\alpha_j}}\right)$$
$$\times \frac{2}{\pi}\int_0^{\pi} \sin^2\theta \prod_{j=1}^{k} \frac{e^{i\theta}\left(1-\frac{e^{i\theta}}{p^{\frac{1}{2}+\alpha_j}}\right)^{-1} - e^{-i\theta}\left(1-\frac{e^{-i\theta}}{p^{\frac{1}{2}+\alpha_j}}\right)^{-1}}{e^{i\theta}-e^{-i\theta}}\, d\theta.$$

A general conjecture is:

$$\sum_{f\in H_2(q)}^{\mathrm{h}} L^k(1/2,f) \sim 2^{k-1}\prod_{\ell=1}^{k-1}\frac{\ell!}{2\ell!} a_k \log^{k(k-1)/2} q\,.$$

A more precise conjecture is

$$\sum_{f\in H_2(q)}^{\mathrm{h}} L_f(1/2)^k = R_k\left(\log \tfrac{q}{4\pi^2}\right) + O(q^{-1/2+\epsilon})$$

as $q\to\infty$, where R_k is a polynomial of degree $k(k-1)/2$ given by the k-fold residue

$$R_k(x) = \frac{(-1)^{k(k-1)/2}2^{k-1}}{k!}\frac{1}{(2\pi i)^k}\oint\cdots\oint$$
$$\frac{H(z_1,\ldots,z_k)\Delta(z_1^2,\ldots,z_k^2)^2}{\prod_{j=1}^k z_j^{2k-1}} e^{\frac{x}{2}\sum_{j=1}^k z_j}\, dz_1\ldots dz_k,$$

where

$$H(\alpha_1,\dots,\alpha_k) = A_k(\alpha_1,\dots,\alpha_k)\prod_{j=1}^{k}\Gamma(1+\alpha_j)\prod_{1\le i<j\le k}\zeta(1+\alpha_i+\alpha_j)$$

Maass Forms

There is another kind of cusp form associated with the group $\Gamma_0(q)$. This is a function $f(z)$ which is real analytic on the upper half-plane. It transforms like a weight 0 cusp form and is an eigenfunction of the Laplace operator:

$$\Delta := y^2\left(\frac{\partial^2}{\partial x^2}+\frac{\partial^2}{\partial y^2}\right).$$

It has a Fourier expansion as a linear combination of terms $e(nx)$ in which the dependence on y is expressed through K-Bessel functions. The prototype for these is given by the Eisenstein series (for the full modular group)

$$E(z,s) = \sum_{\gamma\in\Gamma_\infty\backslash\Gamma_0(1)} y(\gamma z)^s = \sum_{(c,d)=1}\frac{y^s}{|cz+d|^{2s}}$$

where $y(z)$ denotes the imaginary part of z and where Γ_∞ is the group which fixes ∞, i.e. the group of matrices $\begin{pmatrix}1 & b\\ 0 & 1\end{pmatrix}$ for integer b. This is not a cusp form (because it doesn't vanish at iy as $y\to\infty$.) However, its Fourier expansion is similar to that of the Maass cusp forms for which no explicit construction is known (apart from some forms with eigenvalue 1/4). Let

$$\theta(s) := \pi^{-s}\Gamma(s)\zeta(2s) = \theta(1-s).$$

Then $\theta(s)E(z,s) =$

$$\theta(s)y^s+\theta(1-s)y^{1-s}+4y^{1/2}\sum_{n=1}^{\infty}\sum_{ab=n}(a/b)^{s-1/2}K_{s-1/2}(2\pi ny)\cos(2\pi nx).$$

Since $\theta(s)$, $\sum_{ab=n}(a/b)^{s-1/2}$ and $K_{s-1/2}(2\pi ny)$ are all invariant under $s\to 1-s$, we see that $\theta(s)E(z,s)=\theta(1-s)E(z,1-s)$.

A Maass form f with eigenvalue $\lambda = 1/2+\kappa^2$ satisfies $(\Delta+\lambda)f=0$ and has Fourier expansion

$$f(z) = y^{1/2}\sum_{n=1}^{\infty}\lambda_f(n)K_{i\kappa}(2\pi ny)\cos(2\pi nx)$$

for an even Maass form and

$$f(z) = y^{1/2} \sum_{n=1}^{\infty} \lambda_f(n) K_{i\kappa}(2\pi ny) \sin(2\pi nx)$$

for an odd Maass form.

The L-function $L_f(s) = \sum_{n=1}^{\infty} \lambda_f(n) N^{-s}$ associated with a Maass from is entire, has an Euler product, and satisfies the functional equation

$$\xi_f(s) := \pi^{-s} \Gamma((s + i\kappa)/2) \Gamma((s - i\kappa)/2) L_f(s) = \xi_f(1 - s)$$

for even Maass forms and

$$\xi_f(s) := \pi^{-s} \Gamma((s + 1 + i\kappa)/2) \Gamma((s + 1 - i\kappa)/2) L_f(s) = \xi_f(1 - s)$$

for odd Maass forms.

Selberg's trace formula provides us with a kind of Weyl law for the number of Maass forms with eigenvalue less than a given quantity.

Ramanujan's conjecture for Maass forms is that $|\lambda_f(p)| \leq 2$. However, this has not yet been proven. The best result is $\lambda_f(p) \ll p^{1/9}$.

Moments

Motohashi [M1] and Ivic [Iv] have computed moments of these L-series at the central critical points. These have been achieved for the first through fourth moments and agree with random matrix conjectures.

Quadratic Twists of Modular L-Functions

In this section we give a specific example of what we mean by these quadratic twists. This example should be sufficient to allow the reader to understand a more general situation. Let

$$L_{11}(s) = \sum_{n=1}^{\infty} \frac{\lambda_n}{n^{1/2+s}}$$

be the L-function of conductor 11 of the elliptic curve

$$y^2 + y = x^3 - x^2.$$

The coefficients λ_n are obtained from cusp form of weight two and level 11 given by

$$\sum_{n=1}^{\infty} \lambda_n q^n = q \prod_{n=1}^{\infty} (1 - q^n)^2 (1 - q^{11n})^2.$$

Expanding the right hand side using Euler's pentagonal theorem provides an efficient means to compute the λ_n's.

$L_{11}(s)$ satisfies an even functional equation (i.e. its sign is +1)

$$\left(\frac{11^{1/2}}{2\pi}\right)^s \Gamma(s+1/2)L_{11}(s) = \left(\frac{11^{1/2}}{2\pi}\right)^{1-s} \Gamma(3/2-s)L_{11}(1-s),$$

and $L_{11}(s)$ may be written as a product over primes

$$L_{11}(s) = \frac{1}{1-11^{-s-1/2}} \prod_{p\neq 11} \frac{1}{1-\lambda_p p^{-s-1/2}+p^{-2s}}.$$

Consider now quadratic twists of $L_{11}(s)$,

$$L_{11}(s,\chi_d) = \sum_{n=1}^{\infty} \frac{\lambda_n}{n^{1/2+s}}\chi_d(n).$$

with $(d,11)=1$. $L_{11}(s,\chi_d)$ satisfies the functional equation

$$L_{11}(s,\chi_d) = \chi_d(-11)\frac{\Gamma(3/2-s)}{\Gamma(s+1/2)}\left(\frac{2\pi}{N^{1/2}}\right)^{2s-1} |d|^{2(1/2-s)} L_{11}(1-s,\chi_d).$$

We wish to look at moments of $L_{11}(1/2,\chi_d)$ but only for those $L(s,\chi_d)$ that have an even functional equation, i.e. $\chi_d(-11)=1$. We further only look at $d<0$ since in that case a theorem of Kohnen and Zagier enables us to easily gather numerical data for $L_{11}(1/2,\chi_d)$ with which to check our conjecture.

When $d<0$, $\chi_d(-1)=-1$, hence, in order to have an even functional equation, we require that $\chi_d(11)=-1$, i.e. $d = 2,6,7,8,10 \mod 11$. The sum over fundamental discriminants is

$$\sum_{\substack{-D<d<0\\ d=2,6,7,8,10 \mod 11}}^{\star} L_{11}(1/2,\chi_d)^k$$

$$= \sum_{\substack{-D<d<0\\ d=2,6,7,8,10 \mod 11}}^{\star} \mathcal{D}_k\left(\log\left(\frac{|d|11^{1/2}}{2\pi}\right)\right) + O(D^{\frac{1}{2}+\epsilon})$$

where $\mathcal{D}_k$ is the polynomial of degree $k(k-1)/2$ given by the k-fold residue

$$\mathcal{D}_k(x) = \frac{(-1)^{k(k-1)/2}2^k}{k!}\frac{1}{(2\pi i)^k}\oint\cdots\oint \frac{R_{11}(z_1,\ldots,z_k)\Delta(z_1^2,\ldots,z_k^2)^2}{\prod_{j=1}^{k} z_j^{2k-1}} e^{x\sum_{j=1}^k z_j}\, dz_1\ldots dz_k,$$

where

$$R_{11}(z_1,\ldots,z_k) = A_k(z_1,\ldots,z_k)\prod_{j=1}^{k}\left(\frac{\Gamma(1+z_j)}{\Gamma(1-z_j)}\right)^{\frac{1}{2}} \prod_{1\le i<j\le k} \zeta(1+z_i+z_j),$$

and A_k is the Euler product which is absolutely convergent for $\sum_{j=1}^k |z_j| < 1/2$,

$$A_k(z_1,\ldots,z_k) = \prod_p R_{11,p}(z_1,\ldots,z_k) \prod_{1\le i<j\le k}\left(1-\frac{1}{p^{1+z_i+z_j}}\right)$$

with, for $p \neq 11$, $R_{11,p} = \left(1+\frac{1}{p}\right)^{-1} \times$

$$\left(\frac{1}{p}+\frac{1}{2}\left(\prod_{j=1}^k \frac{1}{1-\lambda_p p^{-1-z_j}+p^{-1-2z_j}} + \prod_{j=1}^k \frac{1}{1+\lambda_p p^{-1-z_j}+p^{-1-2z_j}}\right)\right)$$

and

$$R_{11,11} = \prod_{j=1}^k \frac{1}{1+11^{-1-z_j}}.$$

The local factor $R_{11,11}$ differs slightly from (4.3.13) because we are restricting ourselves to $\chi_d(11) = -1$, therefore only one term appears.

In [CFKRS] we compare moments computed numerically with moments estimated by our conjecture. The two agree to within the accuracy we have for the moment polynomial coefficients. We believe that if one were to compute the coefficients to higher accuracy, one would see an even better agreement with the data.

While one can compute $L_{11}(1/2, \chi_d)$ using standard techniques, one can in our case exploit a theorem of Kohnen and Zagier which relates $L_{11}(1/2, \chi_d)$, for fundamental discriminants $d < 0$, $d = 2, 6, 7, 8, 10 \mod 11$, to the coefficients $c_{11}(|d|)$ of a weight 3/2 modular form

$$L_{11}(1/2,\chi_d) = \kappa_{11} c_{11}(|d|)^2/\sqrt{d}$$

where κ_{11} is a constant. The weight 3/2 form in question was determined by Rodriguez-Villegas (private communication)

$$\sum_{n=1}^{\infty} c_{11}(n)q^n = (\theta_1(q)-\theta_2(q))/2$$
$$= -q^3+q^4+q^{11}+q^{12}-q^{15}-2q^{16}-q^{20}\ldots$$

where

$$\theta_1(q) = \sum_{\substack{(x,y,z)\in\mathbb{Z}^3\\ x=y \mod 2}} q^{x^2+11y^2+11z^2} = 1+2q^4+2q^{11}+4q^{12}+4q^{15}+2q^{16}+\ldots$$

and

$$\theta_2(q) = \sum_{\substack{(x,y,z)\in\mathbb{Z}^3\\ x=y \mod 3\\ y=z \mod 2}} q^{(x^2+11y^2+33z^2)/3} = 1+2q^3+2q^{12}+6q^{15}+6q^{16}+\ldots.$$

This was used to compute the $c_{11}(|d|)$'s for $d < 85,000,000$; the numerical evidence supports the conjectures.

Moments

The first moment of these twists, for a general element of $H_k(q)$ has been evaluated by Murty and Murty [MM] and by Bump, Friedberg, and Hoffstein [BFH]. Iwaniec [Iwa] has given a particularly elegant proof of the first moment.

4.5 Symmetric Square L-functions

Recall that the Euler product for a level q modular form has the shape

$$L_f(s) = \prod_p \left(1 - \frac{\alpha_f(p)}{p^s}\right)^{-1} \left(1 - \frac{\alpha'_f(p)}{p^s}\right)^{-1}.$$

We can form the symmetric square L-function associated to f as

$$L_f(\text{sym}^2, s) = \prod_p \left(1 - \frac{\alpha_f^2(p)}{p^s}\right)^{-1} \left(1 - \frac{\alpha_f(p)\alpha'_f(p)}{p^s}\right)^{-1} \left(1 - \frac{\alpha'_f(p)^2}{p^s}\right)^{-1}.$$

Note that this L-function has a degree three Euler product associated with it. Shimura proved that this is an entire function which satisfies the functional equation

$$\begin{aligned}\xi_f(\text{sym}^2, s) :&= \pi^{-3s/2} q^s \Gamma(s/2)\Gamma((s+k-1)/2)\Gamma((s+k)/2) L_f(\text{sym}^2, s) \\ &= \xi_f(\text{sym}^2, 1-s).\end{aligned}$$

1-Level Density

The average spacing for all the zeros of all the $L_f(\text{sym}^2, s)$ with $f \in H_k(1)$ up to a fixed height t_0 is asymptotic to $2\pi/\log(k^2)$. Let ϕ be a test function which is even and rapidly decaying. Iwaniec, Luo, and Sarnak [ILS] proved that if the support of $\hat{\phi}$ is contained in $(-3/2, 3/2)$, then (for fixed q)

$$\begin{aligned}&\lim_{K\to\infty} \sum_{k=2}^{K} \frac{1}{|H_k(q)|} \sum_{f\in H_k(q), L_f(\text{sym}^2, 1/2+i\gamma_f)=0} \phi\left(\frac{\gamma_f \log k^2 q^2}{2\pi}\right) \\ &\quad = \int_{-\infty}^{\infty} \phi(x) W(\text{Sp})(x)\, dx\,.\end{aligned}$$

Moments

The first moment for symmetric square L-functions can be evaluated asymptotically, but so far not the second, see Iwaniec and Michel [IM]. The symmetry type of this family is symplectic and the difficulty of achieving the second moment over this family is like the fourth moment of quadratic L-functions.

4.6 Convolution L-functions

Given two cuspidal L-functions

$$L_f(s) = \prod_p \left(1 - \frac{\alpha_f(p)}{p^s}\right)^{-1} \left(1 - \frac{\alpha'_f(p)}{p^s}\right)^{-1}$$

where $f \in H_k(q_1)$ and

$$L_g(s) = \prod_p \left(1 - \frac{\beta_g(p)}{p^s}\right)^{-1} \left(1 - \frac{\beta'_g(p)}{p^s}\right)^{-1}$$

where $g \in H_\ell(q_2)$ with $(q_1, q_2) = 1$ we form the convolution L-function

$$L_{f\times g}(s) = \prod_p \left(1 - \frac{\alpha_f(p)\beta_g(p)}{p^s}\right)^{-1} \left(1 - \frac{\alpha_f(p)\beta'_g(p)}{p^s}\right)^{-1} \times \\ \left(1 - \frac{\alpha'_f(p)\beta_g(p)}{p^s}\right)^{-1} \left(1 - \frac{\alpha'_f(p)\beta'_g(p)}{p^s}\right)^{-1}.$$

If $f \neq g$, then this L-function is entire – an Euler product of degree 4 – and satisfies the functional equation

$$\begin{aligned} \xi_{f\times g}(s) :&= (2\pi)^{-2s}(q_1q_2)^s \Gamma(s + (|k-\ell|)/2)\Gamma(s - 1 + (k+\ell-1)/2)L_{f\times g}(s) \\ &= \pm\xi_{f\times g}(1-s). \end{aligned}$$

Moments

Kowalski, Michel, and Vanderkam have successfully computed all of the main terms (see (http://www.math.univ-montp2.fr/m̃ichel/publi.html [22]) in the second moment of the convolution L-function at the central point for a fixed g and f varying over $H_k(q)$ for large prime q. The leading term is of size $q \log^3 q$ in accordance with the expectation that this is an orthogonal family.

5 Other Directions

Here we describe a few things that there weren't room for in the notes; or things that are still being developed.

5.1 Integrals of Ratios of Zeta-Functions

In 1994 Farmer [F] made the conjecture that

$$\frac{1}{T}\int_0^T \frac{\zeta(s+a)\zeta(1-s+b)}{\zeta(s+u)\zeta(1-s+v)}\,dt \sim 1 + \frac{T^{-u+v}-1}{(u+v)(a+b)}.$$

Here, $s = 1/2 + ir$ and the real parts of u and v are positive. Farmer showed that this conjecture implies Montgomery's pair-correlation conjecture. He later made a conjecture for an integral of a ratio of three zetas over 3 zetas.

After a lecture at MSRI in June 1999 Martin Zirnbauer and Stephen Nonnemacher proposed to Farmer and colleagues a way to generalize this conjecture to the integral of a ratio with any number of zeta-factors in the numerator and denominator. Their suggestion was based on an analogous moment for characteristic polynomials.

They have a method to compute exact formulas for the average over $\mathrm{U}(N)$, $\mathrm{Sp}(N)$, or $\mathrm{SO}(2N)$ of such a ratio. The method involves representation theory – supersymmetry and Weyl's formula for the characters of representations.

5.2 Mollifiers

For many applications in number theory one needs to compute moments of 'mollified' L-functions. Examples include

$$\int_0^T |\zeta(1/2+it)|^2 \left| \sum_{n\le y} \frac{\mu(n)\log y/n}{n^{1/2+it}} \right|^2 dt.$$

This example is relevant to proofs that a positive proportion of zeros of $\zeta(s)$ are on the one-half line. Levinson used an asymptotic formula for this with the length y of the mollifier taken to be $y = T^{1/2-\epsilon}$ to show that at least one-third of the zeros are on the critical line; Conrey used such a moment with $y = T^{4/7-\epsilon}$ to prove that at least two-fifths of the zeros are on the critical line.

Farmer's conjecture from the previous section can be used to prove an asymptotic formula for this mollified mean-square with an arbitrary length. Such a formula would imply that almost all of the zeros are on the critical line.

In [CF] formulas are given for the mollified mean square of L-functions from three different families. The asymptotic formula are conjectured to depend only on the symmetry type of the family and not on the family itself.

The more general conjectures for ratios could be used to conjecture formulas for any moment of an L-function times a mollifier (not just the second moment).

5.3 Connections with Primes in Short Intervals

Montgomery and Goldston showed that Montgomery's Pair Correlation conjecture is equivalent to an assertion about a second moment of primes in short intervals.

The method relates both quantities to an asymptotic formula for

$$\int_0^T \left| \frac{\zeta'(1/2 + a + it)}{\zeta(1/2 + a + it)} \right|^2 dt$$

with small a with positive real part.

It should be mentioned that Bogolmony and Keating showed how a heuristic beginning with the Hardy-Littlewood conjectures for pairs of primes leads to all of the CUE n-correlation statistics for the zeros of the Riemann zeta-function.

5.4 Distribution of Zeros of Derivatives

The Riemann Hypothesis is equivalent to the assertion that all of the zeros of $\zeta'(s)$ have real parts greater than or equal to 1/2. The question of the distribution of the real parts of these zeros arises in Levinson's method. The proper scaling at a height T is $1/\log T$. So the question is to determine the distribution function

$$d(\alpha) := \lim_{T \to \infty} \#\{\rho' = \beta' + i\gamma' : 0 < \gamma' < T, \beta' < 1/2 + \alpha/\log T\}.$$

The analogous question for CUE has to do with the distribution of the zeros of $Z'(U, s)$ inside the unit circle on the scale of $1/N$. Francesco Mezzadri [Mez] has made some progress on this question but has not solved it completely.

A similar question arises about the distribution of zeros of $\xi'(s)$ on the critical line.

5.5 Moments of Derivatives

The asymptotics of the moments

$$\int_0^T |\zeta'(1/2 + it)|^{2k} \, dt$$

can be conjectured for integral k (see [Hu]) but so far not for non-integral k and, of course, the analogous question for moments of derivatives of characteristic polynomials.

5.6 Lower Order Terms for Non-Integral Moments of L-Functions

We [CFKRS1] have conjectures for all of the main terms for the $2k$-th moment of L-functions in families. The leading main term is an analytic function of k; similarly for the second main term, third, and so on for any fixed term. However, we don't yet have an analytic expression for all of the terms taken together.

5.7 Extremely Large Values

We would like to know the size of the largest values of $|\zeta(1/2+it)|$. These could presumably be deduced from our conjectured formula for the $2k$-th moment by taking k extremely large. However, we have been unable so far to determine the large k asymptotics for the conjecture when we include all of the lower order terms. See [CGo], [Hu], and [U].

5.8 Distribution of Small Values

Random Matrix Theory does seem to predict very well the small and intermediate size values of L-functions. These models can be used [CKRS] together with an appropriate discretization to predict the frequency of vanishing to order two within certain families of the central values of the L-functions in the family. This prediction is especially of interest with regard to elliptic curves and their ranks. However, we don't seem to be able to use Random matrix Theory to predict vanishing to order three.

References

[BFH2] Bump, Daniel; Friedberg, Solomon; Hoffstein, Jeffrey. Nonvanishing theorems for L-functions of modular forms and their derivatives. Invent. Math. 102 (1990), no. 3, 543–618.

[BFH1] Bump, Daniel; Friedberg, Solomon; Hoffstein, Jeffrey. A nonvanishing theorem for derivatives of automorphic L-functions with applications to elliptic curves. Bull. Amer. Math. Soc. (N.S.) 21 (1989), no. 1, 89-93.

[C] Conrey, J. Brian: L-functions and random matrices. In: Mathematics unlimited—2001 and beyond. Springer, Berlin Heidelberg New York (2001) (arXiv math.NT/0005300).

[CKRS] Conrey, J. B.; Farmer, D. W.; Keating, J. P.; Rubinstein, M. O.; Snaith, N. C.: On the frequency of vanishing of quadratic twists of modular L-functions. In: Number theory for the millennium, I Urbana, IL, 2000, A K Peters, Natick, MA (2002).

[CFKRS1] Conrey, J. B.; Farmer, D. W.; Keating, J. P.; Rubinstein, M. O.; Snaith, N. C.; Integral moments of L-functions. AIM Preprint 2002–6, arXiv math.NT/0206018.

[CFKRS2] Conrey, J. B.; Farmer, D. W.; Keating, J. P.; Rubinstein, M. O.; Snaith, N. C.; Autocorrelation of Random Matrix Polynomials. AIM Preprint 2002–10, arXiv math-ph/0208007.

[CGh] Conrey, J. B.; Ghosh, A. A conjecture for the sixth power moment of the Riemann zeta-function. Internat. Math. Res. Notices 1998, no. 15, 775–780.

[CGo] Conrey, J. B.; Gonek, S. M. High moments of the Riemann zeta-function. Duke Math. J. 107 (2001), no. 3, 577–604.

[Dei] Deift, P. A.: Orthogonal polynomials and random matrices: a Riemann-Hilbert approach. Courant Lecture Notes in Mathematics, 3. New York University, Courant Institute of Mathematical Sciences, New York. American Mathematical Society, Providence, RI (1999).

[F] Farmer, David W. Long mollifiers of the Riemann zeta-function. Mathematika 40 (1993), no. 1, 71–87.

[GGM] Goldston, D. A.; Gonek, S. M.; Montgomery, H. L. Mean values of the logarithmic derivative of the Riemann zeta-function with applications to primes in short intervals. J. Reine Angew. Math. 537 (2001), 105–126.

[GM] Goldston, Daniel A.; Montgomery, Hugh L. Pair correlation of zeros and primes in short intervals. In: Analytic number theory and Diophantine problems (Stillwater, OK, 1984), 183–203, Progr. Math., 70, Birkhäuser Boston, Boston, MA, 1987.

[H-B1] Heath-Brown, D. R. The fourth power moment of the Riemann zeta function. Proc. London Math. Soc. (3) 38 (1979), no. 3, 385–422.

[H-B2] Heath-Brown, D. R. The fourth power mean of Dirichlet's L-functions. Analysis 1 (1981), no. 1, 25–32.

[H-B3] Heath-Brown, D. R. An asymptotic series for the mean value of Dirichlet L-functions. Comment. Math. Helv. 56 (1981), no. 1, 148–161.

[H-B4] Heath-Brown, D. R. A mean value estimate for real character sums. Acta Arith. 72 (1995), no. 3, 235–275.

[Hu] Hughes, Christopher Paul; On the characteristic polynomial of a random unitary matrix and the Riemann zeta function, Thesis, University of Bristol, 2001.

[Hu1] Hughes, C. P. Random matrix theory and discrete moments of the Riemann zeta function, AIM Preprint 2002–17, arXiv math.NT/0207236.

[Iv] Ivic, Aleksandr; On the moments of Hecke series at central points. Preprint arXiv math.NT/0210337.

[IvJut] Ivic, Aleksandr and Jutila, Matti; On the moments of Hecke series at central points II. Preprint arXiv math.NT/0305178.

[Iwa] Iwaniec, Henryk. On the order of vanishing of modular L-functions at the critical point. Sém. Théor. Nombres Bordeaux (2) 2 (1990), no. 2, 365–376.

[Iwa1] Iwaniec, Henryk: Introduction to the spectral theory of automorphic forms. Biblioteca de la Revista Matemtica Iberoamericana. [Library of the Revista Matemtica Iberoamericana]*Revista Matemtica Iberoamericana, Madrid* (1995).

[Iwa2] Iwaniec, Henryk: Topics in classical automorphic forms. Graduate Studies in Mathematics, 17. American Mathematical Society, Providence, RI (1997).

[ILS] Iwaniec, Henryk; Luo, Wenzhi; Sarnak, Peter. Low lying zeros of families of L-functions. Inst. Hautes Études Sci. Publ. Math. No. 91, (2000), 55–131 (2001).

[IM] Iwaniec, H.; Michel, P. The second moment of the symmetric square L-functions. Ann. Acad. Sci. Fenn. Math. 26 (2001), no. 2, 465–482.

[Jut] Jutila, M. On the mean value of $L(\frac{1}{2}, \chi)$ for real characters. Analysis 1 (1981), no. 2, 149–161.

[KS1] Katz, Nicholas M. and Sarnak, Peter: Zeroes of zeta functions and symmetry. Bull. Amer. Math. Soc. (N.S.) 36 (1999), no. 1, 1–26.

[KS2] Katz, Nicholas M. and Sarnak, Peter: Random matrices, Frobenius eigenvalues, and monodromy. American Mathematical Society Colloquium Publications, 45. American Mathematical Society, Providence, RI (1999).

[KSn1] Keating, J. P.; Snaith, N. C. Random matrix theory and L-functions at $s = 1/2$. Comm. Math. Phys. 214 (2000), no. 1, 91–110.

[KSn2] Keating, J. P.; Snaith, N. C. Random matrix theory and $\zeta(1/2 + it)$. Comm. Math. Phys. 214 (2000), no. 1, 57–89.

[KSn3] Keating, J. P.; Snaith, N. C. Random matrices and L-functions. J. Phys. A 36 (2003), no. 12, 2859–2851.

[Meh] Mehta, Madan Lal: Random matrices. Second edition. Academic Press, Inc., Boston, MA (1991).

[Mez] Mezzadri, F. Random matrix theory and the zeros of $\zeta'(s)$. AIM Preprint 2002–9, arXiv math-ph/0207044.

[M1] Motohashi, Yōichi. Spectral mean values of Maass waveform L-functions. J. Number Theory 42 (1992), no. 3, 258–284.

[M2] Motohashi, Yoichi. Spectral theory of the Riemann zeta-function. Cambridge Tracts in Mathematics, 127. Cambridge University Press, Cambridge, 1997. x+228 pp. ISBN: 0-521-44520-5

[MM] Murty, M. Ram; Murty, V. Kumar Mean values of derivatives of modular L-series. Ann. of Math. (2) 133 (1991), no. 3, 447–475.

[Odl] Odlyzko A. M.: http://www.dtc.umn.edu/odlyzko/

[Oz] Özlük, Ali E. On the pair correlation of zeros of Dirichlet L-functions. Number theory (Banff, AB, 1988), 471–476, de Gruyter, Berlin, 1990.

[RS] Rudnick, Zeév; Sarnak, Peter. Zeros of principal L-functions and random matrix theory. A celebration of John F. Nash, Jr. Duke Math. J. 81 (1996), no. 2, 269–322.

[So] Soundararajan, K. Nonvanishing of quadratic Dirichlet L-functions at $s = \frac{1}{2}$. Ann. of Math. (2) 152 (2000), no. 2, 447–488.

[Tit] Titchmarsh, E. C.: The theory of the Riemann zeta-function. Second edition. Edited and with a preface by D. R. Heath-Brown. The Clarendon Press, Oxford University Press, Oxford (1986).

[U] Ulmer, Douglas Elliptic curves with large rank over function fields. Ann. of Math. (2) 155 (2002), no. 1, 295–315.

Energy Level Statistics, Lattice Point Problems, and Almost Modular Functions

Jens Marklof

School of Mathematics, University of Bristol, Bristol BS8 1TW, U.K.
j.marklof@bristol.ac.uk

Summary. One of the central aims in quantum chaos is to classify quantum systems according to universal statistical properties. It has been conjectured that the energy levels of generic integrable quantum systems have the same statistical properties as random numbers from a Poisson process (Berry & Tabor 1977), and chaotic quantum systems the same as eigenvalues of random matrices from suitably chosen ensembles (Bohigas, Giannoni & Schmit 1984). I review some recent developments concerning simple classes of integrable systems, where the study of eigenvalue correlations leads to subtle lattice point counting problems which, in some instances, can be solved by ergodic theoretic techniques. In a special example (the so-called "boxed oscillator") energy level statistics are related to the statistical distribution of the fractional parts of the sequence $n^2\alpha$. We will see that the error term of this distribution can be identified with an *almost modular function*, and that the fluctuations of the error term are governed by a general limit theorem for such functions.

1 Introduction

The classification of quantum systems according to universal statistical properties is one of the central objectives in quantum chaos. The topic is discussed in detail in Eugene Bogomolny's lectures [7] and I will here concentrate on a special class of quantum systems whose level statistics can be understood in terms of lattice point counting problems. Let us consider a Hamiltonian with discrete energy spectrum $\lambda_1 \leq \lambda_2 \leq \ldots \to \infty$. We assume that the number of levels (counted with multiplicity) grows asymptotically as

$$\#\{j : \lambda_j \leq \lambda\} \sim \overline{N}(\lambda) \quad (\lambda \to \infty) \tag{1.1}$$

where $\overline{N}(\lambda) = c\lambda^\gamma$ with constants $c > 0$, $\gamma \geq 1$. To investigate its statistical properties it is convenient to rescale the sequence by setting $X_j = \overline{N}(\lambda_j)$ which yields mean density $= 1$, i.e.,

$$\#\{j : X_j \leq X\} \sim X \quad (X \to \infty). \tag{1.2}$$

The central conjecture, put forward by Berry and Tabor in 1977 [1], is that if the Hamiltonian is classically integrable (and sufficiently "generic") then the X_j have the same local statistical properties as independent random variables from a Poisson process. This means that

$$\mathcal{N}(T, L) := \#\{j : T \leq X_j \leq T + L\}, \tag{1.3}$$

the number of X_j in a randomly shifted interval $[T, T+L]$ of fixed length L, is distributed according to the Poisson law $\frac{L^k}{k!}\,\mathrm{e}^{-L}$. More precisely, let $\rho : \mathbb{R}_{>0} \to \mathbb{R}_{\geq 0}$ be a continuous probability density with compact support, and define the family of probability densities ρ_X with $X \in \mathbb{R}_{\geq 1}$ by $\rho_X(T) = X^{-1}\rho(TX^{-1})$. The assertion is now that $\mathcal{N}(T, L)$ has a Poisson limit distribution, if T is distributed according to ρ_X and $X \to \infty$. That is, for any bounded function $g : \mathbb{Z}_{\geq 0} \to \mathbb{C}$ we have

$$\int_0^\infty g\big(\mathcal{N}(T, L)\big)\, \rho_X(T)\, dT \to \sum_{k=0}^\infty g(k) \frac{L^k}{k!}\, \mathrm{e}^{-L}\,. \tag{1.4}$$

This is in contrast to chaotic systems where the spectral statistics are expected to follow those of random matrix ensembles.

The central idea behind the Berry-Tabor conjecture is that the energy levels of an integrable Hamiltonian are in semiclassical approximation given by the EBK quantization

$$\lambda_j(\hbar) \sim H(\hbar(\boldsymbol{m} + \boldsymbol{\alpha})), \qquad \hbar \to 0, \tag{1.5}$$

where $H(\boldsymbol{I})$ is the classical Hamiltonian in the action variables; $\boldsymbol{m}$ runs over integer lattice points and $\boldsymbol{\alpha}$ is a fixed vector determined by topological data such as Maslov indices. One case where this approximation can be controlled sufficiently well to study spectral correlations is when H is the negative Laplacian $-\Delta$ on surfaces with integrable geodesic flow. For examples in the case of surfaces of revolution (with some technical assumptions) one has [10, 11]

$$\lambda_j = F(m_1, m_2 + \tfrac{1}{2}), \quad (m_1, m_2) \in \mathbb{Z}^2, \quad |m_1| \leq m_2, \tag{1.6}$$

where $F(\boldsymbol{x}) = F_2(\boldsymbol{x}) + F_0(\boldsymbol{x}) + O(\|\boldsymbol{x}\|^{-1})$, $\|\boldsymbol{x}\| \to \infty$, and F_2, F_0 are smooth homogeneous functions of degree 2 and 0, respectively. Note that in the case of the Laplacian the semiclassical limit $\hbar \to 0$ is equivalent to the high energy limit $j \to \infty$.

Sinai [42] and Major [17] proved the Poisson limit theorem (1.4) for generic F in a certain function space. A "generic" function has, however, level curves $F(\boldsymbol{x}) = 1$ which are not twice differentiable. Advances towards a proof of the Poisson conjecture for systems with analytic F, such as the Laplacian on surfaces with integrable geodesic flow, have been made only recently. Sarnak [38] showed that the pair correlation statistics are Poisson for the eigenvalues of tori with a generic flat metric (we shall see below that pair correlation or two-point statistics correspond to the variance of the distribution of $\mathcal{N}(T, L)$). The eigenvalues of a flat torus are given by positive definite binary quadratic forms $\alpha m^2 + \beta mn + \gamma n^2$ ($m, n \in \mathbb{Z}$), and "generic" refers to a choice of (α, β, γ) in a set of full Lebesgue measure. These results were extended by VanderKam to tori of arbitrary dimension [43] and also to higher-order correlation functions [44]. Eskin, Margulis and Mozes [12] strengthened considerably Sarnak's result by giving explicit diophantine conditions on (α, β, γ) under which the pair correlation statistics of two-dimensional flat tori is Poisson. It is interesting to note, however, that the fluctuations of the spectral form factor (the Fourier transform of the pair correlation density) are in this case not consistent with the Poisson model [18].

Berry and Tabor point out that there are many examples of integrable systems which violate their general conjecture, and that hence the Poisson distribution should only be expected for "generic" systems. One of the most interesting counter examples is the multi-dimensional harmonic oscillator whose eigenvalues are given by the values of the linear form $\boldsymbol{\omega} \cdot \boldsymbol{m}$ at lattice points $\boldsymbol{m} \in \mathbb{N}^k$; see Berry and Tabor's original work [1], and subsequent papers by Pandey, Bohigas, Giannoni and Ramaswamy [30, 31], Bleher [2, 3], Mazel and Sinai [29], Greenman [13, 14], and myself [21].

In the present paper we focus on two special classes of integrable systems. The first example is the k-dimensional standard torus $\mathbb{T}^k$ threaded by flux lines, where the question of energy level statistics corresponds to counting lattice points in thin spherical shells centered at $\boldsymbol{\alpha}$. It was first studied in connection with the Berry-Tabor conjecture by Cheng, Lebowitz and Major [8, 9]. In Sects. 2 and 3 I will review recent results on the pair correlation statistics [24, 25], which were announced in [22, 23].

The second example is the "boxed oscillator", i.e., a particle constrained by a box in x-direction and by a harmonic potential in the y-direction, so that

$H = -\partial_x^2 - \partial_y^2 + \omega^2 y^2$. In this case the eigenvalue correlations are closely related with the local statistics of the fractional parts of the sequence $n^2\alpha$, which were studied by Sinai [41], Pellegrinotti [32], Rudnick, Sarnak and Zaharescu [35, 36, 45], and Zelditch [46]. In Sects. 4 and 5 I will discuss joint work with Strömbergsson [28], which relates the pair correlation problem for $n^2\alpha$ to a natural equidistribution problem in hyperbolic geometry.

It is crucial in the Poisson limit theorem (1.4) that L is kept fixed. If L increases (sufficiently slowly) with T then the left-hand-side is expected to converge to a Gaussian distribution, see Bleher's review [5] for a detailed discussion. (In a recent paper [16], Hughes and Rudnick prove a central limit theorem for lattice points in annuli.) If, on the other hand, L grows sufficiently fast with T (e.g. $L = T$) the limiting distribution (provided it exists) is typically non-universal. In the case when the eigenvalues are given by values of positive definite binary quadratic forms (or more general functions homogeneous of degree two) the work of Heath-Brown [15] and Bleher [4, 5] shows that the limit distribution can be described in terms of almost periodic functions. Bleher and Bourgain obtained a similar result for the multidimensional torus threaded by flux lines, under certain diophantine conditions on the flux strength [6].

In the case of the boxed oscillator, we will see in Sect. 6 that, rather than almost periodic functions, almost *modular* functions will describe the distribution of the error term. This last section is based on the papers [26, 27].

2 Torus Threaded by Flux Lines and Lattice Points in Thin Spherical Shells

The quantum mechanics of a free particle on a k-dimensional torus threaded by flux lines of strength $\boldsymbol{\alpha} = (\alpha_1, \ldots, \alpha_k)$ is described by the Hamiltonian

$$H = \sum_j \left(\frac{1}{2\pi \mathrm{i}} \frac{\partial}{\partial x_j} - \alpha_j \right)^2 \tag{2.1}$$

acting on periodic functions φ, i.e., $\varphi(\boldsymbol{x} + \boldsymbol{l}) = \varphi(\boldsymbol{x})$, for all $\boldsymbol{l} \in \mathbb{Z}^k$. The eigenfunctions of H are $\varphi_{\boldsymbol{m}}(\boldsymbol{x}) = e(\boldsymbol{m} \cdot \boldsymbol{x})$, where $\boldsymbol{m} = (m_1, \ldots, m_k) \in \mathbb{Z}^k$, and its eigenvalues $0 \leq \lambda_1 \leq \lambda_2 \leq \cdots \to \infty$ are given by

$$\|\boldsymbol{m} - \boldsymbol{\alpha}\|^2 = (m_1 - \alpha_1)^2 + \cdots + (m_k - \alpha_k)^2. \tag{2.2}$$

Geometrically, the eigenvalues of H thus correspond to squared radii of spheres with center $\boldsymbol{\alpha}$ which contain at least one lattice point $\boldsymbol{m} \in \mathbb{Z}^k$; the multiplicity of the eigenvalue corresponds to the number of lattice points on the sphere. Since the number of lattice points in a ball of large radius is approximately its volume, we find that (1.1) holds with $\overline{N}(\lambda) = B_k \lambda^{k/2}$ where B_k is the volume of the unit ball. According to the Berry-Tabor conjecture we expect the rescaled sequence $X_j = B_k \|\boldsymbol{m}-\boldsymbol{\alpha}\|^k$ to satisfy the Poisson limit theorem (1.4), at least for "generic" choices of $\boldsymbol{\alpha}$. Hence, in geometric terms, the conjecture says that the number of lattice points inside a random spherical shell with fixed volume L, whose inner sphere encloses a ball of volume T (randomly distributed with law ρ_X), has a Poisson limit distribution as $X \to \infty$.

As a first step towards a proof of the conjecture we shall here show that the second moment of $\mathcal{N}(T,L)$, the *number variance*

$$\Sigma^2(X,L) := \frac{1}{X}\int_0^\infty \{\mathcal{N}(T,L) - L\}^2 \rho\left(\frac{T}{X}\right) dT, \tag{2.3}$$

converges indeed to the variance of the Poisson distribution, which is L. Note in the above definition of $\Sigma^2(X,L)$ that, in view of (1.2), the expectation value of $\mathcal{N}(T,L)$ is asymptotically

$$\frac{1}{X}\int_0^\infty \mathcal{N}(T,L)\,\rho\left(\frac{T}{X}\right) dT \to L. \tag{2.4}$$

As we shall see the set of "generic" $\boldsymbol{\alpha}$ can be characterized by an explicit diophantine condition which is in fact satisfied by a set of $\boldsymbol{\alpha}$ of full Lebesgue measure.

The vector $\boldsymbol{\alpha} = (\alpha_1,\ldots,\alpha_k) \in \mathbb{R}^k$ is said to be *diophantine of type* κ, if there exists a constant $C > 0$ such that

$$\max_j \left|\alpha_j - \frac{m_j}{q}\right| > \frac{C}{q^\kappa} \tag{2.5}$$

for all $m_1,\ldots,m_k,q \in \mathbb{Z}$, $q > 0$. The smallest possible value for κ is $\kappa = 1+\frac{1}{k}$. In this case $\boldsymbol{\alpha}$ is called *badly approximable*. Examples of badly approximable vectors are $\boldsymbol{\alpha}$ such that the components of $(\boldsymbol{\alpha},1)$ form a basis of a real algebraic number field of the degree $k+1$ ([39], Theorem 6F). On the other hand, for any $\kappa > 1+\frac{1}{k}$, the set of diophantine vectors of type κ is of full Lebesgue measure ([39], Theorem 6G).

Theorem 1 (Poisson limit of the number variance). *Suppose* $\boldsymbol{\alpha}$ *is diophantine of type* $\kappa < \frac{k-1}{k-2}$ *and the components of the vector* $(\boldsymbol{\alpha},1) \in \mathbb{R}^{k+1}$ *are linearly independent over* $\mathbb{Q}$. *Then, for every* $L > 0$,

$$\lim_{X\to\infty} \Sigma^2(X,L) = L. \tag{2.6}$$

This theorem is a corollary of a more general statement on the convergence of the *pair correlation density* of the X_j, which is proved in [24, 25]. For any $\psi \in \mathrm{C}_0(\mathbb{R}_{>0}\times\mathbb{R}_{>0}\times\mathbb{R})$ (i.e., continuous and of compact support) let us define the pair correlation function

$$R_2(\psi,\lambda) = \frac{1}{B_k \lambda^{k/2}} \sum_{i,j=1}^{\infty} \psi\left(\frac{\lambda_i}{\lambda}, \frac{\lambda_j}{\lambda}, \lambda^{k/2-1}(\lambda_i - \lambda_j)\right). \tag{2.7}$$

We then have the following statement (Theorem 2.2, [25]).

Theorem 2 (Poisson limit of pair correlation). *Let $\psi \in \mathrm{C}_0(\mathbb{R}_{>0} \times \mathbb{R}_{>0} \times \mathbb{R})$. Suppose the components of $(\boldsymbol{\alpha}, 1) \in \mathbb{R}^{k+1}$ are linearly independent over $\mathbb{Q}$, and assume $\boldsymbol{\alpha}$ is diophantine of type $\kappa < \frac{k-1}{k-2}$. Then*

$$\lim_{\lambda \to \infty} R_2(\psi, \lambda) = \frac{k}{2} \int_0^\infty \psi(r, r, 0) r^{k/2-1} dr + \frac{k^2}{4} B_k \int_{\mathbb{R}} \int_0^\infty \psi(r, r, s) r^{k-2} dr\, ds. \tag{2.8}$$

To see more clearly what this theorem says about the distribution of the rescaled sequence X_j, let us put

$$\tilde{R}_2(\psi, X) = \frac{1}{X} \sum_{i,j=1}^{\infty} \psi\Big(\frac{X_i}{X}, \frac{X_j}{X}, X_i - X_j\Big). \tag{2.9}$$

The map

$$\omega : \mathbb{R}_{>0} \times \mathbb{R}_{>0} \times \mathbb{R} \to \mathbb{R}_{>0} \times \mathbb{R}_{>0} \times \mathbb{R}, \qquad \begin{pmatrix} r_1 \\ r_2 \\ s \end{pmatrix} \mapsto \begin{pmatrix} B_k r_1^{k/2} \\ B_k r_2^{k/2} \\ B_k R(r_1, r_2) s \end{pmatrix} \tag{2.10}$$

with $R(r_1, r_2) = (r_1^{k/2} - r_2^{k/2})/(r_1 - r_2)$ is invertible, continuous and in particular maps compact sets to compact sets. We may therefore choose as a suitable test function in Theorem 2 the function $\psi = \tilde{\psi} \circ \omega$, for any $\tilde{\psi} \in \mathrm{C}_0(\mathbb{R}_{>0} \times \mathbb{R}_{>0} \times \mathbb{R})$. So

$$\psi(r_1, r_2, s) = \tilde{\psi}(B_k r_1^{k/2}, B_k r_2^{k/2}, B_k R(r_1, r_2) s). \tag{2.11}$$

After a simple change of variables this shows that Theorem 2 is equivalent to the statement that (under the same conditions on $\boldsymbol{\alpha}$) *for any $\tilde{\psi} \in \mathrm{C}_0(\mathbb{R}_{>0} \times \mathbb{R}_{>0} \times \mathbb{R})$ we have*

$$\lim_{X \to \infty} \tilde{R}_2(\tilde{\psi}, X) = \int_0^\infty \tilde{\psi}(r, r, 0)\, dr + \int_{\mathbb{R}} \int_0^\infty \tilde{\psi}(r, r, s)\, dr\, ds. \tag{2.12}$$

The first term represents the asymptotic contribution of the diagonal terms $(i = j)$ in the sum, while the second asserts that the spacings $X_i - X_j$ (for $i \neq j$) are uniformly distributed, as one would expect from independent random variables with constant mean spacing. We will show in Appendix A that Theorem 1 follows in fact from (2.12) for a special choice of test function $\tilde{\psi}$.

The diophantine conditions in the above theorems are in fact sharp; there are diophantine vectors $\boldsymbol{\alpha}$ of type $\kappa = \frac{k-1}{k-2}$ such that the components of $(\boldsymbol{\alpha}, 1) \in \mathbb{R}^{k+1}$ are linearly independent over $\mathbb{Q}$, for which the conclusion of the theorems do not hold. Such $\boldsymbol{\alpha}$ are of the form $\boldsymbol{\alpha} = (\alpha_1, \ldots, \alpha_k)$ where $(\alpha_1, \ldots, \alpha_{k-2}) \in \mathbb{R}^{k-2}$ is badly approximable by rationals (i.e., diophantine of type $\frac{k-1}{k-2} = 1+\frac{1}{k-2}$) and $(\alpha_{k-1}, \alpha_k) \in \mathbb{R}^2$ are very well approximable vectors which form a set of second Baire category in $\mathbb{R}^2$. (A set of second category is a set which cannot be represented as a countable union of nowhere dense sets.) The idea here is that the pair correlation function diverges at a logarithmic rate for $\boldsymbol{\alpha}$ with $(\alpha_{k-1}, \alpha_k) \in \mathbb{Q}^2$, which is still felt by well approximable (α_{k-1}, α_k); see [24, 25] for details. (Note that the set C in Theorem 1.7 [25] is wrongly characterized as a second category subset in $\mathbb{R}^k$, since we impose diophantine conditions. C is only a dense subset in $\mathbb{R}^k$.)

3 Theta Sums and Unipotent Flows

Let us firstly note that it is sufficient (see [25] for details) to prove Theorem 2 for pair correlation functions of the form

$$R_2(\psi_1, \psi_2, h, \lambda) = \frac{1}{B_k \lambda^{k/2}} \sum_{i,j=1}^{\infty} \psi_1\big(\frac{\lambda_i}{\lambda}\big)\psi_2\big(\frac{\lambda_j}{\lambda}\big)\hat{h}\big(\lambda^{k/2-1}(\lambda_i - \lambda_j)\big), \tag{3.1}$$

Here $\psi_1, \psi_2 \in \mathcal{S}(\mathbb{R}_{\geq 0})$ are real-valued, and $\mathcal{S}(\mathbb{R}_{\geq 0})$ denotes the Schwartz class of infinitely differentiable functions of the half line $\mathbb{R}_{\geq 0}$ which, as well as their derivatives, decrease rapidly at $+\infty$. $\hat{h}$ is the Fourier transform of a compactly supported function $h \in \mathrm{C}_0(\mathbb{R})$, $\hat{h}(s) = \int_{\mathbb{R}} h(u) e(\frac{1}{2}us)\, du$ with the shorthand $e(z) := \mathrm{e}^{2\pi \mathrm{i} z}$.

A short calculation shows that $R_2(\psi_1, \psi_2, h, \lambda)$ can be written as an integral over a product of theta sums,

$$\begin{aligned} R_2(\psi_1, \psi_2, h, \lambda) &= \frac{1}{B_k} v^{k/2-1} \times \\ &\times \int_{\mathbb{R}} \Theta_f\left(u + \mathrm{i}\frac{1}{\lambda}, 0; \begin{pmatrix} \mathbf{0} \\ \boldsymbol{\alpha} \end{pmatrix}\right) \overline{\Theta_g\left(u + \mathrm{i}\frac{1}{\lambda}, 0; \begin{pmatrix} \mathbf{0} \\ \boldsymbol{\alpha} \end{pmatrix}\right)}\, h(v^{k/2-1}u)\, du, \end{aligned} \tag{3.2}$$

for the choice of functions $f(\boldsymbol{w}) = \psi_1(\|\boldsymbol{w}\|^2)$ and $g(\boldsymbol{w}) = \psi_2(\|\boldsymbol{w}\|^2)$. Here the theta sum Θ_f is defined for any Schwartz function $f \in \mathcal{S}(\mathbb{R}^k)$ by

$$\Theta_f(\tau, \phi; \boldsymbol{\xi}) = v^{k/4} \sum_{\boldsymbol{m} \in \mathbb{Z}^k} f_\phi((\boldsymbol{m} - \boldsymbol{y})v^{1/2})\, e(\tfrac{1}{2}\|\boldsymbol{m} - \boldsymbol{y}\|^2 u + \boldsymbol{m} \cdot \boldsymbol{x}), \tag{3.3}$$

where

$$\tau = u + \mathrm{i}v, \ (u \in \mathbb{R},\ v \in \mathbb{R}_{>0}), \quad \phi \in \mathbb{R}, \quad \boldsymbol{\xi} = \begin{pmatrix} \boldsymbol{x} \\ \boldsymbol{y} \end{pmatrix} \ (\boldsymbol{x}, \boldsymbol{y} \in \mathbb{R}^k). \tag{3.4}$$

The family of functions f_ϕ is an extension of $f =: f_\phi\big|_{\phi=0}$ defined by

$$f_\phi(\boldsymbol{w}) = \int_{\mathbb{R}^k} G_\phi(\boldsymbol{w}, \boldsymbol{w}') f(\boldsymbol{w}')\, dw', \tag{3.5}$$

with the integral kernel

$$G_\phi(\boldsymbol{w},\boldsymbol{w}') = e(-k\sigma_\phi/8)|\sin\phi|^{-k/2} e\left[\frac{\frac{1}{2}(\|\boldsymbol{w}\|^2+\|\boldsymbol{w}'\|^2)\cos\phi - \boldsymbol{w}\cdot\boldsymbol{w}'}{\sin\phi}\right], \tag{3.6}$$

where $\sigma_\phi = 2\nu+1$ when $\nu\pi < \phi < (\nu+1)\pi$, $\nu\in\mathbb{Z}$. The operators $U^\phi : f \mapsto f_\phi$ are unitary, and in particular $U^0 = \mathrm{id}$.

The idea behind the introduction of the extra variables ϕ and $\boldsymbol{x}$ is that the product $\Theta_f\overline{\Theta_g}$ can be identified with a function on the finite volume homogeneous space $\mathcal{M} = \Gamma\backslash G^k$ where $G^k = \mathrm{SL}(2,\mathbb{R})\ltimes\mathbb{R}^{2k}$ and Γ is a lattice in G^k. The multiplication law for G^k is $(M;\boldsymbol{\xi})(M';\boldsymbol{\xi}') = (MM';\boldsymbol{\xi}+M\boldsymbol{\xi}')$ where $M,M'\in\mathrm{SL}(2,\mathbb{R})$ and $\boldsymbol{\xi},\boldsymbol{\xi}'\in\mathbb{R}^{2k}$; the action of $\mathrm{SL}(2,\mathbb{R})$ on $\mathbb{R}^{2k}$ is defined by

$$M\boldsymbol{\xi} = \begin{pmatrix} a\boldsymbol{x}+b\boldsymbol{y} \\ c\boldsymbol{x}+d\boldsymbol{y}\end{pmatrix}, \quad M = \begin{pmatrix} a & b \\ c & d\end{pmatrix}, \quad \boldsymbol{\xi} = \begin{pmatrix}\boldsymbol{x}\\ \boldsymbol{y}\end{pmatrix}. \tag{3.7}$$

The connection between $M\in\mathrm{SL}(2,\mathbb{R})$ and the variables $\tau = u+\mathrm{i}v$, ϕ used above is given by the Iwasawa decomposition

$$M = \begin{pmatrix} 1 & u \\ 0 & 1\end{pmatrix}\begin{pmatrix} v^{1/2} & 0 \\ 0 & v^{-1/2}\end{pmatrix}\begin{pmatrix}\cos\phi & -\sin\phi \\ \sin\phi & \cos\phi\end{pmatrix}. \tag{3.8}$$

The first of the two crucial ingredients in the proof of the Poisson limit of the pair correlation functions is following equidistribution theorem [24, 25] whose proof in turn uses Ratner's classification of ergodic measures invariant under a unipotent flow [33, 34]. The following theorem may be viewed (strictly speaking only in the case $\sigma = 0$) as a special case of Shah's Theorem 1.4 [40]; for a proof see [24] ($\sigma = 0$) and [25] ($\sigma > 0$).

Theorem 3 (Equidistribution of translates of unipotent orbits). *Let Γ be a subgroup of $\mathrm{SL}(2,\mathbb{Z})\ltimes\mathbb{Z}^{2k}$ of finite index, and assume the components of the vector $(\boldsymbol{y},1)\in\mathbb{R}^{k+1}$ are linearly independent over $\mathbb{Q}$. Let h be a continuous function $\mathbb{R}\to\mathbb{R}_{\geq 0}$ with compact support. Then, for any bounded continuous function F on $\Gamma\backslash G^k$ and any $\sigma\geq 0$, we have*

$$\lim_{v\to 0} v^\sigma \int_{\mathbb{R}} F\left(u+\mathrm{i}v, 0; \begin{pmatrix}\boldsymbol{0}\\ \boldsymbol{y}\end{pmatrix}\right) h(v^\sigma u)\, du = \frac{1}{\mu(\Gamma\backslash G^k)}\int_{\Gamma\backslash G^k} F\, d\mu \int_{\mathbb{R}} h(w)\, dw \tag{3.9}$$

where μ is the Haar measure of G^k.

The dynamical interpretation of the above average is the following. Let us define the flows $\Psi^u, \Phi^t : \Gamma\backslash G^k \to \Gamma\backslash G^k$ by right translation with

$$\Psi_0^u = \left(\begin{pmatrix} 1 & u \\ 0 & 1\end{pmatrix};\boldsymbol{0}\right), \qquad \Phi_0^t = \left(\begin{pmatrix} \mathrm{e}^{-t/2} & 0 \\ 0 & \mathrm{e}^{t/2}\end{pmatrix};\boldsymbol{0}\right), \tag{3.10}$$

respectively. Then

$$\Gamma\left(u+\mathrm{i}\,\mathrm{e}^{-t},0;\begin{pmatrix}\mathbf{0}\\ \boldsymbol{y}\end{pmatrix}\right)=\Gamma g_0\Psi_0^u\Phi_0^t=\Phi^t\circ\Psi^u(\Gamma g_0),\;\; g_0:=\left(1;\begin{pmatrix}\mathbf{0}\\ \boldsymbol{y}\end{pmatrix}\right) \quad (3.11)$$

and we can thus view the integral for $t=0$ as an integral along an orbit of the unipotent flow Ψ^u which includes (at time $u=0$) the point g_0; for $t>0$ we obtain a translate by Φ^t of the above orbit which, by Theorem 3, eventually becomes equidistributed in $\Gamma\backslash G^k$.

The integral on the right-hand-side of the above equidistribution theorem can be worked out explicitly for $F=\Theta_f\overline{\Theta_g}$ and yields precisely the first term in Theorem 2. The problem is that F is not a bounded function. To prove convergence in this case we need to ensure that the translated orbit stays sufficiently far away from the singularities of F; this is achieved by imposing diophantine conditions on $\boldsymbol{y}$. The only exception is a small piece of the orbit at $u=0$ which runs into the singularity and produces an additional contribution, which in fact yields the second term in Theorem 2.

4 The Boxed Oscillator, Lattice Points in Thin Parabolic Strips, and Distribution Modulo One

The Hamiltonian of the boxed oscillator is $H=-\frac{\partial^2}{\partial x^2}-\frac{\partial^2}{\partial y^2}+\omega^2y^2$, where we assume Dirichlet boundary conditions at $x=0,\ell$. Its eigenvalues are $E_j=(\pi/\ell)^2n^2+(2m+1)\omega$, for $n=1,2,3,\ldots$ and $m=0,1,2,\ldots$. Up to overall additive and multiplicative constants these can be written as $\lambda_j=n^2\alpha+m$ with $\alpha=(\pi/\ell)^2/2\omega$. The eigenvalue number is asymptotically $\#\{j:\lambda_j\leq\lambda\}\sim c\lambda^{3/2}$ where $c=\operatorname{meas}\{x,y\geq 0,\alpha x^2+y\leq 1\}=\frac{2}{3\sqrt{\alpha}}$.

The statistical properties of the sequence λ_j are directly related to those of $n^2\alpha$ mod 1. For consider those $\lambda_j=n^2\alpha+m$, which fall into the interval $[\lambda,\lambda+1)$, for some fixed $\lambda>0$. Clearly for every $n=1,2,\ldots$ such that $n^2\alpha<\lambda+1$ there exists a unique $m=0,1,2,\ldots$ such that $\lambda_j\in[\lambda,\lambda+1)$. The values of λ_j in this interval are thus in one-to-one correspondence with $n^2\alpha$ mod 1, $n=1,\ldots,N<\sqrt{(\lambda+1)/\alpha}$. The distribution of the λ_j in small random intervals can therefore be understood in terms of the distribution of $n^2\alpha$ mod 1 in small (i.e. of size of the order of $1/N$) random intervals of the unit circle. Let $[\xi,\xi+N^{-1}\sigma]+\mathbb{Z}$ be such an interval where ξ is uniformly distributed on the unit circle; define the analogue of the counting function (1.3) by

$$\mathcal{N}(N,\xi,\sigma)=\#\{n=1,\ldots,N:n^2\alpha\in[\xi,\xi+N^{-1}\sigma]+\mathbb{Z}\}. \quad (4.1)$$

In view of the Berry-Tabor conjecture we expect that—for generic α— this number is Poisson distributed as $N\to\infty$, i.e., for any bounded function $g:\mathbb{Z}_{\geq 0}\to\mathbb{C}$,

$$\int_0^1 g(\mathcal{N}(N,\xi,\sigma))\,d\xi\to\sum_{k=0}^{\infty}g(k)\frac{\sigma^k}{k!}\,\mathrm{e}^{-\sigma}. \quad (4.2)$$

The best result we have in this direction is again for the number variance

$$\Sigma^2(N,\sigma) := \int_0^1 \{\mathcal{N}(N,\xi,\sigma) - \sigma\}^2 \, d\xi, \tag{4.3}$$

which can be shown to converge to the Poisson limit for almost all α.

Theorem 4 (Poisson limit of the number variance). *There is a set $P \subset \mathbb{R}$ of full Lebsgue measure such that, for every $\alpha \in P$ and every $\sigma > 0$,*

$$\lim_{N\to\infty} \Sigma^2(N,\sigma) = \sigma. \tag{4.4}$$

As for Theorem 1 above, this theorem follows from the Poisson distribution of the more general pair correlation function

$$R_2(\psi,N) = \frac{1}{N} \sum_{j,k=1}^{N} \sum_{\nu\in\mathbb{Z}} \psi(N(j^2\alpha - k^2\alpha + \nu)) \tag{4.5}$$

where $\psi \in \mathrm{C}_0(\mathbb{R})$, continuous and with compact support. The following theorem is proved by Rudnick and Sarnak [35] by averaging $R_2(\psi,N)$ and its square over α and using a variant of the Borel-Cantelli argument.

Theorem 5 (Poisson limit of pair correlation). *There is a set $P \subset \mathbb{R}$ of full Lebsgue measure such that, for every $\alpha \in P$ and every $\psi \in \mathrm{C}_0(\mathbb{R})$, we have*

$$\lim_{N\to\infty} R_2(\psi,N) = \psi(0) + \int_{\mathbb{R}} \psi(x)\, dx. \tag{4.6}$$

The number variance is in this case in fact identical to the pair correlation function, i.e., $\Sigma^2(N,\sigma) = R_2(\psi,N) - \sigma^2$ for the choice $\psi(x) = \max\{\sigma - |x|, 0\}$, see Appendix B.

5 On $n^2\alpha$ mod 1 and the Equidistribution of Kronecker Sequences Along Closed Horocycles

In view of Theorems 1 and 2, one would like to give a more explicit characterization (in terms of diophantine conditions) for the set of α for which $n^2\alpha$ mod 1 is Poisson distributed. Would, for instance, the assertion in Theorem 5 hold for $\alpha = \sqrt{2}$? Motivated by the affirmative answer in the case of the pair correlation problem for quadratic forms discussed in the previous section, the idea is to look for an equidistribution problem involving unipotent orbits, which can be employed to understand the pair correlation densities of $n^2\alpha$ mod 1. To this end, consider a pair correlation function with smooth weighting,

$$R_2(f,h,N) = \frac{1}{N} \sum_{j,k\in\mathbb{Z}} \sum_{\nu\in\mathbb{Z}} f\left(\frac{j}{N}\right) f\left(\frac{k}{N}\right) \hat{h}(N(j^2\alpha - k^2\alpha + \nu)) \tag{5.1}$$

where $f \in \mathrm{C}_0^\infty(\mathbb{R})$, $h \in \mathrm{C}_0(\mathbb{R})$ with Fourier transform

$$\hat{h}(s) = \int_{\mathbb{R}} h(u) e(\tfrac{1}{2}us)\, du = O(|s|^{-2}) \tag{5.2}$$

for $s \to \infty$. Applying Poisson summation to the ν-sum, we obtain

$$R_2(f,h,N) = \frac{1}{N} \sum_{m\in\mathbb{Z}} h\left(\frac{m}{N}\right) \left|\Theta_f(m\alpha + \mathrm{i}N^{-2}, 0)\right|^2 \tag{5.3}$$

where $\Theta_f(\tau,\phi)$ is the theta sum (3.3) for dimension $k=1$ at $\boldsymbol{\xi} = \mathbf{0}$, i.e.,

$$\Theta_f(\tau,\phi) = v^{1/4} \sum_{n\in\mathbb{Z}^k} f_\phi(nv^{1/2})\, e(\tfrac{1}{2}n^2 u). \tag{5.4}$$

The pair correlation function may thus be viewed as a special case of averages of the form

$$\frac{1}{M} \sum_{m=1}^{M} F(m\alpha + \mathrm{i}v, 0) \tag{5.5}$$

as $M \to \infty$ and $v \to 0$, where F is a continuous function on $\mathcal{M} = \Gamma \backslash \mathrm{SL}(2,\mathbb{R})$, where Γ is a lattice in $\mathrm{SL}(2,\mathbb{R})$ which contains the parabolic subgroup $\{\left(\begin{smallmatrix} 1 & j \\ 0 & 1 \end{smallmatrix}\right) : j \in \mathbb{Z}\}$; we will also assume for simplicity that $-1 \in \Gamma$. In particular, for $\Gamma = \Gamma_\theta$, the invariance group of $|\Theta_f|$ (the "theta group"), one can show [28] that if for some fixed $\alpha \in \mathbb{R}$ and $F = |\Theta_f|^2$ we have

$$\lim_{M\to\infty} \frac{1}{M} \sum_{m=1}^{M} F(m\alpha + \mathrm{i}v, 0) = \frac{1}{\mu(\mathcal{M})} \int_{\mathcal{M}} F\, d\mu, \quad v = M^{-2} \to 0, \tag{5.6}$$

then the limiting pair correlation density of $n^2\alpha$ mod 1 is Poisson.

The equidistribution theorem (5.6) we are here interested in combines two classical equidistribution problems. The first is the equidistribution of long closed horocycles [37], i.e.,

$$\lim_{y\to 0} \int_0^1 F(u + \mathrm{i}v, 0)\, du = \frac{1}{\mu(\mathcal{M})} \int_{\mathcal{M}} F\, d\mu, \tag{5.7}$$

for any sufficiently nice test function F, e.g., bounded continuous. The second is the distribution of the Kronecker sequence $\alpha, 2\alpha, 3\alpha \ldots, M\alpha$ mod 1, which is well known to be equidistributed as $M \to \infty$ for all irrational α; that is

$$\lim_{M\to\infty} \frac{1}{M} \sum_{m=1}^{M} F(m\alpha + \mathrm{i}v, 0) = \int_0^1 F(u + \mathrm{i}v, 0)\, du \tag{5.8}$$

for fixed $v > 0$. Taking both limits $M \to \infty$, $v \to 0$ simultaneously requires a careful analysis. Of particular interest is the case when the number M of points on the horocycle grows slower than the length of the horocycle, v^{-1}. In this case the problem is that the mean distance between the points on the horocycles grows as $v \to 0$. It seems therefore difficult to show that any possible limit measure is invariant under some unipotent action, and hence Ratner's theorem cannot be applied (in the present approach, that is). The proof of the following theorem uses instead methods from spectral analysis [28].

Theorem 6 (Equidistribution of $m\alpha \bmod 1$ along closed horocycles). *Let Γ be a lattice in $\mathrm{SL}(2,\mathbb{R})$ as described above. Fix $\nu > 0$. Then there is a set $P = P(\Gamma,\nu) \subset \mathbb{R}$ of full Lebesgue measure such that for any $\alpha \in P$, any bounded continuous function $F : \mathcal{M} \to \mathrm{C}$, and any constants $0 < C_1 < C_2$, we have*

$$\frac{1}{M}\sum_{m=1}^{M} F(m\alpha + \mathrm{i}v, 0) \to \frac{1}{\mu(\mathcal{M})}\int_{\mathcal{M}} F\, d\mu \tag{5.9}$$

uniformly as $M \to \infty$ and $C_1 M^{-\nu} \le v \le C_2 M^{-\nu}$.

This theorem holds in fact for a larger class of test functions F which are continuous but unbounded, and which allow the choice $F = |\Theta_f|^2$. Theorem 6 thus implies Rudnick & Sarnak's result [35] that the pair correlation density of $n^2\alpha \bmod 1$ is Poisson for almost all α.

If $\Gamma = \mathrm{SL}(2,\mathbb{Z})$ or a congruence subgroup, and we increase the number of points on the horocycles sufficiently fast (i.e., ν is chosen sufficiently small) we are able to prove equidistribution under explicit diophantine conditions. The best possible result is obtained under the assumption that the Fourier coefficients of the eigenfunctions on the Laplacian on $\Gamma\backslash\mathfrak{H}$ ($\mathfrak{H}$ denotes the complex upper half plane) are almost bounded; this hypothesis is usually referred to as the *Ramanujan conjecture for Maass wave forms.*

Theorem 7 (Equidistribution of $m\alpha \bmod 1$ along closed horocycles). *Let Γ be a congruence subgroup of $\mathrm{SL}(2,\mathbb{Z})$ and assume the Ramanujan conjecture for Maass waveforms on $\Gamma\backslash\mathfrak{H}$ holds. Let $\alpha \in \mathbb{R}$ be of type $\kappa \ge 2$, and fix $\nu < \min\{2, \frac{2}{\kappa-2}\}$. Then for any bounded continuous function $F : \mathcal{M} \to \mathrm{C}$, and any constant $C_1 > 0$, we have*

$$\frac{1}{M}\sum_{m=1}^{M} F(m\alpha + \mathrm{i}v, 0) \to \frac{1}{\mu(\mathcal{M})}\int_{\mathcal{M}} F\, d\mu \tag{5.10}$$

uniformly as $M \to \infty$, $v \to 0$ so long as $v \ge C_1 M^{-\nu}$.

This statement is proved in [28]. If $\kappa \ge 3$, then $\nu < \frac{2}{\kappa-2}$ is in fact the best possible restriction on ν, in the sense that there are otherwise counter examples for which the assertion of the theorem is wrong [28]. Thus, in contrast with the equidistribution theorem for unipotent flows (Theorem 3), we must impose diophantine conditions even in the case of bounded test functions.

It would be very interesting to extend Theorem 7 to $\nu = 2$, which, as mentioned above, is the case relevant to the pair correlation problem. Note that the theta group Γ_θ is a congruence subgroup of $\mathrm{SL}(2,\mathbb{Z})$.

6 Distribution Modulo One and Almost Modular Functions

In the previous section we have presented some evidence that the distribution of $n^2\alpha$ mod 1 in intervals of size $1/N$ is described by a Poisson distribution. In the same vein (as mentioned in the introduction) it can be expected that a central limit theorem holds for slightly larger intervals. Let us here consider the case when the interval size is macroscopic. For any fixed interval $[\xi, \xi+\eta]$, $0 < \eta < 1$, we are interested in the counting function

$$\mathcal{N}_\alpha(N,\xi,\eta) = \#\{n = 1,\dots,N : n^2\alpha \in [\xi,\xi+\eta] + \mathbb{Z}\}. \tag{6.1}$$

For irrational α, the sequence $n^2\alpha$ is equidistributed mod 1, which means that $\mathcal{N}(N,\xi,\eta) \sim N\eta$ as $N \to \infty$. The *error term* is thus

$$E_\alpha(N,\xi,\eta) = \mathcal{N}_\alpha(N,\xi,\eta) - N\eta. \tag{6.2}$$

There are two possibilities to study the fluctuations of this function. Fix the interval and take α to be uniformly distributed in $[0,1)$, or fix α and take ξ to be uniformly distributed in $[0,1)$ with η fixed as usual (note, however, that this time we consider large intervals compared with the mean separation $1/N$). In the first case we have the following statement.

Theorem 8 (Limit theorem for the error term). *For α uniformly distributed in $[0,1)$, $N^{-1/2}E_\alpha(N,\xi,\eta)$ has a limit distribution as $N \to \infty$. That is, there exists a probability measure $\nu_{\xi,\eta}$ on $\mathbb{R}$ such that, for any bounded continuous function $g : \mathbb{R} \to \mathbb{R}$, we have*

$$\lim_{N\to\infty} \int_0^1 g\big(N^{-1/2}E_\alpha(N,\xi,\eta)\big)d\alpha = \int_{\mathbb{R}} g(w)\,\nu_{\xi,\eta}(dw). \tag{6.3}$$

Furthermore, $\nu_{\xi,\eta}$ is even.

This is a special case of Theorem 2.1 in [26], which also provides an explicit formula for the variance of the limit distribution. To sketch the proof of Theorem 8 let us write

$$E_\alpha(N,\xi,\eta) = \sum_{n=1}^{N} \psi(n^2\alpha) - N\int_0^1 \psi(t)\,dt, \tag{6.4}$$

where ψ is the characteristic function of $[\xi, \xi+\eta]$. ψ could in fact be a more general real- or complex-valued function; we will only require that its Fourier coefficients

$$\widehat{\psi}_k = \int_0^1 \psi(t)\, e(-kt)\, dt. \tag{6.5}$$

satisfy

$$\widehat{\psi}_0 = 0, \tag{6.6}$$

and that there are constants $\beta > 1/2$ and $C(\psi) > 0$ such that

$$|\widehat{\psi}_k| \leq \frac{C(\psi)}{|k|^\beta} \tag{6.7}$$

for all $k \neq 0$. Fourier expansion (which converges only in the L^2 sense) yields

$$E_\alpha(N, \xi, \eta) = \sum_{k \neq 0} \widehat{\psi}_k \left\{ \sum_{n=1}^{N} e(kn^2 x) \right\}. \tag{6.8}$$

It is known [19] that the theta sums inside the curly brackets individually have a limit distribution, as $N \to \infty$. This result follows from the observation (cf. previous sections) that theta sums can be identified with functions on the metaplectic cover of $SL(2, \mathbb{R})$ which are invariant under certain subgroups of finite index in the metaplectic analogue of $SL(2, \mathbb{Z})$. The limit theorem is then a direct consequence for the equidistribution of long closed horocycles on the metaplectic cover [20].

One can show that the truncated Fourier expansion

$$E_\alpha^{(K)}(N, \xi, \eta) = \sum_{0 < |k| \leq K} \widehat{\psi}_k \left\{ \sum_{n=1}^{N} e(kn^2 x) \right\} \tag{6.9}$$

can as well be identified with functions on the metaplectic cover of $SL(2, \mathbb{R})$, where the index of the invariance subgroup is still finite but becomes large with increasing K. Following the same steps as in [19] one can therefore show that $E_\alpha^{(K)}(N, \xi, \eta)$ satisfies the limit theorem, Theorem 8.

The variance of the difference $E_\alpha(N, \xi, \eta) - E_\alpha^{(K)}(N, \xi, \eta)$ is, uniformly in $N \gg 1$, arbitrarily small for K sufficiently large; hence the distributions of $E_\alpha(N, \xi, \eta)$ and $E_\alpha^{(K)}(N, \xi, \eta)$ are arbitrarily close for K large. Theorem 8 follows now from standard probabilistic arguments.

The fact that each approximation $E_\alpha^{(K)}(N, \xi, \eta)$ is a modular function, but $E_\alpha(N, \xi, \eta)$ is not (in general), suggests the name *almost modular function*, in close analogy with almost *periodic* functions in the sense of Besicovitch. The above arguments can in fact be extended to general classes of almost modular functions, which are characterized by the approximability (with respect to a certain L^p norm) by modular functions invariant under congruence subgroups of large index [26].

Another interesting example of an almost modular function is the logarithm of

$$\prod_{n=1}^{\infty}(1-e(n^2 z)), \tag{6.10}$$

which is studied in [27]. Its limit distribution in the complex plane is in fact rotation-invariant.

Acknowledgement

I would like to thank Bernard Julia, Pierre Moussa and Pierre Vanhove for organizing this very inspiring meeting at Les Houches, and Andreas Strömbergsson for his collaboration in the joint work described in Sect. 5. The research presented in this publication has been supported by an EPSRC Advanced Research Fellowship, the Nuffield Foundation (Grant NAL/00351/G), the Royal Society (Grant 22355), and the EC Research Training Network (Mathematical Aspects of Quantum Chaos) HPRN-CT-2000-00103.

A Proof of Theorem 1

Because of (1.2) we have, for large X,

$$\Sigma^2(X,L) \sim \frac{1}{X}\int_0^{\infty} \mathcal{N}(T,L)^2 \rho\Big(\frac{T}{X}\Big)\, dT - L^2. \tag{A.1}$$

Expand

$$\mathcal{N}(T,L) = \sum_j \chi_1\Big(\frac{X_j - T}{L}\Big), \tag{A.2}$$

where χ_1 is the indicator function of the interval $[0,1]$. This yields

$$\Sigma^2(X,L) + L^2 \sim \frac{1}{X}\sum_{i,j}\int_{-\infty}^{\infty} \chi_1\Big(\frac{X_i - T}{L}\Big)\chi_1\Big(\frac{X_j - T}{L}\Big)\rho\Big(\frac{T}{X}\Big)\, dT. \tag{A.3}$$

We have replaced 0 in the lower limit by $-\infty$, which is permitted since ρ is supported on the positive half line. Substitute T by $T + \frac{1}{2}(X_i + X_j)$, and the right hand side becomes

$$\frac{1}{X}\sum_{i,j}\int_{-\infty}^{\infty} \chi_1\Big(\frac{\frac{1}{2}(X_i - X_j) - T}{L}\Big)\times \\ \times \chi_1\Big(\frac{\frac{1}{2}(X_j - X_j) - T}{L}\Big)\rho\Big(\frac{\frac{1}{2}(X_i + X_j) + T}{X}\Big)\, dT. \tag{A.4}$$

The integration in T is restricted by the inequalities

$$0 \leq \frac{1}{2}(X_i - X_j) - T \leq L, \qquad 0 \leq \frac{1}{2}(X_j - X_i) - T \leq L, \tag{A.5}$$

which imply $0 \leq -T \leq L$, so T is bounded. Therefore, by the continuity of ρ,

$$\rho\Big(\frac{\frac{1}{2}(X_i + X_j) + T}{X}\Big) \sim \rho\Big(\frac{\frac{1}{2}(X_i + X_j)}{X}\Big), \tag{A.6}$$

and it is sensible to write (A.4) as

$$\frac{1}{X}\sum_{i,j} \rho\Big(\frac{\frac{1}{2}(X_i + X_j)}{X}\Big) W(X_i - X_j) + \text{error term} \tag{A.7}$$

where

$$W(s) := \int_{-\infty}^{\infty} \chi_1\Big(\frac{T + \frac{1}{2}s}{L}\Big)\chi_1\Big(\frac{T - \frac{1}{2}s}{L}\Big)\, dT = \max\{L - |s|, 0\}. \tag{A.8}$$

Since the function $\psi(r_1, r_2, s) = \rho(\frac{1}{2}(r_1 + r_2))W(s)$ is continuous and has compact support, (2.12) yields

$$\lim_{X\to\infty} \Sigma^2(X, L) + L^2 = \int_0^\infty \psi(r, r, 0)\, dr + \int_{\mathbb{R}} \int_0^\infty \psi(r, r, s)\, dr\, ds = L + L^2, \tag{A.9}$$

which proves Theorem 1, provided the above error term is indeed small. To investigate this, note that

$$|\text{error term}| \leq \frac{1}{X}\sum_{i,j} \tilde{\rho}\Big(\frac{\frac{1}{2}(X_i + X_j)}{X}\Big) W(X_i - X_j) \tag{A.10}$$

where $\tilde{\rho}$ is a continuous function with compact support such that

$$\sup_{0\leq -T\leq L} \Big|\rho\Big(r + \frac{T}{X}\Big) - \rho(r)\Big| \leq \tilde{\rho}(r) \tag{A.11}$$

for all r. It is evident that for any given $\epsilon > 0$ we can find a function $\tilde{\rho}$ meeting this requirement for all X large enough and satisfying in addition $\int_0^\infty \tilde{\rho}(r)dr < \epsilon$. By (2.12) the right hand side of (A.10) converges to

$$(L^2 + L)\int_0^\infty \tilde{\rho}(r)dr < (L^2 + L)\epsilon \tag{A.12}$$

which means that the error term is smaller than any $\epsilon > 0$, hence zero. □

B Proof of Theorem 4

We have

$$\begin{aligned}
&\Sigma^2(N,\sigma)+\sigma^2\\
&=\sum_{j,k=1}^{\infty}\sum_{\nu,\nu'\in\mathbb{Z}}\int_0^1\left\{\chi_{[0,\sigma]}\big(N(j^2\alpha+\xi+\nu)\big)\chi_{[0,\sigma]}\big(N(k^2\alpha+\xi+\nu')\big)\right\}d\xi\\
&\quad=\sum_{j,k=1}^{\infty}\sum_{\nu\in\mathbb{Z}}\int_{\mathbb{R}}\left\{\chi_{[0,\sigma]}\big(N(j^2\alpha+\xi+\nu)\big)\chi_{[0,\sigma]}\big(N(k^2\alpha+\xi)\big)\right\}d\xi\\
&\quad=\frac{1}{N}\sum_{j,k=1}^{\infty}\sum_{\nu\in\mathbb{Z}}\int_{\mathbb{R}}\left\{\chi_{[0,\sigma]}\big(N(j^2\alpha-k^2\alpha+\nu)+\xi\big)\chi_{[0,\sigma]}(\xi)\right\}d\xi \qquad \text{(B.1)}
\end{aligned}$$

and thus $\Sigma^2(N,\sigma)+\sigma^2 = R_2(\psi,N)$ for $\psi(x)=\int_{\mathbb{R}}\left\{\chi_{[0,\sigma]}(x+\xi)\chi_{[0,\sigma]}(\xi)\right\}d\xi = \max\{\sigma-|x|,0\}$. Since $\psi\in \mathrm{C}_0(\mathbb{R})$, and $\psi(0)+\int_{\mathbb{R}}\psi(x)\,dx=\sigma+\sigma^2$, Theorem 4 is thus indeed a special case of Theorem 5. □

References

1. M.V. Berry and M. Tabor, Level clustering in the regular spectrum, *Proc. Roy. Soc.* A **356** (1977) 375-394.
2. P.M. Bleher, The energy level spacing for two harmonic oscillators with golden mean ratio of frequencies, *J. Statist. Phys.* **61** (1990) 869-876.
3. P.M. Bleher, The energy level spacing for two harmonic oscillators with generic ratio of frequencies, *J. Statist. Phys.* **63** (1991) 261-283.
4. P.M. Bleher, On the distribution of the number of lattice points inside a family of convex ovals, *Duke Math. J.* **67** (1992) 461-481.
5. P.M. Bleher, Trace formula for quantum integrable systems, lattice point problem, and small divisors, in: D. Hejhal et al. (eds.), *Emerging Applications of Number Theory*, IMA Volumes in Mathematics and its Applications **109** (Springer, New York, 1999) pp. 1-38.
6. P.M. Bleher and J. Bourgain, Distribution of the error term for the number of lattice points inside a shifted ball, *Analytic number theory*, Vol. 1 (Allerton Park, IL, 1995), 141-153, *Progr. Math.* **138** (Birkhäuser, Boston, 1996).
7. E. Bogomolny, *Quantum and arithmetic chaos*, Proceedings of the Les Houches Winter School "Frontiers in Number Theory, Physics and Geometry", 2003.
8. Z. Cheng and J.L. Lebowitz, Statistics of energy levels in integrable quantum systems, *Phys. Rev.* A **44** (1991) 3399-3402.
9. Z. Cheng, J.L. Lebowitz and P. Major, On the number of lattice points between two enlarged and randomly shifted copies of an oval, *Probab. Theory Related Fields* **100** (1994) 253-268.
10. Y. Colin de Verdière, Quasi-modes sur les varietes Riemanniennes, *Invent. Math.* **43** (1977) 15-52.
11. Y. Colin de Verdière, Spectre conjoint d'opérateurs pseudo-différentiels qui commutent. II. Le cas intégrable. *Math. Z.* **171** (1980) 51-73.

12. A. Eskin, G. Margulis and S. Mozes, Quadratic forms of signature (2,2) and eigenvalue spacings on rectangular 2-tori, preprint 1998.
13. C. Greenman, The generic spacing distribution of the two-dimensional harmonic oscillator, *J. Phys.* A **29** (1996) 4065-4081.
14. C. Greenman, Is the level spacing distribution of the infinite-dimensional harmonic oscillator that of a Poisson process? *J. Phys. A* **30** (1997) 927-936.
15. D.R. Heath-Brown, The distribution and moments of the error term in the Dirichlet divisor problem, *Acta Arith.* **60** (1992) 389-415.
16. C. Hughes and Z. Rudnick, On the distribution of lattice points in thin annuli, *Int. Math. Res. Not.* **13** (2004) 637-658.
17. P. Major, Poisson law for the number of lattice points in a random strip with finite area, *Prob. Theo. Rel. Fields* **92** (1992) 423-464.
18. J. Marklof, Spectral form factors of rectangle billiards, *Comm. Math. Phys.* **199** (1998) 169-202.
19. J. Marklof, Limit theorems for theta sums, *Duke Math. J.* **97** (1999) 127-153.
20. J. Marklof, Theta sums, Eisenstein series, and the semiclassical dynamics of a precessing spin, in: D. Hejhal et al. (eds.), *Emerging Applications of Number Theory*, IMA Vol. Math. Appl. **109** (Springer, New York, 1999) 405-450.
21. J. Marklof, The n-point correlations between values of a linear form, with an appendix by Z. Rudnick, *Ergod. Th. Dyn. Sys.* **20** (2000) 1127-1172.
22. J. Marklof, The Berry-Tabor conjecture, *Proceedings of the 3rd European Congress of Mathematics*, Barcelona 2000, Progress in Mathematics Vol. 202 (Birkhäuser, Basel, 2001) 421-427.
23. J. Marklof, Level spacing statistics and integrable dynamics, *XIIIth International Congress on Mathematical Physics*, London 2000 (International Press, Boston, 2001) 359-363.
24. J. Marklof, Pair correlation densities of inhomogeneous quadratic forms, *Ann. of Math.* (2) **158** (2003) 419-471.
25. J. Marklof, Pair correlation densities of inhomogeneous quadratic forms II, *Duke Math. J.* **115** (2002) 409-434; Correction, *ibid.* **120** (2003) 227-228.
26. J. Marklof, Almost modular functions and the distribution of n^2x modulo one, *Int. Math. Res. Not.* **39** (2003) 2131-2151.
27. J. Marklof, Holomorphic almost modular forms, *Bull. London Math. Soc.* **36** (2004) 647-655.
28. J. Marklof and A. Strömbergsson, Equidistribution of Kronecker sequences along closed horocycles, *Geom. Funct. Anal.* **13** (2003) 1239-1280.
29. A.E. Mazel and Ya.G. Sinai, A limiting distribution connected with fractional parts of linear forms, in: S. Albeverio et al. (ed.), *Ideas and methods in mathematical analysis, stochastics and applications*, Vol. 1 (1992) 220-229.
30. A. Pandey, O. Bohigas and M.-J. Giannoni, Level repulsion in the spectrum of two-dimensional harmonic oscillators, *J. Phys.* A **22** (1989) 4083-4088.
31. A. Pandey and R. Ramaswamy, Level spacings for harmonic-oscillator systems, *Phys. Rev.* A **43** (1991) 4237-4243.
32. A. Pellegrinotti, Evidence for the Poisson distribution for quasi-energies in the quantum kicked-rotator model, *J. Statist. Phys.* **53** (1988) 1327-1336.
33. M. Ratner, On Raghunathan's measure conjecture, *Ann. of Math.* (2) **134** (1991) 545-607.
34. M. Ratner, Raghunathan's topological conjecture and distributions of unipotent flows, *Duke Math. J.* **63** (1991) 235-280.

35. Z. Rudnick and P. Sarnak, The pair correlation function of fractional parts of polynomials, *Comm. Math. Phys.* **194** (1998) 61-70.
36. Z. Rudnick, P. Sarnak and A. Zaharescu, The distribution of spacings between the fractional parts of $n^2\alpha$, *Invent. Math.* **145** (2001) 37-57.
37. P. Sarnak, Asymptotic behavior of periodic orbits of the horocycle flow and Eisenstein series, *Comm. Pure Appl. Math.* **34** (1981) 719-739.
38. P. Sarnak, Values at integers of binary quadratic forms, *Harmonic Analysis and Number Theory* (Montreal, PQ, 1996), 181-203, CMS Conf. Proc. **21**, Amer. Math. Soc., Providence, RI, 1997.
39. W.M. Schmidt, *Approximation to algebraic numbers*, Série des Conférences de l'Union Mathématique Internationale, No. 2. Monographie No. 19 de l'Enseignement Mathématique, Geneva, 1972. (Also in Enseignement Math. (2) **17** (1971), 187-253.)
40. N.A. Shah, Limit distributions of expanding translates of certain orbits on homogeneous spaces, *Proc. Indian Acad. Sci., Math. Sci.* **106** (1996) 105-125.
41. Ya. G. Sinai, The absence of the Poisson distribution for spacings between quasi-energies in the quantum kicked-rotator model. *Phys.* D **33** (1988) 314-316.
42. Ya.G. Sinai, Poisson distribution in a geometrical problem, *Adv. Sov. Math., AMS Publ.* **3** (1991) 199-215.
43. J.M. VanderKam, Pair correlation of four-dimensional flat tori, *Duke Math. J.* **97** (1999) 413-438.
44. J.M. VanderKam, Correlations of eigenvalues on multi-dimensional flat tori, *Comm. Math. Phys.* **210** (2000) 203-223.
45. A. Zaharescu, Correlation of fractional parts of $n^2\alpha$, *Forum Math.* **15** (2003) 1-21.
46. S. Zelditch, Level spacings for integrable quantum maps in genus zero, *Comm. Math. Phys.* **196** (1998) 289-318, Addendum: "Level spacings for integrable quantum maps in genus zero", *ibid.*, 319-329.

Arithmetic Quantum Chaos of Maass Waveforms

H. Then

Abteilung Theoretische Physik, Universität Ulm, Albert-Einstein-Allee 11, 89069 Ulm
holger.then@uni-ulm.de

Summary. We compute numerically eigenvalues and eigenfunctions of the quantum Hamiltonian that describes the quantum mechanics of a point particle moving freely in a particular three-dimensional hyperbolic space of finite volume and investigate the distribution of the eigenvalues.

1 Introduction

The distribution of the eigenvalues of a quantum Hamiltonian is a central subject that is studied in quantum chaos. There are some generally accepted conjectures about the nearest-neighbor spacing distributions of the eigenvalues.

Unless otherwise stated we use the following assumptions: The quantum mechanical system is desymmetrized with respect to all its unitary symmetries, and whenever we examine the distribution of the eigenvalues we regard them on the scale of the mean level spacings. Moreover, it is generically believed that after desymmetrization a generic quantum Hamiltonian possesses no degenerate eigenvalues.

Conjecture 1 (Berry, Tabor [1]). If the corresponding classical system is integrable, the eigenvalues behave like independent random variables and the distribution of the nearest-neighbor spacings is close to the Poisson distribution, i.e. there is no level repulsion.

Conjecture 2 (Bohigas, Giannoni, Schmit [2, 3]). If the corresponding classical system is chaotic, the eigenvalues are distributed like the eigenvalues of hermitian random matrices [4]. The corresponding ensembles depend only on the symmetries of the system:

- For chaotic systems without time-reversal invariance the distribution of the eigenvalues should be close to the distribution of the Gaussian Unitary Ensemble (GUE) which is characterized by a quadratic level repulsion.
- For chaotic systems with time-reversal invariance and integer spin the distribution of the eigenvalues should be close to the distribution of the Gaussian Orthogonal Ensemble (GOE) which is characterized by a linear level repulsion.
- For chaotic systems with time-reversal invariance and half-integer spin the distribution of the eigenvalues should be close to the distribution of the Gaussian Symplectic Ensemble (GSE) which is characterized by a quartic level repulsion.

These conjectures are very well confirmed by numerical calculations, but several exceptions are known. Here are two examples:

Exception 1 *The harmonic oscillator is classically integrable, but its spectrum is equidistant.*

Exception 2 *The geodesic motion on surfaces with constant negative curvature provides a prime example for classical chaos. In some cases, however, the nearest-neighbor distribution of the eigenvalues of the Laplacian on these surfaces appears to be Poissonian.*

"A strange arithmetical structure of chaos" in the case of surfaces of constant negative curvature that are generated by arithmetic fundamental groups was discovered by Aurich and Steiner [5], see also Aurich, Bogomolny, and Steiner [6]. Deviations from the expected GOE-behaviour in the case of a particular arithmetic surface were numerically observed by Bohigas, Giannoni, and Schmit [3] and by Aurich and Steiner [7]. Computations coming out in [7, 8] showed, however, that the level statistics on 30 generic (i.e. non-arithmetic) surfaces were in nice agreement with the expected random-matrix theory prediction in accordance with conjecture 2. This has led Bogomolny, Georgeot, Giannoni, and Schmit [9], Bolte, Steil, and Steiner [10], and Sarnak [11] to introduce the concept of arithmetic quantum chaos.

Conjecture 3 (Arithmetic Quantum Chaos). On surfaces of constant negative curvature that are generated by arithmetic fundamental groups, the distribution of the eigenvalues of the quantum Hamiltonian are close to the Poisson distribution. Due to level clustering small spacings occur comparably often.

We compute numerically the eigenvalues and eigenfunctions of the Laplacian that describes the quantum mechanics of a point particle moving freely in the non-integrable three-dimensional hyperbolic space of constant negative curvature generated by the Picard group. The Picard group is arithmetic and we find that our results are in accordance with the conjecture of arithmetic quantum chaos.

For the definition of an arithmetic group we refer the reader to [12].

2 Preliminaries: The Modular Group

For simplicity we first introduce the topology and geometry of the two-dimensional surface of constant negative curvature that is generated by the modular group [13]. It will then be easy to carry over to the three-dimensional space of constant negative curvature that is generated by the Picard group.

The construction begins with the upper half-plane,

$$\mathcal{H} = \{(x, y) \in \mathbb{R}^2; \quad y > 0\},$$

equipped with the hyperbolic metric of constant negative curvature

$$ds^2 = \frac{dx^2 + dy^2}{y^2}.$$

A free particle on the upper half-plane moves along geodesics , which are straight lines and semicircles perpendicular to the x-axis, respectively, see figure 1.

Expressing a point $(x, y) \in \mathcal{H}$ as a complex number $z = x + iy$, all isometries of the hyperbolic metric are given by the group of linear fractional transformations,

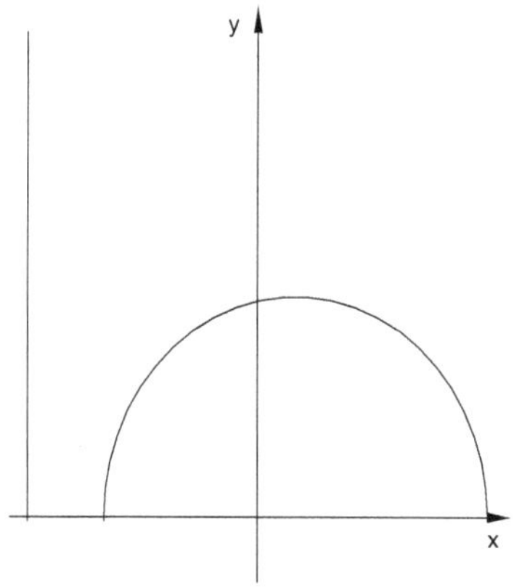

Fig. 1. Geodesics in the upper half-plane of constant negative curvature

$$z \mapsto \gamma z = \frac{az+b}{cz+d}; \quad a,b,c,d \in \mathbb{R}, \quad ad-bc=1,$$

which is isomorphic to the group of matrices

$$\gamma = \begin{pmatrix} a & b \\ c & d \end{pmatrix} \in \mathrm{SL}(2,\mathbb{R}),$$

up to a common sign of the matrix entries,

$$\mathrm{SL}(2,\mathbb{R})/\{\pm 1\} = \mathrm{PSL}(2,\mathbb{R}).$$

In analogy to the concept of a fundamental cell in a regular lattice of a crystal we can introduce a fundamental domain of a discrete group $\Gamma \subset \mathrm{PSL}(2,\mathbb{R})$.

Definition 1. *A fundamental domain of the discrete group Γ is an open subset $\mathcal{F} \subset \mathcal{H}$ with the following conditions: The closure of $\mathcal{F}$ meets each orbit $\Gamma z = \{\gamma z;\ \gamma \in \Gamma\}$ at least once, $\mathcal{F}$ meets each orbit Γz at most once, and the boundary of $\mathcal{F}$ has Lebesgue measure zero.*

If we choose the group Γ to be the modular group,

$$\Gamma = \mathrm{PSL}(2,\mathbb{Z}),$$

which is generated by a translation and an inversion,

$$\begin{pmatrix} 1 & 1 \\ 0 & 1 \end{pmatrix} : z \mapsto z+1,$$
$$\begin{pmatrix} 0 & -1 \\ 1 & 0 \end{pmatrix} : z \mapsto -z^{-1},$$

the fundamental domain of standard shape is

$$\mathcal{F} = \{z = x + \mathrm{i}y \in \mathcal{H}; \quad -\frac{1}{2} < x < \frac{1}{2}, \quad |z| > 1\},$$

see figure 2. The isometric copies of the fundamental domain $\gamma\mathcal{F}$, $\gamma \in \Gamma$, tessellate the upper half-plane completely without any overlap or gap, see figure 3.

Identifying the fundamental domain $\mathcal{F}$ and parts of its boundary with all its isometric copies $\gamma\mathcal{F}$, $\forall\gamma \in \Gamma$, defines the topology to be the quotient space $\Gamma\backslash\mathcal{H}$. The quotient space $\Gamma\backslash\mathcal{H}$ can also be thought of as the fundamental domain $\mathcal{F}$ with its faces glued according to the elements of the group Γ, see figure 4.

Any function being defined on the upper half-plane which is invariant under linear fractional transformations,

$$f(z) = f(\gamma z) \quad \forall\gamma \in \Gamma,$$

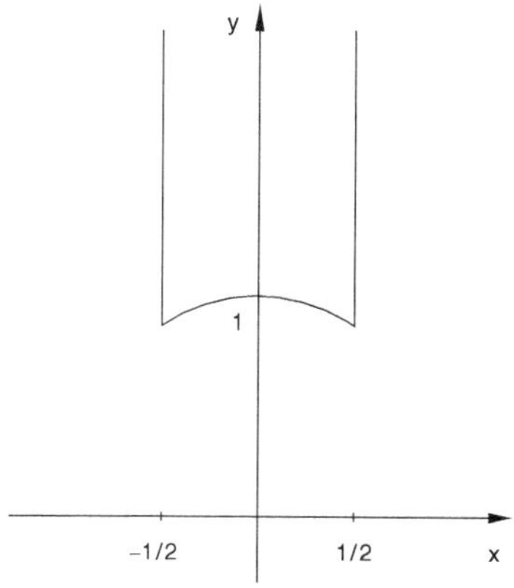

Fig. 2. The fundamental domain of the modular group

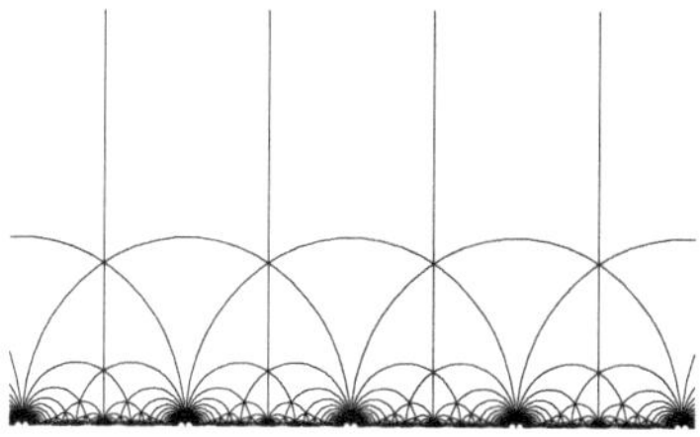

Fig. 3. The upper half-plane tessellated with isometric copies of the fundamental domain

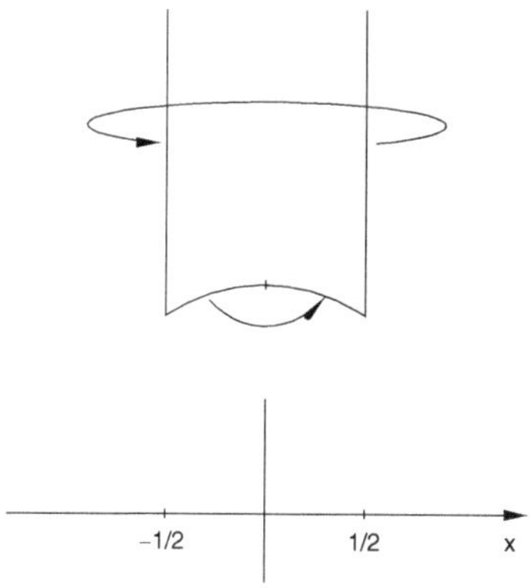

Fig. 4. Identifying the faces of the fundamental domain according to the elements of the modular group

can be identified with a function living on the quotient space $\Gamma\backslash\mathcal{H}$. A function on the quotient space is tantamount to a function on the fundamental domain with periodic boundary conditions. Vice versa, any function being defined on the quotient space can be identified with an automorphic function, $f(z) = f(\gamma z)$, $\forall\gamma\in\Gamma$, living on the upper half-plane.

With the hyperbolic metric the quotient space $\Gamma\backslash\mathcal{H}$ inherits the structure of an orbifold. An orbifold locally looks like a manifold, with the exception that it is allowed to have elliptic fixed-points.

The orbifold of the modular group has one parabolic and two elliptic fixed-points,

$$z = \mathrm{i}\infty, \quad z = \mathrm{i}, \quad \text{and} \quad z = \frac{1}{2} + \mathrm{i}\frac{\sqrt{3}}{2}.$$

The parabolic one fixes a cusp at $z = \mathrm{i}\infty$ which is invariant under the parabolic element

$$\begin{pmatrix} 1 & 1 \\ 0 & 1 \end{pmatrix}.$$

Hence the orbifold of the modular group is non-compact. The volume element corresponding to the hyperbolic metric reads

$$d\mu = \frac{dxdy}{y^2},$$

such that the volume of the orbifold $\Gamma\backslash\mathcal{H}$ is finite,

$$\mathrm{vol}(\Gamma\backslash\mathcal{H}) = \frac{\pi}{3}.$$

Scaling the units such that $\hbar = 1$ and $2m = 1$, the stationary Schrödinger equation which describes the quantum mechanics of a point particle moving freely in the orbifold $\Gamma\backslash\mathcal{H}$ becomes

$$(\Delta + \lambda)f(z) = 0,$$

where the hyperbolic Laplacian is given by

$$\Delta = y^2\left(\frac{\partial^2}{\partial x^2} + \frac{\partial^2}{\partial y^2}\right)$$

and λ is the scaled energy. We can relate the eigenvalue problem defined on the orbifold $\Gamma\backslash\mathcal{H}$ to the eigenvalue problem defined on the upper half-space, with the eigenfunctions being subject to the automorphy condition relative to the discrete group Γ,

$$f(\gamma z) = f(z) \quad \forall\gamma\in\Gamma.$$

In order to avoid solutions that grow exponentially in the cusp, we impose the boundary condition

$$f(z) = O(y^{\kappa}) \quad \text{for} \quad z \to \mathrm{i}\infty$$

where κ is some positive constant.

The solutions of this eigenvalue problem can be identified with Maass waveforms [14]. The identification is worthwhile, since much is known about Maass waveforms from number theory and harmonic analysis which will simplify their computation, see e.g. [15, 16, 17, 18, 19, 20, 13, 21, 22, 23].

3 The Picard Group

In the three-dimensional case one considers the upper half-space,

$$\mathcal{H} = \{(x_0, x_1, y) \in \mathbb{R}^3; \quad y > 0\}$$

equipped with the hyperbolic metric

$$ds^2 = \frac{dx_0^2 + dx_1^2 + dy^2}{y^2}.$$

The geodesics of a particle moving freely in the upper half-space are straight lines and semicircles perpendicular to the x_0-x_1-plane, respectively, see figure 5.

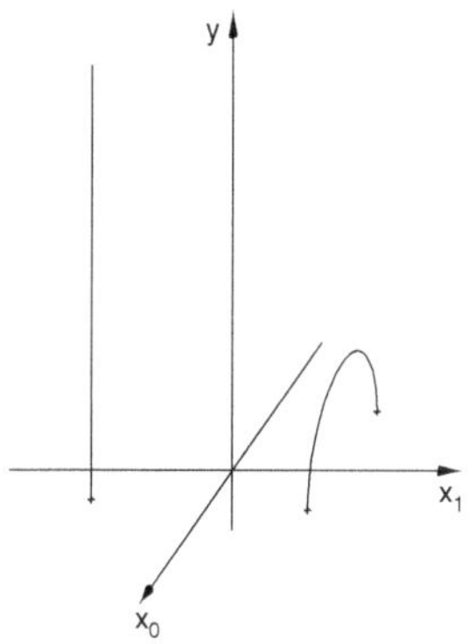

Fig. 5. Geodesics in the upper half-space of constant negative curvature

Expressing any point $(x_0, x_1, y) \in \mathcal{H}$ as a Hamilton quaternion, $z = x_0 + \mathrm{i}x_1 + \mathrm{j}y$, with the multiplication defined by $\mathrm{i}^2 = -1$, $\mathrm{j}^2 = -1$, $\mathrm{ij} + \mathrm{ji} = 0$, all motions in the upper half-space are given by linear fractional transformations

$$z \mapsto \gamma z = (az+b)(cz+d)^{-1}; \quad a,b,c,d \in \mathbb{C}, \quad ad-bc=1.$$

The group of these transformations is isomorphic to the group of matrices

$$\gamma = \begin{pmatrix} a & b \\ c & d \end{pmatrix} \in \mathrm{SL}(2,\mathbb{C})$$

up to a common sign of the matrix entries,

$$\mathrm{SL}(2,\mathbb{C})/\{\pm 1\} = \mathrm{PSL}(2,\mathbb{C}).$$

The motions provided by the elements of $\mathrm{PSL}(2,\mathbb{C})$ exhaust all orientation preserving isometries of the hyperbolic metric on $\mathcal{H}$.

Remark 1. If one wants to avoid using quaternions, the point $(x_0, x_1, y) \in \mathcal{H}$ can be expressed by $(x,y) \in \mathbb{C} \times \mathbb{R}$ with $x = x_0 + \mathrm{i}x_1$ and $y > 0$. But then the linear fractional transformations look somewhat more complicated,

$$(x,y) \mapsto \gamma(x,y) = \Big(\frac{(ax+b)(\bar{c}\bar{x}+\bar{d}) + a\bar{c}y^2}{|cx+d|^2+|cy|^2}, \frac{y}{|cx+d|^2+|cy|^2}\Big).$$

In order to keep the notation simple we hence use quaternions.

We now choose the discrete group $\Gamma \subset \mathrm{PSL}(2,\mathbb{C})$ generated by the cosets of three elements,

$$\begin{pmatrix} 1 & 1 \\ 0 & 1 \end{pmatrix}, \quad \begin{pmatrix} 1 & \mathrm{i} \\ 0 & 1 \end{pmatrix}, \quad \begin{pmatrix} 0 & -1 \\ 1 & 0 \end{pmatrix},$$

which yield two translations and one inversion,

$$z \mapsto z+1, \quad z \mapsto z+\mathrm{i}, \quad z \mapsto -z^{-1}.$$

This group Γ is called the Picard group. The three motions generating Γ, together with the coset of the element

$$\begin{pmatrix} \mathrm{i} & 0 \\ 0 & -\mathrm{i} \end{pmatrix}$$

that is isomorphic to the symmetry

$$z = x + \mathrm{j}y \mapsto \mathrm{i}z\mathrm{i} = -x + \mathrm{j}y,$$

can be used to construct the fundamental domain of standard shape

$$\mathcal{F} = \{z = x_0 + \mathrm{i}x_1 + \mathrm{j}y \in \mathcal{H}; \quad -\frac{1}{2} < x_0 < \frac{1}{2}, \quad 0 < x_1 < \frac{1}{2}, \quad |z| > 1\},$$

see figure 6. Identifying the faces of the fundamental domain according to the elements of the group Γ leads to a realization of the quotient space $\Gamma \backslash \mathcal{H}$, see figure 7.

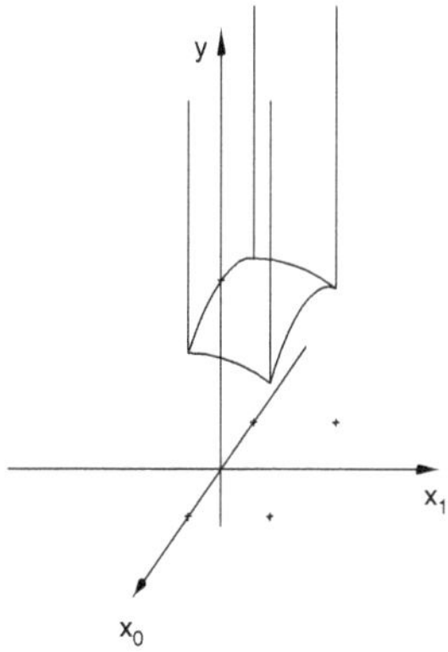

Fig. 6. The fundamental domain of the Picard group

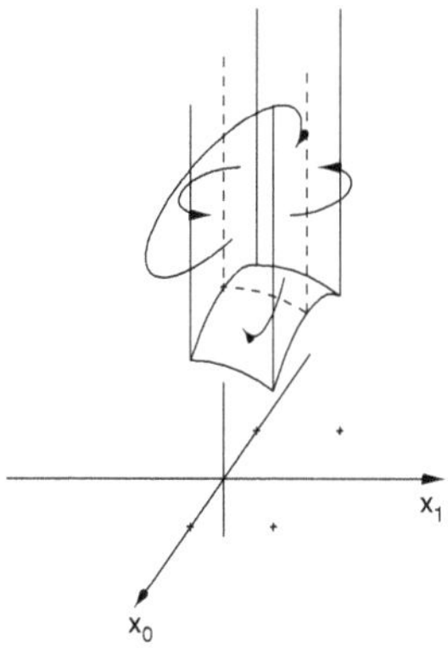

Fig. 7. Identifying the faces of the fundamental domain according to the elements of the Picard group

With the hyperbolic metric the quotient space $\Gamma\backslash\mathcal{H}$ inherits the structure of an orbifold that has one parabolic and four elliptic fixed-points,

$$z = \mathrm{j}\infty, \quad z = \mathrm{j}, \quad z = \frac{1}{2} + \mathrm{j}\sqrt{\frac{3}{4}}, \quad z = \frac{1}{2} + \mathrm{i}\frac{1}{2} + \mathrm{j}\sqrt{\frac{1}{2}}, \quad z = \mathrm{i}\frac{1}{2} + \mathrm{j}\sqrt{\frac{3}{4}}.$$

The parabolic fixed-point corresponds to a cusp at $z = \mathrm{j}\infty$ that is invariant under the parabolic elements

$$\begin{pmatrix} 1 & 1 \\ 0 & 1 \end{pmatrix}, \quad \text{and} \quad \begin{pmatrix} 1 & \mathrm{i} \\ 0 & 1 \end{pmatrix}.$$

The volume element deriving from the hyperbolic metric reads

$$d\mu = \frac{dx_0 dx_1 dy}{y^3},$$

such that the volume of the non-compact orbifold $\Gamma\backslash\mathcal{H}$ is finite [24],

$$\mathrm{vol}(\Gamma\backslash\mathcal{H}) = \frac{\zeta_K(2)}{4\pi^2} \simeq 0.305$$

where

$$\zeta_K(s) = \frac{1}{4} \sum_{\nu\in\mathbb{Z}[\mathrm{i}]-\{0\}} (\nu\bar{\nu})^{-s}, \quad \Re s > 1,$$

is the Dedekind zeta function.

We are interested in the eigenfunctions of the Laplacian,

$$\Delta = y^2\Big(\frac{\partial^2}{\partial x_0^2} + \frac{\partial^2}{\partial x_1^2} + \frac{\partial^2}{\partial y^2}\Big) - y\frac{\partial}{\partial y},$$

which determine the quantum mechanics of a particle moving freely in the orbifold $\Gamma\backslash\mathcal{H}$. As in the preceding section we identify the solutions with Maass waveforms [25].

Since the Maass waveforms are automorphic, and therefore periodic in x_0 and x_1, it follows that they can be expanded into a Fourier series,

$$f(z) = u(y) + \sum_{\beta\in\mathbb{Z}[\mathrm{i}]-\{0\}} a_\beta y K_{\mathrm{i}r}(2\pi|\beta|y)\,\mathrm{e}^{2\pi\mathrm{i}\Re\beta x}, \tag{3.1}$$

where

$$u(y) = \begin{cases} b_0 y^{1+\mathrm{i}r} + b_1 y^{1-\mathrm{i}r} & \text{if } r \neq 0, \\ b_2 y + b_3 y \ln y & \text{if } r = 0. \end{cases}$$

$K_{\mathrm{i}r}(x)$ is the K-Bessel function whose order is connected with the eigenvalue λ by

$$\lambda = r^2 + 1.$$

If a Maass waveform vanishes in the cusp,

$$\lim_{z\to\mathrm{j}\infty} f(z) = 0,$$

it is called a Maass cusp form. Maass cusp forms are square integrable over the fundamental domain, $\langle f, f\rangle < \infty$, where

$$\langle f, g\rangle = \int_{\Gamma\backslash\mathcal{H}} \bar{f} g \, d\mu$$

is the Petersson scalar product.

According to the Roelcke-Selberg spectral resolution of the Laplacian [16, 17], its spectrum contains both a discrete and a continuous part. The discrete part is spanned by the constant eigenfunction f_0 and a countable number of Maass cusp forms $f_1, f_2, f_3, \ldots$ which we take to be ordered with increasing eigenvalues, $0 = \lambda_0 < \lambda_1 \leq \lambda_2 \leq \lambda_3 \leq \ldots$. The continuous part of the spectrum $\lambda \geq 1$ is spanned by the Eisenstein series $E(x, 1 + \mathrm{i}r)$ which are known analytically [26, 27]. The Fourier coefficients of the functions $\Lambda_K(1 + \mathrm{i}r)E(x, 1 + \mathrm{i}r)$ are given by

$$b_0 = \Lambda_K(1 + \mathrm{i}r), \quad b_1 = \Lambda_K(1 - \mathrm{i}r), \quad a_\beta = 2 \sum_{\substack{\lambda,\mu \in \mathbb{Z}[\mathrm{i}] \\ \lambda\mu = \beta}} \left|\frac{\lambda}{\mu}\right|^{\mathrm{i}r},$$

where

$$\Lambda_K(s) = 4\pi^{-s}\Gamma(s)\zeta_K(s)$$

has an analytic continuation into the complex plane except for a pole at $s = 1$.

Normalizing the Maass cusp forms according to

$$\langle f_n, f_n \rangle = 1,$$

we can expand any square integrable function $\phi \in L^2(\Gamma \backslash \mathcal{H})$ in terms of Maass waveforms, [28],

$$\phi(z) = \sum_{n \geq 0} \langle f_n, \phi \rangle f_n(z) + \frac{1}{2\pi \mathrm{i}} \int_{\Re s = 1} \langle E(\cdot, s), \phi \rangle E(z, s)\, ds.$$

The eigenvalues and their associated Maass cusp forms are not known analytically. Thus, one has to approximate them numerically. Previous calculations of eigenvalues for the Picard group can be found in [29, 30, 31, 32]. By making use of the Hecke operators [29, 33] and the multiplicative relations among the coefficients, Steil [32] obtained a non-linear system of equations which allowed him to compute 2545 consecutive eigenvalues. We extend these computations with the use of Hejhal's algorithm [34].

4 Hejhal's Algorithm

Hejhal found a linear stable algorithm for computing Maass waveforms together with their eigenvalues which he used for groups acting on the two-dimensional hyperbolic plane [34], see also [35, 36]. We make use of this algorithm which is based on the Fourier expansion and the automorphy condition. We apply it for the Picard group acting on the three-dimensional hyperbolic space. For the Picard group no small eigenvalues $0 < \lambda = r^2 + 1 < 1$ exist [37]. Therefore, r is real and the term $u(y)$ in the Fourier expansion of Maass cusp forms vanishes. Due to the exponential decay of the K-Bessel function for large arguments (A.1) and the polynomial bound of the coefficients [25],

$$a_\beta = O(|\beta|), \quad |\beta| \to \infty,$$

the absolutely convergent Fourier expansion can be truncated,

$$f(z) = \sum_{\substack{\beta \in \mathbb{Z}[\mathrm{i}]-\{0\} \\ |\beta| \leq M}} a_\beta y K_{\mathrm{i}r}(2\pi|\beta|y)\, \mathrm{e}^{2\pi\mathrm{i}\Re\beta x} + [[\varepsilon]], \tag{4.1}$$

if we bound y from below. Given $\varepsilon > 0$, r, and y, we determine the smallest $M = M(\varepsilon, r, y)$ such that the inequalities

$$2\pi M y \geq r \quad \text{and} \quad K_{\mathrm{i}r}(2\pi M y) \leq \varepsilon \max_x (K_{\mathrm{i}r}(x))$$

hold. Larger y allow smaller M. In all truncated terms,

$$[[\varepsilon]] = \sum_{\substack{\beta \in \mathbb{Z}[\mathrm{i}]-\{0\} \\ |\beta| > M}} a_\beta y K_{\mathrm{i}r}(2\pi|\beta|y)\, \mathrm{e}^{2\pi\mathrm{i}\Re\beta x},$$

the K-Bessel function decays exponentially in $|\beta|$, and already the K-Bessel function of the first truncated summand is smaller than ε times most of the K-Bessel functions in the sum of (4.1). Thus, the error $[[\varepsilon]]$ does at most marginally exceed ε. The reason why $[[\varepsilon]]$ can exceed ε somewhat is due to the possibility that the summands in (4.1) cancel each other, or that the coefficients in the truncated terms are larger than in (4.1). By a finite two-dimensional Fourier transformation the Fourier expansion (4.1) is solved for its coefficients

$$a_\gamma y K_{\mathrm{i}r}(2\pi|\gamma|y) = \frac{1}{(2Q)^2} \sum_{x \in \mathbb{X}[\mathrm{i}]} f(x + \mathrm{j}y)\, \mathrm{e}^{-2\pi\mathrm{i}\Re\gamma x} + [[\varepsilon]], \tag{4.2}$$

where $\mathbb{X}[\mathrm{i}]$ is a two-dimensional equally distributed set of $(2Q)^2$ numbers,

$$\mathbb{X}[\mathrm{i}] = \{\frac{k_0 + \mathrm{i}k_1}{2Q}; \quad k_i = -Q + \tfrac{1}{2}, -Q + \tfrac{3}{2}, \ldots, Q - \tfrac{3}{2}, Q - \tfrac{1}{2}, \quad i = 0, 1\},$$

with $2Q > M + |\gamma|$.

By automorphy we have

$$f(z) = f(z^*),$$

where z^* is the Γ-pullback of the point z into the fundamental domain $\mathcal{F}$,

$$z^* = \gamma z, \quad \gamma \in \Gamma, \quad z^* \in \mathcal{F}.$$

Thus, a Maass cusp form can be approximated by

$$f(x + \mathrm{j}y) = f(x^* + \mathrm{j}y^*) = \sum_{\substack{\beta \in \mathbb{Z}[\mathrm{i}]-\{0\} \\ |\beta| \leq M_0}} a_\beta y^* K_{\mathrm{i}r}(2\pi|\beta|y^*)\, \mathrm{e}^{2\pi\mathrm{i}\Re\beta x^*} + [[\varepsilon]], \tag{4.3}$$

where y^* is always larger or equal than the height y_0 of the lowest points of the fundamental domain $\mathcal{F}$,

$$y_0 = \min_{z \in \mathcal{F}}(y) = \frac{1}{\sqrt{2}},$$

allowing us to replace $M(\varepsilon, r, y)$ by $M_0 = M(\varepsilon, r, y_0)$.

Choosing y smaller than y_0 the Γ-pullback $z \mapsto z^*$ of any point into the fundamental domain $\mathcal{F}$ makes at least once use of the inversion $z \mapsto -z^{-1}$, possibly together with the translations $z \mapsto z+1$ and $z \mapsto z+\mathrm{i}$. This is called implicit automorphy , since it guarantees the invariance $f(z) = f(-z^{-1})$. The conditions $f(z) = f(z+1)$ and $f(z) = f(z+\mathrm{i})$ are automatically satisfied due to the Fourier expansion.

Making use of the implicit automorphy by replacing $f(x+\mathrm{j}y)$ in (4.2) with the right-hand side of (4.3) gives

$$\begin{aligned} &a_\gamma y K_{\mathrm{i}r}(2\pi|\gamma|y) \\ &= \frac{1}{(2Q)^2} \sum_{x \in \mathbb{X}[\mathrm{i}]} \sum_{\substack{\beta \in \mathbb{Z}[\mathrm{i}]-\{0\} \\ |\beta| \le M_0}} a_\beta y^* K_{\mathrm{i}r}(2\pi|\beta|y^*)\, \mathrm{e}^{2\pi\mathrm{i}\Re\beta x^*}\, \mathrm{e}^{-2\pi\mathrm{i}\Re\gamma x} + [[2\varepsilon]], \end{aligned} \quad (4.4)$$

which is the central identity in the algorithm.

The symmetry in the Picard group and the symmetries of the fundamental domain imply that the Maass waveforms fall into four symmetry classes [32] named **D**, **G**, **C**, and **H**, satisfying

$$\begin{aligned} \mathbf{D}: &\quad f(x+\mathrm{j}y) = f(\mathrm{i}x+\mathrm{j}y) = f(-\bar{x}+\mathrm{j}y), \\ \mathbf{G}: &\quad f(x+\mathrm{j}y) = f(\mathrm{i}x+\mathrm{j}y) = -f(-\bar{x}+\mathrm{j}y), \\ \mathbf{C}: &\quad f(x+\mathrm{j}y) = -f(\mathrm{i}x+\mathrm{j}y) = f(-\bar{x}+\mathrm{j}y), \\ \mathbf{H}: &\quad f(x+\mathrm{j}y) = -f(\mathrm{i}x+\mathrm{j}y) = -f(-\bar{x}+\mathrm{j}y), \end{aligned}$$

respectively, see figure 8, from which the symmetry relations among the coefficients follow,

$$\begin{aligned} \mathbf{D}: &\quad a_\beta = a_{\mathrm{i}\beta} = a_{\bar{\beta}}, \\ \mathbf{G}: &\quad a_\beta = a_{\mathrm{i}\beta} = -a_{\bar{\beta}}, \\ \mathbf{C}: &\quad a_\beta = -a_{\mathrm{i}\beta} = a_{\bar{\beta}}, \\ \mathbf{H}: &\quad a_\beta = -a_{\mathrm{i}\beta} = -a_{\bar{\beta}}. \end{aligned}$$

Defining

$$\mathrm{cs}(\beta, x) = \sum_{\sigma \in \mathbb{S}_\beta} s_{\sigma\beta}\, \mathrm{e}^{2\pi\mathrm{i}\Re\sigma x},$$

where $s_{\sigma\beta}$ is given by

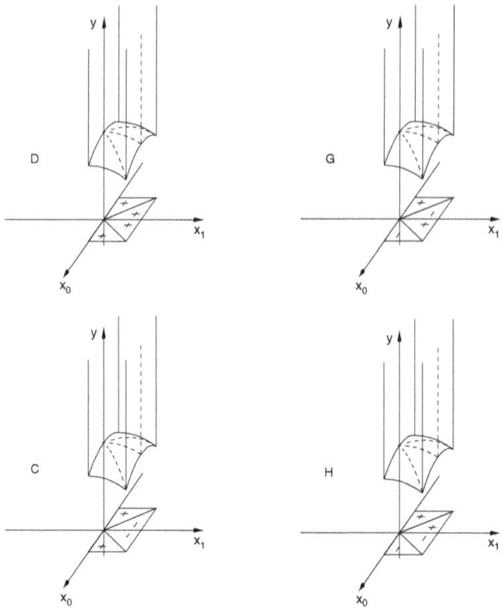

Fig. 8. The symmetries **D**, **G**, **C**, and **H** from top left to bottom right

$$a_\sigma = s_{\sigma\beta} a_\beta$$

and

$$\sigma \in \$_\beta = \begin{cases} \{\beta, \mathrm{i}\beta, -\beta, -\mathrm{i}\beta, \bar\beta, \mathrm{i}\bar\beta, -\bar\beta, -\mathrm{i}\bar\beta\} & \text{if } \bar\beta \notin \{\beta, \mathrm{i}\beta, -\beta, -\mathrm{i}\beta\}, \\ \{\beta, \mathrm{i}\beta, -\beta, -\mathrm{i}\beta\} & \text{else,} \end{cases}$$

the Fourier expansion (3.1) of the Maass waveforms can be written

$$f(z) = u(y) + \sum_{\beta \in \tilde{\mathbb{Z}}[\mathrm{i}] - \{0\}} a_\beta y K_{\mathrm{i}r}(2\pi|\beta|y)\,\mathrm{cs}(\beta, x),$$

where the tilde operator on a set of numbers is defined such that

$$\tilde{\mathbb{X}} \subset \mathbb{X}, \quad \bigcup_{x \in \tilde{\mathbb{X}}} \$_x = \mathbb{X}, \quad \text{and} \quad \bigcap_{x \in \tilde{\mathbb{X}}} \$_x = \emptyset$$

holds.

Forgetting about the error $[[2\varepsilon]]$ the set of equations (4.4) can be written as

$$\sum_{\substack{\beta \in \tilde{\mathbb{Z}}[\mathrm{i}] - \{0\} \\ |\beta| \le M_0}} V_{\gamma\beta}(r, y) a_\beta = 0, \quad \gamma \in \tilde{\mathbb{Z}}[\mathrm{i}] - \{0\}, \quad |\gamma| \le M_0, \tag{4.5}$$

where the matrix $V = (V_{\gamma\beta})$ is given by

$$V_{\gamma\beta}(r,y) = \#\{\sigma \in \mathbb{S}_\gamma\} y K_{\mathrm{i}r}(2\pi|\gamma|y)\delta_{\gamma\beta} - \frac{1}{(2Q)^2} \sum_{x\in\mathbb{X}[\mathrm{i}]} y^* K_{\mathrm{i}r}(2\pi|\beta|y^*)\,\mathrm{cs}(\beta,x^*)\,\mathrm{cs}(\gamma,-x).$$

Since $y < y_0$ can always be chosen such that $K_{\mathrm{i}r}(2\pi|\gamma|y)$ is not too small, the diagonal terms in the matrix V do not vanish for large $|\gamma|$ and the matrix is well conditioned.

We are now looking for the non-trivial solutions of (4.5) for $1 \le |\gamma| \le M_0$ that simultaneously give the eigenvalues $\lambda = r^2 + 1$ and the coefficients a_β. Trivial solutions are avoided by setting one of the coefficients equal to one, $a_\alpha = 1$. Here we choose α to be 1, $2+\mathrm{i}$, 1, and $1+\mathrm{i}$, for the symmetry classes **D**, **G**, **C**, and **H**, respectively.

Since the eigenvalues are unknown we discretize the r axis and solve for each r value on this grid the inhomogeneous system of equations

$$\sum_{\substack{\beta\in\tilde{\mathbb{Z}}[\mathrm{i}]-\{0,\alpha\}\\ |\beta|\le M_0}} V_{\gamma\beta}(r,y^{\#1})a_\beta = -V_{\gamma\alpha}(r,y^{\#1}), \quad 1 \le |\gamma| \le M_0, \tag{4.6}$$

where $y^{\#1} < y_0$ is chosen such that $K_{\mathrm{i}r}(2\pi|\gamma|y^{\#1})$ is not too small for $1 \le |\gamma| \le M_0$. A good value to try for $y^{\#1}$ is given by $2\pi M_0 y^{\#1} = r$.

It is important to check whether

$$g_\gamma = \sum_{\substack{\beta\in\tilde{\mathbb{Z}}[\mathrm{i}]-\{0\}\\ |\beta|\le M_0}} V_{\gamma\beta}(r,y^{\#2})a_\beta, \quad 1 \le |\gamma| \le M_0,$$

vanishes where $y^{\#2}$ is another y value independent of $y^{\#1}$. Only if all g_γ vanish simultaneously the solution of (4.6) is independent of y. In this case $\lambda = r^2+1$ is an eigenvalue and the a_β's are the coefficients of the Fourier expansion of the corresponding Maass cusp form.

The probability to find an r value such that all g_γ vanish simultaneously is zero, because the discrete eigenvalues are of measure zero in the real numbers. Therefore, we make use of the intermediate value theorem where we look for simultaneous sign changes in g_γ. Once we have found them in at least half of the g_γ's we have found an interval which contains an eigenvalue with high probability. By some bisection and interpolation we can see if this interval really contains an eigenvalue, and by nesting up the interval until its size tends to zero we obtain the eigenvalue.

In order not to miss eigenvalues which lie close together nor to waste CPU time with a too fine grid, we use the adaptive r grid introduced in [38].

5 Eigenvalues for the Picard Group

We have found 13950 eigenvalues of the Laplacian for the Picard group in the interval $1 < \lambda = r^2 + 1 \le 19601$. 4115 of them belong to eigenfunctions of the symmetry class **D**, 2805 to **G**, 3715 to **C**, and 3315 to **H**. The smallest eigenvalue is $\lambda = r^2 + 1$ with $r = 6.6221193402528$ which is in agreement with the lower bound $\lambda > \frac{2\pi^2}{3}$ [37]. Table 1 shows the first few eigenvalues of each symmetry class. They agree with those of Steil [32] up to five decimal places. We next regard the statistics of the eigenvalues. First, we compare the output of our algorithm with Weyl's law and higher order corrections drawn from [39]. This serves as a check whether we have found all eigenvalues. We then find it necessary to correct one of the terms in [39] numerically. Finally, we regard the spectral fluctuations and find that the nearest-neighbor spacing distribution closely resembles that of a Poisson random process as predicted by [9, 10, 11] and previously observed by [32].

In the first step we consider the level counting function

$$N(r) = \#\{ i \mid r_i \leq r \}$$

and split it into two parts

$$N(r) = \bar{N}(r) + N_{fluc}(r).$$

Here $\bar{N}$ is a smooth function describing the average increase in the number of levels and N_{fluc} describes the fluctuations around the mean such that

$$\lim_{R\to\infty} \frac{1}{R} \int_1^R N_{fluc}(r) dr = 0.$$

The average increase in the number of levels is given by Weyl's law [40, 41] and higher order corrections have been calculated by Matthies [39]. She obtained

$$\bar{N}(r) = \frac{\mathrm{vol}(\mathcal{F})}{6\pi^2} r^3 + a_2 r \log r + a_3 r + a_4 \tag{5.1}$$

with the constants

$$\begin{aligned} a_2 &= -\frac{3}{2\pi}, \\ a_3 &= \frac{1}{\pi}\left[\frac{13}{16}\log 2 + \frac{7}{4}\log\pi - \log\Gamma(\tfrac{1}{4}) + \frac{2}{9}\log(2+\sqrt{3}) + \frac{3}{2}\right], \\ a_4 &= -\frac{1}{2}. \end{aligned}$$

We compare our results for $N(r)$ with (5.1) by defining

$$N_{fluc}(r) = N(r) - \bar{N}(r). \tag{5.2}$$

N_{fluc} fluctuates around zero or a negative integer whose absolute value gives the number of missing eigenvalues, see figure 9. Unfortunately, our algorithm does not find all eigenvalues in one single run. In the first run it finds about 97% of the eigenvalues. Apart from very few exceptions the remaining eigenvalues are found in the third run. To be more specific, we plotted N_{fluc} decreased by $\frac{1}{2}$, because $N(r) - \bar{N}(r)$ is approximately $\frac{1}{2}$ whenever $\lambda = r^2 + 1$

Table 1. The first few eigenvalues of the Laplacian for the Picard group. Listed is r. related to the eigenvalues via $\lambda = r^2 + 1$.

D	G	C	H
8.55525104		6.62211934	
11.10856737		10.18079978	
12.86991062		12.11527484	12.11527484
14.07966049		12.87936900	
15.34827764		14.14833073	
15.89184204		14.95244267	14.95244267
17.33640443		16.20759420	
17.45131992	17.45131992	16.99496892	16.99496892
17.77664065		17.86305643	17.86305643
19.06739052		18.24391070	
19.22290266		18.83298996	
19.41119126		19.43054310	19.43054310
20.00754583		20.30030720	20.30030720
20.70798880	20.70798880	20.60686743	
20.81526852		21.37966055	21.37966055
21.42887079		21.44245892	
22.12230276		21.83248972	21.83248972
22.63055256		22.58475297	22.58475297
22.96230105	22.96230105	22.85429195	
23.49617692		23.49768305	23.49768305
23.52784503		23.84275866	
23.88978413	23.88978413	23.89515755	23.89515755
24.34601664		24.42133829	24.42133829
24.57501426		25.03278076	25.03278076
24.70045917		25.42905483	
25.47067539		25.77588591	25.77588591
25.50724616		26.03903968	
25.72392169	25.72392169	26.12361823	26.12361823

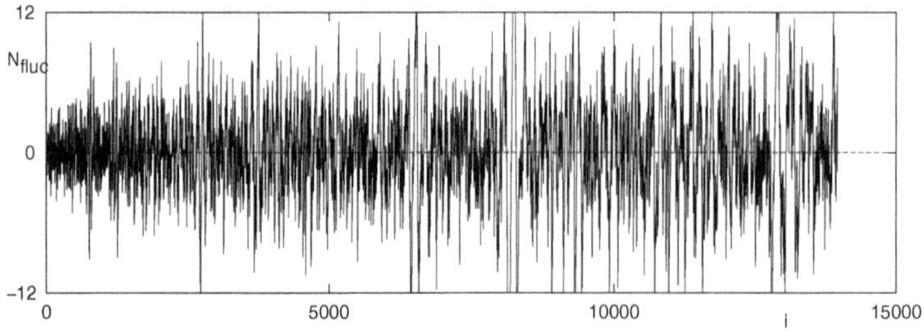

Fig. 9. $N_{fluc}(r_i)$ as a function of i fluctuating around zero

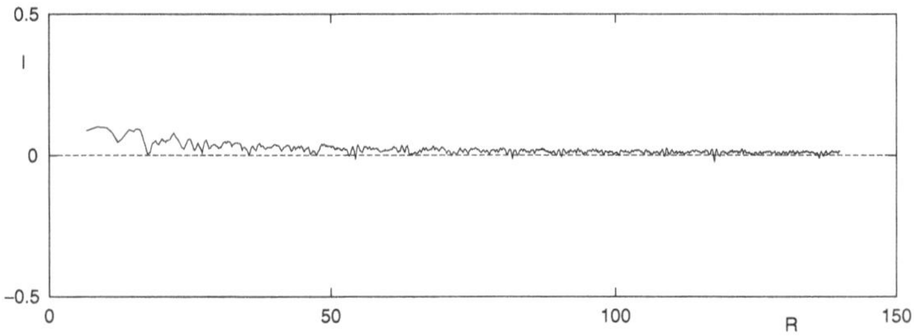

Fig. 10. I as a function of R showing that $I \xrightarrow{R\to\infty} 0$

is an eigenvalue. Furthermore, we took the eigenvalue $\lambda = 0$ into account. We remark that we never find more eigenvalues than predicted by (5.1). A plot indicating that N_{fluc} fluctuates around zero is shown in figure 10 where we plotted the integral

$$I(R) = \frac{1}{R}\int_1^R N_{fluc}(r)dr. \tag{5.3}$$

So far, everything seems to be consistent. Taking the desymmetrized spectra into account (5.1) is modified [39]

$$\bar{N}(r) = \frac{\mathrm{vol}(\mathcal{F})}{24\pi^2}r^3 + b_1 r^2 + b_2 r \log r + b_3 r + b_4 \tag{5.4}$$

with the constants depending on the symmetry class. For the symmetry class **D** the constants are given in [39] as

$$\begin{aligned} b_1 &= \tfrac{1}{24}, \\ b_2 &= -\tfrac{13}{8\pi}, \\ b_3 &= \tfrac{1}{4\pi}\big[-\tfrac{11}{16}\log 2 + \tfrac{19}{4}\log\pi - \log\Gamma(\tfrac{1}{4}) \\ &\qquad + \tfrac{2}{9}\log(2+\sqrt{3}) + \tfrac{1}{4}\log(3+2\sqrt{2}) + \tfrac{13}{2}\big], \\ b_4 &= -\tfrac{47}{72}. \end{aligned}$$

For **G**

$$\begin{aligned}
b_1 &= -\tfrac{1}{24},\\
b_2 &= \tfrac{3}{8\pi},\\
b_3 &= \tfrac{1}{4\pi}\big[\tfrac{37}{16}\log 2 + \tfrac{3}{4}\log\pi - \log\Gamma(\tfrac{1}{4})\\
&\qquad + \tfrac{2}{9}\log(2+\sqrt{3}) + \tfrac{1}{4}\log(3+2\sqrt{2}) - \tfrac{3}{2}\big],\\
b_4 &= -\tfrac{25}{72}.
\end{aligned}$$

For **C**

$$\begin{aligned}
b_1 &= \tfrac{1}{96},\\
b_2 &= -\tfrac{1}{8\pi},\\
b_3 &= \tfrac{1}{4\pi}\big[\tfrac{5}{16}\log 2 + \tfrac{3}{4}\log\pi - \log\Gamma(\tfrac{1}{4})\\
&\qquad + \tfrac{2}{9}\log(2+\sqrt{3}) - \tfrac{1}{4}\log(3+2\sqrt{2}) + \tfrac{1}{2}\big],\\
b_4 &= \tfrac{125}{576}.
\end{aligned}$$

And for **H**

$$\begin{aligned}
b_1 &= -\tfrac{1}{96},\\
b_2 &= -\tfrac{1}{8\pi},\\
b_3 &= \tfrac{1}{4\pi}\big[\tfrac{21}{16}\log 2 + \tfrac{3}{4}\log\pi - \log\Gamma(\tfrac{1}{4})\\
&\qquad + \tfrac{2}{9}\log(2+\sqrt{3}) - \tfrac{1}{4}\log(3+2\sqrt{2}) + \tfrac{1}{2}\big],\\
b_4 &= \tfrac{163}{576}.
\end{aligned}$$

Let $\{r_i\}$ be a sequence related to the consecutive eigenvalues $\lambda = r^2 + 1$. If we plot $N_{fluc}(r_i)$ as a function of i for the desymmetrized spectra we obtain small deviations which can hardly be seen in figure 11. But if we plot the integral (5.3) we see that N_{fluc} does not really fluctuate around zero. Instead, in figure 12 we see systematic deviations, but the discrepancy is much less than one eigenvalue for each symmetry class. Since the number of eigenvalues is integer-valued we do not assume that we have found less or too many eigenvalues. Therefore, we fit the constants b_1, b_2, b_3, b_4 in (5.4) and obtain new constants for each of the symmetry classes. Since the integrals $I(R)$ in figure 12 show a linear behavior, the constants b_1 and b_2 seem to be correct. We thus only change the constants b_3 and b_4 by fitting them numerically. For the symmetry class **D** the new constants are

$$\begin{aligned}
b_3 &= 0.8639... &\qquad& \text{instead of } b_3 = 0.8679...,\\
b_4 &= -0.288... && \text{instead of } b_4 = -0.653....
\end{aligned}$$

For **G**

$$\begin{aligned}
b_3 &= 0.0285... &\qquad& \text{instead of } b_3 = 0.0324...,\\
b_4 &= -0.184... && \text{instead of } b_4 = -0.347... .
\end{aligned}$$

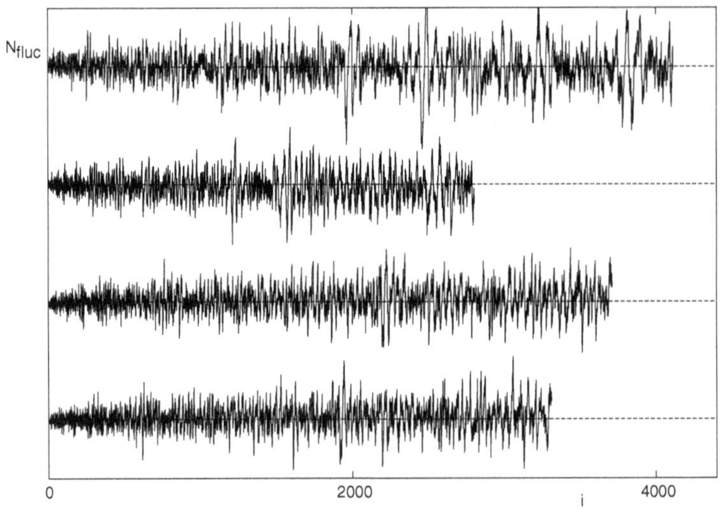

Fig. 11. $N_{fluc}(r_i)$ as a function of i for each symmetry class. The symmetry classes are **D**, **G**, **C**, **H** from top to bottom

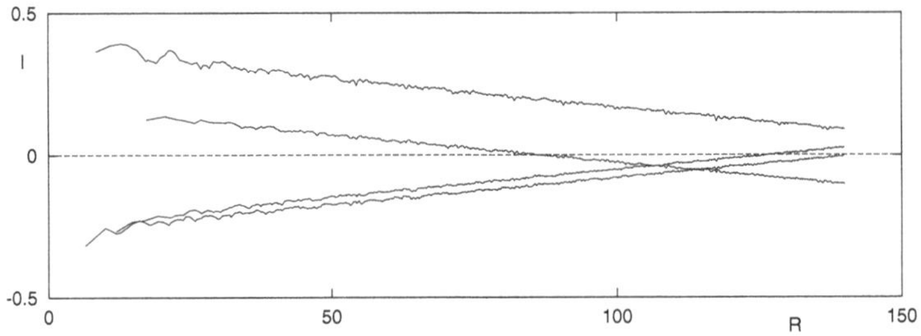

Fig. 12. I as a function of R showing the systematic deviations from $I \xrightarrow{R\to\infty} 0$. Each curve belongs to one of the symmetry classes **D**, **G**, **C**, **H**

For **C**

$$b_3 = 0.0150... \qquad \text{instead of } b_3 = 0.0111...,$$
$$b_4 = -0.062... \qquad \text{instead of } b_4 = 0.217... .$$

And **H**

$$b_3 = 0.0702... \qquad \text{instead of } b_3 = 0.0662...,$$
$$b_4 = 0.034... \qquad \text{instead of } b_4 = 0.283... .$$

In figure 13 we present the integral (5.3) with the corrected constants.

Now we are able to regard the spectral fluctuations . We unfold the spectrum,

$$x_i = \bar{N}(r_i),$$

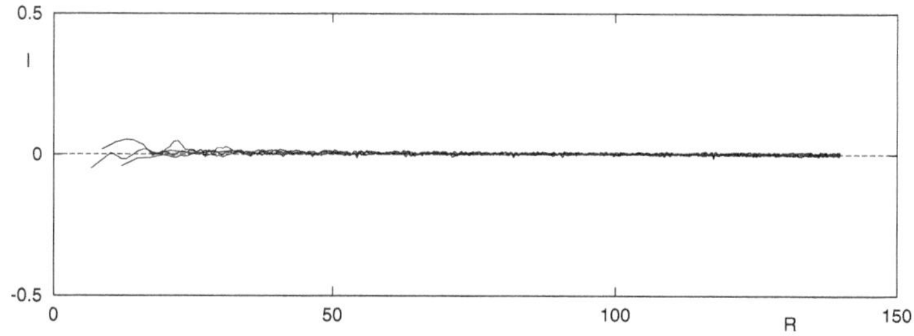

Fig. 13. I as a function of R with the corrected constants. Each curve belongs to one of the symmetry classes. The curves are quite indistinguishable from 0

in order to obtain rescaled eigenvalues x_i with a unit mean density. Then

$$s_i = x_{i+1} - x_i$$

defines the sequence of nearest-neighbor level spacings which has a mean value of 1 as $i \to \infty$. We find that the spacing distribution comes close to that of a Poisson random process,

$$P_{\text{Poisson}}(s) = \mathrm{e}^{-s},$$

see figures 14 to 17, as opposed to that of a Gaussian orthogonal ensemble of random matrix theory,

$$P_{\text{GOE}}(s) \simeq \frac{\pi}{2} s\, \mathrm{e}^{-\frac{\pi}{4}s^2} .$$

The integrated distribution,

$$\mathcal{I}(s) = \int_0^s P(t)\, dt,$$

showing the fraction of spacings up to a given length is also shown in figures 14 to 17. The spacing distributions of the desymmetrized spectra are in accordance with the conjecture of arithmetic quantum chaos. Also in agreement with the conjecture is that we have not found any degenerate eigenvalues within each symmetry class. But taking the eigenvalues of all four symmetry classes together systematic degeneracies occur due to the following:

Theorem 1 (Steil [32]). *If $\lambda = r^2 + 1$ is an eigenvalue corresponding to an eigenfunction of the symmetry class* **G** *resp.* **H** *then there exists an eigenfunction of the symmetry class* **D** *resp.* **C** *corresponding to the same eigenvalue.*

These degeneracies were first observed by Huntebrinker [30] and later explained by Steil [32] with the use of the Hecke operators [29, 33]. The Hecke operators are defined by

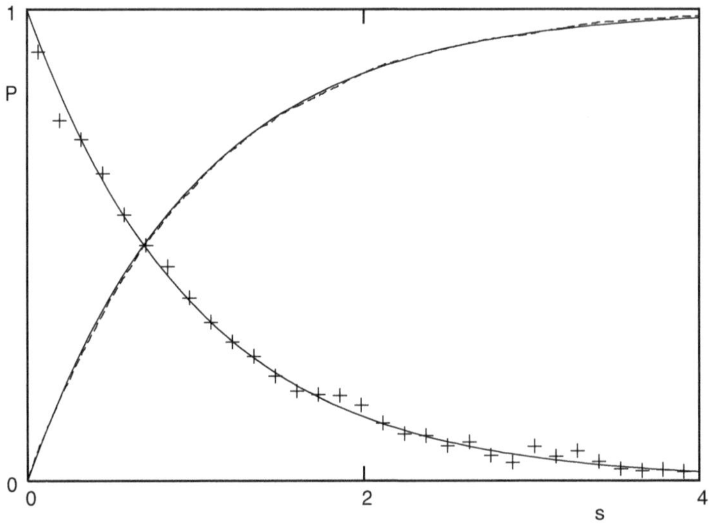

Fig. 14. Level spacing distribution for the symmetry class **D**. The abscissa displays the spacings s. The dashed curve starting at the origin is the integrated distribution. For comparison, the full curves show a Poisson distribution

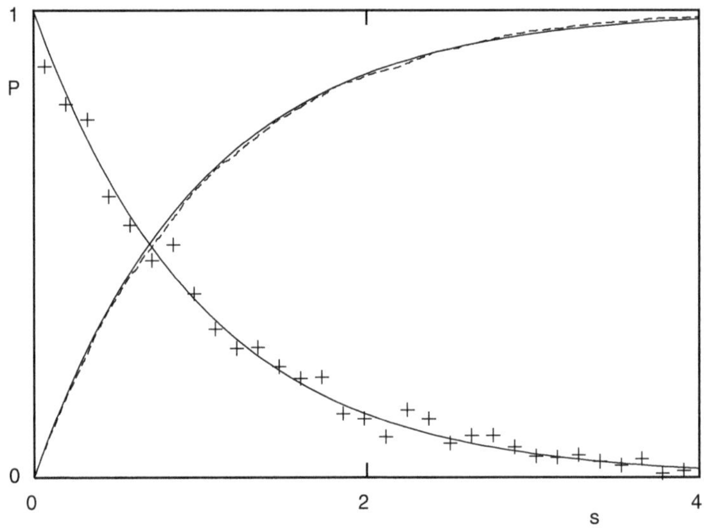

Fig. 15. Level spacing distribution for the symmetry class **G**

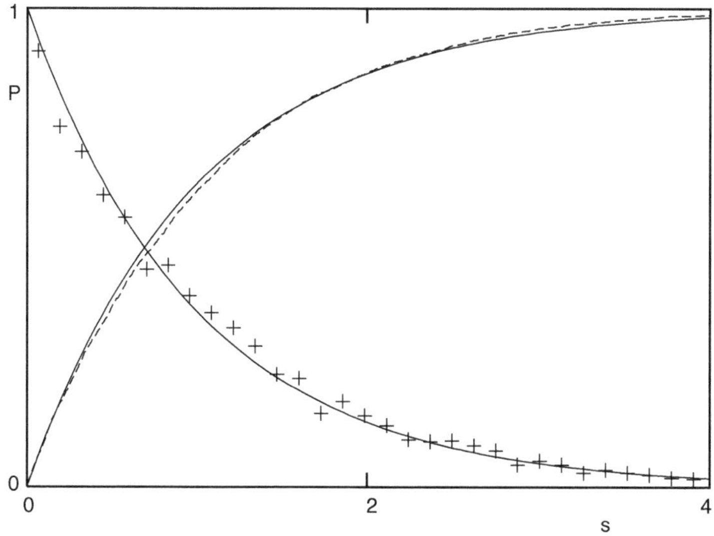

Fig. 16. Level spacing distribution for the symmetry class **C**

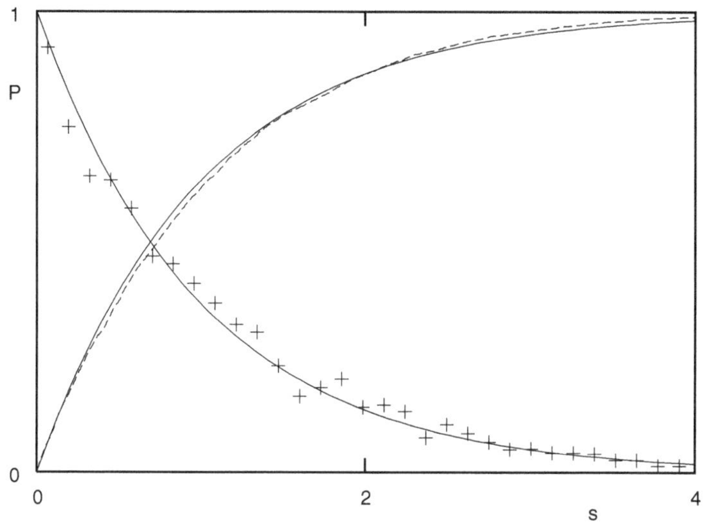

Fig. 17. Level spacing distribution for the symmetry class **H**

$$\mathrm{T}_\gamma\, g(z) = \frac{1}{|\gamma|} \sum_{\substack{a,b,d \in \mathbb{Z}[\mathrm{i}]-\{0\} \\ ad=\gamma \\ b(\mathrm{mod}\ d) \\ \Re d>0,\ \Im d \geq 0.}} g\big((ad)^{-\frac{1}{2}}(az+b)(d)^{-1}(ad)^{\frac{1}{2}}\big), \quad \gamma \in \mathbb{Z}[\mathrm{i}] - \{0\}.$$

They are self adjoint operators which commute with the Laplacian and among each other. One can therefore simultaneously diagonalize these operators. The corresponding Maass cusp forms are then called Hecke eigenfunctions. The eigenvalue equation of the Hecke operators reads

$$\mathrm{T}_\gamma\, g_r(z) = t_\gamma g_r(z), \quad \gamma \in \mathbb{Z}[\mathrm{i}] - \{0\},$$

where each Hecke eigenfunction is either identical to a Maass cusp form with a given symmetry or to a superposition of Maass cusp forms corresponding to the same eigenvalue $\lambda = r^2 + 1$, but to different symmetry classes,

$$g_r(z) = \sum_{\substack{n \in \mathbb{N} \\ (\Delta + (r^2+1)) f_n(z) = 0}} c_n f_n(z).$$

The Hecke operators are multiplicative,

$$\mathrm{T}_\gamma\, \mathrm{T}_\beta\, g_r(z) = \sum_{d|(\gamma,\beta)} \mathrm{T}_{\frac{\gamma\beta}{d^2}}\, g_r(z),$$

and the Hecke eigenvalues are connected to the Fourier coefficients,

$$b_\gamma = b_1 t_\gamma, \quad \gamma \in \mathbb{Z}[\mathrm{i}] - \{0\},$$

where the Fourier coefficients b_γ of the Hecke eigenfunctions are given by

$$b_\gamma = \sum_n c_n a_{\gamma,n}$$

and the index n at the Fourier coefficients of the Maass cusp forms $a_\gamma = a_{\gamma,n}$ means that they belong to the Fourier expansion of the n-th Maass cusp form $f_n(z)$.

Lemma 1 (Steil [32]). *If $g_r(z)$ is a Hecke eigenfunction that does not vanish identically, then:*

(i) Its first Fourier coefficient is never zero, $b_1 \neq 0$.

(ii) A Hecke eigenfunction cannot be of symmetry class **G** *or* **H**.

(iii) Hecke eigenfunctions can always be desymmetrized such that they fall either into the symmetry class $\mathbf{D} \cup \mathbf{G}$ *or* $\mathbf{C} \cup \mathbf{H}$.

Proof (Steil's theorem). Let $f_n(z)$ be a Maass cusp form of the symmetry class **G** or **H**. Due to Steil's lemma, it cannot be a Hecke eigenfunction. Since one can diagonalize the Laplacian and the Hecke operators simultaneously, there have to exist linearly independent Maass cusp forms f_{n+k}, $k = 0, \dots, K$ corresponding to the same eigenvalue $\lambda = r^2 + 1$ such that

$$\sum_{k=0}^{K} c_{n+k} f_{n+k}(z) = g_r(z)$$

is a Hecke eigenfunction.

At least one of these Maass cusp forms has to be of the symmetry class **D** or **C** in order that

$$b_1 = \sum_{k=0}^{K} c_{n+k} a_{1,n+k}$$

does not vanish.

Since the Hecke eigenfunctions can be desymmetrized such that they fall into the symmetry class $\mathbf{D} \cup \mathbf{G}$ resp. $\mathbf{C} \cup \mathbf{H}$ they are a superposition of either Maass cusp forms of the symmetry classes **D** and **G** or of Maass cusp forms of the symmetry classes **C** and **H**. Therefore, if $f_n(z)$ is of symmetry class **G** one of the f_{n+k}, $k = 1, \dots, K$ is of symmetry class **D**, and if $f_n(z)$ is of symmetry class **H** one of the f_{n+k}, $k = 1, \dots, K$ is of symmetry class **C**.

Based on our numerical results we now conjecture the following:

Conjecture 4. Taking all four symmetry classes together, there are no degenerate eigenvalues other than those explained by Steil's theorem. Furthermore, the degenerate eigenvalues which are explained by Steil's theorem occur only in pairs of two degenerate eigenvalues. They never occur in pairs of three or more degenerate eigenvalues.

Maass cusp forms of the symmetry classes **G** and **H** indeed occur. On the one hand we have found a number of them numerically. On the other hand, Weyl's law also explains their existence. Due to this the number of eigenvalues whose corresponding Maass cusp forms belong to a specific symmetry class is in leading order independent of the choice of the symmetry class. Weyl's law together with Steil's theorem lead to the following:

Conjecture 5. The sequence of non-degenerate eigenvalues in the spectrum of the Laplacian for the Picard group is of density zero.

This means that as $\lambda \to \infty$

$$\frac{\#\{\text{non-degenerate eigenvalues} \le \lambda\}}{\#\{\text{degenerate eigenvalues} \le \lambda\}} \to 0.$$

Table 1 looks as if it would contradict this conjecture. But this is due to the fact that only the first few eigenvalues are listed. In table 2 we list some consecutive large eigenvalues where we can see a better agreement with the conjecture.

Table 2. Some consecutive large eigenvalues of the Laplacian for the Picard group. Listed is r, related to the eigenvalues via $\lambda = r^2 + 1$.

D	G	C	H
139.65419675	139.65419675	139.66399548	139.66399548
139.65434417	139.65434417	139.66785333	139.66785333
139.65783548	139.65783548	139.66922266	139.66922266
139.66104047	139.66104047	139.67870460	139.67870460
139.67694018		139.68234200	139.68234200
139.68162707	139.68162707	139.68424704	139.68424704
139.68657976		139.69369972	139.69369972
139.71803029	139.71803029	139.69413379	139.69413379
139.72166907	139.72166906	139.69657741	139.69657741
139.78322452	139.78322452	139.73723373	139.73723373
139.81928622	139.81928622	139.73828541	139.73828541
139.81985670	139.81985670	139.74467774	139.74467774
139.82826034	139.82826034	139.75178180	139.75178180
139.84250751		139.75260292	139.75260292
139.87781072	139.87781072	139.79620628	139.79620628
139.87805540		139.80138072	139.80138072
139.88211647	139.88211647	139.81243991	139.81243991
139.91782003	139.91782003	139.81312982	139.81312982
139.91893517		139.82871870	139.82871870
139.92397167	139.92397167	139.86401372	139.86401372
139.92721861	139.92721861	139.86461581	139.86461581
139.93117207	139.93117207	139.89407865	139.89407865
139.93149277	139.93149277	139.89914777	139.89914777
139.94067283		139.90090849	139.90090849
139.94396890	139.94396890	139.91635302	139.91635302
139.95074070		139.94071729	139.94071729
139.95124805	139.95124805	139.95080198	139.95080198
139.99098324	139.99098324	139.97043676	139.97043676

6 Summary

Our principal goal was to test the conjecture of arithmetic quantum chaos numerically in one example. For this purpose we have chosen a point particle moving freely in the three-dimensional and negatively curved quotient space of the Picard group. Identifying the solutions of the stationary Schrödinger equation with Maass waveforms allowed us to use Hejhal's algorithm to compute the eigenfunctions and eigenvalues numerically. Having computed 13950 eigenvalues (and eigenfunctions), which exceeds all previous computations in non-integrable three-dimensional systems, we demonstrated that our numerical results are in accordance with the conjecture of arithmetic quantum chaos. Within each symmetry class we do not find any degenerate eigenvalues, but taking all four symmetry classes together, almost all eigenvalues become degenerate in the limit of large eigenvalues $\lambda \to \infty$. This behaviour was explained by the interplay of the symmetries with the Hecke-operators.

7 Acknowledgments

The help of Ralf Aurich, Jens Bolte, Dennis A. Hejhal and Frank Steiner is gratefully acknowledged. The author was supported by the Deutsche Forschungsgemeinschaft under the contract no. DFG Ste 241/16-1. Part of the work was done while I was a member of Dennis A. Hejhal's group in Uppsala (Sweden) supported by the European Commission Research Training Network HPRN-CT-2000-00103. The computations were run on the Universitäts-Rechenzentrum Ulm.

A The K-Bessel Function

The K-Bessel function is defined by

$$K_{\mathrm{i}r}(x) = \int_0^\infty \mathrm{e}^{-x\cosh t}\cos(rt)\,dt, \quad \Re x > 0, \quad r \in \mathbb{C},$$

see Watson [42], and is real for real arguments x and real or imaginary order $\mathrm{i}r$. It solves the modified Bessel differential equation

$$x^2u''(x) + xu'(x) - (x^2 - r^2)u(x) = 0,$$

and decays exponentially for large arguments

$$K_{\mathrm{i}r}(x) \sim \sqrt{\frac{\pi}{2x}}\,\mathrm{e}^{-x} \quad \text{for } x \to \infty. \tag{A.1}$$

A second linearly independent solution of the modified Bessel differential equation is the I-Bessel function

$$I_{\mathrm{i}r}(x) = (\frac{x}{2})^{\mathrm{i}r} \sum_{k=0}^{\infty} \frac{(\frac{x}{2})^{2k}}{k!\Gamma(\mathrm{i}r + k + 1)},$$

which grows exponentially for large arguments

$$I_{\mathrm{i}r}(x) \sim \sqrt{\frac{1}{2\pi x}}\, \mathrm{e}^{x} \quad \text{for } x \to \infty.$$

The K-Bessel function decreases exponentially when r increases. This can be compensated by multiplication with the factor $\mathrm{e}^{\frac{\pi r}{2}}$.

In order to compute the K-Bessel function numerically for small or moderate imaginary order we use its continued fraction representation which follows from the Miller algorithm [43].

References

1. M. V. Berry and M. Tabor. Closed orbits and the regular bound spectrum. *Proc. Roy. Soc. London Ser. A*, 349:101–123, 1976.
2. O. Bohigas, M.-J. Giannoni, and C. Schmit. Characterization of chaotic quantum spectra and universality of level fluctuation laws. *Phys. Rev. Lett.*, 52:1–4, 1984.
3. O. Bohigas, M.-J. Giannoni, and C. Schmit. Spectral fluctuations, random matrix theories and chaotic motion. Stochastic processes in classical and quantum systems. *Lecture Notes in Phys.*, 262:118–138, 1986.
4. M. L. Mehta. *Random matrices.* Academic Press, second edition, 1991.
5. R. Aurich and F. Steiner. On the periodic orbits of a strongly chaotic system. *Physica D*, 32:451–460, 1988.
6. R. Aurich, E. B. Bogomolny, and F. Steiner. Periodic orbits on the regular hyperbolic octagon. *Physica D*, 48:91–101, 1991.
7. R. Aurich and F. Steiner. Periodic-orbit sum rules for the Hadamard-Gutzwiller model. *Physica D*, 39:169–193, 1989.
8. R. Aurich and F. Steiner. Energy-level statistics of the Hadamard-Gutzwiller ensemble. *Physica D*, 43:155–180, 1990.
9. E. B. Bogomolny, B. Georgeot, M.-J. Giannoni, and C. Schmit. Chaotic billiards generated by arithmetic groups. *Phys. Rev. Lett.*, 69:1477–1480, 1992.
10. J. Bolte, G. Steil, and F. Steiner. Arithmetical chaos and violation of universality in energy level statistics. *Phys. Rev. Lett.*, 69:2188–2191, 1992.
11. P. Sarnak. Arithmetic quantum chaos. *Israel Math. Conf. Proc.*, 8:183–236, 1995.
12. A. Borel. *Introduction aux groupes arithmétiques. (French).* Hermann, 1969.
13. A. Terras. *Harmonic Analysis on Symmetric Spaces and Applications*, volume 1. Springer, 1985.
14. H. Maaß. Über eine neue Art von nichtanalytischen automorphen Funktionen und die Bestimmung Dirichletscher Reihen durch Funktionalgleichungen. (German). *Math. Ann.*, 121:141–183, 1949.
15. A. Selberg. Harmonic analysis and discontinuous groups in weakly symmetric Riemannian spaces with applications to Dirichlet series. *J. Indian Math. Soc.*, 20:47–87, 1956.

16. W. Roelcke. Das Eigenwertproblem der automorphen Formen in der hyperbolischen Ebene, I. (German). *Math. Ann.*, 167:292–337, 1966.
17. W. Roelcke. Das Eigenwertproblem der automorphen Formen in der hyperbolischen Ebene. II. (German). *Math. Ann.*, 168:261–324, 1967.
18. L. D. Faddeev. Expansion in eigenfunctions of the Laplace operator on the fundamental domain of a discrete group on the Lobačevskiĭ plane. *Transactions of the Moscow Math. Soc.*, 17:357–386, 1967.
19. G. Shimura. *Introduction to the Arithmetic Theory of Automorphic Functions.* Princeton Univ. Press, 1971.
20. D. A. Hejhal. *The Selberg trace formula for* PSL$(2,\mathbb{R})$. Lecture Notes in Math. 1001. Springer, 1983.
21. T. Miyake. *Modular forms.* Springer, 1989.
22. A. B. Venkov. *Spectral Theory of Automorphic Functions and Its Applications.* Kluwer Academic Publishers, 1990.
23. H. Iwaniec. *Introduction to the Spectral Theory of Automorphic Forms.* Revista Matemática Iberoamericana, 1995.
24. G. Humbert. Sur la mesure des classes d'Hermite de discriminant donné dans un corps quadratique imaginaire, et sur certaines volumes non euclidiens. (French). *C. R. Acad. Sci. Paris*, 169:448–454, 1919.
25. H. Maaß. Automorphe Funktionen von mehreren Veränderlichen und Dirichletsche Reihen. (German). *Abh. Math. Semin. Univ. Hamb.*, 16:72–100, 1949.
26. T. Kubota. *Elementary Theory of Eisenstein Series.* Kodansha, Tokyo and Halsted Press, 1973.
27. J. Elstrodt, F. Grunewald, and J. Mennicke. Eisenstein series on three-dimensional hyperbolic space and imaginary quadratic number fields. *J. Reine Angew. Math.*, 360:160–213, 1985.
28. J. Elstrodt, F. Grunewald, and J. Mennicke. *Groups Acting on Hyperbolic Space.* Springer, 1998.
29. M. N. Smotrov and V. V. Golovčanskiĭ. Small eigenvalues of the Laplacian on $\Gamma\backslash H_3$ for $\Gamma = PSL_2(\mathbb{Z}[i])$. *Preprint*, 91-040, Bielefeld 1991.
30. W. Huntebrinker. Numerical computation of eigenvalues of the Laplace-Beltrami operator on three-dimensional hyperbolic spaces by finite-element methods. *Diss. Summ. Math.*, 1:29–36, 1996.
31. F. Grunewald and W. Huntebrinker. A numerical study of eigenvalues of the hyperbolic Laplacian for polyhedra with one cusp. *Experiment. Math.*, 5:57–80, 1996.
32. G. Steil. Eigenvalues of the Laplacian for Bianchi groups. In D. A. Hejhal, J. Friedman, M. C. Gutzwiller, and A. M. Odlyzko, editors, *Emerging applications of number theory*, IMA Series No. 109, pages 617–641. Springer, 1999.
33. D. Heitkamp. Hecke-Theorie zur SL$(2;\mathfrak{o})$. (German). *Schriftenreihe des Mathematischen Instituts der Universität Münster, 3. Serie*, 5, 1992.
34. D. A. Hejhal. On eigenfunctions of the Laplacian for Hecke triangle groups. In D. A. Hejhal, J. Friedman, M. C. Gutzwiller, and A. M. Odlyzko, editors, *Emerging applications of number theory*, IMA Series No. 109, pages 291–315. Springer, 1999.
35. B. Selander and A. Strömbergsson. Sextic coverings of genus two which are branched at three points. *UUDM report 2002:16*, Uppsala 2002.
36. H. Avelin. On the deformation of cusp forms. (Licentiate Thesis). *UUDM report 2003:8*, Uppsala 2003.

37. K. Stramm. Kleine Eigenwerte des Laplace-Operators zu Kongruenzgruppen. (German). *Schriftenreihe des Mathematischen Instituts der Universität Münster, 3. Serie*, 11, 1994.
38. H. Then. Maaß cusp forms for large eigenvalues. *Math. Comp.*, 74:363-381, 2005.
39. C. Matthies. *Picards Billard. Ein Modell für Arithmetisches Quantenchaos in drei Dimensionen. (German).* PhD thesis, Universität Hamburg, 1995.
40. H. Weyl. Das asymptotische Verteilungsgesetz der Eigenwerte linearer partieller Differentialgleichungen. (German). *Math. Ann.*, 71:441–479, 1912.
41. V. G. Avakumović. Über die Eigenfunktionen auf geschlossenen Riemannschen Mannigfaltigkeiten. (German). *Math. Z.*, 65:327–344, 1956.
42. G. N. Watson. *A treatise on the theory of Bessel functions.* Cambridge University Press, 1944.
43. N. M. Temme. On the numerical evaluation of the modified Bessel function of the third kind. *J. Comput. Phys.*, 19:324–337, 1975.

Large N Expansion for Normal and Complex Matrix Ensembles

P. Wiegmann[1] and A. Zabrodin[2]

[1] James Frank Institute and Enrico Fermi Institute of the University of Chicago, 5640 S.Ellis Avenue, Chicago, IL 60637, USA and Landau Institute for Theoretical Physics, Moscow, Russia
[2] Institute of Biochemical Physics, Kosygina str. 4, 119991 Moscow, Russia and ITEP, Bol. Cheremushkinskaya str. 25, 117259 Moscow, Russia

Summary. We present the first two leading terms of the $1/N$ (genus) expansion of the free energy for ensembles of normal and complex random matrices. The results are expressed through the support of eigenvalues (assumed to be a connected domain in the complex plane). In particular, the subleading (genus-1) term is given by the regularized determinant of the Laplace operator in the complementary domain with the Dirichlet boundary conditions. An explicit expression of the genus expansion through harmonic moments of the domain gives some new representations of the mathematical objects related to the Dirichlet boundary problem, conformal analysis and spectral geometry.

1 Introduction

Ensembles of random matrices have numerous important applications in physics and mathematics ranging from energy levels of nuclei to number theory. An important information is encoded in the $1/N$ expansion (N is the size of the matrix) of different expectation values in the ensemble. Many relevant references can be found in [1] .

In this paper we discuss $1/N$-expansion in statistical ensembles of normal and complex matrices. A matrix M is called normal if it commutes with its Hermitian conjugate: $[M, M^\dagger] = 0$, so both matrices can be diagonalized simultaneously. Eigenvalues of a normal matrix are complex numbers. The statistical weight

$$e^{\frac{1}{\hbar}\operatorname{tr} W(M)} d\mu(M)$$

of the normal matrix ensemble is specified by a potential function $W(M)$ (which depends on both M and $M^\dagger$). Here $\hbar$ is a parameter, and the measure $d\mu$ of integration over normal matrices is induced from the flat metric on the space of all complex matrices.

Along the standard procedure of integration over angle variables [2], one passes to the joint probability distribution of eigenvalues $z_1, \ldots, z_N$. The partition function is then given by the integral

$$Z_N = \frac{1}{(2\pi^3\hbar)^{N/2} N!} \int |\Delta_N(z_i)|^2 \prod_{j=1}^{N} e^{\frac{1}{\hbar}W(z_j)} d^2 z_j \tag{1.1}$$

Here $\Delta_N(z_i) = \prod_{i>j}^{N}(z_i - z_j)$ is the Vandermonde determinant and $d^2 z \equiv dx\, dy$ for $z = x + iy$. The N-dependent normalization factor is put here for further convenience.

The model of normal matrices was introduced in [3]. This model is the particular $\beta = 1$ case of a more general one, referred to as 2D Coulomb gas with the joint probability distribution $|\Delta_N(z_i)|^{2\beta} \prod_{j=1}^{N} e^{\frac{1}{\hbar}W(z_j)} d^2 z_j$.

For the potential of the form

$$W(z) = -z\bar{z} + V(z) + \overline{V(z)} \tag{1.2}$$

where $V(z)$ is an analytic function in some region of the complex plane (say, a polynomial), the normal matrix model is equivalent to the ensemble of all complex matrices with the same potential. It generalizes the Gaussian Ginibre-Girko ensemble [4]. When passing to the integral over eigenvalues, the partition function for complex matrices differs from the one for normal matrices by a normalization factor only [2]. Both models are then reduced to the 2D Coulomb gas (with $\beta = 1$) in the external potential. Note also a formal similarity with the model of two Hermitian matrices. Its partition function is given by the same formula (1.1), with the potential (1.2), but z_i and $\bar{z}_i$ are to be regarded as two independent *real* integration variables, with $d^2 z_i$ being understood as $dz_i d\bar{z}_i$.

It appears that the potential of the form (1.2) is most important for applications [5]. In the main part of the paper, we concentrate on this case, so one may, in this context, ignore the difference between the normal and complex ensembles, taking the 2D Coulomb gas partition function as a starting point. Physical applications of this model include the quantum Hall effect, the Saffman-Taylor viscous fingering and, conjecturally, more general growth problems which are mathematically described as a random evolution in the moduli space of complex curves. Recently, the normal matrix model was shown [6] to be closely related to the matrix quantum mechanics, and, therefore, to the $c = 1$ string theory.

In addition to this it appears that the large N limit of the normal or complex random matrices admits a natural geometric interpretation relevant to the 2D inverse potential problem, the Dirichlet boundary problem and to spectral geometry of planar domains. In this paper we concentrate on calculation of the $1/N$ expansion of the free energy, $F \propto \log Z_N$, and on its algebro-geometric meaning, leaving physical aspects for future publications.

The large N limit also implies the limit $\hbar \to 0$, so that $\hbar N$ is finite and fixed. We prefer to work with the equivalent $\hbar$-expansion, rather than with the $1/N$ expansion, thus emphasizing its semiclassical nature. The free energy of the Hermitian, two-Hermitian, normal and complex matrix ensembles with the potential (1.2) has an $\hbar$-expansion of the form $\log Z_N = \sum_{g \geq 0} \hbar^{2g-2} F_g$, where g-th term is associated with the contribution of diagrams with Euler characteristics $2 - 2g$, in the perturbative expansion of the free energy. Here we discuss the first two terms, F_0 and F_1:

$$F = \hbar^2 \log Z_N = F_0 + \hbar^2 F_1 + O(\hbar^4) \tag{1.3}$$

The leading term, F_0, is the contribution of planar diagrams, and F_1 is commonly referred to as genus 1 correction.

When N becomes large new macroscopic structures emerge. Invoking a physical analogy, one may say that the gas of eigenvalues segregates into "phases" with zero and non-zero density separated by a very narrow interface. The domain D in the complex plane where the density is non-zero is called the support of eigenvalues (it may consist of several disconnected domains). The density at any point outside it is exponentially small as $N \to \infty$.

The leading contribution to the free energy, the F_0 term in (1.3), is basically the Coulomb energy of particles confined in the domain D. For the potential of the form (1.2) it is the tau-function of curves introduced in [7]. It encodes solutions to archetypal problems of complex analysis and potential theory in planar domains.

Here we review these results and also compute the genus-1 correction to the free energy. The latter is identified with the free energy of a free bosonic field in the domain D^c which is complementary to the support of eigenvalues, i.e., in the domain where the mean density vanishes:

$$F_1 = -\frac{1}{2} \log \det(-\Delta_{\mathsf{D}^c}) \tag{1.4}$$

Here $\det(-\Delta_{\mathsf{D^c}})$ is a properly regularized determinant of the Laplace operator in $\mathsf{D^c}$ with Dirichlet boundary conditions. This suggests interesting links to spectral geometry of planar domains.

The genus expansion in the Hermitian random matrix model beyond the leading order has been obtained in the seminal paper [8]. In [9], the genus-1 correction was interpreted in terms of bosonic field theory on a hyperelliptic Riemann surface. The genus 1 correction to free energy of the model of two Hermitian matrices with polynomial potential was found only recently [10].

2 The Planar Large N Limit

In this section we briefly recall the large N limit technique. This material is standard since early days of random matrix models (see., e.g., [11]). An appealing feature of the model of normal or complex matrices is a nice geometric interpretation and a direct relation to the inverse potential problem in two dimensions.

As was already mentioned, the parameter $\hbar$ tends to zero simultaneously with $N \to \infty$ in such a way that $t_0 = N\hbar$ is kept finite and fixed. Using the Coulomb gas analogy, one may say that the leading contribution to the free energy is equal to the extremal value of the energy

$$\mathcal{E} = \sum_{i \neq j} \log |z_i - z_j| + \frac{1}{\hbar} \sum_i W(z_i) \tag{2.1}$$

Equilibrium positions of charges are given by the extremum of the plasma energy: $\partial_{z_i}\mathcal{E} = \partial_{\bar{z}_i}\mathcal{E} = 0$.

Consider the 2D Coulomb potential $\Phi(z) = -\hbar \sum_i \log|z - z_i|^2$ created by the charges. Writing it as

$$\Phi(z) = -\int \log|z - \zeta|^2 \rho(z)\, d^2 z$$

where

$$\rho(z) = -\frac{1}{4\pi}\Delta\Phi(z) = \hbar \sum_i \delta^{(2)}(z - z_i) \tag{2.2}$$

is the microscopic density of eigenvalues (a sum of two-dimensional delta-functions), we assume that Φ in the limit can be treated as a continuous function. It is normalized as $\int \Delta\Phi(z) d^2 z = -4\pi t_0$. Let Φ_0 be this function for the equilibrium configuration of charges, then

$$\partial_z(\Phi_0(z) - W(z)) = \partial_{\bar{z}}(\Phi_0(z) - W(z)) = 0 \tag{2.3}$$

with the understanding that this equation holds only for z belonging to a domain (or domains) where the density is nonzero. Applying $\partial_{\bar{z}}$ to the both sides, we see that the equilibrium density, $\rho_0(z)$, is equal to $-\frac{1}{4\pi}\Delta W(z)$ in some domain D (the support of eigenvalues) and zero otherwise:

$$\rho_0(z) = \begin{cases} \sigma & z \in \mathsf{D} \\ 0 & z \in \mathsf{D}^c \end{cases} \quad \text{and} \quad \Phi_0(z) = -\int_{\mathsf{D}} \log|z-\zeta|^2 \sigma \, d^2\zeta$$

Here $\mathsf{D}^c = \mathbf{C} \setminus \mathsf{D}$ is the domain complimentary to the support of eigenvalues and

$$\sigma = -\frac{1}{4\pi}\Delta W(z,\bar{z}) \tag{2.4}$$

In this and in the next section we assume the special form of the potential (1.2). Then $\rho_0 = 1/\pi$ in the domain D.

The shape of D is determined by the function $V(z)$. Let us assume, without loss of generality, that $0 \in \mathsf{D}$ and parametrize $V(z)$ by Taylor coefficients at the origin:

$$V(z) = \sum_{k\geq 1} t_k z^k \tag{2.5}$$

The parameters t_k (coupling constants of the matrix model) are in general complex numbers. Multiplying (2.3) by z^{-k} and integrating over the boundary of D, we conclude that the domain D is such that $-\pi k t_k$'s are moments of its complement, D^c, with respect to the functions z^{-k}:

$$t_k = -\frac{1}{\pi k}\int_{\mathsf{D}^c} z^{-k} d^2 z = \frac{1}{2\pi i k}\oint_{\partial\mathsf{D}} z^{-k}\bar{z}\, dz \tag{2.6}$$

Besides, from the normalization condition we know that the area of D is equal to πt_0. To find the shape of the domain from its moments and area is the subject of the inverse potential problem. These data determine it uniquely, at least locally.

Here we assume that D is a connected domain. For example, in the potential $W = -z\bar{z}$ the eigenvalues uniformly fill the disk of radius $\sqrt{\hbar N}$. Small perturbations of the potential slightly disturb the circular shape.

In what follows, we need some functions associated with the domain D, or rather with its complement, D^c. The basic one is a univalent conformal map from the exterior of the unit disk onto the domain D^c. Such a map exists by virtue of the Riemann mapping theorem. Let U be the unit disk and U^c its complement, i.e., the exterior of the unit disk. Consider the conformal map $z(w)$ from U^c onto D^c normalized so that $z(\infty) = \infty$ and $r = \lim_{w\to\infty} z(w)/w$ is real, then the map is unique. In general, the Laurent expansion of the function $z(w)$ around infinity is

$$z(w) = rw + \sum_{k\geq 0} u_k w^{-k} \tag{2.7}$$

The real number r is called the (external) conformal radius of D. Since the map is conformal, all zeros and poles of the derivative $z'(w) \equiv \partial_w z(w)$ are inside the unit circle. We also need the function $\bar{z}(w)$ given by the Laurent series (2.7) with complex conjugate coefficients and the Green function of the Dirichlet boundary problem in D^c. In terms of the conformal map, the latter is given by the explicit formula

$$G(z, z') = \log \left| \frac{w(z) - w(z')}{w(z)\overline{w(z')} - 1} \right| \tag{2.8}$$

Here $w(z)$ is the conformal map from D^c onto U^c inverse to the $z(w)$.

3 The Leading Term of the Free Energy

The leading contribution to the free energy is the value of the Coulomb energy (2.1) (multiplied by $\hbar^2$) for the extremal configuration of charges:

$$F_0 = \int_{\mathsf{D}}\int_{\mathsf{D}} \log|z - z'|\, \sigma(z)\sigma(z') d^2z\, d^2z' + \int_{\mathsf{D}} W(z, \bar{z})\sigma d^2 z$$

The integrated version of the extremum condition (2.3) tells us that $\Phi_0(z) - W(z) = \text{const}$ for any $z \in \mathsf{D}$. The constant can be found from the same equality at $z = 0$, and we obtain F_0 as an explicit functional of the domain D:

$$F_0 = -\int_{\mathsf{D}}\int_{\mathsf{D}} \log\left|\frac{1}{z} - \frac{1}{z'}\right| \sigma(z)\sigma(z') d^2z\, d^2z' \tag{3.1}$$

For the special potential of the form (1.2), when $\sigma = 1$, the free energy is to be regarded as a function of t_0 and the coupling constants t_k.

Properties of F_0 immediately follow from known correlation functions of the model in the planar large N limit. See [5, 7] for normal and complex matrices and [12] for similar results in the context of the Hermitian 2-matrix model. Some of these correlation functions previously appeared in studies of thermal fluctuations in classical confined Coulomb plasma [13]. Integrable structures associated with F_0 were studied in [14, 7, 15, 16]. Here is the list of main properties of F_0 for the most important case $\sigma = 1/\pi$.

- *1-st order derivatives:*

$$\frac{\partial F_0}{\partial t_k} = \frac{1}{\pi}\int_{\mathsf{D}} z^k\, d^2 z\,, \quad k \geq 1,$$
$$\frac{\partial F_0}{\partial t_0} = \frac{1}{\pi}\int_{\mathsf{D}} \log|z|^2\, d^2 z \tag{3.2}$$

can be combined in the generating formula

$$\mathcal{D}(z) F_0 = \frac{1}{\pi}\int_{\mathsf{D}} \log|z^{-1} - \zeta^{-1}|^2 d^2\zeta\,, \quad z \in \mathsf{D}^c\,, \tag{3.3}$$

where

$$\mathcal{D}(z) = \frac{\partial}{\partial t_0} + \sum_{k\geq 1} \frac{1}{k}\left(z^{-k}\frac{\partial}{\partial t_k} + \bar{z}^{-k}\frac{\partial}{\partial \bar{t}_k}\right) \tag{3.4}$$

Since the derivatives of F_0 with respect to the moments t_k are moments of the complimentary domain, this function formally solves the 2D inverse potential problem.

- *2-nd order derivatives:* for $z, z' \in \mathsf{D}^c$ we have

$$\mathcal{D}(z)\mathcal{D}(z')F_0 = 2G(z,z') - \log\left|\frac{1}{z} - \frac{1}{z'}\right|^2 \tag{3.5}$$

where $G(z,z')$ is the Green function of the Dirichlet boundary problem in D^c (2.8). Note that the logarithmic singularity of the Green function at $z = z'$ cancels by the second term in the right hand side. In a particular case when both z, z' tend to infinity, we get a simple formula for the conformal radius:

$$\partial_{t_0}^2 F_0 = 2\log r \tag{3.6}$$

- *3-d order derivatives.* The generating formula reads [15]:

$$\mathcal{D}(a)\mathcal{D}(b)\mathcal{D}(c)F_0 = -\frac{1}{2\pi}\oint_{\partial\mathsf{D}} \partial_n G(a,\xi)\partial_n G(b,\xi)\partial_n G(c,\xi)|d\xi| \tag{3.7}$$

An important corollary of this formula and eq. (3.5) is the complete symmetry of the expression $\mathcal{D}(a)G(b,c)$ with respect to all permutations of the points a, b, c. Another corollary of (3.7) is the following *residue formula* valid for $j, k, l \geq 0$ [16]:

$$\frac{\partial^3 F_0}{\partial t_j \partial t_k \partial t_l} = \frac{1}{2\pi i}\oint_{|w|=1} \frac{h_j(w)h_k(w)h_l(w)}{z'(w)\bar{z}'(w^{-1})}\frac{dw}{w} \tag{3.8}$$

Here $h_j(w)$ are polynomials in w of degree j:

$$h_j(w) = w\frac{d}{dw}\left[(z^j(w))_+\right] \quad \text{for } j \geq 1 \text{ and } \quad h_0(w) = 1\,,$$

where $(...)_+$ is the positive degree part of the Laurent series. The notation $\bar{z}'(w^{-1})$ means the derivative $d\bar{z}(u)/du$ taken at the point $u = w^{-1}$. This formula is especially useful when $z'(w)$ is a rational function, then the integral is reduced to a finite sum of residues. We use this below.

- *Dispersionless Hirota equations.* The function F_0 obeys an infinite number of non-linear differential equations which are combined into the integrable hierarchy of dispersionless Hirota's equations. See [7, 15] for details.
- *WDVV equations.* Suppose $V(z)$ is a polynomial of m-th degree, i.e., $t_k = 0$ for all $k > m$. On this subspace of parameters, F_0 obeys the system of Witten-Dijkgraaf-Verlinde-Verlinde (WDVV) equations

$$\mathsf{F}_i \mathsf{F}_j^{-1} \mathsf{F}_k = \mathsf{F}_k \mathsf{F}_j^{-1} \mathsf{F}_i \quad \text{for all } 0 \le i, j, k \le m-1 \tag{3.9}$$

where F_i is the m by m matrix with matrix elements $(\mathsf{F}_i)_{jk} = \frac{\partial^3 F_0}{\partial t_i \partial t_j \partial t_k}$. See [16] for details.

To conclude: F_0 is a "master function" which generates objects of complex analysis in planar simply-connected domains. The full free energy of the matrix ensemble, F, may be regarded as its "quantization".

4 The Genus 1 Correction to the Free Energy

4.1 The Result for F_1

We now describe the result for the genus-1 correction F_1. We start with the special potential (1.2). Then F_1 is expressed entirely in terms of the metric on the U^{c} induced from the standard flat metric on the z-plane by the conformal map: $dz\, d\bar{z} = e^{2\phi(w)} dw\, d\bar{w}$. Here

$$\phi(w) = \log |z'(w)| \tag{4.1}$$

and $z(w)$ is the conformal map $\mathsf{U}^{\mathsf{c}} \to \mathsf{D}^{\mathsf{c}}$ (2.7). The derivation of this formula and its extension to a general potential is outlined in Section 5.

We found that

$$F_1 = -\frac{1}{24\pi} \oint_{|w|=1} (\phi \partial_n \phi + 2\phi) \, |dw| \tag{4.2}$$

Here ∂_n is the normal derivative, with the normal vector pointing outside the unit circle. The derivation of this formula is outlined, for a more general model, in Section 5.

Since $\phi(w)$ is harmonic in U^{c}, we may rewrite the r.h.s. of (4.2) as

$$F_1 = \frac{1}{24\pi} \int_{|w|>1} |\nabla \phi|^2 d^2 w - \frac{1}{12\pi} \oint_{|w|=1} \phi \, |dw|$$

Here we recognize the formula for the regularized determinant of the Laplace operator $\Delta_{\mathsf{D}^{\mathsf{c}}} = 4\partial_z \partial_{\bar{z}}$ in D^{c} with Dirichlet conditions on the boundary. The first term is the bulk contribution first found by Polyakov [17] (for a metric induced by a conformal map it reduces to a boundary integral), while the second term, computed in [18], is a net boundary term (see also Section 1 of [19]):

$$F_1 = -\frac{1}{2} \log \det (-\Delta_{\mathsf{D}^{\mathsf{c}}}) \tag{4.3}$$

In the particular case $W(z) = -z\bar{z}$ we get $F_1 = -\frac{1}{12} \log t_0$ that coincides with the result of [20] obtained by a direct calculation.

The appearance of elements of quantum field theory in a curved space is not accidental. A field-theoretical derivation of this result will be given elsewhere.

4.2 Rational Case

Before explaining the origin of the explicit formula for F_1 we write it in yet another suggestive form. Consider a domain such that $z'(w)$ is a rational function:

$$z'(w) = r \prod_{i=0}^{m-1} \frac{w - a_i}{w - b_i}$$

All the points a_i and b_i must be inside the unit circle, otherwise the map $z(w)$ is not conformal. On the unit circle we have $|dw| = \frac{dw}{iw}$ and $\phi(w) = \frac{1}{2}(\log z'(w) + \log \bar{z}'(w^{-1}))$, where the first and the second term (the Schwarz reflection) are analytic outside and inside it, respectively. (Recall that our notation $\bar{z}'(w^{-1})$ means $d\bar{z}(u)/du$ at the point $u = w^{-1}$.) Plugging this into (4.2), we get:

$$\begin{aligned} F_1 = &-\frac{1}{24\pi i}\oint_{|w|=1} \log z'(w)\left[\frac{1}{2}\partial_w \log z'(w) + \frac{1}{w}\right] dw - \\ &-\frac{1}{24\pi i}\oint_{|w|=1} \log \bar{z}'(w^{-1})\frac{dw}{w} - \frac{1}{48\pi i}\oint_{|w|=1} \log \bar{z}'(w^{-1})\frac{z''(w)}{z'(w)} dw \end{aligned}$$

The integrals can be calculated by taking residues either outside or inside the unit circle. The poles are at ∞, at 0, and at the points a_i and b_i. The result is

$$F_1 = -\frac{1}{24}\left(\log r^4 + \sum_{z'(a_i)=0} \log \bar{z}'(a_i^{-1}) - \sum_{z'(b_i)=\infty} \log \bar{z}'(b_i^{-1})\right) \tag{4.4}$$

If the potential $V(z)$ is polynomial, $V(z) = \sum_{k=1}^{m} t_k z^k$, i.e., $t_k = 0$ as $k > m$ for some $m > 0$, then the series for the conformal map $z(w)$ truncates: $z(w) = rw + \sum_{l=0}^{m-1} u_l w^{-l}$ and

$$z'(w) = r \prod_{i=0}^{m-1} (1 - a_i w^{-1})$$

is a polynomial in w^{-1} (all poles b_i of $z'(w)$ merge at the origin). Then the last sum in (4.4) becomes $m \log r$ and the formula (4.4) gives

$$F_1 = -\frac{1}{24}\log\left(r^4 \prod_{z'(a_j)=0} \frac{\bar{z}'(a_j^{-1})}{r}\right) = -\frac{1}{24}\log\left(r^4 \prod_{i,\,j=0}^{m-1} (1 - \bar{a}_i a_j)\right) \tag{4.5}$$

This formula is essentially identical to the genus-1 correction to the free energy of the Hermitian 2-matrix model with a polynomial potential recently computed by Eynard [10].

4.3 Determinant Representation of F_1 for Polynomial Potentials

For polynomial potentials the genus-1 correction enjoys an interesting determinant representation.

Set

$$D_m := \det\left(\frac{\partial^3 F_0}{\partial t_0 \partial t_j \partial t_k}\right)_{0\le j,k\le m-1}$$

Using the residue formula (3.8) we compute:

$$D_m = \frac{1}{(2\pi i)^m}\oint_{|w_0|=1}\frac{dw_0}{w_0}\cdots\oint_{|w_{m-1}|=1}\frac{dw_{m-1}}{w_{m-1}}\,\frac{\det\left[h_j(w_j)h_k(w_j)\right]}{\prod_{l=0}^{m-1} z'(w_l)\bar{z}'(w_l^{-1})} \tag{4.6}$$

Clearly, the determinant in the numerator can be substituted by $\frac{1}{m}\det^2(h_j(w_k))$ and $\det\left[h_j(w_k)\right] = (m-1)!\, r^{\frac{1}{2}m(m-1)}\Delta_m(w_i)$, where $\Delta_m(w_i)$ is the Vandermonde determinant. Each integral in (4.6) is given by the sum of residues at the points a_i inside the unit circle (the residues at $w_i = 0$ vanish). Computing the residues and summing over all permutations of the points a_i, we get:

$$D_m = (-1)^{\frac{1}{2}m(m-1)}((m-1)!)^2\, r^{m(m-3)}\,\frac{\prod_j a_j^{m-1}}{\prod_{i,j}(1-a_j\bar{a}_i)} \tag{4.7}$$

As is seen from (2.6), the last non-zero coefficient of $V(z)$ equals $t_m = \frac{\bar{u}_{m-1}}{mr^{m-1}}$. (We regard it as a fixed parameter.) Therefore, $\prod_{i=1}^m a_i = (-1)^m m(m-1)r^{m-2}\bar{t}_m$, and we represent F_1 (4.5) in the form

$$F_1 = \frac{1}{24}\log D_m - \frac{1}{12}(m^2-3m+3)\log r - \frac{1}{24}(m-1)\log \bar{t}_m + \text{const} \tag{4.8}$$

where const is a numerical constant. Recalling (3.6), we see that F_1, for models with polynomial potentials of degree m, is expressed through derivatives of F_0:

$$F_1 = \frac{1}{24}\log \det_{m\times m}\left(\frac{\partial^3 F_0}{\partial t_0\partial t_j\partial t_k}\right) - \frac{1}{24}(m^2-3m+3)\frac{\partial^2 F_0}{\partial t_0^2} - \frac{1}{24}(m-1)\log\bar{t}_m + \text{const} \tag{4.9}$$

where j, k run from 0 to $m-1$.

Similar determinant formulas are known for genus-1 corrections to free energy in topological field theories [21].

5 F_1 from Loop Equation

The standard (and powerful) method to obtain $1/N$-expansions in matrix models is to use invariance of the partition function under changes of matrix integration variables. In the 2D Coulomb gas formalism, this reduces to invariance of the partition function (1.1) under diffeomorphisms

$$z_i \longrightarrow z_i + \epsilon(z_i, \bar{z}_i)\,, \quad \bar{z}_i \longrightarrow z_i + \bar{\epsilon}(z_i, \bar{z}_i)$$

The invariance of the partition function in the first order in ϵ results in the identity

$$\sum_i \int \partial_{z_i} \left(\epsilon(z_i, \bar{z}_i)\, e^{\mathcal{E}} \right) \prod_j d^2 z_j = 0 \tag{5.1}$$

for any function $\epsilon(z, \bar{z})$. One may read it as Ward identity obeyed by correlation functions of the model. For historical reasons, it is called the loop equation. Since correlation functions are variational derivatives of the free energy with respect to the potential, the loop equation is an implicit functional relation for the free energy.

5.1 Loop Equation in General Normal Matrix Model

A closed loop equation does not emerge for the special potential (1.2). It can be written only for the ensemble of normal matrices with a general potential W in (1.1). Let it be of the form

$$W(z) = -z\bar{z} + V(z) + \overline{V(z)} + U(z)$$

where U is only assumed to have a regular Taylor expansion around the origin starting from cubic terms.

Choosing $\epsilon(z_i, \bar{z}_i) = (z - z_i)^{-1}$ and plugging it into (5.1) with $\mathcal{E}$ given in (2.1), one is able to rewrite (5.1) as a relation between correlation functions of the field

$$\Phi(z) = -\hbar \,\mathrm{tr}\, \log \left[(z - M)(\bar{z} - M^\dagger) \right] = -\hbar \sum_i \log |z - z_i|^2$$

or $\partial\Phi(z) = -\hbar \mathrm{tr}(z - M)^{-1}$ (here and below $\partial \equiv \partial_z$). Note that $\partial\Phi(z)$ is trace of the resolvent of the matrix M and $\Delta\Phi(z) = -4\pi\rho(z)$, where ρ is the density of eigenvalues. After some simple rearrangings, the loop equation following from (5.1) acquires the form

$$\frac{1}{2\pi} \int \frac{\partial W(\zeta)\, \langle \Delta\Phi(\zeta) \rangle}{z - \zeta}\, d^2\zeta = \Big\langle (\partial\Phi(z))^2 \Big\rangle + \hbar \Big\langle \partial^2 \Phi(z) \Big\rangle \tag{5.2}$$

(For any symmetric function $f(\{z_i\})$, the correlation function $\langle f \rangle$ is defined as the integral $\int f(\{z_i\}) |\Delta_N(z_i)|^2 \prod_j e^{\frac{1}{\hbar} W(z_j)} d^2 z_j$ with a normalization factor such that $\langle 1 \rangle = 1$.) This relation is exact for any finite N. Supplemented by the relation

$$\Big\langle \partial\Phi(z) \Big\rangle = -\frac{t_0}{z} + \partial_z \mathcal{D}(z) F \tag{5.3}$$

(also exact) which directly follows from the definitions of the free energy and the field Φ, the loop equation allows one to find the $\hbar$-expansion of the free energy.

5.2 Expanding the Loop Equation

The $\hbar$-expansion of the free energy for the general normal matrix model is more complicated than the one discussed in the previous sections. It contains all powers of $\hbar$, not only even:

$$\hbar^2 \log Z_N = F_0 + \hbar F_{1/2} + \hbar^2 F_1 + O(\hbar^3) \tag{5.4}$$

so it hardly has a direct topological interpretation. Accordingly, $\hbar$-expansions of mean values and other correlation functions are expansions in $\hbar$ rather than $\hbar^2$. In particular,

$$\langle \Phi(z) \rangle = \Phi_0(z) + \hbar \Phi_{1/2}(z) + \hbar^2 \Phi_1(z) + O(\hbar^3) \tag{5.5}$$

We proceed by expanding the loop equation in powers of $\hbar$. In the leading order, the second term in the r.h.s. vanishes, and $\bar{z}$-derivative of the both sides gives:

$$(\partial W(z) - \partial \Phi_0(z)) \, \Delta \Phi_0(z) = 0 \tag{5.6}$$

This just means that for $z \in \mathsf{D}$ the extremum condition (2.3) is satisfied and $\Delta \Phi_0(z) = 0$ otherwise. Inside D, the leading term of the mean density, $\rho_0(z)$, is given by $\rho_0(z) = \sigma(z)$, where $\sigma(z)$ is defined in (2.4). (Note that the function σ is defined by this formula everywhere in the complex plane, and does not depend on the shape of D, while ρ_0 coincides with σ in D and equals 0 in D^{c}.) For potentials of the form (1.2), $\sigma(z) = 1/\pi$.

Being developed into a series in $\hbar$, the loop equation gives an iterative procedure to find the coefficients $\Phi_i(z)$. We need the following results on the correlation functions for the general normal matrix ensemble (see [5]):

$$\Big\langle \partial \Phi(z) \Big\rangle = \int_{\mathsf{D}} \frac{\sigma(\zeta) d^2 \zeta}{\zeta - z} + O(\hbar) \tag{5.7}$$

$$\begin{aligned} \Big\langle \Phi(z_1) \Phi(z_2) \Big\rangle_{\mathrm{conn}} &= 2\hbar^2 \Big(G(z_1, z_2) - G(z_1, \infty) - G(\infty, z_2) \\ &\quad - \log \frac{|z_1 - z_2|}{r} \Big) + O(\hbar^3) \end{aligned} \tag{5.8}$$

where the connected correlation function is defined as $\langle fg \rangle_{\mathrm{conn}} = \langle fg \rangle - \langle f \rangle \langle g \rangle$. Note that the function (5.8) has no singularity at coinciding points $z_1, z_2 \in \mathsf{D}^{\mathsf{c}}$. Merging the points, we get:

$$\Big\langle (\partial \Phi(z))^2 \Big\rangle_{\mathrm{conn}} = \frac{\hbar^2}{6} \{w; \, z\} + O(\hbar^3) \tag{5.9}$$

where

$$\{w;\, z\} = \frac{w'''(z)}{w'(z)} - \frac{3}{2}\left(\frac{w''(z)}{w'(z)}\right)^2$$

is the Schwarzian derivative of the conformal map $w(z)$.

After these preparations, further steps are straightforward. Terms of order $\hbar$ and $\hbar^2$ of the loop equation give:

$$\begin{aligned} &\frac{1}{2\pi}\int L(z,\zeta)\, \Delta\Phi_{1/2}(\zeta)\, d^2\zeta = -\partial^2\Phi_0(z) \\ &\frac{1}{2\pi}\int L(z,\zeta)\, \Delta\Phi_1(\zeta)\, d^2\zeta = -\left[\left(\partial\Phi_{1/2}(z)\right)^2 + \partial^2\Phi_{1/2}(z)\right] - \frac{1}{6}\{w;\, z\} \end{aligned} \tag{5.10}$$

where the kernel of the integral operator in the l.h.s. is

$$L(z,\zeta) = \frac{\partial W(\zeta) - \partial\Phi_0(z)}{\zeta - z} \tag{5.11}$$

It should be noted that the $\hbar$-expansion of the loop equation may break down for $z \in \mathsf{D}$. This is mainly because the correlator $\langle\Phi(z)\Phi(z')\rangle$, when the two points are close to each other and belong to the support of eigenvalues, is not given by eq. (5.8). At the same time, for our purpose we need this correlator just on very small distances, when the two points merge. Naively, for $z, z' \in \mathsf{D}$ the correlator diverges as $z' \to z$. This means that its short-distance behaviour is in fact of a different order in $\hbar$ and must be calculated separately. Fortunately, this problem can be avoided by restricting the equations to D^c, where no divergency emerges on any scale and one may think that the short-distance behaviour of correlation functions is still given by eq. (5.8). (However, we understand that this argument is not rigorous and need to be justified by an actual calculation of correlation functions at small scales.) Hereafter, z in (5.10) is assumed to be outside the support of eigenvalues, i.e., the equations should be solved for $z \in \mathsf{D}^c$. In this region the functions $\Phi_k(z)$ are harmonic.

From (5.3) we see that

$$\partial_z\mathcal{D}(z)F_{1/2} = \partial_z\Phi_{1/2}(z)\,, \qquad \partial_z\mathcal{D}(z)F_1 = \partial_z\Phi_1(z) \tag{5.12}$$

The strategy is to find Φ_k's from (5.10) and then "to integrate" them to get F_k's, i.e., to find a functional F_k such that it obeys (5.12). A general method to find the "derivative" $\mathcal{D}(z)$ of any proper functional of the domain D^c is proposed in [15].

An important remark is in order. Suppose we restrict ourselves to the class of models with potentials of the form (1.2) (i.e., with $\sigma(z) = 1/\pi$), like in previous sections. Applying $\partial_z\mathcal{D}(z)$ to the functional (4.2), that is F_1 in this case, we obtain a wrong answer for $\partial_z\Phi_1(z)$, which does not obey the loop equation (5.10)! This seemingly contradicts eqs. (5.12) and so explains why one has to deal with the arbitrary potential. The matter is simply that there are functionals such that they vanish for potentials with $\sigma(z) = 1/\pi$ but their "derivatives", $\partial_z\mathcal{D}(z)$, do not. They do contribute to Φ_1 and restore the right answer.

5.3 Free Energy of the General Model

Skipping further details, we present the results for the general model of normal matrices.

The answer for F_0 is familiar [5]. It is given by (3.1). The first correction, $F_{1/2}$, is

$$F_{1/2} = -\int_{\mathsf{D}} \sigma(z) \log \sqrt{\pi\sigma(z)}\, d^2 z \tag{5.13}$$

To write down the full answer for F_1 in a compact form, we need to introduce, along with the $\phi(w)$ (4.1), another function,

$$\chi(z) = \log \sqrt{\pi\sigma(z)} \tag{5.14}$$

and the function $\chi^H(z)$ defined in the domain D^{c}. It is a harmonic function in D^{c} with the boundary value $\chi(z)$. The function χ^H is the solution to the Dirichlet boundary problem: $\chi^H(z) = -\frac{1}{2\pi}\int_{\partial\mathsf{D}} \partial_n G(z,\xi)\chi(\xi)|d\xi|$. The explicit formula for F_1 reads:

$$F_1 = \frac{1}{24\pi}\left[\int_{|w|>1} |\nabla(\phi+\chi)|^2\, d^2 w - 2\oint_{|w|=1} (\phi+\chi)\,|dw| - \int_{\mathsf{C}} |\nabla\chi|^2\, d^2 w\right] +$$
$$+\frac{1}{8\pi}\left[\int_{\mathsf{D}} |\nabla\chi|^2 d^2 z - \oint_{\partial\mathsf{D}} \chi\partial_n\chi^H\,|dz| - \frac{1}{2}\int_{\mathsf{D}} \Delta\chi d^2 z\right] \tag{5.15}$$

where χ in the first three integrals is treated as a function of w through $\chi = \chi(z(w))$.

The r.h.s. of this formula is naturally decomposed into two parts having completely different nature, the "quantum" and "classical" parts: $F_1 = F_1^{(\mathrm{q})} + F_1^{(\mathrm{cl})}$. The (most interesting) quantum part can be again represented through the regularized determinant of the Laplace operator in D^{c} with Dirichlet boundary conditions. However, now the Laplacian should be taken in conformal metric with the conformal factor $e^{2\chi(z)}$. Equivalently, on the exterior of the unit circle the Laplacian should be taken in the metric with the conformal factor $e^{2\phi(w)+2\chi(z(w))}$; we see that ϕ and χ do enter as the sum $\phi+\chi$ in the first line. More precisely, the formula for regularized determinants of Laplace operators in domains with boundary known in the literature (eq. (4.42) in [18]) allows us to identify

$$F_1^{(\mathrm{q})} = \frac{1}{2}\log\frac{\det\left(-e^{-2\chi}\Delta_{\mathsf{C}}\right)}{\det\left(-e^{-2\chi}\Delta_{\mathsf{D}^{\mathsf{c}}}\right)} \tag{5.16}$$

The classical part comes from "classical" (though of order $\hbar$) corrections to the shape of the support of eigenvalues, which always exist unless $\sigma(z)$ is a constant (see below). It is essentially given by $F_1^{(\mathrm{cl})} = \lim_{\hbar\to 0}\left(\frac{1}{2\hbar^2}\langle(\mathrm{tr}\,\chi(M))^2\rangle_{\mathrm{conn}}\right)$.

Different terms of the $\hbar$-expansion gain a clear interpretation in terms of the collective field theory of the normal matrix model, in the spirit of [22]. In this context, it is natural to start with the general Coulomb gas model with arbitrary β. The generalized loop equation

$$\frac{1}{2\pi}\int \frac{\partial W(\zeta)\,\langle \Delta\Phi(\zeta)\rangle}{z-\zeta}\,d^2\zeta = \beta\Big\langle (\partial\Phi(z))^2\Big\rangle + (2-\beta)\hbar\Big\langle \partial^2\Phi(z)\Big\rangle \tag{5.17}$$

can be understood as the conformal Ward identity for the collective theory. This allows one to find the effective action in the form

$$\begin{aligned} S &= S_0 + S_1 \\ S_0 &= \beta \int\!\!\int \rho(z)\log|z-\zeta|\rho(\zeta)\,d^2z d^2\zeta + \int W(z)\rho(z)\,d^2z \\ S_1 &= -\Big(1-\frac{\beta}{2}\Big)\hbar\int \rho(z)\log\rho(z)\,d^2z \end{aligned} \tag{5.18}$$

The second term, S_1, is a combination of the short range part $-\frac{\beta}{2}\rho\log\rho$ and the entropy $\rho\log\rho$. (See [23, 24], where similar actions for unitary and Hermitian matrix ensembles were discussed.)

This action suggests to rearrange the $\hbar$-expansion of the free energy (5.4) and write it in the "topological" form $F=\sum_{g\geq 0}\hbar^{2g}F_g$, where each term has its own expansion

$$F_g = F_g^{(0)} + \sum_{n\geq 1}\hbar_\beta^n F_g^{(n)} \tag{5.19}$$

Here $\hbar_\beta \equiv (2-\beta)\hbar$ is regarded as an independent parameter. The equilibrium density of charges, ρ_0, is determined by $\delta S/\delta\rho = 0$ which leads to the Liouville-like equation

$$-\frac{\hbar_\beta}{8\pi}\Delta\log\rho_0(z) + \beta\rho_0(z) = \sigma(z) \tag{5.20}$$

in the bulk. For $\beta \neq 2$ the first term generates corrections to the shape of the support of eigenvalues. The classical free energy is $F_0 = F_0^{(0)} + \hbar_\beta F_0^{(1)} + \hbar_\beta^2 F_0^{(2)} + O(\hbar_\beta^3)$. In particular we see that $F_{1/2}$ given in (5.13) is in fact $F_0^{(1)}$ while the "classical" part $F_1^{(\mathrm{cl})}$ of (5.15) is $F_0^{(2)}$. The "quantum" part is then $F_1^{(\mathrm{q})} = F_1^{(0)}$.

The collective field theory approach to the normal and complex matrix ensembles will be presented elsewhere.

Acknowledgments

We thank A.Cappelli, L.Chekhov, B.Dubrovin, V.Kazakov, I.Kostov, Yu.Makeenko, A.Marshakov, M.Mineev-Weinstein and R.Theodorescu for useful discussions. P.W. would like to thank INFN and the University of Florence, Italy, for hospitality. A.Z. is grateful to B.Julia for opportunity to present these results at the Les Houches Spring School in March 2003. P. W. was supported by the NSF MRSEC Program under DMR-0213745, NSF DMR-0220198 and Humboldt foundation. The work of A.Z. was supported in part by RFBR grant 03-02-17373, by grant for support of scientific schools NSh-1999.2003.2 and by Federal Program of the Russian ministry of industry, science and technology 40.052.1.1.1112. The work of both authors was partially supported by the NATO grant PST.CLG.978817.

References

1. P.J.Forrester, N.C.Snaith and J.J.M.Verbaarschot, *Developments in random matrix theory*, J. Phys. A: Math. Gen. **36** (2003) R1-R10, an introductory review for the special issue "Random Matrix Theory"
2. M.L.Mehta, *Random matrices*, 2nd edition, Academic Press, NY, 1991
3. L.-L.Chau and Y.Yu, Phys. Lett. **167A** (1992) 452; L.-L.Chau and O.Zaboronsky, Commun. Math. Phys. **196** (1998) 203-247, e-print archive: hep-th/9711091
4. J.Ginibre, J. Math. Phys. **6** (1965) 440; V.Girko, Theor. Prob. Appl. **29** (1985) 694
5. P.Wiegmann and A.Zabrodin, J. Phys. A: Math. Gen. **36** (2003) 3411-3424; e-print archive: hep-th/0210159; A.Zabrodin, *New applications of non-hermitian random matrices*, to be published, e-print archive: cond-mat/0210331
6. S.Alexandrov, V.Kazakov and I.Kostov, *2D string theory as normal matrix model*, e-print archive: hep-th/0302106
7. I.Kostov, I.Krichever, M.Mineev-Weinstein, P.Wiegmann and A.Zabrodin, *τ-function for analytic curves*, in: Random Matrices and Their Applications (MSRI publications, vol. 40), ed. P.Bleher and A.Its (Cambridge: Canbridge Academic Press), 285-299; e-print archive: hep-th/0005259
8. J.Ambjorn, L.Chekhov, C.F.Kristjansen and Yu.Makeenko, Nucl. Phys. **B404** (1993) 127-172; Erratum: ibid. **B449** (1995) 681, e-print archive: hep-th/9302014
9. I.Kostov, *Conformal Field Theory Techniques in Random Matrix models*, e-print archive: hep-th/9907060
10. B. Eynard, *Large N expansion of the 2-matrix model*, JHEP 0301 (2003) 051, e-print archive: hep-th/0210047; B.Eynard, *Large N expansion of the 2-matrix model, multicut case*, e-print archive: hep-th/0307052
11. E.Brezin, C.Itzykson, G.Parisi and J.B.Zuber, Commun. Math. Phys. **59** (1978) 35-51
12. M.Bertola, *Free energy of the two-matrix model/dToda tau-function*, e-print archive: hep-th/0306184
13. A.Alastuey and B.Jancovici, J. Stat. Phys. **34** (1984) 557; B.Jancovici, J. Stat. Phys. **80** (1995) 445; P.J.Forrester, Phys. Rep. **301** (1998) 235-270
14. M.Mineev-Weinstein, P.Wiegmann and A.Zabrodin, Phys. Rev. Lett. **84** (2000) 5106-5109, e-print archive: nlin.SI/0001007; P.Wiegmann and A.Zabrodin, Commun. Math. Phys. **213** (2000) 523-538, e-print archive: hep-th/9909147

15. A.Marshakov, P.Wiegmann and A.Zabrodin, Commun. Math. Phys. **227** (2002) 131-153; e-print archive: hep-th/0109048
16. A.Boyarsky, A.Marshakov, O.Ruchayskiy, P.Wiegmann and A.Zabrodin, Phys. Lett. **B515** (2001) 483-492; e-print archive: hep-th/0105260
17. A.Polyakov, Phys. Lett. **B103** (1981) 207-210
18. O.Alvarez, Nucl. Phys. **B216** (1983) 125-184; P.Di Vecchia, B.Durhuus, P.Olesen and J.Petersen, Nucl.Phys. **B207** (1982) 77; J.Ambjorn, B.Durhuus, J.Frölich and P.Orland, Nucl.Phys. **B270** (1986) 457
19. B.Osgood, R.Phillips and P.Sarnak, J. Func. Anal. **80** (1988) 148-211
20. P. Di Francesko, M.Gaudin, C.Itzykson and F.Lesage, Int. J. Mod. Phys. **A9** (1994) 4257-4351
21. B.Dubrovin and Y.Zhang, Commun. Math. Phys. **198** (1998) 311-361, e-print archive: hep-th/9712232
22. A.Jevicki and B.Sakita, Nucl. Phys. **B165** (1980) 511; A.Jevicki and B.Sakita, Nucl. Phys. **B185** (1981) 89; A.Jevicki, *Collective field theory and Schwinger-Dyson equations in matrix models*, preprint Brown-HET-777, Proceedings of the meeting "Symmetries, quarks and strings" held at the City College of New York, Oct. 1-2, 1990
23. F.Dyson, *Statistical theory of the energy levels of complex systems*, part II, J. Math. Phys. **3** (1962) 157
24. O.Lechtenfeld, *Semiclassical approach to finite-N matrix models*, e-print archive: hep-th/9112045; S.Ben-Menahem, $D = 0$ *matrix model as conjugate field theory*, preprint SLAC-PUB-5377, February 1992

Symmetries Arising from Free Probability Theory

Dan Voiculescu

Department of Mathematics
University of California at Berkeley
Berkeley, CA 94720-3840
dvv@math.berkeley.edu

1 Introduction

We discuss two types of symmetry structures.

On the large N limit of a system of independent Gaussian random matrices there is a Lie algebra action, which is the free analogue of the volume-preserving vector fields for a Gaussian measure [7]. The second symmetry is the coalgebra of the free difference-quotient and its remarkable duality feature [6]. It provides the natural framework to analyze an operator with respect to an algebra of operators playing the role of scalars, when these scalars and the operator are in the most noncommutative situation. Note that in combinatorics, for classical (commutative) one-variable polynomials the corresponding coalgebra structure was studied by G.C. Rota, apparently without realizing the selfduality.

We also included background comments on free probability and the connection of random matrices to the von Neumann algebras of free groups (see [5], [8] and references given there). Based on the free difference-quotient coalgebra and on examples ([1], [2]) we emphasize the role of generalized resolvents in the study of random matrices in the free probability approach.

We have resisted the temptation to include in these notes the $U(n,1)$ action on the full Fock space, which is paired with the Lie superalgebra action of $gl(n|1)$ and which appeared in our first free probability paper ([3]), but has not been sufficiently clarified.

No operator algebra knowledge is presumed in this informal presentation, nor in the background material [5] and [8].

Acknowledgements

Most of the work on these notes was done while the author held an International Blaise Pascal Research Chair from the State and Ile de France Region, managed by the Fondation de l'École Normale Supérieure and visiting the Institut de Mathématiques de Jussieu. He was also supported in part by NSF Grant DMS-0079945.

2 Free Probability

Free probability theory is noncommutative probability theory plus free independence.

The basic object in noncommutative probability theory is an algebra A over the complex numbers $\mathbb{C}$, with unit $1 \in A$ and a linear map $\varphi : A \to \mathbb{C}$, so that $\varphi(1) = 1$. The elements $a \in A$ will be called noncommutative random variables and $\varphi(a)$ is the expectation of a (This is like the observables of quantum physics).

If $\alpha = (a_i)_{i \in I} \subset A$ is a family of noncommutative random variables, their joint distribution will be described by the collection of noncommutative moments $\varphi(a_{i_1} \dots a_{i_p})$. A better way to organize the information contained in these numbers is to consider the polynomials in noncommutative indeterminate indexed by I, $\mathbb{C}\langle X_i | i \in I\rangle$ and the linear maps $\Psi_\alpha : \mathbb{C}\langle X_i | i \in I\rangle \to \mathbb{C}$ defined by

$$\Psi_\alpha(P(X_i | i \in I)) = \varphi(P(a_i | i \in I)) \,.$$

In the case of a single random variable $a \in A$, if its moments $\varphi(a^n)$ are equal to the moments of a probability measure μ on $\mathbb{R}$, we will also call μ the distribution of α.

What distinguishes free probability theory, is the definition of independence. A family $(A_i)_{i \in I} \subset A, 1 \in A_i$ of subalgebras in A is **freely independent** if

$$\varphi(a_1 \dots a_n) = 0$$

whenever $\varphi(a_j) = 0,\ 1 \leq j \leq n$ and $a_j \in A_{i_j}, 1 \leq j \leq n,\ i_j \neq i_{j+1}$.

Note that in the above definition, it is only required that consecutive a_j's be in different A_{i_j}'s (for instance it is possible that $i_1 = i_3$ as long as this index $\neq i_2$). The free independence requirement reminds of the conditions defining a nontrivial reduced word in a free product of groups (from which the condition was derived).

Sets of noncommutative random variables $(\omega_i)_{i \in I}$, $\omega_i \in A$ are freely independent if the algebras A_i generated by $\{1\} \cup \omega_i$ are freely independent. Like in the case of classical independence, knowledge of the joint distribution for the variables in each ω_i, together with the free independence of the ω_i's completely determines the joint distribution of all variables in $\omega = \cup_{i \in I} \omega_i$.

Starting from this input, a free probability, parallel to a quite large part of the basic classical probability theory, expressible in terms of expectations of numerical random variables, has emerged.

A salient feature of this parallelism is the correspondence of basic laws. The free analogues of the Gauss, Poisson and Cauchy laws are, respectively, the Wigner semicircle law, the Pastur–Marchenko distribution and the same Cauchy law (a fixed point). This means for instance, that in the central limit theorem for freely independent variables, the limit distribution is the semicircle law.

An inventory of the free parallel includes much more.

Addition and multiplication of free random variables yield nonlinear convolution operations. Free convolutions, both additive and multiplicative, can be computed using the linearizing R and S transformations (which perform the function of the logarithm of the Fourier transform and respectively of the Mellin transform). With this machinery, freely infinitely divisible laws and free stable laws have been classified. These distributions yield noncommutative processes with free increments (additive or multiplicative) the Markovian properties of which have been studied.

A more recent entry is free entropy. For an n-tuple of noncommutative variables this provides the analogue of the Shannon entropy of continuous variables (differential entropy). For one variable, the classical quantity

$$-\int_{\mathbb{R}} p(t) \log p(t) \mathrm{d}t$$

($p(t)$ the density of the distribution) is replaced in the free context by

$$c + \int\int \log |s-t| p(s)p(t) \mathrm{d}s\mathrm{d}t$$

which up to a constant c is the negative of the logarithmic energy of the distribution. The general n-variable case is a long story, which cannot be compressed in a few lines (see the survey [8]).

There is also a combinatorial side to free probability discovered by R. Speicher: it can be seen as replacing in the calculus of cumulants the lattice of all partitions of $\{1,\dots,n\}$ by the lattice of non-crossing partitions of $\{1,\dots,n\}$. Non-crossing means there are no $a < b < c < d$ so that $\{a,c\}$ and $\{b,d\}$ are contained in two different sets of the partition. The non-crossing partitions correspond to planar diagrams in physics.

Where does free independence occur? In general, it is the independence relation for quantities with the highest degree of noncommutativity and there are three basic free probability contexts

- convolution operators on free products of groups
- creation and annihilation operators on the full Fock space
- the large N limit of random matrices

The first two models are actually not so different, since the Boltzmann full Fock space has a basis indexed by a free semigroup, i.e. we are dealing with a free product of semigroups.

From the perspective of quantum statistics embodied by the 3 Fock spaces (bosonic, fermionic and full) free probability may be viewed as mathematics corresponding to the full Boltzmann statistic. It is often said that the full Fock space is somewhat "unphysical", there is however no reason to consider it as "unmathematical".

To conclude this section, I would like to mention a few contributors to this area (references to their work and further names are in my St. Flour notes [5] and in the survey article [8]: P. Biane, R. Speicher, D. Shlyakhtenko, K. Dykema, A. Nica, U. Haagerup, L.Ge, F. Radulescu, H. Bercovici, A. Guionnet, T. Cabanal-Duvillard, M. Anshelevich. Also, via random matrices there have been connections to physics models (papers by I.M. Singer, M. Douglas, D. Gross, R. Gopakumar, P. Zinn-Justin, S.G. Rajeev and others).

3 Operator Algebras

Our noncommutative probability framework (M,τ) will have some extra features. $M \subset B(\mathcal{H})$ will be a selfadjoint algebra of bounded operators on a Hilbert space, i. e. $I \in M$, $T \in M \Rightarrow T^* \in M$ and if $T_1, T_2 \in M$, $\lambda \in \mathbb{C}$, then $T_1+T_2 \in M$, $T_1T_2 \in M$, $\lambda T_1 \in M$. The expectation functional $\tau : M \to \mathbb{C}$ will be a trace-state, which in our case means there is a unit vector $h \in \mathcal{H}$, $||h|| = 1$, so that $\tau(T) = (Th,h)$ and the trace condition $\tau(T_1T_2) = \tau(T_2T_1)$ is satisfied if $T_1, T_2 \in M$.

M is a von Neumann algebra if we additionally require M to be weakly closed, i.e. for any net $(T_i)_{i\in I} \in M$ and $T \in B(\mathcal{H})$, so that if (T_ih,k) converges to (Th,k) for all $h,k \in \mathcal{H}$, then $T \in M$. If $T = T^* \in M$ and M is a von Neumann algebra, then $f(T) \in M$ if $f : \mathbb{R} \to \mathbb{R}$ is a Borel function.

Example 1. Let G be a group, $l^2(G)$ the Hilbert space with orthonormal basis $(e_g)_{g\in G}$ and λ the left regular representation of G on $l^2(G)$ given by

$\lambda(g)e_h = e_{gh}$. The linear span of $\lambda(G)$ is a selfadjoint algebra of operators on $l^2(G)$ and its weak closure will be denoted by $L(G)$. On $L(G)$ there is the von Neumann trace-state

$$\tau(\sum_{g\in G} c_g\lambda(g)) = c_e = (\sum_{g\in G} c_g\lambda(g)e_e, e_e) .$$

Example 2. Let $(T_1(N),\dots,T_n(N))$ be an n-tuple of selfadjoint $N\times N$ matrices with joint distribution the probability measure $\mu_N \in \mathrm{Prob}(\mathfrak{M}_{sa}(N))^n$, where $\mathfrak{M}_{sa}(N)$ denotes the selfadjoint $N\times N$ matrices. Provided the limits make sense (and some boundedness is assumed) there is (M,τ) generated by $T_j = T_j^*, i\le j\le n$ so that

$$\lim_{N\to\infty} N^{-1}\mathrm{ETr}(T_{i_1}(N)\dots T_{i_p}(N)) = \tau(T_{i_1}\cdots T_{i_p})$$

for all $i\le i_1,\dots,i_p\le n, p\in\mathbb{N}$ (here E is the expectation).

In essence this is an instance of the Gelfand–Naimark–Segal construction, which roughly amounts to the following. The noncommutative polynomials $\mathbb{C}\langle X_1,\dots,X_n\rangle$ can be endowed with an involution so that

$$(cX_{i_1}\dots X_{i_p})^* = \bar{c}X_{i_p}\dots X_{i_1}$$

and one defines an inner product

$$(P,Q) = \lim_{N\to\infty} N^{-1}\mathrm{ETr}(P(T_1(N),\dots,T_n(N))Q^*(T_1(N),\dots,T_n(N))) .$$

Let $\mathcal{H}$ be the Hilbert space obtained from $\mathbb{C}\langle X_1,\dots,X_n\rangle$. The left action of the ring $\mathbb{C}\langle X_1,\dots,X_n\rangle$ on itself turns $\mathbb{C}\langle X_1,\dots,X_n\rangle$ into an algebra of operators on $\mathcal{H}$, the weak closure of which will be M. The X_j's give rise to the operators T_j and $\tau(\cdot) = (\cdot 1, 1)$.

4 Two Derivatives

Let (M,τ) be a von Neumann algebra with trace-state τ and $T_j = T_j^*, 1\le j\le n$, a n-tuple of elements generating M. Throughout the rest of this paper we will assume $T_1,\dots,T_n$ do not satisfy any algebraic relation, i.e. there is no $P\in\mathbb{C}\langle X_1,\dots,X_n\rangle, P\ne 0$ so that $P(T_1,\dots,T_n)=0$.

To define the derivatives, it will be convenient to deal with a slightly more general situation. We replace $\mathbb{C}\langle T_1,\dots,T_n\rangle$ by $B\langle X\rangle$, the polynomials in X with "noncommutative coefficients in B", i.e. the monomials are $b_0Xb_1X\dots b_n$ and the only relation between the coefficients B and X is $X = 1X = X1$. Note that if B is the algebra generated by $1, T_1,\dots,T_{j-1},T_{j+1},\dots,T_n$ and $X = T_j$ then $\mathbb{C}\langle T_1,\dots,T_n\rangle = B\langle X\rangle$.

The **free difference-quotient**

$$\partial_{X:B} : B\langle X\rangle \to B\langle X\rangle \otimes B\langle X\rangle$$

is defined by

$$\partial_{X:B} b_0 X b_1 X \dots b_n = \sum_{1\le j\le n} b_0 X \dots b_{j-1} \otimes b_j X \dots b_n \, .$$

This means: for every X in $b_0Xb_1X\dots b_n$ there is a term in the formula where that X has been replaced by $\otimes$.

To state the characteristic property of $\partial_{X:B}$, let us return to $B\langle X\rangle \subset M$. If $K \in M$, let $m_K : B\langle X\rangle \otimes B\langle X\rangle \to M$ be the linear map so that $m_K(P_1 \otimes P_2) = P_1 K P_2$. If $P \in B\langle X\rangle$, we have

$$\frac{\mathrm{d}}{\mathrm{d}\varepsilon} P(X + \varepsilon K)|_{\varepsilon=0} = m_K(\partial_{X:B} P)$$

thus $\partial_{X:B} P$ can be viewed as the differential at X of the map $M \to M$ which takes $m \in M$ to $P(m) \in M$.

The other derivative we shall consider is the **cyclic derivative** of Rota–Sagan–Stein:

$$\delta_{X:B} : B\langle X\rangle \to B\langle X\rangle$$

which is given by

$$\delta_{X:B} b_0 X b_1 X \dots b_n = \sum_{1\le j\le n} b_j X \dots b_n b_0 X \dots b_{j-1} \, .$$

It can also be described as being the free difference-quotient $\partial_{X:B}$ followed by multiplication after the two factors in the tensor product have been permuted.

The characteristic property of $\delta_{X:B}$ is that

$$\frac{\mathrm{d}}{\mathrm{d}\varepsilon} \tau(P(X + \varepsilon K))|_{\varepsilon=0} = \tau((\delta_{X:B} P) K)$$

where $P \in B\langle X\rangle, K \in M$. This means $\delta_{X:B}$ is the gradient of the map

$$M \ni m \to \tau(P(m)) \in \mathbb{C} \, .$$

In case $B = \mathbb{C}\langle T_1, \dots, T_{j-1}, T_{j+1}, \dots, T_n\rangle$ and $X = T_j$, we denote $\partial_{X:B}$ by ∂_j and $\delta_{X:B}$ respectively by δ_j and we refer to them as the partial free-difference quotient and respectively the partial cyclic derivative w.r.t. T_j.

In case $B = \mathbb{C}$, $B\langle X\rangle = \mathbb{C}[X]$ (commutative polynomials) and $\partial_{X:\mathbb{C}}$ can be identified with the classical difference quotient

$$P \to \frac{P(X) - P(Y)}{X - Y}$$

after identifying $\mathbb{C}[X]\otimes\mathbb{C}[X]$ with $\mathbb{C}[X,Y]$. Also $\delta_{X:\mathbb{C}}$ identifies with the usual derivative $P \to P'$.

5 Infinitesimal Automorphisms of M

An automorphism of M is a linear bijection $\alpha : M \to M$ so that $\alpha(x^*) = \alpha(x)^*$, $\alpha(xy) = \alpha(x)\alpha(y)$, $\alpha(1) = 1$, $\tau \circ \alpha = \tau$. Loosely speaking, the infinitesimal version of these will be derivations D_K which are densely defined and which, under suitable circumstances will exponentiate to one-parameter groups of automorphisms. Here $K = (K_1, \ldots, K_n) \in (M_{sa})^n$, M is generated by $T_1, \ldots, T_n$ and D_K is defined on $\mathbb{C}\langle T_1, \ldots, T_n\rangle$ by

$$D_K P = \sum_{1 \le j \le n} m_{K_j} \partial_j P$$

(or equivalently)

$$= \frac{\mathrm{d}}{\mathrm{d}\varepsilon} P(T_1 + \varepsilon K_1, \ldots, T_n + \varepsilon K_n)|_{\varepsilon=0} \,.$$

The key condition on D_K is to preserve the trace-state

$$\tau(D_K P) = 0$$

which is

$$0 = \frac{\mathrm{d}}{\mathrm{d}\varepsilon} \tau(P(T_1 + \varepsilon K_1, \ldots, T_n + \varepsilon K_n))|_{\varepsilon=0} = \sum_{1 \le j \le n} \tau((\delta_j P) K_j) = 0 \,.$$

This means

$$K \in (\delta \mathbb{C}\langle T_1, \ldots, T_n\rangle)^{\perp}$$

where δ denotes the cyclic gradient

$$\delta P = (\delta_1 P, \ldots, \delta_n P)$$

and the orthogonal $\perp$ is with respect to the scalar product

$$((X_1, \ldots, X_n), (Y_1, \ldots, Y_n)) = \sum_{1 \le j \le n} \tau(X_j Y_j)$$

on $(M_{sa})^n$.

The requirement on D_K to be the generator of a one-parameter group of automorphisms involves analytic vectors. This appears to be related to the K_j's being "sufficiently analytic" functions of the T_j's. A **sufficient condition** is

$$K_j \in \mathbb{C}\langle T_1, \ldots, T_n\rangle, \quad 1 \le j \le n \,.$$

The derivations D_K are best understood in the framework of the **Lie algebra of noncommutative vector fields**

$$\mathrm{Vect}\, \mathbb{C}\langle T_1, \ldots, T_n\rangle = (\mathbb{C}\langle T_1, \ldots, T_n\rangle_{sa})^n$$

with the bracket

$$[P,Q] = (D_P Q_j - D_Q P_j)_{1 \le j \le n}$$

(the complexification of which means dropping the selfadjointness requirement on the components). The τ-preserving noncommutative vector fields form a Lie subalgebra

$$\text{Vect } \mathbb{C}\langle T_1, \dots, T_n | \tau \rangle = \text{Vect } \mathbb{C}\langle T_1, \dots, T_n \rangle \cap (\delta \mathbb{C}\langle T_1, \dots, T_n \rangle_{sa})^{\perp} .$$

In general, the orthogonal of the cyclic gradients may not contain a dense set of polynomial elements. Therefore, we will say **that $(T_1, \dots, T_n)$ is regular (or equivalently that Vect $\mathbb{C}\langle T_1, \dots, T_n | \tau \rangle$ is infinitesimally rich) if**

$$\delta \mathbb{C}\langle T_1, \dots, T_n \rangle_{sa} + \text{Vect } \mathbb{C}\langle T_1, \dots, T_n | \tau \rangle$$

is dense in Vect $\mathbb{C}\langle T_1, \dots, T_n \rangle$ w.r.t. the Hilbert space norm corresponding to $((X_j)_{1 \le j \le n}, (Y_j)_{1 \le j \le n}) = \sum_{1 \le j \le n} \tau(X_j Y_j)$.

Since $T_1, \dots, T_n$ generate M, the map

$$\text{Aut}(M) \ni \alpha \to (\alpha(T_1), \dots, \alpha(T_n)) \in (M_{sa})^n$$

is a bijection of Aut(M) onto the automorphic orbit of $(T_1, \dots, T_n)$ and regularity means that the cyclic gradients (which are normal to the orbit) and Vect $\mathbb{C}\langle T_1, \dots, T_n | \tau \rangle$ (which are tangent to the orbit) span a dense set, in particular that Vect $\mathbb{C}\langle T_1, \dots, T_n | \tau \rangle$ be dense in the tangent space to the orbit at $(T_1, \dots, T_n)$. Of course, since the orbit is not really a manifold, this is only a rough picture.

6 Symmetries of the Large N Limit of the n-tuple of Gaussian Random Matrices

It seems that the best conceptual description of the large N limit of a n-tuple of $N \times N$ hermitian random matrices is in the form outlined in Example 2 of Section 2, i.e. a von Neumann algebra with trace-state (M, τ) and a generating n-tuple $(T_1, \dots, T_n)$ of selfadjoint elements. (The unstructured collection of noncommutative moments is a rather non-illuminating presentation of the same information).

Let $T_j(N), 1 \le j \le n$, be independent hermitian random matrices with real and imaginary parts of the entries i.i.d. $(0, 1/N)$ Gaussian (the entries not bound by the hermiticity requirement). **We shall refer to this n-tuple as the n-tuple of Gaussian random matrices.**

Fact. a) The large N limit of the Gaussian n-tuple of random matrices $T_j(N), 1 \le j \le n$ can be realized by the operators $S_j = l_j + l_j^*$ $(1 \le j \le n)$ where $l_j \xi = e_j \otimes \xi$ are left creation operators on the full Fock space

$$\mathcal{T}(\mathbb{C}^n) = \mathbb{C}1 \oplus \bigoplus_{k\geq 1} (\mathbb{C}^n)^{\otimes k}$$

($e_j, 1 \leq j \leq n$ being the canonical basis of $\mathbb{C}^n$) and $\tau(T) = (T1, 1)$.

b) The von Neumann algebra generated by $S_1, \ldots, S_N$ is isomorphic to $L(F_n)$ where F_n is the free group on generators $g_1, \ldots, g_n$. the isomorphism can be given by putting $\lambda(g_j)$ in correspondence with $f(S_j)$ where $f : \mathbb{R} \to \{z \in \mathbb{C} \mid |z| = 1\}$ is a Borel function, injective on $[-2, 2]$, so that $f_*\mu =$ *Haar measure*, where μ is the semicircular measure with density $\frac{1}{\pi}\sqrt{4 - t^2}\chi_{[-2,2]}$.

Note that the vacuum expectation $(\cdot 1, 1)$ is a pure state on the algebra of all operators on $\mathcal{T}\mathbb{C}^n$, however, it is a trace-state on the algebra of $S_1, \ldots, S_n$. The commutant of this algebra is generated by the operators $D_j = r_j + r_j^*$, where $r_j\xi = \xi \otimes e_j$ are the right creation operators. In free probability theory $S_1, \ldots, S_n$ is called a semicircular system and it is the free analogue of a system of n i.i.d. centered Gaussian random variables

The construction of infinitesimal automorphisms of the von Neumann algebra $L(F_n)$ using cyclic gradients w.r.t the semicircular generator $S_1, \ldots, S_n$ enjoys very good properties and can be given in explicit form.

Fact. a) We have

$$\text{Vect } \mathbb{C}\langle S_1, \ldots, S_n | \tau\rangle + \delta\mathbb{C}\langle T_1, \ldots, T_n\rangle_{sa} = \text{Vect } \mathbb{C}\langle S_1, \ldots, S_n\rangle \,.$$

In particular Vect $\mathbb{C}\langle S_1, \ldots, S_n|\tau\rangle$ **is infinitesimally rich and defines derivations which exponentiate to one-parameter groups of automorphisms.**

b) The n-tuples of elements

$$F_I = (\delta_{i_0,j} P_{k_0-1}(S_{i_0}) P_{k_1}(S_{i_1}) \ldots P_{k_p}(S_{i_p}) - \delta_{i_p j} P_0(S_{i_0}) \ldots P_{k_p-1}(S_{i_p}))_{1\leq j\leq n}$$

where I runs over systems of indices

$$I = (\underbrace{i_0, \ldots, i_0}_{k_0 times}, \ldots, \underbrace{i_p, \ldots, i_p}_{k_p times}) \;\; k_r > 0 \;\; (0 \leq r \leq p), \sum_{0\leq r\leq p} k_r \geq 2, i_r \neq i_{r+1}$$

span the complexification of Vect $\mathbb{C}\langle S_1, \ldots, S_n|\tau\rangle$. **Here the P_j's are the Chebyshev polynomials which are the orthogonal polynomials for the semicircular measure with density $\frac{1}{\pi}\sqrt{4 - t^2}\chi_{[-2,2]}$.**

7 The von Neumann Algebras of Free Groups

The connection with the large N limit of Gaussian random matrices has been the key to several other important applications of free probability to the study of the $L(F_n)$. This section contains some brief speculations about the interest in these von Neumann algebras.

Infinite-dimensional von Neumann algebras (M, τ) with scalar center $Z(M) = \mathbb{C}I$, are called II_1-factors. The orthogonal projection operators $P = P^* = P^2 \in M$ determine a wonderful infinite-dimensional geometry of linear subspaces, with dimensions $\tau(P) \in [0, 1]$ discovered by John von Neumann.

The most fundamental II_1-factor is the so-called hyperfinite or injective II_1 factor (it is also the smallest). By a deep theorem of A. Connes all $L(G)$, where G is a countable amenable group with infinite conjugacy classes, are isomorphic to the hyperfinite II_1-factor. The free group factors, $L(F_n)$ are not isomorphic to the hyperfinite II_1-factor and for several reasons, which have emerged from free probability, they seem to represent, after the hyperfinite, the next important class of II_1-factors.

One source of interest in the $L(F_n)$ is the following conjecture about large N limits of random multimatrix models.

"(Not too un-)Reasonable Guess": the von Neumann algebras arising from large N limits of random multimatrix models given by densities

$$c_N \exp(-N \mathrm{Tr} P(A_1, \ldots, A_N)) \mathrm{d}A_1 \ldots \mathrm{d}A_n$$

on hermitian matrices are in the $L(F_n)$ family.

8 Duality for the Coalgebra of the Free Difference-Quotient

We go back to the general situation of $B\langle X\rangle \subset (M, \tau)$ considered in Sect. 4.

The free difference-quotient satisfies the coassociativity equation:

$$(\partial_{X:B} \otimes id) \circ \partial_{X:B} = (\mathrm{id} \otimes \partial_{X:B}) \circ \partial_{X:B}$$

This makes $(B\langle X\rangle, \partial_{X:B})$ a coalgebra and moreover $\partial_{X:B}$ is a derivation of $B\langle X\rangle$ into the $B\langle X\rangle$-bimodule $B\langle X\rangle \otimes B\langle X\rangle$. This can be written

$$\partial \circ \mu = (\mathrm{id} \otimes \mu) \circ (\partial \otimes \mathrm{id}) + (\mu \otimes \mathrm{id}) \circ (\mathrm{id} \otimes \partial) \qquad (\star)$$

In our functional analysis context, under good conditions one can pass to a closure of the unbounded operator $\partial_{X:B}$ and this closure will be defined also for inverses of elements of $B\langle X\rangle$ or matrix elements in $\mathfrak{M}_n(B\langle X\rangle)$. Having all these inverses at hand will make things much nicer, as will soon be visible. First let us emphasize a remarkable feature of the structure.

Selfduality: if (A, μ, ∂) is an algebra (A, μ) and ∂ is a comultiplication which is also a derivation, then the same also holds for the dual $(A^\bullet, \partial^\bullet, \mu^\bullet)$ (where $A^\bullet$ denotes the dual vector space etc.)

This is quite easy to see, taking the dual of the derivation relation ($\star$), we get:

$$\mu^\bullet \circ \partial^\bullet = (\partial^\bullet \otimes \mathrm{id}) \circ (\mathrm{id} \otimes \mu^\bullet) + (\mathrm{id} \otimes \partial^\bullet) \circ (\mu^\bullet \otimes \mathrm{id})$$

which coresponds to replacing (μ, ∂) by $(\partial^\bullet, \mu^\bullet)$.

The significance of duality becomes clearer in the functional analysis context when A **contains matrix elements of matricial B-resolvents**

$$((b_{ij})_{1\le j\le n} - (X\delta_{ij})_{1\le j\le n})^{-1}$$

What happens is that, **roughly speaking, under some invertibility conditions, the corepresentations, i.e. $\beta \in \mathfrak{M}_n(A)$ satisfying**

$$(\mathrm{id}_{\mathfrak{M}_n} \otimes \partial_{X:B})\beta = \beta \otimes_{\mathfrak{M}_n} \beta$$

turn out to be the matricial B-resolvents. Very roughly, the dual object should then be obtained via a map

$$A^\bullet \ni \varphi \to \bigoplus_{\beta \text{ corepresentation}} (\mathrm{id}_{\mathfrak{M}_n} \otimes \varphi)(\beta) \in \bigoplus_{\beta} \mathfrak{M}_{\dim\beta} \,.$$

Simplest Example of Duality: Let μ be a compactly supported probability measure on $\mathbb{R}$, $M = L^\infty(\mathbb{R}, d\mu), \tau = \mu$ and $A \subset M$, the algebra of rational functions with poles outside supp μ. In this 1-dimensional situation, it suffices to restrict the consideration of matricial resolvents to 1×1 matrices.

The map into the dual object of $L^1(\mathbb{R}, d\mu) \subset A^\bullet$ is then given by the Cauchy transform

$$L^1(\mathbb{R}, \mathrm{d}\mu) \ni f \to C(f)$$

where $C(f)(z) = \int f(t)(z-t)^{-1}\mathrm{d}t$ is defined on $\mathbb{C} \setminus \mathrm{supp}\,\mu$. The multiplication $\#$ in the dual (when defined) is such that

$$C(f_1 \# f_2)(z) = C(f_1)(z)C(f_2)(z)$$

and the comultiplication derivation corresponds to the difference-quotient

$$\frac{C(f)(z_1) - C(f)(z_2)}{z_1 - z_2}$$

Thus the dual object will be an algebra of analytic functions on $(\mathbb{C} \cup \{\infty\}) \setminus$ supp μ vanishing at ∞, endowed with the usual multiplication and the comultiplication defined by the difference quotient. Note that on this road there are delicate analytic aspects, with the flavor of analytic capacity theory.

For the general version of this duality, involving matricial B-resolvents, see [9].

9 Matricial B – Resolvents

The initial application of the coalgebra of $\partial_{X:B}$ in free probability has been to analytic subordination related to free Markovianity. One of the simplest examples ($B = \mathbb{C}$) is provided by a pair of hermitian random matrices $T(N), D(N)$ where $D(N)$ is diagonal deterministic and $T(N)$ has a distribution which is rotation invariant. Assuming $T(N)$ and $T(N) + D(N)$ have limit distribution of eigenvalues μ and ν probability measures on $\mathbb{R}$, let $G_\mu(z), G_\nu(z)$ be their Cauchy transforms. We then have analytic subordination $G_\mu \circ u = G_\nu$, where u is an analytic function mapping the upper half-plane to itself and ∞ to ∞. This has analytic consequences on the smoothness of ν given μ. We had obtained some results in this direction using free convolution, which were then shown by Biane using also some combinatorics, to be the consequence of a subordination result for resolvents and which, then in turn, we generalized and simplified showing that certain conditional expectations become homomorphisms for the dual multiplication of a difference-quotient (see the references in [6]).

Actually the matricial B-resolvents appear as a key ingredient in analyzing a selfadjoint operator w.r.t. totally noncommuting "operator-scalars". Also, the free probability machinery has a generalization to the B-valued context. We would like to conclude this discussion by mentioning some important random matrix work related to free probability where the use of these resolvents has been essential.

Fact (Haagerup - Thorbjoernsen, [1]). **If $T_1(N), \dots, T_n(N)$ is the n-tuple of i.i.d. Gaussian random matrices, then almost surely**

$$\lim_{N\to\infty} ||P(T_1(N), \dots, T_n(N))|| = ||P(S_1, \dots, S_n)||$$

where P is a noncommutative polynomial and $S_1, \dots, S_n$ is the semicircular n-tuple.

This result appears as a sweeping generalization of the largest eigenvalue results for one Gaussian matrix.

The other instance is provided by **D.Shlyakhtenko's work [2] on the limit distribution of eigenvalues of so-called generalized Gaussian random band matrices**

$$T(N) = (g_{mn} K(\frac{m}{N}, \frac{n}{N}))_{1 \le m,n \le N}$$

here $(g_{mn})_{1\le m,n\le N}$ is an i.i.d. Gaussian hermitian random matrix and $K : [0,1]^2 \to \mathbb{C}$ is a continuous hermitian kernel. In this case, B is a commutative algebra arising as the limit of diagonal matrices.

In both cases the final result no longer involves the generalized resolvents, but these are essential on the way to that result.

References

[1] Haagerup, U., and Thorbjornsen, S.: A new application of random matrices: $Ext(C_r^*(F_2))$ is not a group (preprint 2002).

[2] Shlyakhtenko, D.:Random Gaussian band matrices and freeness with amalgamation. International Math. Res. Notices, **20**, 1013–1025 (1996)

[3] Voiculescu, D.: Symmetries of some reduced free product C^*-algebras. In: Operator Algebras and their Connections with Topology and Ergodic Theory, Lecture Notes in Math., 1132. Springer, Berlin Heildelberg New York (1985)

[4] Voiculescu, D.:Operations on certain non-commuting operator-valued random variables. Astérisque, 227–241 (1995)

[5] Voiculescu, D.: Lectures on free probability theory. In: École d'Été de Probabilités de Saint-Flour XXVIII–1998, Lecture Notes in Math., 1738. Springer, Berlin Heildelberg New York (2000)

[6] Voiculescu, D.: The coalgebra of the free difference quotient in free probability. International Math. Res. Notices, No.2, 79–106 (2000)

[7] Voiculescu, D.: Cyclomorphy. International Math. Res. Notices, No.6, 299–332 (2002)

[8] Voiculescu, D.: Free entropy. Bull. London Math. Soc. **34**, 257–278 (2002)

[9] Voiculescu, D.: Free analysis questions I: Duality for the coalgebra of $\partial_{X:B}$. International Math. Res. Notices **16**, 793–822 (2004)

Universality and Randomness for the Graphs and Metric Spaces*

A. M. Vershik†

St.Petersburg Mathematical Institute of Russian Academy of Science
Fontanka 27, St.Petersburg, 191011
Russia
vershik@pdmi.ras.ru

1 Universal and Random Graphs

In this section we will show the existence of a denumerable, non-directed graph which is universal in the category of all countable, non-directed graphs and homogeneous (with respect to its finite subgraphs). Moreover, we will see that such a graph is determined unique up to isomorphism. This definition was done by R. Rado. We will give the criteria of universality using incidence matrix. Then we consider the random graphs or probability measures on the set of graphs. We generalize the theorem by P. Erdös and A. Rényi that the random graph is universal graph with probability 1.[1]

---

*A detailed version of this talk was published in "Russian Mathematical Survey", v.59, (2004) No.2

†Partially supported by CRDF RUM1-2622 and Russian Fund 05-01-00899 and Poincare Inst.(Paris).

[1]References added in galley-proof:

We will describe a non-directed countable graph by a pair $\Gamma = (\Gamma, \Gamma^{(1)})$, where Γ is the (countable) set of vertices, and $\Gamma^{(1)} \subseteq \Gamma \times \Gamma$ the set of edges. Any subset $S \subseteq \Gamma$ determines a subgraph $(S, \Gamma^{(1)}|_{S\times S})$ of Γ, which we will again denote by S.

Given two graphs Γ_i, $i = 1, 2$. A mapping $\iota : \Gamma_1 \to \Gamma_2$ is called a mono-(iso)morphism, if it is one-to-one (bijective, resp.) and preserves the structure of the graph Γ_1:

$$(x, y) \in \Gamma_1^{(1)} \text{ if and only if } (\iota x, \iota y) \in \Gamma_2^{(1)}.$$

In other words, Γ_1 is isomorphic to the subgraph $\iota(\Gamma_1)$ of Γ_2. Every isomorphism of Γ_1 onto itself is called automorphism.

Definition 1. *A (countable, non-directed) graph Γ is said to be*

(1) universal, if to any finite (non-directed) graph F there exists a mono-morphism $\iota : F \to \Gamma$. In other words, any finite non-directed graph is isomorphic to a subgraph of Γ.
(2) homogeneous, if to any finite subgraphs F_1, F_2 of Γ, and isomorphism $\iota : F_1 \to F_2$, there exists an automorphism $\iota' : \Gamma \to \Gamma$ which extends ι.

1.1 Construction of the Universal Graph

In the sequel, we will show that universality and homogeneity is equivalent to the following extension property:

Definition 2. *A graph Γ is said to have the* extension property, *if the following holds: Given any finite graph F, and a one-point extension F' of F. If there is a mono-morphism $\iota : F \to \Gamma$, then ι can be extended to an mono-morphism $\iota' : F' \to \Gamma$, i.e. the following diagram commutes:*

$$\begin{array}{ccc} F & \xrightarrow{\iota\ (mono\text{-}morphism)} & \Gamma \\ {\scriptstyle id}\downarrow & & {\scriptstyle id}\downarrow \\ F' & \xrightarrow{\iota'\ (mono\text{-}morphism)} & \Gamma \end{array}.$$

In other words, the graph Γ satisfies the extension property if and only if to any finite subgraph F of Γ, and any (exterior) one-point extension $F' = F \cup \{a\}$ of F, there exists a point $\alpha \in \Gamma$ such that for $x \in F$

$$(\alpha, x) \in \Gamma^{(1)} \text{ if and only if } (a, x) \in F'^{(1)}.$$

The extension property allows us to extend locally defined isomorphisms to global isomorphisms via a so-called *back and forth* argument:

1. A. Vershik. Random metric spaces and universality. Russian Math. Surveys 59, No. 2, 259–295 (2004).
2. P. Cameron A. Vershik. Some isometry groups ofthe Urysohn space. Annal. Pure apl. Logic. t. 1703 1–9 (2006).
3. V. Pestov. Dynamics of Infionite-dimensional Groups and Ramsey-type Phenomena. Impa 25 Coloquio Brasiliero de Mathematics. 1–241 (2005).

Lemma 1. *Suppose Γ, $\tilde{\Gamma}$ are countable, non-directed graphs which have the extension property, and $F \subseteq \Gamma$ finite. Then any mono-morphism $\iota : F \to \tilde{\Gamma}$ can be extended to an isomorphism $\iota' : \Gamma \to \tilde{\Gamma}$.*

Proof. Write $\Gamma = \{x_1, x_2, \ldots\}$ and $\tilde{\Gamma} = \{y_1, y_2, \ldots\}$. Denote $F_1 = F$, $G_1 = \iota(F_1)$, and $\iota_1 = \iota : F_1 \to G_1$. Define $G_2 = G_1 \cup \{y_1\}$. By the extension property, $\iota_1^{-1} : G_1 \to F_1$ extends to a mono-morphism $\iota_2^{-1} : G_2 \to F_2$ onto a subset $F_2 \supseteq F_1$. Now define $F_3 = F_2 \cup \{x_1\}$. Again there is a mono-morphism $\iota_3 : F_3 \to G_3$ onto a subset $G_3 \supseteq G_2$. Take $G_4 = G_3 \cup \{y_2\}$. Continuing in the same manner as above, we can construct inductively subgraphs

$$F_1 \subseteq F_2 \subseteq \cdots \subseteq F_n \subseteq \cdots \subseteq \Gamma$$

with

$$F_n \supseteq \{x_k : 1 \le k \le \frac{n-1}{2}\},$$

subgraphs

$$G_1 \subseteq G_2 \subseteq \cdots \subseteq G_n \subseteq \cdots \subseteq \tilde{\Gamma}$$

satisfying

$$G_n \supseteq \{y_k : 1 \le k \le \frac{n}{2}\},$$

and isomorphisms $\iota_n : F_n \to G_n$,

$$\iota = \iota_1 \subseteq \iota_2 \subseteq \cdots \subseteq \iota_n \subseteq \cdots$$

The mapping $\iota_\infty = \bigcup_{n=1}^{\infty} \iota_n$ maps $\bigcup_{n=1}^{\infty} F_n = \Gamma$ isomorphically onto $\bigcup_{n=1}^{\infty} G_n = \tilde{\Gamma}$. □

Theorem 1. *A countable, non-directed graph is universal and homogeneous if and only if it has the extension property. Two such graphs are isomorphic.*

Proof. Any universal and homogeneous graph Γ has the extension property: By universality there is an embedding $\kappa : F_2 \to M$. Define $i : \kappa(F_1) \to \iota(M_1)$, $i = \iota \circ \kappa^{-1}$. Now, choose an automorphism $J : M \to M$ which extends i. The mapping $J \circ \kappa : F_2 \to M$ is an extension of ι.

Conversely, if Γ has the extension property, then it is obviously universal. Moreover, if $\iota : F \to F'$ is an isomorphism between finite subgraphs of Γ, then Lemma 1 (setting $\Gamma = \tilde{\Gamma}$) shows that ι can be extended to a global automorphism $\iota' : \Gamma \to \Gamma$. Hence Γ is homogeneous.

Suppose $\tilde{\Gamma}$ is another countable graph which satisfies the extension property. To show that Γ and $\tilde{\Gamma}$ are isomorphic, choose any $a \in \Gamma$, $\tilde{a} \in \tilde{\Gamma}$, and apply Lemma 1 to the mono-morphism $\iota : \{a\} \to \tilde{\Gamma}$, which maps a to $\tilde{a}$. □

Remark 1. A non-directed countable graph Γ which has the extension property is universal in the category of all countable non-directed graphs, i.e. every countable, non-directed graph Γ' can be embedded into Γ.

Consider a countable, non-directed graph $(\Gamma, \Gamma^{(1)})$, $\Gamma = \{x_1, x_2, x_2, \ldots\}$. We define its incidence matrix $m_\Gamma = (\varepsilon_{ij})_{i,j=1}^{\infty} \in \{0,1\}^{\mathbb{N}\times\mathbb{N}}$ by

$$\varepsilon_{ij} = \begin{cases} 1 & \text{if } (x_i, x_j) \in \Gamma^{(1)} \\ 0 & \text{otherwise} \end{cases}. \tag{1.1}$$

Of course the definition of m_Γ depends on the order of the x_i.

Definition 3. *An incidence matrix* $(\varepsilon_{ij})_{i,j=1}^{\infty}$ *is said to be* universal, *if for every* $n \in \mathbb{N}$

$$\{(\varepsilon_{1k}, \varepsilon_{2k}, \ldots, \varepsilon_{nk}) : k > n\} = \{0,1\}^n. \tag{1.2}$$

In other words, every finite binary word $(\varepsilon_1, \varepsilon_2, \ldots, \varepsilon_n)$ *occurs infinitely often:*

$$(\varepsilon_{1k}, \varepsilon_{2k}, \ldots, \varepsilon_{nk}) = (\varepsilon_1, \varepsilon_2, \ldots, \varepsilon_n) \text{ for infinitely many } k > n. \tag{1.3}$$

The set of all universal matrices is denoted by $\mathcal{U}$.

Universal matrices are connected with universal graphs as follows:

Theorem 2. *A countable non-directed graph* Γ *is universal and homogeneous, if its incidence matrix* m_Γ *is universal.*

Proof. Write $\Gamma = \{x_1, x_2, x_3, \ldots\}$, and let m_Γ be its incidence matrix defined by 1.1. By Theorem 1, we have to show that Γ satisfies the extension property if and only if m_Γ is universal. But this is obvious, since Definition 1.2 is only a reformulation of the extension property. □

To show the existence of a universal and homogeneous graph, we construct a universal incidence matrix $(\varepsilon_{ij})_{i,j=1}^{\infty}$ as follows: Choose an enumeration $\{w^{(1)}, w^{(2)}, w^{(3)}, \ldots\}$ of all finite words with alphabet $\{0,1\}$, such that every finite word occurs infinitely often. Extend these finite words $w^{(i)}$ arbitrarily to infinite words $\tilde{w}^{(i)} \in \{0,1\}^{\mathbb{N}}$. Now define ε_{ij}, $i<j$, as

$$(\varepsilon_{1,n+1}, \varepsilon_{2,n+1}, \ldots, \varepsilon_{n,n+1}) = \pi_n \tilde{w}^{(n)}, \qquad n \in \mathbb{N}$$

where π_n is the projection onto the first n coordinates. Further define $\varepsilon_{ii} = 0$, and $\varepsilon_{ij} = \varepsilon_{ji}$ for $i > j$. This already defines an incidence matrix, such that (1.3) is satisfied.

We thus have proved

Theorem 3. *There exists a countable, non-directed graph which is universal and homogeneous. By Theorem 1 this graph is determined uniquely up to isomorphism.*

Since such a universal and homogeneous graph is unique (up to isometry), we simply speak of *the* universal graph Γ_U.

Suppose $x, y \in \Gamma_U$ are such that $(x,y) \notin \Gamma_U^{(1)}$. By the extension property, there is a point $z \in \Gamma_U$ such that both $(x,z), (z,y) \in \Gamma_U^{(1)}$. Thus the graph metric

$$d(x,y) = \min \# \text{ edges in } \gamma,$$

where the min is taken over all paths which join x and y, can only attain the values

$$d(x,y) = \begin{cases} 0 & \text{if } x = y \\ 1 & \text{for } (x,y) \in \Gamma^{(1)} \\ 2 & \text{otherwise} \end{cases}.$$

1.2 Action of the Group s^∞ and the Set of Universal Matrices

We denote by $\mathcal{M} \subseteq \{0,1\}^{\mathbb{N}\times\mathbb{N}}$ the set of all incidence matrices, equipped with the topology that inherits from the product topology on $\{0,1\}^{\mathbb{N}\times\mathbb{N}}$, and by $\mathcal{U}$ the subset of all universal incidence matrices. The space $\mathcal{M}$ is compact and metrizable, and $\mathcal{U}$ is a closed subset of $\mathcal{M}$.

Let denote by S^∞ the group of all permutations on the naturals $\mathbb{N}$. Equipped with the topology of point-wise convergence, and choosing a proper metric, for example the usual metric for point-wise convergence

$$d(g_1,g_2) = \sum 1/2^n \frac{|g_1(n) - g_2(n)|}{1 + |g_1(n) - g_2(n)|},$$

S^∞ is a complete separable metric space (=polish space), which is not locally compact. Moreover, *this topology makes S^∞ to a non-locally compact, polish group.* All finite permutations form a dense subgroup S_∞ of S^∞.

For any $g \in S^\infty$ and infinite matrix $(a_{ij})_{i,j=1}^\infty$ we define

$$T_g(a_{ij}) = g\,(a_{ij})\,g^{-1} = (a_{g(i),g^{-1}(j)})_{i,j=1}^\infty. \tag{1.4}$$

Clearly T_g leaves the set $\mathcal{M}$ of incidence matrices invariant. Via $g \mapsto T_g$, the group S^∞ (or S_∞) acts on $\mathcal{M}$. Note that for any finite permutation $g \in S_\infty$, $T_g : \mathcal{M} \to \mathcal{M}$ is a homeomorphism, whereas for any infinite permutation it is only a Borel-automorphism.

Consider a graph Γ with incidence matrix m_Γ: Another (countable, non-directed) graph $\tilde{\Gamma}$ is isomorphic to Γ if and only if

$$m_{\tilde{\Gamma}} = g m_\Gamma g^{-1}$$

for some $g \in S^\infty$, or equivalently if the orbits $S^\infty m_\Gamma$ and $S^\infty m_{\tilde{\Gamma}}$ coincide. In this sense we may regard the group $\operatorname{Aut}\Gamma$ of all automorphisms of Γ as a (closed) subgroup of S^∞.

PROBLEM

How does the group $\operatorname{Aut}(\Gamma_U)$ of automorphisms of the universal graph Γ_U look like?

It is known that this group is simple (see [2]) Very interesting question whether this (uncountable) group has unitary representations which do not extend to the whole group S^∞

As we saw $\mathcal{U}$ is S^∞-invariant subset of $\mathcal{M}$. Moreover, since any two universal graphs are isomorphic, S^∞ *acts transitively on the set* $\mathcal{U}$ *of universal matrices*, i.e. if $m_1, m_2 \in \mathcal{U}$, then there is a $g \in S^\infty$ such that

$$m_2 = g\, m_1\, g^{-1}.$$

By the universality condition, it is clear that *for any universal matrix* m*, its orbit* $S^\infty m$ *is dense in* $\mathcal{M}$.

Theorem 4. *The set* $\mathcal{U}$ *of all universal incidence matrices is a dense* G_δ*-subset of* $\mathcal{M}$.

Proof. For any finite binary word $w = (\varepsilon_1, \varepsilon_2, \ldots, \varepsilon_n)$, the set $M_w = \{(\varepsilon_{ij})_{i,j=1}^\infty \in \mathcal{M} : (1.3) \text{ holds }\}$ equals

$$\bigcap_{m>n} \bigcup_{k \geq m} \{(\varepsilon_{ij})_{i,j=1}^\infty \in \mathcal{M} : (\varepsilon_{1k}, \varepsilon_{2k}, \ldots, \varepsilon_{nk}) = (\varepsilon_1, \varepsilon_2, \ldots, \varepsilon_n)\}$$

is a G_δ set in $\mathcal{M}$. Therefore, as the intersection of countably many M_w, the set $\mathcal{U}$ is also G_δ.

For any universal incidence matrix m, the orbit $S_\infty M$ is dense in $\mathcal{M}$. Since $\mathcal{U}$ is S_∞-invariant, this completes the proof of the theorem. $\square$

1.3 Random Graphs

Now we will define the probability measures on the set of all countable graphs and in particular give a probabilistic proof of the existence of the universal graph Γ_U. That fact that the simplest such probability measures (independent entries) gives with probability one the universal graph is due to P. Erdös and A. Rényi ([4] see also [2]).

This proof will also imply the existence S^∞-invariant measures which are concentrated on the set $\mathcal{U}$ of universal incidence matrices.

We first give an intuitive description. Starting with a single point (one-point graph), we successively define random graphs Γ_n, $n = 1, 2, 3, \ldots$ in the following manner: To a given n-point graph $\Gamma_n = \{x_1, x_2, \ldots, x_n\}$ we add a $n+1$-th point x_{n+1} and independently set edges in $\Gamma_{n+1} = \Gamma_n \cup \{x_{n+1}\}$ between the new point x_n and the old points $x \in \Gamma_n$ according to

$$\begin{aligned} \text{Prob}[(x_{n+1}, x_m) \in \Gamma_{n+1}] &= p \\ \text{Prob}[(x_{n+1}, x_m) \notin \Gamma_{n+1}] &= 1 - p, \end{aligned}$$

where $0 < p < 1$ is a fixed number. Then, with probability one, the so obtained graph is the universal graph Γ_U:

Theorem 5 (P. Erdos, A. Renyi). *Suppose* $\{\xi_{ij} : i < j\}$ *is an array of independent identically distributed random variables with*

$$P[\xi_{ij} = 1] = p,$$
$$P[\xi_{ij} = 0] = 1 - p,$$

where $0 < p < 1$. *Set* $\xi_{ii} = 0$ *and* $\xi_{ij} = \xi_{ji}$ *for* $i > j$. *Then, with probability one,* $(\xi_{ij})_{i,j=1}^{\infty}$ *is a universal incidence matrix.*

Proof. We only have to show that, with probability one, every binary word $(\varepsilon_1, \varepsilon_2, \dots, \varepsilon_n) \in \{0,1\}^n$ occurs infinitely often:

$$P[(\xi_{11}, \xi_{2k}, \dots, \xi_{nk}) = (\varepsilon_1, \varepsilon_2, \dots, \varepsilon_n) \text{ for infinitely many } k > n] = 1.$$

But this is evident since all the ξ_{ij}, $i < j$, are independent, and since for $k > n$

$$0 < P[(\xi_{1k}, \xi_{2k}, \dots, \xi_{nk}) = (\varepsilon_1, \varepsilon_2, \dots, \varepsilon_n)] < 1,$$

which follows from $0 < p < 1$. □

Remark 2. The measures which were defined are invariant and ergodic with respect to the NW-shift $\sigma_{NW} : \mathcal{M} \to \mathcal{M}$ defined by

$$\sigma_{NW}(a_{i,j})_{i,j=1}^{\infty} = (a_{i+1,j+1})_{i,j=1}^{\infty}. \tag{1.5}$$

There are lot of measures (besides those above) which are invariant and ergodic with respect to the group S_∞ which acts on the set of symmetric $0-1$ matrices and concentrated on the subset of the universal matrices (see [13]). So we have a strong version of so called Kolmogorov's effect: there are uncountable many invariant ergodic measures for the transitive action of non-locally compact group S_∞ .

Moreover we can say that the set of the probability measures on the space of symmetric $0-1$ matrices which have as support the set of universal matrices is everywhere dense G_δ-set in the weak topology of the space of measures.

In another words for the generic probability measure on the set of countable graphs a universal graph has probability 1.

2 The Urysohn Metric Space and Random Metric Spaces

In his last works [10], Paul S. Urysohn (1898–1924) gave a concrete construction of a universal separable metric space which is now known as "Urysohn Space". Different to other universal separable metric spaces (for example $C([0,1])$ with the sup-norm, as was proven by Banach and Mazur), the Urysohn space is more special, since it is homogeneous in the sense of Definition 4 and the space $C([0,1])$ is not homogeneous. Moreover it turns out that, up to isometric equivalence, there is only one universal and homogeneous space – Urysohn space $\mathfrak{U}$.

In this section we give an explicit construction of the Urysohn space $\mathfrak{U}$ which is in the spirit of our construction of the universal graph in Section 1, i.e. it is based on the property that isometries from finite metric spaces F into $\mathfrak{U}$ can be extended to arbitrary one-point extensions of F (see also [12]). Beside universality, this extension property also implies homogeneity and the uniqueness of the Urysohn space. Another analogue to the case (Gurarij spaces (see [8]), or the Poulsen Simplex (see [7]) will be considered elsewhere. For different constructions of the Urysohn space, see [6] or [10].

Definition 4. *A polish space* (U, d) *is said to be*

(1) universal, if to any finite metric space (F, d_F) *there exists an isometry* $\iota : (F, d_F) \to (U, d)$.
(2) homogeneous , if to any finite subsets F_1, F_2 *of* U*, and bijective isometry* $\iota : F_1 \to F_2$*, there exists an automorphism (=bijective isometry)* $\iota' : U \to U$ *which extends* ι.
(3) a Urysohn space, if it is universal and homogeneous.

Note that any separable metric space (M, d) can be embedded isometrically into a separable Banach space $(B, \|\,.\,\|)$: For example, choose any point $x_0 \in M$ and consider the mapping

$$\iota : M \to C_b(M), \qquad x \mapsto \iota(x) = d(x,\,.\,) - d(x_0,\,.\,),$$

where $C_b(M)$ is the space of all bounded continuous functions on M, equipped with the sup-norm $\|\,.\,\|_\infty$. The triangular inequality

$$|d(x, z) - d(y, z)| \leq d(x, y), \qquad x, y, z \in K,$$

implies that $\|\iota(x) - \iota(y)\|_\infty = d(x, y)$, and setting $z = y$ in the formula above shows that $\|\iota(x) - \iota(y)\|_\infty = d(x, y)$. Thus ι is an isometric embedding of M into $C_b(M)$. Set $B =$ closed linear span of ιM, $\|\,.\,\| = \|\,.\,\|_\infty$. This is a Banach space and by the separability of M, B is also separable.

This shows that any universal (separable) Banach space (for example $C[0, 1]$ with the sup-norm, see [3] or other textbooks on topology) is also a universal separable metric space. But every known model of a universal Banach space is not homogeneous in the sense of Definition 4.

2.1 Extending Isometries

In the sequel we will show that the Urysohn property can be characterized by certain extension properties of isometries. This enables us to see that Urysohn spaces are uniquely determined up to isomorphism.

Definition 5. *Let* (M_1, d_1), (M_2, d_2) *be metric spaces and* $\varepsilon > 0$*. A mapping* $\iota : M_1 \to M_2$ *is said to be an* ε*-isometry, if*

$$\|d_1(x, y) - d_2(x, y)\| \leq \varepsilon, \qquad x, y \in M_1.$$

Definition 6. *Suppose $\varepsilon \geq 0$. A metric space (M, d) is said to have the ε-extension property, if the following holds: Given any finite metric space (F, d_F), and a one-point extension $(F', d_{F'})$ of (F, d_F). If there is an isometric mapping $\iota : F \to M$, then ι can be extended to an ε-isometry $\iota' : F' \to M$, i.e. the following diagram commutes:*

$$\begin{array}{ccc} F & \xrightarrow{\iota \quad (isometry)} & M \\ {\scriptstyle id}\downarrow & & {\scriptstyle id}\downarrow \\ F' & \xrightarrow{\iota' \quad (\varepsilon-isometry)} & M \end{array} .$$

If $\varepsilon = 0$, we simply speak of the extension property.

In other words, (M, d) satisfies the ε-extension property if and only if to any finite subspace $F \subseteq M$, and any (exterior) one-point metric extension $(F' = F \cup \{a\}, d_{F'})$ of F, there exists a point $a_\varepsilon \in M$ such that

$$|d(a_\varepsilon, x) - d_{F'}(a, x)| \leq \varepsilon, \qquad x \in F.$$

The extension property allows us to extend locally defined isometries to global isometries via a so-called *back and forth* argument:

Lemma 2. *Suppose (M, d), $(\tilde{M}, \tilde{d})$ are polish spaces which have the extension property, $F \subseteq M$ finite. Then any isometric mapping $\iota : F \to \tilde{M}$ can be extended to a bijective isometry $\iota' : M \to \tilde{M}$.*

Proof. Choose countable dense subsets $\{x_1, x_2, \ldots\} \subseteq M$ and $\{y_1, y_2, \ldots\} \subseteq \tilde{M}$.

Denote $F_1 = F$, $G_1 = \iota(F_1)$, and $\iota_1 = \iota : F_1 \to G_1$. Define $G_2 = G_1 \cup \{y_1\}$. By the extension property, $\iota_1^{-1} : G_1 \to F_1$ extends to an isometry $\iota_2^{-1} : G_2 \to F_2$ onto a subset $F_2 \supseteq F_1$. Now define $F_3 = F_2 \cup \{x_1\}$. Again there is an isometry $\iota_3 : F_3 \to G_3$ onto a subset $G_3 \supseteq G_2$. Take $G_4 = G_3 \cup \{y_2\}$. Continuing in the same manner as above, we can construct inductively subsets

$$F_1 \subseteq F_2 \subseteq \cdots \subseteq F_n \subseteq \cdots \subseteq M$$

with

$$F_n \supseteq \{x_k : 1 \leq k \leq \frac{n-1}{2}\},$$

subsets

$$G_1 \subseteq G_2 \subseteq \cdots \subseteq G_n \subseteq \cdots \subseteq \tilde{M}$$

satisfying

$$G_n \supseteq \{y_k : 1 \leq k \leq \frac{n}{2}\},$$

and isometric bijections $\iota_n : F_n \to G_n$,

$$\iota = \iota_1 \subseteq \iota_2 \subseteq \cdots \subseteq \iota_n \subseteq \cdots$$

The mapping $\iota_\infty = \bigcup_{n=1}^\infty \iota_n$ maps $F_\infty = \bigcup_{n=1}^\infty F_n$ isometrically onto $G_\infty = \bigcup_{n=1}^\infty G_n$. Since F_∞ and G_∞ are dense, and since the metrics on M and $\tilde{M}$ are complete we may extend ι_∞ to an bijective isometry $\iota' : M \to \tilde{M}$. □

Theorem 6. *A polish space is a Urysohn space if and only if it has the extension property. Two Urysohn spaces are isometrically isomorphic.*

Proof. A Urysohn space (M, d) has the finite extension property: By universality there is an isometry $\kappa : F_2 \to M$. Define $i : \kappa(F_1) \to \iota(M_1)$, $i = \iota \circ \kappa^{-1}$. Now, choose an automorphism $J : M \to M$ which extends i. The mapping $J \circ \kappa : F_2 \to M$ is an extension of ι.

On the other hand, if (M, d) has the extension property, then (M, d) is obviously universal. Moreover, if $\iota : F_1 \to F_2$ is an isometry between finite subspaces of M, then Lemma 2 (setting $M = \tilde{M}$) shows the existence of an automorphism $\iota' : M \to M$ which extends ι.

The uniqueness of two Urysohn spaces (M, d),$(\tilde{M}, \tilde{d})$ is now obvious: Choose any two points $x_1 \in M$, $x_2 \in \tilde{M}$ and consider the isometry $\iota : x_1 \mapsto x_2$. Since both spaces posses the extension property, Lemma 2 allows us to extend ι to an isomorphism $\iota' : M \to \tilde{M}$. □

Remark 3. The existence of a polish space which is Urysohn is shown in Section 2.2. Since such a space is unique up to isometry, we will simply speak of *the* Urysohn space $(\mathfrak{U}, d)$.

Note that if a polish space (M, d) satisfies the extension property, then it is also universal in the category of polish spaces, i.e. every polish space (P, d) can be embedded isometrically into M: Choose a countable dense subset $\{x_1, x_2, \ldots\} \subseteq M'$, and denote $F_n = \{x_1, \ldots, x_n\}$. There exist exists isometric mappings $\iota_n : F_n \to M$,

$$\iota_1 \subseteq \iota_2 \subseteq \cdots \subseteq \iota_n \subseteq \cdots$$

The mapping $\iota_\infty = \bigcup_{n=1}^\infty \iota_n$ maps $F_\infty = \bigcup_{n=1}^\infty F_n$ isometrically into M. Since F_∞ is dense in P, and since the metric on M is are complete, we may extend ι_∞ to an isometry $\iota : P \to M$.

Lemma 3. *Suppose a complete metric space (M, d) satisfies the ε-extension property for all $\varepsilon > 0$. Then (M, d) has the extension property for compact metric spaces: Given any compact metric space (C, d_C), and any one-point (or polish) extension $(C', d_{C'})$ of (C, d_C). If there is an isometric mapping $\iota : C \to M$, then ι can be extended to an isometry $\iota' : C' \to M$, i.e. the following diagram commutes:*

$$\begin{array}{ccc} C & \xrightarrow{\ \iota \quad (isometry)\ } & M \\ {\scriptstyle id}\downarrow & & {\scriptstyle id}\downarrow \\ C' & \xrightarrow{\ \iota' \quad (\varepsilon-isometry)\ } & M \end{array}$$

Proof. First of all, note that for arbitrary $\varepsilon > 0$ the ε-extension property holds also for compact metric spaces: Given any compact metric space (C, d_C), one-point extension $(C' = C \cup \{a\}, d_{C'})$, and isometry $\iota : C \to M$. Because C is compact, we may choose a finite set $F \subseteq C$ such that

$$\min_{y \in F} d(\iota x, \iota y) \leq \varepsilon/2, \qquad x \in C.$$

By the assumption on M, there exists an $\varepsilon/2$-extension of the restriction of ι to F, which maps a to $x_a \in M$, say. Now the function

$$\iota' : C \cup \{a\} \to M, \begin{cases} \iota' = \iota & \text{on } C \\ a \mapsto x_a \end{cases}$$

is an ε-extension of $\iota : C \to M$, as one easily can verify.

To show the extension property, we have to find a point $\alpha \in M$ such that

$$d(\alpha, \iota(x)) = d_{C'}(a, x), \qquad x \in C. \tag{2.1}$$

Choose ε_n, such that $\sum_{n=0}^{\infty} \varepsilon_n < \infty$. We will define inductively a sequence of points $x_n \in M$, $n = 0, 1, 2, \ldots$, such that

$$|d(x_n, \iota(x)) - d_{C'}(a, x)| \leq \varepsilon_n, \qquad x \in C, \tag{2.2}$$

and such that the sequence (x_n) is Cauchy in M. Then, since M is complete, there exists a limit α which certainly satisfies (2.1).

Since M satisfies the ε-extension property for $\varepsilon = \varepsilon_0$, we can find a point $x_0 \in M$ which satisfies (2.2) with $n = 0$. Define the other x_n, $n \geq 1$, by applying the following induction step:

Given a point x_n which satisfies (2.2). Define an (exterior) one-point extension $(C \cup \{x_n, b\}, d')$ of $(C \cup \{x_0\}, d)$ by:

$$\begin{aligned} d'(x, y) &= d(x, y) & x, y \in C \cup \{x_n\} \\ d'(x, b) &= d_{C'}(x, a) & x \in C \\ d'(x_n, b) &= \varepsilon_n \end{aligned}$$

In fact, by (2.2), d' satisfies the triangular inequality (if in addition $\varepsilon_n \leq \inf_{x \in C} d_{C'}(x, a)$, which we certainly may assume) as one can easily verify. Since M satisfies the ε-extension property for $\varepsilon = \varepsilon_{n+1}$, we can find a point $x_{n+1} \in M$ such that

$$|d(x_{n+1}, \iota(x)) - d_{C'}(a, x)| \leq \varepsilon_{n+1}, \qquad x \in C,$$

and

$$d(x_n, x_{n+1}) \leq d'(x_n, b) + \varepsilon_{n+1} = \varepsilon_n + \varepsilon_{n+1}.$$

By construction, 2.2 is satisfied, and since $\sum \varepsilon_n < \infty$, the latter inequality shows that (x_n) is a Cauchy sequence, which completes the proof. □

As an immediate consequence of Lemma 3, any polish space (M, d) which satisfies the extension property (for finite metric spaces) also satisfies the extension property for compact metric spaces. Repeating the proof of Lemma 2 verbatim we may conclude:

Lemma 4. *Suppose (M, d), $(\tilde{M}, \tilde{d})$ are polish spaces which have the extension property, $C \subseteq M$ compact. Then any isometric mapping $\iota : C \to \tilde{M}$ can be extended to a bijective isometry $\iota' : M \to \tilde{M}$.*

In particular, this shows that the Urysohn space is homogeneous with respect to its compact subspaces. Hence the Urysohn space $\mathcal{U}$ is a "good" universal object in the category of all compact metric spaces: every compact metric space (C, d_C) can be embedded isometrically into $\mathcal{U}$, and this embedding is unique up to isometry, i.e. if $\iota_1 : C \to U$, $\iota_2 : C \to U$ are two such isometries, then there exists an automorphism $J : \mathcal{U} \to \mathcal{U}$ such that

$$\iota_1 = J \circ \iota_2.$$

Lemma 4 is not true assuming C to be closed, even countable (see [1], [5]).

The following Theorem which will be important for our construction of the Urysohn space (See subsection 2.2):

Theorem 7. *A polish space is Urysohn if and only if it satisfies the ε-extension property for every $\varepsilon > 0$.*

Proof. Combine Lemma 3 and Theorem 6. □

2.2 Construction, Universal Matrices

Before show the existence of a Urysohn space, we need some preparations.

Suppose (M, d) is an infinite polish space. Choose a countable dense subset $D = \{x_1, x_2, x_3, \ldots\}$ and define a matrix $r = (r_{ij})_{i,j=1}^{\infty}$ by

$$r_{i,j} = d(x_i, x_j). \tag{2.3}$$

Any matrix obtained in such a way is called a *(infinite) distance matrix.* The set of all infinite distance matrices is denoted by $\mathcal{R}$. We will consider $\mathcal{R}$ as topological space equipped with the topology that inherits from the product topology on $\mathbb{R}^{\mathbb{N}\times\mathbb{N}}$. $\mathcal{R}$ is a closed convex cone in $\mathbb{R}^{\mathbb{N}\times\mathbb{N}}$, in particular it is a polish space.

Similarly we say that a finite matrix $(r_{i,j})_{i,j=1}^{n}$ is a *(finite) distance matrix,* if there exists a finite metric space $(\{x_1, x_2, \ldots, x_n\}, d)$ such that (2.3) holds. The set of all n-dimensional distance matrices, equipped with the product topology, is denoted by $\mathcal{R}_n$. If we define a (continuous) projection $\pi_n : \mathcal{R}_{n+1} \to \mathcal{R}_n$ by

$$\pi_n (r_{ij})_{i,j=1}^{n+1} = (r_{ij})_{i,j=1}^{n},$$

then we can regard $\mathcal{R}$ as the inverse limit of the cones $\mathcal{R}_n$:

$$\mathbb{R}^+ \overset{\pi_1}{\leftarrow} \mathcal{R}_2 \overset{\pi_2}{\leftarrow} \cdots \overset{\pi_{n-1}}{\leftarrow} \mathcal{R}_n \overset{\pi_n}{\leftarrow} \mathcal{R}_{n+1} \overset{\pi_{n+1}}{\leftarrow} \cdots$$

Given a n-dimensional distance matrix $r_n = (r_{ij})_{i,j=1}^n$, which corresponds to $F = (\{x_1, \ldots, x_n\}, d)$. A vector $b = (b_1, \ldots, b_n)$ is called *admissible*, if the matrix

$$r_n^b = \begin{pmatrix} r_{11} & r_{12} & \cdots & r_{1n} & b_1 \\ r_{21} & r_{22} & \cdots & r_{2n} & b_2 \\ \vdots & \vdots & \ddots & \vdots & \vdots \\ r_{n1} & r_{n2} & \cdots & r_{nn} & b_n \\ b_1 & b_2 & \cdots & b_n & 0 \end{pmatrix}$$

is also a distance matrix. In other words, b is admissible if there exists a one-point extension $(F \cup \{x\}, d')$ of (F, d), such that

$$d'(x, x_i) = b_i, \qquad i = 1, \ldots, n.$$

The set of all admissible vectors is then denoted by $\operatorname{Adm}(r_n)$. It determines all one-point extensions of (F, d) up to isomorphism. Note that

$$\operatorname{Adm}(r_n) = \{(b_i)_{i=1}^n \in \mathbb{R}^n : b_i - b_j \le r_{ij} \le b_i + b_j\}, \tag{2.4}$$

hence $\operatorname{Adm}(r_n)$ is a closed convex cone $\subseteq \mathbb{R}^n$.

Lemma 5. *Suppose $M = (\{x_1, x_2, x_3, \ldots\}, d)$ is countable metric space, and consider the subspaces $F_n = (\{x_1, \ldots, x_n\}, d)$, with $r_n = (d(x_i, x_j))_{i,j=1}^n$ as their corresponding distance matrices. Then*

$$\mathbb{R}^+ \overset{\pi_1}{\leftarrow} \operatorname{Adm}(r_2) \overset{\pi_2}{\leftarrow} \cdots \overset{\pi_{n-1}}{\leftarrow} \operatorname{Adm}(r_n) \overset{\pi_n}{\leftarrow} \operatorname{Adm}(r_{n+1}) \overset{\pi_{n+1}}{\leftarrow} \cdots,$$

where $\pi_n : \mathbb{R}^{n+1} \to \mathbb{R}^n$ is the projection onto the first n coordinates.

Proof. It is clear that $\pi_{n+1} \operatorname{Adm}(r_{n+1}) \subseteq \operatorname{Adm}(r_n)$. To prove the opposite inclusion, it suffices to show that to any two vectors $a, b \in \operatorname{Adm}(r_n)$ we can construct a two-point extension $M = (F_n \cup \{y_a, y_b\}, d)$ of F_n with $d(x_i, y_a) = a_i$ and $d(x_i, y_b) = b_i$, $i = 1, \ldots, n$. The only distance left to choose is $\alpha = d(y_a, y_b)$. The validity of the triangular inequality for d is equivalent to

$$|a_i - b_i| \le \alpha \le a_i + b_i, \qquad 1 \le i \le n,$$

which is only possible if

$$\max_{1 \le i \le n} |a_i - b_i| \le \min_{1 \le i \le n} a_i + b_i.$$

But the latter inequality is true since both vectors a and b are admissible: There exists a one-point extension $(F_n \cup \{z_a\}, d_a)$ of F_n with

$$d_a(z_a, x_i) = a_i, \qquad 1 \le i \le n,$$

and a one-point extension $(F_n \cup \{z_b\}, d_b)$ of F_n with

$$d_b(z_b, x_i) = b_i, \qquad 1 \le i \le n.$$

Now, using the triangular inequality for both metrics, we have

$$a_i - b_i = d_a(z_a, x_i) - d_b(z_b, x_i) \le d_a(z_a, x_j) + d(x_i, x_j) - d_b(z_b, x_i) \le$$
$$\le d_a(z_a, x_j) + d_b(z_b, x_j) = a_j + b_j.$$

Since i and j where arbitrary, this implies the desired inequality. □

Remark. The above lemma shows that the cones $\mathrm{Adm}(r_n)$ are consistent under projections. *They are also consistent under permutations:* If g is any permutation on the set $\{1, 2, \dots n\}$ then we have

$$g\,\mathrm{Adm}(r_n) = \mathrm{Adm}(g r_n g^{-1}),$$

where for vectors $x = (x_1, x_2, \dots, x_n)$, its permutation gx is defined by $gx = (x_{g(1)}, x_{g(2)}, \dots, x_{g(n)})$.

Definition 7. *A distance matrix $r = (r_{ij})_{i,j=1}^{\infty}$ is said to be* universal, *if for every $n \in \mathbb{N}$*

$$\textit{closure}\ \{(r_{1k}, r_{2k}, \dots, r_{nk}) : k = n, n+1, \dots\} = \mathrm{Adm}((r_{ij})_{i,j=1}^{n}). \qquad (2.5)$$

It is called weakly universal, *if $S_\infty r$ is dense in $\mathcal{R}$. The set of all universal matrices is denoted by $\mathcal{M}$.*

Universal matrices are connected with Urysohn spaces:

Theorem 8. *Suppose (M, d) is a polish space, $(x_n)_{n=1}^{\infty}$ a dense sequence. The matrix $r = (d(x_i, x_j))_{i,j=1}^{\infty}$ is universal if and only if M is Urysohn.*

Proof. Suppose M is an Urysohn space. Fix $n \in \mathbb{N}$, $r_n = (d(x_i, x_j))_{i,j=1}^{n}$ and choose an admissible vector $(a_1, a_2, \dots, a_n) \in \mathrm{Adm}(r_n)$. By the extension property, there is an $x \in M$ such that

$$d(x, x_i) = a_i, \qquad i = 1, 2, \dots, n.$$

Since the x_k are dense in M, the property (2.5) is evident.

Conversely, suppose the distance matrix $(d(x_i, x_j))$ is universal. Choose a finite subset $F = \{y_1, y_2, \dots, y_n\} \subseteq M$, and an admissible vector $(a_1, a_2, \dots, a_n) \in \mathrm{Adm}(r_n)$, where $r_n = (d(y_i, y_j))_{i,j=1}^{n}$. Fix $\varepsilon > 0$. Since the x_k are dense, and since (2.5) holds, there are points $x_{k_1}, \dots, x_{k_n}$ and $x = x_{k_{n+1}}$ such that

$$|x_{k_i} - y_i| < \varepsilon/2, \qquad i = 1, 2, \dots, n$$

and

$$|d(x, x_{k_i}) - a_i| < \varepsilon/2, \qquad i = 1, 2, \ldots, n.$$

Therefore $|d(x, y_i) - a_i| < \varepsilon$, $i = 1, \ldots, n$, which proves the ε-extension property of M. Since $\varepsilon > 0$ was chosen arbitrarily, we can conclude by Lemma 3 that M has the extension property, hence it is a universal and homogeneous space. □

It is not difficult to show that *the distance matrix* $r = (d(x_i, x_j))_{i,j=1}^{\infty}$ *is weakly universal, if and only if* (M, d) *is weakly universal in the following sense: To any finite metric space* (F, d_F), *and any* $\varepsilon > 0$, *there exists an* ε*-isometry* ι *which maps* F *into* M. Thus every universal matrix is also weakly universal. The converse is not true: there exists weakly universal matrices (spaces), which are not universal matrices (spaces, resp.). For example consider a sequence of metric spaces

$$(F_1, d_1), (F_2, d_2), (F_3, d_3), \ldots,$$

such that their corresponding distance matrices

$$r_1, r_2, r_3, \ldots$$

enumerate all rational distance matrices of all dimensions. Now consider the disjoint sum (M, d) of the spaces (F_i, d_i), i.e. $M = \bigcup_i F_i$ and

$$d(x, y) = \begin{cases} d_i(x, y) & \text{for } x, y \in F_i \\ \frac{\operatorname{diam}(F_i) + \operatorname{diam}(F_j)}{2} & \text{for } x \in F_i, y \in F_j \end{cases},$$

where $\operatorname{diam}(F_i) = \max_{x,y \in F_i} d_i(x, y)$. The space (M, d) is complete, since it consists of isolated points only, separable, by construction weakly universal but not every finite metric space can be embedded into M.

Before we can start with the construction of the Urysohn space, we need another auxiliary lemma, which proof we leave to the reader.

Lemma 6. *Suppose* A_n,A_m *are convex cones in* $\mathbb{R}^n$, $\mathbb{R}^m$ *respectively* ($n > m$), *such that* $\pi A_n = A_m$, *where* $\pi : \mathbb{R}^n \to \mathbb{R}^m$ *is the projection onto the first* m *coordinates. If* $(a_i)_{i=1}^{\infty}$ *is a sequence dense in* A_m, *then we can find vectors* $(a'_i)_{i=1}^{\infty}$ *dense in* A_n *such that*

$$\pi a'_i = a_i, \qquad i \in \mathbb{N}.$$

Now we are able to construct a Urysohn space, or equivalently a universal matrix: To do so, we define a metric d on the set of naturals $\mathbb{N}$ as follows. Choose a sequence $(a_n^{(1)})_{n=2}^{\infty}$ dense in $\mathbb{R}_+$ and define

$$d(1, n) = a_n^{(1)}, \qquad i = 2, 3, \ldots$$

Note that d already defines a metric on the two-point set $\{1, 2\}$. Set $r_1 = (0)$, and let r_2 be the distance matrix which corresponds to the two-point space

$(\{1,2\},d)$. Since the projection $\pi : \mathbb{R}^2 \to \mathbb{R}$, $(x_1,x_2) \to x_1$ maps $\mathrm{Adm}(r_2)$ onto $\mathrm{Adm}(r_1) = \mathbb{R}_+$, Lemma 6 shows that there is a sequence of numbers $(a_n^{(2)})_{n=3}^\infty$ such that

$$\mathrm{Adm}(r_2) = \text{closure}\,\{\begin{pmatrix} a_n^{(1)} \\ a_n^{(2)} \end{pmatrix} : n \geq 3\}.$$

Define

$$d(2,n) = a_n^{(2)}, \qquad n = 3,4,\dots$$

This defines already a metric on the three-point set $\{1,2,3\}$. Now let r_3 be the distance matrix corresponding to $(\{1,2,3\},d)$. Again, we can find a sequence of non-negative numbers $(a_n^{(3)})_{n=4}^\infty$ such that

$$\mathrm{Adm}(r_3) = \text{closure}\,\{\begin{pmatrix} a_n^{(1)} \\ a_n^{(2)} \\ a_n^{(3)} \end{pmatrix} : i \geq 4\}.$$

Define

$$d(3,n) = a_n^{(2)}, \qquad n \geq 4.$$

Continuing in this manner, we obtain a metric d on $\mathbb{N}$ such that by construction the matrix $(d(i,j))_{i,j=1}^\infty$ is universal. Therefore the completion of $(\mathbb{N},d)$ is a Urysohn space. We have proved the main theorem of this section:

Theorem 9 (Urysohn). *There exists a polish space $(\mathfrak{U},d)$ which is universal and homogeneous. By Theorem 6 this space is determined uniquely up to isomorphism.*

Remark. For $\alpha > 0$, restricting ourselves to bounded distance matrices $r \in [0,\alpha]^{\mathbb{N}\times\mathbb{N}}$, one similarly can construct a homogeneous polish space $\mathfrak{U}_\alpha$ of diameter α which is universal in the following sense: *Every polish space of diameter $\leq \alpha$ can be embedded isometrically into $\mathfrak{U}_\alpha$*. Such a space is also unique up to isometric equivalence.

2.3 Some Properties of the Urysohn Space

We now consider topological properties of the Urysohn space.

Theorem 10. *Any continuous map $f : C \to \mathfrak{U}$ from a compact metric space (C,d_C) into the Urysohn space $(\mathfrak{U},d)$ is contractible (=homotopic to the constant map).*

Proof. Consider the image $f(C)$ as a subspace of $\mathfrak{U}$. This is a compact metric space and we therefore can find a isometric embedding $\iota : f(C) \to B$ into a separable Banach space $(B,\|\,.\,\|)$ (as was shown at the beginning of this section). By the extension property, the isometry $\iota^{-1} : \iota(f(C)) \to C$ can be extended to an isometry $J : B \to \mathfrak{U}$:

$$\begin{array}{ccc} (C, d_C) & \xrightarrow{\;f\;} & (\mathfrak{U}, d) \\ {\scriptstyle g=\iota\circ f}\Big\downarrow & & \Big\downarrow{\scriptstyle id} \\ (B, \|\,.\,\|) & \xrightarrow{\;J\;} & (\mathfrak{U}, d) \end{array}$$

Since the mapping $g : C \to B$ is contractible, f is also contractible. □

Theorem 11. *The Urysohn space $(\mathfrak{U}, d)$ is path-wise connected. Moreover all homotopy groups π_n (and therefore all homology groups) are trivial.*

Proof. That $\mathfrak{U}$ is path-wise connected is an immediate consequence of the extension property: The mapping which maps two different numbers $a, b \in \mathbb{R}$ to two different points $x, y \in \mathfrak{U}$ is an isometry, provided we have chosen the proper norm on $\mathbb{R}$, and can therefore be extended to an isometric embedding of the whole interval $[a, b]$ into $\mathfrak{U}$.

By Theorem 10, any continuous mapping $f : S^n \to \mathfrak{U}$ from the n-dimensional sphere into $\mathfrak{U}$ is contractible, hence all homotopy groups of U are trivial. □

PROBLEM

Is Urysohn space $\mathfrak{U}$ contractible or not?

Consider the group $\mathrm{Aut}(\mathfrak{U})$ of isometries of the Urysohn space, equipped with the topology of point-wise convergence. In [11], the author shows that $\mathrm{Aut}(\mathfrak{U})$ is a universal topological group with a countable base, i.e. every Hausdorff topological group with a countable base is isomorphic to a subgroup of $\mathrm{Aut}(\mathfrak{U})$. Apart from that, not very much is known about this group.

PROBLEM

To describe algebraic or topological properties of the Isometry Group $\mathrm{Aut}(\mathfrak{U})$ of the Urysohn space $\mathfrak{U}$.

We continue with miscellaneous considerations about the Urysohn space.

Theorem 12. *Consider the product $\mathfrak{U}^2 = \mathfrak{U} \times \mathfrak{U}$ equipped with the maximum metric $d_\infty((x, y), (x', y')) = \max\{d(x, x'), d(y, y')\}$. The product $\mathfrak{U}^2$ is separable and universal, but not isometrically isomorphic to $\mathfrak{U}$*

Proof. We show that $(\mathfrak{U}^2, d_\infty)$ does not satisfy the extension property: Choose a subspace $F_1 = \{x_1, x_2, x_3\} \subseteq \mathfrak{U}$ with distance matrix

$$r_{F_1} = \begin{pmatrix} 0 & 1 & 1/4 \\ 1 & 0 & 1 \\ 1/4 & 1 & 0 \end{pmatrix}.$$

and a subspace $F_2 = \{y_1, y_2, y_3\} \subseteq \mathfrak{U}$ with distance matrix

$$r_{F_2} = \begin{pmatrix} 0 & 1/4 & 1 \\ 1/4 & 0 & 1 \\ 1 & 1 & 0 \end{pmatrix}.$$

Let $F \subseteq \mathfrak{U}^2$ be the subspace which consists of the points (x_i, y_i), $i = 1, 2, 3$. All edges of this triangle have length one, i.e. the distance matrix of F is

$$r_F = \begin{pmatrix} 0 & 1 & 1 \\ 1 & 0 & 1 \\ 1 & 1 & 0 \end{pmatrix}.$$

The vector $(1/2, 1/2, 1) \in \mathrm{Adm}(r_F)$, but, by equation (2.4), any vector $(b_1, b_2, b_3) \in \mathrm{Adm}(r_{F_j})$, $j = 1, 2$, with $b_j \leq 1/2$ must have $b_3 < 1$. Thus there is no point $(a, b) \in \mathfrak{U}^2$ such that

$$(d((a, b), (x_i, y_i)))_{i=1}^3 = (1/2, 1/2, 1).$$

□

Remark 4. The same argument as in the proof of Theorem 12 also works out for the product $\mathfrak{U} \times [0, 1]^2$ (equipped with the maximum metric). Hence $\mathfrak{U} \times [0, 1]^2$ and therefore $\mathfrak{U} \times [0, 1]$ (also equipped with the maximum metric) is not isometrically isomorphic to $\mathfrak{U}$.

For any $\alpha > 0$, $(\mathfrak{U}, \alpha d)$ also has the extension property and is therefore isometrically isomorphic to $(\mathfrak{U}, d)$: Suppose $\iota : C \to \mathfrak{U}$ is an isometric embedding of a compact space (C, d_C) into $(\mathfrak{U}, \alpha d)$, and suppose (C', d'_C) is a one-point extension of (C, d_C). By

$$\alpha d(\iota x, \iota y) = d_C(x, y), \qquad x, y \in C$$

the same ι embeds $(C, \alpha^{-1} d_C)$ isometrically into $(\mathfrak{U}, d)$. We thus can find an isometry $\iota' : (C', \alpha^{-1} d_{C'}) \to (\mathfrak{U}, d)$ which extends ι. By the same argument as before, ι' is also an isometric embedding of $(C', d_{C'})$ into $(\mathfrak{U}, \alpha d)$.

2.4 The Set $\mathcal{U}$ of Universal Matrices

As for incidence matrices in Section 1, the group S^∞ of permutations on $\mathbb{N}$ acts via formula (1.4) also on the cone of all distance matrices. Unlike to the case of graphs may not regard $\mathrm{Aut}(\mathfrak{U})$ as a subgroup of S^∞.

Theorem 13. *The set $\mathcal{U}$ of all universal distance matrices is S_∞-invariant, i.e. $g\mathcal{M}g^{-1} = \mathcal{M}$ for every $g \in S_\infty$. Moreover, it is a dense G_δ-subset of $\mathcal{R}$.*

Proof. From the remark which follows Lemma 5 it is obvious that a distance matrix $r = (r_{ij})_{i,j=1}^\infty$ is universal if and only if for every finite subset $\{k_1, k_2, \ldots, k_n\} \subseteq \mathbb{N}$,

$$\mathrm{Adm}((r_{k_i k_j})_{i,j=1}^n) = \text{closure}\ \{(r_{k_1 k}, r_{k_2 k}, \ldots, r_{k_n k}) : k = n, n+1, \ldots\}.$$

Thus $\mathcal{M}$ is S_∞-invariant.

By the above construction, there is at least one universal matrix $r \in \mathcal{R}$. Its orbit $S_\infty r$ dense in $\mathcal{R}$, and consists of universal matrices only. Hence $\mathcal{U}$ is dense in $\mathcal{R}$. For fixed $n, m \in \mathbb{N}$ and $\alpha > 0$, the function

$$\varepsilon_{n,m,\alpha} : r = (r_{ij}) \mapsto \sup_{a \in \mathrm{Adm}(\pi_n r), \|a\| \leq \alpha} \ \min_{n<k\leq m} \|a - (r_{ik})_{i=1}^n\|,$$

where $\|\cdot\|$ is the maximum norm on $\mathbb{R}^n$, is a continuous map $\mathcal{R} \to \mathbb{R}$. By definition,

$$\mathcal{U} = \bigcap_{k\in N} \bigcap_{n\in\mathbb{N}} \bigcap_{\alpha\in\mathbb{N}} \bigcup_{m>n} [\varepsilon_{n,m,\alpha} < \frac{1}{k}]$$

is therefore a G_δ-set. □

We will now give a probabilistic proof of the existence of the Urysohn space in the spirit of Erdös Theorem 5. This procedure will also give an idea of a random metric space. To emphasize the principle idea we only give an intuitive but not formally rigorous presentation:

Starting with a single point we successively construct a random sequence (M_n, d_n) of n-point metric spaces (or n-dimensional distance matrices r_n) ,

$$(M_0, d_0) \subseteq (M_1, d_1) \subseteq (M_2, d_2) \subseteq \ldots \subseteq (M_n, d_n) \subseteq \ldots,$$

as follows: to a given n-point metric space M_n (with distance matrix $r_n \in \mathcal{R}_n$) we randomly add a $n+1$-th point choosing the distances between the new and the previous points (= admissible vector $\in \mathrm{Adm}(r_n)$) according to a certain (conditional) probability distribution μ_{r_n} (this is a probability measure on $\mathrm{Adm}(r_n)$).

To ensure that, with probability one, the so obtained space is Urysohn (or equivalently, its distance matrix is universal), we choose the distributions μ_r as follows: On the one-dimensional admissible cone $\mathbb{R}^+$ we take any fixed measure $\mu_\emptyset$ which supports the whole half-line (i.e. to any open subset B of $\mathbb{R}^+$, $\mu_\emptyset(B) > 0$). Now we continue inductively via the following procedure:

Suppose we had constructed all the measures μ_{r_n} for all one-dimensional distance matrices $r_n \in \mathcal{R}_n$ such that its support $\mathrm{supp}\, \mu_{r_n}$ is always the whole admissible cone $\mathrm{Adm}(r_n)$. For any $n+1$-dimensional distance matrix $r_{n+1} \in \mathcal{R}_{n+1}$, recall that

$$\mathrm{Adm}(\pi_n r_{n+1}) \stackrel{\pi_n}{\longleftarrow} \mathrm{Adm}(r_{n+1}),$$

where the projection is onto. Now in $\mathrm{Adm}(r_{n+1})$ we take any measure $\mu_{r_{n+1}}$ such that $\pi_n \mu_{r_{n+1}} = \mu_{r_n}$ and

$$\mathrm{supp}\, \mu_{r_{n+1}} = \mathrm{Adm}(r_{n+1})$$

Such a measure does exist, but we won't give the detailed construction here.

On the cone $\mathcal{R}_1$ we put a probability measure μ_1, $\mu_1 = \mu_\emptyset$. Using the measures μ_r, we define probability measures μ_n on $\mathcal{R}_n$ via

$$\mu_n(\,.\,|\pi_{n-1}r) = \mu_{\pi_{n-1}r}(\tilde{\pi}_{n-1}\,.\,),$$

where $\tilde{\pi}_{n-1}(r_{ij})_{i,j=1}^n = (r_{1,n}, r_{2,n}, \dots, r_{n-1,n})$. Note that $\tilde{\pi}_{n-1}$ maps each matrix $r \in \mathcal{R}_n$ into the the admissible cone $\mathrm{Adm}(\pi_{n-1}r)$, hence this definition makes sense.

The so constructed family μ_n is consistent under projection,

$$\pi_m \mu_n = \mu_m, \qquad m \leq n,$$

hence it defines a probability measure μ on $\mathcal{R}$. By construction the probability that the random distance matrix is universal equals one, thus

$$\mu(\mathcal{U}) = 1.$$

Remark 5. To assure that, with probability one, the random distance matrix is universal, it is certainly not necessary that the conditional distributions are consistent under projections, i.e.

$$\mu_{\pi_m r} = \pi_m \mu_r, \qquad r \in \mathcal{R}_n, m \leq n,$$

as assumed in the discussion above (see [12]). Anyway, note that the measure μ in the above construction is generically not S_∞-invariant.

At the end of this section we gather

- (cf. Theorem 9) There exists polish space $(\mathfrak{U}, d)$ which is Urysohn, i.e. universal and homogeneous. This space is unique up to isometric equivalence.
- (cf. Theorem 11) The Urysohn space $(\mathfrak{U}, d)$ is path-wise connected. All homotopy groups π_n of $(\mathfrak{U}, d)$, and therefore all homology groups, are trivial.
- Open question: Is $(\mathfrak{U}, d)$ contractible?
- Open question: Find familiar spaces to which $(\mathfrak{U}, d)$ is homeomorphic.
- (cf. Theorem 8) A space is Urysohn if and only if its distance matrix of any (= some) everywhere dense countable subset is universal.
- (cf. Theorem 13) The set of all universal matrices is dense G_δ subset of $\mathcal{R}$, which is invariant under finite permutations.

2.5 Random Metric Space

Here we will give a sketch of the construction of the random metric space. For this we will define a probability measure on the space of distance matrices which is concentrated on the subset of universal matrices. This construction is analogue to the same construction of the random graph but more complicate because it is not possible to use independence of entries (triangle inequality is the obstruction to that). Choose some measure γ on the half line $\mathbf{R}_+$ which has full support (e.g. Gaussian measure on the half line) and let entries

$$r_{1,2}, \dots r_{1,n}, \dots$$

are i.i.d with distribution γ. We will define the conditional distribution of the element of distance matrix $r_{m,n}, (n > m)$ under the condition that all the elements with indices (i, k) are fixed where $1 \leq i \leq m-1, 1 \leq j \leq n$

Then the conditional distribution of the element $r_{m.n}$ is uniform distribution on the interval $[a, b]$ where

$$a = \max_{i=1,\dots m-1} |r_{i,m} - r_{i,n}|; \mathrm{b} = \min_{i=1\dots m-1} |r_{i,m} + r_{i,n}|$$

Lemma 2.10 shows that $a \leq b$ so interval $[a, b]$ is not empty and we can by induction to define the distributions of all entries of the distance matrix. Remark that distribution of the element $r_{m,n}$ actually depends on the distribution of $r_{i,m}$ and $r_{i,n}$ for $i = 1, \dots m-1$ only. As result we had define a probability measure μ_γ on the space of distance matrices with parameter – measure γ on the half-line.

Theorem 14. *For any absolutely continuous measure γ on the half line* $\mathbf{R}$ *with positive density the set of universal distance matrices has measure 1 with respect to the measure μ_γ. In another word. Let r is a random distance matrix with distribution mu_γ in the space of all distance matrices; then a metric space which is completion of the set of naturals* $\mathbf{N}$ *under the distance matrix r is the Urysohn space with probability 1.*

The proof uses the law of large numbers in the special situation but we will not prove the theorem here. As in the case of graphs we emphasize that the last fact is valid for very reach set of measures on the space of distance matrices, – we gave only one simple example which is in a sense similar to the example with independent entries for the graphs. Namely, as in paragraph 1 for the case of graph, the set of probability measures on the space of distance matrices for which subset of universal matrices has a full measure is again everywhere dense G_δ subset of the space of all probability measures with weak topology.

Acknowledgements

Author is grateful to U. Haboeck and Yu. Yakubovich for the help in the preparation of the manuscript.

References

1. S. A. Bogatyĭ, *Metrically homogeneous spaces. (russian. russian summary)*, Uspekhi Mat. Nauk **57** (2002), no. 2(344), 3–22, translation in Russian Math. Surveys 57 (2002), no. 2, 221–240.
2. P. J. Cameron, *The random graph*, The mathematics of Paul Erdös, II, 333–351, Algorithms Combin., 14, Springer, Berlin, 1997.
3. J. Dugundji, *Topology*, Allyn and Bacon, Boston, Mass., 1966.
4. P. Erdös and A. Rényi, *Asymetric graphs*, Acta. Math. Acad. Sci. Hungar **14** (1963), 295–315.
5. G. E. Huhunaišvili, *On a property of Uryson's universal metric space.(russian)*, Dokl. Akad. Nauk SSSR (N.S.) **101** (1955), 607–610.
6. M. Katětov, *On universal metric spaces*, General topology and its relations to modern analysis and algebra, VI (Prague,1986), pp. 323–330, Heldermann, Berlin, 1988.
7. J. Lindenstrauss, G. Olsen, and Y. Sternfeld, *The Poulsen simplex*, Ann. Inst. Fourier (Grenoble) **28** (1978), no. 1,vi, 91–114.
8. W. Lusky, *The Gurarij spaces are unique*, Arch. Math. (Basel) **27** (1976), no. 6, 627–635.
9. R. R. Phelps, *Lectures on Choquet's Theorem*, Van Nostrand Mathematical Studies, vol. 7, Van Nostrand Company, Canada, 1966.
10. P. S. Urysohn, *Sur un espace metrique universel*, Bull. Sci. Math. **51** (1927), 1–38.
11. V. V. Uspenskij, *On the group of isometries of the Urysohn universal metric space*, Comment. Math. Univ. Carolin. **31** (1990), no. 1, 181–182.
12. A. M. Vershik, *Random and universal metric spaces*, Preprint of the Erwin Schrödinger Institute, No. 1234; to appear in Fundamental Mathematics Today, publ. by MCCMI (Moscow), 2003.
13. A. M. Vershik, *Classification of measureable functions of several arguments and invariantly distributed random matrices*, Funct. Anal. Appl. **36** (2002), no. 2, 12–27
14. A. M. Vershik, *Random matrix space is Urysohn space*, Dokl. Russian Acad. of Sci. **386** (2002), no. 6.

Part II

Zeta Functions

From Physics to Number Theory Via Noncommutative Geometry

Alain Connes[1] and Matilde Marcolli[2]

[1] Collège de France, 3, rue Ulm, F-75005 Paris, France
I.H.E.S. 35 route de Chartres F-91440 Bures-sur-Yvette, France
Department of Mathematics, Vanderbilt University, TN-37240, USA
connes@ihes.fr, alain@connes.org
[2] Max–Planck Institut für Mathematik, Vivatsgasse 7,
D-53111 Bonn, Germany, marcolli@mpim-bonn.mpg.de

Introduction

e volta nostra poppa nel mattino,
de' remi facemmo ali al folle volo
—— *Dante, Inf. XXVI 124-125*

Several recent results reveal a surprising connection between modular forms and noncommutative geometry.

The first occurrence came from the classification of noncommutative three spheres, [C–DuboisViolette-I] [C–DuboisViolette-II]. Hard computations with the noncommutative analog of the Jacobian involving the ninth power of the Dedekind eta function were necessary in order to analyze the relation between such spheres and noncommutative nilmanifolds. Another occurrence can be seen in the computation of the explicit cyclic cohomology Chern character of a spectral triple on $SU_q(2)$ [C–02]. Another surprise came recently from a remarkable action of the Hopf algebra of transverse geometry of foliations of codimension one on the space of lattices modulo Hecke correspondences, described in the framework of noncommutative geometry, using a modular Hecke algebra obtained as the cross product of modular forms by the action of Hecke correspondences [C–Moscovici-I] [C–Moscovici-II]. This action determines a differentiable structure on this noncommutative space, related to the Rankin–Cohen brackets of modular forms, and shows their compatibility with Hecke operators. Another instance where properties of modular forms can be recast in the context of noncommutative geometry can be found in the theory of modular symbols and Mellin transforms of cusp forms of weight two, which can be recovered from the geometry of the moduli space of Morita equivalence classes of noncommutative tori viewed as boundary of the modular curve [Manin–M].

The theory of modular Hecke algebras, the spectral realization of zeros of L-functions, and the arithmetic properties of KMS states in quantum statistical mechanics combine into a unique general picture based on the noncommutative geometry of the space of commensurability classes of $\mathbb{Q}$-lattices. This theme will be explored in depth in our forthcoming book [C–M-1]. In this paper we concentrate on the arithmetic properties of the spaces of commensurability classes of 1 and 2-dimensional $\mathbb{Q}$-lattices up to scaling.

An n-dimensional $\mathbb{Q}$-lattice consists of an ordinary lattice Λ in $\mathbb{R}^n$ and a homomorphism

$$\phi : \mathbb{Q}^n/\mathbb{Z}^n \to \mathbb{Q}\Lambda/\Lambda.$$

Two such $\mathbb{Q}$-lattices are *commensurable* if and only if the corresponding lattices are commensurable and the maps agree modulo the sum of the lattices.

The description of the spaces of commensurability classes of $\mathbb{Q}$-lattices via noncommutative geometry yields two quantum systems related by a duality. The first system is of quantum statistical mechanical nature, with the algebra of coordinates parameterizing commensurability classes of $\mathbb{Q}$-lattices modulo scaling and with a time evolution with eigenvalues given by the index of pairs of commensurable $\mathbb{Q}$-lattices. There is a symmetry group acting on the system, in general by *endomorphisms*. It is this symmetry that is spontaneously broken at low temperatures, where the system exhibits distinct phases parameterized by arithmetic data. We completely analyze the phase transition with spontaneous symmetry breaking in the two-dimensional case, where a new phenomenon appears, namely that there is a second critical temperature, beyond which no equilibrium state survives.

In the "dual system", which corresponds just to commensurability of $\mathbb{Q}$-lattices, the scaling group is acting. In physics language, what emerges is that the zeros of zeta appear as an absorption spectrum of the scaling action in the L^2 space of the space of commensurability classes of $\mathbb{Q}$-lattices as in [C–99]. While the zeros of zeta and L-functions appear at the critical temperature, the analysis of the low temperature equilibrium states concentrates on the subspace

$$\mathrm{GL}_n(\mathbb{Q})\backslash\mathrm{GL}_n(\mathbb{A})$$

of *invertible* $\mathbb{Q}$-lattices, which as is well known plays a central role in the theory of automorphic forms.

While, at first sight, at least in the 1-dimensional case, it would seem easy to classify commensurability classes of $\mathbb{Q}$-lattices, we shall see that ordinary geometric tools fail because of the ergodic nature of the equivalence relation.

Such quotients are fundamentally of "quantum nature", in that, even though they are sets in the ordinary sense, it is impossible to distinguish points by any finite (or countable) collection of invariants. Noncommutative geometry is specifically designed to handle such quantum spaces by encoding them by algebras of non-commuting coordinates and extending the techniques of ordinary geometry using the tools of functional analysis, noncommutative algebra, and quantum physics.

Direct attempts to define function spaces for such quotients lead to invariants that are of a cohomological nature. For instance, let the fundamental group Γ of a Riemann surface act on the boundary $\mathbb{P}^1(\mathbb{R})$ of its universal cover identified with the Poincaré disk. The space

$$L^\infty(\Gamma\backslash\mathbb{P}^1(\mathbb{R})) := L^\infty(\mathbb{P}^1(\mathbb{R}))^\Gamma$$

is in natural correspondence with global sections of the sheaf of (real parts of) holomorphic functions on the Riemann surface, as boundary values. More generally, the cyclic cohomology of the noncommutative algebra of coordinates on such quotients is obtained by applying derived functors to these naive functorial definition of function spaces.

In the case of 1-dimensional $\mathbb{Q}$-lattices, the states at zero temperature are related to the Kronecker–Weber construction of the maximal abelian extension $\mathbb{Q}^{ab}$. In fact, in this case the quantum statistical mechanical system is the one constructed in [Bost–C], which has underlying geometric space X_1 parameterizing commensurability classes of 1-dimensional $\mathbb{Q}$-lattices modulo scaling by $\mathbb{R}_+^*$. The corresponding algebra of coordinates is a Hecke algebra for an almost normal pair of solvable groups. The regular representation is of type III_1 and determines the time evolution of the system, which has the set of $\log(p)$, p a prime number, as set of basic frequencies. The system has an action of the idèle class group modulo the connected component of identity as a group of symmetries. This induces a Galois action on the ground states of the system at zero temperature. When raising the temperature the system has a phase transition, with a unique equilibrium state above the critical temperature. The Riemann zeta function appears as the partition function of the system, as in [Julia].

Each equivalence class of $\mathbb{Q}$-lattices determines an irreducible covariant representation, where the Hamiltonian is implemented by minus the log of the covolume. For a general class, this is not bounded below. It is so, however, in the case of equivalence classes of *invertible* $\mathbb{Q}$-lattices, *i.e.* where the labelling of torsion points is one to one. These classes then define positive energy representations and corresponding KMS states for all temperatures below critical.

In the 2-dimensional case, as the temperature lowers, the system settles down on these invertible $\mathbb{Q}$-lattices, so that the zero temperature space is commutative and is given by the Shimura variety

$$\mathrm{GL}_2(\mathbb{Q})\backslash \mathrm{GL}_2(\mathbb{A})/\mathbb{C}^*.$$

The action of the symmetry group, which in this case is nonabelian and isomorphic to $\mathbb{Q}^*\backslash \mathrm{GL}_2(\mathbb{A}_f)$, is more subtle due to the presence of inner automorphisms and the necessary use of the formalism of superselection sectors. Moreover, its effect on the zero temperature states is not obtained directly but is induced by the action at non-zero temperature, which involves the full noncommutative system. The quotient $\mathrm{GL}_2(\mathbb{Q})\backslash(M_2(\mathbb{A}_f)\times \mathrm{GL}_2(\mathbb{R}))/\mathbb{C}^*$ and the space of 2–dimensional $\mathbb{Q}$-lattices modulo commensurability and scaling are the same, hence the corresponding algebras are Morita equivalent. However, it is preferable to work with the second description, since, by taking the classical quotient by the action of the subgroup $\mathrm{SL}_2(\mathbb{Z})$, it reduces the group part in the cross product to the classical Hecke algebra.

The GL_2 system has an arithmetic structure provided by a *rational* subalgebra, given by a natural condition on the coefficients of the q-series. We show that it is a Hecke algebra of modular functions, closely related to the modular Hecke algebra of [C–Moscovici-I], [C–Moscovici-II]. The symmetry group acts on the values of ground states on this rational subalgebra as the automorphism group of the modular field.

Evaluation of a generic ground state φ of the system on the rational subalgebra generates an embedded copy of the modular field in $\mathbb{C}$ and there exists a unique isomorphism of the symmetry group of the system with the Galois group of the embedded modular field, which intertwines the Galois action on the image with the symmetries of the system,

$$\theta(\sigma)\circ\varphi = \varphi\circ\sigma.$$

The relation between this GL_2 system and class field theory is being investigated in ongoing work [C–M–Ramachandran].

The arithmetic structure is inherited by the dual of the GL_2 system and enriches the structure of the noncommutative space of commensurability classes of 2-dimensional $\mathbb{Q}$-lattices to that of a "noncommutative arithmetic variety".

The dual of the GL_1 system, under the duality obtained by taking the cross product by the time evolution, corresponds to the space of commensurability classes of 1-dimensional $\mathbb{Q}$-lattices, not considered up to scaling. This corresponds geometrically to the total space $\mathcal{L}$ of a principal $\mathbb{R}_+^*$ bundle over the base X_1, and determines a natural scaling action of $\mathbb{R}_+^*$. The space $\mathcal{L}$ is described by the quotient

$$\mathcal{L} = \mathrm{GL}_1(\mathbb{Q})\backslash\mathbb{A}^{\cdot},$$

where $\mathbb{A}^{\cdot}$ denotes the set of adèles with nonzero archimedean component. The corresponding algebra of coordinates is Morita equivalent to $C(X_1) \rtimes_{\sigma_t} \mathbb{R}$.

Any approach to a spectral realization of the zeros of zeta through the quantization of a classical dynamical system faces the problem of obtaining the leading term in the Riemann counting function for the number of zeros of imaginary part less than E as a volume in phase space. The solution [C–99] of this issue is achieved in a remarkably simple way, by the scaling action of $\mathbb{R}_+^*$ on the phase space of the real line $\mathbb{R}$, and will be dealt with in our forthcoming book [C–M-1].

In particular, this shows that the space $\mathcal{L}$ requires a further compactification at the archimedean place, obtained by replacing the quotient $\mathcal{L} = \mathrm{GL}_1(\mathbb{Q})\backslash\mathbb{A}^{\cdot}$ by $\overline{\mathcal{L}} = \mathrm{GL}_1(\mathbb{Q})\backslash\mathbb{A}$ *i.e.* dropping the non vanishing of the archimedean component. This compactification has an analog for the GL_2 case, given by the noncommutative boundary of modular curves considered in [Manin–M], which corresponds to replacing $\mathrm{GL}_2(\mathbb{R})$ by $M_2(\mathbb{R})$ at the archimedean place, and is related to class field theory for real quadratic fields through Manin's real multiplication program [Manin].

The space $\overline{\mathcal{L}}$ appears as the configuration space for a quantum field theory, where the degrees of freedom are parameterized by prime numbers, including infinity. When only finitely many degrees of freedom are considered, and in particular only the place at infinity, the semiclassical approximation exhibits the main terms in the asymptotic formula for the number of zeros of the Riemann zeta function.

The zeros of zeta appear as an absorption spectrum, namely as lacunae in a continuous spectrum, where the width of the absorption lines depends on the presence of a cutoff. The full idèles class group appears as symmetries of the system and L-functions with Grössencharakter replace the Riemann zeta function in nontrivial sectors.

From the point of view of quantum field theory, the field configurations are given by adèles, whose space $\mathbb{A}$ is then divided by the action of the gauge group $\mathrm{GL}_1(\mathbb{Q})$. As mentioned above, the quotient space is essentially the same as the space $\mathcal{L}$ of commensurability classes of 1-dimensional $\mathbb{Q}$-lattices. The $\log(p)$ appear as periods of the orbits of the scaling action. The Lefschetz formula for the scaling action recovers the Riemann–Weil explicit formula as a semi-classical approximation. The exact quantum calculation for finitely many degrees of freedom confirms this result. The difficulty in extending this calculation to the global case lies in the quantum field theoretic problem of passing to infinitely many degrees of freedom. These aspects will be discussed in [C–M-1].

The dual system $\mathcal{L}$ can be interpreted physically as a "universal scaling system", since it exhibits the continuous renormalization group flow and its relation with the discrete scaling by powers of primes. For the primes two and three, this discrete scaling manifests itself in acoustic systems, as is well known in western classical music, where the two scalings correspond, respectively, to passing to the octave (frequency ratio of 2) and transposition (the perfect fifth is the frequency ratio 3/2), with the approximate value $\log(3)/\log(2) \sim 19/12$ responsible for the difference between the "circulating temperament" of the Well Tempered Clavier and the "equal temperament" of XIX century music. It is precisely the irrationality of $\log(3)/\log(2)$ which is responsible for the noncommutative nature of the quotient corresponding to the three places $\{2, 3, \infty\}$.

The main features of the dual systems in the GL_1 case are summarized in the following table:

Quantum statistical mechanics	Quantum field theory
Commensurability classes of $\mathbb{Q}$-lattices modulo scaling	Commensurability classes of $\mathbb{Q}$-lattices
$A = C^*(\mathbb{Q}/\mathbb{Z}) \rtimes \mathbb{N}^\times$	$A \rtimes_{\sigma_t} \mathbb{R}$
Time evolution σ_t	Energy scaling $U(\lambda)$, $\lambda \in \mathbb{R}_+^*$
$\{\log p\}$ as frequencies	$\{\log p\}$ as periods of orbits
Arithmetic rescaling μ_n	Renormalization group flow $\mu\partial_\mu$
Symmetry group $\hat{\mathbb{Z}}^*$ as Galois action on $T = 0$ states	Idèles class group as gauge group
System at zero temperature	$\mathrm{GL}_n(\mathbb{Q})\backslash\mathrm{GL}_n(\mathbb{A})$
System at critical temperature (Riemann's ζ as partition function)	Spectral realization (Zeros of ζ as absorption spectrum)
Type III_1	Type II_∞

There is a similar duality (and table) in the GL_2 case, where part of the picture remains to be clarified. The relation with the modular Hecke algebra of [C–Moscovici-I], [C–Moscovici-II] is more natural in the dual system where modular forms with non-zero weight are naturally present.

The fact that the KMS state at critical temperature can be expressed as a noncommutative residue (Dixmier trace) shows that the system at critical temperature should be analyzed with tools from quantum field theory and renormalization. The key role of the continuous renormalization group flow as a symmetry of the dual system $\mathcal{L}$ and its similarity with a Galois group at the archimedean place brings us to Chapter II of this work which deals with the relation between renormalization and motivic Galois theory and whose content will now be briefly described.

The mathematical theory of renormalization in QFT developed in [C–Kreimer-I] [C–Kreimer-II] shows in geometric terms that the procedure of perturbative renormalization can be described as the Birkhoff decomposition

$$\gamma(z) = \gamma_-(z)^{-1}\,\gamma_+(z) \tag{0.1}$$

on the projective line of complexified dimensions z of the loop $\gamma(z) \in G$ given by the unrenormalized theory. The $\gamma_-(z)$ side of the Birkhoff decomposition yields the counterterms and the $\gamma_+(z)$ side evaluated at the critical dimension gives the renormalized value of the theory. The group G is the group of "diffeographisms" of the physical theory based on the Hopf algebra of Feynman graphs. It contains the renormalization group as a natural 1-parameter subgroup.

Moreover, two types of considerations motivated in [C–01] the expectation of relating concretely the renormalization group to a Galois group. On the one hand, it was shown in [C–00] that the classification of approximately finite factors provides a nontrivial Brauer theory for central simple algebras over $\mathbb{C}$, and an archimedean analog of the module of central simple algebras over nonarchimedean fields. The relation of Brauer theory to the Galois group is via the construction of central simple algebras as cross products of a field by a group of automorphisms. This was realized for type II_1 in [C–DuboisViolette-II], as the cross product of the field of elliptic functions by an automorphism given by translation on the elliptic curve. The results on the GL_2 system suggest the possibility of an analogous construction for type III_1 factors using the modular field. On the other hand, the coupling constants g of the fundamental interactions (electromagnetic, weak and strong) are not really constants but depend on the energy scale μ and are therefore functions $g(\mu)$. Thus, high energy physics implicitly extends the "field of constants", passing from the field of scalars $\mathbb{C}$ to a field of functions containing all the $g(\mu)$. On this field, the renormalization group provides the corresponding theory of ambiguity. These considerations suggest the idea that the renormalization group should be related to a still mysterious Galois theory at the archimedean place. We will return to discuss issues related to Galois theory and arithmetic geometry at the archimedean place in [C–M-1].

In the treatment of renormalization that we present in this paper, we realize concretely a Galois interpretation of the renormalization group, in the context of motivic Galois theory. In fact, we show that perturbative renormalization, in the dimensional regularization (Dim-Reg) and minimal substraction scheme, is governed by a universal "motivic Galois group" U, which is independent of the physical theory and acts on the set of dimensionless coupling constants of physical theories, through a map to the corresponding group G of diffeographisms, which in turn maps to formal diffeomorphisms, as shown in [C–Kreimer-II]. The natural appearance of the "motivic Galois group" in the context of renormalization confirms a suggestion made by Cartier in [Cartier], that in the Connes–Kreimer theory of perturbative renormalization one should find a hidden "cosmic Galois group" closely related in structure to the Grothendieck–Teichmüller group.

The starting point for the relation of perturbative renormalization to motivic Galois theory is a form of the 't Hooft relations, given by the scattering formula

$$\gamma_-(z) = \lim_{t\to\infty} e^{-t\left(\frac{\beta}{z}+Z_0\right)}\, e^{tZ_0} \tag{0.2}$$

proved in [C–Kreimer-II], which expresses the counterterms in the Birkhoff decomposition (0.1) through the residues of graphs.

When this formula is expressed more explicitly in terms of the time ordered exponential of physicists (also known as expansional in mathematical terminology), the resulting expansional can be recognized as the solution of a differential system. This step of passing from Birkhoff decomposition of loops to a class of differential equations suggests the presence of an underlying Riemann–Hilbert correspondence. This, in general, establishes an equivalence between a category of differential systems with singularities and certain representation theoretic data. In our setting, the appropriate class of differential systems is identified via a geometric reformulation of the main properties of the loops $\gamma_\mu(z)$ of the unrenormalized theories.

We consider as base space a punctured disk Δ^*, which is the space of complexified dimensions around the dimension D of space-time, and a principal $\mathbb{G}_m$-bundle B over Δ^*, whose fibers account for the arbitrariness in the normalization of integration in complexified dimension $z \in \Delta^*$. The $\mathbb{G}_m$-action corresponds to the rescaling $\hbar\,\partial/\partial\hbar$. For G the group of diffeographisms of a given theory, we then consider G-valued flat connection on B, which are *equisingular*. The equisingularity condition translates in geometric terms the physical fact that the counterterms are independent of the additional choice of a unit of mass μ. An equisingular flat G-valued connection on B is $\mathbb{G}_m$-invariant, singular on the fiber over zero, and such that the equivalence class of the singularity of the pullback of the connection by a section of the principal $\mathbb{G}_m$-bundle only depends on the value of the section at the origin.

The classification of equivalence classes of such differential systems can then be obtained in the form of a Riemann–Hilbert correspondence, by considering the category of flat equisingular vector bundles. These can be organized into a neutral Tannakian category with a natural fiber functor to the category of vector spaces. The Tannakian category obtained this way is equivalent to the category of finite dimensional representations of the affine group scheme $U^* = U \rtimes \mathbb{G}_m$, which is uniquely determined by this property and universal with respect to the set of physical theories.

We construct a specific "universal singular frame" on principal U-bundles over B. When using in this frame the dimensional regularization technique of QFT, all divergences disappear and one obtains a finite theory, which only depends upon the choice of a local trivialization for the principal $\mathbb{G}_m$-bundle B. The coefficients of the universal singular frame, written out in the expansional form, are the same rational numbers that appear as coefficients in the local index formula of Connes–Moscovici [C–Moscovici-0].

In particular, representations $U^* \to G^* = G \rtimes \mathbb{G}_m$ classify flat equisingular G-valued differential systems for G the diffeographisms group of a given physical theory, while U^* is universal and independent of the physical theory. More explicitly, U^* is the semi-direct product by its grading of the graded pro-unipotent Lie group U whose Lie algebra is the free graded Lie algebra

$$\mathcal{F}(1,2,3,\cdots)_\bullet$$

generated by elements e_{-n} of degree n, $n > 0$. The way it maps to the diffeographism group G is by mapping the generator e_{-n} to the n-th graded piece of the β-function, viewed as an element in $\mathrm{Lie}(G)$. In particular, it follows that the renormalization group (whose infinitesimal generator is the element β) can be lifted canonically to a 1-parameter subgroup of the universal group U^*.

Closely related group schemes appear in motivic Galois theory and U^* is for instance abstractly (but non-canonically) isomorphic to the motivic Galois group $G_{\mathcal{M}_T}(\mathcal{O})$ ([Deligne-Goncharov], [Goncharov]) of the scheme $S_4 = \mathrm{Spec}(\mathcal{O})$ of 4-cyclotomic integers, $\mathcal{O} = \mathbb{Z}[i][\frac{1}{2}]$. This suggests an intriguing relation between renormalization and mixed Tate motives.

These facts altogether indicate that the divergences of Quantum Field Theory, far from just being an unwanted nuisance, are a clear sign of the presence of totally unexpected symmetries of geometric origin. This shows, in particular, that one should understand how the universal singular frame "renormalizes" the geometry of space-time using Dim-Reg and the minimal subtraction scheme.

The structure of this work is organized as follows.

- The first Chapter is dedicated to the quantum statistical mechanical system of $\mathbb{Q}$-lattices, in the cases of dimension one and two, and its behavior at zero temperature.
- The second Chapter is dedicated to the theory of renormalization of [C–Kreimer-I], [C–Kreimer-II], the Riemann-Hilbert problem and the relation with motivic Galois theory, according to the results announced in [C–M].

References

[Bost–C] J.B. Bost, A. Connes, *Hecke algebras, Type III factors and phase transitions with spontaneous symmetry breaking in number theory*, Selecta Math. (New Series) Vol.1 (1995) N.3, 411–457.

[Cartier] P. Cartier, *A mad day's work: from Grothendieck to Connes and Kontsevich. The evolution of concepts of space and symmetry*, in "Les relations entre les mathématiques et la physique théorique", 23–42, Inst. Hautes Études Sci., Bures-sur-Yvette, 1998. (English translation in Bull. Amer. Math. Soc. (N.S.) 38 (2001), no. 4, 389–408).

[C–99] A. Connes, *Trace formula in Noncommutative Geometry and the zeros of the Riemann zeta function.* Selecta Mathematica. (New Series) Vol.5 (1999) 29–106.

[C–00] A. Connes, *Noncommutative Geometry and the Riemann Zeta Function*, Mathematics: Frontiers and perspectives, IMU 2000 volume.

[C–01] A. Connes, *Symétries Galoisiennes et Renormalisation*, in "Poincaré Seminar 2002: Vacuum Energy-Renormalization", Progress in Mathematical Physics, V. 30, Birkhauser 2003.

[C–02] A. Connes, *Cyclic cohomology, Quantum group Symmetries and the Local Index Formula for* $SU_q(2)$.(2002), Math QA/0209142.

[C–DuboisViolette-I] A. Connes, M. Dubois-Violette, *Noncommutative finite-dimensional manifolds. I. spherical manifolds and related examples*, Comm. Math. Phys. Vol.230 (2002) N.3, 539–579.

[C–DuboisViolette-II] A. Connes, M. Dubois-Violette, *Moduli space and structure of noncommutative 3-spheres*, Lett. Math. Phys., Vol.66 (2003) N.1-2, 91–121.

[C–Kreimer-I] A. Connes and D. Kreimer, *Renormalization in quantum field theory and the Riemann-Hilbert problem. I. The Hopf algebra structure of graphs and the main theorem.* Comm. Math. Phys. 210 (2000), N.1, 249–273.

[C–Kreimer-II] A. Connes and D. Kreimer, *Renormalization in quantum field theory and the Riemann-Hilbert problem. II. The* β*-function, diffeomorphisms and the renormalization group.* Comm. Math. Phys. 216 (2001), N.1, 215–241.

[C–M] A. Connes, M. Marcolli, *Renormalization and motivic Galois theory.* (2004), Math NT/0409306.

[C–M-1] A. Connes, M. Marcolli, *Noncommutative Geometry from Quantum Physics to Motives.* Book in preparation.

[C–M–Ramachandran] A. Connes, M. Marcolli, N. Ramachandran, *KMS states and complex multiplication*, in preparation.

[C–Moscovici-0] A. Connes, H. Moscovici, *The local index formula in noncommutative geometry, GAFA* **5** (1995), 174-243.

[C–Moscovici-I] A. Connes, H. Moscovici, *Modular Hecke algebras and their Hopf symmetry*, Moscow Math. Journal, Vol.4 (2004) N.1, 67–109.

[C–Moscovici-II] A. Connes, H. Moscovici, *Rankin-Cohen Brackets and the Hopf Algebra of Transverse Geometry*, Moscow Math. Journal, Vol.4 (2004) N.1, 111–130.

[Deligne-Goncharov] P. Deligne, A.B. Goncharov *Groupes fondamentaux motiviques de Tate mixte*, preprint, math.NT/0302267

[Goncharov] A. Goncharov, *Multiple polylogarithms and mixed Tate motives*, preprint, math.AG/0103059.

[Julia] B. Julia, *Statistical theory of numbers*, in Number Theory and Physics, J.-M. Luck, P. Moussa and M. Waldschmidt (Eds.), Springer Verlag, Berlin, 1990.

[Manin] Yu.I. Manin, *Real multiplication and noncommutative geometry (ein Alterstraum).* in "The legacy of Niels Henrik Abel", pp.685–727, Springer, 2004.

[Manin–M] Yu.I. Manin, M. Marcolli, *Continued fractions, modular symbols, and noncommutative geometry*, Selecta Mathematica (New Series) Vol.8 (2002) N.3, 475–520.

Quantum Statistical Mechanics of $\mathbb{Q}$-Lattices

1 Introduction

In this chapter we shall start by giving a geometric interpretation in terms of the space of commensurability classes of 1-dimensional $\mathbb{Q}$-lattices of the quantum statistical dynamical system (BC [5]). This system exhibits the relation between the phenomenon of spontaneous symmetry breaking and number theory. Its dual system obtained by taking the cross product by the time evolution is basic in the spectral interpretation of zeros of zeta.

Since $\mathbb{Q}$-lattices and commensurability continue to make sense in dimension n, we shall obtain an analogous system in higher dimension and in particular we derive a complete picture of the system in dimension $n = 2$. This shows two distinct phase transitions with arithmetic spontaneous symmetry breaking.

In the initial model of BC ([5]) the partition function is the Riemann zeta function. Equilibrium states are characterized by the KMS-condition. While at large temperature there is only one equilibrium state, when the temperature gets smaller than the critical temperature, the equilibrium states are no longer unique but fall in distinct phases parameterized by number theoretic data. The pure phases are parameterized by the various embeddings of the cyclotomic field $\mathbb{Q}^{ab}$ in $\mathbb{C}$.

The physical observables of the BC system form a C^*-algebra endowed with a natural time evolution σ_t. This algebra is interpreted here as the algebra of noncommuting coordinates on the space of commensurability classes of 1-dimensional $\mathbb{Q}$-lattices up to scaling by $\mathbb{R}_+^*$.

What is remarkable about the ground states of this system is that, when evaluated on the rational observables of the system, they only affect values that are algebraic numbers. These span the maximal abelian extension of $\mathbb{Q}$. Moreover, the class field theory isomorphism intertwines the two actions of the idèles class group, as symmetry group of the system, and of the Galois group, as permutations of the expectation values of the rational observables. That the latter action preserves positivity is a rare property of states. We abstract this property as a definition of "fabulous[3] states", in the more general context of arbitrary number fields and review recent developments in the direction of extending this result to other number fields.

We present a new approach, based on the construction of an analog of the BC system in the GL_2 case. Its relation to the complex multiplication case of the Hilbert 12th problem will be discussed specifically in ongoing work of the two authors with N. Ramachandran [13].

The C^*-algebra of observables in the GL_2-system describes the noncommutative space of commensurability classes of $\mathbb{Q}$-lattices in $\mathbb{C}$ up to scaling by $\mathbb{C}^*$.

---

[3]This terminology is inspired from John Conway's talk on "fabulous" groups.

A $\mathbb{Q}$-lattice in $\mathbb{C}$ is a pair (Λ, ϕ) where $\Lambda \subset \mathbb{C}$ is a lattice while

$$\phi : \mathbb{Q}^2/\mathbb{Z}^2 \longrightarrow \mathbb{Q}\Lambda/\Lambda$$

is a homomorphism of abelian groups (not necessarily invertible). Two $\mathbb{Q}$-lattices (Λ_j, ϕ_j) are commensurable iff the lattices Λ_j are commensurable (*i.e.* $\mathbb{Q}\Lambda_1 = \mathbb{Q}\Lambda_2$) and the maps ϕ_j are equal modulo $\Lambda_1 + \Lambda_2$. The time evolution corresponds to the ratio of covolumes of pairs of commensurable $\mathbb{Q}$-lattices. The group

$$S = \mathbb{Q}^* \backslash \mathrm{GL}_2(\mathbb{A}_f)$$

quotient of the finite adèlic group of GL_2 by the multiplicative group $\mathbb{Q}^*$ acts as symmetries of the system, and the action is implemented by endomorphisms, as in the theory of superselection sectors of Doplicher-Haag-Roberts ([16]).

It is this symmetry which is spontaneously broken below the critical temperature $T = \frac{1}{2}$. The partition function of the GL_2 system is $\zeta(\beta)\zeta(\beta - 1)$, for $\beta = 1/T$, and the system exhibits three distinct phases, with two phase transitions at $T = \frac{1}{2}$ and at $T = 1$. At low temperatures ($T < \frac{1}{2}$) the pure phases are parameterized by the set

$$\mathrm{GL}_2(\mathbb{Q}) \backslash \mathrm{GL}_2(\mathbb{A})/\mathbb{C}^*$$

of classes of invertible $\mathbb{Q}$-lattices (up to scaling). The equilibrium states of the "crystalline phase" merge as $T \to 1/2$ from below, as the system passes to a "liquid phase", while at higher temperatures ($T \geq 1$) there are no KMS states.

The subalgebra of *rational* observables turns out to be intimately related to the modular Hecke algebra introduced in (Connes-Moscovici [10]) where its surprising relation with transverse geometry of foliations is analyzed. We show that the KMS states at zero temperature when evaluated on the rational observables generate a specialization of the modular function field F. Moreover, as in the BC system the state intertwines the two actions of the group S, as symmetry group of the system, and as permutations of the expectation values of the rational observables by the Galois group of the modular field, identified with S by Shimura's theorem ([51]).

We shall first explain the general framework of quantum statistical mechanics, in terms of C^*-algebras and KMS states. Noncommutative algebras concretely represented in Hilbert space inherit a canonical time evolution, which allows for phenomena of phase transition and spontaneous symmetry breaking for KMS states at different temperatures.

There are a number of important nuances between the abelian BC case and the higher dimensional non-abelian cases. For instance, in the abelian case, the subfield of $\mathbb{C}$ generated by the image of the rational subalgebra under an extremal KMS_∞ state does not depend on the choice of the state and the intertwining between the symmetry and the Galois actions is also independent of the state. This no longer holds in the non-abelian case, because of the presence of inner automorphisms of the symmetry group S.

Moreover, in the GL_2 case, the action of S on the extremal KMS_∞ states is not transitive, and the corresponding invariant of the orbit of a state φ under S is the subfield $F_\varphi \subset \mathbb{C}$, which is the specialization of the modular field given by evaluation at the point in the upper half plane parameterizing the ground state φ.

Another important nuance is that the algebra A is no longer unital while $\mathcal{A}_\mathbb{Q}$ is a subalgebra of the algebra of unbounded multipliers of A. Just as an ordinary function need not be bounded to be integrable, so states can be evaluated on unbounded multipliers. In our case, the rational subalgebra $\mathcal{A}_\mathbb{Q}$ is not self-adjoint.

Finally the action of the symmetry group on the ground states is obtained via the action on states at positive temperature. Given a ground state, one warms it up below the critical temperature and acts on it by endomorphisms. When taking the limit to zero temperature of the resulting state, one obtains the corresponding transformed ground state. In our framework, the correct notion of ground states is given by a stronger form of the KMS_∞ condition, where we also require that these are weak limits of KMS_β states for $\beta \to \infty$.

We then consider the "dual" system of the GL_2-system, which describes the space of commensurability classes of 2-dimensional $\mathbb{Q}$-lattices (not up to scaling). The corresponding algebra is closely related to the modular Hecke algebra of [10]. As in the 1-dimensional case, where the corresponding space is compactified by removing the non-zero condition for the archimedean component of the adèle, the compactification of the two-dimensional system amounts to replacing the archimedean component $\mathrm{GL}_2(\mathbb{R})$ with matrices $M_2(\mathbb{R})$. This corresponds to the noncommutative compactification of modular curves considered in [37]. In terms of $\mathbb{Q}$-lattices this corresponds to degenerations to pseudo-lattices, as in [34].

It is desirable to have a concrete physical (experimental) system realizing the BC symmetry breaking phenomenon (as suggested in [43]). In fact, we shall show that the explicit presentation of the BC algebra not only exhibits a strong analogy with phase states, as in the theory of optical coherence, but it also involves an action on them of a discrete scaling group, acting by integral multiplication of frequencies.

Acknowledgements

We are very grateful to Niranjan Ramachandran for many extremely useful conversations on class field theory and KMS states, that motivated the GL_2 system described here, whose relation to the theory of complex multiplication is being investigated in [13]. We thank Marcelo Laca for giving us an extensive update on the further developments on [5]. We benefited from visits of the first author to MPI and of the second author to IHES and we thank both institutions for their hospitality. The second author is partially supported by a Sofja Kovalevskaya Award of the Humboldt Foundation and the German Government.

2 Quantum Statistical Mechanics

In classical statistical mechanics a state is a probability measure μ on the phase space that assigns to each observable f an expectation value, in the form of an average

$$\int f\, d\mu. \tag{2.1}$$

In particular for a Hamiltonian system, the Gibbs canonical ensemble is a measure defined in terms of the Hamiltonian and the symplectic structure on the phase space. It depends on a parameter β, which is an inverse temperature, $\beta = 1/kT$ with k the Boltzmann constant. The Gibbs measure is given by

$$d\mu_G = \frac{1}{Z} e^{-\beta H} d\mu_{Liouville}, \tag{2.2}$$

normalized by $Z = \int e^{-\beta H} d\mu_{Liouville}$.

When passing to infinitely many degrees of freedom, where the interesting phenomena of phase transitions and symmetry breaking happen, the definition of the Gibbs states becomes more involved (*cf.* [45]). In the quantum mechanical framework, the analog of the Gibbs condition is given by the KMS condition at inverse temperature β ([17]). This is simpler in formulation than its classical counterpart, as it relies only on the involutive algebra A of observables and its time evolution $\sigma_t \in \mathrm{Aut}(A)$, and does not involve any additional structure like the symplectic structure or the approximation by regions of finite volume.

In fact, quantum mechanically, the observables form a C^*-algebra A, the Hamiltonian is the infinitesimal generator of the (pointwise norm continuous) one parameter group of automorphisms $\sigma_t \in \mathrm{Aut}(A)$, and the analog of a probability measure, assigning to every observable a certain average, is given by a *state*.

Definition 2.1 *A state on a C^*-algebra A is a linear form on A such that*

$$\varphi(1) = 1\,, \quad \varphi(a^* a) \geqq 0 \qquad \forall a \in A\,. \tag{2.3}$$

When the C^*-algebra A is non unital the condition $\varphi(1) = 1$ is replaced by $||\varphi|| = 1$ where

$$||\varphi|| := \sup_{x \in A, ||x|| \leq 1} |\varphi(x)|\,. \tag{2.4}$$

Such states are restrictions of states on the unital C^*-algebra $\tilde{A}$ obtained by adjoining a unit.

The evaluation $\varphi(a)$ gives the expectation value of the observable a in the statistical state φ. The Gibbs relation between a thermal state at inverse temperature $\beta = \frac{1}{kT}$ and the time evolution

$$\sigma_t \in \mathrm{Aut}(A) \tag{2.5}$$

is encoded by the KMS condition ([17]) which reads

$$\forall a, b \in A\,,\ \exists F \text{ bounded holomorphic in the strip } \{z \mid \operatorname{Im} z \in [0, \beta]\} \tag{2.6}$$

$$F(t) = \varphi(a\,\sigma_t(b)) \qquad F(t + i\beta) = \varphi(\sigma_t(b)a) \quad \forall t \in \mathbb{R}.$$

In the case of a system with finitely many quantum degrees of freedom, the algebra of observables is the algebra of operators in a Hilbert space $\mathcal{H}$ and the time evolution is given by $\sigma_t(a) = e^{itH}\, a\, e^{-itH}$, where H is a positive self-adjoint operator such that $\exp(-\beta H)$ is trace class for any $\beta > 0$. For such a system, the analog of (2.2) is

$$\varphi(a) = \frac{1}{Z} \operatorname{Tr}\left(a\, e^{-\beta H}\right) \qquad \forall a \in A, \tag{2.7}$$

with the normalization factor $Z = \operatorname{Tr}(\exp(-\beta H))$. It is easy to see that (2.7) satisfies the KMS condition (2.6) at inverse temperature β.

In the nonunital case, the KMS condition is defined in the same way by (2.6). Let $M(A)$ be the multiplier algebra of A and let $B \subset M(A)$ be the C^*-subalgebra of elements $x \in M(A)$ such that $t \mapsto \sigma_t(x)$ is norm continuous.

Proposition 2.2 *Any state φ on A admits a canonical extension to a state still noted φ on the multiplier algebra $M(A)$ of A. The canonical extension of a KMS state still satisfies the KMS condition on B.*

Proof. For the first statement we refer to [42]. The proof of the second statement illustrates a general density argument, where the continuity of $t \mapsto \sigma_t(x)$ is used to control the uniform continuity in the closed strip, in order to apply the Montel theorem of normal families. Indeed, by weak density of A in $M(A)$, one obtains a sequence of holomorphic functions, but one only controls their uniform continuity on smooth elements of B. □

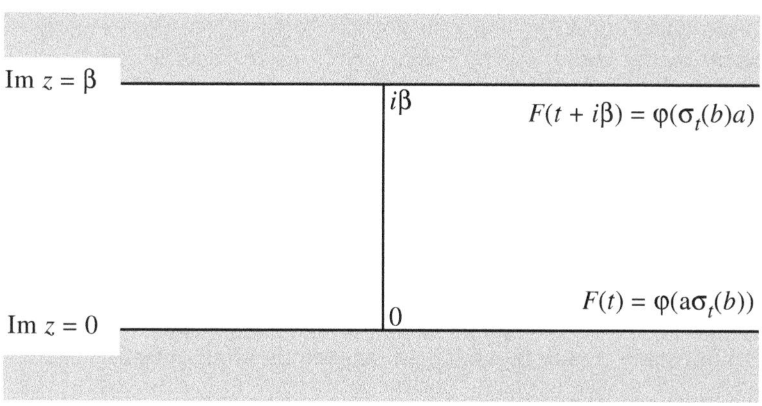

Fig. 1. The KMS condition

As we shall see, it will also be useful to extend whenever possible the integration provided by a state to unbounded multipliers of A.

In the unital case, for any given value of β, the set Σ_β of KMS_β states on A forms a convex compact Choquet simplex (possibly empty and in general infinite dimensional). In the nonunital case, given a σ_t-invariant subalgebra C of B, the set $\Sigma_\beta(C)$ of KMS_β states on C should be viewed as a compactification of the set of KMS_β states on A. The restriction from C to A maps $\Sigma_\beta(C)$ to KMS_β positive linear forms on A of norm less than or equal to one (quasi-states).

The typical pattern for a system with a single phase transition is that this simplex consists of a single point for $\beta \leqq \beta_c$ *i.e.* when the temperature is larger than the critical temperature T_c, and is non-trivial (of some higher dimension in general) when the temperature lowers. Systems can exhibit a more complex pattern of multiple phase transitions, where no KMS state exists above a certain temperature. The GL_2 system, which is the main object of study in this paper, will actually exhibit this more elaborate behavior.

We refer to the books ([6], [16]) for the general discussion of KMS states and phase transitions. The main technical point is that for finite β a β-KMS state is extremal iff the corresponding GNS representation is factorial. The decomposition into extremal β-KMS states is then the primary decomposition for a given β-KMS state.

At 0 temperature ($\beta = \infty$) the interesting notion is that of weak limit of β-KMS states for $\beta \to \infty$. It is true that such states satisfy a weak form of the KMS condition. This can be formulated by saying that, for all $a, b \in A$, the function

$$F(t) = \varphi(a\, \sigma_t(b))$$

extends to a bounded holomorphic function in the upper half plane $\mathbb{H}$. This implies that, in the Hilbert space of the GNS representation of φ (*i.e.* the completion of A in the inner product $\varphi(a^* b)$), the generator H of the one-parameter group σ_t is a positive operator (positive energy condition). However, this condition is too weak in general to be interesting, as one sees by taking the trivial evolution ($\sigma_t = \mathrm{id}\,, \quad \forall t \in \mathbb{R}$). In this case any state fulfills it, while weak limits of β-KMS states are automatically tracial states. Thus, we shall define $\Sigma_{\beta=\infty}$ as the set of weak limit points of the sets Σ_β of β-KMS states for $\beta \to \infty$.

The framework for spontaneous symmetry breaking ([16]) involves a (compact) group of automorphisms $G \subset \mathrm{Aut}(A)$ of A commuting with the time evolution,

$$\sigma_t\, \alpha_g = \alpha_g\, \sigma_t \qquad \forall\, g \in G\,,\ t \in \mathbb{R}\,. \tag{2.8}$$

The group G is the symmetry group of the system, and the choice of an equilibrium state φ may break it to a smaller subgroup given by the isotropy group of φ

$$G_\varphi = \{g \in G\,,\ g\varphi = \varphi\}. \tag{2.9}$$

The group G acts on Σ_β for any β, hence on its extreme points $\mathcal{E}(\Sigma_\beta) = \mathcal{E}_\beta$.

The unitary group $\mathcal{U}$ of the fixed point algebra of σ_t acts by inner automorphisms of the dynamical system (A, σ_t): for $u \in \mathcal{U}$,

$$(\mathrm{Ad}u)\,(x) := u\,x\,u^*\,, \quad \forall x \in A\,.$$

These *inner* automorphisms of (A, σ_t) act trivially on KMS_β states, as one checks using the KMS condition. This gives us the freedom to wipe out the group $\mathrm{Int}(A, \sigma_t)$ of inner symmetries and to define an action *modulo inner* of a group G on the system (A, σ_t) as a map

$$\alpha : G \to \mathrm{Aut}(A, \sigma_t)$$

fulfilling the condition

$$\alpha(g_1 g_2)\,\alpha(g_2)^{-1}\,\alpha(g_1)^{-1} \in \mathrm{Int}(A, \sigma_t)\,, \quad \forall g_j \in G\,.$$

Such an action gives an action of the group G on the set Σ_β of KMS_β states since the ambiguity coming from $\mathrm{Int}(A, \sigma_t)$ disappears in the action on Σ_β. In fact there is one more generalization of the above obvious notion of symmetries that we shall crucially need later – it involves actions by endomorphisms. This type of symmetry plays a key role in the theory of superselection sectors developed by Doplicher-Haag-Roberts (cf.[16], Chapter IV).

Definition 2.3 *An* endomorphism ρ *of the dynamical system* (A, σ_t) *is a* $*$-*homomorphism* $\rho : A \to A$ *commuting with* σ_t.

It follows then that $\rho(1) = e$ is an idempotent fixed by σ_t. Given a KMS_β state φ the equality

$$\rho^*(\varphi) := Z^{-1}\,\varphi \circ \rho\,, \quad Z = \phi(e)$$

gives a KMS_β state, provided that $\varphi(e) \neq 0$. Exactly as above for unitaries, consider an isometry

$$u \in A\,, \quad u^*\,u = 1$$

which is an eigenvector for σ_t, *i.e.* that fulfills, for some $\lambda \in \mathbb{R}_+^*$ $(\lambda \geq 1)$, the condition

$$\sigma_t(u) = \lambda^{it}\,u\,, \quad \forall t \in \mathbb{R}\,.$$

Then u defines an *inner* endomorphism $\mathrm{Ad}u$ of the dynamical system (A, σ_t) by the equality

$$(\mathrm{Ad}u)\,(x) := u\,x\,u^*\,, \quad \forall x \in A\,,$$

and one obtains the following.

Proposition 2.4 *The inner endomorphisms of the dynamical system* (A, σ_t) *act trivially on the set of* KMS_β *states,*

$$(\mathrm{Ad}u)^*(\varphi) = \varphi\,, \quad \forall \varphi \in \Sigma_\beta\,.$$

Proof. The KMS_β condition shows that $\varphi(u\,u^*) = \lambda^{-\beta} > 0$ so that $(\mathrm{Ad}u)^*(\varphi)$ is well defined. The same KMS_β condition applied to the pair $(x\,u^*,\, u)$ shows that $(\mathrm{Ad}u)^*(\varphi) = \varphi$. □

At 0 temperature ($\beta = \infty$) it is no longer true that the endomorphisms act directly on the set Σ_∞ of KMS_∞ states, but one can use their action on KMS_β-states together with the "warming up" operation. This is defined as the map

$$W_\beta(\varphi)(x) = Z^{-1}\,\mathrm{Trace}(\pi(x)\,e^{-\beta H}), \quad \forall x \in A, \tag{2.10}$$

where H is the positive energy Hamiltonian, implementing the time evolution in the representation π associated to the KMS_∞ state φ and $Z = \mathrm{Trace}(e^{-\beta H})$. Typically, W_β gives a bijection

$$W_\beta : \Sigma_\infty \to \Sigma_\beta,$$

for β larger than critical. Using the bijection W_β, one can transfer the action back to zero temperature states.

Another property of KMS states that we shall need later is the following functoriality. Namely, besides the obvious functoriality under pullback, discussed above, KMS states push forward under equivariant surjections, modulo normalization.

Proposition 2.5 *Let (A, σ_t) be a C^*-dynamical system (A separable) and J a norm closed two sided ideal of A globally invariant under σ_t. Let u_n be a quasi central approximate unit for J. For any KMS_β-state φ on (A, σ_t) the following sequence converges and defines a KMS_β positive linear form on $(A/J, \sigma_t)$,*

$$\psi(x) = \lim_{n\to\infty} \varphi((1-u_n)\,x)\,, \quad \forall x \in A\,.$$

Proof. Let A'' be the double dual of A and $p \in A''$ the central open projection corresponding to the ideal J (cf. [42]). By construction the u_n converge weakly to p (cf. [42] 3.12.14) so the convergence follows as well as the positivity of ψ. By construction ψ vanishes on J. To get the KMS_β condition one applies (2.6) with $a = (1-u_n)\,x$, $b = y$ where y is a smooth element in A. Then one gets a bounded uniformly continuous sequence $F_n(z)$ of holomorphic functions in the strip $\{z \,|\, \mathrm{Im}\, z \in [0, \beta]\}$ with

$$F_n(t) = \varphi((1-u_n)\,x\,\sigma_t(y)) \qquad F_n(t+i\beta) = \varphi(\sigma_t(y)(1-u_n)\,x) \quad \forall t \in \mathbb{R}.$$

Using the Montel theorem on normal families and the quasi-central property of u_n one gets the KMS_β condition for ψ. □

3 $\mathbb{Q}^{ab}$ and KMS States

We shall now describe an explicit system (*cf.* [4], [5]) that will make contact between the general framework above and arithmetic. The algebra $\mathcal{A}$ of this system is defined over the rationals,

$$\mathcal{A} = \mathcal{A}_{\mathbb{Q}} \otimes_{\mathbb{Q}} \mathbb{C}\,, \tag{3.1}$$

where $\mathcal{A}_{\mathbb{Q}}$ is a $\mathbb{Q}$-algebra and is of countable (infinite) dimension as a vector space over $\mathbb{Q}$. The algebra $\mathcal{A}$ has a C^*-completion A and a natural time evolution σ_t.

To any vacuum state $\varphi \in \mathcal{E}_\infty$ for (A, σ_t) we attach the $\mathbb{Q}$-vector space of complex numbers,

$$V_\varphi := \{\varphi(a)\,;\; a \in \mathcal{A}_{\mathbb{Q}}\} \tag{3.2}$$

that is of countable dimension over $\mathbb{Q}$. It turns out that V_φ is included in algebraic numbers, so that one can act on these numbers by the Galois group

$$\mathrm{Gal}\,(\overline{\mathbb{Q}}/\mathbb{Q})\,. \tag{3.3}$$

The symmetry group G is the inverse limit with the profinite topology

$$G = \hat{\mathbb{Z}}^* = \varprojlim_n \mathrm{GL}_1(\mathbb{Z}/n\mathbb{Z}). \tag{3.4}$$

This can also be described as the quotient of the idèle class group of $\mathbb{Q}$ by the connected component of the identity,

$$G = \mathrm{GL}_1(\mathbb{Q})\backslash \mathrm{GL}_1(\mathbb{A})/\mathbb{R}_+^* = C_{\mathbb{Q}}/D_{\mathbb{Q}}\,. \tag{3.5}$$

Here $\mathbb{A} = \mathbb{A}_{\mathbb{Q}}$ denotes the adèles of $\mathbb{Q}$, namely $\mathbb{A} = \mathbb{A}_f \times \mathbb{R}$, with $\mathbb{A}_f = \hat{\mathbb{Z}} \otimes \mathbb{Q}$.

The following amazing fact holds:

For any $\varphi \in \mathcal{E}_\infty$ and any $\gamma \in \mathrm{Gal}(\overline{\mathbb{Q}}/\mathbb{Q})$, the composition (3.6)

$\gamma \circ \varphi$ defined on $\mathcal{A}_{\mathbb{Q}}$ does extend to a *state* on $\mathcal{A}$.

What is "unreasonable" in this property defining "fabulous" states is that, though elements

$$\gamma \in \mathrm{Gal}(\overline{\mathbb{Q}}/\mathbb{Q}) \tag{3.7}$$

extend to automorphisms of $\mathbb{C}$, these are extremely discontinuous and not even Lebesgue measurable (except for $z \mapsto \bar{z}$), and certainly do not preserve positivity.

It follows from (3.6) that the composition $\varphi \mapsto \gamma \circ \varphi$ defines uniquely an action of $\mathrm{Gal}(\overline{\mathbb{Q}}/\mathbb{Q})$ on $\mathcal{E}_\infty$ and the equation

$$\gamma \circ \varphi = \varphi \circ g \tag{3.8}$$

gives a relation between Galois automorphisms and elements of G, *i.e.* idèle classes (3.5), which is in fact the class field theory isomorphism $C_{\mathbb{Q}}/D_{\mathbb{Q}} \cong \mathrm{Gal}(\mathbb{Q}^{ab}/\mathbb{Q})$.

Let us now concretely describe our system, consisting of the algebra $\mathcal{A}$ (defined over $\mathbb{Q}$) and of the time evolution σ_t.

The main conceptual steps involved in the construction of this algebra are:

- The construction, due to Hecke, of convolution algebras associated to double cosets on algebraic groups over the rational numbers;
- The existence of a canonical time evolution on a von Neumann algebra.

More concretely, while Hecke was considering the case of GL_2, where Hecke operators appear in the convolution algebra associated to the almost normal subgroup $\mathrm{GL}_2(\mathbb{Z}) \subset \mathrm{GL}_2(\mathbb{Q})$, the BC system arises from the Hecke algebra associated to the corresponding pair of parabolic subgroups.

Indeed, let P be the algebraic group "$ax+b$", *i.e.* the functor which to any abelian ring R assigns the group P_R of 2 by 2 matrices over R of the form

$$P_R = \left\{ \begin{bmatrix} 1 & b \\ 0 & a \end{bmatrix} \; ; \; a, b \in R\,, \; a \text{ invertible} \right\} . \tag{3.9}$$

By construction $P_{\mathbb{Z}}^+ \subset P_{\mathbb{Q}}^+$ is an inclusion $\Gamma_0 \subset \Gamma$ of countable groups, where P_R^+ denotes the restriction to $a > 0$. This inclusion fulfills the following commensurability condition:

$$\text{The orbits of the left action of } \Gamma_0 \text{ on } \Gamma/\Gamma_0 \text{ are all } \textit{finite}. \tag{3.10}$$

For obvious reasons the same holds for orbits of Γ_0 acting on the right on $\Gamma_0\backslash\Gamma$.

The Hecke algebra $\mathcal{A}_{\mathbb{Q}} = \mathcal{H}_{\mathbb{Q}}(\Gamma, \Gamma_0)$ is by definition the convolution algebra of functions of finite support

$$f : \Gamma_0\backslash\Gamma \to \mathbb{Q}\,, \tag{3.11}$$

which fulfill the Γ_0-invariance condition

$$f(\gamma\gamma_0) = f(\gamma) \qquad \forall \gamma \in \Gamma\,, \; \gamma_0 \in \Gamma_0 \tag{3.12}$$

so that f is really defined on $\Gamma_0\backslash\Gamma/\Gamma_0$. The convolution product is then given by

$$(f_1 * f_2)(\gamma) = \sum_{\Gamma_0\backslash\Gamma} f_1(\gamma\gamma_1^{-1}) f_2(\gamma_1)\,. \tag{3.13}$$

The time evolution appears from the analysis of the *regular representation* of the pair (Γ, Γ_0). It is trivial when Γ_0 is normal, or in the original case of Hecke, but it becomes interesting in the parabolic case, due to the lack of unimodularity of the parabolic group, as will become clear in the following.

The regular representation

$$(\pi(f)\xi)(\gamma) = \sum_{\Gamma_0\backslash\Gamma} f(\gamma\gamma_1^{-1})\xi(\gamma_1) \tag{3.14}$$

in the Hilbert space

$$\mathcal{H} = \ell^2(\Gamma_0 \backslash \Gamma) \tag{3.15}$$

extends to the complexification

$$\mathcal{A}_{\mathbb{C}} = \mathcal{A}_{\mathbb{Q}} \otimes_{\mathbb{Q}} \mathbb{C} \tag{3.16}$$

of the above algebra, which inherits from this representation the involution $a \mapsto a^*$, uniquely defined so that $\pi(a^*) = \pi(a)^*$ (the Hilbert space adjoint), namely

$$f^*(\gamma) := \overline{f(\gamma^{-1})} \qquad \forall \gamma \in \Gamma_0 \backslash \Gamma / \Gamma_0. \tag{3.17}$$

It happens that the time evolution (*cf.* [54]) of the von Neumann algebra generated by $\mathcal{A}$ in the regular representation restricts to the dense subalgebra $\mathcal{A}$. This implies that there is a uniquely determined time evolution $\sigma_t \in \mathrm{Aut}(\mathcal{A})$, such that the state φ_1 given by

$$\varphi_1(f) = \langle \pi(f) \varepsilon_e, \varepsilon_e \rangle \tag{3.18}$$

is a KMS_1 state *i.e.* a KMS state at inverse temperature $\beta = 1$. Here ε_e is the cyclic and separating vector for the regular representation given by the left coset $\{\Gamma_0\} \in \Gamma_0 \backslash \Gamma$.

Explicitly, one gets the following formula for the time evolution:

$$\sigma_t(f)(\gamma) = \left(\frac{L(\gamma)}{R(\gamma)}\right)^{-it} f(\gamma) \qquad \forall \gamma \in \Gamma_0 \backslash \Gamma / \Gamma_0 \,, \tag{3.19}$$

where the integer valued functions L and R on the double coset space are given respectively by

$$L(\gamma) = \text{ Cardinality of left } \Gamma_0 \text{ orbit of } \gamma \text{ in } \Gamma/\Gamma_0 \,, \quad R(\gamma) = L(\gamma^{-1}) \,. \tag{3.20}$$

Besides the conceptual description given above, the algebra $\mathcal{A}_{\mathbb{Q}}$ also has a useful explicit presentation in terms of generators and relations (*cf.* [5] §4, Prop.18). We recall it here, in the slightly simplified version of [21], Prop.24.

Proposition 3.1 *The algebra $\mathcal{A}_{\mathbb{Q}}$ is generated by elements μ_n, $n \in \mathbb{N}^\times$ and $e(r)$, for $r \in \mathbb{Q}/\mathbb{Z}$, satisfying the relations*

- $\mu_n^* \mu_n = 1$, *for all* $n \in \mathbb{N}^\times$,
- $\mu_k \mu_n = \mu_{kn}$, *for all* $k, n \in \mathbb{N}^\times$,
- $e(0) = 1$, $e(r)^* = e(-r)$, *and* $e(r)e(s) = e(r+s)$, *for all* $r, s \in \mathbb{Q}/\mathbb{Z}$,
- *For all* $n \in \mathbb{N}^\times$ *and all* $r \in \mathbb{Q}/\mathbb{Z}$,

$$\mu_n \, e(r) \, \mu_n^* = \frac{1}{n} \sum_{ns=r} e(s). \tag{3.21}$$

In this form the time evolution preserves pointwise the subalgebra $R_{\mathbb{Q}} = \mathbb{Q}[\mathbb{Q}/\mathbb{Z}]$ generated by the $e(r)$ and acts on the μ_n's as

$$\sigma_t(\mu_n) = n^{it}\,\mu_n.$$

The Hecke algebra considered above admits an automorphism α, $\alpha^2 = 1$ whose fixed point algebra is the Hecke algebra of the pair $P_{\mathbb{Z}} \subset P_{\mathbb{Q}}$. The latter admits an equivalent description[4], from the pair

$$(P_R, P_{\mathbb{A}_f}),$$

where R is the maximal compact subring of the ring of finite adèles

$$\mathbb{A}_f = \prod_{\text{res}} \mathbb{Q}_p\,. \tag{3.22}$$

This adèlic description displays, as a natural symmetry group, the quotient G of the idèle class group of $\mathbb{Q}$ by the connected component of identity (3.5).

Let $\overline{\mathbb{Q}}$ be an algebraic closure of $\mathbb{Q}$ and $\mathbb{Q}^{ab} \subset \overline{\mathbb{Q}}$ be the maximal abelian extension of $\mathbb{Q}$. Let $r \mapsto \zeta_r$ be a (non-canonical) isomorphism of $\mathbb{Q}/\mathbb{Z}$ with the multiplicative group of roots of unity inside $\mathbb{Q}^{ab}$.

We can now state the basic result that gives content to the relation between phase transition and arithmetic (BC [5]):

Theorem 3.2 *1. For $0 < \beta \leqq 1$ there exists a unique* KMS_β *state φ_β for the above system. Its restriction to $R_{\mathbb{Q}} = \mathbb{Q}[\mathbb{Q}/\mathbb{Z}] \subset \mathcal{A}$ is given by*

$$\varphi_\beta\left(e(a/b)\right) = b^{-\beta} \prod_{p\,\text{prime},\, p|b} \left(\frac{1 - p^{\beta-1}}{1 - p^{-1}}\right). \tag{3.23}$$

2. For $\beta > 1$ the extreme KMS_β *states are parameterized by embeddings $\rho : \mathbb{Q}^{ab} \to \mathbb{C}$ and*

$$\varphi_{\beta,\rho}\left(e(a/b)\right) = Z(\beta)^{-1} \sum_{n=1}^{\infty} n^{-\beta} \rho\left(\zeta_{a/b}^n\right), \tag{3.24}$$

where the partition function $Z(\beta) = \zeta(\beta)$ is the Riemann zeta function.

3. For $\beta = \infty$, the Galois group $\mathrm{Gal}(\overline{\mathbb{Q}}/\mathbb{Q})$ *acts by composition on $\mathcal{E}_\infty$. The action factors through the abelianization* $\mathrm{Gal}(\mathbb{Q}^{ab}/\mathbb{Q})$*, and the class field theory isomorphism $\theta : G \to \mathrm{Gal}(\mathbb{Q}^{ab}/\mathbb{Q})$ intertwines the actions,*

$$\alpha \circ \varphi = \varphi \circ \theta^{-1}(\alpha)\,, \qquad \alpha \in \mathrm{Gal}(\mathbb{Q}^{ab}/\mathbb{Q})\,.$$

4 Further Developments

The main theorem of class field theory provides a classification of finite abelian extensions of a local or global field K in terms of subgroups of a locally compact abelian group canonically associated to the field. This is the multiplicative group $K^* = \mathrm{GL}_1(K)$ in the local nonarchimedean case, while in the global

[4]This procedure holds more generally (*cf.* [47] [48]) for arbitrary almost normal inclusions (Γ_0, Γ).

case it is the quotient C_K/D_K of the idèle class group C_K by the connected component of the identity. The construction of the group C_K is at the origin of the theory of idèles and adèles.

Hilbert's 12th problem can be formulated as the question of providing an explicit set of generators of the maximal abelian extension K^{ab} of a number field K, inside an algebraic closure $\bar{K}$, and of the action of the Galois group $\mathrm{Gal}(K^{ab}/K)$. The typical example where this is achieved, which motivated Hilbert's formulation of the explicit class field theory problem, is the Kronecker–Weber case: the construction of the maximal abelian extension of $\mathbb{Q}$. In this case the torsion points of $\mathbb{C}^*$ (roots of unity) generate $\mathbb{Q}^{ab} \subset \mathbb{C}$.

Remarkably, the only other case for number fields where this program has been carried out completely is that of imaginary quadratic fields, where the construction relies on the theory of elliptic curves with complex multiplication (*cf. e.g.* [53]). Generalizations to other number fields involve other remarkable problems in number theory like the Stark conjectures. Recent work of Manin [34] [35] suggests a close relation between the real quadratic case and noncommutative geometry.

To better appreciate the technical difficulties underlying any attempt to address the Hilbert 12th problem of explicit class field theory via the BC approach, in view of the problem of fabulous states that we shall formulate in §5, we first summarize briefly the state of the art (to this moment and to our knowledge) in the study of C^*-dynamical systems with phase transitions associated to number fields.

Some progress from the original BC paper followed in various directions, and some extensions of the BC construction to other global fields (number fields and function fields) were obtained. Harari and Leichtnam [19] produced a C^*-dynamical system with phase transition for function fields and algebraic number fields. A localization is used in order to deal with lack of unique factorization into primes. In the number field case, one replaces the ring $\mathcal{O}$ of integers of K by the principal ring $\mathcal{O}_S$ obtained by inverting a suitable finite set of prime ideals. The construction is based on the inclusion $\mathcal{O}_S \rtimes 1 \subset K \rtimes K_+^*$, where K_+^* is the subgroup of K^* generated by the generators of prime ideals of $\mathcal{O}_S$. The symmetry group G of (3.5) is replaced by the group

$$G = \hat{\mathcal{O}}_S^* = \mathrm{GL}_1(\hat{\mathcal{O}}_S),$$

with $\hat{\mathcal{O}}_S$ the profinite completion of the ring $\mathcal{O}_S$. There is a group homomorphism $s : G \to C_K/D_K$, but it is in general neither injective nor surjective, hence, even in the case of imaginary quadratic fields, the construction does not capture the action of the Galois group $\mathrm{Gal}(K^{ab}/K)$, except in the very special class number one case.

P. Cohen gave in [8] a construction of a C^*-dynamical system associated to a number field K, which has spontaneous symmetry breaking and recovers the full Dedekind zeta function as partition function. The main point of her approach is to involve the semigroup of all ideals rather than just the principal ideals used in other approaches as the replacement of the semi-group of positive integers involved in BC. Still, the group of symmetries is $G = \hat{\mathcal{O}}^*$ and not the desired C_K/D_K.

Typically, the extensions of the number field K obtained via these constructions are given by roots of unity, hence they do not recover the maximal abelian extension.

The Hecke algebra of the inclusion $\mathcal{O} \rtimes 1 \subset K \rtimes K^*$ for an arbitrary algebraic number field K was considered by Arledge, Laca, and Raeburn in [1], where they discuss its structure and representations, but not the problem of KMS states.

Further results on this Hecke algebra have been announced by Laca and van Frankenhuysen [27]: they obtain some general results on the structure and representations for all number fields, while they analyze the structure of KMS states only for the class number one case. In this case, their announced result is that there are enough ground states to support a transitive free action of $\mathrm{Gal}(K^{ab}/K)$ (up to a copy of $\{\pm 1\}$ for each real embedding). However, it appears that the construction does not give embeddings of K^{ab} as actual values of the ground states on the Hecke algebra over K, hence it does not seem suitable to treat the class field theory problem of providing explicit generators of K^{ab}.

The structure of the Hecke algebra of the inclusion $\mathcal{O} \rtimes \mathcal{O}^* \subset K \rtimes K^*$ was further clarified by Laca and Larsen in [23], using a decomposition of the Hecke algebra of a semidirect product as the cross product of the Hecke algebra of an intermediate (smaller) inclusion by an action of a semigroup.

The original BC algebra was also studied in much greater details in several following papers. It was proved by Brenken in [7] and by Laca and Raeburn in [25] that the BC algebra can be written as a semigroup cross product. Brenken also discusses the case of Hecke algebras from number fields of the type considered in [25, 1].

Laca then re-derived the original BC result from the point of view of semigroup cross products in [21]. This allows for significant simplifications of the argument in the case of $\beta > 1$, by looking at the conditional expectations and the KMS condition at the level of the "predual" (semigroup) dynamical system. A further simplification of the original phase transition theorem of BC was given by Neshveyev in [40], via a direct argument for ergodicity, which implies uniqueness of the KMS states for $0 \leq \beta \leq 1$.

The BC algebra can also be realized as a full corner in the cross product of the finite adèles by the multiplicative rationals, as was shown by Laca in [22], by dilating the semigroup action to a minimal full group action. Laca and Raeburn used the dilation results of [22] to calculate explicitly the primitive (and maximal) ideal spaces of the BC algebra as well as of the cross product of the full adèles by the action of the multiplicative rationals.

Using the cross product description of the BC algebra, Leichtnam and Nistor computed Hochschild, cyclic, and periodic cyclic homology groups of the BC algebra, by computing the corresponding groups for the C^*-dynamical system algebras arising from the action of $\mathbb{Q}^*$ on the adèles of $\mathbb{Q}$. The calculation for the BC algebra then follows by taking an increasing sequence of smooth subalgebras and an inductive limit over certain Morita equivalent subalgebras.

Further results related to aspects of the BC construction and generalizations can be found in [3], [15], [24], [29], [30], [55].

5 Fabulous States

Given a number field K, we let $\mathbb{A}_K$ denote the adèles of K and $J_K = \mathrm{GL}_1(\mathbb{A}_K)$ be the group of idèles of K. We write C_K for the group of idèles classes $C_K = J_K/K^*$ and D_K for the connected component of the identity in C_K.

If we remain close to the spirit of the Hilbert 12th problem, we can formulate a general question, aimed at extending the results of [5] to other number fields K. Given a number field K, with a choice of an embedding $K \subset \mathbb{C}$, the "problem of fabulous states" consists in constructing a C^*-dynamical system (A, σ_t) and an "arithmetic" subalgebra $\mathcal{A}$, which satisfy the following properties:

1. The idèles class group $G = C_K/D_K$ acts by symmetries on (A, σ_t) preserving the subalgebra $\mathcal{A}$.
2. The states $\varphi \in \mathcal{E}_\infty$, evaluated on elements of $\mathcal{A}$, satisfy:
   - $\varphi(a) \in \bar{K}$, the algebraic closure of K in $\mathbb{C}$;
   - the elements of $\{\varphi(a) : a \in \mathcal{A}\}$, for $\varphi \in \mathcal{E}_\infty$ generate K^{ab}.
3. The class field theory isomorphism
$$\theta : C_K/D_K \xrightarrow{\simeq} \mathrm{Gal}(K^{ab}/K)$$
   intertwines the actions,
$$\alpha \circ \varphi = \varphi \circ \theta^{-1}(\alpha), \tag{5.1}$$
   for all $\alpha \in \mathrm{Gal}(K^{ab}/K)$ and for all $\varphi \in \mathcal{E}_\infty$.

Notice that, with this formulation, the problem of the construction of fabulous states is intimately related to Hilbert's 12th problem. This question will be pursued in [13].

We shall construct here a system which is the analog of the BC system for $\mathrm{GL}_2(\mathbb{Q})$ instead of $\mathrm{GL}_1(\mathbb{Q})$. This will extend the results of [5] to this non-abelian GL_2 case and will exhibit many new features which have no counterpart in the abelian case. Our construction involves the explicit description of the automorphism group of the modular field, [51]. The construction of fabulous states for imaginary quadratic fields, which will be investigated with N. Ramachandran in [13], involves specializing the GL_2 system to a subsystem compatible with complex multiplication in a given imaginary quadratic field.

The construction of the GL_2 system gives a C^*-dynamical system (A, σ_t) and an involutive subalgebra $\mathcal{A}_{\mathbb{Q}}$ defined over $\mathbb{Q}$, satisfying the following properties:

- The quotient group $S := \mathbb{Q}^*\backslash \mathrm{GL}_2(\mathbb{A}_f)$ of the finite adèlic group of GL_2 acts as symmetries of the dynamical system (A, σ_t) preserving the subalgebra $\mathcal{A}_{\mathbb{Q}}$.
- For generic $\varphi \in \mathcal{E}_\infty$, the values $\{\varphi(a) \in \mathbb{C} : a \in \mathcal{A}_{\mathbb{Q}}\}$ generate a subfield $F_\varphi \subset \mathbb{C}$ which is an extension of $\mathbb{Q}$ of transcendence degree 1.
- For generic $\varphi \in \mathcal{E}_\infty$, there exists an isomorphism

$$\theta : S \xrightarrow{\simeq} \mathrm{Gal}(F_\varphi/\mathbb{Q})$$

  which intertwines the actions

$$\alpha \circ \varphi = \varphi \circ \theta^{-1}(\alpha), \tag{5.2}$$

  for all $\alpha \in \mathrm{Gal}(F_\varphi/\mathbb{Q})$.

There are a number of important nuances between the abelian case above and the non-abelian one. For instance, in the abelian case the field generated by $\varphi(\mathcal{A})$ does not depend on the choice of $\varphi \in \mathcal{E}_\infty$ and the isomorphism θ is also independent of φ. This no longer holds in the non-abelian case, as is clear from the presence of inner automorphisms of the symmetry group S. Also, in the latter case, the action of S on $\mathcal{E}_\infty$ is not transitive and the corresponding invariant of the orbit of φ under S is the subfield $F_\varphi \subset \mathbb{C}$. Another important nuance is that the algebra A is no longer unital while $\mathcal{A}_{\mathbb{Q}}$ is an algebra of unbounded multipliers of A. Finally, the symmetries require the full framework of endomorphisms as explained above in §2.

6 The subalgebra $\mathcal{A}_{\mathbb{Q}}$ and Eisenstein Series

In this section we shall recast the BC algebra in terms of the trigonometric analog of the Eisenstein series, following the analogy developed by Eisenstein and Kronecker between trigonometric and elliptic functions, as outlined by A.Weil in [56].

This will be done by first giving a geometric interpretation in terms of $\mathbb{Q}$-lattices of the noncommutative space X whose algebra of continuous functions $C(X)$ is the BC C^*-algebra. The space X is by construction the quotient of the Pontrjagin dual of the abelian group $\mathbb{Q}/\mathbb{Z}$ by the equivalence relation generated by the action by multiplication of the semi-group $\mathbb{N}^\times$.

Let

$$R = \prod_p \mathbb{Z}_p$$

be the compact ring product of the rings $\mathbb{Z}_p$ of p-adic integers. It is the maximal compact subring of the locally compact ring of finite adèles

$$\mathbb{A}_f = \prod_{\mathrm{res}} \mathbb{Q}_p$$

We recall the following standard fact

Proposition 6.1 • *The inclusion $\mathbb{Q} \subset \mathbb{A}_f$ gives an isomorphism of abelian groups*

$$\mathbb{Q}/\mathbb{Z} = \mathbb{A}_f/R\,.$$

• *The following map is an isomorphism of compact rings*

$$j:\ R \to \mathrm{Hom}(\mathbb{Q}/\mathbb{Z},\, \mathbb{Q}/\mathbb{Z})\,, \quad j(a)(x) = a\,x\,, \quad \forall x \in \mathbb{A}_f/R\,, \quad \forall a \in R\,.$$

We shall use j from now on to identify R with $\mathrm{Hom}(\mathbb{Q}/\mathbb{Z},\, \mathbb{Q}/\mathbb{Z})$. Note that by construction $\mathrm{Hom}(\mathbb{Q}/\mathbb{Z},\, \mathbb{Q}/\mathbb{Z})$ is endowed with the topology of pointwise convergence. It is identified with $\varprojlim \mathbb{Z}/N\mathbb{Z}$ using the restriction to N-torsion elements.

For every $r \in \mathbb{Q}/\mathbb{Z}$ one gets a function $e(r) \in C(R)$ by,

$$e(r)(\rho) := \exp 2\pi i \rho(r) \quad \forall \rho \in \mathrm{Hom}(\mathbb{Q}/\mathbb{Z},\, \mathbb{Q}/\mathbb{Z})$$

and this gives the identification of R with the Pontrjagin dual of $\mathbb{Q}/\mathbb{Z}$ and of $C(R)$ with the group C^*-algebra $C^*(\mathbb{Q}/\mathbb{Z})$.

One can then describe the BC C^*-algebra as the cross product of $C(R)$ by the semigroup action of $\mathbb{N}^\times$ as follows. For each integer $n \in \mathbb{N}^\times$ we let $nR \subset R$ be the range of multiplication by n. It is an open and closed subset of R whose characteristic function π_n is a projection $\pi_n \in C(R)$. One has by construction

$$\pi_n\,\pi_m = \pi_{n\vee m}\,, \quad \forall n\,,\ m \in \mathbb{N}^\times$$

where $n \vee m$ denotes the lowest common multiple of n and m.

The semigroup action of $\mathbb{N}^\times$ on $C(R)$ corresponds to the isomorphism

$$\alpha_n(f)(\rho) := f(n^{-1}\,\rho)\,, \quad \forall \rho \in nR\,. \tag{6.1}$$

of $C(R)$ with the reduced algebra $C(R)_{\pi_n}$ of $C(R)$ by the projection π_n. In the BC algebra one has

$$\mu_n\, f\, \mu_n^* = \alpha_n(f)\,, \quad \forall f \in C(R). \tag{6.2}$$

There is an equivalent description of the BC algebra in terms of the étale groupoid G of pairs (r, ρ), where $r \in \mathbb{Q}_+^*$, $\rho \in R$ and $r\,\rho \in R$. The composition in G is given by

$$(r_1, \rho_1) \circ (r_2, \rho_2) = (r_1\, r_2, \rho_2)\,, \quad \text{if } r_2\,\rho_2 = \rho_1\,, \tag{6.3}$$

and the convolution of functions by

$$f_1 * f_2(r, \rho) := \sum f_1(rs^{-1}, s\,\rho)\, f_2(s, \rho)\,, \tag{6.4}$$

while the adjoint of f is

$$f^*(r, \rho) := \overline{f(r^{-1}, r\,\rho)}\,. \tag{6.5}$$

All of this is implicit in ([5]) and has been amply described in the subsequent papers mentioned in §4. In the description above, μ_n is given by the function $\mu_n(r,\rho)$ which vanishes unless $r = n$ and is equal to 1 for $r = n$. The time evolution is given by

$$\sigma_t(f)(r,\rho) := r^{it} f(r,\rho), \quad \forall f \in C^*(G). \tag{6.6}$$

We shall now describe a geometric interpretation of this groupoid G in terms of commensurability of $\mathbb{Q}$-lattices. In particular, it will pave the way to the generalization of the BC system to higher dimensions. The basic simple geometric objects are $\mathbb{Q}$-lattices in $\mathbb{R}^n$, defined as follows.

Definition 6.2 *A $\mathbb{Q}$-lattice in $\mathbb{R}^n$ is a pair (Λ,ϕ), with Λ a lattice in $\mathbb{R}^n$, and $\phi : \mathbb{Q}^n/\mathbb{Z}^n \longrightarrow \mathbb{Q}\Lambda/\Lambda$ an homomorphism of abelian groups.*

Two lattices Λ_j in $\mathbb{R}^n$ are commensurable iff their intersection $\Lambda_1 \cap \Lambda_2$ is of finite index in Λ_j. Their sum $\Lambda = \Lambda_1 + \Lambda_2$ is then a lattice and, given two homomorphisms of abelian groups $\phi_j : \mathbb{Q}^n/\mathbb{Z}^n \longrightarrow \mathbb{Q}\Lambda_j/\Lambda_j$, the difference $\phi_1 - \phi_2$ is well defined modulo $\Lambda = \Lambda_1 + \Lambda_2$.

Notice that in Definition 6.2 the homomorphism ϕ, in general, is not an isomorphism.

Definition 6.3 *A $\mathbb{Q}$-lattice (Λ,ϕ) is* invertible *if the map ϕ is an isomorphism of abelian groups.*

We consider a natural equivalence relation on the set of $\mathbb{Q}$-lattices defined as follows.

Proposition 6.4 *The following defines an equivalence relation called* commensurability *between $\mathbb{Q}$-lattices: (Λ_1,ϕ_1), (Λ_2,ϕ_2) are commensurable iff Λ_j are commensurable and $\phi_1 - \phi_2 = 0$ modulo $\Lambda = \Lambda_1 + \Lambda_2$.*

Proof. Indeed, let (Λ_j,ϕ_j) be three $\mathbb{Q}$-lattices and assume commensurability between the pairs $(1,2)$ and $(2,3)$. Then the lattices Λ_j are commensurable and are of finite index in $\Lambda = \Lambda_1 + \Lambda_2 + \Lambda_3$. One has $\phi_1 - \phi_2 = 0$ modulo Λ, $\phi_2 - \phi_3 = 0$ modulo Λ and thus $\phi_1 - \phi_3 = 0$ modulo Λ. But $\Lambda' = \Lambda_1 + \Lambda_3$ is of finite index in Λ and thus $\phi_1 - \phi_3$ gives a group homomorphism

$$\mathbb{Q}^n/\mathbb{Z}^n \longrightarrow \Lambda/\Lambda'$$

which is zero since $\mathbb{Q}^n/\mathbb{Z}^n$ is infinitely divisible and Λ/Λ' is finite. This shows that $\phi_1 - \phi_3 = 0$ modulo $\Lambda' = \Lambda_1 + \Lambda_3$ and hence that the pair $(1,3)$ is commensurable. □

Notice that every $\mathbb{Q}$-lattice in $\mathbb{R}$ is uniquely of the form

$$(\Lambda,\phi) = (\lambda\,\mathbb{Z}, \lambda\,\rho), \quad \lambda > 0, \tag{6.7}$$

with $\rho \in \mathrm{Hom}(\mathbb{Q}/\mathbb{Z}, \mathbb{Q}/\mathbb{Z}) = R$.

Proposition 6.5 *The map*

$$\gamma(r,\rho) = ((r^{-1}\mathbb{Z},\,\rho)\,,(\mathbb{Z},\rho))\,,\quad \forall (r,\rho)\in G\,,$$

defines an isomorphism of locally compact étale groupoids between G and the quotient $\mathcal{R}\,/\,\mathbb{R}_+^*$ of the equivalence relation $\mathcal{R}$ of commensurability on the space of $\mathbb{Q}$-lattices in $\mathbb{R}$ by the natural scaling action of $\mathbb{R}_+^*$.

Proof. First since $r\,\rho \in R$ the pair $(r^{-1}\mathbb{Z},\,\rho) = r^{-1}(\mathbb{Z},\,r\,\rho)$ is a $\mathbb{Q}$-lattice and is commensurable to $(\mathbb{Z},\rho)$. Thus, the map γ is well defined. Using the identification (6.7), we see that the restriction of γ to the objects $G^{(0)}$ of G is an isomorphism of R with the quotient of the space of $\mathbb{Q}$-lattices in $\mathbb{R}$ by the natural scaling action of $\mathbb{R}_+^*$. The freeness of this action shows that the quotient $\mathcal{R}\,/\,\mathbb{R}_+^*$ is still a groupoid, and one has

$$\gamma(r_1,\rho_1)\circ\gamma(r_2,\rho_2) = \gamma(r_1\,r_2,\rho_2)\quad \text{if } r_2\,\rho_2 = \rho_1\,.$$

Finally, up to scaling, every element of $\mathcal{R}$ is of the form

$$((r^{-1}\,\mathbb{Z},\,r^{-1}\,\rho')\,,(\mathbb{Z},\rho))$$

where both ρ' and ρ are in R and $r = \frac{a}{b} \in \mathbb{Q}_+^*$. Moreover since $r^{-1}\,\rho' = \rho$ modulo $\frac{1}{a}\mathbb{Z}$ one gets $a\rho - b\rho' = 0$ and $r^{-1}\,\rho' = \rho$. Thus γ is surjective and is an isomorphism. □

This geometric description of the BC algebra allows us to generate in a natural manner a rational subalgebra which will generalize to the two dimensional case. In particular the algebra $C(R)$ can be viewed as the algebra of homogeneous functions of "weight 0" on the space of $\mathbb{Q}$-lattices for the natural scaling action of the multiplicative group $\mathbb{R}_+^*$ where weight k means

$$f(\lambda\,\Lambda,\lambda\,\phi) = \lambda^{-k}\,f(\Lambda,\phi)\,,\quad \forall\lambda\in\mathbb{R}_+^*\,.$$

We let the function $c(\Lambda)$ be the multiple of the covolume $|\Lambda|$ of the lattice, specified by

$$2\,\pi\,i\,c(\mathbb{Z}) = 1 \tag{6.8}$$

The function c is homogeneous of weight -1 on the space of $\mathbb{Q}$-lattices. For $a \in \mathbb{Q}/\mathbb{Z}$, we then define a function $e_{1,a}$ of weight 0 by

$$e_{1,a}(\Lambda,\phi) = c(\Lambda)\sum_{y\in\Lambda+\phi(a)} y^{-1}\,, \tag{6.9}$$

where one uses Eisenstein summation *i.e.* $\lim_{N\to\infty}\sum_{-N}^{N}$ when $\phi(a)\neq 0$ and one lets $e_{1,a}(\Lambda,\phi) = 0$ when $\phi(a) = 0$.

The main result of this section is the following

Theorem 6.6 • *The $e_{1,a}\,, a\in\mathbb{Q}/\mathbb{Z}$ generate $\mathbb{Q}[\mathbb{Q}/\mathbb{Z}]$.*

- *The rational algebra $\mathcal{A}_{\mathbb{Q}}$ is the subalgebra of $A = C^*(G)$ generated by the $e_{1,a}$, $a \in \mathbb{Q}/\mathbb{Z}$ and the μ_n, μ_n^*.*

We define more generally for each weight $k \in \mathbb{N}$ and each $a \in \mathbb{Q}/\mathbb{Z}$ a function $\epsilon_{k,a}$ on the space of $\mathbb{Q}$-lattices in $\mathbb{R}$ by

$$\epsilon_{k,a}(\Lambda, \phi) = \sum_{y \in \Lambda + \phi(a)} y^{-k} . \tag{6.10}$$

This is well defined provided $\phi(a) \neq 0$. For $\phi(a) = 0$ we let

$$\epsilon_{k,a}(\Lambda, \phi) = \lambda_k \, c(\Lambda)^{-k}, \tag{6.11}$$

where we shall fix the constants λ_k below in (6.14). The function $\epsilon_{k,a}$ has weight k *i.e.* it satisfies the homogeneity condition

$$\epsilon_{k,a}(\lambda \Lambda, \lambda \phi) = \lambda^{-k} \epsilon_{k,a}(\Lambda, \phi) , \quad \forall \lambda \in \mathbb{R}_+^* .$$

When $a = \frac{b}{N}$ the function $\epsilon_{k,a}$ has level N in that it only uses the restriction ϕ_N of ϕ to N-torsion points of $\mathbb{Q}/\mathbb{Z}$,

$$\phi_N : \frac{1}{N}\mathbb{Z}/\mathbb{Z} \longrightarrow \frac{1}{N}\Lambda/\Lambda.$$

The products

$$e_{k,a} := c^k \, \epsilon_{k,a} \tag{6.12}$$

are of weight 0 and satisfy two types of relations.

The first relations are multiplicative and express $e_{k,a}$ as a polynomial in $e_{1,a}$,

$$e_{k,a} = P_k(e_{1,a}) \tag{6.13}$$

where the P_k are the polynomials with rational coefficients uniquely determined by the equalities

$$P_1(u) = u \, , \quad P_{k+1}(u) = \frac{1}{k}(u^2 - \frac{1}{4}) \, \partial_u P_k(u) \, .$$

This follows for $\phi(a) \notin \Lambda$ from the elementary formulas for the trigonometric analog of the Eisenstein series ([56] Chapter II). Since $e_{1,a}(\Lambda, \phi) = 0$ is the natural choice for $\phi(a) \in \Lambda$, the validity of (6.13) uniquely dictates the choice of the normalization constants λ_k of (6.11). One gets

$$\lambda_k = P_k(0) = (2^k - 1) \, \gamma_k \, , \tag{6.14}$$

where $\gamma_k = 0$ for odd k and $\gamma_{2j} = (-1)^j \, \dfrac{B_j}{(2j)!}$ with $B_j \in \mathbb{Q}$ the Bernoulli numbers. Equivalently,

$$\frac{x}{e^x - 1} = 1 - \frac{x}{2} - \sum_1^{\infty} \gamma_{2j} \, x^{2j}.$$

One can express the $e_{k,a}$ as $\mathbb{Q}$-linear combinations of the generators $e(r)$. We view $e(r)$ as the function on $\mathbb{Q}$-lattices which assigns to $(\Lambda, \phi) = (\lambda\,\mathbb{Z}, \lambda\,\rho)$, $\lambda > 0$, the value

$$e(r)(\Lambda, \phi) := \exp 2\pi i \rho(r).$$

One then has

Lemma 6.7 *Let $a \in \mathbb{Q}/\mathbb{Z}$, and $n > 0$ with $n\,a = 0$. Then*

$$e_{1,a} = \sum_{k=1}^{n-1} (\frac{k}{n} - \frac{1}{2})\, e(k\,a). \tag{6.15}$$

Proof. We evaluate both sides on $(\Lambda, \phi) = (\lambda\,\mathbb{Z}, \lambda\,\rho)$, $\lambda > 0$. Both sides only depend on the restriction $x \mapsto c\,x$ of ρ to n-torsion elements of $\mathbb{Q}/\mathbb{Z}$ which we write as multiplication by $c \in \mathbb{Z}/n\mathbb{Z}$. Let $a = \frac{b}{n}$. If $bc = 0(n)$ then $\phi(a) = 0$ and both sides vanish since $e(k\,a)(\Lambda, \phi) = \exp 2\pi i(\frac{kbc}{n}) = 1$ for all k. If $bc \neq 0(n)$ then $\phi(a) \neq 0$ and the left side is $\frac{1}{2}(U+1)/(U-1)$ where $U = \exp 2\pi i \frac{bc}{n}$, $U^n = 1$, $U \neq 1$. The right hand side is

$$\sum_{k=1}^{n-1} (\frac{k}{n} - \frac{1}{2})\, U^k\,,$$

which gives $\frac{1}{2}(U+1)$ after multiplication by $U - 1$. □

This last equality shows that $e_{1,a}$ is (one half of) the Cayley transform of $e(a)$ with care taken where $e(a) - 1$ fails to be invertible. In particular while $e(a)$ is unitary, $e_{1,a}$ is skew-adjoint,

$$e_{1,a}^* = -e_{1,a}\,.$$

We say that a $\mathbb{Q}$-lattice (Λ, ϕ) is divisible by an integer $n \in \mathbb{N}$ when $\phi_n = 0$. We let π_n be the characteristic function of the set of $\mathbb{Q}$-lattices divisible by n. It corresponds to the characteristic function of $n\,R \subset R$.

Let $N > 0$ and (Λ, ϕ) a $\mathbb{Q}$-lattice with $\phi_N(a) = c\,a$ for $c \in \mathbb{Z}/N\mathbb{Z}$. The order of the kernel of ϕ_N is $m = \gcd(N, c)$. Also a divisor $b|N$ divides (Λ, ϕ) iff it divides c. Thus for any function f on $\mathbb{N}^*$ one has

$$\sum_{b|N} f(b)\, \pi_b(\Lambda, \phi) = \sum_{b|\gcd(N,c)} f(b)\,,$$

which allows one to express any function of the order $m = \gcd(N, c)$ of the kernel of ϕ_N in terms of the projections π_b, $b|N$. In order to obtain the function $m \mapsto m^j$ we let

$$f_j(n) := \sum_{d|n} \mu(d)(n/d)^j\,,$$

where μ is the Möbius function so that

$$f_j(n) = n^j \prod_{p\,\mathrm{prime},\, p|n} (1 - p^{-j}) .$$

Notice that f_1 is the Euler totient function and that the ratio $f_{-\beta+1}/f_1$ gives the r.h.s. of (3.23) in Theorem 3.2.
The Möbius inversion formula gives

$$\sum_{b|N} f_j(b)\, \pi_b(\Lambda, \phi) = m^j , \quad m = \gcd(N, c) . \tag{6.16}$$

We can now write division relations fulfilled by the functions (6.12).

Lemma 6.8 *Let $N > 0$ then*

$$\sum_{N\, a=0} e_{k,a} = \gamma_k \sum_{d|N} ((2^k - 2)\, f_1(d) + N^k f_{-k+1}(d))\, \pi_d. \tag{6.17}$$

Proof. For a given $\mathbb{Q}$-lattice (Λ, ϕ) with $\mathrm{Ker}\, \phi_N$ of order $m|N$, $N = m\, d$, the result follows from

$$\sum_{N\, a=0} \epsilon_{k,a}(\Lambda, \phi) = m \sum_{y \in \frac{1}{d}\Lambda \setminus \Lambda} y^{-k} + m\, (2^k - 1)\, \gamma_k\, c^{-k}(\Lambda) = m\, (d^k + 2^k - 2) \gamma_k\, c^{-k}$$

together with (6.16) applied for $j = 1$ and $j = 1 - k$. □

The semigroup action of $\mathbb{N}^\times$ is given on functions of $\mathbb{Q}$-lattices by the endomorphisms

$$\alpha_n(f)(\Lambda,\, \phi) := f(n\, \Lambda, \phi) , \quad \forall (\Lambda, \phi) \in \pi_n , \tag{6.18}$$

while $\alpha_n(f)(\Lambda,\, \phi) = 0$ outside π_n. This semigroup action preserves the rational subalgebra $\mathcal{B}_{\mathbb{Q}}$ generated by the $e_{1,a}\, , a \in \mathbb{Q}/\mathbb{Z}\, ,$ since one has

$$\alpha_n(e_{k,a}) = \pi_n\, e_{k,a/n}\, , \tag{6.19}$$

(independently of the choice of the solution $b = a/n$ of $n\, b = a$) and we shall now show that the projections π_n belong to $\mathcal{B}_{\mathbb{Q}}$.

Proof of Theorem 6.6
Using (6.17) one can express π_n as a rational linear combination of the $e_{k,a}$, with k even, but special care is needed when n is a power of two. The coefficient of $\gamma_k\, \pi_N$ in (6.17), when $N = p^b$ is a prime power, is given by $(2^k - 2)(p - 1)p^{b-1} - p^b(p^{k-1} - 1)$, which does not vanish unless $p = 2$, and is $-p^{b-1}(2 - 3p + p^2)$ for $k = 2$. Thus, one can express π_N as a linear combination of the $e_{2,a}$ by induction on b. For $p = 2$, $N = 2^b$, $b > 1$ the coefficient of $\gamma_k\, \pi_N$ in (6.17) is zero but the coefficient of $\gamma_k\, \pi_{N/2}$ is $-2^{b-2}(2^k - 1)(2^k - 2) \neq 0$ for k even. This allows one to express π_N as a linear combination of the $e_{2,a}$ by induction on b. Thus, for instance, π_2 is given by

$$\pi_2 = 3 + 2 \sum_{4\,a=0} e_{2,a}.$$

In general, π_{2^n} involves $\sum_{2^{n+1}\,a=0} e_{2,a}$.

Since for relatively prime integers n, m one has $\pi_{nm} = \pi_n \, \pi_m$, we see that the algebra $\mathcal{B}_{\mathbb{Q}}$ generated over $\mathbb{Q}$ by the $e_{1,a}$ contains all the projections π_n. In order to show that $\mathcal{B}_{\mathbb{Q}}$ contains the $e(r)$ it is enough to show that for any prime power $N = p^b$ it contains $e(\frac{1}{N})$. This is proved by induction on b. Multiplying (6.17) by $1 - \pi_p$ and using $(1 - \pi_p)\, \pi_{p^l} = 0$ for $l > 0$ we get the equalities

$$(1 - \pi_p) \sum_{N\,a=0} e_{k,a} = (N^k + 2^k - 2)\, \gamma_k \, (1 - \pi_p)\,.$$

Let then $z(j) = (1 - \pi_p)\, e_{1,\frac{j}{N}}$. The above relations together with(6.13) show that in the reduced algebra $(\mathcal{B}_{\mathbb{Q}})_{1-\pi_p}$ one has, for all k,

$$\sum_{j=1}^{N-1} P_k(z(j)) = (N^k - 1)\, \gamma_k\,.$$

Thus, for $j \in \{1, ..., N-1\}$, the symmetric functions of the $z(j)$ are fixed rational numbers σ_h. In particular $z = z(1)$ fulfills

$$Q(z) = z^{N-1} + \sum_1^{N-1} (-1)^h \, \sigma_h \, z^{N-1-h} = 0$$

and $\pm \frac{1}{2}$ is not a root of this equation, whose roots are the $\dfrac{1}{2i} \cot(\dfrac{\pi\, j}{N})$. This allows us, using the companion matrix of Q, to express the Cayley transform of $2z$ as a polynomial with rational coefficients,

$$\frac{2z + 1}{2z - 1} = \sum_0^{N-2} \alpha_n \, z^n.$$

One then has

$$\sum_0^{N-2} \alpha_n \, z^n = (1 - \pi_p)\, e(\frac{1}{N}),$$

where the left-hand side belongs to $\mathcal{B}_{\mathbb{Q}}$ by construction. Now $\pi_p \, e(\frac{1}{N})$ is equal to $\alpha_p(e(\frac{p}{N}))$. It follows from the induction hypothesis on b, $(N = p^b)$, that $e(\frac{p}{N}) \in \mathcal{B}_{\mathbb{Q}}$ and therefore using (6.19) that $\alpha_p(e(\frac{p}{N})) \in \mathcal{B}_{\mathbb{Q}}$. Thus, we get $e(\frac{1}{N}) \in \mathcal{B}_{\mathbb{Q}}$ as required. This proves the first part. To get the second notice that the cross product by $\mathbb{N}^\times$ is obtained by adjoining to the rational group ring of $\mathbb{Q}/\mathbb{Z}$ the isometries μ_n and their adjoints μ_n^* with the relation

$$\mu_n\, f\, \mu_n^* = \alpha_n(f)\,, \quad \forall f \in \mathbb{Q}[\mathbb{Q}/\mathbb{Z}]\,,$$

which gives the rational algebra $\mathcal{A}_{\mathbb{Q}}$. □

It is not true, however, that the division relations (6.17) combined with the multiplicative relations (6.13) suffice to present the algebra. In particular there are more elaborate division relations which we did not need in the above proof. In order to formulate them, we let for $d|N$, $\pi(N,d)$ be the projection belonging to the algebra generated by the π_b, $b|N$, and corresponding to the subset

$$\gcd(N,\, (\Lambda,\phi)) = N/d$$

so that

$$\pi(N,d) = \pi_{N/d} \prod_{k|d} (1 - \pi_{k\,N/d}),$$

where the product is over non trivial divisors $k \neq 1$ of d.

Proposition 6.9 *The $e_{k,a}$, $a \in \mathbb{Q}/\mathbb{Z}$, k odd, fulfill for any $x \in \mathbb{Q}/\mathbb{Z}$ and any integer N the relation*

$$\frac{1}{N} \sum_{N\,a=0} e_{k,x+a} = \sum_{d|N} \pi(N,d)\, d^{\,k-1}\, e_{k,d\,x}.$$

Proof. To prove this, let (Λ,ϕ) be such that $\gcd(N,\, (\Lambda,\phi)) = N/d = m$ and assume by homogeneity that $\Lambda = \mathbb{Z}$. Then when a ranges through the $\frac{j}{N}$, $j \in \{0,...,N-1\}$, the $\phi(a)$ range m-times through the $\frac{j}{d}$, $j \in \{0,...,d-1\}$. Thus the left-hand side of (6.9) gives m-times

$$c(\mathbb{Z})^k \sum_{j=0}^{d-1} \sum_{y \in \mathbb{Z}+\phi(x)+\frac{j}{d}} y^{-k} = c(\mathbb{Z})^k\, d^{\,k} \sum_{y \in \mathbb{Z}+\phi(d\,x)} y^{-k}.$$

This is clear when $y = 0$ does not appear in the sums involved. When it does one has, for $\epsilon \notin \frac{\mathbb{Z}}{d}$,

$$\sum_{j=0}^{d-1} \sum_{y \in \mathbb{Z}+\phi(x)+\frac{j}{d}} (y+\epsilon)^{-k} = d^{\,k} \sum_{y \in \mathbb{Z}+\phi(d\,x)} (y + d\,\epsilon)^{-k}.$$

Subtracting the pole part on both sides and equating the finite values gives the desired equality, since for odd k the value of $\epsilon_{k,a}(\Lambda,\phi)$ for $\phi(a) = 0$ can be written as the finite value of

$$\sum_{y \in \Lambda+\phi(a)} (y+\epsilon)^{-k}.$$

For even k this no longer holds and the finite value $\gamma_k c(\Lambda)^k$ is replaced by $(2^k - 1)\,\gamma_k c(\Lambda)^k$. Thus when $\phi(d\,x) \in \mathbb{Z}$ one gets an additional term which is best taken care of by multiplying the right hand side in Proposition 6.9 by $(1-\pi_{\delta(d\,x)})$, with $\delta(y)$ the order of y in $\mathbb{Q}/\mathbb{Z}$, and adding corresponding terms to the formula, which becomes

$$\frac{1}{N}\sum_{N\,a=0} e_{k,x+a} = \sum_{d|N} \pi(N,d)(1-\pi_{\delta(d\,x)})\,d^{k-1}\,e_{k,d\,x} \tag{6.20}$$
$$+\;\gamma_k \sum_{d|N} (d^{k-1} + d^{-1}(2^k-2))\,\pi(N,d)\,\pi_{\delta(d\,x)}$$

These relations are more elaborate than the division relations for trigonometric functions. They restrict to the latter on the subset of invertible $\mathbb{Q}$-lattices, for which all π_n, $n \neq 1$ are zero and the only non-zero term in the r.h.s. is the term in $d = N$. The standard discussion of Eisenstein series in higher dimension is restricted to invertible $\mathbb{Q}$-lattices, but in our case the construction of the endomorphisms implemented by the μ_n requires the above extension to non-invertible $\mathbb{Q}$-lattices. We shall now proceed to do it in dimension 2.

7 The Determinant Part of the GL_2-System

As we recalled in the previous sections, the algebra of the 1-dimensional system can be described as the semigroup cross product

$$C(R) \rtimes \mathbb{N}^\times .$$

Thus, one may wish to follow a similar approach for the 2-dimensional case, by replacing $C(R)$ by $C(M_2(R))$ and the semigroup action of $\mathbb{N}^\times$ by the semigroup action of $M_2(\mathbb{Z})^+$. Such construction can be carried out, as we discuss in this section, and it corresponds to the "determinant part" of the GL_2 system. It is useful to analyze what happens in this case first, before we discuss the full GL_2-system in the next section. In fact, this will show quite clearly where some important technical issues arise.

For instance, just as in the case of the BC algebra, where the time evolution acts on the isometries μ_n by n^{it} and leaves the elements of $C(R)$ fixed, the time evolution here is given by $\mathrm{Det}\,(m)^{it}$ on the isometries implementing the semigroup action of $m \in M_2(\mathbb{Z})^+$, while leaving $C(M_2(R))$ pointwise fixed. In this case, however, the vacuum state of the corresponding Hamiltonian is highly degenerate, because of the presence of the $\mathrm{SL}_2(\mathbb{Z})$ symmetry. This implies that the partition function and the KMS states below critical temperature can only be defined via the type II_1 trace Trace_Γ.

This issue will be taken care more naturally in the full GL_2-system, by first taking the classical quotient by $\Gamma = \mathrm{SL}_2(\mathbb{Z})$ on the space $M_2(R) \times \mathbb{H}$. This will resolve the degeneracy of the vacuum state and the counting of modes of the Hamiltonian will be on the coset classes $\Gamma\backslash M_2(\mathbb{Z})^+$.

The whole discussion of this section extends to $\mathrm{GL}(n)$ for arbitrary n and we shall briefly indicate how this is done, but we stick to $n = 2$ for definiteness.

We start with the action of the semigroup

$$M_2(\mathbb{Z})^+ = \{m \in M_2(\mathbb{Z}), \mathrm{Det}\,(m) > 0\} = \mathrm{GL}_2^+(\mathbb{Q}) \cap M_2(R) \tag{7.1}$$

on the compact space $M_2(R)$, given by left multiplication

$$\rho \mapsto m\,\rho, \tag{7.2}$$

where the product $m\rho$ takes place in $M_2(R)$ using the natural homomorphism

$$M_2(\mathbb{Z})^+ \to M_2(R), \tag{7.3}$$

which is the extension to two by two matrices of the inclusion of the ring $\mathbb{Z}$ in $\hat{\mathbb{Z}} = R$. The relevant C^*-algebra is the semi-group cross product

$$A = C(M_2(R)) \rtimes M_2(\mathbb{Z})^+. \tag{7.4}$$

It can be viewed as the C^*-algebra $C^*(G_2)$ of the étale groupoid G_2 of pairs (r, ρ), with $r \in \mathrm{GL}_2^+(\mathbb{Q})$, $\rho \in M_2(R)$ and $r\,\rho \in M_2(R)$, where the product takes place in $M_2(\mathbb{A}_f)$. The composition in G_2 is given by

$$(r_1, \rho_1) \circ (r_2, \rho_2) = (r_1\, r_2, \rho_2)\,, \quad \text{if } r_2\, \rho_2 = \rho_1$$

and the convolution of functions by

$$f_1 * f_2(r, \rho) := \sum f_1(rs^{-1}, s\,\rho)\, f_2(s, \rho)\,, \tag{7.5}$$

while the adjoint of f is

$$f^*(r, \rho) := \overline{f(r^{-1}, r\,\rho)} \tag{7.6}$$

(*cf.* the analogous expressions (6.3), (6.4), (6.5) in the 1-dimensional case).

A homomorphism $G_2 \to H$ of the groupoid G_2 to an abelian group H determines a *dual action* of the Pontrjagin dual of H on the algebra of G_2, as in the case of the time evolution σ_t, with $H = \mathbb{R}_+^*$ and its dual identified with $\mathbb{R}$. We shall use the same term "dual action" for nonabelian H.

The main structure is given by the *dual action* of $\mathrm{GL}_2^+(\mathbb{R})$ corresponding to the groupoid homomorphism j

$$j\,:\, G_2 \to \mathrm{GL}_2^+(\mathbb{R})\,, \quad j(r, \rho) = r \tag{7.7}$$

obtained from the inclusion $\mathrm{GL}_2^+(\mathbb{Q}) \subset \mathrm{GL}_2^+(\mathbb{R})$. As a derived piece of structure one gets the one parameter group of automorphisms $\sigma_t \in \mathrm{Aut}(A)$ which is dual to the determinant of the homomorphism j,

$$\sigma_t(f)(r, \rho) := \mathrm{Det}(r)^{it}\, f(r, \rho)\,, \quad \forall f \in A\,. \tag{7.8}$$

The obtained C^*-dynamical system (A, σ_t) only involves $\mathrm{Det} \circ j$ and it does not fully correspond to the BC system. We shall make use of the full dual action of $\mathrm{GL}_2^+(\mathbb{R})$ later in the construction of the full GL_2 system.

The algebra $C(M_2(R))$ embeds as a $*$-subalgebra of A. The analogs of the isometries μ_n, $n \in \mathbb{N}^\times$ are the isometries μ_m, $m \in M_2(\mathbb{Z})^+$ given by

$$\mu_m(m,\rho) = 1\,, \quad \mu_m(r,\rho) = 0\,, \quad \forall r \neq m\,.$$

The range $\mu_m\,\mu_m^*$ of μ_m is the projection given by the characteristic function of the subset $P_m = m\,M_2(R) \subset M_2(R)$. It depends only on the lattice $L = m(\mathbb{Z}^2) \subset \mathbb{Z}^2$. Indeed, if $m, m' \in M_2(\mathbb{Z})^+$ fulfill $m(\mathbb{Z}^2) = m'(\mathbb{Z}^2)$, then $m' = m\gamma$ for some $\gamma \in \Gamma$, hence $m\,M_2(R) = m'\,M_2(R)$. Thus, we shall label this analog of the π_n by lattices

$$L \subset \mathbb{Z}^2 \mapsto \pi_L \in C(M_2(R))\,, \tag{7.9}$$

where π_L is the characteristic function of P_m, for any m such that $m(\mathbb{Z}^2) = L$. The algebra generated by the π_L is then governed by

$$\pi_L\,\pi_{L'} = \pi_{L\cap L'}\,, \qquad \pi_{\mathbb{Z}^2} = 1\,. \tag{7.10}$$

In fact, the complete rules are better expressed in terms of partial isometries $\mu(g,L)$, with $g \in \mathrm{GL}_2^+(\mathbb{Q})$, $L \subset \mathbb{Z}^2$ a lattice, and $g(L) \subset \mathbb{Z}^2$, satisfying

$$\mu(g,L)(g,\rho) = \pi_L(\rho)\,, \quad \mu(g,L)(r,\rho) = 0\,, \quad \forall r \neq g\,.$$

One has

$$\mu(g_1,L_1)\,\mu(g_2,L_2) = \mu(g_1g_2, g_2^{-1}(L_1)\cap L_2)\,, \tag{7.11}$$

and

$$\mu(g,L)^* = \mu(g^{-1}, g(L))\,. \tag{7.12}$$

The $\mu(g,L)$ generate the semi-group C^*-subalgebra $C^*(M_2(\mathbb{Z})^+) \subset A$ and together with $C(M_2(R))$ they generate A. The additional relations are

$$f\,\mu(g,L) = \mu(g,L)\,f^g\,, \quad \forall f \in C(M_2(R))\,, \quad g \in \mathrm{GL}_2^+(\mathbb{Q}), \tag{7.13}$$

where $f^g(y) := f(gy)$ whenever gy makes sense.

The action of $\mathrm{GL}_2(R)$ on $M_2(R)$ by right multiplication commutes with the semi-group action (7.2) of $M_2(\mathbb{Z})^+$ and with the time evolution σ_t. They define symmetries

$$\alpha_\theta \in \mathrm{Aut}(A,\sigma)\,.$$

Thus, we have a C^*-dynamical system with a compact group of symmetries. The following results show how to construct KMS$_\beta$-states for $\beta > 2$. We first describe a specific positive energy representation of the C^*-dynamical sub-system $(C^*(M_2(\mathbb{Z})^+), \sigma_t)$. We let $\mathcal{H} = \ell^2(M_2(\mathbb{Z})^+)$ with canonical basis ε_m, $m \in M_2(\mathbb{Z})^+$. We define $\pi(\mu(g,L))$ as the partial isometry in $\mathcal{H}$ with initial domain given by the span of

$$\varepsilon_m\,, \quad m = \begin{bmatrix} m_{11} & m_{12} \\ m_{21} & m_{22} \end{bmatrix}, \quad (m_{11}, m_{21}) \in L\,,\ (m_{12}, m_{22}) \in L\,, \tag{7.14}$$

i.e. matrices m whose columns belong to the lattice $L \subset \mathbb{Z}^2$. On this domain we define the action of $\pi(\mu(g,L))$ by

$$\pi(\mu(g,L))\,\varepsilon_m = \varepsilon_{gm}\,. \tag{7.15}$$

Notice that the columns of gm belong to gL.

Proposition 7.1 *1) π is an involutive representation of $C^*(M_2(\mathbb{Z})^+)$ in $\mathcal{H}$.*
2) The Hamiltonian H given by $H\varepsilon_m = \log \mathrm{Det}(m)\,\varepsilon_m$ is positive and implements the time evolution σ_t:

$$\pi(\sigma_t(x)) = e^{itH}\,\pi(x)\,e^{-itH} \qquad \forall x \in C^*(M_2(\mathbb{Z})^+)\,.$$

3) $\Gamma = \mathrm{SL}_2(\mathbb{Z})$ acts on the right in $\mathcal{H}$ by

$$\rho(\gamma)\,\epsilon_m := \epsilon_{m\,\gamma^{-1}}\,, \quad \forall \gamma \in \Gamma\,, \quad m \in M_2(\mathbb{Z})^+\,.$$

and this action commutes with $\pi(C^*(M_2(\mathbb{Z})^+)$.

Proof. The map $m \mapsto gm$ is injective so that $\pi(\mu(g,L))$ is a partial isometry. Its range is the set of $h \in M_2(\mathbb{Z})^+$ of the form $g\,m$ where $\mathrm{Det}(m) > 0$ and the columns of m are in L. This means that $\mathrm{Det}(h) > 0$ and the columns of h are in $gL \subset \mathbb{Z}^2$. This shows that

$$\pi(\mu(g,L))^* = \pi(\mu(g^{-1}, gL)), \tag{7.16}$$

so that π is involutive on these elements.

Then the support of $\pi(\mu(g_1,L_1))\,\pi(\mu(g_2,L_2))$ is formed by the ε_m with columns of m in L_2, such that the columns of $g_2 m$ are in L_1. This is the same as the support of $\pi(\mu(g_1 g_2, g_2^{-1} L_1 \cap L_2))$ and the two partial isometries agree there. Thus, we get

$$\pi(\mu(g_1,L_1)\,\mu(g_2,L_2)) = \pi(\mu(g_1,L_1))\,\pi(\mu(g_2,L_2))\,. \tag{7.17}$$

Next, using (7.15) we see that

$$H\,\pi(\mu(g,L)) - \pi(\mu(g,L))\,H = \log(\mathrm{Det}\,g)\,\pi(\mu(g,L)), \tag{7.18}$$

since both sides vanish on the kernel while on the support one can use the multiplicativity of Det.

Now $\Gamma = \mathrm{SL}_2(\mathbb{Z})$ acts on the right in $\mathcal{H}$ by

$$\rho(\gamma)\,\epsilon_m := \epsilon_{m\,\gamma^{-1}}\,, \quad \forall \gamma \in \Gamma\,, \quad m \in M_2(\mathbb{Z})^+ \tag{7.19}$$

and this action commutes by construction with the algebra $\pi(C^*(M_2(\mathbb{Z})^+)$.
□

The image $\rho(C^*(\Gamma))$ generates a type II_1 factor in $\mathcal{H}$, hence one can evaluate the corresponding trace Trace_Γ on any element of its commutant. We let

$$\varphi_\beta(x) := \mathrm{Trace}_\Gamma(\pi(x)\, e^{-\beta H}), \quad \forall x \in C^*(M_2(\mathbb{Z})^+) \tag{7.20}$$

and we define the normalization factor by

$$Z(\beta) = \mathrm{Trace}_\Gamma(e^{-\beta H}). \tag{7.21}$$

We then have the following:

Lemma 7.2 *1) The normalization factor $Z(\beta)$ is given by*

$$Z(\beta) = \zeta(\beta)\,\zeta(\beta-1),$$

where ζ is the Riemann ζ-function.

2) For all $\beta > 2$, $Z^{-1}\,\varphi_\beta$ is a KMS_β state on $C^*(M_2(\mathbb{Z})^+)$.

Proof. Any sublattice $L \subset \mathbb{Z}^2$ is uniquely of the form $L = \begin{bmatrix} a & b \\ 0 & d \end{bmatrix} \mathbb{Z}^2$, where $a, d \geq 1$, $0 \leq b < d$ (*cf.* [50] p. 161). Thus, the type II_1 dimension of the action of Γ in the subspace of $\mathcal{H}$ spanned by the ε_m with $\mathrm{Det}\, m = N$ is the same as the cardinality of the quotient of $\{m \in M_2(\mathbb{Z})^+, \mathrm{Det}\, m = N\}$ by Γ acting on the right. This is equal to the cardinality of the set of matrices $\begin{bmatrix} a & b \\ 0 & d \end{bmatrix}$ as above with determinant $= N$. This gives $\sigma_1(N) = \sum_{d|N} d$. Thus, $Z(\beta)$ is given by

$$\sum_{N=1}^{\infty} \frac{\sigma_1(N)}{N^\beta} = \zeta(\beta)\,\zeta(\beta-1). \tag{7.22}$$

One checks the KMS_β-property of φ_β using the trace property of Trace_Γ together with the second equality in Proposition 7.1. □

Proposition 7.3 *1) For any $\theta \in \mathrm{GL}_2(R)$ the formula*

$$\pi_\theta(f)\,\epsilon_m := f(m\,\theta)\,\epsilon_m, \quad \forall m \in M_2(\mathbb{Z})^+$$

extends the representation π to an involutive representation π_θ of the cross product $A = C(M_2(R)) \rtimes M_2(\mathbb{Z})^+$ in $\mathcal{H}$.

2) Let $f \in C(M_2(\mathbb{Z}/N\mathbb{Z})) \subset C(M_2(R))$. Then $\pi_\theta(f) \in \rho(\Gamma_N)'$ where Γ_N is the congruence subgroup of level N.

3) For each $\beta > 2$ the formula

$$\psi_\beta(x) := \mathrm{Lim}_{\mathrm{N}\to\infty} \quad Z_N^{-1}\, \mathrm{Trace}_{\Gamma_N}(\pi_\theta(x)\, e^{-\beta H}), \quad \forall x \in A$$

defines a KMS_β state on A, where $Z_N := \mathrm{Trace}_{\Gamma_N}(e^{-\beta H})$.

Proof. 1) The invertibility of θ shows that letting f_L be the characteristic function of P_L one has $\pi_\theta(f_L) = \pi_L$ independently of θ. Indeed $f_L(m\,\theta) = 1$ iff $m\,\theta \in P_L$ and this holds iff $m(\mathbb{Z}^2) \subset L$.

To check (7.13) one uses

$$f(gm\,\theta) = f^g(m\,\theta))\,, \quad \forall g \in \mathrm{GL}_2^+(\mathbb{Q})\,.$$

2) Let $p_N : M_2(R) \to M_2(\mathbb{Z}/N\mathbb{Z})$ be the canonical projection. It is a ring homomorphism. Let then $f = h \circ p_N$ where h is a function on $M_2(\mathbb{Z}/N\mathbb{Z})$. One has, for any $\gamma \in \Gamma_N$,

$$\pi_\theta(f)\,\rho(\gamma)\,\epsilon_m = \pi_\theta(f)\,\epsilon_{m\,\gamma^{-1}} =$$
$$f(m\,\gamma^{-1}\theta)\,\epsilon_{m\,\gamma^{-1}} = h(p_N(m\,\gamma^{-1}\theta))\,\epsilon_{m\,\gamma^{-1}}\,.$$

The equality $p_N(\gamma) = 1$ shows that

$$p_N(m\,\gamma^{-1}\theta) = p_N(m)\,p_N(\gamma^{-1})\,p_N(\theta) = p_N(m)\,p_N(\theta) = p_N(m\,\theta),$$

hence

$$\pi_\theta(f)\,\rho(\gamma)\,\epsilon_m = \rho(\gamma)\,\pi_\theta(f)\,\epsilon_m\,.$$

3) One uses 2) to show that, for all N and $f \in C(M_2(\mathbb{Z}/N\mathbb{Z})) \subset C(M_2(R))$, the products $f\,\mu(g,L)$ belong to the commutant of $\rho(\Gamma_N)$. Since Γ_N has finite index in Γ it follows that for $\beta > 2$ one has $Z_N := \mathrm{Trace}_{\Gamma_N}(e^{-\beta H}) < \infty$. Thus, the limit defining $\psi_\beta(x)$ makes sense on a norm dense subalgebra of A and extends to a state on A by uniform continuity. One checks the KMS_β condition on the dense subalgebra in the same way as above. □

It is not difficult to extend the above discussion to arbitrary n using ([51]). The normalization factor is then given by

$$Z(\beta) = \prod_0^{n-1} \zeta(\beta - k)\,.$$

What happens, however, is that the states ψ_β only depend on the determinant of θ. This shows that the above construction should be extended to involve not only the one-parameter group σ_t but in fact the whole dual action given by the groupoid homomorphism (7.7).

Definition 7.4 *Given a groupoid G and a homomorphism $j : G \to H$ to a group H, the "cross product" groupoid $G \times_j H$ is defined as the product $G \times H$ with units $G^{(0)} \times H$, range and source maps*

$$r(\gamma, \alpha) := (r(\gamma), j(\gamma)\,\alpha)\,, \quad s(\gamma,\alpha) := (s(\gamma), \alpha)$$

and composition

$$(\gamma_1, \alpha_1) \circ (\gamma_2, \alpha_2) := (\gamma_1 \circ \gamma_2, \alpha_2)\,.$$

In our case this cross product $\tilde{G}_2 = G_2 \times_j \mathrm{GL}_2^+(\mathbb{R})$ corresponds to the groupoid of the partially defined action of $\mathrm{GL}_2^+(\mathbb{Q})$ on the locally compact space Z_0 of pairs $(\rho, \alpha) \in M_2(R) \times \mathrm{GL}_2^+(\mathbb{R})$ given by

$$g(\rho,\alpha) := (g\,\rho, g\,\alpha)\,, \quad \forall g \in \mathrm{GL}_2^+(\mathbb{Q})\,, \quad g\,\rho \in M_2(R)\,.$$

Since the subgroup $\Gamma \subset \mathrm{GL}_2^+(\mathbb{Q})$ acts freely and properly by translation on $\mathrm{GL}_2^+(\mathbb{R})$, one obtains a Morita equivalent groupoid S_2 by dividing $\tilde{G}_2$ by the following action of $\Gamma \times \Gamma$:

$$(\gamma_1,\gamma_2).(g,\rho,\alpha) := (\gamma_1\, g\gamma_2^{-1},\, \gamma_2\rho, \gamma_2\alpha). \tag{7.23}$$

The space of units $S_2^{(0)}$ is the quotient $\Gamma\backslash Z_0$. We let $p : Z_0 \to \Gamma\backslash Z_0$ be the quotient map. The range and source maps are given by

$$r(g,\rho,\alpha) := p(g\rho, g\alpha)\,, \quad s(g,\rho,\alpha) := p(\rho,\alpha)$$

and the composition is given by

$$(g_1,\rho_1,\alpha_1)\circ(g_2,\rho_2,\alpha_2) = (g_1 g_2, \rho_2, \alpha_2),$$

which passes to $\Gamma \times \Gamma$-orbits.

8 Commensurability of $\mathbb{Q}$-Lattices in $\mathbb{C}$ and the Full GL_2-System

We shall now describe the full GL_2 C^*-dynamical system (A, σ_t). It is obtained from the system of the previous section by taking a cross product with the dual action of $\mathrm{GL}_2^+(\mathbb{R})$ *i.e.* from the groupoid S_2 that we just described. It admits an equivalent and more geometric description in terms of the notion of *commensurability* between $\mathbb{Q}$-lattices developed in section 1.6 above and we shall follow both points of view. The C^*-algebra A is a Hecke algebra, which is a variant of the *modular Hecke algebra* defined in ([10]). Recall from section 1.6:

Definition 8.1 *1) A $\mathbb{Q}$-lattice in $\mathbb{C}$ is a pair (Λ, ϕ), with Λ a lattice in $\mathbb{C}$, and $\phi : \mathbb{Q}^2/\mathbb{Z}^2 \longrightarrow \mathbb{Q}\,\Lambda/\Lambda$ an homomorphism of abelian groups.*

2) Two $\mathbb{Q}$-lattices (Λ_j, ϕ_j) are commensurable iff Λ_j are commensurable and $\phi_1 - \phi_2 = 0$ modulo $\Lambda = \Lambda_1 + \Lambda_2$.

This is an equivalence relation $\mathcal{R}$ between $\mathbb{Q}$-lattices (Proposition 6.4). We use the basis $\{e_1 = 1, e_2 = -i\}$ of the $\mathbb{R}$-vector space $\mathbb{C}$ to let $\mathrm{GL}_2^+(\mathbb{R})$ act on $\mathbb{C}$ as $\mathbb{R}$-linear transformations. We let

$$\Lambda_0 := \mathbb{Z}\, e_1 + \mathbb{Z}\, e_2 = \mathbb{Z} + i\mathbb{Z}$$

Also we view $\rho \in M_2(R)$ as the homomorphism

$$\rho : \mathbb{Q}^2/\mathbb{Z}^2 \longrightarrow \mathbb{Q}.\Lambda_0/\Lambda_0\,, \quad \rho(a) = \rho_1(a)e_1 + \rho_2(a)e_2\,.$$

Proposition 8.2 *The map*

$$\gamma(g,\rho,\alpha) = ((\alpha^{-1} g^{-1} \Lambda_0,\, \alpha^{-1}\rho), (\alpha^{-1}\Lambda_0,\, \alpha^{-1}\rho)), \quad \forall (g,\rho,\alpha) \in S_2$$

defines an isomorphism of locally compact étale groupoids between S_2 and the equivalence relation $\mathcal{R}$ of commensurability on the space of $\mathbb{Q}$-lattices in $\mathbb{C}$.

Proof. The proof is the same as for Proposition 6.5. □

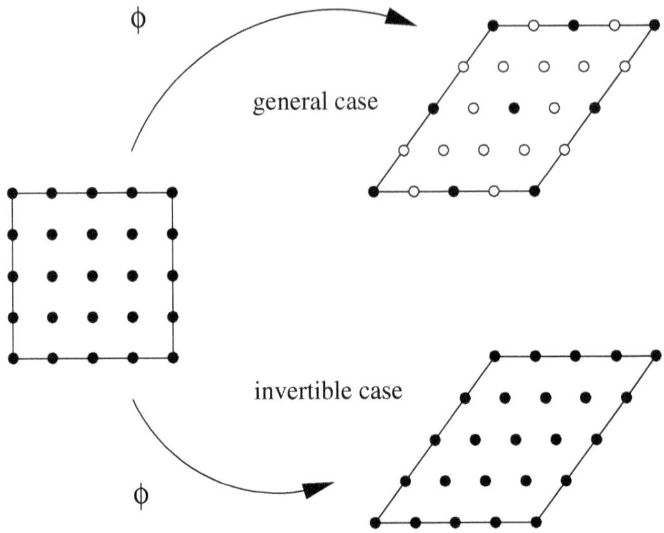

Fig. 2. $\mathbb{Q}$-Lattices in $\mathbb{C}$

We shall now describe the quotient of $S_2 \sim \mathcal{R}$ by the natural scaling action of $\mathbb{C}^*$. We view $\mathbb{C}^*$ as a subgroup of $\mathrm{GL}_2^+(\mathbb{R})$ by the map

$$\lambda = a + i\,b \in \mathbb{C}^* \mapsto \begin{bmatrix} a & b \\ -b & a \end{bmatrix} \in \mathrm{GL}_2^+(\mathbb{R}) \tag{8.1}$$

and identify the quotient $\mathrm{GL}_2^+(\mathbb{R})/\mathbb{C}^*$ with $\mathbb{H}$ by the map

$$\alpha \in \mathrm{GL}_2^+(\mathbb{R}) \mapsto \tau = \alpha(i) \in \mathbb{H}. \tag{8.2}$$

Given a pair (Λ_j, ϕ_j) of commensurable $\mathbb{Q}$-lattices and a non zero complex number $\lambda \in \mathbb{C}^*$ the pair $(\lambda\Lambda_j, \lambda\phi_j)$ is still a pair of commensurable $\mathbb{Q}$-lattices. Moreover, one has

$$\gamma(g,\rho,\alpha\,\lambda^{-1}) = \lambda\,\gamma(g,\rho,\alpha), \quad \forall \lambda \in \mathbb{C}^*. \tag{8.3}$$

The scaling action of $\mathbb{C}^*$ on $\mathbb{Q}$-lattices in $\mathbb{C}$ is not free, since the lattice Λ_0 for instance is invariant under multiplication by i. It follows that the quotient

$S_2/\mathbb{C}^* \sim \mathcal{R}/\mathbb{C}^*$ is not a groupoid. One can nevertheless define its convolution algebra in a straightforward manner by restricting the convolution product on $S_2 \sim \mathcal{R}$ to functions which are homogeneous of *weight* 0, where weight k means

$$f(g, \rho, \alpha\,\lambda) = \lambda^k\, f(g, \rho, \alpha)\,, \quad \forall \lambda \in \mathbb{C}^*\,. \tag{8.4}$$

Let

$$Y = M_2(R) \times \mathbb{H}, \tag{8.5}$$

endowed with the natural action of $\mathrm{GL}_2^+(\mathbb{Q})$ by

$$\gamma \cdot (\rho, \tau) = \left(\gamma\,\rho,\ \frac{a\tau + b}{c\tau + d}\right), \tag{8.6}$$

for $\gamma = \begin{bmatrix} a & b \\ c & d \end{bmatrix} \in \mathrm{GL}_2^+(\mathbb{Q})$ and $(\rho, \tau) \in Y$. Let then

$$Z \subset \Gamma\backslash\mathrm{GL}_2^+(\mathbb{Q}) \times_\Gamma Y \tag{8.7}$$

be the locally compact space quotient of $\{(g, y) \in \mathrm{GL}_2^+(\mathbb{Q}) \times Y\,,\ g\,y \in Y\}$ by the following action of $\Gamma \times \Gamma$:

$$(g, y) \mapsto (\gamma_1\, g\, \gamma_2^{-1}, \gamma_2\, y)\,, \quad \forall \gamma_j \in \Gamma\,.$$

The natural lift of the quotient map (8.2), together with proposition 8.2, first gives the identification of the quotient of Y by Γ with the space of $\mathbb{Q}$-lattices in $\mathbb{C}$ up to scaling, realized by the map

$$\theta : \Gamma\backslash Y \to (\text{Space of } \mathbb{Q}\text{-lattices in } \mathbb{C})/\mathbb{C}^* = X\,, \tag{8.8}$$

$$\theta(\rho, \tau) = (\Lambda, \phi)\,, \quad \Lambda = \mathbb{Z} + \mathbb{Z}\tau\,, \quad \phi(x) = \rho_1(x) - \tau\rho_2(x)\,.$$

It also gives the isomorphism $\theta : S_2/\mathbb{C}^* = Z \to \mathcal{R}/\mathbb{C}^*$,

$$\theta(g, y) = (\lambda\theta(gy), \theta(y)), \tag{8.9}$$

where $\lambda = \mathrm{Det}(g)^{-1}(c\tau + d)$ for $\gamma = \begin{bmatrix} a & b \\ c & d \end{bmatrix} \in \mathrm{GL}_2^+(\mathbb{Q})$ and $y = (\rho, \tau) \in Y$.

We let $\mathcal{A} = C_c(Z)$ be the space of continuous functions with compact support on Z. We view elements $f \in \mathcal{A}$ as functions on $\mathrm{GL}_2^+(\mathbb{Q}) \times Y$ such that

$$f(\gamma\, g, y) = f(g, y) \quad f(g\,\gamma, y) = f(g, \gamma\, y)\,, \quad \forall \gamma \in \Gamma\,,\ g \in \mathrm{GL}_2^+(\mathbb{Q})\,,\ y \in Y\,.$$

This does not imply that $f(g, y)$ only depends on the orbit $\Gamma.y$ but that it only depends on the orbit of y under the congruence subgroup $\Gamma \cap g^{-1}\Gamma\, g$.

We define the convolution product of two such functions by

$$(f_1 * f_2)(g, y) := \sum_{h \in \Gamma\backslash\mathrm{GL}_2^+(\mathbb{Q}),\, hy \in Y} f_1(gh^{-1}, hy)\, f_2(h, y) \tag{8.10}$$

and the adjoint by

$$f^*(g,y) := \overline{f(g^{-1}, g\,y)}. \tag{8.11}$$

Notice that these rules combine (3.13) and (7.5).

For any $x \in X$ we let $c(x)$ be the commensurability class of x. It is a countable subset of X and we want to define a natural representation in $l^2(c(x))$. We let p be the quotient map from Y to X. Let $y \in Y$ with $p(y) = x$ be an element in the preimage of x. Let

$$G_y = \{g \in \mathrm{GL}_2^+(\mathbb{Q}) \,|\, gy \in Y\}.$$

The natural map $g \in G_y \mapsto p(gy) \in X$ is a surjection from $\Gamma \backslash G_y$ to $c(x)$ but it fails to be injective in degenerate cases such as $y = (0, \tau)$ with $\tau \in \mathbb{H}$ a complex multiplication point (*cf.* Lemma 8.8). This corresponds to the phenomenon of holonomy in the context of foliations ([11]). To handle it one defines the representation π_y directly in the Hilbert space $\mathcal{H}_y = l^2(\Gamma \backslash G_y)$ of left Γ-invariant functions on G_y by

$$(\pi_y(f)\,\xi)(g) := \sum_{h \in \Gamma \backslash G_y} f(gh^{-1}, hy)\,\xi(h)\,, \quad \forall g \in G_y\,, \tag{8.12}$$

for $f \in \mathcal{A}$ and $\xi \in \mathcal{H}_y$.

Proposition 8.3 *1) The vector space $\mathcal{A}$ endowed with the product $*$ and the adjoint $f \mapsto f^*$ is an involutive algebra.*

2) For any $y \in Y$, π_y defines a unitary representation of $\mathcal{A}$ in $\mathcal{H}_y$ whose unitary equivalence class only depends on $x = p(y)$.

3) The completion of $\mathcal{A}$ for the norm given by

$$||f|| := \mathrm{Sup}_{y \in Y}\, ||\pi_y(f)||$$

is a C^*-algebra.

The proof of (1) and (2) is similar to ([10], Proposition 2). Using the compactness of the support of f, one shows that the supremum is finite for any $f \in \mathcal{A}$ (cf. [11]). □

Remark 8.4 The locally compact space Z of (8.7) is not a groupoid, due to the torsion elements in Γ, which give nontrivial isotropy under scaling, for the square and equilateral lattices. Nonetheless, Proposition 8.3 yields a well defined C^*-algebra. This can be viewed as a subalgebra of the C^*-algebra of the groupoid obtained by replacing Γ by its commutator subgroup in the definition of S_2 as in (7.23).

We let σ_t be the one parameter group of automorphisms of A given by

$$\sigma_t(f)(g,y) = (\mathrm{Det}\, g)^{it}\, f(g,y)\,. \tag{8.13}$$

Notice that since X is not compact (but still locally compact) the C^*-algebra A does not have a unit, hence the discussion of Proposition 2.2 applies.

The one parameter group σ_{2t} (8.13) is the modular automorphism group associated to the regular representation of $\mathcal{A}$. To obtain the latter we endow $X = \Gamma\backslash Y$ with the measure

$$dy = d\rho \times d\mu(\tau),$$

where $d\rho = \prod d\rho_{ij}$ is the normalized Haar measure of the additive compact group $M_2(R)$ and $d\mu(\tau)$ is the Riemannian volume form in $\mathbb{H}$ for the Poincaré metric, normalized so that $\mu(\Gamma\backslash\mathbb{H}) = 1$. We then get the following result.

Proposition 8.5 *The expression*

$$\varphi(f) = \int_X f(1,y)\, dy\,. \tag{8.14}$$

defines a state on A, which is a KMS_2 state for the one parameter group σ_t.

Proof. At the measure theory level, the quotient $X = \Gamma\backslash Y$ is the total space over $\Gamma\backslash\mathbb{H}$ of a bundle with fiber the probability space $M_2(R)/\{\pm 1\}$, thus the total mass $\int_X dy = 1$. One gets

$$\varphi(f^* * f) = \int_X \sum_{h\in\Gamma\backslash \mathrm{GL}_2^+(\mathbb{Q}),\, hy\in Y} \overline{f(h,y)} f(h,y)\, dy\,, \quad \forall f \in \mathcal{A}\,,$$

which suffices to get the Hilbert space $\mathcal{H}$ of the regular representation and the cyclic vector ξ implementing the state φ, which corresponds to

$$\xi(g,y) = 0\,, \quad \forall g \notin \Gamma\,, \quad \xi(1,y) = 1\,, \quad \forall y \in X\,.$$

The measure $d\rho$ is the product of the additive Haar measures on column vectors, hence one gets

$$d(g\,\rho) = (\mathrm{Det}\, g)^{-2}\, d\rho\,, \quad \forall g \in \mathrm{GL}_2^+(\mathbb{Q})\,.$$

Let us prove that φ is a KMS_2 state. The above equality shows that, for any compactly supported continuous function α on $\Gamma\backslash\mathrm{GL}_2^+(\mathbb{Q}) \times_\Gamma Y$, one has

$$\int_X \sum_{h\in\Gamma\backslash \mathrm{GL}_2^+(\mathbb{Q})} \alpha(h,y)\, dy = \int_X \sum_{k\in\Gamma\backslash \mathrm{GL}_2^+(\mathbb{Q})} \alpha(k^{-1}, k\,y)\, (\mathrm{Det}\, k)^{-2}\, dy\,. \tag{8.15}$$

Let then $f_j \in \mathcal{A}$ and define $\alpha(h,y) = 0$ unless $hy \in Y$ while otherwise

$$\alpha(h,y) = f_1(h^{-1}, h\,y)\, f_2(h,y) (\mathrm{Det}\, h)^{it-2}.$$

The l.h.s. of (8.15) is then equal to $\varphi(f_1\,\sigma_z(f_2))$ for $z = t+2i$. The r.h.s. of (8.15) gives $\varphi(\sigma_t(f_2)\,f_1)$ and (8.15) gives the desired equality $\varphi(f_1\,\sigma_{t+2i}(f_2)) = \varphi(\sigma_t(f_2)\,f_1)$. □

We can now state the main result on the analysis of KMS states on the C^*-dynamical system (A,σ_t). Recall that a $\mathbb{Q}$-lattice $l = (\Lambda,\phi)$ is invertible if ϕ is an isomorphism (Definition 6.3). We have the following result.

Theorem 8.6 *1) For each invertible $\mathbb{Q}$-lattice $l = (\Lambda,\phi)$, the representation π_l is a positive energy representation of the C^*-dynamical system (A,σ_t).*

2) For $\beta > 2$ and $l = (\Lambda,\phi)$ an invertible $\mathbb{Q}$-lattice, the formula

$$\varphi_{\beta,l}(f) = Z^{-1} \sum_{\Gamma\backslash M_2(\mathbb{Z})^+} f(1, m\,\rho, m(\tau))\,\mathrm{Det}(m)^{-\beta}\,,$$

defines an extremal KMS$_\beta$ state $\varphi_{\beta,l}$ on (A,σ_t), where $Z = \zeta(\beta)\,\zeta(\beta-1)$ is the partition function.

3) For $\beta > 2$ the map $l \mapsto \varphi_{\beta,l}$ is a bijection from the space of invertible $\mathbb{Q}$-lattices (up to scaling) to the space $\mathcal{E}_\beta$ of extremal KMS$_\beta$ states on (A,σ_t).

The proof of 1) reflects the following fact, which in essence shows that the invertible $\mathbb{Q}$-lattices are *ground states* for our system.

Lemma 8.7 *1) Let $s : \mathbb{Q}^2/\mathbb{Z}^2 \to \mathbb{Q}^2$ be a section of the projection $\pi : \mathbb{Q}^2 \to \mathbb{Q}^2/\mathbb{Z}^2$. Then the set of $s(a+b) - s(a) - s(b)\,,\ a,b \in \mathbb{Q}^2/\mathbb{Z}^2$, generates $\mathbb{Z}^2$.*

2) Let $l = (\Lambda,\phi)$ be an invertible $\mathbb{Q}$-lattice and $l' = (\Lambda',\phi')$ be commensurable with l. Then $\Lambda \subset \Lambda'$.

Proof. 1) Let $L \subset \mathbb{Z}^2$ be the subgroup generated by the $s(a+b) - s(a) - s(b)\,,\ a,b \in \mathbb{Q}^2/\mathbb{Z}^2$. If $L \neq \mathbb{Z}^2$ we can assume, after a change of basis, that for some prime number p one has $L \subset p\,\mathbb{Z} \oplus \mathbb{Z}$. Restricting s to the p-torsion elements of $\mathbb{Q}^2/\mathbb{Z}^2$ and multiplying it by p, we get a morphism of groups

$$\mathbb{Z}/p\,\mathbb{Z} \oplus \mathbb{Z}/p\,\mathbb{Z} \to \mathbb{Z}/p^2\mathbb{Z} \oplus \mathbb{Z}/p\,\mathbb{Z},$$

which is a section of the projection

$$\mathbb{Z}/p^2\mathbb{Z} \oplus \mathbb{Z}/p\,\mathbb{Z} \to \mathbb{Z}/p\,\mathbb{Z} \oplus \mathbb{Z}/p\,\mathbb{Z}.$$

This gives a contradiction, since the group $\mathbb{Z}/p^2\mathbb{Z} \oplus \mathbb{Z}/p\,\mathbb{Z}$ contains elements of order p^2.

2) Let s (resp. s') be a lift of ϕ modulo Λ (resp. of ϕ' modulo Λ'). Since $\phi - \phi' = 0$ modulo $\Lambda'' = \Lambda + \Lambda'$ one has $s(a) - s'(a) \in \Lambda + \Lambda'$ for all $a \in \mathbb{Q}^2/\mathbb{Z}^2$. This allows one to correct s modulo Λ and s' modulo Λ' so that $s = s'$. Then for any $a,b \in \mathbb{Q}^2/\mathbb{Z}^2$ one has $s(a+b) - s(a) - s(b) \in \Lambda \cap \Lambda'$ and the first part of the lemma together with the invertibility of ϕ show that $\Lambda \cap \Lambda' = \Lambda$. □

Given $y \in Y$ we let H_y be the diagonal operator in $\mathcal{H}_y$ given by

$$(H_y \, \xi)(h) := \log(\mathrm{Det}(h)) \; \xi(h) \,, \quad \forall h \in G_y \tag{8.16}$$

It implements the one parameter group σ_t *i.e.*

$$\pi_y(\sigma_t(x)) = e^{itH_y} \, \pi_y(x) \, e^{-itH_y} \,, \quad \forall x \in A \,. \tag{8.17}$$

In general the operator H_y is not positive but when the lattice $l = (\Lambda, \phi) = \theta(p(y))$ is invertible one has

$$\mathrm{Det}(h) \in \mathbb{N}^* \,, \quad \forall h \in G_y,$$

hence $H_y \geq 0$. This proves the first part of the theorem. The basis of the Hilbert space $\mathcal{H}_y$ is then labeled by the lattices Λ' containing Λ and the operator H_y is diagonal with eigenvalues the logarithms of the orders $\Lambda' : \Lambda$. Equivalently, one can label the orthonormal basis ϵ_m by the coset space $\Gamma \backslash M_2(\mathbb{Z})^+$. Thus, the same counting as in the previous section (cf.[51]) shows that

$$Z = \mathrm{Trace}(e^{-\beta H_y}) = \zeta(\beta) \, \zeta(\beta - 1)$$

and in particular that it is finite for $\beta > 2$. The KMS$_\beta$ property of the functional

$$\varphi_{\beta,l}(f) = Z^{-1} \, \mathrm{Trace}(\pi_y(f) \; e^{-\beta H_y})$$

then follows from (8.17). One has, using (8.12) for $y = (\rho, \tau) \in Y$,

$$\langle \pi_y(f)(\epsilon_m), \, \epsilon_m \rangle \; = \; f(1, m \, \rho, m(\tau)) \,,$$

hence we get the following formula for $\varphi_{\beta,l}$:

$$\varphi_{\beta,l}(f) = Z^{-1} \sum_{\Gamma \backslash M_2(\mathbb{Z})^+} f(1, m \, \rho, m(\tau)) \, \mathrm{Det}(m)^{-\beta} \,. \tag{8.18}$$

Finally, the irreducibility of the representation π_y follows as in [11] p.562 using the absence of holonomy for invertible $\mathbb{Q}$-lattices. This completes the proof of 2) of Theorem 8.6.

In order to prove 3) of Theorem 8.6 we shall proceed in two steps. The first shows (Proposition 8.10 below) that KMS$_\beta$ states are given by measures on the space X of $\mathbb{Q}$-lattices (up to scaling). The second shows that when $\beta > 2$ this measure is carried by the commensurability classes of invertible $\mathbb{Q}$-lattices.

We first describe the stabilizers of the action of $\mathrm{GL}_2^+(\mathbb{Q})$ on the space of $\mathbb{Q}$-lattices in $\mathbb{C}$.

Lemma 8.8 *Let $g \in \mathrm{GL}_2^+(\mathbb{Q})$, $g \neq 1$ and $y \in Y$, $y = (\rho, \tau)$ such that $gy = y$. Then $\rho = 0$. Moreover $g \in \mathbb{Q}^* \subset \mathrm{GL}_2^+(\mathbb{Q})$ unless τ is an imaginary quadratic number in which case $g \in K^* \subset \mathrm{GL}_2^+(\mathbb{Q})$ where $K = \mathbb{Q}(\tau)$ is the corresponding quadratic field.*

Proof. Let $g = \begin{bmatrix} a & b \\ c & d \end{bmatrix}$, then $g(\tau) = \tau$ means $a\tau + b = \tau\,(c\tau + d)$. If $c \neq 0$ this shows that τ is an imaginary quadratic number. Let $K = \mathbb{Q}(\tau)$ be the corresponding field and let $\{\tau, 1\}$ be the natural basis of K over $\mathbb{Q}$. Then the multiplication by $(c\tau + d)$ is given by the transpose of the matrix g. Since $g \neq 1$ and K is a field we get

$$g - 1 = \begin{bmatrix} a-1 & b \\ c & d-1 \end{bmatrix} \in \mathrm{GL}_2(\mathbb{Q})$$

and thus $(g-1)\rho = 0$ implies $\rho = 0$. If $c = 0$ then $a\tau + b = \tau\,(c\tau + d)$ implies $a = d$, $b = 0$ so that $g \in \mathbb{Q}^* \subset \mathrm{GL}_2^+(\mathbb{Q})$. Since $g \neq 1$ one gets $\rho = 0$. □

We let $X = \Gamma\backslash Y$ be the quotient and $p : Y \to X$ the quotient map. We let F be the closed subset of X, $F = p(\{(0,\tau); \tau \in \mathbb{H}\})$.

Lemma 8.9 *Let $g \in \mathrm{GL}_2^+(\mathbb{Q})$, $g \notin \Gamma$ and $x \in X$, $x \notin F$. There exists a neighborhood V of x, such that*

$$p(g\,p^{-1}(V)) \cap V = \emptyset$$

Proof. Let $\Gamma_0 = \Gamma \cap g^{-1}\Gamma g$ and $X_0 = \Gamma_0\backslash Y$. For $x_0 \in X_0$ the projections $p_1(x_0) = p(y)$ and $p_2(x_0) = p(gy)$ are independent of the representative $y \in Y$. Moreover if $p_1(x_0) \notin F$ then $p_1(x_0) \neq p_2(x_0)$ by Lemma 8.8. By construction Γ_0 is of finite index n in Γ and the fiber $p_1^{-1}(x)$ has at most n elements. Let then $W \subset X_0$ be a compact neighborhood of $p_1^{-1}(x)$ in X_0 such that $x \notin p_2(W)$. For $z \in V \subset X$ sufficiently close to x one has $p_1^{-1}(z) \subset W$ and thus $p_2(p_1^{-1}(z)) \subset V^c$ which gives the result. □

We can now prove the following.

Proposition 8.10 *Let $\beta > 0$ and φ a KMS$_\beta$ state on (A, σ_t). Then there exists a probability measure μ on $X = \Gamma\backslash Y$ such that*

$$\varphi(f) = \int_X f(1,x)\,d\mu(x)\,, \quad \forall f \in A\,.$$

Proof. Let $g_0 \in \mathrm{GL}_2^+(\mathbb{Q})$ and $f \in C_c(Z)$ such that

$$f(g,y) = 0\,, \quad \forall g \notin \Gamma\, g_0\, \Gamma\,.$$

Since any element of $C_c(Z)$ is a finite linear combination of such functions, it is enough to show that $\varphi(f) = 0$ provided $g_0 \notin \Gamma$. Let $h_n \in C_c(X)$, $0 \leq h_n \leq 1$ with support disjoint from F and converging pointwise to 1 in the complement of F. Let $u_n \in A$ be given by

$$u_n(1,y) := h_n(y)\,, \quad u_n(g,y) = 0\,, \quad \forall g \notin \Gamma\,.$$

The formula

$$\Phi(f)(g,\tau) := f(g,0,\tau) \quad \forall f \in A \tag{8.19}$$

defines a homomorphism of (A,σ_t) to the C^* dynamical system (B,σ_t) obtained by specialization to $\rho = 0$, with convolution product

$$f_1 * f_2(\rho,\tau) = \sum_h f_1(gh^{-1}, h(\tau)) f_2(h,\tau),$$

where now we have no restriction on the summation, as in [10].

For each $n \in \mathbb{N}^*$ we let

$$\mu_{[n]}(g,y) = 1 \text{ if } g \in \Gamma.[n]\,, \quad \mu_{[n]}(g,y) = 0 \text{ if } g \notin \Gamma.[n]\,. \tag{8.20}$$

One has $\mu_{[n]}^* \, \mu_{[n]} = 1$ and $\sigma_t(\mu_{[n]}) = n^{2it} \, \mu_{[n]}\,, \quad \forall t \in \mathbb{R}$. Moreover, the range $\pi(n) = \mu_{[n]} \, \mu_{[n]}^*$ of $\mu_{[n]}$ is the characteristic function of the set of $\mathbb{Q}$-lattices that are divisible by n, *i.e.* those of the form $(\Lambda, n\,\phi)$.

Let $\nu_{[n]} = \Phi(\mu_{[n]})$. These are unitary multipliers of B. Since they are eigenvectors for σ_t, the system (B,σ_t) has no non-zero KMS_β positive functional. This shows that the pushforward of φ by Φ vanishes and by Proposition 2.5 that, with the notation introduced above,

$$\varphi(f) = \lim_n \varphi(f * u_n)\,.$$

Thus, since $(f * u_n)(g,y) = f(g,y)\,h_n(y)$, we can assume that $f(g,y) = 0$ unless $p(y) \in K$, where $K \subset X$ is a compact subset disjoint from F. Let $x \in K$ and V as in lemma 8.9 and let $h \in C_c(V)$. Then, upon applying the KMS_β condition (2.6) to the pair a,b with $a = f$ and

$$b(1,y) := h(y)\,, \quad b(g,y) = 0\,, \quad \forall g \notin \Gamma\,,$$

one gets $\varphi(b * f) = \varphi(f * b)$. One has $(b * f)(g,y) = h(gy)\, f(g,y)$. Applying this to $f * b$ instead of f and using $h(gy)\,h(y) = 0\,, \quad \forall y \in X$ we get $\varphi(f * b^2) = 0$ and $\varphi(f) = 0$, using a partition of unity on K. □

We let Det be the continuous map from $M_2(R)$ to R given by the determinant

$$\begin{bmatrix} a & b \\ c & d \end{bmatrix} \in M_2(R) \mapsto a\,d - b\,c \in R\,.$$

For each $n \in \mathbb{N}^*$, the composition $\pi_n \circ \mathrm{Det}$ defines a projection $\pi'(n)$, which is the characteristic function of the set of $\mathbb{Q}$-lattices whose determinant is divisible by n. If a $\mathbb{Q}$-lattice is divisible by n its determinant is divisible by n^2 and one controls divisibility using the following family of projections $\pi_p(k,l)$. Given a prime p and a pair (k,l) of integers $k \le l$, we let

$$\pi_p(k,l) := (\pi(p^k) - \pi(p^{k+1}))\,(\pi'(p^{k+l}) - \pi'(p^{k+l+1}))\,. \tag{8.21}$$

This corresponds, when working modulo $N = p^b, b > l$, to matrices in the double class of

$$\begin{bmatrix} a & 0 \\ 0 & d \end{bmatrix}\,, \quad v_p(a) = k\,, \quad v_p(d) = l\,,$$

where v_p is the p-adic valuation.

Lemma 8.11 • *Let φ be a KMS$_\beta$ state on (A, σ_t). Then, for any prime p and pair (k,l) of integers $k < l$, one has*

$$\varphi(\pi_p(k,l)) = p^{-(k+l)\beta}\, p^{l-k}\, (1+p^{-1})\,(1-p^{-\beta})\,(1-p^{1-\beta})$$

while for $k = l$ one has

$$\varphi(\pi_p(l,l)) = p^{-2l\beta}\,(1-p^{-\beta})\,(1-p^{1-\beta})\,.$$

- *For distinct primes p_j one has*

$$\varphi(\prod \pi_{p_j}(k_j,l_j)) = \prod \varphi(\pi_{p_j}(k_j,l_j)).$$

Proof. For each $n \in \mathbb{N}^*$ we let $\nu_n \in M(A)$ be given by

$$\nu_n(g,y) = 1\,, \quad \forall g \in \Gamma \begin{bmatrix} n & 0 \\ 0 & 1 \end{bmatrix} \Gamma\,, \quad \nu_n(g,y) = 0 \quad \text{otherwise}\,.$$

One has $\sigma_t(\nu_n) = n^{it}\,\nu_n\,,\quad \forall t \in \mathbb{R}$. The double class $\Gamma \begin{bmatrix} n & 0 \\ 0 & 1 \end{bmatrix} \Gamma$ is the union of the left Γ-cosets of the matrices $\begin{bmatrix} a & b \\ 0 & d \end{bmatrix}$ where $a\,d = n$ and $\gcd(a,b,d) = 1$. The number of these left cosets is

$$\omega(n) := n \prod_{p\,\mathrm{prime},\, p|n} (1+p^{-1})$$

and

$$\nu_n^* * \nu_n(1,y) = \omega(n)\,, \quad \forall y \in Y. \tag{8.22}$$

One has

$$\nu_n * \nu_n^*(1,y) = \sum_{h \in \Gamma\backslash \mathrm{GL}_2^+(\mathbb{Q}),\, hy \in Y} \nu_n(h^{-1},hy)^2\,.$$

With $y = (\rho,\tau)$, the r.h.s. is independent of τ and only depends upon the $\mathrm{SL}_2(\mathbb{Z}/n\mathbb{Z}) - \mathrm{GL}_2(\mathbb{Z}/n\mathbb{Z})$ double class of $\rho_n = p_n(\rho) \in M_2(\mathbb{Z}/n\mathbb{Z})$.

Let us assume that $n = p^l$ is a prime power. We can assume that $\rho_n = p_n(\rho)$ is of the form

$$\rho_n = \begin{bmatrix} p^a & 0 \\ 0 & p^b \end{bmatrix}\,, \quad 0 \le a \le b \le l\,.$$

We need to count the number $\omega(a,b)$ of left Γ-cosets $\Gamma\,h_j$ in the double class $\Gamma \begin{bmatrix} n^{-1} & 0 \\ 0 & 1 \end{bmatrix} \Gamma$ such that $h_j y \in Y$ *i.e.* $h_j \rho \in M_2(R)$. A full set of representatives of the double class is given by $h_j = (\alpha_j^t)^{-1}$ where the α_j are

$$\alpha_0 = \begin{bmatrix} n & 0 \\ 0 & 1 \end{bmatrix} \quad \alpha(s) = \begin{bmatrix} 1 & s \\ 0 & n \end{bmatrix}\,, s \in \{0, 1, \ldots, n-1\}$$

and for $x \in \{1, 2, ..., l-1\}$, $s \in \mathbb{Z}/p^{l-x}\mathbb{Z}$ prime to p

$$\alpha(x,s) = \begin{bmatrix} p^x & s \\ 0 & p^{l-x} \end{bmatrix}.$$

The counting gives

- $\omega(a,b) = 0$ if $b < l$.
- $\omega(a,b) = p^a$ if $a < l, b \geq l$.
- $\omega(a,b) = p^l(1+p^{-1})$ if $a \geq l$.

Let $e_p(i,j)$, $(i \leq j)$ be the projection corresponding to $a \geq i$, $b \geq j$. Then for $i < j$ one has

$$\pi_p(i,j) = e_p(i,j) - e_p(i+1,j) - e_p(i,j+1) + e_p(i+1,j+1) \qquad (8.23)$$

while

$$\pi_p(j,j) = e_p(j,j) - e_p(j,j+1). \qquad (8.24)$$

The computation above gives

$$\nu_n * \nu_n^*(1,y) = p^l(1+p^{-1})e_p(l,l) + \sum_0^{l-1} p^k \left(e_p(k,l) - e_p(k+1,l)\right), \quad (8.25)$$

where we omit the variable y in the r.h.s.

Let φ be a KMS$_\beta$ state, and $\sigma(k,l) := \varphi(e_p(k,l))$. Then, applying the KMS$_\beta$ condition to the pair $(\mu_{[p]}f, \mu_{[p]}^*)$ for $f \in C(X)$, one gets

$$\sigma(k,l) = p^{-2k\beta}\,\sigma(0,l-k).$$

Let $\sigma(k) = \sigma(0,k)$. Upon applying the KMS$_\beta$ condition to (ν_n, ν_n^*), one gets

$$p^l(1+p^{-1})p^{-l\beta} = p^l(1+p^{-1})p^{-2l\beta} + \sum_0^{l-1} p^k\left(p^{-2k\beta}\sigma(l-k) - p^{-2(k+1)\beta}\sigma(l-k-1)\right.$$

Since $\sigma(0) = 1$, this determines the $\sigma(n)$ by induction on n and gives

$$\sigma(n) = a\,p^{n(1-\beta)} + (1-a)\,p^{-2n\beta},$$

with

$$a = (1+p)\frac{p^\beta - 1}{p^{1+\beta} - 1}.$$

Combined with (8.23) and (8.24), this gives the required formulas for $\varphi(\pi_p(k,l))$ and the first part of the lemma follows.

To get the second part, one proceeds by induction on the number m of primes p_j. The function $f = \prod_1^{m-1} \pi_{p_j}(k_j, l_j)$ fulfills

$$f(h\,y) = f(y), \quad \forall y \in Y, \quad \forall h \in \Gamma \begin{bmatrix} n^{-1} & 0 \\ 0 & 1 \end{bmatrix} \Gamma,$$

where $n = p_m^l$. Thus, when applying the KMS_β condition to $(\nu_n f, \nu_n^*)$, the above computation applies with no change to give the result. □

Let us now complete the proof of 3) of Theorem 8.6. Let φ be a KMS_β state. Proposition 8.10 shows that there is a probability measure μ on X such that

$$\varphi(f) = \int_X f(1,x)\, d\mu(x)\,, \quad \forall f \in A\,.$$

With $y = (\rho, \tau) \in X$, Lemma 8.11 shows that the probability $\varphi(e_p(1,1)) = \sigma(1,1)$ that a prime p divides ρ is $p^{-2\beta}$. Since the series $\sum p^{-2\beta}$ converges ($\beta > \frac{1}{2}$ would suffice here), it follows (*cf.* [44] Thm. 1.41) that, for almost all $y \in X$, ρ is only divisible by a finite number of primes. Next, again by Lemma 8.11, the probability that the determinant of ρ is divisible by p is

$$\varphi(e_p(0,1)) = \sigma(1) = (1+p)p^{-\beta} - p^{1-2\beta}.$$

For $\beta > 2$ the corresponding series $\sum((1+p)p^{-\beta} - p^{1-2\beta})$ is convergent. Thus, we conclude that with probability one

$$\rho_p \in \mathrm{GL}_2(\mathbb{Z}_p)\,, \quad \text{for almost all } p.$$

Moreover, since $\sum \varphi(\pi_p(k,l)) = 1$, one gets with probability one

$$\rho_p \in \mathrm{GL}_2(\mathbb{Q}_p)\,, \quad \forall p\,.$$

In other words, the measure μ gives measure one to finite idèles. (Notice that finite idèles form a Borel subset which is not closed.) However, when ρ is a finite idèle the corresponding $\mathbb{Q}$-lattice is commensurable to a unique invertible $\mathbb{Q}$-lattice. Then the KMS_β condition shows that the measure μ is entirely determined by its restriction to invertible $\mathbb{Q}$-lattices, so that, for some probability measure ν,

$$\varphi = \int \varphi_{\beta,l}\, d\nu(l).$$

It follows that the Choquet simplex of extremal KMS_β states is the space of probability measures on the locally compact space

$$\mathrm{GL}_2(\mathbb{Q})\backslash \mathrm{GL}_2(\mathbb{A})/\mathbb{C}^*$$

of invertible $\mathbb{Q}$-lattices [5] and its extreme points are the $\varphi_{\beta,l}$. □

In fact Lemma 8.11 admits the following corollary:

Corollary 8.12 *For $\beta \leq 1$ there is no KMS_β state on (A, σ_t).*

---

[5] *cf. e.g.* [39] for the standard identification of the set of invertible $\mathbb{Q}$-lattices with the above double quotient

Proof. Indeed the value of $\varphi(\pi_p(k,l))$ provided by the lemma is strictly negative for $\beta < 1$ and vanishes for $\beta = 1$. In the latter case this shows that the measure μ is supported by $\{0\} \times \mathbb{H} \subset Y$ and one checks that no such measure fulfills the KMS condition for $\beta = 1$. □

In fact the measure provided by Lemma 8.11 allows us to construct a specific KMS_β state on (A, σ_t) for $1 < \beta \leq 2$. We shall analyze this range of values in Chapter III in connection with the renormalization group.

To get some feeling about what happens when $\beta \to 2$ from above, we shall show that, on functions f which are independent of τ, the states $\varphi_{\beta,l}$ converge weakly to the KMS_2 state φ of (8.14), independently of the choice of the invertible $\mathbb{Q}$-lattice l. Namely, we have

$$\varphi_{\beta,l}(f) \to \int_{M_2(R)} f(a)\, da\,.$$

Using the density of functions of the form $f \circ p_N$ among left Γ-invariant continuous functions on $M_2(R)$, this follows from:

Lemma 8.13 *For $N \in \mathbb{N}$, let $\Gamma(N)$ be the congruence subgroup of level N and*

$$Z_\beta = \sum_{\Gamma(N)\backslash M_2(\mathbb{Z})^+} \mathrm{Det}(m)^{-\beta}\,.$$

When $\beta \to 2$ one has, for any function f on $M_2(\mathbb{Z}/N\mathbb{Z})$,

$$Z_\beta^{-1} \sum_{\Gamma(N)\backslash M_2(\mathbb{Z})^+} f(p_N(m))\, \mathrm{Det}(m)^{-\beta} \to N^{-4} \sum_{M_2(\mathbb{Z}/N\mathbb{Z})} f(a)\,.$$

Proof. For $x \in M_2(\mathbb{Z}/N\mathbb{Z})$ we let

$$h(x) = \lim_{\beta \to 2} Z_\beta^{-1} \sum_{m \in \Gamma(N)\backslash M_2(\mathbb{Z})^+,\, p_N(m)=x} \mathrm{Det}(m)^{-\beta}$$

be the limit of the above expression, with f the characteristic function of the subset $\{x\} \subset M_2(\mathbb{Z}/N\mathbb{Z})$. We want to show that

$$h(x) = N^{-4}\,, \quad \forall x \in M_2(\mathbb{Z}/N\mathbb{Z})\,. \tag{8.26}$$

Since p_N is a surjection $SL_2(\mathbb{Z}) \to SL_2(\mathbb{Z}/N\mathbb{Z})$ and $\Gamma(N)$ a normal subgroup of Γ, one gets

$$h(\gamma_1 x \gamma_2) = h(x)\,, \quad \forall \gamma_j \in SL_2(\mathbb{Z}/N\mathbb{Z})\,. \tag{8.27}$$

Thus, to prove (8.26) we can assume that x is a diagonal matrix

$$x = \begin{bmatrix} n & 0 \\ 0 & n\,\ell \end{bmatrix} \in M_2(\mathbb{Z}/N\mathbb{Z})\,.$$

Dividing both n and N by their g.c.d. k does not affect the validity of (8.26), since all $m \in \Gamma(N)\backslash M_2(\mathbb{Z})^+$ with $p_N(m) = x$ are of the form $k\,m'$, while $\mathrm{Det}(m)^{-\beta} = k^{-2\beta}\,\mathrm{Det}(m')^{-\beta}$. This shows that (8.26) holds for $n = 0$ and allows us to assume that n is coprime to N. Let then r be the g.c.d. of ℓ and N. One can then assume that $x = \begin{bmatrix} n & 0 \\ 0 & n'\,r \end{bmatrix}$, with $r|N$ and with n and n' coprime to N. Let $\Delta \subset SL_2(\mathbb{Z}/N\mathbb{Z})$ be the diagonal subgroup. The left coset $\Delta\, x \subset M_2(\mathbb{Z}/N\mathbb{Z})$ only depends on r and the residue $\delta \in (\mathbb{Z}/N'\mathbb{Z})^*$ of $n\,n'$ modulo $N' = N/r$. It is the set of all diagonal matrices of the form

$$y = \begin{bmatrix} n_1 & 0 \\ 0 & n_2\,r \end{bmatrix}, \quad n_1 \in (\mathbb{Z}/N\mathbb{Z})^*, \quad n_1\,n_2 = \delta\;(N').$$

Let $\Gamma_\Delta(N) \subset \Gamma$ be the inverse image of Δ by p_N. By (8.27) h is constant on $\Delta\, x$, hence

$$h(x) = \lim_{\beta\to 2} Z_\beta^{-1} \sum_{m\in\Gamma_\Delta(N)\backslash M_2(\mathbb{Z})^+,\, p_N(m)\in\Delta\, x} \mathrm{Det}(m)^{-\beta}. \tag{8.28}$$

In each left coset $m \in \Gamma_\Delta(N)\backslash M_2(\mathbb{Z})^+$ with $p_N(m) \in \Delta\, x$ one can find a unique triangular matrix $\begin{bmatrix} a & b \\ 0 & d \end{bmatrix}$ with $a > 0$ coprime to N, $d > 0$ divisible by r, $a\,d/r = \delta\,(N')$ and $b = N\,b'$ with $0 \leq b' < d$. Thus, we can rewrite (8.28) as

$$h(x) = \lim_{\beta\to 2} Z_\beta^{-1} \sum_Y d\,(ad)^{-\beta}, \tag{8.29}$$

where Y is the set of pairs of positive integers (a,d) such that

$$p_N(a) \in (\mathbb{Z}/N\mathbb{Z})^*, \quad r|d, \quad a\,d = r\,\delta\,(N).$$

To prove (8.26) we assume first that $r < N$ and write $N = N_1\,N_2$, where N_1 is coprime to N' and N_2 has the same prime factors as N'. One has $r = N_1\,r_2$, with $r_2|N_2$. An element of $\mathbb{Z}/N_2\mathbb{Z}$ is invertible iff its image in $\mathbb{Z}/N'\mathbb{Z}$ is invertible. To prove (8.26) it is enough to show that, for any of the r_2 lifts $\delta_2 \in \mathbb{Z}/N_2\mathbb{Z}$ of δ, one has

$$\lim_{\beta\to 2} Z_\beta^{-1} \sum_{Y'} d\,(ad)^{-\beta} = N_1\,N^{-4}, \tag{8.30}$$

where Y' is the set of pairs of positive integers (a,d) such that

$$p_N(a) \in (\mathbb{Z}/N\mathbb{Z})^*, \quad p_{N_2}(a\,d) = \delta_2.$$

We let 1_N be the trivial Dirichlet character modulo N. Then when $\mathcal{X}_{N_2}$ varies among Dirichlet characters modulo N_2 one has

$$\sum_{Y'} d\,(ad)^{-\beta} = \varphi(N_2)^{-1} \sum \mathcal{X}_{N_2}(\delta_2)^{-1}\, L(1_{N_1} \times \mathcal{X}_{N_2}, \beta)\, L(\mathcal{X}_{N_2}, \beta - 1),$$

where φ is the Euler totient function. Only the trivial character $\mathcal{X}_{N_2} = 1_{N_2}$ contributes to the limit (8.30), since the other L-functions are regular at 1. Moreover, the residue of $L(1_{N_2}, \beta-1)$ at $\beta = 2$ is equal to $\frac{\varphi(N_2)}{N_2}$ so that, when $\beta \to 2$, we have

$$\sum_{Y'} d\,(ad)^{-\beta} \sim N_2^{-1}\, L(1_N, 2)\, (\beta-2)^{-1}\,.$$

By construction one has

$$Z_\beta \sim |\Gamma : \Gamma(N)|\, \zeta(2)\, (\beta-2)^{-1}\,,$$

where the order of the quotient group $\Gamma : \Gamma(N)$ is $N^3 \prod_{p|N} (1-p^{-2})$ ([51]). Since

$$L(1_N, s) = \prod_{p|N} (1-p^{-s})\, \zeta(s)$$

one gets (8.30). A similar argument handles the case $r = N$. □

The states $\varphi_{\beta,l}$ converge when $\beta \to \infty$ and their limits restrict to $C_c(X) \subset A$ as characters given by evaluation at l:

$$\varphi_{\infty,l}(f) = f(l)\,, \quad \forall f \in C_c(X).$$

These characters are all distinct and we thus get a bijection of the space

$$\mathrm{GL}_2(\mathbb{Q})\backslash \mathrm{GL}_2(\mathbb{A})/\mathbb{C}^*$$

of invertible $\mathbb{Q}$-lattices with the space $\mathcal{E}_\infty$ of extremal KMS$_\infty$ states.

We shall now describe the natural symmetry group of the above system, from an action of the quotient group S

$$S := \mathbb{Q}^* \backslash \mathrm{GL}_2(\mathbb{A}_f)$$

as symmetries of our dynamical system. Here the finite adèlic group of GL_2 is given by

$$\mathrm{GL}_2(\mathbb{A}_f) = \prod_{\mathrm{res}} \mathrm{GL}_2(\mathbb{Q}_p),$$

where in the restricted product the p-component lies in $\mathrm{GL}_2(\mathbb{Z}_p)$ for all but finitely many p's. It satisfies

$$\mathrm{GL}_2(\mathbb{A}_f) = \mathrm{GL}_2^+(\mathbb{Q})\, \mathrm{GL}_2(R).$$

The action of the subgroup

$$\mathrm{GL}_2(R) \subset S$$

is defined in a straightforward manner using the following right action of $\mathrm{GL}_2(R)$ on $\mathbb{Q}$-lattices:

$$(\Lambda, \phi).\gamma = (\Lambda, \phi \circ \gamma), \quad \forall \gamma \in \mathrm{GL}_2(R).$$

By construction this action preserves the commensurability relation for pairs of $\mathbb{Q}$-lattices and preserves the value of the ratio of covolumes for such pairs. We can view it as the action

$$(\rho, \tau).\gamma = (\rho \circ \gamma, \tau)$$

of $\mathrm{GL}_2(R)$ on $Y = M_2(R) \times \mathbb{H}$, which commutes with the left action of $\mathrm{GL}_2^+(\mathbb{Q})$. Thus, this action defines automorphisms of the dynamical system (A, σ_t) by

$$\theta_\gamma(f)(g, y) := f(g, y.\gamma), \quad \forall f \in \mathcal{A}, \gamma \in \mathrm{GL}_2(R),$$

and one has

$$\theta_{\gamma_1}\, \theta_{\gamma_2} = \theta_{\gamma_1 \gamma_2}, \quad \forall \gamma_j \in \mathrm{GL}_2(R).$$

The complementary action of $\mathrm{GL}_2^+(\mathbb{Q})$ is more subtle and is given by endomorphisms of the dynamical system (A, σ_t), following Definition 2.3.

For $m \in M_2(\mathbb{Z})^+$, let $\tilde{m} = \mathrm{Det}(m)\, m^{-1} \in M_2(\mathbb{Z})^+$. The range R_m of the map $\rho \to \rho\, \tilde{m}$ only depends on $L = m(\mathbb{Z}^2)$. Indeed if $m_j \in M_2(\mathbb{Z})^+$ fulfill $m_1(\mathbb{Z}^2) = m_2(\mathbb{Z}^2)$ then $m_2 = m_1 \gamma$ for some $\gamma \in \Gamma$, hence $M_2(R)\, \tilde{m}_1 = M_2(R)\, \tilde{m}_2$. Let then

$$e_L \in C(X) \tag{8.31}$$

be the characteristic function of $\Gamma \backslash (R_m \times \mathbb{H}) \subset \Gamma \backslash (M_2(R) \times \mathbb{H})$, for any m such that $m(\mathbb{Z}^2) = L$. Equivalently, it is the characteristic function of the open and closed subset $E_L \subset X$ of $\mathbb{Q}$-lattices of the form $(\Lambda, \phi \circ \tilde{m})$. One has

$$e_L\, e_{L'} = e_{L \cap L'}, \qquad e_{\mathbb{Z}^2} = 1.. \tag{8.32}$$

For $l = (\Lambda, \phi) \in E_L \subset X$ and $m \in M_2(\mathbb{Z})^+$, $m(\mathbb{Z}^2) = L$ we let

$$l \circ \tilde{m}^{-1} := (\Lambda, \phi \circ \tilde{m}^{-1}) \in X.$$

This map preserves commensurability of $\mathbb{Q}$-lattices. On Y it is given by

$$(\rho, \tau) \circ \tilde{m}^{-1} := (\rho \circ \tilde{m}^{-1}, \tau), \quad \forall (\rho, \tau) \in R_m \times \mathbb{H}$$

and it commutes with the left action of $\mathrm{GL}_2^+(\mathbb{Q})$. The formula

$$\theta_m(f)(g, y) := f(g, y \circ \tilde{m}^{-1}), \quad \forall y \in R_m \times \mathbb{H}, \tag{8.33}$$

extended by $\theta_m(f)(g, y) = 0$ for $y \notin R_m \times \mathbb{H}$, defines an endomorphism θ_m of A that commutes with the time evolution σ_t. Notice that $\theta_m(1) = e_L \in M(A)$ is a multiplier of A and that θ_m lands in the reduced algebra A_{e_L}, so that (8.33) is unambiguous. Thus one obtains an action of the semigroup $M_2(\mathbb{Z})^+$ by endomorphisms of the dynamical system (A, σ_t), fulfilling Definition 2.3.

Proposition 8.14 *The above actions of the group* $\mathrm{GL}_2(R) \subset S$ *and of the semigroup* $M_2(\mathbb{Z})^+ \subset S$ *assemble to an action of the group* $S = \mathbb{Q}^* \backslash \mathrm{GL}_2(\mathbb{A}_f)$ *as symmetries of the dynamical system* (A, σ_t).

Proof. The construction above applies to give an action by endomorphisms of the semigroup $\mathrm{GL}_2(\mathbb{A}_f) \cap M_2(R)$, which contains both $\mathrm{GL}_2(R)$ and $M_2(\mathbb{Z})^+$. It remains to show that the sub-semigroup $\mathbb{N}^\times \subset M_2(\mathbb{Z})^+$ acts by inner endomorphisms of (A, σ_t). Indeed for any $n \in \mathbb{N}^*$, the endomorphism $\theta_{[n]}$ (where $[n] = \begin{bmatrix} n & 0 \\ 0 & n \end{bmatrix} \in M_2(\mathbb{Z})^+$) is inner and implemented by the multiplier $\mu_{[n]} \in M(A)$ which was defined in (8.20) above *i.e.* one has

$$\theta_n(f) = \mu_{[n]}\, f\, \mu_{[n]}^*\,, \quad \forall f \in A\,.$$

□

9 The Subalgebra $\mathcal{A}_{\mathbb{Q}}$ and the Modular Field

The strategy outlined in §6 allows us to find, using Eisenstein series, a suitable *arithmetic* subalgebra $\mathcal{A}_{\mathbb{Q}}$ of the algebra of unbounded multipliers of the basic Hecke C^*-algebra A of the previous section. The extremal KMS_∞ states $\varphi \in \mathcal{E}_\infty$ extend to $\mathcal{A}_{\mathbb{Q}}$ and the image $\varphi(\mathcal{A}_{\mathbb{Q}})$ generates, in the generic case, a specialization $F_\varphi \subset \mathbb{C}$ of the modular field F. The state φ will then intertwine the symmetry group S of the system (A, σ_t) with the Galois group of the modular field *i.e.* we shall show that there exists an isomorphism θ of S with $\mathrm{Gal}(F_\varphi/\mathbb{Q})$ such that

$$\alpha \circ \varphi = \varphi \circ \theta^{-1}(\alpha)\,, \quad \forall \alpha \in \mathrm{Gal}(F_\varphi/\mathbb{Q})\,. \tag{9.1}$$

Let us first define $\mathcal{A}_{\mathbb{Q}}$ directly without any reference to Eisenstein series and check directly its algebraic properties. We let $Z \subset \Gamma\backslash \mathrm{GL}_2^+(\mathbb{Q}) \times_\Gamma Y$ be as above and $f \in C(Z)$ be a function with *finite support* in the variable $g \in \Gamma\backslash\mathrm{GL}_2^+(\mathbb{Q})$. Such an f defines an unbounded multiplier of the C^*-algebra A with the product given as above by

$$(f_1 * f_2)(g, y) := \sum_{h \in \Gamma\backslash\mathrm{GL}_2^+(\mathbb{Q}), hy \in Y} f_1(gh^{-1}, hy)\, f_2(h, y).$$

One has $Y = M_2(R) \times \mathbb{H}$ and we write $f(g, y) = f(g, \rho, z)$, with $(g, \rho, z) \in \mathrm{GL}_2^+(\mathbb{Q}) \times M_2(R) \times \mathbb{H}$. In order to define the *arithmetic* elements $f \in \mathcal{A}_{\mathbb{Q}}$ we first look at the way f depends on $\rho \in M_2(R)$. Let as above $p_N : M_2(R) \to M_2(\mathbb{Z}/N\mathbb{Z})$ be the canonical projection. It is a ring homomorphism. We say that f has level N iff $f(g, \rho, z)$ only depends upon $(g, p_N(\rho), z) \in \mathrm{GL}_2^+(\mathbb{Q}) \times M_2(\mathbb{Z}/N\mathbb{Z}) \times \mathbb{H}$. Then specifying f amounts to assigning the finitely many continuous functions $f_{g,m} \in C(\mathbb{H})$ with $m \in M_2(\mathbb{Z}/N\mathbb{Z})$ and

$$f(g,\rho,z) = f_{g,p_N(\rho)}(z).$$

The invariance condition

$$f(g\,\gamma,y) = f(g,\gamma\,y)\,, \quad \forall \gamma \in \Gamma\,,\; g \in \mathrm{GL}_2^+(\mathbb{Q})\,,\; y \in Y \tag{9.2}$$

then shows that

$$f_{g,m}|\gamma = f_{g,m}\,, \quad \forall \gamma \in \Gamma(N) \cap g^{-1}\Gamma g,$$

with standard notations for congruence subgroups and for the slash operation in weight 0 (*cf.* (9.29)).

We denote by F the field of modular functions which are rational over $\mathbb{Q}^{ab}$, *i.e.* the union of the fields F_N of modular functions of level N rational over $\mathbb{Q}(e^{2\pi i/N})$. Its elements are modular functions $h(\tau)$ whose $q^{\frac{1}{N}}$-expansion has all its coefficients in $\mathbb{Q}(e^{2\pi i/N})$ (*cf.* [51]).

The first requirement for arithmetic elements is that

$$f_{g,m} \in F \quad \forall (g,m). \tag{9.3}$$

This condition alone, however, is not sufficient. In fact, the modular field F_N of level N contains (*cf.* [51]) a primitive N-th root of 1. Thus, the condition (9.3) alone allows the algebra $\mathcal{A}_{\mathbb{Q}}$ to contain the cyclotomic field $\mathbb{Q}^{ab} \subset \mathbb{C}$, but this would prevent the existence of "fabulous states", because the "fabulous" property would not be compatible with $\mathbb{C}$-linearity. We shall then impose an additional condition, which forces the spectrum of the corresponding elements of $\mathcal{A}_{\mathbb{Q}}$ to contain all Galois conjugates of such a root, so that no such element can be a scalar. This is, in effect, a consistency condition on the roots of unity that appear in the coefficients of the q-series, when ρ is multiplied on the left by a diagonal matrix.

Consider elements $g \in \mathrm{GL}_2^+(\mathbb{Q})$ and $\alpha \in \mathrm{GL}_2(\mathbb{Z}/N\mathbb{Z})$, respectively of the form

$$g = r\begin{bmatrix} n & 0 \\ 0 & 1 \end{bmatrix} \quad \text{and} \quad \alpha = \begin{bmatrix} k & 0 \\ 0 & 1 \end{bmatrix}, \tag{9.4}$$

with k prime to N and $n|N$.

Definition 9.1 *We shall say that f of level N_0 is* arithmetic *($f \in \mathcal{A}_{\mathbb{Q}}$) iff for any multiple N of N_0 and any pair (g,α) as in (9.4) we have $f_{g,m} \in F_N$ for all $m \in M_2(\mathbb{Z}/N\mathbb{Z})$ and the q-series of $f_{g,\alpha m}$ is obtained from the q-series for $f_{g,m}$ by raising to the power k the roots of unity that appear as coefficients.*

The arithmetic subalgebra $\mathcal{A}_{\mathbb{Q}}$ enriches the structure of the noncommutative space to that of a "noncommutative arithmetic variety". As we shall prove in Theorem 9.5, a generic ground state φ of the system, when evaluated on $\mathcal{A}_{\mathbb{Q}}$ generates an embedded copy F_φ of the modular field in $\mathbb{C}$. Moreover, there exists a unique isomorphism $\theta = \theta_\varphi$ of the symmetry group S of the system with $\mathrm{Gal}(F_\varphi/\mathbb{Q})$, such that

$$\theta(\sigma) \circ \varphi = \varphi \circ \sigma\,, \quad \forall \sigma \in S\,.$$

A first step towards this result is to show that the arithmeticity condition is equivalent to a covariance property under left multiplication of ρ by elements $\alpha \in \mathrm{GL}_2(R)$, in terms of Galois automorphisms. The condition is always satisfied for $\alpha \in \mathrm{SL}_2(R)$.

For each $g \in \mathrm{GL}_2(\mathbb{A}_f)$, we let $\mathrm{Gal}(g) \in \mathrm{Aut}(F)$ be its natural action on F, written in a covariant way so that

$$\mathrm{Gal}(g_1\, g_2) = \mathrm{Gal}(g_1) \circ \mathrm{Gal}(g_2).$$

With the standard contravariant notation $f \mapsto f^g$ (*cf. e.g.* [28]) we let, for all $f \in F$,

$$\mathrm{Gal}(g)(f) := f^{\tilde{g}}\,, \quad \tilde{g} = \mathrm{Det}(g)\, g^{-1}\,. \tag{9.5}$$

Lemma 9.2 *For any $\alpha \in \mathrm{SL}_2(R)$ one has*

$$f_{g,\alpha\, m} = \mathrm{Gal}(\alpha) f_{g',m},$$

where $g\,\alpha = \alpha'\, g'$ is the decomposition of $g\,\alpha$ as a product in $\mathrm{GL}_2(R).\mathrm{GL}_2(\mathbb{Q})$.

Proof. Notice that the decomposition $\alpha'\, g'$ is not unique, but the left invariance

$$f(\gamma\, g', \rho, \tau) = f(g', \rho, \tau)\,, \quad \forall \gamma \in \Gamma$$

shows that the above condition is well defined. Let $p_N : M_2(R) \to M_2(\mathbb{Z}/N\mathbb{Z})$ be the projection. Then $f_{g,\alpha\, m} = f_{g,p_N(\alpha)p_N(m)}$, for f of level N. Let $\gamma \in \Gamma$ be such that $p_N(\gamma) = p_N(\alpha)$. Then

$$f_{g,\alpha m}(\tau) = f(g, \gamma m, \tau) = f(g\gamma, m, \gamma^{-1}(\tau)).$$

Thus, for $g' = g\gamma$, one obtains the required condition. □

Lemma 9.3 *A function f is in $\mathcal{A}_{\mathbb{Q}}$ iff condition* (9.3) *is satisfied and*

$$f_{g,\alpha\, m} = \mathrm{Gal}(\alpha) f_{g',m}, \quad \forall \alpha \in \mathrm{GL}_2(R), \tag{9.6}$$

where $g\,\alpha = \alpha'\, g'$ is the decomposition of $g\,\alpha$ as a product in $\mathrm{GL}_2(R).\mathrm{GL}_2(\mathbb{Q})$.

Proof. By Lemma 9.2, the only nontrivial part of the covariance condition (9.6) is the case of diagonal matrices $\delta = \begin{bmatrix} u & 0 \\ 0 & 1 \end{bmatrix}$ with $u \in \mathrm{GL}_1(R)$.

To prove (9.6) we can assume that $g = g_0\, \gamma$ with g_0 diagonal as in (9.4) and $\gamma \in \Gamma$. Let then $\gamma\, \alpha = \delta\, \alpha_1$ with δ as above and $\alpha_1 \in \mathrm{SL}_2(R)$. One has

$$f_{g_0,\delta \alpha_1 m} = \mathrm{Gal}(\delta)\, f_{g_0,\alpha_1 m}$$

by Definition 9.1 since $\mathrm{Gal}(\delta)$ is given by raising the roots of unity that appear as coefficients of the q-expansion to the power k where u is the residue of k modulo N (*cf.* [51] (6.2.1) p.141). One then has

$$f_{g,\alpha\, m} = \mathrm{Gal}(\gamma^{-1}) f_{g_0,\gamma\alpha m} = \mathrm{Gal}(\gamma^{-1}\delta)\, f_{g_0,\alpha_1 m}$$

and by Lemma 9.2, with $g_0\alpha_1 = \alpha_1' g_0'$ we get

$$f_{g,\alpha\, m} = \mathrm{Gal}(\gamma^{-1}\delta\alpha_1)\, f_{g_0',m} = \mathrm{Gal}(\alpha)\, f_{g_0',m}$$

Moreover

$$g\,\alpha = g_0\,\gamma\,\alpha = g_0\,\delta\,\alpha_1 = \delta\,g_0\,\alpha_1 = \delta\,\alpha_1'\,g_0'$$

which shows that $g' = g_0'$

One checks similarly that the converse holds. □

Proposition 9.4 *$\mathcal{A}_{\mathbb{Q}}$ is a subalgebra of the algebra of unbounded multipliers of A, globally invariant under the action of the symmetry group S.*

Proof. For each *generic* value of $\tau \in \mathbb{H}$ the evaluation map

$$h \in F \mapsto I_\tau(h) = h(\tau) \in \mathbb{C}$$

gives an isomorphism of F with a subfield $F_\tau \subset \mathbb{C}$ and a corresponding action Gal_τ of $\mathrm{GL}_2(\mathbb{A}_f)$ by automorphisms of F_τ, such that

$$\mathrm{Gal}_\tau(g)(I_\tau(h)) = I_\tau(\mathrm{Gal}(g)(h)). \tag{9.7}$$

We first rewrite the product as

$$(f_1 * f_2)(g,\rho,\tau) = \sum_{g_1\in\Gamma\backslash\mathrm{GL}_2^+(\mathbb{Q}),\, g_1\rho\in M_2(R)} f_1(gg_1^{-1}, g_1\rho, g_1(\tau))\, f_2(g_1,\rho,\tau)\,.$$

The proof that $(f_1 * f_2)_{g,m} \in F$ is the same as in Proposition 2 of ([10]). It remains to be shown that condition (9.6) is stable under convolution. Thus we let $\alpha \in \mathrm{GL}_2(R)$ and we want to show that $f_1 * f_2$ fulfills (9.6). We let $g' \in \mathrm{GL}_2(\mathbb{Q})$ and $\beta \in \mathrm{GL}_2(R)$ with $g\,\alpha = \beta\, g'$. By definition, one has

$$(f_1{*}f_2)(g,\alpha\,\rho,\tau) = \sum_{g_1\in\Gamma\backslash\mathrm{GL}_2^+(\mathbb{Q}),\, g_1\alpha\rho\in M_2(R)} f_1(gg_1^{-1}, g_1\,\alpha\,\rho, g_1(\tau))\, f_2(g_1,\alpha\,\rho,\tau).$$

We let $g_1\alpha = \alpha' g_1'$ be the decomposition of $g_1\alpha$, and use 9.6 to write the r.h.s. as

$$(f_1 * f_2)(g,\alpha\,\rho,\tau) = \sum f_1(gg_1^{-1}, g_1\,\alpha\,\rho, g_1(\tau))\, \mathrm{Gal}_\tau(\alpha)(f_2(g_1',\rho,\tau))\,,$$

with Gal_τ as in (9.7). The result then follows from the equality

$$f_1(gg_1^{-1}, g_1\,\alpha\,\rho, g_1(\tau)) = \mathrm{Gal}_\tau(\alpha)(f_1(g'g_1'^{-1}, g_1'\,\rho, g_1'(\tau))), \tag{9.8}$$

which we now prove. The equality $g_1\alpha = \alpha' g_1'$ together with (9.6) shows that

$$f_1(gg_1^{-1}, g_1\,\alpha\,\rho, g_1(\tau)) = \mathrm{Gal}_{g_1(\tau)}(\alpha')(f_1(g'g_1'^{-1}, g_1'\,\rho, g_1(\tau)))\,,$$

using $g g_1^{-1}\alpha' = g g_1^{-1}(g_1 \alpha g_1^{'-1}) = g\alpha g_1^{'-1} = \beta g' g_1^{'-1}$.

For any $h \in F$, one has

$$I_{g_1(\tau)}(\mathrm{Gal}(\alpha')h) = I_{g_1(\tau)}(\mathrm{Gal}(g_1)\mathrm{Gal}(\alpha)\mathrm{Gal}(g_1^{'-1})(h))$$

and, by construction of the Galois action [51],

$$I_{g_1(\tau)} \circ \mathrm{Gal}(g_1) = I_\tau,$$

so that in fact

$$\mathrm{Gal}_{g_1(\tau)}(\alpha')I_{g_1(\tau)}(h) = \mathrm{Gal}_\tau(\alpha)I_{g_1'(\tau)}(h)).$$

This proves (9.8) and it shows that $\mathcal{A}_{\mathbb{Q}}$ is a subalgebra of the algebra of unbounded multipliers of A. To prove the invariance under S is straightforward, since the endomorphisms are all acting on the ρ variable by right multiplication, which does not interfere with condition (9.6). □

In fact, modulo the nuance between "forms" and functions, the above algebra $\mathcal{A}_{\mathbb{Q}}$ is intimately related to the modular Hecke algebra of [10].

We can now state the main result extending Theorem 3.2 to the two dimensional case.

Theorem 9.5 *Let $l = (\rho, \tau)$ be a generic invertible $\mathbb{Q}$-lattice and $\varphi_l \in \mathcal{E}_\infty$ be the corresponding KMS$_\infty$ state. The image $\varphi_l(\mathcal{A}_{\mathbb{Q}}) \subset \mathbb{C}$ generates the specialization $F_\tau \subset \mathbb{C}$ of the modular field F obtained for the modulus τ. The action of the symmetry group S of the dynamical system (A, σ_t) is intertwined by φ with the Galois group of the modular field F_τ by the formula*

$$\varphi \circ \alpha = \mathrm{Gal}_\tau(\rho\alpha\rho^{-1}) \circ \varphi.$$

Proof. We first need to exhibit enough elements of $\mathcal{A}_{\mathbb{Q}}$. Let us first deal with functions $f(g, \rho, \tau)$ which vanish except when $g \in \Gamma$. By construction these are functions on the space X of $\mathbb{Q}$-lattices

$$X = (\text{Space of } \mathbb{Q}\text{-lattices in } \mathbb{C})/\mathbb{C}^* \sim \Gamma\backslash(M_2(R) \times \mathbb{H}). \tag{9.9}$$

To obtain such elements of $\mathcal{A}_{\mathbb{Q}}$ we start with Eisenstein series and view them as functions on the space of $\mathbb{Q}$-lattices. Recall that to a pair $(\rho, \tau) \in Y$ we associate the $\mathbb{Q}$-lattice $(\Lambda, \phi) = \theta(\rho, \tau)$ by

$$\Lambda = \mathbb{Z} + \tau\mathbb{Z}, \ \phi(a) = \rho_1(a) - \tau\rho_2(a) \in \mathbb{Q}\Lambda/\Lambda, \tag{9.10}$$

where $\rho_j(a) = \sum \rho_{jk}(a_k) \in \mathbb{Q}/\mathbb{Z}$, for $a = (a_1, a_2) \in (\mathbb{Q}/\mathbb{Z})^2$. The Eisenstein series are given by

$$E_{2k,a}(\rho, \tau) = \pi^{-2k} \sum_{y \in \Lambda + \phi(a)} y^{-2k}. \tag{9.11}$$

This is undefined when $\phi(a) \in \Lambda$, but we shall easily deal with that point below. For $k=1$ we let

$$X_a(\rho,\tau) = \pi^{-2}\left(\sum_{y\in\Lambda+\phi(a)} y^{-2} - {\sum_{y\in\Lambda}}' y^{-2}\right) \tag{9.12}$$

when $\phi(a) \notin \Lambda$ and $X_a(\rho,\tau) = 0$ if $\phi(a) \in \Lambda$. This is just the evaluation of the Weierstrass $\wp$-function on $\phi(a)$.

For $\gamma \in \Gamma = \mathrm{SL}_2(\mathbb{Z})$ we have $X_a(\gamma\,\rho, \gamma\,\tau) = (c\tau + d)^2\, X_a(\rho,\tau)$, which shows that the function $c(\tau)X_a$ is Γ-invariant on Y, where

$$c(\tau) = -2^7\,3^5\,\frac{g_2 g_3}{\Delta} \tag{9.13}$$

has weight -2 and no pole in $\mathbb{H}$. We use c as we used the covolume in the 1-dimensional case, to pass to modular functions. This corresponds in weight 2 to passing from division values of the Weierstrass $\wp$-function to the Fricke functions (*cf.* [28] §6.2)

$$f_v(\tau) = -2^7 3^5\,\frac{g_2\, g_3}{\Delta}\,\wp(\lambda(v,\tau)), \tag{9.14}$$

where $v = (v_1, v_2) \in (\mathbb{Q}/\mathbb{Z})^2$ and $\lambda(v,\tau) := v_1\,\tau + v_2$. Here g_2, g_3 are the coefficients giving the elliptic curve $E_\tau = \mathbb{C}/\Lambda$ in Weierstrass form,

$$y^2 = 4x^3 - g_2 x - g_3,$$

with discriminant $\Delta = g_2^3 - 27 g_3^2$. One has (up to powers of π)

$$g_2 = 60\, e_4\,, \quad g_3 = 140\, e_6,$$

where one defines the standard modular forms of even weight $k \in 2\,\mathbb{N}$ as

$$e_k(\Lambda) := \pi^{-k} \sum_{y\in\Lambda\setminus\{0\}} y^{-k}$$

with q-expansion ($q = e^{2\pi i \tau}$)

$$e_k = \frac{2^k}{k!} B_{\frac{k}{2}} + (-1)^{k/2}\frac{2^{k+1}}{(k-1)!}\sum_1^\infty \sigma_{k-1}(N)\, q^N,$$

where the B_n are the Bernoulli numbers and $\sigma_n(N)$ is the sum of d^n over the divisors d of N. The e_{2n} for $n \geq 2$ are in the ring $\mathbb{Q}[e_4, e_6]$ (cf [56]) thanks to the relation

$$\frac{1}{3}(m-3)(4m^2-1)e_{2m} = \sum_2^{m-2} (2r-1)(2m-2r-1)e_{2r}e_{2m-2r}. \tag{9.15}$$

Notice that $\mathcal{X}_a := c X_a \in C(M_2(R) \times \mathbb{H}) = C(Y)$ is a continuous function on Y. The continuity of X_a as a function of ρ comes from the fact that it only involves the restriction $\rho_N \in M_2(\mathbb{Z}/N\mathbb{Z})$ of ρ to N-torsion elements a with $Na = 0$.

We view $\mathcal{X}_a$ as a function on Z by

$$\mathcal{X}_a(\gamma, \rho, \tau) := \mathcal{X}_a(\rho, \tau), \quad \forall \gamma \in \Gamma,$$

while it vanishes for $\gamma \notin \Gamma$. Let us show that $\mathcal{X}_a \in \mathcal{A}_{\mathbb{Q}}$. Since the Fricke functions belong to the modular field F we only need to check (9.6). For $\alpha \in \mathrm{GL}_2(R)$ and generic τ we want to show that

$$\mathcal{X}_a(\alpha\, \rho, \tau) = \mathrm{Gal}_\tau(\alpha)\mathcal{X}_a(\rho, \tau).$$

If $\rho(a) = 0$ both sides vanish, otherwise they are both given by Fricke functions f_v, $f_{v'}$, corresponding respectively to the labels (using (9.10))

$$v = s\,\alpha\,\rho(a)\,, \quad v' = s\,\rho(a)\,, \quad s = \begin{bmatrix} 0 & -1 \\ 1 & 0 \end{bmatrix}.$$

Thus, $v' = s\,\alpha\,s^{-1}(v) = (s^{-1}\,\alpha^t\,s)^t(v)$ and the result follows from (9.5) and the equality

$$\tilde{\alpha} = \mathrm{Det}(\alpha)\,\alpha^{-1} = s^{-1}\,\alpha^t\,s,$$

with the Galois group $\mathrm{GL}_2(\mathbb{Z}/n\mathbb{Z})/\pm 1$ of the modular field F_n over $\mathbb{Q}(j)$ acting on the Fricke functions by permutation of their labels:

$$f_v^{\sigma(u)} = f_{u^t v}\,, \quad \forall u \in \mathrm{GL}_2(\mathbb{Z}/n\mathbb{Z}).$$

This shows that $\mathcal{X}_a \in \mathcal{A}_{\mathbb{Q}}$ and it suffices to show that, with the notation of the theorem, $\varphi_l(\mathcal{A}_{\mathbb{Q}})$ generates F_τ, since the modular field F is the field generated over $\mathbb{Q}$ by all the Fricke functions. It already contains $\mathbb{Q}(j)$ at level 2 and it contains in fact $\mathbb{Q}^{ab}(j)$.

Let us now display elements $T_{r_1,r_2} \in \mathcal{A}_{\mathbb{Q}}$, $r_j \in \mathbb{Q}_+^*$, $r_1 | r_2$, associated to the classical Hecke correspondences. We let $C_{r_1,r_2} \subset \Gamma\backslash\mathrm{GL}_2(\mathbb{Q})^+$ be the finite subset given by the double class of $\begin{bmatrix} r_1 & 0 \\ 0 & r_2 \end{bmatrix}$ in $\Gamma\backslash\mathrm{GL}_2(\mathbb{Q})^+/\Gamma$. We then define

$$T_{r_1,r_2}(g, \rho, \tau) = 1 \text{ if } g \in C_{r_1,r_2}\,,\ g\rho \in M_2(R)\,,\ T_{r_1,r_2}(g, \rho, \tau) = 0 \text{ otherwise}.$$

One needs to check (9.6), but if $g\,\alpha = \alpha'\,g'$ is the decomposition of $g\,\alpha$ as a product in $\mathrm{GL}_2(R).\mathrm{GL}_2(\mathbb{Q})^+$, then g' belongs to the double coset of $g \in \Gamma\backslash\mathrm{GL}_2(\mathbb{Q})^+/\Gamma$, which gives the required invariance. It is not true that the $T_{r_1,r_2} \in \mathcal{A}_{\mathbb{Q}}$ fulfill the relations of the Hecke algebra $\mathcal{H}(\mathrm{GL}_2(\mathbb{Q})^+, \Gamma)$ of double cosets, but this holds when r_1, r_2 are restricted to vary among positive *integers*. To see this one checks that the map

$$\tau(f)(g, y) := f(g) \quad \text{if} \quad g \in M_2(\mathbb{Z})^+\,, \quad \tau(f)(g, y) = 0 \quad \text{otherwise}$$

defines an isomorphism

$$\tau : \mathcal{H}(M_2(\mathbb{Z})^+, \Gamma) \to \mathcal{A}_{\mathbb{Q}} \tag{9.16}$$

of the standard Hecke algebra $\mathcal{H}(M_2(\mathbb{Z})^+, \Gamma)$ of Γ-biinvariant functions (with Γ-finite support) on $M_2(\mathbb{Z})^+$ with a subalgebra $\mathcal{H} \subset \mathcal{A}_{\mathbb{Q}}$. Notice that it is only because the condition $h\,y \in Y$ of definition (8.10) is now automatically satisfied that τ is a homomorphism.

Let us now show the intertwining equality

$$\varphi_l \circ \alpha = \mathrm{Gal}_\tau(\rho\,\alpha\,\rho^{-1}) \circ \varphi_l\,, \quad \forall \alpha \in S\,. \tag{9.17}$$

One has

$$\varphi_l(f) = f(1, \rho, \tau)\,, \quad \forall f \in \mathcal{A}_{\mathbb{Q}}$$

It is enough to prove (9.17) for $\alpha \in \mathrm{GL}_2(R)$ and for $\alpha \in \mathrm{GL}_2(\mathbb{Q})$.

For $\alpha \in \mathrm{GL}_2(R)$, the state $\varphi_l \circ \alpha$ is given simply by

$$(\varphi_l \circ \alpha)(f) = f(1, \rho\,\alpha, \tau)\,, \quad \forall f \in \mathcal{A}_{\mathbb{Q}}\,,$$

and using (9.6) one gets (9.17) in that case.

Let $m \in M_2^+(\mathbb{Z})$. Then the state $\varphi_l \circ m$ is more tricky to obtain, since it is not the straight composition but the 0-temperature limit of the states obtained by composition of the KMS_β state $\varphi_{l,\beta}$ with the endomorphism θ_m defined in (8.33). Indeed, the range of θ_m is the reduced algebra by the projection e_L, with $L = m(\mathbb{Z}^2)$, on which any of the zero temperature states vanishes identically.

Let us first show that for finite β we have

$$\varphi_{l,\beta} \circ \theta_m = \varphi_{l,\beta}(e_L)\,\varphi_{l',\beta}, \tag{9.18}$$

where $L = m(\mathbb{Z}^2)$ and l' is given by

$$l' = (\rho', {m'}^{-1}(\tau))\,, \quad \rho\,m = m'\,\rho' \in M_2^+(\mathbb{Z}).\mathrm{GL}_2(R)\,. \tag{9.19}$$

By (8.18) we have

$$\varphi_{\beta,l}(\theta_m(f)) = Z^{-1} \sum_{\Gamma\backslash M_2(\mathbb{Z})^+} f(1, \mu\,\rho\,\tilde{m}^{-1}, \mu(\tau))\,\mathrm{Det}(\mu)^{-\beta}\,,$$

where $\mu \in M_2(\mathbb{Z})^+$ is subject to the condition $\mu\,\rho\,\tilde{m}^{-1} \in M_2(R)$. The other values of μ a priori involved in the summation (8.18) do not contribute, since they correspond to the orthogonal of the support of $\theta_m(f)$.

One has $\mathrm{Det}(m) = \mathrm{Det}(m')$ by construction, hence

$$\rho\,\tilde{m}^{-1} = \rho\,m\,\mathrm{Det}(m)^{-1} = \mathrm{Det}(m')^{-1}\,m'\,\rho' = \tilde{m'}^{-1}\,\rho'\,.$$

Therefore the condition $\mu\,\rho\,\tilde{m}^{-1} \in M_2(R)$ holds iff $\mu = \nu\,\tilde{m'}$ for some $\nu \in M_2(\mathbb{Z})^+$. Thus, since $\mathrm{Det}(\mu) = \mathrm{Det}(\nu) \cdot \mathrm{Det}(\tilde{m}')$, we can rewrite, up to multiplication by a scalar,

$$\varphi_{\beta,l}(\theta_m(f)) = Z'^{-1} \sum_{\Gamma\backslash M_2(\mathbb{Z})^+} f(1, \nu\,\rho', \nu\,\tilde{m}'(\tau))\,\mathrm{Det}(\nu)^{-\beta}\,.$$

This proves (9.18). It remains to show that on $\mathcal{A}_{\mathbb{Q}}$ we have

$$\varphi_{l'}(f) = \mathrm{Gal}_\tau(\rho\, m\, \rho^{-1}) \circ \varphi_l(f)\,, \quad \forall f \in \mathcal{A}_{\mathbb{Q}}\,.$$

Both sides only involve the values of f on invertible $\mathbb{Q}$-lattices, and there, by (9.6) one has

$$f(1,\,\alpha,\tau) = \mathrm{Gal}_\tau(\alpha) f(1,\,1,\tau)\,, \quad \forall \alpha \in \mathrm{GL}(2,R)\,.$$

Thus, we obtain

$$\varphi_{l'}(f) = f(1,\rho',\,m'^{-1}(\tau)) = I_{m'^{-1}(\tau)}(\mathrm{Gal}(\rho')f) = I_\tau(\mathrm{Gal}(m'\,\rho')f)\,.$$

Since $m'\,\rho' = \rho\, m$, this gives $\mathrm{Gal}_\tau(\rho\, m\, \rho^{-1}) \circ \varphi_l(f)$ as required. □

We shall now work out the algebraic relations fulfilled by the $\mathcal{X}_a$ as extensions of the division formulas of elliptic functions.

We first work with lattice functions of some weight k, or equivalently with forms $f(g,y)\,dy^{k/2}$, and then multiply them by a suitable factor to make them homogeneous of weight 0 under scaling. The functions of weight 2 are the generators, the higher weight ones will be obtained from them by universal formulas with modular forms as coefficients.

The powers X_a^m of the function X_a are then expressed as universal polynomials with coefficients in the ring $\mathbb{Q}\,[e_4,e_6]$ in the following weight $2k$ functions $(k>1)$:

$$E_{2k,a}(\rho,\tau) = \pi^{-2k} \sum_{y\in\Lambda+\phi(a)} y^{-2k}. \tag{9.20}$$

These fulfill by construction ([56]) the relations

$$\begin{aligned} E_{2m,a} &= X_a(E_{2m-2,a} - e_{2m-2}) + \left(1 - \binom{2m}{2}\right) e_{2m} \\ &\quad - \sum_{1}^{m-2} \binom{2k+1}{2k} e_{2k+2}(E_{2m-2k-2,a} - e_{2m-2k-2})\,. \end{aligned} \tag{9.21}$$

These relations dictate the value of $E_{2k}(\rho,\tau)$ when $\varphi(a) \in \Lambda$: one gets

$$E_{2k}(\rho,\tau) = \nu_{2k}(\tau) \qquad \text{if } \varphi(a)\in\Lambda, \tag{9.22}$$

where ν_{2k} is a modular form of weight $2k$ obtained by induction from (9.21), with X_a replaced by 0 and E_{2m} by ν_{2m}. One has $\nu_{2k} \in \mathbb{Q}\,[e_4,e_6]$ and the first values are

$$\nu_4 = -5\,e_4\,, \quad \nu_6 = -14\,e_6\,, \quad \nu_8 = \frac{45}{7}\,e_4^2\,, \ldots \tag{9.23}$$

We shall now write the important algebraic relations between the functions X_a, which extend the division relations of elliptic functions from invertible $\mathbb{Q}$-lattices to arbitrary ones.

In order to work out the division formulas for the Eisenstein series $E_{2m,a}$ we need to control the image of $\left(\frac{1}{N}\mathbb{Z}\right)^2 = \frac{1}{N}\mathbb{Z}^2$ under an arbitrary element $\rho \in M_2(R)$. This is done as follows using the projections π_L defined in (7.9).

Lemma 9.6 *Let $N \in \mathbb{N}^*$, and $\rho \in M_2(R)$. There exists a smallest lattice $L \subset \mathbb{Z}^2$ with $L \supset N\mathbb{Z}^2$, such that $\pi_L(\rho) = 1$. One has*

$$\rho(\frac{1}{N}\mathbb{Z}^2) = \frac{1}{N}L\,.$$

Proof. There are finitely many lattices L with $N\mathbb{Z}^2 \subset L \subset \mathbb{Z}^2$. Thus, the intersection of those L for which $\pi_L(\rho) = 1$ is still a lattice and fulfills $\pi_L(\rho) = 1$ by (7.10). Let L be this lattice, and let us show that $\rho(\frac{1}{N}\mathbb{Z}^2) \subset \frac{1}{N}L$. Let $m \in M_2(\mathbb{Z})^+$ be such that $m(\mathbb{Z}^2) = L$. Then $\pi_L(\rho) = 1$ implies that $\rho = m\mu$ for some $\mu \in M_2(R)$. Thus $\rho(\frac{1}{N}\mathbb{Z}^2) \subset m(\frac{1}{N}\mathbb{Z}^2) = \frac{1}{N}L$. Conversely let $L' \subset \mathbb{Z}^2$ be defined by $\rho(\frac{1}{N}\mathbb{Z}^2) = \frac{1}{N}L'$. We need to show that $\pi_{L'}(\rho) = 1$, *i.e.* that there exists $m' \in M_2(\mathbb{Z})^+$ such that $L' = m'(\mathbb{Z}^2)$ and $m'^{-1}\rho \in M_2(R)$. Replacing ρ by $\gamma_1\rho\gamma_2$ for $\gamma_j \in \mathrm{SL}_2(\mathbb{Z})$ does not change the problem, hence we can use this freedom to assume that the restriction of ρ to $\left(\frac{1}{N}\mathbb{Z}\right)^2$ is of the form

$$\rho_N = \begin{bmatrix} d_1 & 0 \\ 0 & d_2 \end{bmatrix}, \quad d_j \mid N\,,\ d_1 \mid d_2\,. \tag{9.24}$$

One then takes $m' = \begin{bmatrix} d_1 & 0 \\ 0 & d_2 \end{bmatrix} \in M_2(\mathbb{Z})^+$ and checks that $L' = m'(\mathbb{Z}^2)$ while $m'^{-1}\rho$ belongs to $M_2(R)$. □

Given an integer $N > 1$, we let S_N be the set of lattices

$$N\,\mathbb{Z}^2 \subset L \subset \mathbb{Z}^2, \tag{9.25}$$

which is the same as the set of subgroups of $(\mathbb{Z}/N\mathbb{Z})^2$. For each $L \in S_N$ we define a projection $\pi(N,L)$ by

$$\pi(N,L) = \pi_L \prod_{L' \in S_N,\, L' \subsetneqq L} (1 - \pi_{L'})\,. \tag{9.26}$$

By Lemma 9.6 the range of $\pi(N,L)$ is exactly the set of $\rho \in M_2(R)$ such that

$$\rho(\frac{1}{N}\mathbb{Z}^2) = \frac{1}{N}L\,. \tag{9.27}$$

The general form of the division relations is as follows.

Proposition 9.7 *There exists canonical modular forms $\omega_{N,L,k}$ of level N and weight $2k$, such that for all k and $(\rho,\tau)\in Y$ they satisfy*

$$\sum_{Na=0} X_a^k(\rho,\tau) = \sum_{L\in S_N} \pi(N,L)(\rho)\,\omega_{N,L,k}(\tau)\,.$$

In fact, we shall give explicit formulas for the $\omega_{N,L,k}$ and show in particular that

$$\omega_{N,L,k}(\gamma\,\tau) = (c\tau+d)^{2k}\,\omega_{N,\gamma^{-1}\,L,k}(\tau),$$

which implies that $\omega_{N,L,k}$ is of level N.

We prove it for $k=1$ and then proceed by induction on k. The division formulas in weight 2 involve the 1-cocycle on the group $\mathrm{GL}_2^+(\mathbb{Q})$ with values in Eisenstein series of weight 2 given in terms of the Dedekind η-function by (cf. [10])

$$\mu_\gamma(\tau) \;=\; \frac{1}{12\pi i}\,\frac{d}{d\tau}\log\frac{\Delta|\gamma}{\Delta} \;=\; \frac{1}{2\pi i}\,\frac{d}{d\tau}\log\frac{\eta^4|\gamma}{\eta^4}\,, \tag{9.28}$$

where we used the standard 'slash operator' notation for the action of $\mathrm{GL}_2^+(\mathbb{R})$ on functions on the upper half plane:

$$f|_k\,\alpha\,(z) = \mathrm{Det}\,(\alpha)^{k/2}\,f(\alpha\cdot z)\,j(\alpha,z)^{-k}\,, \tag{9.29}$$

$$\alpha = \begin{bmatrix} a & b\\ c & d\end{bmatrix} \in \mathrm{GL}_2^+(\mathbb{R}),\quad \alpha\cdot z = \frac{az+b}{cz+d}\quad \text{and}\quad j(\alpha,z) \;=\; cz+d\,.$$

Since $\mu_\gamma=0$ for $\gamma\in\Gamma$, the cocycle property

$$\mu_{\gamma_1\cdot\gamma_2} \;=\; \mu_{\gamma_1}|\gamma_2 \;+\; \mu_{\gamma_2} \tag{9.30}$$

shows that, for $m\in M_2(\mathbb{Z})^+$, the value of $\mu_{m^{-1}}$ only depends upon the lattice $L=m(\mathbb{Z}^2)$. We shall denote it by μ_L.

Lemma 9.8 *For any integer N, the X_a, $a\in\mathbb{Q}/\mathbb{Z}$, fulfill the relation*

$$\sum_{N\,a=0} X_a \;=\; N^2\,\sum_{L\in S_N}\pi(N,L)\;\mu_L$$

By construction the projections $\pi(N,L)$, $L\in S_N$ form a partition of unity,

$$\sum_{L\in S_N}\pi(N,L) = 1\,.$$

Thus to prove the lemma it is enough to evaluate both sides on $\rho\in\pi(N,L)$.

We can moreover use the equality

$$\mu_{\gamma^{-1}L} = \mu_L|\gamma\,,\quad \forall\gamma\in\Gamma$$

to assume that L and ρ_N are of the form

$$L = \begin{bmatrix} d_1 & 0 \\ 0 & d_2 \end{bmatrix} \mathbb{Z}^2 , \qquad \rho_N = \begin{bmatrix} d_1 & 0 \\ 0 & d_2 \end{bmatrix} , \quad d_j \mid N \, , \; d_1 \mid d_2 \, .$$

Let $d_2 = n\, d_1$. The order of the kernel of ρ_N is $d_1\, d_2$ and the computation of $\sum_{N\, a=0} X_a(\rho, \tau)$ gives

$$-N^2 e_2(\tau) \, + \, d_1\, d_2\, N^2 \sum_{(a,b)\in\mathbb{Z}^2\setminus\{0\}} (a\, d_1 - b\, d_2 \tau)^{-2}$$

which gives $N^2(n\, e_2(n\, \tau) - e_2(\tau)) = N^2 \mu_L$.

This proves the proposition for $k = 1$. Let us proceed by induction using (9.21) to express X_a^k as $E_{2k,a}$ plus a polynomial of degree $< k$ in X_a with coefficients in $\mathbb{Q}[e_4, e_6]$. Thus, we only need to prove the equality

$$\sum_{N\, a=0} E_{2k,a} = \sum_{L\in S_N} \pi(N, L) \; \alpha_{N,L,k},$$

where the modular forms $\alpha_{N,L,k}$ are given explicitly as

$$\alpha_{N,L,k} \; = \; N^{2k} d^2 n^{k+1} e_{2k} | m^{-1} \; - \mathrm{Det}(m) \; (e_{2k} \, - \nu_{2k}),$$

with $m \in M_2(\mathbb{Z})^+$, $m(\mathbb{Z}^2) = L$, and (d, dn) the elementary divisor of L.

The proof is obtained as above by evaluating both sides on arbitrary $\rho \in \pi(N, L)$. □

One can rewrite all the above relations in terms of the weight 0 elements

$$\mathcal{X}_a := c\, X_a \, , \quad \mathcal{E}_{2k,a} := c^k \, E_{2k,a} \in \mathcal{A}_{\mathbb{Q}}.$$

In particular, the two basic modular functions $c^2\, e_4$ and $c^3\, e_6$ are replaced by

$$c^2\, e_4 = \frac{1}{5}\, j\, (j - 1728)\, , \quad c^3\, e_6 = -\frac{2}{35}\, j\, (j - 1728)^2.$$

We can now rewrite the relations (9.21) in terms of universal polynomials

$$P_n \in \mathbb{Q}(j)[X],$$

which express the generators $\mathcal{E}_{2k,a}$ in terms of $\mathcal{X}_a$ by

$$\mathcal{E}_{2k,a} = P_k(\mathcal{X}_a).$$

In fact, from (9.21) we see that the coefficients of P_k are themselves polynomials in j rather than rational fractions, so that

$$P_n \in \mathbb{Q}[j, X].$$

The first ones are given by

$$P_2 = X^2 - j(j - 1728)\, , \quad P_3 = X^3 - \frac{9}{5}\, X j(j - 1728) + \frac{4}{5} j(j - 1728)^2\, , \quad \cdots$$

10 The Noncommutative Boundary of Modular Curves

We shall explain in this section how to combine the dual of the GL_2-system described above with the idea, originally developed in the work of Connes–Douglas–Schwarz [12] and Manin–Marcolli [37], of enlarging the boundary of modular curves with a noncommutative space that accounts for the degeneration of elliptic curves to noncommutative tori.

The GL_2-system described in the previous sections admits a "dual" system obtained by considering $\mathbb{Q}$-lattices up to commensurability but no longer up to scaling. Equivalently this corresponds to taking the cross product of the GL_2-system by the action of the Pontrjagin dual of $\mathbb{C}^*$, which combines the time evolution σ_t with an action by the group $\mathbb{Z}$ of integral weights of modular forms. The resulting space is the total space of the natural $\mathbb{C}^*$-bundle.

In adélic terms this "dual" noncommutative space $\mathcal{L}_2$ is described as follows,

Proposition 10.1 *There is a canonical bijection from the space of* $GL_2(\mathbb{Q})$-*orbits of the left action of* $GL_2(\mathbb{Q})$ *on* $M_2(\mathbb{A}_f) \times GL_2(\mathbb{R})$ *to the space* $\mathcal{L}_2$ *of commensurability classes of two-dimensional* $\mathbb{Q}$-*lattices.*

Proof. The space of $GL_2(\mathbb{Q})$ orbits on $M_2(\mathbb{A}_f) \times GL_2(\mathbb{R})$ is the same as the space of $GL_2^+(\mathbb{Q})$ orbits on $M_2(R) \times GL_2^+(\mathbb{R})$. □

By the results of the previous sections, the classical space obtained by considering the zero temperature limit of the quantum statistical mechanical system describing commensurability classes of 2-dimensional $\mathbb{Q}$-lattices up to scaling is the Shimura variety that represents the projective limit of all the modular curves

$$GL_2(\mathbb{Q})\backslash GL_2(\mathbb{A})/\mathbb{C}^* . \tag{10.1}$$

Usually, the Shimura variety is constructed as the projective limit of the

$$\Gamma'\backslash GL_2(\mathbb{R})^+/\mathbb{C}^*$$

over congruence subgroups $\Gamma' \subset \Gamma$. This gives a connected component in (10.1). The other components play a crucial role in the present context, in that the existence of several connected components allows for non-constant solutions of the equation $\zeta^n = 1$. Moreover all the components are permuted by the Galois covariance property of the arithmetic elements of the GL_2 system.

The total space of the natural $\mathbb{C}^*$-bundle, *i.e.* the quotient

$$GL_2(\mathbb{Q})\backslash GL_2(\mathbb{A}), \tag{10.2}$$

is the space of *invertible* 2-dimensional $\mathbb{Q}$-lattices (not up to scaling).

In the GL_1 case, the analog of (10.2), *i.e.* the space of idèle classes

$$GL_1(\mathbb{Q})\backslash GL_1(\mathbb{A}),$$

is compactified by first considering the noncommutative space of commensurability classes of $\mathbb{Q}$-lattices not up to scaling

$$\mathcal{L} = \mathrm{GL}_1(\mathbb{Q})\backslash\mathbb{A}^{\cdot},$$

where $\mathbb{A}^{\cdot}$ is the space of adèles with nonzero archimedean component. The next step, which is crucial in obtaining the geometric space underlying the spectral realization of the zeros of the Riemann zeta function, is to add an additional "stratum" that gives the noncommutative space of adèle classes

$$\overline{\mathcal{L}} = \mathrm{GL}_1(\mathbb{Q})\backslash\mathbb{A}\,,$$

which will be analyzed in the next Chapter.

Similarly, in the GL_2 case, the classical space given by the Shimura variety (10.1) is first "compactified" by adding noncommutative "boundary strata" obtained by replacing $\mathrm{GL}_2(\mathbb{A}_f)$ in $\mathrm{GL}_2(\mathbb{A}) = \mathrm{GL}_2(\mathbb{A}_f) \times \mathrm{GL}_2(\mathbb{R})$ by all matrices $M_2(\mathbb{A}_f)$. As boundary stratum of the Shimura variety it corresponds to degenerating the invertible $\mathbb{Q}$-structure ϕ on the lattice to a non-invertible one and yields the notion of $\mathbb{Q}$-lattice. The corresponding space of commensurability classes of $\mathbb{Q}$-lattices up to scaling played a central role in this whole chapter.

The space of commensurability classes of 2-dimensional $\mathbb{Q}$-lattices (not up to scaling) is

$$\mathcal{L}_2 = \mathrm{GL}_2(\mathbb{Q})\backslash(M_2(\mathbb{A}_f) \times \mathrm{GL}_2(\mathbb{R}))\,. \tag{10.3}$$

On $\mathcal{L}_2$ we can consider not just modular functions but all modular forms as functions. One obtains in this way an antihomomorphism of the modular Hecke algebra of level one of [10] (with variable $\alpha \in \mathrm{GL}_2(\mathbb{Q})^+$ restricted to $M_2(\mathbb{Z})^+$) to the algebra of coordinates on $\mathcal{L}_2$.

The further compactification at the archimedean place, corresponding to $\mathcal{L} \hookrightarrow \overline{\mathcal{L}}$ in the GL_1 case, now consists of replacing $\mathrm{GL}_2(\mathbb{R})$ by matrices $M_2(\mathbb{R})$. This corresponds to degenerating the lattices to pseudo-lattices (in the sense of [34]) or in more geometric terms, to a degeneration of elliptic curves to noncommutative tori. It is this part of the "noncommutative compactification" that was considered in [12] and [37].

A $\mathbb{Q}$-pseudolattice in $\mathbb{C}$ is a pair (Λ, ϕ), with $\Lambda = j(\mathbb{Z}^2)$ the image of a homomorphism $j : \mathbb{Z}^2 \to \ell$, with $\ell \subset \mathbb{R}^2 \cong \mathbb{C}$ a real 1-dimensional subspace, and with a group homomorphism

$$\phi : \mathbb{Q}^2/\mathbb{Z}^2 \to \mathbb{Q}\Lambda/\Lambda.$$

The $\mathbb{Q}$-pseudolattice is nondegenerate if j is injective and is invertible if ϕ is invertible.

Proposition 10.2 *Let* $\partial Y := M_2(R) \times \mathbb{P}^1(\mathbb{R})$. *The map*

$$(\rho, \theta) \mapsto (\Lambda, \phi), \qquad \Lambda = \mathbb{Z} + \theta\mathbb{Z}, \quad \phi(x) = \rho_1(x) - \theta\rho_2(x) \tag{10.4}$$

gives an identification

$$\Gamma\backslash\partial Y \simeq (\textit{Space of } \mathbb{Q}\textit{-pseudolattices in } \mathbb{C})/\mathbb{C}^*. \tag{10.5}$$

This space parameterizes the degenerations of 2-dimensional $\mathbb{Q}$*-lattices*

$$\lambda(y) = (\Lambda, \phi) \quad \textit{where} \quad \Lambda = \tilde{h}(\mathbb{Z} + i\,\mathbb{Z}) \quad \textit{and} \quad \phi = \tilde{h} \circ \rho, \tag{10.6}$$

for $y = (\rho, h) \in M_2(R) \times \mathrm{GL}_2(\mathbb{R})$ *and* $\tilde{h} = h^{-1}\mathrm{Det}\,(h)$*, when* $h \in \mathrm{GL}_2(\mathbb{R})$ *degenerates to a non-invertible matrix in* $M_2(\mathbb{R})$*.*

Proof. $\mathbb{Q}$-pseudolattices in $\mathbb{C}$ are of the form

$$\Lambda = \lambda(\mathbb{Z} + \theta\mathbb{Z}), \quad \phi(a) = \lambda\rho_1(a) - \lambda\theta\rho_2(a), \tag{10.7}$$

for $\lambda \in \mathbb{C}^*$ and $\theta \in \mathbb{P}^1(\mathbb{R})$ and $\rho \in M_2(R)$. The action of $\mathbb{C}^*$ multiplies λ, while leaving θ unchanged. This corresponds to changing the 1-dimensional linear subspace of $\mathbb{C}$ containing the pseudo-lattice and rescaling it. The action of $\mathrm{SL}_2(\mathbb{Z})$ on $\mathbb{P}^1(\mathbb{R})$ by fractional linear transformations changes θ. The nondegenerate pseudolattices correspond to the values $\theta \in \mathbb{P}^1(\mathbb{R}) \smallsetminus \mathbb{P}^1(\mathbb{Q})$ and the degenerate pseudolattices to the cusps $\mathbb{P}^1(\mathbb{Q})$.

For $y = (\rho, h) \in M_2(R) \times \mathrm{GL}_2(\mathbb{R})$ consider the $\mathbb{Q}$-lattice (10.6), for $\tilde{h} = h^{-1}\mathrm{Det}\,(h)$. Here we use the basis $\{e_1 = 1, e_2 = -i\}$ of the $\mathbb{R}$-vector space $\mathbb{C}$ to let $\mathrm{GL}_2^+(\mathbb{R})$ act on $\mathbb{C}$ as $\mathbb{R}$-linear transformations. These formulas continue to make sense when $h \in M_2(\mathbb{R})$ and the image $\Lambda = \tilde{h}(\mathbb{Z} + i\,\mathbb{Z})$ is a pseudolattice when the matrix h is no longer invertible.

To see this more explicitly, consider the right action

$$m \mapsto m \cdot z \tag{10.8}$$

of $\mathbb{C}^*$ on $M_2(\mathbb{R})$ determined by the inclusion $\mathbb{C}^* \subset \mathrm{GL}_2(\mathbb{R})$ as in (8.1), The action of $\mathbb{C}^*$ on $M_2(\mathbb{R}) \smallsetminus \{0\}$ is free and proper. The map

$$\rho(\alpha) = \begin{cases} \alpha(i) & (c,d) \neq (0,0) \\ \infty & (c,d) = (0,0) \end{cases} \qquad \text{with} \quad \alpha = \begin{bmatrix} a & b \\ c & d \end{bmatrix} \tag{10.9}$$

defines an isomorphism

$$\rho : (M_2(\mathbb{R}) \smallsetminus \{0\})/\mathbb{C}^* \to \mathbb{P}^1(\mathbb{C}), \tag{10.10}$$

equivariant with respect to the left action of $\mathrm{GL}_2(\mathbb{R})$ on $M_2(\mathbb{R})$ and the action of $\mathrm{GL}_2(\mathbb{R})$ on $\mathbb{P}^1(\mathbb{C})$ by fractional linear transformations. Moreover, this maps $M_2(\mathbb{R})^+$ to the closure of the upper half plane

$$\overline{\mathbb{H}} = \mathbb{H} \cup \mathbb{P}^1(\mathbb{R}). \tag{10.11}$$

The rank one matrices in $M_2(\mathbb{R})$ map to $\mathbb{P}^1(\mathbb{R}) \subset \mathbb{P}^1(\mathbb{C})$. In fact, the isotropy group of $m \in M_2(\mathbb{R})$ is trivial if $m \neq 0$, since $m \cdot z = m$ only has nontrivial solutions for $m = 0$, since $z - 1$ is invertible when nonzero. This shows that $M_2(\mathbb{R})\backslash\{0\}$ is the total space of a principal $\mathbb{C}^*$-bundle. □

Notice that, unlike the case of $\mathbb{Q}$-lattices of (8.8), where the quotient $\Gamma\backslash Y$ can be considered as a classical quotient, here the space $\Gamma\backslash\partial Y$ should be regarded as a noncommutative space with function algebra

$$C(\partial Y) \rtimes \Gamma.$$

The usual algebro–geometric compactification of a modular curve $Y_{\Gamma'} = \Gamma'\backslash\mathbb{H}$, for Γ' a finite index subgroup of Γ, is obtained by adding the cusp points $\Gamma\backslash\mathbb{P}^1(\mathbb{Q})$,

$$X_{\Gamma'} = Y_{\Gamma'} \cup \{\text{ cusps }\} = \Gamma'\backslash(\mathbb{H} \cup \mathbb{P}^1(\mathbb{Q})) . \tag{10.12}$$

Replacing $\mathrm{GL}_2(\mathbb{R})$ by $M_2(\mathbb{R})$ in (10.3) corresponds to replacing the cusp points $\mathbb{P}^1(\mathbb{Q})$ by the full boundary $\mathbb{P}^1(\mathbb{R})$ of $\mathbb{H}$. Since Γ does not act discretely on $\mathbb{P}^1(\mathbb{R})$, the quotient is best described by noncommutative geometry, as the cross product C^*-algebra $C(\mathbb{P}^1(\mathbb{R})) \rtimes \Gamma'$ or, up to Morita equivalence,

$$C(\mathbb{P}^1(\mathbb{R}) \times \mathbb{P}) \rtimes \Gamma, \tag{10.13}$$

with $\mathbb{P}$ the coset space $\mathbb{P} = \Gamma/\Gamma'$.

The noncommutative boundary of modular curves defined this way retains a lot of the arithmetic information of the classical modular curves. Various results of [37] show, from the number theoretic point of view, why the irrational points of $\mathbb{P}^1(\mathbb{R})$ in the boundary of $\mathbb{H}$ should be considered as part of the compactification of modular curves.

The first such result is that the classical definition of modular symbols (*cf.* [36]), as homology classes on modular curves defined by geodesics connecting cusp points, can be generalized to "limiting modular symbols", which are asymptotic cycles determined by geodesics ending at irrational points. The properties of limiting modular symbols are determined by the spectral theory of the Ruelle transfer operator of a dynamical system, which generalizes the Gauss shift of the continued fraction expansion by taking into account the extra datum of the coset space $\mathbb{P}$.

Manin's modular complex (*cf.* [36]) gives a combinatorial presentation of the first homology of modular curves, useful in the explicit computation of the intersection numbers obtained by pairing modular symbols to cusp forms. It is shown in [37] that the modular complex can be recovered canonically from the K-theory of the C^*-algebra (10.13).

Moreover, Mellin transforms of cusp forms of weight two for the congruence subgroups $\Gamma_0(p)$, with p prime, can be obtained by integrating along the boundary $\mathbb{P}^1(\mathbb{R})$ certain "automorphic series" defined in terms of the continued fraction expansion and of modular symbols.

These extensions of the theory of modular symbols to the noncommutative boundary appear to be interesting also in relation to the results of [10], where the pairing with modular symbols is used to give a formal analog of the Godbillon–Vey cocycle and to obtain a rational representative for the Euler class in the group cohomology $H^2(\mathrm{SL}_2(\mathbb{Q}), \mathbb{Q})$.

The fact that the arithmetic information on modular curves is stored in their noncommutative boundary (10.13) is interpreted in [38] as an instance of the physical principle of holography. Noncommutative spaces arising at the boundary of Shimura varieties have been further investigated by Paugam [41] from the point of view of Hodge structures.

This noncommutative boundary stratum of modular curves representing degenerations of lattices to pseudo-lattices has been proposed by Manin ([34] [35]) as a geometric space underlying the explicit class field theory problem for real quadratic fields. In fact, this is the first unsolved case of the Hilbert 12th problem. Manin developed in [34] a theory of real multiplication, where noncommutative tori and pseudolattices should play for real quadratic fields a role parallel to the one that lattices and elliptic curves play in the construction of generators of the maximal abelian extensions of imaginary quadratic fields. The picture that emerges from this "real multiplication program" is that the cases of $\mathbb{Q}$ (Kronecker–Weber) and of both imaginary and real quadratic fields should all have the same underlying geometry, related to different specializations of the GL_2 system. The relation of the GL_2 system and explicit class field theory for imaginary quadratic fields is analyzed in [13].

11 The BC Algebra and Optical Coherence

It is very natural to look for concrete physical realizations of the phase transition exhibited by the BC system. An attempt in this direction has been proposed in [43], in the context of the physical phenomenon of quantum phase locking in lasers.

This interpretation relates the additive generators $e(r)$ of the BC algebra (*cf.* Proposition 3.1) with the quantum phase states, which are a standard tool in the theory of optical coherence (*cf. e.g.* [32]), but it leaves open the interpretation of the generators μ_n. Since on a finite dimensional Hilbert space isometries are automatically unitary, this rules out nontrivial representations of the μ_n in a fixed finite dimensional space.

After recalling the basic framework of phase states and optical coherence, we interpret the action of the μ_n as a "renormalization" procedure, relating the quantum phase states at different scales.

There is a well known analogy (*cf.* [46] §21-3) between the quantum statistical mechanics of systems with phase transitions, such as the ferromagnet or the Bose condensation of superconducting liquid Helium, and the physics of lasers, with the transition to single mode radiation being the analog of "condensation". The role of the inverse temperature β is played in laser physics by the "population inversion" parameter, with critical value at the inversion threshold. The injected signal of the laser acts like the external field responsible for the symmetry breaking mechanism. Given these identifications, one in fact obtains similar forms in the two systems for both thermodynamic potential and statistical distribution. The phase locking phenomenon is also analogous in systems with phase transitions and lasers, with the modes in the laser assuming same phase and amplitude above threshold being the analog of Cooper pairs of electrons acquiring the same energy and phase below critical temperature in superconductors.

In a laser cavity typically many longitudinal modes of the radiation are oscillating simultaneously. For a linewidth $\Delta\nu$ around a frequency ν_0 for the active medium in the cavity of length L and frequency spacing $\delta\nu = c/2L$, the number of oscillating modes is $N = \lceil \Delta\nu/\delta\nu \rceil$ and the field output of the laser is

$$E(x,t) = \sum_{n=-N/2}^{N/2} A_n \exp(-2\pi i \nu_n (t - x/c) + 2\pi i \theta_n), \tag{11.1}$$

with all the beat frequencies between adjacent modes $\nu_n - \nu_{n-1} = \delta\nu$. Due to noise in the cavity all these modes are uncorrelated, with a random distribution of amplitudes A_n and phases θ_n.

A mode locking phenomenon induced by the excited lasing atoms is responsible for the fact that, above the threshold of population inversion, the phases and amplitudes of the frequency modes become locked together. The resulting field

$$E(x,t) = A e^{2\pi i \theta} \exp(-2\pi i \nu_0 (t - x/c)) \left(\frac{\sin(\pi \delta\nu (N+1)(t - x/c))}{\sin(\pi \delta\nu (t - x/c))} \right), \tag{11.2}$$

shows many locked modes behaving like a single longitudinal mode oscillating inside the cavity (*cf.* Figure 3). This phenomenon accounts for the typical narrowness of the laser linewidth and monocromaticity of laser radiation.

Since the interaction of radiation and matter in lasers is essentially a quantum mechanical phenomenon, the mode locking should be modeled by quantum mechanical phase operators corresponding to the resonant interaction

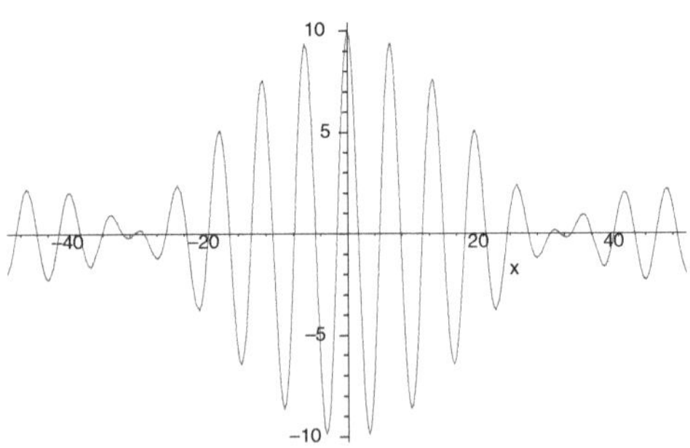

Fig. 3. Output pulse train in lasers above threshold

of many different oscillators. In the quantum theory of radiation one usually describes a single mode by the Hilbert space spanned by the occupation number states $|n\rangle$, with creation and annihilation operators a^* and a that raise and lower the occupation numbers and satisfy the relation $[a, a^*] = 1$. The polar decomposition $a = S\sqrt{N}$ of the annihilation operator is used to define a quantum mechanical phase operator, which is conjugate to the occupation number operator $N = a^* a$. Similar phase operators are used in the modeling of Cooper pairs. This approach to the definition of a quantum phase has the drawback that the one sided shift S is not a unitary operator. This can also be seen in the fact that the inverse Cayley transform of S, which gives the cotangent of the phase, is a symmetric non self-adjoint operator.

The emission of lasers above threshold can be described in terms of coherent state excitations,

$$|\alpha\rangle = \exp(-|\alpha|^2) \sum_n \frac{\alpha^n}{(n!)^{1/2}} |n\rangle,$$

which are eigenfunctions of the annihilation operators,

$$a\,|\alpha\rangle = \alpha\,|\alpha\rangle.$$

These are quantum mechanical analogs of classical electromagnetic waves as in (11.1) (11.2). One can show (*cf. e.g.* [32] §7.4) that the field excitation in a laser approaches a coherent state as the pumping increases to values above the population inversion threshold, with the phase diffusion governed by the equation of motion for quantum mechanical phase states.

The problem in defining a proper quantum phase operator, due to lack of self-adjointness, has been overcome by the following approximation of the basic quantum operators on the Fock space $\mathcal{H}$. One selects a scale, given by a positive integer $N \in \mathbb{N}$ and cuts down $\mathcal{H}$ to a finite dimensional subspace by the phase state projector

$$P_N = \sum_m |\theta_{m,N}\rangle \langle \theta_{m,N}|,$$

where the orthonormal vectors $|\theta_{m,N}\rangle$ in $\mathcal{H}$ are given by

$$|\theta_{m,N}\rangle := \frac{1}{(N+1)^{1/2}} \sum_{n=0}^{N} \exp\left(2\pi i \frac{m\,n}{N+1}\right) |n\rangle. \tag{11.3}$$

These are eigenvectors for the phase operators, that affect discrete values given by roots of unity, replacing a continuously varying phase.

This way, phase and occupation number behave like positions and momenta. An occupation number state has randomly distributed phase and, conversely, a phase state has a uniform distribution of occupation numbers.

We now realize the ground states of the BC system as representations of the algebra $\mathcal{A}$ in the Fock space $\mathcal{H}$ of the physical system described above.

Given an embedding $\rho : \mathbb{Q}^{ab} \to \mathbb{C}$, which determines the choice of a ground state, the generators $e(r)$ and μ_n (*cf.* Proposition 3.1) act as

$$e(a/b)\,|n\rangle = \rho(\zeta_{a/b}^n)\,|n\rangle,$$

$$\mu_k\,|n\rangle = |kn\rangle.$$

In the physical system, the choice of the ground state is determined by the primitive $N+1$-st root of unity

$$\rho(\zeta_{N+1}) = \exp(2\pi i/(N+1)).$$

One can then write (11.3) in the form

$$|\theta_{m,N}\rangle = e\left(\frac{m}{N+1}\right)\cdot v_N. \tag{11.4}$$

where we write v_N for the superposition of the first $N+1$ occupation states

$$v_N := \frac{1}{(N+1)^{1/2}}\sum_{n=0}^{N}|n\rangle.$$

Any choice of a primitive $N+1$-st root of unity would correspond to another ground state, and can be used to define analogous phase states. This construction of phase states brings in a new hidden group of symmetry, which is different from the standard rotation of the phase, and is the Galois group $\mathrm{Gal}(\mathbb{Q}^{ab}/\mathbb{Q})$. This raises the question of whether such symmetries are an artifact of the approximation, or if they truly represent a property of the physical system.

In the BC algebra, the operators μ_n act on the algebra generated by the $e(r)$ by endomorphisms given by

$$\mu_n\,P(e(r_1),\ldots,e(r_k))\,\mu_n^* = \frac{\pi_n}{n^k}\sum_{ns=r}P(e(s_1),\ldots,e(s_k)), \tag{11.5}$$

for an arbitrary polynomial P in k-variables, with $\pi_n = \mu_n\mu_n^*$ and $s = (s_1,\ldots,s_k)$, $r = (r_1,\ldots,r_k)$ in $(\mathbb{Q}/\mathbb{Z})^k$. In particular, this action has the effect of averaging over different choices of the primitive roots.

The averaging on the right hand side of (11.5), involving arbitrary phase observables $P(e(r_1),\ldots,e(r_k))$, has physical meaning as statistical average over the choices of primitive roots. The left hand side implements this averaging as a renormalization group action.

Passing to the limit $N\to\infty$ for the phase states is a delicate process. It is known in the theory of optical coherence (*cf. e.g.* [33] §10) that one can take such limit only after expectation values have been calculated. The analogy between the laser and the ferromagnet suggests that this limiting procedure should be treated as a case of statistical limit, in the sense of [57]. In fact, when analyzing correlations near a phase transition, one needs a mechanism that handles changes of scale. In statistical mechanics, such mechanism exists in the form of a renormalization group, which expresses the fact that different length or energy scales are locally coupled. This is taken care here by the action of (11.5).

References

1. J. Arledge, M. Laca, I. Raeburn, *Semigroup crossed products and Hecke algebras arising from number fields*, Doc. Math. 2 (1997) 115–138.
2. M.W. Binder, *Induced factor representations of discrete groups and their types*, J. Functional Analysis, 115 (1993), 294–312.
3. F. Boca and A. Zaharescu, *Factors of type III and the distribution of prime numbers*, Proc. London Math. Soc. (3) Vol.80 (2000), no. 1, 145–178.
4. J.-B. Bost and A. Connes, *Produits Euleriens et facteurs de type III*, C. R. Acad. Sci. Paris Sér. I Math. Vol.315 (1992), no. 3, 279–284.
5. J.B. Bost, A. Connes, *Hecke algebras, Type III factors and phase transitions with spontaneous symmetry breaking in number theory*, Selecta Math. (New Series) Vol.1 (1995) N.3, 411–457.
6. O. Bratteli, D.W. Robinson. *Operator Algebras and Quantum Statistical Mechanics* I and II, Springer, New York 1979 and 1981.
7. B. Brenken, *Hecke algebras and semigroup crossed product C^*-algebras*, Pacific J. Math., 187 (1999), 241–262.
8. P.B. Cohen, *A C^*-dynamical system with Dedekind zeta partition function and spontaneous symmetry breaking*, Journal de Théorie des Nombres de Bordeaux 11 (1999) 15–30.
9. A. Connes, *Trace formula in noncommutative geometry and the zeros of the Riemann zeta function*, Selecta Math. (New Series) Vol.5 (1999) 29–106.
10. A. Connes, H. Moscovici, *Modular Hecke Algebras and their Hopf Symmetry*, Moscow. Math. Journal, vol.4, n 1, (2004) 67–109.
11. A. Connes, *A survey of foliations and operator algebras*, Proc. Sympos. Pure Math., 38, Amer. Math. Soc., Providence, R.I., (1982) 521–628.
12. A. Connes, M. Douglas, A. Schwarz, *Noncommutative geometry and matrix theory: compactification on tori*, J. High Energy Phys. 1998, no. 2, Paper 3, 35 pp. (electronic).
13. A. Connes, M. Marcolli, N. Ramachandran, *KMS states and complex multiplication*, in preparation.
14. J. Dixmier, *Les C^*-algèbres et leurs Représentations*, Gauthier-Villars, Paris 1964.
15. H. Glöckner and G. Willis, *Topologisation of Hecke pairs and Hecke C^*-algebras*, (preprint), University of Newcastle, Australia (2002).
16. R. Haag, *Local Quantum Physics*, Springer, Berlin 1992.
17. R. Haag, N. M. Hugenholtz, M. Winnink *On the equilibrium states in quantum statistical mechanics*, Comm. Math. Phys. 5 (1967), 215–236.
18. R. Hall, *Hecke C^*-algebras*, PhD Thesis, Pennsylvania State University (1999).
19. D. Harari, E. Leichtnam, *Extension du phénomène de brisure spontanée de symétrie de Bost–Connes au cas des corps globaux quelconques*, Selecta Math. (New Series) Vol.3 (1997) 205–243.

20. B. Julia, *Statistical theory of numbers*, in "Number Theory and Physics", J.-M. Luck, P. Moussa and M. Waldschmidt (Eds.), Springer Verlag, Berlin, 1990.
21. M. Laca, *Semigroups of ∗-endomorphisms, Dirichlet series, and phase transitions*, J. Funct. Anal. 152 (1998) 330–378.
22. M. Laca, *From endomorphisms to automorphisms and back: dilations and full corners*, J. London Math. Soc. (2) 61 (2000), no. 3, 893–904.
23. M. Laca and N. Larsen, *Hecke algebras of semidirect products*, Proc. Amer. Math. Soc. (to appear in 2003).
24. M. Laca and I. Raeburn, *Semigroup crossed products and the Toeplitz algebras of nonabelian groups*, J. Funct. Anal., 139 (1996), 415–440.
25. M. Laca and I. Raeburn, *A semigroup crossed product arising in number theory.* J. London Math. Soc. (2) 59 (1999), no. 1, 330–344.
26. M. Laca and I. Raeburn, *The ideal structure of the Hecke C^*-algebra of Bost and Connes*, Math. Ann. 318 (2000), no. 3, 433–451.
27. M. Laca and M. van Frankenhuysen, *Phase transitions on Hecke C^*-algebras and class field theory* (in preparation).
28. S. Lang, *Elliptic Functions*, (Second Edition), Graduate Texts in Mathematics, Vol.112, Springer-Verlag 1987.
29. N. Larsen, I. Putnam, I. Raeburn, *The two-prime analogue of the Hecke C^*-algebra of Bost and Connes*, Indiana Univ. Math. J. Vol.51 (2002), no. 1, 171–186.
30. N. Larsen and I. Raeburn, *Representations of Hecke algebras and dilations of semigroup cross products*, J. London Math. Soc. (2) **66** (2002), no. 1, 198–212.
31. E. Leichtnam and V. Nistor, *Cross product algebras and the homology of certain p-adic and adelic dynamical systems*, K-Theory, **21** (2000), 1–23.
32. R. Loudon, *The Quantum Theory of Light*, third edition, Oxford University Press, 2000.
33. L. Mandel, E. Wolf, *Optical Coherence and Quantum Optics*, Cambridge University Press, 1995.
34. Yu.I. Manin, *Real Multiplication and noncommutative geometry*, preprint arXiv math.AG/0202109.
35. Yu.I. Manin, *Von Zahlen und Figuren*, preprint arXiv math.AG/0201005.
36. Yu.I. Manin, *Parabolic points and zeta functions of modular curves*, Math. USSR Izvestjia 6 (1972) N.1, 19–64.
37. Yu.I. Manin, M. Marcolli, *Continued fractions, modular symbols, and noncommutative geometry*, Selecta Math. (New Series) Vol.8 (2002) N.3, 475–520.
38. Yu.I. Manin, M. Marcolli, *Holography principle and arithmetic of algebraic curves*, Adv. Theor. Math. Phys. Vol.3 (2001) N.5, 617–650.
39. D. Mumford, *Tata Lectures on Theta. I*, Progress in Mathematics, Vo.28, Birkhäuser, Boston, MA, 1983.
40. S. Neshveyev, *Ergodicity of the action of the positive rationals on the group of finite adeles and the Bost-Connes phase transition theorem*, Proc. Amer. Math. Soc. 130 (2002), no. 10, 2999–3003.
41. F. Paugam *Three examples of Noncommutative moduli spaces and related questions* Preprint 2004.
42. G. Pedersen, *C^*-algebras and their Automorphism Groups*. Academic Press, New York 1979.
43. M. Planat, *Invitation to the "spooky" quantum phase–locking effect and its link to $1/f$ fluctuations*, math-ph/0310082.

44. W. Rudin, *Real and Complex Analysis*, McGraw-Hill, NewYork 1987.
45. D. Ruelle, *Statistical Mechanics.* World Scientific Publishing Co. Inc., River Edge, NJ, 1999. Reprint of the 1989 edition.
46. M. Sargent, M. Scully, W. Lamb, *Laser Physics*, Addison–Wesley, 1974.
47. G. Schlichting, *Polynomidentitäten und Permutationsdarstellungen lokalkompakter Gruppen*, Invent. Math. 55 (1979) N.2, 97–106.
48. G. Schlichting, *Operationen mit periodischen Stabilisatoren*, Arch.Math. 34 (1980) N.2, 97–99.
49. B. Schoeneberg, *Elliptic Modular Functions*, Springer-Verlag, New York - Heidelberg - Berlin, 1974.
50. J.P. Serre, *Cours d'Arithmétique*, Presse Universitaire de France, 1977.
51. G. Shimura, *Arithmetic Theory of Automorphic Functions*, Iwanami Shoten and Princeton 1971.
52. D. Spector, *Supersymmetry and the Möbius inversion function*, Commun. Math. Phys., Vol.127 (1990), 239–252.
53. P. Stevenhagen, *Hilbert's 12th problem, complex multiplication and Shimura reciprocity*, Advanced Studies in Pure Math. 30 (2001) "Class Field Theory – its centenary and prospect" pp. 161–176.
54. M. Takesaki, *Tomita's theory of modular Hilbert algebras and its applications.* Lecture Notes in Math., 28, Springer, 1970.
55. K. Tzanev, *Hecke C*-algebras and amenability.* To appear in JOT.
56. A. Weil, *Elliptic functions according to Eisenstein and Kronecker*, Springer 1976.
57. K.G. Wilson, *The renormalization group: critical phenomena and the Kondo problem*, Rev. Mod. Phys. Vol. 47 (1975) N.4, 773–840.

More Zeta Functions for the Riemann Zeros

André Voros[1]

CEA, Service de Physique Théorique de Saclay (CNRS URA 2306)
F-91191 Gif-sur-Yvette Cedex (France)
E-mail : voros@spht.saclay.cea.fr

Summary. Another family of generalized zeta functions built over the Riemann zeros $\{\rho\}$, namely $\mathscr{Z}(s,x) = \sum_\rho (x-\rho)^{-s}$, has its analytic properties and (countably many) special values listed in explicit detail.

This work is a partial expansion of our first paper [20] on zeta functions built over the *Riemann zeros* $\{\rho\}$, i.e., the nontrivial zeros of the Riemann zeta function $\zeta(s)$. While our oral presentation was more introductory, here we will pursue a fully parallel treatment, begun in [20], for *two* such generalized (i.e., parametric) zeta functions:

[1] Also at: Institut de Mathématiques de Jussieu–Chevaleret (CNRS UMR 7586), Université Paris 7, F-75251 Paris Cedex 05 (France).

$$\mathcal{Z}(\sigma,v) \stackrel{\text{def}}{=} \sum_{k=1}^{\infty} ({\tau_k}^2+v)^{-\sigma} \qquad \text{(and } \mathcal{Z}(\sigma) \stackrel{\text{def}}{=} \mathcal{Z}(\sigma,0)\text{)}, \quad (0.1)$$

$$\mathscr{Z}(s,x) \stackrel{\text{def}}{=} \sum_{\rho} (x-\rho)^{-s} \equiv \sum_{\rho} (\rho+x-1)^{-s} \quad \text{(and } \mathscr{Z}(s) \stackrel{\text{def}}{=} \mathscr{Z}(s,1)\text{)}, \quad (0.2)$$

$$\text{where } \{\rho\} = \{\tfrac{1}{2} \pm \mathrm{i}\tau_k\}_{k=1,2,\ldots} = \{\text{the Riemann zeros}\} \quad (0.3)$$

(or, in a latest extension, the zeros of arithmetic zeta or L-functions [21]).

The two families (0.1) and (0.2) are truly inequivalent except for one function,

$$\mathcal{Z}(\sigma,0) \equiv \mathcal{Z}(\sigma) \equiv (2\cos\pi\sigma)^{-1}\,\mathscr{Z}(2\sigma,\tfrac{1}{2}), \quad (0.4)$$

already considered in [8, Sect. 4 ex. (A)], [4]. Other previous results appear in [11, 15] for the functions $\mathcal{Z}(\sigma,\frac{1}{4})$, in [5, 18] for the family $\{\mathscr{Z}\}$ and earlier [16, 12, 10] for the specific sums $\mathscr{Z}(n) \equiv \sum_\rho \rho^{-n}$ (often denoted σ_n).

In [20], we mainly strived at exhausting explicit results for the family (0.1), handling the family (0.2) in lesser detail. Here we will pursue a fully parallel *explicit* description for the family (0.2), but now based on a *parametric* analytical-continuation formula, (2.1). At the same time we will switch from a Hadamard to a zeta-regularized product formalism, definitely simpler for the family (0.2). This zeta-regularization technique is adapted from spectral theory and quantum mechanics, where it serves to define *spectral* (or *functional*) *determinants* [19, 17]. However, our analysis remains wholly decoupled from any actual spectral meaning whatsoever for the Riemann zeros.

We recapitulate the results of [20] in Sect. 1, but refer to that article for further details. We basically keep the same notations, with (0.2) subsuming the main few changes: the second family used in [20] was $\xi(s,x) \equiv (2\pi)^s\,\mathscr{Z}(s,x)$, and $\mathscr{Z}(n) = \sum_\rho \rho^{-n}$ was formerly $\mathscr{Z}_n$; we also slightly renormalize the function called **D**, cf. (1.27) below. The other essential notations are [1, 7, 3, 6]:

$$\begin{Bmatrix} B_n \\ E_n \end{Bmatrix} : \begin{Bmatrix} \text{Bernoulli} \\ \text{Euler} \end{Bmatrix} \text{numbers;} \qquad B_n(\cdot)\text{: Bernoulli polynomials;}$$

$$\gamma\text{: Euler's constant;} \quad \gamma^{\mathrm{c}}_{n-1}\text{: "Stieltjes \textit{cumulants}", defined by:} \quad (0.5)$$

$$\log\left[s\,\zeta(1+s)\right] \equiv \sum_{n=1}^{\infty} \frac{(-1)^{n-1}}{(n-1)!}\,\gamma^{\mathrm{c}}_{n-1}\,s^n \quad (e.g.,\ \gamma^{\mathrm{c}}_0 = \gamma);$$

the $\gamma^{\mathrm{c}}_{n-1}$ are cumulants [20] for the more classic *Stieltjes constants* γ_{n-1} [1, 12]; see also $\eta_{n-1} \equiv (-1)^n n\,\gamma^{\mathrm{c}}_{n-1}/(n-1)!$ in [2] – notations are not standardized (the so denoted constants and cumulants all truly have degree n, anyway);

$$\Xi(s) \stackrel{\text{def}}{=} s(s-1)\pi^{-s/2}\,\Gamma(s/2)\,\zeta(s), \quad (0.6)$$

which is an entire function, even under $s \longleftrightarrow (1-s)$, normalized to $\Xi(0) = \Xi(1) = 1$, and only keeping the nontrivial zeros of $\zeta(s)$;

$$\beta(s) \stackrel{\text{def}}{=} \sum_{n=0}^{\infty}(-1)^n(2n+1)^{-s} : \text{the Dirichlet } \beta\text{-function}, \tag{0.7}$$

which is a particular L-series of period 4;

$$\zeta(s,a) \stackrel{\text{def}}{=} \sum_{n=0}^{\infty}(n+a)^{-s} : \text{the Hurwitz zeta function}, \tag{0.8}$$

which has a single pole at $s=1$, of polar part $1/(s-1)$, and the special values

$$\zeta(-m,a) = -B_{m+1}(a)/(m+1) \quad (m \in \mathbb{N}), \quad (e.g., \ \zeta(0,a) = \tfrac{1}{2} - a) \tag{0.9}$$

$$\mathrm{FP}_{s=1}\,\zeta(s,a) = -\Gamma'(a)/\Gamma(a) \qquad (\mathrm{FP} \stackrel{\text{def}}{=} \text{finite part at a pole}) \tag{0.10}$$

$$\zeta'(0,a) = \log\left[\Gamma(a)/(2\pi)^{1/2}\right]; \tag{0.11}$$

upon generalized zeta functions as in (0.1), (0.2), (0.11), $'$ will always mean differentiation with respect to the principal variable: the exponent, s or σ.

1 Summary of Previous Results

1.1 Zeta Functions and Zeta-Regularized Products

We first recall some needed results on zeta and infinite-product functions built over certain abstract numerical sequences $\{x_k\}_{k=1,2,\ldots}$ ($0 < x_1 \leq x_2 \leq \cdots$, $x_k \uparrow +\infty$ as in [19]; or $x_k \in \mathbb{C}^*$ with $|x_k| \uparrow \infty$, $|\arg x_k|$ sufficiently bounded as in [17, 14, 9]). Such a sequence is deemed *admissible of order* μ_0 for some $\mu_0 \in (0, +\infty)$ if, essentially, the series

$$Z(s) \stackrel{\text{def}}{=} \sum_{k=1}^{\infty} {x_k}^{-s} \quad \textit{converges in } \{\mathrm{Re}\ s > \mu_0\}, \tag{1.1}$$

and this zeta function $Z(s)$ (analytic for Re $s > \mu_0$) admits a *meromorphic extension to the whole s-plane*, with poles lying in a real sequence $\mu_0 > \mu_1 > \cdots$ ($\mu_n \downarrow -\infty$). The smaller details are better fine-tuned to each context: thus, the zeta functions $\mathcal{Z}$ in (0.1) could be treated earlier [20] using a very low order $\mu_0 < 1$ but *double* poles, which are handled in [14, 9]; now, the functions $\mathscr{Z}$ in (0.2) will require $\mu_0 = 1$ but only *simple* poles, as in [19, 17].

As a consequence of (1.1), the Weierstrass infinite product

$$\Delta(x) \stackrel{\text{def}}{=} \prod_{k=1}^{\infty}\left(1 + \frac{x}{x_k}\right) \exp\left\{\sum_{1 \leq m \leq \mu_0} \frac{1}{m}\left(-\frac{x}{x_k}\right)^m\right\} \tag{1.2}$$

converges $\forall x \in \mathbb{C}$, to an entire function. In the context of the Riemann zeros, the above meromorphic continuation requirements for $Z(s)$ are more easily enforced through a *controlled large-x behavior of* $\log \Delta(x)$ [20]; here we impose

$$\log \Delta(x) \sim \sum_{n=0}^{\infty} (\tilde{a}_{\mu_n} \log x + a_{\mu_n}) x^{\mu_n} \qquad (x \to \infty,\ |\arg x| < \theta) \tag{1.3}$$

uniformly in x for some $\theta > 0$, with $\tilde{a}_{\mu_n} \neq 0$ only for the (finitely many) $\mu_n \in \mathbb{N}$ [19]: this will fit the family $\{\mathscr{Z}\}$, which only features simple poles (any $\mu_n \notin \mathbb{N}$ with $\tilde{a}_{\mu_n} \neq 0$ would give a *double* pole).

At the same time, $\log \Delta(x)$ has a specially simple *Taylor series at* $x = 0$:

$$-\log \Delta(x) = \sum_{m > \mu_0} \frac{Z(m)}{m} (-x)^m \quad (\text{converging for } |x| < \inf_k |x_k|) \tag{1.4}$$

$$= O(|x|^{m_0}) \quad \text{for } m_0 \stackrel{\text{def}}{=} \text{the least integer} > \mu_0.$$

The latter bound and (1.3) allow these Mellin representations for $Z(s)$:

$$\frac{\pi}{s \sin \pi s} Z(s) = \int_0^\infty \log \Delta(y)\, y^{-s-1} \mathrm{d}y \qquad (\mu_0 < \operatorname{Re} s < m_0) \tag{1.5}$$

$$\equiv \cdots \equiv \frac{(-1)^{m_0}\, \Gamma(-s)}{\Gamma(m_0 - s)} \int_0^\infty (\log \Delta)^{(m_0)}(y)\, y^{m_0 - s - 1} \mathrm{d}y. \tag{1.6}$$

[Proof: equations (1.5) and (1.6) are equivalent through integrations by parts; now to verify (1.6), expand $(\log \Delta)^{(m_0)}(y) = (-1)^{m_0 - 1}(m_0 - 1)! \sum_k (y + x_k)^{-m_0}$ and integrate term by term.] Then, repeated integrations by parts, as in [20, Sect. 2.2 and App. A] but pushed further, likewise imply that $Z(s)$ is meromorphic in $\mathbb{C}$, with poles lying in the sequence $\{\mu_n\}$ and polar parts

$$Z(\mu_n + \varepsilon) = \mu_n \left[\pi^{-1} \sin \pi\mu_n\, a_{\mu_n} + \cos \pi\mu_n\, \tilde{a}_{\mu_n}\right] \varepsilon^{-1} + O(1)_{\varepsilon \to 0}\,, \tag{1.7}$$

by specializing formula (23) in [20]. Thus for $Z(s)$, all the poles are *simple*, and $s = 0$ is a *regular* point (as well as all points $s \in -\mathbb{N}$).

All previous results transfer to shifted admissible sequences $\{x + x_k\}$ up to reasonable limitations on the shift parameter x (e.g., $(x + x_k) \notin \mathbb{R}^-\ \forall k$), and hence to the *generalized zeta function* $Z(s, x) \stackrel{\text{def}}{=} \sum_k (x + x_k)^{-s}$. Then, the *zeta-regularized product* $D(x)$ (formally "$\prod_k (x + x_k)$") can be defined as

$$D(x) \stackrel{\text{def}}{=} \exp\left[-Z'(0, x)\right] \qquad (\text{recalling that } ' \equiv \partial/\partial s, \text{ as in (0.11)}). \tag{1.8}$$

It can also be uniquely characterized in several concrete ways [19]. On the one hand, it relates to $\Delta(x)$ through a definite multiplicative factor, trivial in the sense that $D(x)$ stays entire and keeps the same zeros (and order) as $\Delta(x)$:

$$D(x) \equiv \exp\Big[-Z'(0) - \sum_{1 \le m \le \mu_0} \frac{Z_m}{m} (-x)^m\Big]\, \Delta(x), \tag{1.9}$$

$$\text{with} \quad Z_1 = \mathrm{FP}_{s=1} Z(s) \quad (\text{finite part}) \tag{1.10}$$

$$\text{and} \quad Z_m = Z(m) \qquad \text{if } Z(s) \text{ is regular at } m,$$

otherwise Z_m $(m \geq 2)$ is more contrived [19, eq. (4.12)] but unneeded when $\mu_0 = 1$; in which case (1.4), (1.9) and (1.10) finally simplify to

$$-\log D(x) \equiv Z'(0) - [\mathrm{FP}_{s=1} Z(s)]\, x - \log \Delta(x) \tag{1.11}$$

$$= Z'(0) - [\mathrm{FP}_{s=1} Z(s)]\, x + \sum_{m=2}^{\infty} \frac{Z(m)}{m} (-x)^m \quad (|x| < \inf_k |x_k|). \tag{1.12}$$

On the other hand, $\log D(x)$ has a characteristic large-x asymptotic behavior as well: a *generalized Stirling expansion*, of a very specific or "canonical" form,

$$-\log D(x) \sim \sum_{n=0}^{\infty} \hat{a}_{\mu_n} \{x^{\mu_n}\} \qquad (x \to +\infty), \tag{1.13}$$

$$\text{where} \quad \{x^{\mu_n}\} = x^{\mu_n} \qquad \text{for } \mu_n \notin \mathbb{N}$$
$$\{x^{\mu_n}\} = x^{\mu_n}(\log x - C_{\mu_n}) \quad \text{for } \mu_n \in \mathbb{N}, \quad C_0 = 0,\ C_1 = 1$$

(higher C_m [19, eq. (5.1)] are again unneeded when $\mu_0 = 1$); conversely, the constrained form of expansion (1.3) for $\log \Delta(x)$ is implied by (1.9) and (1.13).

A basic feature of the zeta-regularized product prescription is, by construction, its full *invariance under pure translations* $\{x_k\} \mapsto \{x_k + y\}$ (but under no other change of variables in general). As an application, we now express $Z(s, x)$ as a Mellin transform over $\log D$. First, for integer $m > \mu_0$, the formulae (1.4), (1.9), (1.10) shifted by y yield

$$Z(m, y) = \sum_k (y + x_k)^{-m} \equiv -\frac{1}{(m-1)!} \left(-\frac{\mathrm{d}}{\mathrm{d}y}\right)^m \log D(y); \tag{1.14}$$

whereas for $m = 1$, they yield the *finite part* value

$$\mathrm{FP}_{s=1} Z(s, y) = (\log D)'(y). \tag{1.15}$$

Then, since $(\log D)^{(m)} \equiv (\log \Delta)^{(m)}$ for $m > \mu_0$ by (1.9), it follows that (1.14) can be substituted into (1.6) shifted by x, giving

$$Z(s, x) = \frac{(m_0 - 1)!}{\Gamma(s)\, \Gamma(m_0 - s)} \int_0^{\infty} Z(m_0, x + y)\, y^{m_0 - s - 1} \mathrm{d}y \qquad (\mu_0 < \mathrm{Re}\, s < m_0), \tag{1.16}$$

Remarks: – (1.16) actually defines an extension of (1.14) to $m \equiv s$ no longer an integer; – the rightmost pole of $Z(s, x)$ remains $s = \mu_0$ for any x.

The above results will be invoked later for $\mu_0 = 1$, hence $m_0 = 2$; except that we will actually need a formula analogous to (1.16) but *for some* $\mathrm{Re}\, s < 1$: this just requires a couple of integrations by parts upon (1.16), as

$$Z(s,x) = \frac{\sin \pi s}{\pi(1-s)} \int_0^\infty Z(2, x+y)\, y^{1-s} \mathrm{d}y \qquad (1 < \mathrm{Re}\ s < 2) \quad (1.17)$$

$$= -\frac{\sin \pi s}{\pi(1-s)^2} \int_0^\infty \frac{\mathrm{d}}{\mathrm{d}y}[yZ(2,x+y)] y^{1-s} \mathrm{d}y \quad (0 < \mathrm{Re}\ s < 2) \quad (1.18)$$

$$= \frac{\sin \pi s}{\pi(1-s)} \int_0^\infty \tilde{Z}_x(2,y)\, y^{1-s} \mathrm{d}y \qquad (0 < \mathrm{Re}\ s < 1), \quad (1.19)$$

$$\text{with} \quad y\tilde{Z}_x(2,y) \stackrel{\text{def}}{=} yZ(2, x+y) + \tilde{a}_1 \qquad (\text{vanishing at } y = +\infty). \quad (1.20)$$

1.2 The Family $\{\mathcal{Z}(\sigma, v)\}$

We make a digression to recall earlier formulae for that family [20]. The main primary result was the integral representation (72) therein for $\mathcal{Z}(\sigma) \equiv \mathcal{Z}(\sigma, 0)$,

$$\mathcal{Z}(\sigma) = \frac{-\mathbf{Z}(2\sigma) + 2^{2\sigma} \mathrm{e}^{\mp 2\pi \mathrm{i}\sigma}}{2 \cos \pi\sigma} + \frac{\sin \pi\sigma}{\pi} \int_0^{+\mathrm{e}^{\pm \mathrm{i}\varepsilon}\infty} t^{-2\sigma} \frac{\zeta'}{\zeta}\left(\tfrac{1}{2} + t\right) \mathrm{d}t, \quad (1.21)$$

$$\mathbf{Z}(2\sigma) \stackrel{\text{def}}{=} \mathbf{Z}(2\sigma, \tfrac{1}{2}), \qquad \text{where} \qquad (1.22)$$

$$\mathbf{Z}(s,x) \stackrel{\text{def}}{=} \sum_{k=1}^{\infty} (x + 2k)^{-s} \equiv 2^{-s} \zeta(s, 1 + x/2) \qquad (1.23)$$

(admitting real variants [20, eqs. (73)–(74)]); $\mathbf{Z}(2\sigma)$ is the "shadow" zeta function of $\mathcal{Z}(\sigma)$, i.e., the precise counterpart of $\mathcal{Z}(\sigma)$ for the *trivial* zeros of $\zeta(s)$; the more general form $\mathbf{Z}(s,x)$ will enter in Sect. 2, and (1.23) writes it as a variant of the Hurwitz zeta function (0.8).

As shown in [20], (1.21) supplies an explicit analytical continuation of $\mathcal{Z}(\sigma)$ to a meromorphic function in the whole complex σ-plane, plus exhaustive explicit results and special values for $\mathcal{Z}(\sigma)$. Their extension to the full family $\{\mathcal{Z}(\sigma, v)\}$ then follows using the expansion formula (91) in [20], i.e.,

$$\mathcal{Z}(\sigma, v) = \sum_{\ell=0}^{\infty} \frac{\Gamma(1-\sigma)}{\ell!\, \Gamma(1-\sigma-\ell)} \mathcal{Z}(\sigma + \ell)\, v^\ell \qquad (|v| < \tau_1{}^2). \qquad (1.24)$$

The following explicit formulae for $\{\mathcal{Z}(\sigma, v)\}$ resulted [20].

a) the full polar parts (of order 2):

$$\mathcal{Z}(\tfrac{1}{2} - n + \varepsilon, v) = \frac{1}{8\pi} \frac{\Gamma(n + \frac{1}{2})}{n!\, \Gamma(\frac{1}{2})} v^n \varepsilon^{-2} + \mathcal{R}_n(v)\, \varepsilon^{-1} + O(1)_{\varepsilon \to 0} \quad \text{for } n \in \mathbb{N},$$

$$\text{with} \quad \mathcal{R}_n(v) = -\frac{\Gamma(n+1/2)}{n!\, \Gamma(1/2)} \left[\frac{1}{4\pi} \sum_{j=1}^{n} \frac{1}{2j-1} + \frac{\log 2\pi}{4\pi}\right] v^n \qquad (1.25)$$

$$+ \sum_{j=1}^{n} \frac{\Gamma(n+1/2)}{(n-j)!\, \Gamma(j+1/2)} \left[\frac{(-1)^j}{8\pi j} (1 - 2^{1-2j}) B_{2j}\right] v^{n-j};$$

Table 1. Special values of $\mathcal{Z}(\sigma, v)$ (upper half: algebraic, lower half: transcendental [20, Sect. 4]). Notations: see (0.5)–(0.6); m is an integer.

σ	$\mathcal{Z}(\sigma, v) = \sum_{k=1}^{\infty} ({\tau_k}^2 + v)^{-\sigma}$
$-m \le 0$	$\sum_{j=0}^{m} \binom{m}{j} (-1)^j \, 2^{-2j} (1 - \frac{1}{8} E_{2j}) \, v^{m-j}$
0	$7/8$
derivative at 0	$\mathcal{Z}'(0, v) = \frac{1}{4} \log 8\pi - \log \Xi(\frac{1}{2} \pm v^{1/2})$
$+m \ge 1$	$\dfrac{(-1)^{m-1}}{(m-1)!} \dfrac{\mathrm{d}^m}{\mathrm{d}v^m} \log \Xi(\frac{1}{2} \pm v^{1/2})$

b) special values at integer σ, compiled in Table 1; these evaluations can still be pushed further ([20, Table 1] for $v = 0$ and $\frac{1}{4}$; [21, Table 2] for general v).

Finally, as an extra result (useful for comparison with (2.4) below), we now recast the Hadamard product for the Riemann zeta function,

$$\zeta(x) = \frac{\exp(\log 2\pi - 1 - \gamma/2)x}{2(x-1)\,\Gamma(1 + x/2)} \prod_{\rho} (1 - x/\rho) \mathrm{e}^{x/\rho}, \tag{1.26}$$

in terms of *zeta-regularized* factors related to $\mathcal{Z}(\sigma, v)$.

First, the zeta-regularized product underlying the Gamma factor for the trivial zeros, $\Gamma(1+x/2)^{-1}$, i.e. the spectral determinant $\mathbf{D}(x)$ for the sequence $\{2k\}_{k=1,2,\ldots}$, can be specified using (1.8), (1.23), (0.9) and (0.11), as

$$\mathbf{D}(x) = \exp[-\mathbf{Z}'(0, x)] = 2^{-x/2} \pi^{1/2} / \,\Gamma(1 + x/2) \tag{1.27}$$

(warning: the determinant called $\mathbf{D}$ in [20] was normalized differently). Check: $\log \mathbf{D}(x)$ has a large-x asymptotic behavior of the *canonical* form (1.13) for the order $\mu_0 = 1$ (this also being the order of the entire function $\Gamma(1 + x/2)^{-1}$),

$$\log \mathbf{D}(x) \sim -\tfrac{1}{2} x (\log x - 1) - \tfrac{1}{2} \log x \ \left[+ \sum_{1}^{\infty} c_n x^{-n} \right]. \tag{1.28}$$

The other factor in (1.26) essentially contains the function $\Xi(x)$ of (0.6): it can be related to the zeta-regularized product $\mathcal{D}(v)$ for the sequence $\{{\tau_k}^2\}$, which is admissible of order $\mu_0 = \frac{1}{2}$ [20], through (cf. Table 1)

$$\mathcal{D}(v) = \exp[-\,\mathcal{Z}'(0, v)] = (8\pi)^{-1/4} \,\Xi(\tfrac{1}{2} + v^{1/2}). \tag{1.29}$$

The factorization formula (1.26) thus admits a zeta-regularized form as

$$\zeta(\tfrac{1}{2} + t) = (2\pi)^{t/2} \frac{\mathbf{D}(\frac{1}{2} + t) \mathcal{D}(t^2)}{t - \frac{1}{2}} \,. \tag{1.30}$$

This is quite analogous to an earlier decomposition of hyperbolic *Selberg zeta functions* over spectral determinants [19, eq. (7.18)]. In (1.30), the denominator also has the zeta-regularized normalization for an elementary factor; as for the prefactor $(2\pi)^{t/2}$, it corrects for the discrepancy between the zeta-regularizations with respect to t (as in $\mathbf{D}$) and t^2 (in $\mathcal{D}$).

2 The Family $\{\mathscr{Z}(s,x)\}$: Analytical Continuation Formula

Apart from the special values $\mathscr{Z}(n)$, $n \in \mathbb{N}^*$ [16, 12, 10], functions equivalent to $\{\mathscr{Z}(s,x)\}$ of (0.2) were considered first (to our knowledge) by Deninger for $\mathrm{Re}\, x > 1$ [5], then proved by Schröter and Soulé [18] to be meromorphic in the whole s-plane over the larger domain $x \in \Omega \stackrel{\mathrm{def}}{=} \{x \in \mathbb{C} \mid (x+\rho) \notin \mathbb{R}^- \ (\forall \rho)\}$.

2.1 The Primary Result

For the family (0.2), the "shadow" zeta function over the trivial zeros (definable just as before, but now with x as second argument) is just the function $\mathbf{Z}(s,x)$ of (1.23). It governs an integral representation for $\mathscr{Z}(s,x)$ similar to (1.21), but simpler and *now available for all* $x \in \Omega \setminus (-\infty, +1]$:

$$\mathscr{Z}(s,x) = -\mathbf{Z}(s,x) + \frac{1}{(x-1)^s} + \frac{\sin \pi s}{\pi} \int_0^\infty \frac{\zeta'}{\zeta}(x+y)\, y^{-s} \mathrm{d}y \qquad (\mathrm{Re}\, s < 1); \tag{2.1}$$

here, $(x-1)^s$ is given its standard determination in $\mathbb{C} \setminus (-\infty, +1]$; *this* cut is not a singularity for $\mathscr{Z}(s,x)$, indeed the discontinuities across it of the three right-hand side terms in (2.1) can be seen to precisely cancel out when added.

Alternative real forms can be built; a very simple one for real $x > -2$ is

$$\mathscr{Z}(s,x) = -\mathbf{Z}(s,x) + \frac{\sin \pi s}{\pi} \int_0^\infty \left[\frac{\zeta'}{\zeta}(x+y) + \frac{1}{x+y-1}\right] y^{-s} \mathrm{d}y \ (0 < \mathrm{Re}\, s < 1); \tag{2.2}$$

this form only converges in the stated s-plane strip, but contrary to (2.1), it enjoys a well defined $x \to +1$ limit:

$$\mathscr{Z}(s)\ (\equiv \mathscr{Z}(s,1)) = 1 - (1-2^{-s})\,\zeta(s) + \frac{\sin \pi s}{\pi} \int_0^\infty \left[\frac{\zeta'}{\zeta}(1+y) + \frac{1}{y}\right] y^{-s} \mathrm{d}y. \tag{2.3}$$

Equation (2.1) (plus (2.2) for x real) is the new basic result here, extending an earlier formula by Deninger valid only for $\mathrm{Re}\, x > 1$ [5, p. 149]. It is a genuine analog for $\mathscr{Z}(s,x)$ to the Joncquière–Lerch functional relation for $\zeta(s,a)$ [7, Sect. 1.11 (16)], itself generalizing the functional equation of $\zeta(s)$. At $x = \frac{1}{2}$, (2.1) also restores our previous formula (1.21) for $\mathcal{Z}(\sigma)$ by virtue of the relation (0.4). Every explicit consequence that (1.21) implied for $\mathcal{Z}(\sigma)$ alone will extend here *to the whole family* $\mathscr{Z}(s,x)$ solely by (2.1).

2.2 Derivation of the Main Formula (2.1)

As a preliminary step, we transform the Hadamard product (1.26) for $\zeta(s)$ into a zeta-regularized factorization even simpler than (1.30).

We now just factor out the previous "shadow" determinant $\mathbf{D}(x)$, as

$$\zeta(x) \equiv \frac{\mathbf{D}(x)\mathscr{D}(x)}{x-1}, \tag{2.4}$$

$$\mathscr{D}(x) = (x-1)\, 2^{x/2}\pi^{-1/2}\,\Gamma(1+x/2)\zeta(x) \equiv \tfrac{1}{2}\pi^{-1/2}(2\pi)^{x/2}\Xi(x). \tag{2.5}$$

We can then anticipate that this factor $\mathscr{D}(x)$ must be a *zeta-regularized product in* x over the Riemann zeros $\{\rho\}$. Indeed, $\mathscr{D}(x)$ has precisely the $\{\rho\}$ as zeros, *and* $\log \mathscr{D}(x)$ has a large-x expansion of the canonical form (1.13) in x because all other factors present in (2.4) have that property. We will accordingly confirm that $\log \mathscr{D}(x) \equiv -\mathscr{Z}'(0,x)$ below: see (3.5), and earlier in a variant form, [5, thm 3.3] (for $\operatorname{Re} x > 1$) and [18] (for general x).

Now to prove (2.1), we specialize the results of Sect. 1.1 to the sequences $\{-\rho\}$ and $\{2k\}$: both of these *are admissible* of order $\mu_0 = 1$ [5, 18, 9], mainly because $\log \mathbf{D}(x)$, and hence $\log \mathscr{D}(x)$, comply with (1.3) (cf. (1.28), then (2.4)).

Specifically here, the factorization (2.4), together with (1.14) at $m = 2$ (first with $Z = \mathscr{Z}$, then with $Z = \mathbf{Z}$) and with (1.20), entail

$$\tilde{\mathscr{Z}}_x(2,y) \equiv -\tilde{\mathbf{Z}}_x(2,y) + (x+y-1)^{-2} - [\zeta'/\zeta]'(x+y) \tag{2.6}$$

$$\text{with} \quad \tilde{\mathscr{Z}}_x(2,y) \equiv \mathscr{Z}(2,x+y) + \frac{1}{2y}, \quad \tilde{\mathbf{Z}}_x(2,y) \equiv \mathbf{Z}(2,x+y) - \frac{1}{2y}; \tag{2.7}$$

the last line comes from generalized Stirling expansions for $\mathscr{D}$ and $\mathbf{D}$, cf. (1.28). Upon the specific decomposition (2.6), it is allowed to apply the Mellin transformation (1.19) term by term on both sides, at fixed $x \in \Omega \setminus (-\infty, +1]$. Then, the left-hand side yields $\mathscr{Z}(s,x)$; as for the right-hand side, the first term yields $-\mathbf{Z}(s,x)$ by exactly the same argument, the second term trivially evaluates to $(x-1)^{-s}$, and the last term can be subjected to an ultimate integration by parts now valid in the whole half-plane $\{\operatorname{Re} s < 1\}$, using

$$\mathscr{I}_\zeta(s,x) \stackrel{\text{def}}{=} \frac{1}{1-s}\int_0^\infty -\left[\frac{\zeta'}{\zeta}\right]'(x+y)\, y^{1-s}\mathrm{d}y = \int_0^\infty \frac{\zeta'}{\zeta}(x+y)\, y^{-s}\mathrm{d}y; \tag{2.8}$$

all that yields the desired formula (2.1). If the last two terms in (2.6) are kept together instead, (2.2) can be obtained likewise. The structure of the representation (2.1) thus clearly stems from the simple factorization formula (2.4).

3 Explicit Consequences for the Family $\{\mathscr{Z}(s,x)\}$

3.1 Analytical Results (in the s-Variable)

First, (2.1) gives an explicit one-step analytical continuation of $\mathscr{Z}(s,x)$ to the half-plane $\{\operatorname{Re} s < 1\}$. It also implies its analytical continuation in s to all of $\mathbb{C} \setminus \{1\}$, since the Mellin transform $\mathscr{I}_\zeta(s,x)$ of (2.8) is seen (through repeated integrations by parts, using $[\log\zeta]^{(n)}(x) = o(x^{-N})_{x\to+\infty}\ \forall n,\, N$) to be meromorphic in the whole s-plane, and to have only simple poles at $s = 1, 2, \ldots$ with residues

$$\operatorname{Res}_{s=n} \mathscr{I}_\zeta(s,x) = -[\log|\zeta|]^{(n)}(x)/(n-1)! \qquad (x \neq 1), \quad n = 1, 2, \ldots \tag{3.1}$$

(the singularity at $x = 1$ is harmless: see after (3.9), and left part of Table 3). At fixed x, (2.1) and (3.1) imply that $\mathscr{Z}(s,x)$ acquires its polar structure solely from $-\mathbf{Z}(s,x)$: it thus has the only pole $s = 1$, of polar part $-\frac{1}{2}/(s-1)$ [18].

Still for fixed x, the mere substitution into (2.1) of the classic Dirichlet series

$$\frac{\zeta'}{\zeta}(z) = -\sum_{n\geq 2} \frac{\Lambda(n)}{n^z} \quad (\Lambda(n) \stackrel{\text{def}}{=} \log p \text{ if } n = p^r \text{ for some prime } p, \text{ else } 0) \tag{3.2}$$

for $z = x + y$, followed by term-by-term y-integration, yields

$$\mathscr{Z}(s,x) + \mathbf{Z}(s,x) - (x-1)^{-s} \sim -\frac{1}{\Gamma(s)} \sum_{n\geq 2} \frac{\Lambda(n)}{n^x} (\log n)^{s-1}. \tag{3.3}$$

The summation in the right-hand side of (3.3) converges iff the Dirichlet series (3.2) converges uniformly for $y > 0$: i.e., for $\operatorname{Re} x > 1$, where (3.3) becomes an *identity* – written in [5, p. 148], but it is just a particular case of *Weil's explicit formula*, or equivalently of equation (1.1) in [8], again provided $\operatorname{Re} x > 1$. Here, by contrast, (3.3) is meant for *general* fixed x, albeit only as an *asymptotic expansion* (for $s \to -\infty$) if $\operatorname{Re} x \leq 1$: in *this* range (e.g., at the most interesting points $x = 1$ and $\frac{1}{2}$) Weil's explicit formula *breaks down*, so life is harder.

3.2 Special Values for General x

Finally, (2.1) outputs all the special values of $\mathscr{Z}(s,x)$ just by inspection:

$$\mathscr{Z}(-n,x) = -2^n \zeta(-n, 1 + x/2) + (x-1)^n \qquad (n \in \mathbb{N}), \tag{3.4}$$

$$\begin{aligned} \mathscr{Z}'(0,x) &= -\tfrac{1}{2}(\log 2)x + \tfrac{1}{2}\log\pi - \log\Gamma(1+x/2) - \log[(x-1)\zeta(x)] \\ &\equiv -\log\mathscr{D}(x), \end{aligned} \tag{3.5}$$

$$\operatorname{FP}_{s=1} \mathscr{Z}(s,x) = \frac{1}{2}\left(\log 2 + \frac{\Gamma'}{\Gamma}(1+x/2)\right) + \left[\frac{1}{x-1} + \frac{\zeta'}{\zeta}(x)\right] \tag{3.6}$$

$$\equiv (\log\mathscr{D})'(x), \tag{3.7}$$

$$\begin{aligned} \mathscr{Z}(+n,x) &= -2^{-n}\zeta(n, 1+x/2) \\ &\quad + \left[(x-1)^{-n} - \frac{(-1)^n}{(n-1)!}[\log|\zeta|]^{(n)}(x)\right] \quad (n = 2, 3, \ldots) \end{aligned} \tag{3.8}$$

$$\equiv \frac{(-1)^{n-1}}{(n-1)!}(\log\mathscr{D})^{(n)}(x) \qquad (n = 2, 3, \ldots) \tag{3.9}$$

using (0.9)–(0.11), (1.23), (2.5), (3.1); the quantities in square brackets are apparently singular for $x = +1$ but globally extend there by continuity, using the expansion (0.5) with the Stieltjes cumulants.

In particular, (3.5) confirms that the factor $\mathscr{D}(x)$ in (2.4) *is the zeta-regularized product* in x over the sequence of Riemann zeros $\{\rho\}$ – the argument is not circular, because our derivation of the basic formula (2.1) does not rely on that fact but purely on the factorization itself. Thereupon, (3.7), (3.9) simply repeat the general formulae (1.15), (1.14) respectively.

The point $s = 1$ ($= \mu_0$) deserves extra attention. Upon logarithmic differentiation, one Hadamard product formula for $\Xi(x)$ [6, Sects. 1.10, 2.8] directly yields

$$\Xi(x) = \prod_\rho (1 - x/\rho) \quad \Longrightarrow \quad \mathscr{Z}(1, x) \stackrel{\text{def}}{=} \sum_\rho (x - \rho)^{-1} \equiv (\log \Xi)'(x), \quad (3.10)$$

where both product and sum (now only *semiconvergent*) are performed with zeros grouped in symmetrical pairs, as usual. Thus, *in spite of the pole of* $\mathscr{Z}(s, x)$ at $s = 1$, (3.10) yields a *finite* value for $\mathscr{Z}(1, x)$, which however *differs from the finite part* (FP) *of* $\mathscr{Z}(s, x)$ *at* $s = 1$:

$$\mathscr{Z}(1, x) - \mathrm{FP}_{s=1}\, \mathscr{Z}(s, x) \equiv -\tfrac{1}{2} \log 2\pi \quad (3.11)$$

according to (2.5), (3.7). This fixed discrepancy can also be traced to the *nonzero residue of the double pole* in the former zeta function $\mathcal{Z}(\sigma)$, see (3.20) below.

The resulting special values for $\{\mathscr{Z}(s, x)\}$ are fully compiled in Table 2, in their form closest to their analogs for the family $\{\mathcal{Z}(\sigma, v)\}$ in Table 1.

Finally, we state several sets of linear identities imposed upon the values $\mathscr{Z}(n, x)$ purely by the symmetry $(\rho \longleftrightarrow 1 - \rho)$ in (0.2). First:

Table 2. Special values of $\mathscr{Z}(s, x)$ (upper part: algebraic, lower part: transcendental [20, Sect. 4]); see also (3.4)–(3.13′). Notations: see (0.6); $B_{n+1}(\cdot)$: Bernoulli polynomial; n is an integer.

s	$\mathscr{Z}(s, x) = \sum_\rho (x - \rho)^{-s}$
$-n \leq 0$	$2^n B_{n+1}(1 + \frac{x}{2}) \,/\, (n + 1) \,+\, (x - 1)^n$
0	$\frac{1}{2}(x + 3)$
derivative at 0	$\mathscr{Z}'(0, x) = -\frac{1}{2}(\log 2\pi)\, x + \frac{1}{2}(\log 4\pi) - \log \Xi(x)$
finite part at +1	$\mathrm{FP}_{s=1}\, \mathscr{Z}(s, x) = \frac{1}{2} \log 2\pi + (\log \Xi)'(x)$
$+n \geq 1$	$\dfrac{(-1)^{n-1}}{(n-1)!} (\log \Xi)^{(n)}(x)$

$$\mathscr{Z}(n,x) = (-1)^n\, \mathscr{Z}(n,1-x) \qquad \text{for } n=1,2,\ldots. \tag{3.12}$$

Then, these "sum rules" previously known for $x=1$ only [10, eq. (18)]:

$$\mathscr{Z}(k,x) = -\frac{1}{2}\sum_{\ell=k+1}^{\infty}\binom{\ell-1}{k-1}(2x-1)^{\ell-k}\,\mathscr{Z}(\ell,x) \qquad \text{for each } odd\ k\geq 1 \tag{3.13}$$

which actually admit this *more convergent* variant (*unreported elsewhere*),

$$\sum_{\ell=k}^{\infty}\binom{\ell-1}{k-1}(x-\tfrac{1}{2})^{\ell-k}\,\mathscr{Z}(\ell,x) = 0 \qquad \text{for each } odd\ k\geq 1. \tag{3.13$'$}$$

The sum rules (3.13) merely result from the Taylor expansion around $x=\frac{1}{2}$ of $(x-1+\rho)^{-k} = (-1)^k(x-\rho)^{-k}\left[1-(2x-1)/(x-\rho)\right]^{-k}$ followed by summation over the zeros ρ grouped in pairs. But the zeros' symmetry also readily implies

$$\sum_{\rho}(\tfrac{1}{2}-\rho)^{-k}\ \left(\equiv \mathscr{Z}(k,\tfrac{1}{2})\right) \equiv 0 \qquad \text{for each } odd\ k\geq 1, \tag{3.14}$$

and now the expansion of *this* as $0=\sum_\rho (x-\rho)^{-k}\left[1-(x-\frac{1}{2})/(x-\rho)\right]^{-k}$ yields (3.13′). Either set of sum rules recursively allows to eliminate any *finite* subset of odd-k values, in terms of all higher-ℓ values. (In the infinite recursion limit, every odd-k value would end up eliminated in terms of the higher even-ℓ values only, formally as $\mathscr{Z}(2m+1,x) = \sum_{j=m+1}^{\infty} A_{m,j}\,\mathscr{Z}(2j,x)\,(x-\frac{1}{2})^{2j-(2m+1)}$, but *these* have to be *divergent* series (exercise!); only for $x=\frac{1}{2}$ do all odd-k values eliminate truly but trivially, according to (3.14).)

Also linked to the functional equation ($\Xi(s)\equiv\Xi(1-s)$) like (3.12)–(3.14), finite triangular linear relations connect the values $\mathscr{Z}(n,x)$ to the *other* special values $\mathcal{Z}(m,v)$, $m=1,2,\ldots$ at $v\equiv(x-\frac{1}{2})^2$, as derived in [21, Sect. 3.3]:

$$\mathcal{Z}(m,v) = \begin{cases} \displaystyle\sum_{\ell=0}^{m-1}\binom{m+\ell-1}{m-1}(2x-1)^{-m-\ell}\,\mathscr{Z}(m-\ell,x) & (v\neq 0)\\ \tfrac{1}{2}(-1)^m\,\mathscr{Z}(2m,\tfrac{1}{2}) & (v=0)\end{cases}$$

$$\Longleftrightarrow\quad \frac{\mathscr{Z}(n,x)}{n} = \sum_{0\leq\ell\leq n/2}(-1)^\ell\binom{n-\ell}{\ell}(2x-1)^{n-2\ell}\,\frac{\mathcal{Z}(n-\ell,v)}{n-\ell}. \tag{3.15}$$

Remark: the derivation of (3.13) reminds of the v-expansion (1.24) which yielded the general-v properties of $\mathcal{Z}(\sigma,v)$; otherwise, the general-x description of $\mathscr{Z}(s,x)$ is better drawn from the continuation formula (2.1) alone.

3.3 Special Values for $x=1$ and $\frac{1}{2}$

For half-integer x, the values $\zeta(\pm m, 1+x/2)$ which arose in (3.4) and (3.8) can be made slightly more explicit, and even more so for integer x. The most interesting cases are $x=1$ and $\frac{1}{2}$: then, (1.23) implies

$$\mathbf{Z}(s,1) \equiv (1-2^{-s})\,\zeta(s)-1, \qquad \mathbf{Z}(s,\tfrac{1}{2}) \equiv \tfrac{1}{2}\big[(2^s-1)\,\zeta(s)+2^s\beta(s)\big]-2^s, \tag{3.16}$$

and the resulting special values of $\mathscr{Z}(s,1) \equiv \mathscr{Z}(s)$ and $\mathscr{Z}(s,\frac{1}{2})$ form Table 3. They display many relations with the special values of $\mathcal{Z}(\sigma,\frac{1}{4})$ and $\mathcal{Z}(\sigma,0) \equiv \mathcal{Z}(\sigma)$ respectively [20, Table 1], as discussed next.

Table 3. Special values of the functions $\mathscr{Z}(s,x)$ for $x=1$ (see also (3.13′), (3.15), (3.22)), and for $x=\frac{1}{2}$ (see also (3.14), (3.17)–(3.21)). Notations: see (0.5), (0.7); n is an integer. In the bottom line, when n is even, $\zeta(n) \equiv (2\pi)^n|B_n|/(2\,n!)$ while $\beta(n)$ remains elusive.

s	$\mathscr{Z}(s) \equiv \sum_\rho \rho^{-s} \quad [x=1]$	$\mathscr{Z}(s,\frac{1}{2}) \equiv \sum_\rho (\rho-\frac{1}{2})^{-s} \quad [x=\frac{1}{2}]$
$-n<0$	$1-(2^n-1)\frac{B_{n+1}}{n+1}$	$\begin{cases} 2^{-n+1}(1-\frac{1}{8}E_n) & n \text{ even} \\ -\frac{1}{2}(1-2^{-n})\frac{B_{n+1}}{n+1} & n \text{ odd} \end{cases}$
0	2	7/4
derivative at 0	$\mathscr{Z}'(0)=\frac{1}{2}\log 2$	$\mathscr{Z}'(0,\frac{1}{2}) = \log\left[2^{11/4}\pi^{1/2}\,\Gamma(\frac{1}{4})^{-1}\lvert\zeta(\frac{1}{2})\rvert^{-1}\right]$
finite part at +1	$\mathrm{FP}_{s=1}\,\mathscr{Z}(s) = 1-\frac{1}{2}(\log 2-\gamma)$	$\mathrm{FP}_{s=1}\,\mathscr{Z}(s,\frac{1}{2}) = \frac{1}{2}\log 2\pi$
+1	$-\frac{1}{2}\log 4\pi + 1 + \frac{1}{2}\gamma$	0
$+n>1$	$\begin{Bmatrix} 1-(1-2^{-n})\,\zeta(n)+\frac{n}{(n-1)!}\,\gamma^{\mathrm{c}}_{n-1} \\ \equiv \\ 1-(-1)^n 2^{-n}\zeta(n)-\frac{(\log\lvert\zeta\rvert)^{(n)}(0)}{(n-1)!} \end{Bmatrix}$	$\begin{cases} \left.\begin{matrix} -\frac{1}{2}\left[(2^n-1)\,\zeta(n)+2^n\beta(n)\right] \\ +2^{n+1}-\frac{1}{(n-1)!}(\log\lvert\zeta\rvert)^{(n)}(\frac{1}{2}) \end{matrix}\right\} & n \text{ even} \\ 0 & n \text{ odd} \end{cases}$

In the case $x=\frac{1}{2}$ (right-hand column), there is a 1–1 correspondence between the latter explicit results and those for $\mathcal{Z}(\sigma)$ (Sect. 1.2, and [20]), through the relation (0.4):

$$\mathscr{Z}(2m,\tfrac{1}{2}) = 2(-1)^m\,\mathcal{Z}(m) \qquad \text{for } m \in \mathbb{Z} \tag{3.17}$$

$$\mathscr{Z}(1+2m,\tfrac{1}{2}) = (-1)^{m+1}2\pi\,\mathrm{Res}_{\sigma=\frac{1}{2}+m}\,\mathcal{Z}(\sigma) \qquad \text{for } m \in \mathbb{Z}^* \tag{3.18}$$

$$\mathrm{Res}_{s=1}\,\mathscr{Z}(s,\tfrac{1}{2}) = -4\pi\lim_{\varepsilon\to 0}[\varepsilon^2\mathcal{Z}(\tfrac{1}{2}+\varepsilon)] \qquad (=-\tfrac{1}{2}) \tag{3.19}$$

$$\mathrm{FP}_{s=1}\,\mathscr{Z}(s,\tfrac{1}{2}) = -2\pi\,\mathrm{Res}_{\sigma=\frac{1}{2}}\,\mathcal{Z}(\sigma) \qquad (=\tfrac{1}{2}\log 2\pi) \tag{3.20}$$

$$\mathscr{Z}'(0,\tfrac{1}{2}) = \mathcal{Z}'(0). \tag{3.21}$$

For $m=+1,+2,\ldots$, (3.18) restores (3.14) or $\mathscr{Z}(1+2m,\frac{1}{2}) \equiv 0$, just missing the first case $\mathscr{Z}(1,\frac{1}{2})=0$. The latter and (3.20) in turn imply, upon setting $x=\frac{1}{2}$ in the formula (3.11), that the constant discrepancy $[\mathscr{Z}(1,x)-\mathrm{FP}_{s=1}\,\mathscr{Z}(s,x)]$ relates to the *nonzero residue* in the double pole of $\mathcal{Z}(\sigma)$ at $\sigma=\frac{1}{2}$. The actual value $(-(4\pi)^{-1}\log 2\pi)$ of this residue [20], used in (3.20), follows from (1.25) taken at $n=0$, $v=0$.

In the case $x = 1$ (left-hand column), and with $n = 1, 2, \ldots$ henceforth, the special values $\mathscr{Z}(n)$ were already known: for $n = 1$, see [3, ch. 12], [6, Sect. 3.8]; for $n > 1$, we tabulate two equivalent expressions [16, 12, 20] and we refer to [12, Table 5] for numerical values. Furthermore, the $\mathscr{Z}(n)$ satisfy three sets of linear identities:
- the (infinite) sum rules (3.13) or better, (3.13′), specialized to $x = 1$ (remark: [10, eq. (18)] states the $x = 1$ sum rule (3.13) for even indices k as well, but these reduce to finite linear combinations of higher *odd-k* sum rules only);
- the (finite, triangular) relations (3.15) specialized to $x = 1$, $v = \frac{1}{4}$, which then connect the $\mathscr{Z}(n)$ to the other special values $\mathcal{Z}(m, \frac{1}{4})$ [15, 20];
- a similar connection to the sequence $\lambda_n \stackrel{\text{def}}{=} \sum_\rho [1 - (1 - 1/\rho)^n]$ used by *Li's criterion* for the Riemann Hypothesis (i.e., $\lambda_n > 0\ \forall n$ [13]) [10, eq. (27)] [2, thm 2]:

$$\lambda_n = \sum_{j=1}^{n} (-1)^{j+1} \binom{n}{j} \mathscr{Z}(j) \iff \mathscr{Z}(n) = \sum_{j=1}^{n} (-1)^{j+1} \binom{n}{j} \lambda_j . \qquad (3.22)$$

(Note: the λ_n of [13], used here, are n times the λ_n of [10].)

Aside from those $\mathscr{Z}(n)$, $n = 1, 2, \ldots$ and $\mathscr{Z}'(0, x)$ [5, eq. (3.3.1)] [18], the values in Table 3 seem new to us. Remark: the fully explicit $\mathscr{Z}'(0)$ yields the *zeta-regularized product of all the Riemann zeros*: "$\prod_\rho \rho$" $= \mathrm{e}^{-\mathscr{Z}'(0)} = 2^{-1/2}$.

3.4 Concluding Remark

Just as stated for the family (0.1) [20, Sect. 5.5], all of the foregoing analysis straightforwardly extends to zeros of other zeta and L-functions having functional equations similar enough to that of $\zeta(s)$ [21].

References

1. M. Abramowitz and I.A. Stegun, *Handbook of Mathematical Functions*, chap. 23, Dover, New York (1965).
2. E. Bombieri and J.C. Lagarias, *Complements to Li's criterion for the Riemann Hypothesis*, J. Number Theory **77** (1999) 274–287.
3. H. Davenport, *Multiplicative Number Theory* (3rd ed., revised by H.L. Montgomery), Graduate Texts in Mathematics **74**, Springer-Verlag (2000).
4. J. Delsarte, *Formules de Poisson avec reste*, J. Anal. Math. (Jerusalem) **17** (1966) 419–431 (Sect. 7).
5. C. Deninger, *Local L-factors of motives and regularized determinants*, Invent. Math. **107** (1992) 135–150 (thm 3.3 and Sect. 4).
6. H.M. Edwards, *Riemann's Zeta Function*, Academic Press (1974).
7. A. Erdélyi (ed.), *Higher Transcendental Functions* (Bateman Manuscript Project), Vols. I chap. 1 and III chap. 17, McGraw-Hill, New York (1953).

8. A.P. Guinand, *A summation formula in the theory of prime numbers*, Proc. London Math. Soc. Series 2, **50** (1949) 107–119.
9. J. Jorgenson and S. Lang, *Explicit formulas for regularized products and series*, Lecture Notes in Mathematics **1593**, Springer-Verlag (1994) chap. I (and refs. therein).
10. J.B. Keiper, *Power series expansions of Riemann's ξ function*, Math. Comput. **58** (1992) 765–773.
11. N. Kurokawa, *Parabolic components of zeta functions*, Proc. Japan Acad. **64**, Ser. A (1988) 21–24.
12. D.H. Lehmer, *The Sum of Like Powers of the Zeros of the Riemann Zeta Function*, Math. Comput. **50** (1988) 265–273 (and refs. therein).
13. X.-J. Li, *The positivity of a sequence of numbers and the Riemann Hypothesis*, J. Number Theory **65** (1997) 325–333.
14. Yu. Manin, *Lectures on zeta functions and motives (according to Deninger and Kurokawa)*, in: *Columbia University Number Theory Seminar* (New York, 1992), Astérisque **228** (1995) 121–163.
15. Yu.V. Matiyasevich, *A relationship between certain sums over trivial and nontrivial zeros of the Riemann zeta-function*, Mat. Zametki **45** (1989) 65–70 [Translation: Math. Notes (Acad. Sci. USSR) **45** (1989) 131–135.]
16. Y. Matsuoka, *A note on the relation between generalized Euler constants and the zeros of the Riemann zeta function*, J. Fac. Educ. Shinshu Univ. **53** (1985) 81–82, and *A sequence associated with the zeros of the Riemann zeta function*, Tsukuba J. Math. **10** (1986) 249–254.
17. J.R. Quine, S.H. Heydari and R.Y. Song, *Zeta regularized products*, Trans. Amer. Math. Soc. **338** (1993) 213–231.
18. M. Schröter and C. Soulé, *On a result of Deninger concerning Riemann's zeta function*, in: *Motives*, Proc. Symp. Pure Math. **55** Part 1 (1994) 745–747.
19. A. Voros, *Spectral functions, special functions and the Selberg zeta function*, Commun. Math. Phys. **110** (1987) 439–465.
20. A. Voros, *Zeta functions for the Riemann zeros*, Ann. Inst. Fourier, Grenoble **53** (2003) 665–699 [erratum: **54** (2004) 1139].
21. A. Voros, *Zeta functions over zeros of general zeta and L-functions*, in: *Zeta functions, topology and quantum physics* (Proceedings, Osaka, March 2003), T. Aoki, S. Kanemitsu, M. Nakahara and Y. Ohno eds., Developments in Mathematics **14**, Springer-Verlag (2005) 171–196.

Hilbert Spaces of Entire Functions and Dirichlet L-Functions

Jeffrey C. Lagarias

Department of Mathematics, University of Michigan, Ann Arbor,MI 48109-1109 USA. email: lagarias@umich.edu

Summary. We describe connections between the de Branges theory of Hilbert spaces of entire functions and the Riemann hypothesis for Dirichlet L-functions. Assuming the Riemann hypothesis holds for a given L-function, there exists an associated de Branges space with interesting properties, and conversely. This de Branges space comes with an associated self-adjoint operator having as eigenvalues the imaginary parts of the L-function zeros on the critical line, and this operator has an interpretation as a "Hilbert-Polya" generalized differential operator.

1 Introduction

The object of this paper is to formulate a connection between Hilbert spaces of entire functions and the Riemann hypothesis for various L-functions. We note that Louis de Branges has long advocated the applicability of his theory of Hilbert spaces of entire functions to the Riemann hypothesis. He has proposed several approaches to the problem, including [11], [12]. The approach taken here differs from these, as indicated below.

We associate to the Riemann zeta function the entire function

$$E_{\chi_0}(z) := \xi(\frac{1}{2} - iz) + \xi'(\frac{1}{2} - iz)$$

in which $\xi(s) = \frac{1}{2}s(s-1)\pi^{-s/2}\Gamma(\frac{s}{2})\zeta(s)$ is the Riemann ξ-function, and ξ' denotes its derivative with respect to the s-variable. More generally, to each

Dirichlet L-function with a primitive character χ we associate the entire function

$$E_\chi(z) := \xi_\chi(\frac{1}{2} - iz) + \xi_\chi'(\frac{1}{2} - iz),$$

in which $\xi_\chi(s)$ denotes the Dirichlet L-function completed with its archimedean factors, multiplied by a certain constant of modulus one which makes $\xi(s)$ real on the critical line $\Re(s) = \frac{1}{2}$. Our main observation is that for each χ, $E_\chi(z)$ is the structure function of a de Branges space $\mathcal{H}(E_\chi(z))$ if and only if the Riemann hypothesis holds for $L(s,\chi)$. Furthermore it gives a strict de Branges space (defined below) if and only if the Riemann hypothesis holds for $L(s,\chi)$ and all its non-trivial zeros are simple zeros. Thus, assuming the Riemann hypothesis holds, these associated de Branges spaces exist, and we then explore what de Branges's theory implies about such spaces.

We use the fact that the de Branges theory associates to each de Branges space an integral transform which we term here the *de Branges transform.* For the spaces above this transform produces a "Hilbert-Polya" (generalized) differential operator together with self-adjoint boundary conditions that give an eigenvalue interpretation of the zeros of these L-functions. The Riemann hypothesis is interpretable as a positivity property of the coefficient functions of this "Hilbert-Polya" operator. This allows the possibility of approaching the Riemann hypothesis by finding a direct construction of this operator.

We now remark on de Branges's approaches to the Riemann hypotheses taken in [11], [12]. There he formulates theorems that state that any de Branges space $\mathcal{H}(E)$ that satisfies certain hypotheses on the Hilbert space scalar product necessarily has a structure function $E(z)$ that has all its zeros on the horizontal line $\Im(z) = -\frac{1}{2}$. His hope was that these results might apply to the de Branges space $\mathcal{H}(E)$ with structure function $E(z) = \xi(1-iz)$ (more generally $E(z) = \xi_\chi(1-iz)$ for certain Dirichlet L-functions), where the conclusions of his theorems would yield the Riemann hypothesis. These de Branges spaces exist unconditionally. However Conrey and Li [13] have shown that the de Branges spaces with $E(z) = \xi(1-iz)$ and $E(z) = \xi_{\chi_{-4}}(1-iz)$ fail to satisfy the hypotheses of these general theorems.

This paper presents one theorem and then indicates consequences of it. An expanded version of this paper, with additional results, is in preparation [20]. In particular the formulation given here extends to automorphic L-functions, i.e. principal L-functions for GL_n. The research in this paper was done while the author was employed at AT&T Labs-Research, whom he thanks for support.

2 Hilbert Spaces of Entire Functions

We give a brief review of the de Branges theory of Hilbert spaces of entire functions. This review formulates some of de Branges's results in [10] into an operator-theoretic language, using the terminology of canonical differential systems as in Remling [22] or Sakhnovich [23], rather than in terms of integral equations as in de Branges's formulation. In places it makes some simplifying assumptions, and for technically precise statements of the results, see [10] as indicated. The complex variable used is $z = x + iy$ and x, y always denote real variables.

A *structure function* or a *de Branges function* $E(z)$ is an entire function having the property that

$$|E(z)| > |E(\bar{z})| \text{ when } \Im(z) > 0. \tag{2.1}$$

This property implies that $E(z)$ has no zeros in the upper half plane. We say it is a *strict de Branges function* if $E(z)$ has no zeros on the real axis. This class of functions has a long history, see Chapter VII of Levin [21], who uses the term Hermite-Biehler functions, and M. Krein [17, Theorems 9 and 11]. The de Branges theory makes use of a decomposition of the structure function $E(z)$ into entire functions that are pure real and pure imaginary on the real axis, namely

$$E(z) = A(z) - iB(z),$$

given by

$$A(z) = \frac{1}{2}\left(E(z) + \overline{E(\bar{z})}\right),$$

$$B(z) = \frac{1}{2i}\left(E(z) - \overline{E(\bar{z})}\right).$$

A crucial observation of de Branges is that the condition (2.1) implies: *For a de Branges function $E(z) = A(z) - iB(z)$, the functions $A(z)$ and $B(z)$ have only real zeros, and these zeros interlace. If $E(z)$ is a strict de Branges function, then all the zeros are simple zeros* (de Branges [7, Lemma 5]).

One associates to any de Branges function $E(z)$ a de Branges Hilbert space $\mathcal{H}(E)$ of entire functions, as follows. The Hilbert space scalar product is

$$\langle f, g\rangle_E = \int_{-\infty}^{\infty} \frac{f(x)\overline{g(x)}}{|E(x)|^2}dx. \tag{2.2}$$

(conjugate-linear in the second factor). The entire functions $f(z)$ that belong to the space are those which have a finite norm $||f||_E$ and whose growth with respect to $E(z)$ is controlled in the upper half-plane $\mathbb{C}^+ := \{z : \Im(z) > 0\}$ and in the lower half-plane. The growth conditions are that $\frac{f(z)}{E(z)}$ and $\frac{\overline{f(\bar{z})}}{E(z)}$ be of bounded type and nonpositive mean type in $\mathbb{C}^+$. A function $h(z)$ is of *bounded type* if it can be written as a quotient of two bounded analytic functions in $\mathbb{C}^+$ and it is of nonpositive mean type if it grows no faster than $e^{\epsilon y}$ for each $\epsilon > 0$ as $y \to \infty$ on the positive imaginary axis $\{iy \ : y > 0\}$. One can show there always exist such functions, so the space $\mathcal{H}(E)$ is always nontrivial. There are examples where it is a finite-dimensional Hilbert space, but in the cases we will consider here it will always be infinite-dimensional.

A de Branges space $\mathcal{H}(E)$ is a *reproducing kernel Hilbert space*, with a kernel function $K_E(w, z)$ having the property that for each $f(z) \in \mathcal{H}(E)$, there holds

$$f(w) = \langle f(z), K(w, z)\rangle_E \text{ for all } w \in \mathbb{C}.$$

That is, evaluation of a function in $\mathcal{H}(E)$ at the point w is a continuous linear functional on $\mathcal{H}(E)$ and is therefore represented by a scalar product with some function $g_w(z) \in \mathcal{H}(E)$ and we have $K(w, z) := g_w(z)$. (Only values of z on the real axis are used in computing the scalar product.) The reproducing kernel is

$$K(w, z) = \frac{\overline{A(w)}B(z) - A(z)\overline{B(w)}}{\pi(z - \bar{w})}.$$

If we consider the de Branges space to be determined by its reproducing kernel, then there is some freedom in the choice of de Branges functions $E(z)$. For $k \in \mathbb{R}^+$ the function $E_k(z) := kA(z) - \frac{i}{k}B(z)$ gives the same reproducing kernel, for $\lambda \in \mathbb{R}$ so does $E_\lambda(z) := [A(z) + \lambda B(z)] - iB(z)$, and for $0 \leq \theta < 2\pi$ so does $E_\theta(z) = e^{i\theta}E(z) := A_\theta(z) - iB_\theta(z)$. This gives an $SL(2, \mathbb{R})$-action on structure functions that preserve the reproducing kernel. In the case of strict de Branges functions, we can remove this ambiguity by requiring that $E(0) = 1$, and $E'(0) \in i\mathbb{R}$, i.e. $A(0) = 1$, $A'(0) = 0$, and $B(0) = 0$. It proves convenient to only partially remove this ambiguity and call a structure function *normalized* if $E(0) = 1$, with no condition imposed on $A'(0)$.

There is an additional degree of freedom in that one can remove zeros on the real axis from the structure function without changing the de Branges space in an essential way. Indeed if $E(z)$ has a zero on the real axis, at $z = x_0$, say, then the form of the Hilbert space norm in (2.2) shows that every function in $\mathcal{H}(E(z)$ must have a zero at the same location, so we can divide all functions in the space by $z - x_0$ and obtain a new Hilbert space of entire functions having structure function $\mathcal{H}(\frac{E(z)}{z - x_0})$, preserving the Hilbert space inner product. The reproducing kernel changes, with the new reproducing kernel obtained from the old by dividing by $(z - x_0)(\bar{w} - x_0)$. In this way we can in principle reduce to the case of a *strict de Branges space*, one where the structure function $E(z)$ is a strict de Branges function,

There is an abstract theory of de Branges spaces. An *(abstract) de Branges space* is a nonzero Hilbert space $\mathcal{H}$ whose elements are entire functions, such that $\mathcal{H}(E)$ satisfies the axioms:

(H1) Whenever $f(z)$ is in the space and has a non-real zero z_0 then $g(z) := f(z)\frac{z - \bar{z}_0}{z - z_0}$ is in the space and has the same norm as $f(z)$.

(H2) For every nonreal number $w \in \mathbb{C}$, the linear functional on $\mathcal{H}$ defined by $f(z) \mapsto f(w)$ is continuous.

(H3) If $f(z) \in \mathcal{H}$ then $f^*(z) := \overline{f(\bar{z})}$ belongs to $\mathcal{H}$ and has the same norm as $f(z)$.

Two abstract de Branges spaces are *isomorphic* if there is an isometry between them that preserves properties (H1)–(H3). Then any (abstract) de Branges space is isomorphic to some de Branges space $\mathcal{H}(E)$ ([10, Theorem

23]). Each such space is isomorphic to a de Branges space $\mathcal{H}(E(z))$ for which $E(z)$ is a strict de Branges function that is normalized, i.e. $E(0) = 1$.

A de Branges space $\mathcal{H}(E)$ comes with an unbounded operator $(M_z, \mathcal{D}(M_z))$ in which M_z is "multiplication by z" and its domain is

$$\mathcal{D}(M_z) = \{f(z) \in \mathcal{H}(E) : \; zf(z) \in \mathcal{H}(E)\}.$$

This domain is either dense in $\mathcal{H}(E)$ (the "dense" case) or has closure of codimension 1 in $\mathcal{H}(E)$ (the "non-dense" case). We are interested here only in the "dense" case; the property of being "dense" can be read off from properties of $E(z)$ on the real axis. The operator M_z is symmetric and closed (i.e. its graph is closed in $\mathcal{H}(E) \oplus \mathcal{H}(E)$). In the "dense" case the operator has deficiency indices $(1, 1)$, and so has a family of self-adjoint extensions parametrized by the group $U(1) = \{e^{i\theta} : 0 \le \theta < 2\pi\}$.

One interpretation of the de Branges theory is that it supplies a "canonical model" for a particular subset of closed symmetric operators with deficiency indices $(1, 1)$. This class of operators contains the class of "entire operators" introduced by M. Krein (see [15]). The "canonical model" allows various properties of the operator M_z to be read off by inspection.

First, from the normalized structure function $E(z)$ we obtain a description of all self-adjoint extensions of the operator M_z. These extensions all have discrete, simple spectra, as we describe below.

Second, the "canonical model" exhibits a complete chain of nested invariant subspaces for the operator, which consists of subspaces which are themselves de Branges spaces, and which is uniquely with this property. Associated to this chain of invariant subspaces is an integral transform somewhat like the Fourier transform, which we will call here the *de Branges transform*, with a corresponding inverse de Branges transform. The de Branges transform gives an isometry of a de Branges transform Hilbert space $\mathcal{K}(M)$ (defined below) onto the Hilbert space $\mathcal{H}(E)$, with inverse transform going the opposite direction. (See [10, Theorem 44].) The inverse de Branges transform takes the multiplication operator M_z is to a (generalized) linear differential operator [1] D_t acting on a system of 2×1 vectors of functions, whose dependent variable t runs over an interval of the real line $\mathbb{R}$, which can be taken to be $(0, b]$ with b finite, parametrizing the chain of invariant subspaces.

A major theorem of de Branges used in the construction of this transform is the *total ordering theorem* which says if two de Branges spaces $\mathcal{H}(E_1)$ and $\mathcal{H}(E_2)$ are isometrically embedded in a de Branges space $\mathcal{H}(E)$ (i.e. their

[1] More accurately, the de Branges theory uses a 2×2 matrix integral equation in the parameter t. If the integral equation could be differentiated, then one obtains the canonical differential system (2.3) (2.4) given below, see for example Dym [14, p. 396]. The matrix $M(t)$ in (2.4) is related to de Branges's symmetric 2×2 matrix $m(t)$ with entries $(\alpha(t), \beta(t), \gamma(t))$ given in [10, Theorem 38] by $M(t) = \frac{d}{dt}m(t)$. The canonical differential equation formalism works more generally by allowing $M(t)dt = dm(t)$ to be a 2×2 matrix-valued measure, see Remling [22].

reproducing kernels are obtained by restriction) then either $\mathcal{H}(E_1) \subset \mathcal{H}(E_2)$ or $\mathcal{H}(E_2) \subset \mathcal{H}(E_1)$ ([10, Theorem 35]). The order type of the resulting chain of subspaces can be either discrete, where the dimension jumps by 1 at some points, or continuous, or some mixture of discrete and continuous. In any case we can embed such an order type in an interval, and write the family as $\mathcal{H}(E_t)$ where $0 \le t \le b$, say, with $\mathcal{H}(E_{t_1}) \subset \mathcal{H}(E_{t_2})$ if $t_1 < t_2$. In a parametrization by an interval, some values of t correspond to members of the chain, and other values of t do not, being filler values to permit parametrization of the chain by an interval on the real axis. For this de Branges introduces notions of "regular" and "singular" values of a parametrization ([10, p. 136]) in which a "regular" value of t apparently corresponds to belonging to the chain. For the discussion here we shall suppose that we are dealing with a pure continuous case, and also that it is legitimate to differentiate and obtain the canonical system (2.3) below.

For a given normalized strict structure function $E(z)$ it is possible to find a family of normalized strict structure functions of the de Branges chain $\mathcal{H}(E_t)$, denoted by $E_t(z) := A(t,z) - iB(t,z)$ parametrized with $0 < t \le b$, that satisfy [2] a "canonical differential system" (see [23]) for each $z \in \mathbb{C}$,

$$\frac{d}{dt}\begin{bmatrix} A(t) \\ B(t) \end{bmatrix} = zJM(t)\begin{bmatrix} A(t) \\ B(t) \end{bmatrix}, \tag{2.3}$$

in which

$$J = \begin{bmatrix} 0 & -1 \\ 1 & 0 \end{bmatrix} \quad \text{and} \quad M(t) = \begin{bmatrix} \tilde{\alpha}(t) & \tilde{\beta}(t) \\ \tilde{\beta}(t) & \tilde{\gamma}(t) \end{bmatrix}. \tag{2.4}$$

such that at the right endpoint $E_b(z) = E(z)$ for all $z \in \mathbb{C}$, and at the left endpoint satisfies

$$\lim_{t\to 0+} A(t,z) = 1 \quad \text{and} \quad \lim_{t\to 0+} B(t,z) = 0, \tag{2.5}$$

see [10, Theorem 40]. A principal feature is that for (almost all) t the matrix $M(t)$ is real, symmetric and positive semidefinite. The right endpoint is "regular" in a sense defined by de Branges [10, p. 136]. The Hilbert space $\mathcal{K}(M)$ consists of vector-valued functions $[A(t), B(t)]^T$ on an interval, say $[0,b]$, with norm

$$||(f(t), g(t))||_M^2 = \int_0^b [f(t),\ g(t)] M(t) \begin{bmatrix} \overline{f(t)} \\ \overline{g(t)} \end{bmatrix} dt.$$

The *de Branges transform* $\mathcal{T} : \mathcal{K}(M) \to \mathcal{H}(E)$ is:

$$V(t) := (f(t), g(t)) \longmapsto \mathcal{T}(V)(z) := \frac{1}{\pi}\int_0^b [f(t)\ g(t)] M(t) \begin{bmatrix} \overline{A(t,\bar{z})} \\ \overline{B(t,\bar{z})} \end{bmatrix} dt, \tag{2.6}$$

see [10, Theorems 43 and 44]. Note that $\overline{A(t,\bar{z})} = A(t,z)$ and $\overline{B(t,\bar{z})} = B(t,z)$.

---

[2] A a differential equation of the general form (2.3), (2.4) for a fixed z is called a canonical differential equation. A "canonical differential system" is a family of such equations with family parameter z.

The *de Branges direct theorem* asserts that: Any canonical differential system (2.3) with "initial condition" (2.5) with the property that the 2×2 matrix function $M(t)$ is measurable and positive semi-definite symmetric for all $t \in (0, b]$, and which is integrable over the interval, has solutions $\{(A(t,z), B(t,z)) : 0 < t < b, \text{ all } z \in \mathbb{C}\}$ such that each $E(t,z) = A(t,z) - iB(t,z)$ with t constant and $z \in \mathbb{C}$ is a strict, normalized de Branges structure function. This is proved ([10, Theorem 41]) provided some growth conditions are imposed on the coefficients of $M(t)$. These growth conditions characterize those (strict, normalized) de Branges functions belonging to the Polya class. The Polya class consists of those de Branges functions whose modulus is nondecreasing on each vertical line in the upper half-plane, see [10, Sect. 7]. The de Branges functions which are entire functions of exponential type are exactly the class for which the canonical differential equation (2.3) is regular at its left endpoint for each $z \in \mathbb{C}$. For de Branges functions of faster growth this endpoint is singular.

The *de Branges inverse theorem* asserts the following: For any strict normalized de Branges structure function $E(z) = A(z) - iB(z)$ in the Polya class, there exists a set of real coefficient functions $(\tilde{\alpha}(t), \tilde{\beta}(t), \tilde{\gamma}(t))$ such that the real matrix $M(t)$ is positive semidefinite for almost all t in a finite half-open interval $(0, b]$ and the solutions to the canonical differential system (2.3) at the left-endpoint satisfy (2.5) and at the right endpoint $t = b$ have

$$A(b,z) = A(z) \quad \text{and} \quad B(b,z) = B(z). \tag{2.7}$$

In addition the right endpoint is a "regular" value in de Branges's sense. The precise assertion (see [10, Theorem 40]), is an extremely strong inverse spectral theorem which subsumes many known inverse spectral theorems, see Krein [18] and Remling [22, Theorem 7.3]. The canonical differential system (2.3) can be made essentially unique by reparametrizing it to have $Tr(M(t)) \equiv 1$ almost everywhere. With this reparametrization the interval may become infinite to the left, with an endpoint at $-\infty$.

We now describe the self-adjoint extensions of the operator M_z, assuming that we are in the "dense" case. The structure function $E(z)$ specifies two particular self-adjoint extensions associated to $A(z)$ and $B(z)$, respectively. Recall that $A(z)$ and $B(z)$ have real zeros and these zeros interlace. The self-adjoint extension M_z associated to $A(z)$, denoted $\tilde{M}_z$ or $M_z(A)$, has pure discrete simple spectrum located at those zeros of $A(z)$ that have multiplicity exceeding that of $B(z)$ at the same point, and for each such zero ρ an eigenfunction $f_\rho(z) = \frac{A(z)}{(z-\rho)^j}$, where $A(z)$ has a zero of order j at $z = \rho$. The domain $\mathcal{D}(M_z(A)) = \mathcal{D}(M_z) \oplus \mathbb{C}[f_{\rho_0}]$ for any single function f_{ρ_0}. We obtain all self-adjoint extensions of M_z by considering instead $\{A_\theta(z) : 0 \le \theta < 2\pi\}$, obtained from $E_\theta(z) = e^{i\theta}E(z)$. Now suppose that $E(z)$ is a strict de Branges function, in which case $j = 1$ always, and the functions $\{f_\rho(z) = \frac{A(z)}{z-\rho} : A(\rho) = 0\}$ form an orthogonal basis of $\mathcal{H}(E(z))$ ([10, Theorem 22]). This orthogonal basis gives rise to a "summation formula" expressing the

Hilbert space norm of an arbitrary function $f(z) \in \mathcal{H}(E)$,

$$||f(z)||_E^2 = \int_{-\infty}^{\infty} \left| \frac{f(x)}{E(x)} \right|^2 dx = \sum_{\rho} \frac{\pi}{\phi'(\rho)} \left| \frac{f(\rho)}{E(\rho)} \right|^2, \tag{2.8}$$

in which the phase function $\phi(t)$ is given by $E(x) = e^{-i\phi(x)} E_0(x)$, with $E_0(x)$ real-valued.

The operator D_t on the de Branges transform space $\mathcal{K}(M)$ having deficiency indices (1, 1) can be formally written as

$$D_t := M(t)^{-1} J^{-1} \frac{d}{dt} = M(t)^{-1} \begin{bmatrix} 0 & \frac{d}{dt} \\ -\frac{d}{dt} & 0 \end{bmatrix},$$

under the extra assumption that $M(t)$ is invertible everywhere. Theorem 45 of [10] gives a description of the range of the symmetric operator D_t. In the de Branges transform space $\mathcal{K}(M)$ is a corresponding orthogonal basis of eigenfunctions $V_\rho(t) := [A_\rho(t), B_\rho(t)]^T$ of $\tilde{D}_t$ defined indirectly by $\mathcal{M}(V_\rho) = \frac{A(z)}{z-\rho}$, where $A(\rho) = 0$. Expressing members of the Hilbert space $\mathcal{K}(M)$ in terms of this basis gives an "eigenfunction expansion" associated with the de Branges theory.

There are additional aspects to the de Branges theory not covered here. Some of this is to be discussed in [20].

3 de Branges Spaces Associated to Dirichlet L-Functions

Associated to the Riemann zeta function is the Riemann ξ-function, given by

$$\xi(s) = \frac{1}{2} s(s-1) \pi^{-\frac{s}{2}} \Gamma(\frac{s}{2}) \zeta(s).$$

It is an entire function, real on the real axis and on the critical line $\Re(s) = \frac{1}{2}$, satisfies the functional equation $\xi(s) = \xi(1-s)$ and its zeros are exactly the nontrivial zeros of the Riemann zeta function, those in the critical strip $0 < \Re(s) < 1$. We write $\xi(s) := \xi_{\chi_0}(s)$ where χ_0 is the identity Dirichlet character.

We can similarly associate to each Dirichlet L-function $L(s, \chi)$ with a primitive character χ of conductor q a corresponding ξ-function $\xi_\chi(s)$. We define the completed L-function

$$\hat{L}_\chi(s) := (\frac{\pi}{q})^{-\frac{s+k}{2}} \Gamma(\frac{s+k}{2}) L(s, \chi),$$

in which $k = 0$ if $\chi(-1) = 1$ and $k = 1$ if $\chi(-1) = -1$. This is an entire function which satisfies the functional equation

$$\hat{L}_\chi(s) = \epsilon(\chi) \hat{L}_{\bar{\chi}}(1-s),$$

in which $\bar{\chi}$ is the complex conjugate of the character χ and $\epsilon(\chi) := i^k \frac{\tau(\chi)}{q^{\frac{1}{2}}}$ is a constant of absolute value 1, see for example Davenport [6, Chap. 9]. The fact that $\hat{L}_\chi(s)$ transforms under complex conjugation as $\hat{L}_\chi(s) = \overline{\hat{L}_\chi(\bar{s})}$ together with the functional equation implies that $\hat{L}_\chi(s)$ has constant modulus $(\bmod\ \pi)$ on the critical line, i.e. there is a constant $e^{i\theta}$ such that $\hat{L}_\chi(\frac{1}{2}+it) = e^{i\theta} g_\chi(t)$ for some continuous real-valued function $g_\chi(t)$. There remains an ambiguity of a sign in the choice of $e^{i\theta}$ which is removed by requiring that $\xi_\chi(s)$ be positive on the critical line in the upper half-plane just above $s = \frac{1}{2}$. We then define the modified function

$$\xi_\chi(s) := e^{-i\theta} \hat{L}(s, \chi), \tag{3.1}$$

which is real-valued on the critical line, and satisfies the functional equation

$$\xi_\chi(s) = \xi_{\bar{\chi}}(1 - s).$$

The zeros of the function $\xi_\chi(s)$ are exactly the non-trivial zeros of the Dirichlet L-function in the critical strip, counting multiplicities.

In the following result the notation $f'(s)$ denotes the derivative with respect to the s-variable, following standard usage in number theory.

Theorem 1. *For each primitive Dirichlet character χ including the trivial character χ_0, set $E_\chi(z) = A_\chi(z) - iB_\chi(z)$ with*

$$A_\chi(z) = \xi_\chi(\frac{1}{2} - iz), \quad B_\chi(z) = i\ \xi_\chi^{'}(\frac{1}{2} - iz).$$

Then these functions are real on the real axis, and the following holds.

(i) $E_\chi(z)$ is a de Branges function if and only if the Riemann hypothesis holds for $L(s, \chi)$.

(ii) $E_\chi(z)$ is a strict de Branges function if and only if the Riemann hypothesis holds for $L(s, \chi)$ and all its nontrivial zeros are simple zeros.

Proof. The function $A_\chi(z)$ is real on the real axis since $\xi_\chi(s)$ is real on the critical line. Then $B_\chi(z)$ inherits this property under differentiation.

(i) If $E_\chi(z)$ is a de Branges function then by de Branges' lemma both $A_\chi(z)$ and $B_\chi(z)$ have only real zeros, which interlace. The reality of zeros of $A_\chi(z)$ is the Riemann hypothesis for $\xi_\chi(s)$.

Now assume that the Riemann hypothesis holds for $\xi_\chi(s)$. Then $A_\chi(z)$ has real zeros. Since $A_\chi(z)$ is an entire function of order 1 (and infinite type), Laguerre's theorem ([16, Theorem 5.7]) applies to show that $B_\chi(z) = -\frac{d}{dz} A_\chi(z)$ has real zeros and they interlace with those of $A_\chi(z)$.

We show that the Riemann hypothesis for $L(s, \chi)$ implies that

$$\Re\left(\frac{\xi_\chi'(s)}{\xi_\chi(s)}\right) > 0 \ \text{ for } \ \Re(s) > \frac{1}{2}. \tag{3.2}$$

This fact is well known for the Riemann ξ-function, see Lagarias [19]. Starting from the Hadamard product factorization

$$\xi_\chi(s) = e^{A+Bs} \prod_\rho \left(1 - \frac{s}{\rho}\right) e^{\frac{s}{\rho}},$$

and the logarithmic derivative we obtain

$$g_\chi(s) := \frac{\xi_\chi'(s)}{\xi_\chi(s)} = B + \sum_\rho (\frac{1}{s-\rho} + \frac{1}{\rho}) = (B + \sum_\rho \frac{1}{\rho}) + \sum_\rho (\frac{1}{s-\rho}),$$

where the sum is not absolutely convergent in the last equality and must be viewed as taken over $|\rho| < T$ and then letting $T \to \infty$. Taking the real part of this sum, one can check term by term that $\Re(\frac{1}{s-\rho}) > 0$ whenever $\Re(s) > \Re(\rho)$. We also have $\Re(B + \sum_\rho \frac{1}{\rho}) = 0$, which is deduced from the functional equation $g_\chi(s) = -g_\chi(1-s)$. By hypothesis $\Re(\rho) \le \frac{1}{2}$, and (3.2) follows.

Now (3.2) gives $\Im\left(\frac{B_\chi(z)}{A_\chi(z)}\right) = \Im\left(i \frac{\xi'(\frac{1}{2}-iz)}{\xi(\frac{1}{2}-iz)}\right) > 0$ when $\Im(z) > 0$. Then

$$|\frac{B_\chi(z)}{A_\chi(z)} + i| > |\frac{B_\chi(z)}{A_\chi(z)} - i| = |\frac{B_\chi(\bar{z})}{A_\chi(\bar{z})} + i|$$

under the same condition. Thus for $\Im(z) > 0$,

$$\begin{aligned} |E_\chi(z)| &= |-iA_\chi(z)||i + \frac{B_\chi(z)}{A_\chi(z)}| \\ &> |-iA_\chi(\bar{z})||i + \frac{B_\chi(\bar{z})}{A_\chi(\bar{z})}| = |E_\chi(\bar{z})|, \end{aligned}$$

so that $E_\chi(z)$ is a de Branges function.

(ii) This is straightforward. □

Assuming the Riemann hypothesis, it can be shown that the functions $E_\chi(z)$ belong to the Polya class, and that the associated de Branges spaces fall in the "dense" case. We now consider the consequences of having a strict de Branges space, where we can make use of the de Branges transform.

First, the self-adjoint extension $\tilde{M}_z$ of M_z corresponding to the function $A_\chi(z) = \xi_\chi(\frac{1}{2} - iz)$ has a complete orthogonal set of eigenfunctions given by

$$f_\rho(z) := \frac{\xi_\chi(\frac{1}{2} - iz)}{z - \gamma}, \text{ with } \rho = \frac{1}{2} + i\gamma,$$

where ρ runs over all the zeros (assumed simple) of $\xi_\chi(s)$.

Second, the de Branges summation formula applied to this set of orthogonal eigenfunctions gives for all $f(z) \in \mathcal{H}(E_\chi)$, putting $F(\frac{1}{2} - iz) = f(z)$,

$$||f(z)||^2_{E_\chi} = \pi \sum_{\{\rho : \xi_\chi(\rho) = 0\}} \frac{|F(\rho)|^2}{|\xi_\chi'(\rho)|^2},$$

The right side of this formula resembles the spectral side of the "explicit formula" of prime number theory. Viewed this way, the positivity of the Hilbert space norm appears to encode "Weil positivity," compare [1, Sect. 4].

Third, the associated de Branges transform gives an encoding of the Riemann hypothesis plus simplicity of the zeros as a positivity property. To show a given $E_\chi(z)$ is a strict de Branges function, it suffices to show that a corresponding normalized function $E_\chi^N(z) = k_0 E(z)$ (with constant k_0 chosen so that $E_\chi^N(1) = 1$) is a normalized strict de Branges function. The de Branges inverse theorem then says there exists data

$$M(t) = \begin{bmatrix} \tilde{\alpha}(t) & \tilde{\beta}(t) \\ \tilde{\beta}(t) & \tilde{\gamma}(t) \end{bmatrix}.$$

which is real, symmetric and positive semidefinite, and whose canonical differential system, suitably parametrized, produces on the interval $(0, b]$ the function

$$E_\chi^N(z) = A_\chi^N(b, z) - iB_\chi^N(b, z)$$

at its right endpoint. (The $Tr(M) \equiv 1$ reparametrization would necessarily be on an infinite interval $(-\infty, b]$ in this case.) If these coefficient functions are found, then the de Branges direct theorem certifies that $E_\chi^N(z)$ is a strict de Branges function, so that $E_\chi(z)$ is as well, whence $A(b, z) = \xi_\chi(\frac{1}{2} - iz)$ has real simple zeros. Thus the Riemann hypothesis plus simple zeros is encoded as the positive semidefiniteness property of the coefficient matrix $M(t)$ on $(0, b]$. It seems reasonable to expect that for these particular de Branges spaces the matrix $M(t)$ will always be positive definite. We note that the fact that $E_\chi^N(z)$ is not a bandlimited function implies that the canonical differential system for it will necessarily be singular at the left endpoint $t = 0$, with $\gamma(t) \to \infty$ as $t \to 0$.

Fourth, the de Branges transform produces a "Hilbert-Polya" operator, by which we mean a self-adjoint differential operator on a Hilbert space whose eigenvalues encode the zeta zeros. We take the operator $\tilde{D}_t$ to be the self-adjoint extension of the (generalized) differential operator D_t on $\mathcal{K}(M)$ that corresponds to the extension $\tilde{M}_z$ of the de Branges operator M_z under the de Branges transform. It is possible to describe this operator and its domain more concretely. There are particularly interesting forms for it in the case of a real primitive character χ, the self-dual case.

According to the de Branges theory there has been so far no loss of information. That is, if the Riemann hypothesis plus simple zeros holds, then the objects above all exist, if properly interpreted as integral equations rather than differential equations, and conversely. Some inferences on what the coefficient functions of $M(t)$ might look like for the Riemann zeta function case $\mathcal{H}(E_{\chi_0})$ can be obtained by analogy with those of certain Sonine spaces of entire functions, cf. de Branges [8], [9], Burnol [3], [4]. Burnol has also studied some other Hilbert spaces associated to the zeta function and Dirichlet L-functions [2] [5].

4 Conclusions

We have formulated the Riemann hypothesis for Dirichlet L-functions in terms of the existence of particular de Branges spaces. This provides a possible approach to the Riemann hypothesis plus simplicity of the zeta zeros, namely to construct these hypothetical spaces directly in a way that certifies they are de Branges spaces with the correct structure function.

There are at least three ways to construct a de Branges space. The first way is to find a structure function $E(z)$ for the space, and directly prove $E(z)$ has the defining property (2.2). The second way is to obtain the de Branges transform data $\{M(t) : 0 \leq t \leq b\}$, verify that each 2×2 matrix $M(t)$ is real and positive semi-definite symmetric, and integrable over the specified interval. The third way is to construct in some fashion a Hilbert space of entire functions and show directly that it satisfies the axioms (H1)–(H3), without obtaining either the structure function or the de Branges transform. This last approach can sometimes be taken using a weighted Mellin transform, as in de Branges [9] and Burnol [3]. In following the latter two approaches, an additional necessary task is to establish that the resulting de Branges space has the desired structure function $E(z)$.

The usefulness of this reformulation of the Riemann hypothesis will likely depend on whether information coming from number theory, either from arithmetical algebraic geometry, automorphic representations, or from some other source entirely, can be applied to show the existence of these particular (hypothetical) de Branges spaces.

References

1. E. Bombieri and J. C. Lagarias, Complements to Li's criterion for the Riemann hypothesis, J. Number Theory **77** (1999), 274–287.
2. J.-F. Burnol, Sur certains espaces de Hilbert de fonctions entières, liés à transformation de Fourier et aux fonctions L de Dirichlet et de Riemann, C. R. Acad. Sci. Paris Ser. I. Math. **333** (2001), no. 3, 201–206.
3. J.-F. Burnol, Sur les "Espaces de Sonine" associes par de Branges a la transformation de Fourier, C. R. Acad. Sci. Paris, Math. **335** (2002), no. 8, 689–692.
4. J.-F. Burnol, Des Équations de Dirac et de Schrödinger pour la transformation de Fourier, C. R. Acad. Sci. Paris., Math **336** (2003), No. 11, 919–924.
5. J.-F. Burnol, Two complete and minimal systems associated with the zeros of the Riemann zeta function, J. de Theorie des Nombres de Bordeaux **16** (2004), 65–94.
6. H. Davenport, *Multiplicative Number Theory. Second Edition.* Revised by Hugh Montgomery. Springer-Verlag: New York 1980.
7. L. de Branges, Some Hilbert spaces of entire functions, Proc. Amer. Math. Soc. **10** (1959), 840–846.
8. L. de Branges, Symmetry in Hilbert spaces of entire functions, Duke Math. J. **29** (1963), 383–392.
9. L. de Branges, Self-reciprocal functions, J. Math. Anal. Appl. **9** (1964), 433–457.

10. L. de Branges, *Hilbert Spaces of Entire Functions,* Prentice-Hall: Englewood Cliffs, NJ 1968.
11. L. de Branges, The Riemann hypothesis for Hilbert spaces of entire functions, Bull. Amer. Math. Soc., N. S. **15** (1986), 1–17.
12. L. de Branges, The convergence of Euler products, J. Funct. Anal. **107** (1992), 122–210.
13. B. Conrey and X.-J. Li, A note on some positivity conditions related to zeta and L-functions, Internat. Math. Res. Notices **2000**, no. 18, 929–940.
14. H. Dym, An introduction to de Branges spaces of entire functions with applications to differential equations of the Sturm-Liouville type, Advances in Math. **5** (1971), 395–471.
15. M. L. Gorbachuk and V. I. Gorbachuk, *M. G. Krein's Lectures on Entire Operators,* Birkhäuser Verlag, Basel 1997.
16. A. S. B. Holland, *Introduction to the Theory of Entire Functions*, Academic Press: New York 1973.
17. M. G. Krein, Concerning a special class of entire and meromorphic functions, in: N. I. Ahiezer and M. Krein, *Some Questions in the Theory of Moments* (Russian), Gos. Naucno.-Tehn. Izda. Ukrain., Kharkov 1938. (English Translation: Article 6 in: N. I. Ahiezer and M. Krein, *Some Questions in the Theory of Moments*, Trans. of Math. Monographs Vol. 2, AMS: Providence RI, 1962.)
18. M. G. Krein, On a difficult problem in operator theory and its relation to classical analysis, Moscow Math. Soc. jubilee address, 1964. English translation in [15], pp. 199–210.
19. J. C. Lagarias, On a positivity property of the Riemann ξ-function, Acta Arith. **89**, No. 3 (1999), 217–234.
20. J. C. Lagarias, Hilbert spaces of entire functions and automorphic L-functions, paper in preparation.
21. B. Ya. Levin, *Distribution of Zeros of Entire Functions. Revised Edition*, Transl. of Math. Monographs Vol. 5, AMS: Providence, RI 1980.
22. C. Remling, Schrödinger operators and de Branges spaces, J. Funct. Anal. **196** (2002), 323–394.
23. L. A. Sakhnovich, *Spectral theory of canonical differential systems. Method of operator identities*, Operator Theory: Advances and Applications, Vol. 107, Birkhaüser: Basel 1999.

Dynamical Zeta Functions and Closed Orbits for Geodesic and Hyperbolic Flows

Mark Pollicott

Department of Mathematics, Manchester University, Oxford Road, Manchester M13 9PL UK
mp@ma.man.ac.uk

Introduction

In this article we want to give the basic definitions and properties of dynamical zeta functions, and describe a few of their applications. The emphasis is on giving the flavour of the subject rather than a detailed summary.

To fix ideas, let us assume that V is a compact surface with some appropriate Riemannian metric $< \cdot, \cdot >_{TV}$, say. We shall always assume that V has negative curvature at every point on V (although we will not necessarily assume that it has constant negative curvature). In studying geometric properties of manifolds it is sometimes convenient to study the associated geodesic flow. Fortunately, geodesic flows for negatively curved surfaces are important examples of a broader class of flows, namely hyperbolic flows, which are amenable to quite powerful techniques in dynamical systems which have evolved over the last 40 years (from the work of Anosov, Sinai, Ratner, Smale, Bowen, Ruelle, and many others). In particular, it is often (but not always) convenient to introduce simple symbolic models for these flows. The basic hope is that, despite the sacrifice of some of the geometry, we can benefit from being able to apply fairly directly ideas from ergodic theory and what is often colloquially called "Thermodynamic Formalism". Somewhat surprisingly, this method is successful for various classes of problems, including:

(a) Geometric problems (e.g., counting closed geodesics, or equivalently closed orbits for the geodesic flow);
(b) Statistical Properties (e.g., determining rates of mixing for flows); and
(c) Distributional properties (e.g., linear actions associated to the horocycle foliation).

Of course, anyone familiar with the Selberg zeta function for surfaces of constant negative curvature will recognise many of the ideas in (a), for example. The main difference is that instead of using the Selberg trace formula, say, we use transfer operators to study the zeta function. What we lose in elegance (and error terms!) we hope to make up for in the generality of the setting.

In this overview we want to recall a number of the key themes and outline some recent and ongoing developments. The choice of topics reflects the author's idiosyncratic tastes. The results are organised so as to give the illusion of coherence, but are in fact a mixture of older and more recent material. For different accounts and perspectives, the reader is referred to [7], [62]. In particular, nowadays non-symbolic methods are catching up in terms of efficiency in the above areas.

Finally, I would like to express my gratitude to the organisers of the Les Houches School for their invitation to participate.

1 Symbolic Dynamics and Zeta Functions

The familiar geodesic flow for V is a flow ϕ_t $(t \in \mathbb{R})$ defined on the (three dimensional) unit tangent bundle $T_1V = \{(x,v) \in TV : ||v||_{TV} = 1\}$, i.e., those tangent vectors to V having length one with respect to the ambient Riemannian norm. The flow acts in the standard way by moving one tangent vector $v \in T_1V$ to another $v' =: \phi_t x$, using parallel transport [5]. [1] It is the hypothesis of negative curvature ensures that this geodesic flow is a *hyperbolic flow*, i.e., one for which directions transverse to the flow direction (in a natural sense) are either expanding or contracting. [2]

1.1 Sections

The modern use of symbolic dynamics to model hyperbolic systems probably dates back to the work of Adler and Weiss [2], who showed that the famous Arnold CAT map could be modelled by a shift map on the space of sequences from a finite alphabet of symbols. This lead to Sinai's seminal work introducing Markov partitions for more general hyperbolic maps and then Ratner and Bowen's extension to hyperbolic flows [10], [56]. Historically, the use of sequences to model geodesic flows goes back even further to the work of Morse and Hedlund [25] who coded geodesics in terms of generators for the fundamental group.

Step 1 (Discrete maps from flows)

At its most general (and probably least canonical) the coding of orbits for hyperbolic flows $\phi_t : M \to M$ on any compact manifold M starts with a finite number of codimension one sections $T_1, \ldots, T_k$ to the flow. Let $X = \cup_i T_i$ denote the union of the sections. We can consider the discrete Poincaré return map $T : X \to X$, i.e, the map which takes a point x on a section to the point $T(x)$ where its ϕ-orbit next intersects a section. Of course, we need to assume that the sections are chosen so that

(i) every orbit hits the union of the sections infinitely often.

We would also like to consider the map $r : \cup_i T_i \to \mathbb{R}^+$ which gives the time it takes for $x \in X$ to flow to $T(x) \in X$, i.e., $\phi_{r(x)}(x) = T(x)$.

---

[1] More precisely, given any $(x, v) \in M$ we let $\gamma_{(x,v)} : \mathbb{R} \to M$ be the unit speed geodesic with $\gamma_{(x,v)}(0) = x$ and $\dot{\gamma}_{(x,v)}(0) = v$. We define the *geodesic flow* $\phi_t : M \to M$ by $\phi_t(x, v) = (\gamma_{(x,v)}(t), \dot{\gamma}_{(x,v)}(t))$.

[2] For completeness we recall the formal definition, although we won't need it in the sequel. Let M be any C^∞ compact manifold then we call a C^1 flow $\phi_t : M \to M$ *hyperbolic* (or Anosov) if:

(a) the tangent bundle TM has a continuous splitting $T_\Lambda M = E^0 \oplus E^u \oplus E^s$ into $D\phi_t$-invariant sub-bundles E^0 is the one-dimensional bundle tangent to the flow; and there exist $C, \lambda > 0$ such that $||D\phi_t|E^s|| \leq Ce^{-\lambda t}$ for $t \geq 0$ and $||D\phi_{-t}|E^u|| \leq Ce^{-\lambda t}$ for $t \geq 0$;
(b) $\phi_t : M \to M$ is transitive (i.e., there exists a dense orbit); and
(c) the periodic orbits are dense in M.

(More generally, if there is a closed ϕ-invariant set Λ with the above properties then $\phi_t : \Lambda \to \Lambda$ is called a hyperbolic flow.)

Key idea (modulo a slight fudge) There is a natural correspondence between periodic discrete orbits $T^n x = x$ and continuous periodic orbits τ of period $\lambda = \lambda(\tau) > 0$ (i.e., the smallest value such that $\phi_\lambda(x_\tau) = x_\tau$ for all $x_\tau \in \tau$), where

$$\lambda = r(x) + r(Tx) + \ldots + r(T^{n-1}x).$$

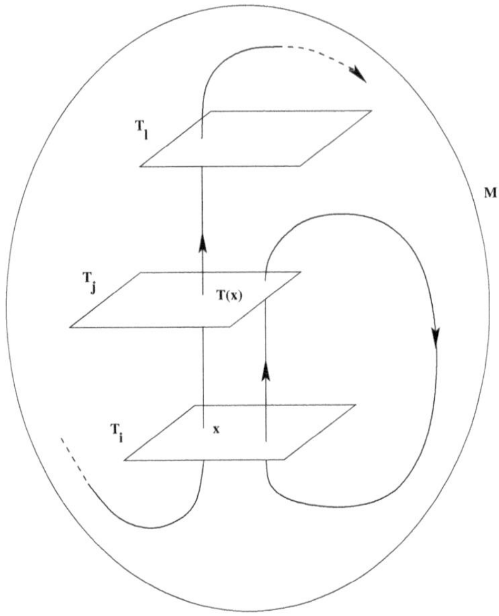

Fig. 1. Transverse (Markov) sections for a hyperbolic flow code a typical orbit and a closed orbit

Like many simple ideas, it is not quite true. There is an additional technical complication because of the closed orbits which pass through the boundaries of sections. However, this is not the typical case and an extra level of technical analysis sorts out this problem [10].

Step 2 (Sequence spaces from the Poincaré map)

The essential idea in symbolic dynamics is that a typical orbit $\{\phi_t(x) : -\infty < t < \infty\}$ will traverse these sections infinitely often (both in forward time and backward time) giving rise to a bi-infinite sequence $(x_n)_{n=-\infty}^{\infty}$ of labels of the sections it traverses [10], [56].

(ii) The sections are chosen to have a Markov property (i.e., essentially that the space Σ of all possible sequences $(x_n)_{n=-\infty}^{\infty}$ is given by a nearest neighbour condition: there exists a $k \times k$ matrix A with entries either 0 or 1 such that the sequence occurs if and only if $A(x_n, x_{n+1}) = 1$).

Alternatively, we can retain a little of the regularity of the functions as follows.

Step 2′ (Expanding maps from the Poincaré map)

Instead of a reducing orbits to sequences, we can replace the invertible Poincaré map by an expanding map (on a smaller space). The basic idea is to remove the contracting direction by identifying the sections X along the stable directions. We can then replace the union of two dimensional sections X by a union of one dimensional intervals Y. The Poincaré map $T : X \to X$ then quotients down to an expanding map $S : Y \to Y$ [59]. Of course, we lose track of the "pasts" of orbits, but for most purposes this is not a real problem.

1.2 An Alternative Approach for Constant Curvature: The Modular Surface and Compact Surfaces

We mentioned that for geodesic flows on surfaces of constant negative curvature there is an alternative method of Hedlund and Hopf to code geodesics. This method was further developed by Adler and Flatto [1] and Series [69]. Again it leads to a C^ω expanding Markov map $T : Y \to Y$. In this case, Y corresponds to the boundary of the universal cover

$$\mathbb{D}^2 = \{z = x + iy \in \mathbb{C} : |z| < 1\}$$

of the surface,i.e., the unit circle. This is divided into a finite number of arcs (actually determined by the sides of a fundamental domain for the surface). The corresponding metric on $\mathbb{D}^2$ is $ds^2 = (dx^2 + dy^2)/(1 - x^2 - y^2)^2$. The side pairs of the fundamental domain correspond to linear fractional transformations which preserve $\mathbb{D}^2$. On the boundary they give rise to expanding interval maps. A geodesic on $\mathbb{D}^2$ is uniquely determined by its two end points on the unit circle. We can associate a function $r : Y \to \mathbb{R}$ by $r(x) = \log |T'(x)|$, then we have the ingredients of the symbolic model.

Example: Modular surface We can consider the geodesic flow on the modular surface. In this case the surface is non-compact, and the difference is that the linear fractional transformation $T : Y \to Y$ is on an infinite number of intervals. However, in this case the transformation T is the well known continued fractional transformation on $[0, 1]$, i.e., $T : [0, 1) \to [0, 1)$ by $Tx = \frac{1}{x}$ (mod 1). The corresponding function $r : I \to \mathbb{R}$ is $r(x) = -2 \log x$, as is easily checked.

In this case the associated transfer operator is very easy to describe. We look at the Banach space B of analytic functions (with a continuous extension to the boundary) on a disk $\{z \in \mathbb{C} : |z - \frac{1}{2}| < \frac{3}{2}\}$. The transfer operator is given by $\mathcal{L}_s h(x) = \sum_{n=1}^{\infty} h\left(\frac{1}{x+n}\right) \frac{1}{(x+n)^{2s}}$ and the determinant $\det(I - \mathcal{L}_s)$ is analytic for $Re(s) > 1$. Using an approach of Ruelle [58], [59], Mayer [40], [41] showed that

(i) $\det(I - \mathcal{L}_s)$ has analytic extension to $\mathbb{C}$; and
(ii) $\zeta(s)$ is related to $\det(I - \mathcal{L}_{s+1})/\det(I - \mathcal{L}_s)$

For the Modular surface (and related surfaces) this special model leads to very elegant connections with functional equations, the Riemann zeta function and Modular functions cf. [38].

2 Zeta Functions, Symbolic Dynamics and Determinants

Let us denote by τ closed orbits for ϕ and let us write $\lambda(\tau) > 0$ for the period, (i.e., given $x \in \tau$ we have $\phi_{\lambda(\tau)}(x) = x$). We shall call τ a primitive closed orbit if $\lambda(\tau)$ is the smallest such value. Let us assume for simplicity a fact which is patently not true (but which has the virtue that it makes a complicated argument into a simple one!) that $r(x) = r(x_0, x_1)$ depends on only two terms in the sequence $x = (x_n)_{n=-\infty}^{\infty} \in \Sigma$. We can then associate to A a weighted $k \times k$ matrix $M_s(i,j) = A(i,j)e^{-sr(i,j)}$, i.e., the entries 1 in A are replaced by values of the exponential of $-sr$ (with $s \in \mathbb{C}$) [45], [44]:

$$\begin{aligned}
\zeta(s) = \prod_{\tau = \substack{\text{prime}\\ \text{orbits}}} \left(1 - e^{-s\lambda(\tau)}\right)^{-1} &= \exp\left(\sum_{\tau = \substack{\text{prime}\\ \text{orbits}}} \sum_{m=1}^{\infty} \frac{(e^{-s\lambda(\tau)})^m}{m} \right) \\
= \exp\left(\sum_{m=1}^{\infty} \sum_{p=1}^{\infty} \sum_{\substack{\text{prime orbits}\\ \{x,\dots,\sigma^{p-1}x\}}} \frac{e^{-sm[r(x_0,x_1)+r(x_1,x_2)+\cdots+r(x_{p-1},x_0)]}}{m} \right) \\
= \exp\left(\sum_{n=1}^{\infty} \sum_{\sigma^n x = x} \frac{e^{-s[r(x_0,x_1)+r(x_1,x_2)+\cdots+r(x_{-1},x_0)]}}{n} \right) \qquad (2.1) \\
= \exp\left(\sum_{n=1}^{\infty} \frac{\operatorname{trace}(M_s^n)}{n} \right) = \frac{1}{\det(I - M_s)}.
\end{aligned}$$

In particular, in this model case we see that $\zeta(s)$ has a (non-zero) meromorphic extension to the entire complex plane. Moreover, the poles are characterised as those values s for which the matrix M_s has 1 as an eigenvalue.

More generally, the function r will be more complicated, but still retains sufficient regularity that the spirit of the above simple argument applies. In the more general setting, the matrix is replaced by a bounded linear operator (the Ruelle transfer operator). [3] The spectrum of this operator is quasi-compact (i.e., aside from isolated eigenvalues of finite multiplicity, the remaining essential spectrum is in a "small" disk). The corresponding result is then in general [58], [44], [49]:

[3] The transfer operator in the context of the Modular surface is the operator $\mathcal{L}_s$ described in the last section

Theorem 2.1 *The zeta function $\zeta(s)$ converges on a half-plane $Re(s) > h$. The zeta function $\zeta(s)$ has a non-zero meromorphic extension to a larger half-plane $Re(s) > h - \epsilon$, for some $\epsilon > 0$.*

There is a simple pole at $s = h$ and, for geodesic flows, there are no other poles on the line $Re(s) = h$.

In the special case of hyperbolic flows with analytic horocycle foliations it is possible to show much more. This includes, for example, constant curvature geodesic flows. This gives much stronger results [59]:

Theorem 2.2 *The zeta function $\zeta(s)$ has a non-zero meromorphic extension to $\mathbb{C}$.*

The proof of this result is similar in spirit, except that from the hypothesis on the foliations the expanding map on the sections (in Case 2′ before) is also C^ω. The transfer operator on analytic functions is trace class and so the determinant makes perfect sense.

If the foliations are not analytic (which is the case for variable curvature surfaces) then slightly less is known [32], [62] and [21].

3 Counting Orbits

We want to mimic the use of the Riemann zeta function in prime number theory, except we want to count closed orbits instead of prime numbers. The aim is to describe that asymptotic behaviour of the number of prime numbers less than x, i.e.,

$$\pi(x) = \#\{p \leq x : p \text{ is a prime }\} \text{ for } x > 0.$$

Notation: We write $f(x) \sim g(x)$ if $\frac{f(x)}{g(x)} \to 1$ as $x \to +\infty$.

In 1896, Hadamard and de la Vallée Poussin independently showed the asymptotic estimate $\pi(x) \sim \frac{x}{\log x}$, as $x \to +\infty$, i.e., the *prime number theorem* [19]. The basic properties of $\pi(x)$ come from the *Riemann zeta function* defined by

$$\zeta_R(s) = \sum_{n=1}^{\infty} \frac{1}{n^s} = \prod_{p=prime} \left(1 - p^{-s}\right)^{-1}.$$

This converges to an analytic non-zero function on the domain $Re(s) > 1$. Moreover, $\zeta_R(s)$ has the following important properties [19]:

(1) $\zeta_R(s)$ has an analytic non-zero extension to a neighbourhood of $Re(s) \geq 1$, *except* for a simple pole at $s = 1$;
(2) $\zeta_R(s)$ has a meromorphic extension to all of $\mathbb{C}$; and $\zeta_R(s)$ and $\zeta_R(1 - s)$ are related by a functional equation.

Property (1) has a direct analogue for most hyperbolic flows (including geodesic flows). We say that a hyperbolic flow is weak mixing if the set of lengths of closed orbits $\{\lambda(\tau) : \tau = \text{ closed orbit}\}$ isn't contained in $a\mathbb{N}$, for some $a > 0$. In particular, any geodesic flow is weak mixing. [4] The following is the analogue of property (1) for the Riemann zeta function.

Theorem 3.1 *Let ϕ be a weak mixing hyperbolic flow. There exists $h > 0$ such that $\zeta(s)$ has an analytic non-zero extension to a neighbourhood of $Re(s) > h$, except for a simple pole at $s = h$.*

The value h is the topological entropy of the geodesic flow.

Unfortunately, property (2) doesn't always have a direct analogue for general hyperbolic flows (although it does for constant curvature geodesic flows). [5] However, since the proof of the prime number theorem only required property (1) for the Riemann zeta function, we expect that something similar will hold for closed orbits of hyperbolic flows. We can denote

$$\pi(T) = \text{Card}\{\tau : \lambda(\tau) \leq T\}, \text{ for } T > 0.$$

The following result is the analogue of the prime number theorem for closed orbits.

Corollary 3.2 *Let $\phi_t : M \to M$ be a weak mixing hyperbolic flow then*

$$\pi(T) \sim \frac{e^{hT}}{hT}, \text{ as } T \to +\infty. \tag{3.1}$$

This was proved, although the details were not published at the time, by Margulis in 1969. This proof did not use zeta functions, but properties of transverse measures for the horocycle foliation. (The proof was reconstructed by Toll in his unpublished Ph.D. thesis from the University of Maryland in 1984.) An alternative proof using zeta functions was given by Parry and Pollicott in [45]. Prior to this Sinai had shown in 1966 that $\lim_{T\to+\infty} \frac{1}{T} \log \pi(T) = h$. For the special case of geodesic flows on surfaces of constant curvature $\kappa = -1$, Huber showed in 1959, using the Selberg trace formula, that $\pi(T) = \text{li}(e^{hT}) + O\left(e^{cT}\right)$ where $\text{li}(x) = \int_2^x \frac{du}{\log u}$ and $c < h$ is actually related to the first non-zero eigenvalue of the Laplacian on the surface.

There are related results for counting geodesic arcs between two given points in place of closed geodesics [54].

3.1 Riemann Hypothesis and Error Terms for Primes

The (still unproved) Riemann hypothesis states that: *Riemann hypothesis* $\zeta(s)$ has all of its (non-trivial) zeros on the line $Re(s) = 1/2$.

---

[4] Although height one suspended flows over hyperbolic diffeomorphisms aren't!

[5] Indeed there are examples (due to Gallavotti) of zeta functions which have logarithmic singularities.

We recall the following:

Notation: We write $f(T) = g(T) + O(h(T))$ if there exists $C > 0$ such that $|f(T) - g(T)| \leq C|h(T)|$.

The Riemann hypothesis would imply that for any $\epsilon > 0$ we can estimate $\pi(x) = \mathrm{li}(x) + O(x^{1/2+\epsilon})$. To date, only smaller non-uniform estimates on the zero free region are known which lead to weaker error terms [19].

3.2 Error Terms for Closed Orbits

It turns out that it is more convenient to replace the principal asymptotic term by $\mathrm{li}(e^{hT}) \sim \frac{e^{hT}}{hT}$, as $T \to +\infty$.

The following result shows that for variable curvature geodesic flows we get exponential error terms (cf. [16] [53]).

Theorem 3.3 *Let $\phi_t : M \to M$ be the geodesic flow for a compact surface with negative curvature. There exists $0 < c < h$, where h again denotes the topological entropy, such that*

$$\pi(T) = \mathrm{li}(e^{hT}) + O\left(e^{cT}\right), \text{ as } T \to +\infty \tag{3.2}$$

Unfortunately, in contrast to the constant curvature case, there is little insight into the value of $c > 0$. The estimate (3.2) extends to counting closed geodesics on compact manifolds of arbitrary dimension providing that the sectional curvature is pinched $-4 \leq \kappa \leq -1$. The following result on $\zeta(s)$ is the main ingredient in the proof of Theorem 3.3.

Proposition 3.4 *For a geodesic flow there exists $c < h$ such that $\zeta(s)$ is analytic in the half-plane $Re(s) > c$, except for a simple pole at $s = h$. Moreover, there exists $0 < \alpha < 1$ such that $\zeta'(h + it)/\zeta(h + it) = O(|t|^{\alpha})$, as $|t| \to +\infty$.*

This can be viewed as an analogue of the classical Riemann Hypothesis for the zeta function for prime numbers. It is well-known for the case of constant negative curvature (using the approach of the Selberg trace formula).

At the level of more general (weak mixing) hyperbolic flows no such result can hold. Indeed, there are very simple examples with poles $\sigma_n + it_n$ for $\zeta(s)$ such that $\sigma_n \nearrow h$ (and $t_n \nearrow \infty$) [47].

3.3 Spatial Distribution of Closed Orbits

Given a geodesic flow $\phi_t : M \to M$, a classical result of Bowen [13] shows that the closed orbits τ are evenly distributed (according to the measure of maximal entropy μ). Consider a Hölder continuous function $g : \Lambda \to \mathbb{R}$, then we can weight a given closed orbit τ by $\lambda_g(\tau) = \int_0^{\lambda(\tau)} g(\phi_t x_\tau)dt$ (for $x_\tau \in \tau$). The following result was originally proved by Bowen [13], with a subsequent proof by Parry [43] using suitably weighted zeta functions.

Theorem 3.5 *Given a geodesic flow $\phi : M \to M$ there exists a probability measure μ such that*

$$\sum_{\lambda(\tau)\leq T} \lambda_g(\tau) / \sum_{\lambda(\tau)\leq T} \lambda(\tau) \to \int g d\mu \text{ as } T \to +\infty.$$

In the case of constant curvature surfaces the measure of maximal entropy is the Liouville measure (i.e., the natural normalised volume).

There are also Central Limit Theorems [57] and Large Deviation Theorems [31] for closed geodesics. In particular, the latter can be viewed as generalisations of Theorem 3.5. More generally, the following result of Kifer is valid for any hyperbolic flow and so, in particular, for the geodesic flow $\phi_t : SV \to SV$. Let μ_τ be the natural invariant measure supported on a closed orbit τ.

Proposition 3.6 *Let $\mathcal{U}$ be an open neighbourhood of the measure of maximal entropy μ in the space $\mathcal{M}$ of all ϕ-invariant probability measures on M. Then*

$$\frac{1}{\pi(T)} \#\{\tau : \lambda(\tau) \leq T \text{ and } \mu_\tau/\lambda(\tau) \notin \mathcal{U}\} = O(e^{-\delta T}),$$

as $T \to +\infty$, where $\delta = \inf_{\nu \in \mathcal{M}-\mathcal{U}}\{h - h(\nu)\} > 0$.

3.4 Homological Distribution of Closed Orbits

By way of motivation, recall the asymptotic behaviour of the number $B(x)$ of integers less than x which can be written as a square or as the sum of two squares, i.e., $B(x) = \#\{1 \leq n \leq x : n = u_1^2 + u_2^2, u_1, u_2 \in \mathbb{Z}\}$ for $x > 0$. Landau [35] showed that $B(x) \sim Kx/(\log x)^{1/2}$, for some $K > 0$, and the same result appears in Ramanujan's famous letter to Hardy in 1913 [8]. The full asymptotic expansion for $B(x)$ has the simple form

$$B(x) = \frac{Kx}{(\log x)^{1/2}} \left(1 + \sum_{n=1}^{N} \frac{\alpha_n}{(\log T)^n} + O\left(\frac{1}{(\log x)^N}\right)\right)$$

for any $N \geq 1$. [23]. The proof of the above asymptotic expansion involves studying the complex function

$$s \mapsto \frac{1}{1-2^{-s}} \prod_q \frac{1}{1-q^{-s}} \prod_r \frac{1}{1-r^{-2s}},$$

where q runs through all primes equal to 1 (mod 4) and r runs through all primes equal to 3 (mod 4). Of course, this differs from the Riemann zeta function only in the factor of 2 in the last exponent, but the result is a singularity of the form $(s-1)^{-1/2}$, rather than a simple pole, which leads to a different asymptotic behaviour.

As usual, we let V denote a compact surface of negative curvature. Let $\alpha \in H_1(V, \mathbb{Z})$ be a fixed element in the first homology. Given a closed geodesic γ we denote by $[\gamma]$ the homology class associated to a closed geodesic V. Let $\pi(T, \alpha)$ be the number of closed geodesics in the homology class α of length at most T, i.e.,

$$\pi(T, \alpha) = \#\{\gamma : l(\gamma) \leq T, \quad [\gamma] = \alpha\}.$$

The following formula was proved independently by Anantharaman [4] and Pollicott and Sharp [55].

Theorem 3.7 *Let $b = \dim(H_1(V, \mathbb{R}))$ be the first Betti number for V. There exist $C_0, C_1, C_2, \ldots$ (with $C_0 > 0$) such that*

$$\pi(T, \alpha) = \frac{e^{hT}}{T^{b/2+1}} \left(\sum_{n=0}^{N} \frac{C_n}{T^n} + O\left(\frac{1}{T^N}\right)\right) \quad \textit{as } T \to +\infty,$$

for any $N > 0$.

The similarity with Landau's result comes from a $(s-1)^{-1/2}$ singularity also appearing in the domain of the corresponding zeta function for $\pi(T, \alpha)$.

For surfaces of constant curvature $\kappa = -1$ this was originally proved by Phillips and Sarnak [46]. Katsuda and Sunada [30], Lalley [33] and Pollicott [50] then each independently showed that for more general surfaces of variable curvature the basic asymptotic formula $\pi(T, \alpha) \sim \frac{e^{hT}}{T^{b/2+1}}$, as $T \to +\infty$, still holds.

Finally, we should remark that there are interesting results on special values of the closely related homological L-functions cf. [20], [22].

3.5 Intersections of Closed Orbits

There are a number of results describing the average number of times a typical closed geodesic intersects itself [48] and [34] [6]. However, we shall describe a more topical result conjectured by Sieber and Richter[71].

Given $0 \leq \theta_1 < \theta_2 \leq 2\pi$, let $i_{\theta_1,\theta_2}(\gamma)$ denote the number of self-intersections of the closed geodesic γ such that the absolute value of the angle of intersection lies in the interval $[\theta_1, \theta_2]$.

Theorem 3.8 *Given $0 \leq \theta_1 < \theta_2 \leq 2\pi$, there exists $I = I(\theta_1, \theta_2)$ and $c < h$ such that, for any $\epsilon > 0$,*

$$\#\left\{\gamma : l(\gamma) \leq T, \frac{i_{\theta_1\theta_2}(\gamma)}{l(\gamma)^2} \in (I-\epsilon, I+\epsilon)\right\} = \mathrm{li}(e^{hT}) + O(e^{cT}).$$

We shall outline the idea of the proof, due to Sharp and the author. Let $\mathcal{F}$ denote the foliation of SV by orbits of the geodesic flow ϕ. Given any

[6] Which also corrects an error in the asymptotic expression in [48]

ϕ-invariant finite measure μ (not necessarily normalised to be a probability measure) we can consider the associated transverse measure $\tilde{\mu}$ for $\mathcal{F}$. The set of such transverse measures $\mathcal{C}$ is usually called the *space of currents*. Let $E = SV \oplus SV - \Delta$ be the Whitney sum of the bundle SV with itself, minus the diagonal $\Delta = \{(x, v, v) : x \in V, v \in S_xV\}$. Let $p : E \to V$ denote the canonical projection. In particular, points of the four dimensional vector bundle E (with two dimensional fibres) consist of triples $\{(x, v, w) : x \in V$ and $v, w \in S_xV\}$. Let $p_1 : E \to SV$ be the projection defined by $p_1(x, v, w) = v$ and let $p_2 : E \to SV$ be defined by $p_2(x, v, w) = w$. Following closely Bonahon's construction [9], we consider the two transverse foliations (with one dimensional leaves) of E given by $\mathcal{F}_1 = p_1^{-1}(\mathcal{F})$ and $\mathcal{F}_2 = p_2^{-1}(\mathcal{F})$. Given $0 \le \theta_1 < \theta_2 \le \pi$, we define the *angular intersection bundle* $E_{\theta_1,\theta_2} \subset E$ by $E_{\theta_1,\theta_2} = \{(x, v, w) \in E : \angle vw \in [\theta_1, \theta_2]\}$, where $0 \le \angle vw \le \pi$ denotes the angle between the two vectors. This is a closed sub-bundle of E.

Given currents $\tilde{\mu}, \tilde{\mu}' \in \mathcal{C}$, we can take the lifts $\widehat{\mu}_1 := p_1^{-1}\tilde{\mu}$ and $\widehat{\mu}_2' := p_2^{-1}\tilde{\mu}'$, which are transverse measures to the foliations $\mathcal{F}_1$ and $\mathcal{F}_2$ for E, respectively. Bonahon defined the *intersection form* $i : \mathcal{C} \times \mathcal{C} \to \mathbb{R}^+$ to be the total mass of the E with respect to the product measure $\widehat{\mu}_1 \times \widehat{\mu}_2'$, i.e., $i(\tilde{\mu}, \tilde{\mu}') = (\widehat{\mu}_1 \times \widehat{\mu}_2')(E)$ [9]. By analogy, we can define an *angular intersection form* $i_{\theta_1,\theta_2} : \mathcal{C} \times \mathcal{C} \to \mathbb{R}^+$ to be the total mass of the E_{θ_1,θ_2} with respect to the product measure $\widehat{\mu}_1 \times \widehat{\mu}_2'$, i.e., $i_{\theta_1,\theta_2}(\tilde{\mu}, \tilde{\mu}') = (\widehat{\mu}_1 \times \widehat{\mu}_2')(E_{\theta_1,\theta_2})$.

In the present context, the large deviation result Proposition 3.6 gives the following estimates.

Lemma 3.9 *Given $\epsilon > 0$, there exists $\delta > 0$ such that*

$$\begin{aligned}\frac{1}{\pi(T)} \quad & \#\{\gamma : l(\gamma) \le T \text{ and } |l(\gamma)^{-2}(\widehat{\mu}_{\gamma,1} \times \widehat{\mu}_{\gamma,2})(E_{\theta_1,\theta_2}) \\ & \qquad -(\widehat{m}_1 \times \widehat{m}_2)(E_{\theta_1,\theta_2})| \ge \epsilon\} \\ &= O(e^{-\delta T}), \text{ as } T \to +\infty.\end{aligned}$$

In particular, we can set $I(\theta_1, \theta_2) := i_{\theta_1,\theta_2}(\tilde{\mu}, \tilde{\mu})$, where μ is the measure of maximal entropy. We deduce that, except for an exceptional set with cardinality of order $O(e^{(h-\delta)T})$, the set of closed geodesics of length at most T satisfy $|l(\gamma)^{-2} i_{\theta_1,\theta_2}(\gamma) - I(\theta_1, \theta_2)| < \epsilon$. Theorem 3.8 then follows easily by applying the asymptotic counting results described in §3.2.

3.6 Decay of Correlations (a Compliment to Counting Orbits)

A closely related problem to that of counting closed orbits is that of decay of correlations. Let $\phi_t : M \to M$ be a weak-mixing hyperbolic flow and let μ again be the measure of maximal entropy (i.e., the measure in Theorem 3.5). The flow ϕ is strong mixing, i.e.,

$$\rho_{F,G}(t) := \int F \circ \phi_t G d\mu - \int F d\mu \int G d\mu \to 0, \text{ for all } F, G \in L^2(X, \mu).$$

(i.e., the "correlation of the flow tends to zero".)

Dolgopyat proved the following result on exponential decay of correlations in the case of geodesic flows on compact negatively curved surfaces [17].

Theorem 3.10 *Let $\phi_t : M \to M$ be the geodesic flow for a surface of variable negative curvature. There exists $\epsilon > 0$ such that for any smooth functions $F, G : M \to \mathbb{R}$ there exists $C > 0$ with $\rho_{F,G}(t) \le Ce^{-\epsilon|t|}$.*

For constant negative curvature surfaces this result can be proved using representation theory [42], [15]. Moreover, there are very few examples of hyperbolic flows for which exponential decay of correlations is known to hold [17].

The complex function used in the study of $\rho_{F,G}(t)$ is its Fourier transform $\widehat{\rho}_{F,G}(s) = \int e^{ist}\rho_{F,G}(t)dt$.

Theorem 3.11 *Let $\phi : M \to M$ be a C^r hyperbolic flow ($r \ge 2$ or $r = \omega$). There is a neighbourhood $\mathcal{V}$ of ϕ (amongst C^r flows on M) such that:*

> *there exists $\epsilon > 0$ such that the associated correlation function $\widehat{\rho}^{(\psi)}(s)$ has a meromorphic extension to a strip $|Im(s)| < \epsilon$, for each $\psi \in \mathcal{V}$ [47]; and whenever $s_i = s_i(\phi)$ is a simple pole for $\widehat{\rho}^{(\phi)}(s)$ in the strip $|Im(s)| < \epsilon$ then the map $\mathcal{V} \ni \psi \mapsto s_i(\psi)$ is C^{r-2} [51].*

Moreover, since the analysis of the Fourier transform also depends on the Ruelle transfer operator there is a direct relationship between the poles of $\hat{\rho}_{F,G}(s)$ and $\zeta_\phi(s)$ (described in [47]). More precisely, s (with $Im(s) < 0$) is a pole for $\widehat{\rho}(s)$ if and only if $h + is$ is a pole for $\zeta(s)$.

If we replace μ by the Liouville measure (or any other suitable Gibbs measure) analogous results hold, with a suitably weighted zeta function.

4 Other Applications of Closed Geodesics

Ruelle's approach to the proof of theorem 2.2 has a number of other applications. Here we recall a couple of our favourites.

4.1 Determinants of the Laplacian

A very interesting object in the case of surfaces V of constant negative curvature $\kappa = -1$ is the (functional) *determinant of the Laplacian*. The Laplacian $\Delta : L^2(V) \to L^2(V)$ is a self-adjoint linear differential operator. Let us write the spectrum of $-\Delta$ as $0 = \lambda_0 < \lambda_1 \le \lambda_2 \le \cdots \nearrow +\infty$ and consider the associated Dirichlet series

$$\eta(s) := \sum_{n=1}^{\infty} \lambda_n^{-s}.$$

This converges for $Re(s) > 1$, as is easily seen using Weyl's theorem. The function $\eta(s)$ has a meromorphic extension to $\mathbb{C}$ and we define the determinant by $\det(\Delta) := \exp(-\eta'(0))$. [7] The function $\det(\Delta)$ depends smoothly on the Riemann metric. There is considerable interest in understanding its critical points [66].

Somewhat surprisingly this quantity can be explicitly expressed in terms of the closed geodesics. The starting point is that is a direct connection between $\det(\Delta)$ and the Selberg zeta function. First we define for each $n \geq 1$ the function

$$a_n := \sum_{|\tau_1|+\cdots+|\tau_r|=n} (-1)^{r+1} \frac{\lambda(\tau_1) + \cdots \lambda(\tau_k)}{(e^{\lambda(\tau_1)} - 1) \cdots (e^{\lambda(\tau_k)} - 1)},$$

where the sum is over collections of closed orbits for the geodesic flow (or, equivalently, closed geodesics) and $|\tau|$ denotes the word length of a corresponding conjugacy class in $\pi_1(V)$ with respect to a suitable choice of generators (i.e., the smallest number of generators needed to write an element in this conjugacy class). The following theorem was proved in [52]. [8]

Theorem 4.1 *We can write* $\det(\Delta) = C(g) \sum_{n=1}^{\infty} a_n$, *where the series is absolutely convergent (and* $|a_n| = O(\theta^{n^2})$*) and* $C(g)$ *is a constant depending only on the genus* g *of the surface* V.

It is also possible to use the zeta functions to describe the dependence of other dynamical invariants, such as entropy [28].

4.2 Computation

It is an interesting problem to get numerical estimates on dynamical properties for interval maps. For example, given an expanding interval maps it might be interesting to estimate the entropy (of the absolutely continuous invariant measure). The "classical" approach to this problem is the Ulam method, in which the map is essentially approximated by a piecewise linear map and the density can be estimated from the eigenvectors of the matrix.

We can now describe a somewhat different method which applies to C^ω expanding maps $T : I \to I$ on an interval I. We can define invariant (signed) measures ν_M defined by

$$\nu_M = \sum_{\substack{(n_1,\ldots,n_m) \\ n_1+\ldots+n_m \leq M}} \frac{(-1)^m}{m!} \sum_{i=1}^{m} \sum_{x \in \mathrm{Fix}(n_i)} \left(\prod_{\substack{j=1 \\ j \neq i}}^{m} \sum_{z \in \mathrm{Fix}(n_j)} r(z, n_j) \right) \frac{\delta_x}{|(T^{n_i})'(x) - 1|}$$

where δ_x is the Dirac measure and the first summation is over ordered m-tuples of positive integers whose sum is not greater than M, where $\mathrm{Fix}(n)$ denotes the set of fixed points of T^n, and where

[7] A particularly nice introduction to this subject is [66].

[8] The title of this article is good humoured reference to the title of the Ph.D. thesis of G. McShane.

$$r(x,n) = \frac{1}{n|(T^n)'(x) - 1|}.$$

The measure ν_M is supported on those periodic points of period at most M, which can easily be computed in practise. Introducing the normalisation constant

$$I_M = \sum_{\substack{(n_1,\ldots,n_m)\\ n_1+\ldots+n_m \le M}} \frac{(-1)^m}{m!} \sum_{i=1}^{m} \sum_{x\in\mathrm{Fix}(n_i)} \left(\prod_{\substack{j=1\\ j\neq i}}^{m} \sum_{z\in\mathrm{Fix}(n_j)} r(z,n_j) \right) \frac{1}{|(T^{n_i})'(x) - 1|},$$

we then define the invariant signed probability measures $\mu_M = I_M^{-1}\nu_M$. For real analytic maps we have the following [27]:

Theorem 4.2 *Let μ be the absolutely continuous T-invariant probability measure. There is a sequence of T-invariant signed probability measures μ_M, supported on the points of period at most M, such that for every C^ω function $g : I \to \mathbb{R}$, there exists $0 < \theta < 1$ and $C > 0$ with*

$$\left| \int g \, d\mu_M - \int g \, d\mu \right| \le C\theta^{M(M+1)/2}.$$

In particular, with the choice $g = \log|T'|$ we have as a corollary that the "Lyapunov exponent" $\lambda_\mu = \int \log|T'| \, d\mu$ can be quickly approximated, i.e.,

$$\left| \int \log|T'| \, d\mu_M - \lambda_\mu \right| \le C\theta^{M(M+1)/2}.$$

Many related ideas appear in the beautiful work of Cvitanovic and his coauthors.

5 Frame Flows

Recently, there has been interest in extending results for hyperbolic flows to partially hyperbolic flows. That is, we allow some transverse directions to the flow that are neither expanding nor contracting. The principle example of such systems are probably the frame flow, which is an extension of the geodesic flow $\phi_t : M \to M$ on the unit tangent bundle, for a manifold V with negative sectional curvatures.

5.1 Frame Flows: Archimedean Version

Let $St_{n+1}(V)$ be the space of (positively oriented) orthonormal $(n+1)$-frames. The frames $St_{n+1}(V)$ form a fibre bundle over M with a natural projection $\pi : St_{n+1}(V) \to M$ which simply forgets all but the first vector in the frame, i.e., $\pi(v_1, \ldots, v_{n+1}) = v_1$. The frame flow $f_t : St_{n+1}(V) \to St_{n+1}(V)$ acts on frames $(v_1, \ldots, v_{n+1}) \in St_{n+1}(V)$ by parallel transporting for time t the frame along the geodesic $\gamma_{v_1} : \mathbb{R} \to V$ satisfying $v_1 = \dot{\gamma}_{v_1}(0)$. In particular, the frame flow semi-conjugates to the geodesic flow, i.e., $\pi f_t = \phi_t \pi$ for all $t \in \mathbb{R}$.

The associated structure group acts on each fibre by rotating the frames about the first vector v_1. In particular, we can identify each fibre $\pi^{-1}(v) \subset St_{n+1}(V)$, for $v \in St_1(V)$, with the compact group $SO(n)$. We can associate to each closed orbit τ a holonomy element $[\tau] \in SO(n-1)$ (defined up to conjugacy). The following is the natural analogue of Theorem 3.5 [44].

Theorem 5.1 *Let $f : SO(n-1) \to \mathbb{R}$ be a continuous function constant on conjugacy classes. Then*

$$\frac{1}{\pi(T)} \sum_{\lambda(\tau) \le T} f([\tau]) \to \int f d\omega, \quad \text{as } T \to +\infty,$$

where ω is the Haar measure on $SO(n-1)$.

The idea of the proof is that we can model the underlying geodesic flow symbolically by a sequence space Σ, etc. But for the frame flow we additionally have an associated map $\Theta : \Sigma \to SO(n-1)$ which essentially measures the "twist" in $SO(n-1)$ along the orbits.

The distribution properties of frame flows on certain non-compact manifolds have been considered in [36]. In this context, there is a particularly interesting connection with Clifford numbers [3].

5.2 Non-Archimedean Version

Let $\mathbb{Q}_p$ denote the p-adic numbers with the usual valuation $|\cdot|_p$. Let $\mathbb{Z}_p = \{x \in \mathbb{Q}_p : |x|_p \le 0\}$ denote the p-adic integers. We can study a natural analogue of the frame flow and geodesic flow for $G = PSL(2, \mathbb{Q}_p)$. The rôle of the hyperbolic upper half plane $\mathbb{H}^2$ in the usual archimedean case is taken here by a regular tree X, say. We recall the basic construction.

Vertices Given any pair of vectors $v_1, v_2 \in \mathbb{Q}_p^2$ one associates a *lattice* $L = v_1\mathbb{Z}_p + v_2\mathbb{Z}_p$. We can define an equivalence relation on lattices: $L \sim L'$ if lattices L, L' are homothetically related (i.e., there exists $\alpha \in \mathbb{Q}_p$ such that $L' = \alpha L$). We take the equivalence classes $[L]$ to be the vertices of the tree X.

Edges Given two vertices (equivalence classes) $[L_1], [L_2]$ we can associate an edge $[L_1] \to [L_2]$ whenever we can find a basis $\{v_1, v_2\}$ for L_1 and $\{\pi v_1, v_2\}$ for L_2, where $\pi = \frac{1}{p}$ is called the *uniformizer.*

Lemma 5.2 *[70] X is a homogeneous tree, with every vertex having $(p+1)$-edges.*

There is a natural action $GL(2, \mathbb{Q}_p) \times X \to X$ on the tree given by $\gamma[v_1\mathbb{Z}_p + v_2\mathbb{Z}_p] = [(\gamma v_1)\mathbb{Z}_p + (\gamma v_2)\mathbb{Z}_p]$. The construction and action is elegantly described by Serre [70]. The frame flow is actually a discrete action defined on the quotient space $\Gamma\backslash X$ of the associated tree X by a lattice Γ and is given as multiplication by $\left(\begin{smallmatrix} 1 & p \\ 0 & \pi \end{smallmatrix}\right)$. If Γ is torsion free then there is a natural shift map on the space of paths $\sigma : \Sigma \to \Sigma$ which plays the role of the geodesic flow. Let $\mathcal{S}$ be the closed multiplicative subgroup of squares in $\mathcal{O}^{\times} = \{x \in \mathbb{Q}_p : |x|_p = 0\}$. There exists a Hölder continuous function $\Theta : \Sigma \to \mathcal{S}$ such that the p-adic frame flow for a lattice Γ corresponds to a simple skew product

$$\begin{aligned} \widehat{\sigma} &: \Sigma \times \mathcal{S} \to \Sigma \times \mathcal{S} \\ \widehat{\sigma}(x, s) &= (\sigma x, \Theta(x)s). \end{aligned} \tag{5.1}$$

Let Γ_n be the set of conjugacy classes of $\gamma \in \Gamma$ with $|\mathrm{tr}\gamma|_p = n$. For each conjugacy class $[\gamma] \in \Gamma_n$, denote by $\sigma([\gamma]) \in \mathcal{S}$ the common value of $p^{|\lambda_\gamma^2|_p}\lambda_\gamma^2$, where λ_γ denotes the maximal eigenvalue. The analogue of Theorem 5.1 is the following result of Ledrappier and Pollicott.

Theorem 5.3 *Eigenvalues of matrices in Γ are uniformly distributed in the sense that for any continuous function ϕ on $\mathcal{S}$, we have:*

$$\lim_{n\to\infty} \frac{1}{p^{2n}} \sum_{[\gamma]\in\Gamma_n} \phi(\sigma([\gamma])) = \int \phi(s) d\omega(s),$$

where ω is the Haar probability measure on $\mathcal{S}$.

This can be viewed as a non-archemidean version of the results in [67]. Moreover, in the particular case that Γ is an arithmetic lattice it is possible to use Deligne's solution of the Ramanujan-Petersson conjecture to get uniform exponential convergence in Theorem 5.3.

References

1. R. Adler and L. Flatto *Geodesic flows, interval maps, and symbolic dynamics* Bull. Amer. Math. Soc. **25** (1991) 229–334
2. R. Adler and B. Weiss *Entropy, a complete metric invariant for automorphisms of the torus* Proc. Nat. Acad. Sci. U.S.A. **57** (1967) 1573–1576
3. L. Ahlfors *Möbius transformations and Clifford numbers* Differential Geometry and Complex Analysis I. Chavel and H.M. Farkas ed. Springer Berlin (1985)
4. N. Anantharaman *Precise counting results for closed orbits of Anosov flows* Ann. Sci. École Norm. Sup. **33** (2000) 33–56
5. D. Anosov *Geodesic flows on closed Riemann manifolds with negative curvature* Proceedings of the Steklov Institute of Mathematics, No. 90 American Mathematical Society Providence, R.I. (1969)
6. M. Babillot and F. Ledrappier *Lalley's theorem on periodic orbits of hyperbolic flows* Ergod. Th. and Dynam. Sys. **18** (1998) 17–39
7. V. Baladi *Periodic orbits and dynamical spectra* Ergod. Th. and Dynam. Sys. **18** (1998) 255–292

8. B. Berndt and R. Rankin *Ramanujan: Letters and commentary.* History of Mathematics, 9 American Mathematical Society Providence, RI (1995)
9. F. Bonahon *Bouts des variétés hyperboliques de dimension* 3 Ann. of Math. **124** (1986) 71–158
10. R. Bowen *Symbolic Dynamics for hyperbolic flows* Amer. J. Math. **95** (1973) 429–460
11. R. Bowen *On Axiom A diffeomorphisms* Reg. Conf. Series, No. 35. Amer. Math. Soc Providence (1978)
12. R. Bowen *Equilibrium states and the ergodic theory of Anosov diffeomorphisms.* Lecture Notes in Mathematics, Vol. 470 Springer (1975)
13. R. Bowen *Periodic orbits for hyperbolic flows* Amer. J. Math. **94** (1972) 1–30
14. L. Carleson and T. Gamelin *Complex dynamics* Universitext: Tracts in Mathematics Springer-Verlag New York (1993
15. P. Collet, H. Epstein and G. Gallavotti *Perturbations of geodesic flows on surfaces of constant negative curvature and their mixing properties* Comm. Math. Phys. **95** (1984) 61–112
16. D. Dolgopyat *On the statistical properties of geodesic flows on negatively curved surfaces* Thesis, Princeton University (1997)
17. D. Dolgopyat *On decay of correlations in Anosov flows* Annals of Math. **147** (1998) 357–390
18. D. Dolgopyat and M. Pollicott *Addendum to: "Periodic orbits and dynamical spectra" by V. Baladi* Ergod. Th. and Dynam. Sys. **18** (1998) 293–301
19. W. Ellison and F. W. Ellison *Prime Numbers* Wiley Paris (1985)
20. D. Fried *The zeta functions of Ruelle and Selberg I* Ann. Sc. Ec. Norm. Sup. **19** (1986) 491–517
21. D. Fried *The flat-trace asymptotics of a uniform system of contractions* Ergod. Th. and Dynam. Sys. **15** (1995) 1061–1071
22. D. Fried *Lefschetz formulas for flows* The Lefschetz centennial conference (Mexico City, 1984) Contemp. Math. Vol 58 III 19–69, Amer. Math. Soc. Providence, R.I. (1987)
23. G. Hardy *Ramanujan: twelve lectures on subjects suggested by his life and work* Chelsea New York (1940)
24. N. Hayden *Meromorphic extension of the zeta function for Axiom A flows* Ergod. Th. and Dynam.Sys. **10** (1990) 347–360
25. G. Hedlund *On the Metrical Transitivity of the Geodesics on Closed Surfaces of Constant Negative Curvature* Ann. of Math. **35** (1934) 787–808
26. H. Huber *Zur analytischen Theorie hyperbolischen Raumformen und Bewegungsgruppen* Math. Ann. **138** (1959) 1–26
27. O. Jenkinson and M. Pollicott *Calculating Hausdorff dimensions of Julia sets and Kleinian limit sets* Amer. J. Math. **124** (2002) 495–545
28. A. Katok, G. Knieper, M. Pollicott and H. Weiss *Differentiability and analyticity of entropy for Anosov and geodesic flows* Invent. Math. **98** (1989) 581–597
29. A. Katsuda and T. Sunada *Homology and closed geodesics in a compact Riemann surface* Amer. J. Math. **110** (1988) 145–155
30. A. Katsuda and T. Sunada *Closed orbits in homology classes* Publ. Math. **71** (1990) 5–32
31. Y. Kifer *Large deviations, averaging and periodic orbits of dynamical systems* Comm. Math. Phys. **162** (1994) 33–46

32. A. Kitaev *Fredholm determinants for hyperbolic diffeomorphisms of finite smoothness* Nonlinearity **12** (1999) 141–179
33. S. Lalley *Closed geodesic in homology classes on surfaces of variable negative curvature* Duke Math, J. **58** (1989) 795–821
34. S. Lalley *Self-intersections of closed geodesics on a negatively curved surface: statistical regularities* Convergence in ergodic theory and probability (Columbus, OH, 1993) Ohio State Univ. Math. Res. Inst. Publ., 5 263–272 de Gruyter Berlin (1996)
35. E. Landau *Über die Einteilung der positiven ganzen Zahlen in vier Klassen nach der Mindeszahl der zu ihrer additiven Zusammensetzung erforderlichen Quadrate* Arch. Math. Phys. **13** 305–312 (1908)
36. F. Ledrappier and M. Pollicott *Ergodic properties of linear actions of* 2×2 *matrices* Duke Math. J. **116** (2003) 353–388
37. F. Ledrappier and M. Pollicott *Distribution results for lattices in* $SL(2, \mathbb{Q}_p)$, *Preprint*
38. J. Lewis, and D. Zagier *Period functions for Maass wave forms, I* Ann. of Math. **153** (2001) 191–258
39. A. Margulis *Certain applications of ergodic theory to the investigation of manifolds of negative curvature* Fun. Anal. Appl. **3** (1969) 89–90
40. D. Mayer *On a* ζ *function related to the continued fraction transformation* Bull.Soc. Math. France **104** (1976) 195–203
41. D. Mayer *The thermodynamic formalism approach to Selberg's zeta function for* $\mathrm{PSL}(2, Z)$ Bull. Amer. Math. Soc. **25** (1991) 55–60
42. C. Moore *Exponential decay of correlation coefficients for geodesic flows* 163–181 Math. Sci. Res. Inst. Publ., 6 Springer New York (1987)
43. W. Parry *Bowen's equidistribution theory and the Dirichlet density theorem* Ergod. Th. and Dynam. Sys. **4** (1984) 117–134
44. W. Parry and M. Pollicott *Zeta functions and the periodic orbit structure of hyperbolic dynamics* Asterisque **187–188** (1990) 1–268
45. W. Parry and M. Pollicott *An analogue of the prime number theorem for closed orbits of Axiom A flows* Annals Math. **118** (1983) 573–591
46. R. Phillips and P. Sarnak *Geodesics in homology classes* Duke Math. J. **55** (1987) 287–297
47. M. Pollicott *On the rate of mixing of Axiom A flows* Invent. Math. **81** (1985) 413–426
48. M. Pollicott *Asymptotic distribution of closed geodesics* Israel J. Math. **52** (1985) 209–224
49. M. Pollicott *Meromorphic extensions of generalized zeta functions* Invent. Math. **85** (1986) 147–164
50. M. Pollicott *Homology and closed geodesics in a compact negatively curved surface,* Amer. J. Math. **113** (1991) 379–385
51. M. Pollicott *Stability of mixing rates for Axiom A attractors* Nonlinearity **16** (2003) 567–578
52. M. Pollicott and A. Rocha *A remarkable formula for the determinant of the Laplacian* Invent. Math **130** (1997) 399–414
53. M. Pollicott and R. Sharp *Exponential error terms for growth functions on negatively curved surfaces* Amer. J. Math **120** (1998) 1019–1042
54. M. Pollicott and R. Sharp *Orbit counting for some discrete groups acting on simply connected manifolds with negative curvature* Invent. Math. **117** (1994) 275–302

55. M. Pollicott and R. Sharp *Asymptotic expansions for closed orbits in homology classes* Geometriae Dedicata **87** (2001) 123–160
56. M. Ratner *Markov partitions for Anosov flows on n-dimensional manifolds* Israel J. Math. **15** (1973) 92–114
57. M. Ratner *The central limit theorem for geodesic flows on n-dimensional manifolds of negative curvature* Israel J. Math. 16 **16** (1973) 181–197
58. D. Ruelle *Generalized zeta-functions for Axiom A basic sets* Bull. Amer. Math. Soc. **82** (1976) 153–156
59. D. Ruelle *Zeta-functions for expanding maps and Anosov flows* Invent. Math. **34** (1976) 231–242
60. D. Ruelle *Thermodynamic Formalism* Addison Wesley New York (1978)
61. D. Ruelle *An extension of the theory of Fredholm determinants* Publ. Math. (IHES) **72** (1990) 175–193
62. D. Ruelle *Dynamical zeta functions: where do they come from and what are they good for?* Mathematical physics, X (Leipzig, 1991) 43–51, Springer Berlin **1992**
63. D. Ruelle *Analytic completion for dynamical zeta functions* Helv. Phys. Acta **66** (1993) 181–191
64. D. Ruelle *Repellers for real analytic maps* Ergod. Th. and Dynam. Sys. **2** (1982) 99–107
65. H. Rugh *The correlation spectrum for hyperbolic analytic maps* Nonlinearity **5** (1992) 1237–1263
66. P. Sarnak *Arithmetic quantum chaos The Schur lectures,8, Bar-Ilan Univ., Ramat Gan* Israel Math. Conf. Proc. **8** (1995) 183–236
67. P. Sarnak and M. Wakayama *Equidistribution of holonomy about closed geodesics.* Duke Math. J. **100** (1999) 1–57
68. A. Selberg *Harmonic analysis and discontinuous groups in weakly symmetric Riemannian spaces with applications to Dirichlet series* J. Indian Math. Soc. **20** (1956) 47–87
69. C. Series *Geometrical Markov coding of geodesics on surfaces of constant negative curvature* Ergodic Theory Dynam. Systems **6** (1986) 601–625
70. S.-P. Serre *Trees* Springer Berlin (1970)
71. M. Sieber and K. Richter *Correlations between periodic orbits and their rôle in spectral statistics* Physica Scripta **T90** (2001) 128–133
72. Y. Sinai *Gibbs measures in ergodic theory* Russ. Math. Surv. **27** (1972) 21–69
73. S. Smale *Differentiable dynamical systems* Bull. Amer. Math. Soc. **73** (1967) 747–817
74. F. Tangerman *Ph. D. thesis Boston University* (1984)

Part III

Dynamical Systems: Interval Exchange, Flat Surfaces, and Small Divisors

Continued Fraction Algorithms for Interval Exchange Maps: an Introduction

Jean-Christophe Yoccoz

Collège de France, 3 Rue d'Ulm, F-75005 Paris, France

1 Introduction

Rotations on the circle $\mathbf{T} = \mathbf{R}/\mathbf{Z}$ are the prototype of quasiperiodic dynamics. They also constitute the starting point in the study of smooth dynamics on the circle, as attested by the concept of rotation number and the celebrated Denjoy theorem. In these two cases, it is important to distinguish the case of rational and irrational rotation number. But, if one is interested in the deeper question of the smoothness of the linearizing map, one has to solve a small divisors problem where the diophantine approximation properties of the irrational rotation number are essential. The classical continuous fraction algorithm generated by the Gauss map $G(x) = \{x^{-1}\}$ (where $x \in (0,1)$ and $\{y\}$ is the fractional part of a real number y) is the natural way to analyze or even define these approximation properties. The modular group $GL(2,\mathbf{Z})$ is here of fundamental importance, viewed as the group of isotopy classes of diffeomorphisms of $\mathbf{T}^2$, where act the linear flows obtained by suspension from rotations.

There is one obvious and classical way to generalize linear flows on the 2-dimensional torus : linear flows on higher dimensional tori. One can still define the classical diophantine approximation properties and obtain KAM-type linearization results. However, we are far from understanding these approximation properties as well as in the classical case, basically because for $n \geq 3$ the group $GL(n, \mathbf{Z})$ is far from hyperbolic and we cannot hope for a continuous fraction algorithm having all the wonderful properties it has for $n = 2$.

A less obvious way to generalize linear flows on the 2-dimensional torus, but one which has received a lot of attention in recent years, is to consider linear flows on compact surfaces of higher genus called translation surfaces. We refer to Zorich's paper in this volume for a precise definition and an introduction to these very natural geometrical structures.

Linear flows on translation surfaces may be obtained as singular suspensions of one-dimensional maps of an interval called interval exchange maps (i.e.m). Such a map is obtained by cutting the interval into d pieces and rearranging the pieces ; when $d = 2$, this is nothing else than a rotation if the endpoints of the interval are identified to get a circle ; for $d = 3$, one is still quite close to rotations (see Section 2.7 below) ; for $d \geq 4$, one can already obtain surfaces of genus ≥ 2. Interval exchange maps (and translation surfaces) occur naturally when analyzing the dynamics of rational polygonal billiards.

An early important result is the proof by Katok-Stepin [4] that almost all i.e.m with $d = 3$ are weakly mixing. Somewhat later, Keane began a systematic study of i.e.m and discovered the right concept of irrationality in this setting ([Ke1]). He also conjectured that almost all i.e.m are uniquely ergodic. One should here beware that minimality is not sufficient to guarantee unique ergodicity, as shown by examples of Keynes-Newton [8], see also [1] and Keane [6]. Keane's conjecture was proved by Masur [11] and Veech [17] independently, see also Kerckhoff [7] and Rees [15]. The key tool developed by Veech, and also considered by Rauzy [14], is a continuous fraction algorithm for i.e.m which has most of the good properties of the classical Gauss map. However, the unique absolutely continuous invariant measure for the elementary step of this algorithm is infinite. In order to be able to apply powerful ergodic-theoretical tools such as Oseledets multiplicative ergodic theorem, one needs an absolutely continuous invariant probability measure ; this was achieved by Zorich [22] by considering an appropriate acceleration of the Rauzy-Veech continuous fraction algorithm.

Our aim in the following is to present the basic facts on the continuous fraction algorithm and its acceleration. After defining precisely interval exchange maps (Section 2), we introduce Keane's condition (Section 3), which guarantees minimality and is exactly the right condition of irrationality to start a continuous fraction algorithm. The basic step of the Rauzy-Veech algorithm is then introduced (Section 4). It appears that unique ergodicity is easily characterized in terms of the algorithm, and we give a proof of the Mazur-Veech theorem (Section 4.4). Next we explain how to suspend i.e.m to obtain linear flows on translation surfaces (Section 5). The continuous fraction algorithm extends to this setting and becomes basically invertible in this context. In the last chapter, we introduce Zorich's accelerated algorithm (Section 6.2) and the absolutely continuous invariant probability measure. However, we stop short of making use of this probability measure and develop the ergodic-theoretic properties of i.e.m and the continuous fraction algorithm. We refer the reader for these to [19, 20, 21, 23, 24, 25, 3].

Coming back to small divisors problems, there does not exist today a KAM-like theory of non linear perturbations of i.e.m. However, as far as the linearized conjugacy equation (also known as the cohomological equation, or the cocycle equation, or the difference equation) is concerned, Forni has obtained [2] fundamental results (in the continuous time setting) which leave some hope that such a theory could exist. Forni solves the cohomological equation (under a finite number of linear conditions) for an unspecified full measure set of i.e.m. In a jointwork with Marmi and Moussa [12], we use the continuous fraction algorithm to formulate an explicit diophantine condition (Roth type i.e.m) of full measure which allows to solve the cohomological equation (with slightly better loss of differentiability than Forni).

One last word of caution : one of the nice properties of the algorithm is its invariance under the basic time-reversal involution. However, the usual notations do not reflect this and lead by forcing an unnatural renormalization to complicated combinatorial formulas. We have thus chosen to depart from the usual notations by adopting from the start notations which are invariant under this fundamental involution. This may cause some trouble but the initial investment should be more than compensated later by simpler combinatorics.

2 Interval Exchange Maps

2.1 An interval exchange map (i.e.m) is determined by combinatorial data on one side, length data on the other side.

The combinatorial data consist of a finite set $\mathcal{A}$ of names for the intervals and of two bijections π_0, π_1 from $\mathcal{A}$ onto $\{1, \ldots, d\}$ (where $d = \#\mathcal{A}$); these indicate in which order the intervals are met between and after the map.

The length data $(\lambda_\alpha)_{\alpha\in\mathcal{A}}$ give the length $\lambda_\alpha > 0$ of the corresponding interval.

More precisely, we set

$$\begin{aligned} I_\alpha &:= [0, \lambda_\alpha) \times \{\alpha\} , \\ \lambda^* &:= \textstyle\sum_{\mathcal{A}} \lambda_\alpha , \\ I &:= [0, \lambda^*) . \end{aligned}$$

We then define, for $\varepsilon = 0, 1$, a bijection j_ε from $\bigsqcup_{\mathcal{A}} I_\alpha$ onto I :

$$j_\varepsilon(x, \alpha) = x + \sum_{\pi_\varepsilon(\beta) < \pi_\varepsilon(\alpha)} \lambda_\beta .$$

The i.e.m T associated to these data is the bijection $T = j_1 \circ j_0^{-1}$ of I.

2.2 If $\mathcal{A}, \pi_0, \pi_1, \lambda_\alpha$ are as above and $X : \mathcal{A}' \to \mathcal{A}$ is a bijection, we can define a new set of data by

$$\pi'_\varepsilon = \pi_\varepsilon \circ X \ , \quad \varepsilon = 0, 1 \ ,$$
$$\lambda'_{\alpha'} = \lambda_{X(\alpha')} \ , \ \alpha' \in \mathcal{A}' \ .$$

Obviously, the "new" i.e.m T' determined by these data is the same, except for names, than the old one. In particular, we could restrict to consider **normalized combinatorial data** characterized by

$$\mathcal{A} = \{1, \ldots, d\} \ , \ \pi_0 = id_{\mathcal{A}} \ .$$

However, this leads later to more complicated formulas in the continuous fraction algorithm because the basic operations on i.e.m do not preserve normalization.

2.3 Given combinatorial data $(\mathcal{A}, \pi_0, \pi_1)$, we set, for $\alpha, \beta \in \mathcal{A}$

$$\Omega_{\alpha,\beta} = \begin{cases} +1 \text{ if } \pi_0(\beta) > \pi_0(\alpha), \pi_1(\beta) < \pi_1(\alpha) \ , \\ -1 \text{ if } \pi_0(\beta) < \pi_0(\alpha), \pi_1(\beta) > \pi_1(\alpha) \ , \\ 0 \quad \text{otherwise} \ . \end{cases}$$

The matrix $\Omega = (\Omega_{\alpha,\beta})_{(\alpha,\beta)\in\mathcal{A}^2}$ is antisymmetric.

Let $(\lambda_\alpha)_{\alpha\in\mathcal{A}}$ be length data, and let T be the associated i.e.m. For $\alpha \in \mathcal{A}, y \in j_0(I_\alpha)$, we have

$$T(y) = y + \delta_\alpha \ ,$$

where the **translation vector** $\delta = (\delta_\alpha)_{\alpha\in\mathcal{A}}$ is related to the **length vector** $\lambda = (\lambda_\alpha)_{\alpha\in\mathcal{A}}$ by :

$$\delta = \Omega\lambda$$

2.4 There is a **canonical involution** $\mathcal{I}$ acting on the set of combinatorial data which exchange π_0 and π_1. For any set $(\lambda_\alpha)_{\alpha\in\mathcal{A}}$ of length data, the interval I_α, I are unchanged, but j_0, j_1 are exchanged and T is replaced by T^{-1}. The matrix Ω is replaced by $-\Omega$ and the translation vector δ by $-\delta$.

Observe that $\mathcal{I}$ does not respect combinatorial normalization.

2.5 In the following, we will always consider only combinatorial data $(\mathcal{A}, \pi_0, \pi_1)$ which are **admissible**, meaning that for all $k = 1, 2, \ldots, d-1$, we have

$$\pi_0^{-1}(\{1, \ldots, k\}) \neq \pi_1^{-1}(\{1, \ldots, k\}) \ .$$

Indeed, if we had $\pi_0^{-1}(\{1, \ldots, k\}) = \pi_1^{-1}(\{1, \ldots, k\})$ for some $k < d$, for any length data $(\lambda_\alpha)_{\alpha\in\mathcal{A}}$, the interval I would decompose into two disjoint invariant subintervals and the study of the dynamics would be reduced to simpler combinatorial data.

2.6 Assume that $\#\mathcal{A} = 2, \mathcal{A} = \{A, B\}$. Without loss of generality, we have $\pi_0(A) = \pi_1(B) = 1, \pi_1(A) = \pi_0(B) = 2$. When we identify $I = [0, \lambda^*)$ with the circle $\mathbf{R}/\lambda^*\mathbf{Z}$, the i.e.m T becomes the rotation by λ_B.

2.7 Assume that $\#\mathcal{A} = 3, \mathcal{A} = \{A, B, C\}$. Without loss of generality, we may also assume that $\pi_0(A) = 1, \pi_0(B) = 2, \pi_0(C) = 3$. Amongst the 6 bijections from $\mathcal{A}$ onto $\{1, 2, 3\}$, there are 3 choices for π_1 giving rise to admissible pairs (π_0, π_1), namely :

(i) $\pi_1(A) = 2,\ \pi_1(B) = 3,\ \pi_1(C) = 1;$
(ii) $\pi_1(A) = 3,\ \pi_1(B) = 1,\ \pi_1(C) = 2;$
(iii) $\pi_1(A) = 3,\ \pi_1(B) = 2,\ \pi_1(C) = 1.$

In case (i) and (ii), we obtain again a rotation on the circle $\mathbf{R}/\lambda^*\mathbf{Z}$ identified to I. In case (iii), consider $\hat{I} = [0, \lambda^* + \lambda_B)$ and $\hat{T} : \hat{I} \to \hat{I}$ defined by

$$\hat{T}(y) = \begin{cases} y + \lambda_C + \lambda_B \text{ for } y \in [0, \lambda_A + \lambda_B) \\ y - \lambda_A - \lambda_B \text{ for } y \in [\lambda_A + \lambda_B, \lambda^* + \lambda_B) \end{cases}$$

Then $\hat{T}$ is an i.e.m on $\hat{I}$. For $y \in [0, \lambda_A)$ or $y \in [\lambda_A + \lambda_B, \lambda^*)$, we have $T(y) = \hat{T}(y)$; for $y \in [\lambda_A, \lambda_A + \lambda_B)$, we have $\hat{T}(y) \notin I$ and $T(y) = \hat{T}^2(y)$. Therefore, T appears as the first return map of $\hat{T}$ in I.

Thus, all i.e.m with $\#\mathcal{A} \leq 3$ are rotations or are closely connected to rotations.

3 The Keane's Property

3.1 Let T be an i.e.m defined by combinatorial data $(\mathcal{A}, \pi_0, \pi_1)$ and length data $\lambda = (\lambda_\alpha)_{\alpha \in \mathcal{A}}$.

DEFINITION – A **connexion** for T is a triple (α, β, m) where $\alpha, \beta \in \mathcal{A}, \pi_0(\beta) > 1$, m is a positive integer and $T^m(j_0(0, \alpha)) = j_0(0, \beta)$.

We say that T has **Keane's property** if there is no connexion for T.

EXERCICE 1 – For $d = 2$, T has Keane's property iff λ_A, λ_B are rationally independent.

EXERCICE 2 – For $d = 3$, in case (i) of 2.7 above, we have $T(y) = y + \lambda_C \mod \lambda^*\mathbf{Z}$.

Show that T has Keane's property iff the two following conditions are satisfied

1. T is an irrational rotation, i.e λ_C/λ^* is irrational ;

2. the points 0 and λ_A are not on the same T-orbit, i.e there are no relations

$$\lambda_A = m\lambda_C + n\lambda^*$$

with $m, n \in \mathbf{Z}$.

3.2 THEOREM – (Keane [Ke1]) *An i.e.m T with the Keane's property is minimal, i.e all orbits are dense.*

Proof – Let T be an i.e.m with the Keane's property.

1. We first show that T has no periodic orbits. Otherwise, there exists $m > 0$ s.t $P_m(T) = \{y, T^m y = y\}$ is non-empty. Then $y^* := \inf P_m(T)$ belongs to $P_m(T)$. If $y^* > 0$, there exists $k \in \{0, \dots m-1\}$ and $\alpha \in \mathcal{A}$ such that $T^k(y^*) = j_0(0,\alpha) > 0$ and (α, α, m) is a connexion. If $y^* = 0, T^{-1}(y^*) = j_0(0,\alpha) > 0$ for some $\alpha \in \mathcal{A}$ and (α, α, m) is again a connexion.
2. Assume now by contradiction that there exists $y \in I$ such that $(T^n(y))_{n \geq 0}$ is not dense in I. Then there exists an half-open interval $J = [y_0, y_1)$ which does not contain any accumulation point of $(T^n(y))_{n \geq 0}$, nor any $j_0(0,\alpha)$. Let D be the finite set consisting of y_0, y_1 and the $j_0(0,\alpha)$; let D^* be the set consisting of the points $\hat{y} \in J$ such that there exists $m > 0$ with $T^m(\hat{y}) \in D$ but $T^l(\hat{y}) \notin J$ for $0 < l < m$. There is a canonical injective map $\hat{y} \mapsto T^m(\hat{y})$ from D^* to D thus D^* is a finite set. Cut J along D^* into half open intervals $J_1, \dots, J_k$.

For each $r \in \{1, \dots, k\}$, there is by Poincaré recurrence a smallest $n_r > 0$ such that $T^{n_r}(J_r) \cap J \neq \emptyset$. But then, by definition of D^*, we must have $T^{n_r}(J_r) \subset J$. We conclude that

$$J^* := \bigcup_{n \geq 0} T^n(J) = \bigcup_r \bigcup_{0 \leq n < n_r} T^n(J_r)$$

is a finite union of half-open intervals, is fully invariant under T (because $J = \bigcup_r T^{n_r}(J_r)$) and does not contain any accumulation point of $(T^n(y))_{n \geq 0}$.

Because λ^* cannot be the only accumulation point of $(T^n(y))_{n \geq 0}$, we cannot have $J^* = I$. Because the combinatorial data are admissible (an obvious consequence of Keane's property), J^* cannot be of the form $[0, \bar{y}), 0 < \bar{y} < \lambda^*$.

Therefore, there exists $y^* \in J^* \cap \partial J^*$ with $y^* > 0$. If $T^l(y^*) \neq j_0(0,\alpha)$ for all $l < 0, \alpha \in \mathcal{A}$, then $T^l(y^*) \in J^* \cap \partial J^*$ for all $l \leq 0$ and y^* is periodic. Similarly, if $T^l(y^*) \neq j_0(0,\alpha)$ for all $l \geq 0, \alpha \in \mathcal{A}$. Both cases are impossible by the first part of the proof. Thus there exists $l_1 < 0, l_2 \geq 0$ and $\alpha_1, \alpha_2 \in \mathcal{A}$ with $T^{l_1}(y^*) = j_0(0,\alpha_1), T^{l_2}(y^*) = j_0(0,\alpha_2)$. Taking l_2 minimal, we have $j_0(0,\alpha_2) > 0$ and $(\alpha_1, \alpha_2, l_2 - l_1)$ is a connexion. □

3.3 Irrationality and Keane's Property

PROPOSITION – (Keane [Ke1]). *If the length data $(\lambda_\alpha)_{\alpha \in \mathcal{A}}$ are rationally independent and the combinatorial data are admissible, then T has Keane's property.*

Proof – Assume on the opposite that there is a connexion (α_0, α_m, m). For $0 < l < m$, let $\alpha_l \in \mathcal{A}$ such that $T^l(j_0(0,\alpha_0)) \in j_0(I_{\alpha_l})$. Denote by $(\delta_\alpha)_{\alpha \in \mathcal{A}}$ the translation vector. We have

$$j_0(0,\alpha_m) - j_0(0,\alpha_0) = \sum_{0 \le l < m} \delta_{\alpha_l}$$

which, in view of 2.3, gives

$$\sum_{\pi_0\alpha < \pi_0\alpha_m} \lambda_\alpha - \sum_{\pi_0\alpha < \pi_0\alpha_0} \lambda_\alpha = \sum_{0 \le l < m} \left(\sum_{\pi_1\alpha < \pi_1\alpha_l} \lambda_\alpha - \sum_{\pi_0\alpha < \pi_0\alpha_l} \lambda_\alpha \right).$$

Setting, for $\alpha \in \mathcal{A}$:

$$n_\alpha := \#\{l \in [0,m), \pi_1(\alpha_l) > \pi_1(\alpha)\} - \#\{l \in (0,m], \pi_0(\alpha_l) > \pi_0(\alpha)\}$$

we obtain $\sum n_\alpha \lambda_\alpha = 0$ and therefore $n_\alpha = 0$ for all $\alpha \in \mathcal{A}$ from rational independence.

Let $\hat{d}$ be the highest value taken by the $\pi_1(\alpha_l), l \in [0,m)$ or the $\pi_0(\alpha_l), l \in (0,m]$. Because the combinatorial data are admissible, there must exists $\hat{\alpha} \in \mathcal{A}$ with $\pi_0(\hat{\alpha}) \ge \hat{d}$ but $\pi_1(\hat{\alpha}) < \hat{d}$. Then $\pi_0(\alpha_l) \le \pi_0(\hat{\alpha})$ for $l \in (0,m]$. As $n_{\hat{\alpha}} = 0$, we must have $\pi_1(\alpha_l) \le \pi_1(\hat{\alpha}) < \hat{d}$ for all $l \in [0,m)$. In a symmetric way, we also prove that $\pi_0(\alpha_l) < \hat{d}$ for all $l \in (0,m]$. This contradicts the definition of $\hat{d}$. □

3.4 A Continuous Version of Interval Exchange Maps

The construction which follows is completely similar to the construction of Denjoy counter examples, i.e C^1 diffeomorphisms of the circle with no periodic orbits and a minimal invariant Cantor set.

Let T be an i.e.m with combinatorial data $(\mathcal{A}, \pi_0, \pi_1)$; for simplicity we assume that T has Keane's property.

For $n \ge 0$, define

$$D_0(n) = \{T^{-n}(j_0(0,\alpha)), \alpha \in \mathcal{A}, \pi_0(\alpha) > 1\} ,$$
$$D_1(n) = \{T^{+n}(j_1(0,\alpha)), \alpha \in \mathcal{A}, \pi_1(\alpha) > 1\} .$$

It follows from the Keane's property that these sets are disjoint from each other and do not contain 0.

Define an atomic measure μ by

$$\mu = \sum_{n \ge 0} \sum_{D_0(n) \sqcup D_1(n)} 2^{-n} \delta_y ,$$

and then increasing maps $i^+, i^- : I \to \mathbf{R}$ by

$$i^-(y) = y + \mu([0,y)) ,$$
$$i^+(y) = y + \mu([0,y]) .$$

We therefore have

$$\begin{aligned} i^+(y) &< i^-(y') && \text{for } y < y' \\ i^+(y) &= i^-(y) && \text{for } y \notin \bigsqcup_{n \geq 0} (D_0(n) \sqcup D_1(n)) , \\ i^+(y) &= i^-(y) + 2^{-n} && \text{for } y \in D_0(n) \sqcup D_1(n) . \end{aligned}$$

We also define

$$\begin{aligned} i^-(\lambda^*) &= \lambda^* + 4(d-1) \\ &= \lim_{y \nearrow \lambda^*} i^{\pm}(y) , \end{aligned}$$

and

$$\begin{aligned} K &= i^-(I) \cup i^+(I) \cup \{i^-(\lambda^*)\} \\ &= \overline{i^-(I)} = \overline{i^+(I)} . \end{aligned}$$

As T is minimal, K is a Cantor set whose gaps are the intervals

$$(i^-(y), i^+(y)), \quad y \in \bigcup_{n \geq 0} \bigcup_{\varepsilon} D_\varepsilon(n) .$$

PROPOSITION – *There is a unique continuous map $\hat{T} : K \to K$ such that $\hat{T} \circ i^+ = i^+ \circ T$ on I. Moreover, $\hat{T}$ is a minimal homeomorphism.*

Proof – $\hat{T}$ is unique because $i^+(I)$ is dense in K. Let us check that $\hat{T}$ is uniformly continuous on $i^+(I)$: if $y < y'$ satisfy $i^+(y') < i^+(y) + 1$, it is easy to check that we have

$$\begin{aligned} \hat{T} \circ i^+(y') - \hat{T} \circ i^+(y) &= i^+(Ty') - i^+(Ty) \\ &< 2(i^+(y') - i^+(y)) . \end{aligned}$$

The first statement of the proposition follows. That $\hat{T}$ is an homeomorphism follows from the observation that our setting gives symmetrical roles to T and T^{-1}. We leave the minimality as an exercice for the reader. □

4 The Continuous Fraction Algorithm

4.1 The Basic Operation (Rauzy [Ra], Veech [V1], [V2])

Let T be an i.e.m defined by combinatorial data $(\mathcal{A}, \pi_0, \pi_1)$ and length data $(\lambda_\alpha)_{\alpha \in \mathcal{A}}$. We assume as always that the combinatorial data are admissible.

We denote by α_0, α_1 the (distinct) elements of $\mathcal{A}$ such that

$$\pi_0(\alpha_0) = \pi_1(\alpha_1) = d .$$

Observe that if $\lambda_{\alpha_0} = \lambda_{\alpha_1}$ the triple $(\alpha_1, \alpha_0, 1)$ is a connexion and T has not the Keane's property.

We now assume that $\lambda_{\alpha_0} \neq \lambda_{\alpha_1}$ and define $\varepsilon \in \{0, 1\}$ by

$$\lambda_{\alpha_\varepsilon} = \max(\lambda_{\alpha_0}, \lambda_{\alpha_1}) .$$

We set

$$\hat{\lambda}^* = \lambda^* - \lambda_{\alpha_{1-\varepsilon}} ,$$
$$\hat{I} = [0, \hat{\lambda}^*) \subset I ,$$

and define $\hat{T} : \hat{I} \to \hat{I}$ to be the first return map of T in $\hat{I}$.

When $\varepsilon = 0$, we have

$$\hat{T}(y) = \begin{cases} T(y) & \text{if } y \notin j_0(I_{\alpha_1}) , \\ T^2(y) & \text{if } y \in j_0(I_{\alpha_1}) . \end{cases}$$

When $\varepsilon = 1$, we have similarly

$$\hat{T}^{-1}(y) = \begin{cases} T^{-1}(y) & \text{if } y \notin j_1(I_{\alpha_0}) , \\ T^{-2}(y) & \text{if } y \in j_1(I_{\alpha_0}) . \end{cases}$$

In both case, it appears that $\hat{T}$ is again an interval exchange map which can be defined using the same alphabet $\mathcal{A}$. The length data for $\hat{T}$ are given by

$$\begin{cases} \hat{\lambda}_\alpha = \lambda_\alpha & \text{if } \alpha \neq \alpha_\varepsilon \\ \hat{\lambda}_{\alpha_\varepsilon} = \lambda_{\alpha_\varepsilon} - \lambda_{\alpha_{1-\varepsilon}} . \end{cases}$$

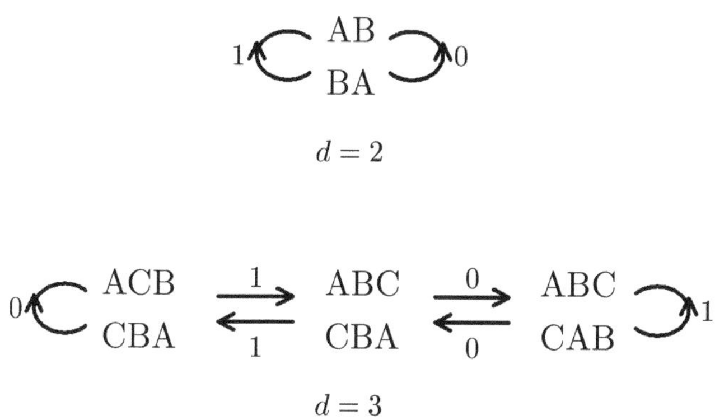

Fig. 1. Rauzy diagrams $d = 2$ and $d = 3$

The combinatorial data $(\hat{\pi}_0, \hat{\pi}_1)$ for $\hat{T}$ are given by $\hat{\pi}_\varepsilon = \pi_\varepsilon$ and

$$\hat{\pi}_{1-\varepsilon}(\alpha) = \begin{cases} \pi_{1-\varepsilon}(\alpha) & \text{if } \pi_{1-\varepsilon}(\alpha) \leq \pi_{1-\varepsilon}(\alpha_\varepsilon) \\ \pi_{1-\varepsilon}(\alpha)+1 & \text{if } \pi_{1-\varepsilon}(\alpha_\varepsilon) < \pi_{1-\varepsilon}(\alpha) < d \\ \pi_{1-\varepsilon}(\alpha_\varepsilon)+1 & \text{if } \pi_{1-\varepsilon}(\alpha) = d \end{cases}$$

We rewrite the relation between old and new length data as

$$\lambda = V\hat{\lambda}\,,$$

where $V = \mathbf{1} + E_{\alpha_\varepsilon \alpha_{1-\varepsilon}}$ has now non negative integer coefficients an belongs to $SL(\mathbf{Z}^\mathcal{A})$.

We also write

$$(\hat{\pi}_0, \hat{\pi}_1) = R_\varepsilon(\pi_0, \pi_1)$$

and observe that these new combinatorial data are admissible.

4.2 Rauzy Diagrams

Let $\mathcal{A}$ be an alphabet. We define an oriented graph, as follows. The vertices are the admissible pairs (π_0, π_1). Each vertex (π_0, π_1) is the starting point of exactly two arrows with endpoints at $R_0(\pi_0, \pi_1)$ and $R_1(\pi_0, \pi_1)$. The arrow connecting (π_0, π_1) to $R_\varepsilon(\pi_0, \pi_1)$ is said to be of **type** ε.

The operations R_0, R_1 are obviously invertible. Therefore each vertex is also the endpoint of exactly two arrows, one of each type.

To each arrow in the graph, we associate a **name** in $\mathcal{A}$: it is the element α_ε such that $\pi_\varepsilon(\alpha_\varepsilon) = d$ (where (π_0, π_1) is the starting point of the arrow and ε is its type). The element $\alpha_{1-\varepsilon}$ will then be called the **secondary name** of this arrow.

A Rauzy diagram is a connected component in this oriented graph.

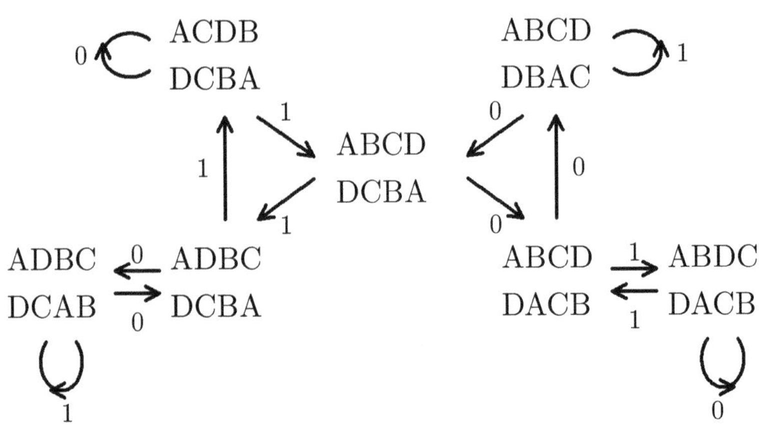

Fig. 2. Rauzy diagram $d = 4$, first case

Obviously, the Rauzy operations R_0, R_1 commute with change of names (cf. 2.2).

Up to change of names, there is only one Rauzy diagram with $d = \#\mathcal{A} = 2$, and one with $d = \#\mathcal{A} = 3$ (see figure 1), where the pair (π_0, π_1) is denoted by the symbol

$$\begin{array}{l} \pi_0^{-1}(1) \ldots \pi_0^{-1}(d) \\ \pi_1^{-1}(1) \ldots \pi_1^{-1}(d)\,. \end{array}$$

For $d = \#\mathcal{A} = 4$, there are 2 distinct Rauzy diagrams (see figures 2 and 3).

In each of these diagrams, the symmetry with respect to the vertical axis corresponds to the action of the canonical involution.

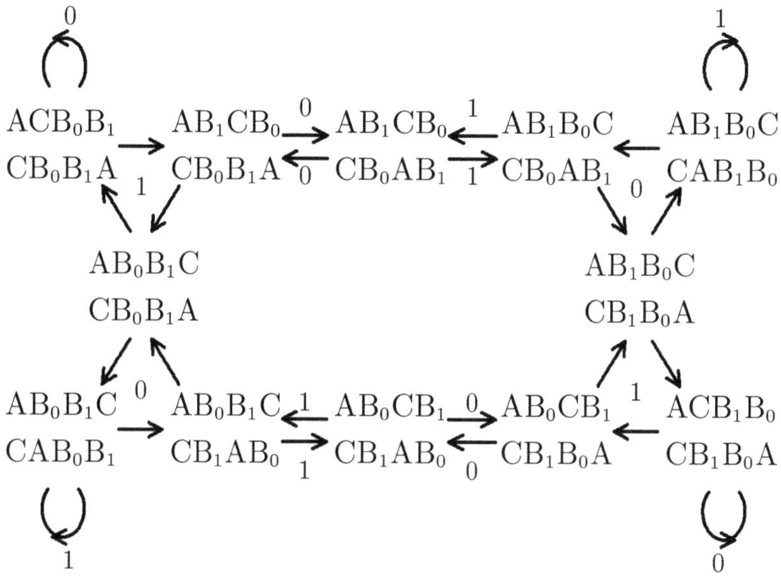

Fig. 3. Rauzy diagram $d = 4$, second case

In the last diagram, there is a further symmetry with respect to the center of the diagram, which corresponds to the exchange of the names B_0, B_1. This is a monodromy phenomenon : to each admissible pair (π_0, π_1), one can associate the permutation $\pi := \pi_1 \circ \pi_0^{-1}$ of $\{1, \ldots, d\}$, which is invariant under change of names. When we identify vertices with the same permutation, we obtain a **reduced Rauzy diagram** and we have a covering map from the Rauzy diagram onto the reduced Rauzy diagram.

In the first three examples above, the covering map is an isomorphism. In the last exemple, the degree of the covering map is 2 and the reduced Rauzy diagram is shown in figure 4, where π is denoted by $(\pi^{-1}(1), \ldots \pi^{-1}(d))$.

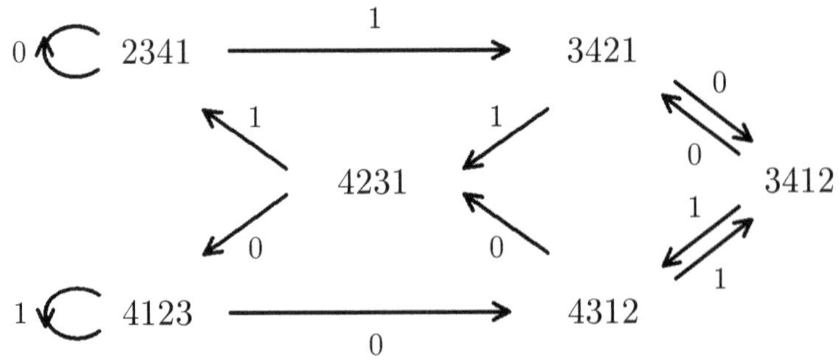

Fig. 4. Reduced Rauzy diagram $d = 4$ (second case)

4.3 Dynamics in Parameter Space

Let $\mathcal{D}$ be a Rauzy diagram on an alphabet $\mathcal{A}$; denote by $V(\mathcal{D})$ the set of vertices of $\mathcal{D}$. For $(\pi_0, \pi_1) \in V(\mathcal{D})$, let

$$\begin{aligned}
\mathcal{C}(\pi_0,\pi_1) &= (\mathbf{R}_+^*)^{\mathcal{A}} \times \{(\pi_0,\pi_1)\} ,\\
\mathcal{C}^*(\pi_0,\pi_1) &= \{((\lambda_\alpha),\pi_0,\pi_1) \in \mathcal{C}(\pi_0,\pi_1), \lambda_{\alpha_0} \neq \lambda_{\alpha_1}\} ,\\
\Delta(\pi_0,\pi_1) &= \{((\lambda_\alpha),\pi_0,\pi_1) \in \mathcal{C}(\pi_0,\pi_1), \Sigma\lambda_\alpha = 1\} ,\\
\Delta^*(\pi_0,\pi_1) &= \Delta(\pi_0,\pi_1) \cap \mathcal{C}^*(\pi_0,\pi_1) .
\end{aligned}$$

For $\varepsilon \in \{0,1\}$, we also write $\Delta^\varepsilon(\pi_0,\pi_1), \mathcal{C}^\varepsilon(\pi_0,\pi_1)$ for the subsets of $\Delta^*(\pi_0,\pi_1), \mathcal{C}^*(\pi_0,\pi_1)$ defined by $\lambda_{\alpha_\varepsilon} > \lambda_{\alpha_{1-\varepsilon}}$.

The basic operation of 4.1 defines a 2 to 1 map from $\mathcal{C}^*(\mathcal{D}) := \sqcup \mathcal{C}^*(\pi_0,\pi_1)$ onto $\mathcal{C}(\mathcal{D}) := \sqcup \mathcal{C}(\pi_0,\pi_1)$; its restriction to $\mathcal{C}^\varepsilon(\pi_0,\pi_1)$ is an isomorphism onto $\mathcal{C}(R_\varepsilon(\pi_0,\pi_1))$ given by the matrix $V = \mathbf{1} + E_{\alpha_\varepsilon \alpha_{1-\varepsilon}}$ of 4.1. We denote this map by $\mathcal{R}$. In other terms, in the context of Section 4.1, we set

$$\mathcal{R}(T) = \hat{T} .$$

Because $\hat{T}$ is a first return map for T, if T has the Keane's property, the same will be true for $\hat{T}$. This means that for such maps we will be able to iterate infinitely many times $\mathcal{R}$.

There is a canonical projection from $\mathcal{C}(\pi_0,\pi_1)$ onto $\Delta(\pi_0,\pi_1)$ which sends $\mathcal{C}^*(\pi_0,\pi_1)$ onto $\Delta^*(\pi_0,\pi_1)$. We define $\Delta(\mathcal{D}) = \sqcup \Delta(\pi_0,\pi_1), \Delta^*(\mathcal{D}) = \sqcup \Delta^*(\pi_0,\pi_1)$, and we get a quotient map which we still denote by $\mathcal{R}$ and which is 2 to 1 from $\Delta^*(\mathcal{D})$ onto $\Delta(\mathcal{D})$.

Let $(\lambda_\alpha)_{\alpha\in\mathcal{A}}, \pi_0, \pi_1) \in \mathcal{C}(\mathcal{D})$ be data defining an i.e.m T ; assume that T satisfies the Keane's property. Iterating $\mathcal{R}$, we get a sequence $(T^{(n)})_{n\geq 0}$ of i.e.m with $T^{(0)} = T$. The data for $T^{(n+1)}$ are related to the data of $T^{(n)}$ by formulas :

$$\begin{aligned}
&(\pi_0^{(n+1)}, \pi_1^{(n+1)}) = R_{\varepsilon_{n+1}}(\pi_0^{(n)}, \pi_1^{(n)}) ,\\
&\lambda^{(n)} = V^{(n+1)} \lambda^{(n+1)} .
\end{aligned}$$

Denote by $\gamma^{(n+1)}$ the arrow in $\mathcal{D}$ which connects the pair $(\pi_0^{(n)}, \pi_1^{(n)})$ to $(\pi_0^{(n+1)}, \pi_1^{(n+1)})$. The sequence $(\gamma^{(n)})_{n>0}$ determines an infinite path in $\mathcal{D}$ starting at $(\pi_0^{(0)}, \pi_1^{(0)})$.

PROPOSITION – *Each name in $\mathcal{A}$ is taken infinitely many times by the sequence $(\gamma^{(n)})_{n>0}$.*

Proof – Let $\mathcal{A}'$ be the set of names which are taken infinitely many times and let $\mathcal{A}'' = \mathcal{A} - \mathcal{A}'$. Replacing T by some $T^{(N)}$, we can assume that names in $\mathcal{A}''$ are not taken at all. Then the length $\lambda_\alpha^{(n)}, \alpha \in \mathcal{A}''$, do no depend on n. But then elements $\alpha \in \mathcal{A}''$ can only appear as secondary names at most finitely many times. Replacing again T by some $T^{(N)}$, we can assume that secondary names are never in $\mathcal{A}''$. Then the sequences $(\pi_\varepsilon^{(n)}(\alpha))_{n>0}$, for $\varepsilon \in \{0,1\}, \alpha \in \mathcal{A}''$, are non decreasing and we can assume (replacing again T by $T^{(N)}$) that they are constant.

We now claim that we must have $\pi_\varepsilon^{(0)}(\alpha'') < \pi_\varepsilon^{(0)}(\alpha')$ for all $\alpha'' \in \mathcal{A}'', \alpha' \in \mathcal{A}'$ and $\varepsilon \in \{0,1\}$. Because the pair $(\pi_0^{(0)}, \pi_1^{(0)})$ is admissible, this implies $\mathcal{A}' = \mathcal{A}$.

To prove the claim, assume that there exist $\alpha' \in \mathcal{A}', \alpha'' \in \mathcal{A}'', \varepsilon \in \{0,1\}$ with $\pi_\varepsilon^{(0)}(\alpha') < \pi_\varepsilon^{(0)}(\alpha'')$.

As $\pi_\varepsilon^{(n)}(\alpha'') = \pi_\varepsilon^{(0)}(\alpha'')$ for all $n \geq 0$, we can never have $\pi_\varepsilon^{(n)}(\alpha') = d$ for some $n > 0$. By definition of $\mathcal{A}'$, there must exist $n \geq 0$ such that $\pi_{1-\varepsilon}^{(n)}(\alpha') = d$; but then $\pi_\varepsilon^{(n+1)}(\alpha'') \neq \pi_\varepsilon^{(0)}(\alpha'')$, which gives a contradiction. □

COROLLARY 1 – *Each type and each secondary name is taken infinitely many times.*

Proof – The first assertion is obvious (we do not need the proposition here). The second follows from the proposition and the following fact : if $\gamma^{(n)}, \gamma^{(n+1)}$ have not the same name, the secondary name of $\gamma^{(n+1)}$ is the (main) name of $\gamma^{(n)}$. □

COROLLARY 2 – *The length of the intervals $I^{(n)}$ goes to 0 as n goes to ∞.*

Proof – All sequences $(\lambda_\alpha^{(n)})_{n\geq 0}$ are non increasing and we want to show that they go to 0. Let $\lambda_\alpha^{(\infty)}$ be the limit. Given $\varepsilon > 0$, let $N \geq 0$ such that $\lambda_\alpha^{(N)} \leq \lambda_\alpha^{(\infty)} + \varepsilon$ for all $\alpha \in \mathcal{A}$. For each $\alpha \in \mathcal{A}$, there exists $n > N$ such that α is a secondary name for $\gamma^{(n)}$; this implies that $\lambda_\alpha^{(\infty)} \leq \lambda_\alpha^{(n)} \leq \varepsilon$ and concludes the proof. □

COROLLARY 3 – *Let T be an i.e.m with admissible combinatorial data which does not have the Keane's property. Then the continuous fraction algorithm stops because at some point the equality $\lambda_{\alpha_0}^{(n)} = \lambda_{\alpha_1}^{(n)}$ (with $\pi_0^{(n)}(\alpha_0) = \pi_1^{(n)}(\alpha_1) = d$) holds.*

Proof – Let (α, β, m) a connexion for $T = T^{(0)}$. We show by infinite descent that the algorithm has to stop. Set $y_0 = j_0((0,\beta))$; set $y_1 = j_1((0,\alpha))$ if $\pi_1(\alpha) \neq 1, y_1 = T(0)$ if $\pi_1(\alpha) = 1$. We have $T^{\bar{m}}(y_1) = y_0$ with $\bar{m} = m - 1$ if $\pi_1(\alpha) \neq 1, \bar{m} = m - 2$ if $\pi_1(\alpha) = 1$, and $\bar{m} \geq 0$ in both cases, with $y_0, y_1 > 0$.

Assume by contradiction that the algorithm never stops. Observe that the proposition and the corollaries 1 and 2 hold, because the Keane's property was not used in their proof. Let n be the largest integer such that

$$|I^{(n)}| > \max(y_0, y_1) ,$$

where $I^{(n)}$ is the domain for $T^{(n)}$. Such an n exists by Corollary 2. If we had $y_0 = y_1$, the equality case would happen at the next step of the basic operation. We therefore have $\bar{m} > 0, y_0 \neq y_1$. Assume for instance that $y_1 > y_0$ (the other case is symmetric). First, because $T^{(n)}$ is the first return map of T into $I^{(n)}$, there exists $0 < \hat{m} \leq \bar{m}$ such that $(T^{(n)})^{\hat{m}}(y_1) = y_0$. Second, by the definition of the basic operation, $y_1' = T^{(n)}(y_1)$ is equal to $j_1(0, \alpha_1)$ at step $n+1$, where $\pi_1^{(n)}(\alpha_1) = d$. Then $(T^{(n)})^{\hat{m}-1}(y_1') = y_0$ and therefore (as $T^{(n+1)}$ is a first return map of $T^{(n)}$) there exists $\bar{m}' \leq \hat{m} - 1 < \bar{m}$ such that $T^{(n+1)}(y_1') = y_0$. We have completed one step of the descent argument, and this concludes the proof. □

COROLLARY 4 – *For each $m \geq 0$, there exists $n > m$ such that the matrix $Q := V^{(m+1)} \ldots V^{(n)}$ satisfies $Q_{\alpha\beta} > 0$ for all $\alpha, \beta \in \mathcal{A}$.*

Proof – Write $Q = Q(n)$. Let $\alpha, \beta \in \mathcal{A}$; if $Q_{\alpha\beta}(n_0) > 0$ for some n_0, then $Q_{\alpha\beta}(n) > 0$ for all $n \geq n_0$: indeed the diagonal terms of the V matrices are equal to 1. It therefore suffices to prove that for all $\alpha, \beta \in \mathcal{A}$ there exists n_0 such that $Q_{\alpha\beta}(n_0) > 0$. Fix $\alpha, \beta \in \mathcal{A}$. If $\alpha = \beta$, we already have $Q_{\alpha\beta}(m+1) = 1$. Assume $\alpha \neq \beta$. Let $n_1 > m$ the smallest integer such that the arrow $\gamma^{(n_1)}$ has name α. Set $\alpha_1 := \alpha$ and let α_2 be the secondary name of $\gamma^{(n_1)}$; we have $Q_{\alpha_1\alpha_i}(n_1) > 0$ for $i = 1, 2$. If $\beta = \alpha_2$, we are done. Otherwise, $d \geq 3$ and there exists a smallest integer $n_1' > n_1$ such that the name of $\gamma^{(n_1')}$ is not α_1 or α_2. There also exists a smallest integer $n_2 > n_1'$ such that the name of $\gamma^{(n_2)}$ is α_1 or α_2. Then, the secondary name α_3 of $\gamma^{(n_2)}$ is the name of $\gamma^{(n_2-1)}$ and therefore is different from α_1 or α_2. We have $V^{(n_2)}_{\alpha_j\alpha_3} = 1$ for some $j \in \{1, 2\}$, and therefore $Q_{\alpha_1\alpha_i}(n_2) > 0$ for $i \in \{1, 2, 3\}$. If $\beta = \alpha_3$ we are done. Otherwise $d \geq 4$ and we define $n_2' > n_2, n_3 > n_2', \alpha_4 \notin \{\alpha_1, \alpha_2, \alpha_3\}$ as above ... At some point we must have $\beta = \alpha_j$. □

COROLLARY 5 – *Define a decreasing sequence of open simplicial cones in $\mathbf{R}^{\mathcal{A}}$ by*

$$\mathcal{C}^{(0)} = (\mathbf{R}_+^*)^{\mathcal{A}}, \mathcal{C}^{(n+1)} = V^{(n+1)}\mathcal{C}^{(n)}$$

and let $\mathcal{C}^{(\infty)} = \cap\, \mathcal{C}^{(n)}$. Then $\mathcal{C}^{(\infty)} \cup \{0\}$ is a closed simplicial cone, of dimension $< d = \#\mathcal{A}$.

Proof – From Corollary 4 it follows that for all $m \geq 0$ there exists $n > m$ such that the closure of $\mathcal{C}^{(n)}$ is contained in $\mathcal{C}^{(m)} \cup \{0\}$. This shows that $\mathcal{C}^{(\infty)} \cup \{0\}$ is closed. For $n \geq 0, \alpha \in \mathcal{A}$, let $e_\alpha^{(n)} = V^{(1)} \ldots V^{(n)}(e_\alpha)$, where $(e_\alpha)_{\alpha \in \mathcal{A}}$ is the canonical base of $\mathbf{R}^{\mathcal{A}}$. Let n_k be an increasing sequence of integers such that $e_\alpha^{(n_k)} \|e_\alpha^{(n_k)}\|^{-1}$ converge towards a limit $e_\alpha^{(\infty)}$ for every $\alpha \in \mathcal{A}$. Then we must have

$$\mathcal{C}^{(\infty)} \cup \{0\} = \{\sum_{\mathcal{A}} t_\alpha e_\alpha^{(\infty)}, t_\alpha \geq 0\} .$$

The limits $e_\alpha^{(\infty)}$ cannot be all distinct, because all coefficients of $V^{(1)} \dots V^{(n)}$ go to ∞ as n goes to ∞ (by Corollary 4), and these matrices are unimodular. Thus $\mathcal{C}^{(\infty)} \cup \{0\}$ is closed, polyhedral of dimension $< d$. Indeed it is simplicial because, as we will see in the next section, it can be interpreted as a cone of invariant measures. □

4.4 Unique Ergodicity and the Continuous Fraction Algorithm

Recall that a transformation is **uniquely ergodic** if it has exactly one invariant probability measure.

For an i.e.m T, (normalized) Lebesgue measure is invariant, hence there should be no other invariant probability measure.

Let T be an i.e.m with the Keane's property. In particular, T is minimal. Therefore, every finite invariant measure μ is continuous and supported by the whole of I. For such a measure, we set

$$H_\mu(x) = \mu([0, x)) .$$

This defines an homeomorphism from I onto $I_\mu := [0, \mu(I))$. Let

$$T_\mu = H_\mu \circ T \circ H_\mu^{-1} .$$

This is a one-to-one transformation of I_μ. Actually, T_μ is immediately seen to be an i.e.m on I_μ, whose combinatorial data are the same as for T, and whose length data $(\lambda_\alpha(\mu))_{\alpha \in \mathcal{A}}$ are given by

$$\lambda_\alpha(\mu) = \mu(j_0(I_\alpha)) = \mu(j_1(I_\alpha)) .$$

Obviously, the image of μ under the conjugacy H_μ is the Lebesgue measure on I_μ.

PROPOSITION – *The map $\mu \mapsto (\lambda_\alpha(\mu))_{\alpha \in \mathcal{A}}$ is a linear homeomorphism from the set of T-invariant finite measures onto the cone $\mathcal{C}^{(\infty)}$ of Corollary 5 In particular, T is uniquely ergodic if and only if $\mathcal{C}^{(\infty)}$ is a ray.*

Proof – The map is obviously linear and continuous ; as T and T_μ are topologically conjugated, T_μ has also the Keane's property ; moreover, the restriction of H_μ to $I^{(n)}$ is an homeomorphism on $I_\mu^{(n)}$ which conjugates $T^{(n)}$ and $T_\mu^{(n)}$. Thus, the length vector $(\lambda_\gamma(\mu))_{\alpha \in \mathcal{A}}$ belongs to $\mathcal{C}^{(n)}$ for every $n \geq 0$ and therefore to $\mathcal{C}^{(\infty)}$. Conversely, let $(\tilde{\lambda}_\alpha)_{\alpha \in \mathcal{A}}$ be a length vector in $\mathcal{C}^{(\infty)}$. Let $\tilde{T}$ be the i.e.m defined by this length vector and the same combinatorial data than T. The continuous fraction algorithm for $\tilde{T}$ never stops (with the same path in the Rauzy diagram than for T), hence $\tilde{T}$ has the Keane's property ; the same is true for the i.e.m. $\tilde{T}_t$ whose length vector is $(1-t)\lambda + t\tilde{\lambda} \in \mathcal{C}^{(\infty)}$. Therefore, for each $t \in [0, 1]$, the points $(T_t^k(0))_{k \geq 0}$ are distinct, form a dense set in I_t and we have

$$T_t^k(0) > T_t^l(0) \Longleftrightarrow T_{t'}^k(0) > T_{t'}^l(0)$$

for all $k, l \geq 0, t, t' \in [0, 1]$. If we set

$$H(T^k(0)) = \tilde{T}^k(0) ,$$

for all $k \geq 0$, the map H extends in a unique way to an homeomorphism from I onto $\tilde{I}$ which conjugates T and $\tilde{T}$. If μ is the image of Lebesgue measure under H^{-1}, then μ is a finite T-invariant measure on I and $\tilde{T} = T_\mu$. □

For $d \leq 3$, interval exchange maps are rotations or first return maps of rotations and thus are uniquely ergodic if minimal. On the other hand, Keane has constructed ([Ke2], see also [KN], [Co]) i.e.m with $d = 4$ which are minimal but not uniquely ergodic. Nevertheless, we have the following fundamental result :

THEOREM – (Mazur [Ma], Veech [V2]) *Let (Q, π_0, π_1) be any admissible combinatorial data. Then, for almost all length data $(\lambda_\alpha)_{\alpha \in \mathcal{A}}$, the associated i.e.m is (minimal and) uniquely ergodic.*

Proof – We will give a slightly simplified version of the proof of Kerchkoff ([Ker]). Let $\mathcal{D}$ be the Rauzy diagram which contains the combinatorial data $(\mathcal{A}, \pi_0, \pi_1)$.

For any finite path $\gamma = (\gamma^{(i)})_{0<i\leq n}$ in $\mathcal{D}$ starting at (π_0, π_1), let $(V^{(i)})_{0<i\leq n}$ be the associated matrices ; let

$$\begin{aligned} Q(\gamma) &= V^{(1)} \dots V^{(n)} \\ \mathcal{C}(\gamma) &= Q(\gamma)[(\mathbf{R}_+^*)^{\mathcal{A}}] \times \{(\pi_0, \pi_1)\} , \\ \Delta(\gamma) &= \mathcal{C}(\gamma) \cap \Delta(\pi_0, \pi_1) . \end{aligned}$$

For $\beta \in \mathcal{A}$, we also write

$$Q_\beta(\gamma) = \sum_\alpha Q_{\alpha\beta}(\gamma) .$$

LEMMA 1 – *We have*

$$\mathrm{vol}_{d-1}(\Delta(\gamma)) = (\prod_\beta Q_\beta(\gamma))^{-1} \; \mathrm{vol}_{d-1}(\Delta(\pi_0, \pi_1)) .$$

Proof – Indeed, $Q(\gamma)$ is unimodular and we have, for $\lambda^{(0)} = Q(\gamma)\lambda^{(n)}$:

$$\sum_\alpha \lambda_\alpha^{(0)} = \sum_\beta Q_\beta(\gamma)\lambda_\beta^{(n)} .$$

□

LEMMA 2 – *Let $C \geq 1$ a constant such that*

$$\max_\alpha Q_\alpha(\gamma) \leq C \min_\alpha Q_\alpha(\gamma) .$$

There exists a constant $c \in (0,1)$, depending only on C and d, and a path γ' extending γ such that

$$\mathrm{vol}_{d-1}(\Delta(\gamma')) \geq c\, \mathrm{vol}_{d-1}(\Delta(\gamma))\,,$$
$$\mathrm{diam}\,(\Delta(\gamma')) \leq (1-c)\, \mathrm{diam}\,(\Delta(\gamma))\,.$$

Proof – Choose a path $\tilde{\gamma}$ starting from the endpoint of γ such that $Q_{\alpha\beta}(\tilde{\gamma}) > 0$ for all $\alpha, \beta \in \mathcal{A}$. We have $Q_{\alpha\beta}(\tilde{\gamma}) \leq C_1$, with C_1 depending only on d. Let $\gamma' = \gamma \star \tilde{\gamma}$. We have, for $\beta \in \mathcal{A}$

$$Q_\beta(\gamma') = \sum_\alpha Q_\alpha(\gamma) Q_{\alpha\beta}(\tilde{\gamma})\,,$$

and thus, by Lemma 1

$$\mathrm{vol}_{d-1}(\Delta(\gamma')) \geq (CC_1 d)^{-d}\, \mathrm{vol}_{d-1}(\Delta(\gamma))\,.$$

It is also clear, considering orthogonal projections on 1-dimensional lines, that we have

$$\mathrm{diam}\,(\Delta(\gamma')) \leq \left(1 - \frac{2}{C_1(d-1)+1}\right) \mathrm{diam}(\Delta(\gamma))\,.$$

□

LEMMA 3 – *Let $(\pi_0^{(n)}, \pi_1^{(n)})$ be the vertex of $\mathcal{D}$ endpoint of γ ; define $\alpha_0, \alpha_1 \in \mathcal{A}$ by $\pi_\varepsilon^{(n)}(\alpha_\varepsilon) = d, \varepsilon = 0, 1$. For $\varepsilon = 0, 1$, let $\Delta^\varepsilon(\gamma)$ be formed of those length data in $\Delta(\gamma)$ for which the $(n+1)^{th}$ arrow has type ε. Then*

$$\mathrm{vol}_{d-1}(\Delta^\varepsilon(\gamma)) = \frac{Q_{\alpha_{1-\varepsilon}}(\gamma)}{Q_{\alpha_0}(\gamma) + Q_{\alpha_1}(\gamma)}\, \mathrm{vol}_{d-1}(\Delta(\gamma))\,.$$

Proof – Clear from Lemma 1. □

Let T be an i.e.m in $\Delta(\gamma)$ satisfying Keane's condition, and let $(\gamma^{(i)}(T))_{i\geq 0}$ be the associated path ; we therefore have $\gamma^{(i)}(T) = \gamma^{(i)}$ for $0 < i \leq n$. Let $(V^{(i)}(T))_{i\geq O}$ be the associated matrices ; define

$$Q(i,T) = V^{(1)}(T) \ldots V^{(i)}(T)\,.$$

Fix $\alpha \in \mathcal{A}$, and define $Q'_\alpha(T) = Q_\alpha(n(\alpha,T),T)$, where $n(\alpha,T)$ is the smallest integer $m > n$ such that the name of $\gamma^{(m)}(T)$ is α (this is well defined by the proposition in 4.3). We then have :

LEMMA 4 – *For any $\mathcal{C} \geq 1$, we have :*

$$\mathrm{vol}_{d-1}(\{T \in \Delta(\gamma), Q'_\alpha(T) \geq \mathcal{C} Q_\alpha(\gamma)\}) \leq \mathcal{C}^{-1}\, \mathrm{vol}_{d-1}(\Delta(\gamma))\,.$$

Proof – We will show the slightly stronger result that the inequality of the lemma holds even after conditioning by the value $\bar{n}$ of $n(\alpha,T) - n$. We show this last result by induction on $\bar{n}$.

We have $\bar{n} = 1$ iff the name of $\gamma^{(n+1)}(T)$ is α ; in this case, we have $Q'_\alpha(T) = Q_\alpha(\gamma)$ and the estimate holds for all $\mathcal{C} \geq 1$.

If $\pi_0^{(n)}(\alpha) < d$ and $\pi_1^{(n)}(\alpha) < d$, we divide $\Delta(\gamma)$ into $\Delta^0(\gamma)$ and $\Delta^1(\gamma)$ and apply the induction hypothesis to both simplices to conclude.

Assume on the other hand that $\pi_0^{(n)}(\alpha) < d, \pi_1^{(n)}(\alpha) = d$; if $\bar{n} > 1$, the name of $\gamma^{(n+1)}(T)$ is the element $\alpha_0 \in \mathcal{A}$ such that $\pi_0^{(n)}(\alpha_0) = d$ and we have

$$Q_\alpha(n+1,T) = Q_\alpha(\gamma) + Q_{\alpha_0}(\gamma) \,,$$

$$\mathrm{vol}_{d-1}(\Delta^0(\gamma)) = \frac{Q_\alpha(\gamma)}{Q_\alpha(n+1,T)} \mathrm{vol}_{d-1}(\Delta(\gamma)) \,,$$

by Lemma 3. We will have $Q'_\alpha(T) \geq Q_\alpha(n+1,T)$. If $1 \leq \mathcal{C} \leq (Q_\alpha(\gamma))^{-1} Q_\alpha(n+1,T)$, the estimate of the lemma holds immediately. For $\mathcal{C} > (Q_\alpha(\gamma))^{-1} Q_\alpha(n+1,T)$, we set

$$\mathcal{C}' = \mathcal{C} Q_\alpha(\gamma)(Q_\alpha(n+1,T))^{-1} \,,$$
$$\gamma' = \gamma \star \gamma^{(n+1)}(T) \,,$$

and use the induction hypothesis to conclude. The case $\pi_0^{(n)}(\alpha) = d > \pi_1^{(n)}(\alpha)$ is symmetric. □

LEMMA 5 – *Let $C_0 \geq 1$ a constant and a non trivial non empty subset $\mathcal{A}_0 \subset \mathcal{A}$, $\mathcal{A}_0 \neq \mathcal{A}$, such that*

$$\max_{\alpha \in \mathcal{A}_0} Q_\alpha(\gamma) \leq C_0 \min_{\alpha \in \mathcal{A}_0} Q_\alpha(\gamma) \,,$$

$$\max_{\alpha \in \mathcal{A}} Q_\alpha(\gamma) \leq \max_{\alpha \in \mathcal{A}_0} Q_\alpha(\gamma) \,.$$

There exist a constant $C_1 \geq 1$, a constant $c_1 \in (0,1)$, depending only on C_0 and d, and paths $(\gamma(l))_{1 \leq l \leq L}$ extending γ such that

(i) the simplices $\Delta(\gamma(l))$ have disjoint interiors and

$$\mathrm{vol}_{d-1}(\sqcup \Delta(\gamma(l))) \geq c_1 \,\mathrm{vol}_{d-1}\,(\Delta(\gamma)) \,;$$

(ii) for every $l \in [1, L]$, there exists a subset $\mathcal{A}_l$ of $\mathcal{A}$ strictly larger than $\mathcal{A}_0$ such that

$$\max_{\mathcal{A}_l} Q_\alpha(\gamma(l)) \leq C_1 \min_{\mathcal{A}_l} Q_\alpha(\gamma(l)),$$

$$\max_{\mathcal{A}} Q_\alpha(\gamma(l)) \leq \max_{\mathcal{A}_l} Q_\alpha(\gamma(l)) \,.$$

Proof – We first extend γ to a path $\tilde{\gamma}$ such that the name of the last arrow of $\tilde{\gamma}$ does not belong to $\mathcal{A}_0$; we can do this having

$$\max_{\mathcal{A}} Q_\alpha(\tilde{\gamma}) \le C_1' \max_{\mathcal{A}} Q_\alpha(\gamma),$$

$$\mathrm{vol}_{d-1}(\Delta(\tilde{\gamma})) \ge c_1' \ \mathrm{vol}_{d-1}(\Delta(\gamma)) \ ,$$

C_1', c_1' depend only on d.

We then apply Lemma 4, for every $\alpha \in \mathcal{A}_0$, to $\tilde{\gamma}$ with $C = 2\#\mathcal{A}_0$. We obtain that the volume of those $T \in \Delta(\tilde{\gamma})$ for which $Q'_\alpha(T) \le 2\#\mathcal{A}_0 Q_\alpha(\tilde{\gamma})$ for every $\alpha \in \mathcal{A}_0$ is at least half the volume of $\Delta(\tilde{\gamma})$. For such a T, let $m > \tilde{n} =$ length $(\tilde{\gamma})$ the smallest integer such that the name $\bar{\alpha}$ of $\gamma^{(m)}(T)$ belongs to $\mathcal{A}_0$. We define (for those T) a finite path $\gamma(T)$ as follows :

1. If for some $\tilde{m} \in (\tilde{n}, m)$, some $\alpha \in \mathcal{A} - \mathcal{A}_0$, we have

$$Q_\alpha(\tilde{m}, T) \ge \max Q_\alpha(\tilde{\gamma}) \ ,$$

   we let $\gamma(T) = (\gamma^{(i)}(T))_{0 \le i \le \tilde{m}}$, where $\tilde{m}$ is the smallest such integer.
2. Otherwise, $\gamma(T) = (\gamma^{(i)}(T))_{0 \le i \le m}$.
   We select finitely many such $T_1, \ldots, T_L$ such that, setting $\gamma(l) = \gamma(T_l)$, we have

$$\mathrm{vol}\ (\cup \Delta(\gamma(l))) \ge \frac{1}{4}\ \mathrm{vol}\ \Delta(\tilde{\gamma})$$

   and the $\Delta(\gamma(l))$ have disjoint interiors. Let $l \in [1, L]$; if T_l is as in case a), we take $\mathcal{A}_l$ to be the union of $\mathcal{A}_0$ and all $\alpha \in \mathcal{A} - \mathcal{A}_0$ satisfying $Q_\alpha(\tilde{m}, T_l) \ge \max_\alpha Q_\alpha(\tilde{\gamma})$. If T_l is as in case b), by definition of m, the name β of $\gamma^{(m-1)}(T_l)$ does not belong to $\mathcal{A}_0$ and we have

$$Q_\beta^{(m)}(T_l) = Q_\alpha^{(m-1)}(T_l) + Q_\beta^{(m-1)}(T_l) \ ,$$

   where α is the name of $\gamma^{(m-1)}(T_l)$. It follows that

$$Q_\beta^{(m)}(T_l) \ge C_0^{-1} \max_{\alpha \in \mathcal{A}_0} Q_\alpha(\gamma) \ .$$

   We take $\mathcal{A}_l = \mathcal{A}_0 \cup \{\beta\}$ in this case. We obtain the conclusions of the lemma with $c_1 = \frac{1}{4} c_1'$ and $C_1 = C_0(1 + 2(\#\mathcal{A}_0)C_1')$.

□

Iterating Lemma 5, we obtain

LEMMA 6 – *There exists a constant C, depending only on d, and paths $(\gamma(l))_{1 \le l \le L}$ extending γ such that*

(i) *the simplices $\Delta(\gamma(l))$ have disjoint interiors and*

$$\mathrm{vol}_{d-1}(\sqcup \Delta(\gamma(l))) \ge C^{-1}\ \mathrm{vol}(\Delta(\gamma)) \ ;$$

(ii) *for every $1 \le l \le L$, we have*

$$\max_\alpha\ Q_\alpha(\gamma(l))) \le C\ \min_\alpha\ (Q_\alpha(\gamma(l))) \ .$$

□

The proof of the theorem is now clear : for almost every i.e.m T, with associated path $(\gamma^{(i)})_{i>0}$, it follows from Lemma 6 that there are infinitely many integers n_k such that the path $(\gamma^{(i)})_{0<i\leq n_k}$ satisfy the hypothesis of Lemma 2. It follows then from Lemma 2 that the intersection of the simplices $\Delta((\gamma^{(i)})_{0<i\leq n})$ is reduced to a point. □

5 Suspension of i.e.m

5.1 Suspension Data

Let $(\mathcal{A}, \pi_0, \pi_1)$ be admissible combinatorial data, and let T be an i.e.m of this combinatorial type, determined by length data $(\lambda_\alpha)_{\alpha\in\mathcal{A}}$.

We will construct a Riemann surface with a flow which can be considered as a suspension of T. In order to do this, we need data which we call **suspension data**.

We will identify $\mathbf{R}^2$ with $\mathbf{C}$. Consider a family $\tau = (\tau_\alpha)_{\alpha\in\mathcal{A}} \in \mathbf{R}^{\mathcal{A}}$. To this family we associate

$$\zeta_\alpha = \lambda_\alpha + i\tau_\alpha \ , \quad \alpha \in \mathcal{A}$$

$$\xi^\varepsilon_\alpha = \sum_{\pi_\varepsilon \beta \leq \pi_\varepsilon \alpha} \zeta_\beta \ , \ \alpha \in \mathcal{A} \ , \ \varepsilon \in \{0,1\} \ .$$

We always have $\xi^0_{\alpha_0} = \xi^1_{\alpha_1}$, where as before $\pi_\varepsilon(\alpha_\varepsilon) = d$. We say that τ defines suspension data if the following inequalities hold :

$$Im\xi^0_\alpha > 0 \text{ for all } \alpha \in \mathcal{A}, \alpha \neq \alpha_0 \ ,$$

$$Im\xi^1_\alpha < 0 \text{ for all } \alpha \in \mathcal{A}, \alpha \neq \alpha_1 \ .$$

We also set

$$\theta_\alpha = \xi^1_\alpha - \xi^0_\alpha, \alpha \in \mathcal{A} \ .$$

We then have

$$\theta = \Omega\zeta \ ,$$

$$Re\theta = \delta \ ,$$

and define $h = -Im\theta = -\Omega\tau$.

One has $h_\alpha > 0$ for all $\alpha \in \mathcal{A}$, because of the formula

$$\theta_\alpha = (\xi^1_\alpha - \zeta_\alpha) - (\xi^0_\alpha - \zeta_\alpha) \ .$$

One has also

$$Im\ \xi^0_{\alpha_0} = Im\ \xi^1_{\alpha_1} \in [-h_{\alpha_1}, h_{\alpha_0}]\ .$$

5.2 Construction of a Riemann Surface

Let $(\mathcal{A}, \pi_0, \pi_1)$ and $(\zeta_\alpha = \lambda_\alpha + i\tau_\alpha)_{\alpha\in\mathcal{A}}$ as above. For $\alpha \in \mathcal{A}$, consider the rectangles in $\mathbf{C} = \mathbf{R}^2$:

$$R^0_\alpha = (Re\xi^0_\alpha - \lambda_\alpha, Re\xi^0_\alpha) \times [0, h_\alpha]\ ,$$

$$R^1_\alpha = (Re\xi^1_\alpha - \lambda_\alpha, Re\xi^1_\alpha) \times [-h_\alpha, 0]\ ,$$

and the segments

$$S^0_\alpha = \{Re\xi^0_\alpha\} \times [0, Im\xi^0_\alpha)\ ,\ \alpha \neq \alpha_0$$

$$S^1_\alpha = \{Re\xi^1_\alpha\} \times (Im\xi^1_\alpha, 0]\ ,\ \alpha \neq \alpha_1\ .$$

Let also $S^0_{\alpha_0} = S^1_{\alpha_1}$ be the half-open vertical segment $[\lambda^*, \xi^0_{\alpha_0}) = [\lambda^*, \xi^1_{\alpha_1})$. Define then

$$R_\zeta = \bigcup_\varepsilon \bigcup_\alpha R^\varepsilon_\alpha \bigcup_\varepsilon \bigcup_\alpha S^\varepsilon_\alpha\ .$$

The translation by θ_α sends R^0_α onto R^1_α. If $\xi^0_{\alpha_0} = \xi^1_{\alpha_1} = 0, S^0_{\alpha_0} = S^1_{\alpha_1}$ is empty, $\xi^0_{\alpha_1}$ is the top right corner of $R^0_{\alpha_1}$ and $\xi^1_{\alpha_0}$ is the bottom right corner of $R^1_{\alpha_0}$. If $\xi^0_{\alpha_0} = \xi^1_{\alpha_1} > 0$, the translation by θ_{α_1} sends the top part $\tilde{S}^0_{\alpha_1} = \{Re\xi^0_{\alpha_1}\} \times [h_{\alpha_1}, Im\xi^0_{\alpha_1})$ of $S^0_{\alpha_1}$ onto $S^1_{\alpha_1}$. If $\xi^0_{\alpha_0} = \xi^1_{\alpha_1} < 0$, the translation by θ_{α_0} sends $S^0_{\alpha_0}$ onto the bottom part $\tilde{S}^1_{\alpha_0} = \{Re\xi^1_{\alpha_0}\} \times (Im\xi^1_{\alpha_0}, -h_{\alpha_0}]$ of $S^1_{\alpha_0}$.

We use these translations to identify in R_ζ each R^0_α to each R^1_α, and $S^0_{\alpha_0} = S^1_{\alpha_1}$ (if non empty) to either $\tilde{S}^0_{\alpha_1}$ or $\tilde{S}^1_{\alpha_0}$. Denote by M^*_ζ the topological space obtained from R_ζ by these identifications.

Observe that M^*_ζ inherits from $\mathbf{C}$ the structure of a Riemann surface, and also a nowhere vanishing holomorphic 1-form ω (given by dz) and a vertical vector field (given by $\frac{\partial}{\partial y}$).

5.3 Compactification of M^*_ζ

Let $\bar{\mathcal{A}}$ be the set with $2d-2$ elements of pairs (α, L) and (α, R), except that we identify $(\alpha_0, R) = (\alpha_1, R)$ and $(\alpha'_0, L) = (\alpha'_1, L)$, where $\pi_\varepsilon(\alpha_\varepsilon) = d, \pi_\varepsilon(\alpha'_\varepsilon) = 1$.

Let σ be the permutation of $\bar{\mathcal{A}}$ defined by

$$\sigma(\alpha, R) = (\beta_0, L)\ ,$$

$$\sigma(\alpha, L) = (\beta_1, R)\ ,$$

with $\pi_0(\beta_0) = \pi_0(\alpha) + 1, \pi_1(\beta_1) = \pi_1(\alpha) - 1$; in particular, we have

$$\sigma(\alpha_0, R) = (\pi_0^{-1}(\pi_0(\alpha_1) + 1), L) \ ,$$

$$\sigma(\alpha_1', L) = (\pi_1^{-1}(\pi_1(\alpha_0') - 1), R) \ .$$

The permutation describes which half planes are met when one winds around an end of M_ζ^*. Denote by $\sum$ the set of cycles of σ. To each $c \in \Sigma$ is associated in a one-to-one correspondance an end q_c of M_ζ^*. From the local structure around q_c, it is clear that the compactification $M_\zeta = M_\zeta^* \bigcup_{\Sigma} \{q_c\}$ will be a compact Riemann surface, with the set of marked points $\{q_c\} = M_\zeta - M_\zeta^*$ in canonical correspondence with Σ. Moreover, the 1-form ω extends to a holomorphic 1-form on M_ζ ; the length of a cycle c is an even number $2n_c$; the corresponding marked point q_c is a zero of ω of order $n_c - 1$.

Let $\nu = \#\Sigma$, and let g be the genus of M_ζ. We have

$$d - 1 = \Sigma n_c$$

$$2g - 2 = \Sigma(n_c - 1)$$

hence

$$d = 2g + \nu - 1 \ .$$

Example : Suppose that π_0, π_1 satisfy

$$\pi_0(\alpha) + \pi_1(\alpha) = d + 1, \ \text{for all } \alpha \in \mathcal{A}$$

If d is even, there is only 1 cycle ; we have $d = 2g$ and the only zero of ω has order $2g - 2$. If d is odd, there are two cycles of equal length $d - 1$; we have $d = 2g + 1$, and each of the two zeros of ω has order $g - 1$.

The vertical vector field on M_ζ^* does not extend (continuously) to M_ζ when $g > 1$, unless one slows it near the marked points (which we will not do here). Nevertheless, it can be considered as a suspension of T : starting from a point $(x, 0)$ on the bottom side of R_α^0, one flows up till reaching the top side where the point (x, h_α) is identified with the point $(x + \delta_\alpha, 0) = (T(x), 0)$ in the top side of R_α^1. The return time is h_α. The vector field is not complete, as some orbits reach marked points in finite time.

5.4 The Basic Operation of the Algorithm for Suspensions

Let $(\mathcal{A}, \pi_0, \pi_1)$ and $(\zeta_\alpha = \lambda_\alpha + i\tau_\alpha)_{\alpha \in \mathcal{A}}$ as above. Construct R_ζ, M_ζ as in 5.2, 5.3. With $\pi_\varepsilon(\alpha_\varepsilon) = d$ as above, assume that

$$\lambda_{\alpha_0} \neq \lambda_{\alpha_1} \ ..$$

Then the formula $\lambda_{\alpha_\varepsilon} = \max(\lambda_{\alpha_0}, \lambda_{\alpha_1})$ defines uniquely $\varepsilon \in \{0, 1\}$ and determines uniquely the basic step of the continuous fraction algorithm ; this step produces new combinatorial data $(\mathcal{A}, \hat{\pi}_0, \hat{\pi}_1)$ and length data $(\hat{\lambda}_\alpha)_{\alpha \in \mathcal{A}}$ given by

$$\begin{cases} \hat{\lambda}_\alpha & = \lambda_\alpha \ , \quad \alpha \neq \alpha_\varepsilon \\ \hat{\lambda}_{\alpha_\varepsilon} & = \lambda_{\alpha_\varepsilon} - \lambda_{\alpha_{1-\varepsilon}} \end{cases}$$

For suspension data, we just define in the same way

$$\begin{cases} \hat{\zeta}_\alpha & = \zeta_\alpha \ , \quad \alpha \neq \alpha_\varepsilon \\ \hat{\zeta}_{\alpha_\varepsilon} & = \zeta_{\alpha_\varepsilon} - \zeta_{\alpha_{1-\varepsilon}} \end{cases}$$

This has a nice representation in terms of the corresponding regions $R_\zeta, R_{\hat{\zeta}}$. One cuts from R_ζ the part where $x > \hat{\lambda}^* = \lambda^* - \lambda_{\alpha_\varepsilon}$; it is made of $R^{1-\varepsilon}_{\alpha_{1-\varepsilon}}$ and a right part of $R^{\varepsilon}_{\alpha_\varepsilon}$. We glue back $R^{1-\varepsilon}_{\alpha_{1-\varepsilon}}$ to the free horizontal side of $R^{1-\varepsilon}_{\alpha_\varepsilon}$, and the right part of $R^{\varepsilon}_{\alpha_\varepsilon}$ to $R^{\varepsilon}_{\alpha_{1-\varepsilon}}$: see figure 5.

It is easy to check that the new suspension data satisfy the inequalities required in 5.1 ; if for instance $\varepsilon = 0$, one has

$$\hat{\xi}^0_\alpha = \xi^0_\alpha \quad , \alpha \neq \alpha_0$$

with $\hat{\pi}_0 = \pi_0$ on one hand and

$$\begin{aligned} \hat{\xi}^1_\alpha &= \xi^1_\alpha, \qquad \alpha \neq \alpha_0, \alpha_1 \\ \hat{\xi}^1_{\alpha_1} &= \xi^1_{\alpha_0}, \\ \hat{\xi}^1_{\alpha_0} &= \xi^1_{\alpha_0} - \zeta_{\alpha_1}. \end{aligned}$$

The last formula gives

$$\begin{aligned} -\hat{\xi}^1_{\alpha_0} &= \zeta_{\alpha_1} - \xi^1_{\alpha_0} \\ &= \zeta_{\alpha_1} - \xi^0_{\alpha_0} - \theta_{\alpha_0} \\ &= \zeta_{\alpha_1} - \xi^1_{\alpha_1} - \theta_{\alpha_0} \\ &= -\xi^1_{\tilde{\alpha}_1} - \theta_{\alpha_0} \ , \end{aligned}$$

with $\pi_1(\tilde{\alpha}_1) = d - 1$. We therefore have

$$-Im\hat{\xi}^1_{\alpha_0} = -Im\xi^1_{\tilde{\alpha}_1} + h_{\alpha_1} > 0 \ .$$

We also see that (still with $\varepsilon = 0$), if $\hat{\alpha}_1 \in \mathcal{A}$ is such that $\hat{\pi}_1(\hat{\alpha}_1) = d$ (we have $\hat{\alpha}_1 = \tilde{\alpha}_1$ if $\tilde{\alpha}_1 \neq \alpha_0, \hat{\alpha}_1 = \alpha_1$ if $\tilde{\alpha}_1 = \alpha_0$), one has

$$Im\hat{\xi}^1_{\hat{\alpha}_1} = Im\xi^1_{\tilde{\alpha}_1} < 0$$

Conversely, given $(\mathcal{A}, \pi_0, \pi_1)$ and $(\zeta_\alpha = \lambda_\alpha + i\tau_\alpha)_{\alpha \in \mathcal{A}}$ as above, assume that

$$Im\xi^0_{\alpha_0} = Im\xi^1_{\alpha_1} \neq 0 \ ,$$

and define ε as 0 if $Im\xi^1_{\alpha_1} < 0$, 1 if $Im\xi^0_{\alpha_0} > 0$. Set

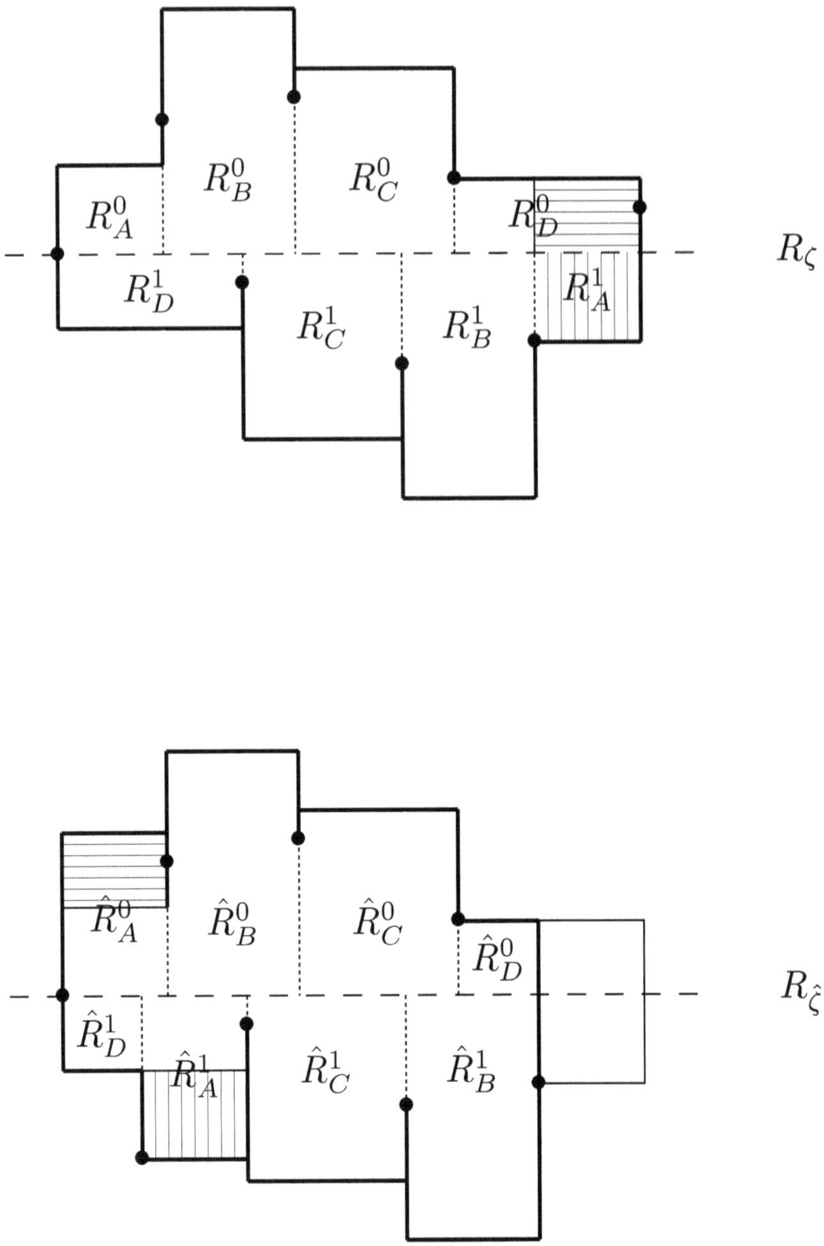

Fig. 5. The Rauzy-Veech operation for suspensions

$$\begin{cases} \hat{\zeta}_\alpha = \zeta_\alpha \text{ for } \alpha \neq \alpha_\varepsilon \ , \\ \hat{\zeta}_{\alpha_\varepsilon} = \zeta_{\alpha_\varepsilon} + \zeta_{\alpha_{1-\varepsilon}} \ , \end{cases}$$

and define appropriately new combinatorial data ; this operation is the inverse of the one above. Thus the dynamics of the continuous fraction algorithm at the level of suspensions is invertible (on a full measure set) and can be viewed as the natural extension of the dynamics at the level of i.e.m.

It is clear that the Riemann surfaces $M_\zeta, M_{\hat{\zeta}}$ are canonically isomorphic, and the isomorphism respects the holomorphic 1-form and the vertical vector field.

5.5 Cohomological Interpretation of Ω

Consider the following homology classes :

- $c_\alpha \in H_1(M_\zeta, \Sigma, \mathbf{Z})$ is defined by a path in R_ζ joining $\xi_\alpha^0 - \zeta_\alpha$ to ξ_α^0 (or by a path joining $\xi_\alpha^1 - \zeta_\alpha$ to ξ_α^1) ;
- $c_\alpha^* \in H_1(M_\zeta - \Sigma, \mathbf{Z})$ is defined by a path in R_ζ joining the center of R_α^0 to the center of R_α^1.

Then $(c_\alpha)_{\alpha\in\mathcal{A}}$ is a basis of $H_1(M_\zeta, \Sigma, \mathbf{Z})$, and $(c_\alpha^*)_{\alpha\in\mathcal{A}}$ is a basis of $H_1(M_\zeta - \Sigma, \mathbf{Z})$.

For the intersection pairing on $H_1(M_\zeta - \Sigma, \mathbf{Z}) \times H_1(M_\zeta, \Sigma, \mathbf{Z}), (c_\alpha^*)$ and (c_α) are dual bases.

We have canonical maps

$$H_1(M_\zeta - \Sigma, \mathbf{Z}) \to H_1(M_\zeta, \mathbf{Z}) \to H_1(M_\zeta, \Sigma, \mathbf{Z})$$

where the first map is surjective and the second injective ; the image of c_α^* in $H_1(M_\zeta, \Sigma, \mathbf{Z})$ is equal to $\sum_\beta \Omega_{\alpha\beta}\, c_\beta$.

The 1-form ω determines a cohomology class $[\omega]$ in $H^1(M_\zeta, \Sigma, \mathbf{C})$: we have

$$\int_{c_\alpha} \omega = \zeta_\alpha$$

We have the dual sequence

$$H^1(M_\zeta, \Sigma, \mathbf{C}) \to H^1(M_\zeta, \mathbf{C}) \to H^1(M_\zeta - \Sigma, \mathbf{C})$$

where the first map is surjective and the second injective. The image of $[\omega]$ in $H^1(M_\zeta - \Sigma, \mathbf{C})$ satisfies

$$\int_{c_\alpha^*} \omega = \theta_\alpha = (\Omega\zeta)_\alpha \ .$$

Thus Ω is the matrix of the composition

$$H^1(M_\zeta, \Sigma, \mathbf{C}) \to H^1(M_\zeta - \Sigma, \mathbf{C}) .$$

The image of Ω is equal to the image of $H^1(M_\zeta, \mathbf{C})$ into $H^1(M_\zeta - \Sigma, \mathbf{C})$.

When one performs the basic operation of the continuous fraction algorithm and one identifies M_ζ with $M_{\hat{\zeta}}$, the relation between the old and new bases is given by

$$\begin{cases} \hat{c}_\alpha = c_\alpha \quad \text{if } \alpha \neq \alpha_\varepsilon , \\ \hat{c}_{\alpha_\varepsilon} = c_{\alpha_\varepsilon} - c_{\alpha_{1-\varepsilon}} \end{cases}$$

$$\begin{cases} \hat{c}^*_\alpha = c^*_\alpha \quad \text{if } \alpha \neq \alpha_{1-\varepsilon} \\ \hat{c}^*_{\alpha_{1-\varepsilon}} = c^*_{\alpha_{1-\varepsilon}} + c^*_{\alpha_\varepsilon} \end{cases}$$

At the cohomological level, we have an isomorphism of $H^1(M_\zeta, \mathbf{C})$ given by

$$\begin{cases} \hat{\theta}_\alpha = \theta_\alpha \quad \text{if } \alpha \neq \alpha_{1-\varepsilon} \\ \hat{\theta}_{\alpha_{1-\varepsilon}} = \theta_{\alpha_{1-\varepsilon}} + \theta_{\alpha_\varepsilon} \end{cases}$$

(these formulas determine an isomorphism from $Im\Omega$ onto $Im\hat{\Omega}$). This is the discrete version of the so-called Kontsevich-Zorich cocycle.

5.6 The Teichmüller Flow

Fix combinatorial data $(\mathcal{A}, \pi_0, \pi_1)$. Given length data (λ_α) and suspension data (τ_α), one defines for $t \in \mathbf{R}$

$$U^t(\lambda, \tau) = (e^{t/2}\lambda, e^{-t/2}\tau)$$

This flow is called the Teichmüller flow. Observe that the conditions on the length data ($\lambda_\alpha > 0$) and on the suspension data (cf. 4.1) are preserved under the flow.

It is also obvious that the flow commutes with the basic operation of the continuous fraction algorithm. In particular, the inequality $\lambda_{\alpha_\varepsilon} > \lambda_{\alpha_{1-\varepsilon}}$ is preserved.

The surface M_ζ is canonically equipped with an area form (coming from $\mathbf{C}$) for which its area is

$$A := \text{area } (M_\zeta) = \sum_{\alpha \in \mathcal{A}} \lambda_\alpha h_\alpha .$$

The area is preserved by the Teichmüller flow, and also by the basic operation of the continuous fraction algorithm.

The Lebesgue measure $d\lambda\, d\tau$ on the domain of $\mathbf{R}^{\mathcal{A}} \times \mathbf{R}^{\mathcal{A}}$ defined by the restrictions on length and suspension data is preserved by the Teichmüller flow, and by the basic operation of the continuous fraction algorithm.

6 Invariant Measures

6.1 The Case $d = 2$

We have seen in 2.6 that i.e.m in this case are just rotations on the circle.

Let (λ_A, λ_B) be the length data. The basic step of the continuous fraction algorithm sends these data on $(\lambda_A - \lambda_B, \lambda_B)$ (resp. $(\lambda_A, \lambda_B - \lambda_A)$) if $\lambda_A > \lambda_B$ (resp. $\lambda_A < \lambda_B$). Set $x = \lambda_B/\lambda_A$ if $\lambda_B < \lambda_A, x = \lambda_A/\lambda_B$ if $\lambda_A < \lambda_B$. We obtain the well-known map

$$g(x) = \begin{cases} \frac{x}{1-x} & \text{for } \; 0 < x < 1/2 \\ g(1-x) = \frac{1-x}{x} & \text{for } \; 1/2 < x < 1 \,, \end{cases}$$

with a parabolic fixed point at 0. This map has $\frac{dx}{x}$ as a unique (up to a multiplicative constant) invariant measure absolutely continuous w.r.t Lebesgue measure, but this measure is infinite !

Instead, the Gauss map

$$G(x) = \{x^{-1}\}$$

has $\frac{dx}{1+x}$ as a unique (up to a multiplicative constant) invariant measure absolutely continuous w.r.t Lebesgue measure, but the density is now analytic on $[0, 1]$.

The map G is related to g as follows : we have $G(x) = g^n(x)$, where n is the smallest integer > 0 such that $g^{n-1}(x) \in [1/2, 1)$.

For a general Rauzy diagram (with admissible combinatorial data), Veech has shown ([V2]) that there exists a unique (up to a multiplicative constant) measure absolutely continuous w.r.t Lebesgue measure which is invariant under the normalized continuous fraction algorithm. But again, this measure is infinite.

Following Zorich, it is however possible to accelerate the Rauzy-Veech algorithm, concatenating several successive steps in a single one (as the Gauss map does). For the new algorithm, there will exist an invariant absolutely continuous probability measure, which is very useful for ergodic - theoretic considerations.

6.2 The Accelerated Algorithm ([Z1])

Let $(\mathcal{A}, \pi_0, \pi_1)$ be admissible combinatorial data and $(\lambda_\alpha)_{\alpha \in \mathcal{A}}$ be length data. Assume for simplicity that the i.e.m T defined by these data satisfies the Keane property.

The continuous fraction algorithm applied to T gives an infinite path in the Rauzy diagram of $(\mathcal{A}, \pi_0, \pi_1)$, starting at the vertex (π_0, π_1), that we denote by $(\gamma_n(T))_{n>0}$. To each arrow γ_n is associated a type (0 or 1) and a name (a letter in $\mathcal{A}$) ; it is obvious from the definitions of type and name that γ_n, γ_{n+1} have the same type iff they have the same name. We also know that each name is taken infinitely many times (proposition in 4.3); the same assertion for types is actually obvious.

In the accelerated algorithm, one performs in a single step the consecutive steps of the (slow) algorithm for which the associated arrows have the same type (or name).

Assume for instance that $\lambda_{\alpha_0} > \lambda_{\alpha_1}$. Write $\pi_1(\alpha_0) = d - \bar{d} < d$ and $\pi_1(\alpha_1^{(i)}) = d - i$ for $0 \leq i < \bar{d}$. The accelerated algorithm makes the following "euclidean division" : one substracts from λ_{α_0} in turn $\lambda_{\alpha_1^{(0)}}, \lambda_{\alpha_1^{(1)}}, \ldots \lambda_{\alpha_1^{(\bar{d}-1)}}, \lambda_{\alpha_1^{(0)}}, \lambda_{\alpha_1^{(1)}} \ldots$ stopping just before the result becomes negative. This is a **single** step for the accelerated algorithm. When $\bar{d} = 1$, for instance when $d = 2$, it just amounts to ordinary euclidean division with remainder.

We can extend the definition of the accelerated algorithm at the level of suspension data. Recall that at this level, the dynamics of the slow algorithm are essentially invertible (i.e modulo a set of codimension one). The dynamics of the accelerated algorithm is a first return map of the dynamics of the slow one. Indeed, for fixed combinatorial data $(\mathcal{A}, \pi_0, \pi_1)$, the simplicial cone of length data is divided into the two simplicial subcones $\{\lambda_{\alpha_0} > \lambda_{\alpha_1}\}$ and $\{\lambda_{\alpha_1} > \lambda_{\alpha_0}\}$ according to the type 0 or 1 of the basic step. On the other hand, we have seen in 5.4 that the polyhedral cone of suspension data is divided into $\{Im\ \xi^1_{\alpha_1} < 0\}$ and $\{Im\ \xi^0_{\alpha_0} > 0\}$ according to the type 0 or 1 of the prior basic step.

Therefore, we set

$$\begin{aligned} \mathcal{Z}_0 &= \{\lambda_{\alpha_0} > \lambda_{\alpha_1}\ ,\ Im\ \xi^0_{\alpha_0} > 0\}\ , \\ \mathcal{Z}_1 &= \{\lambda_{\alpha_1} > \lambda_{\alpha_0}\ ,\ Im\ \xi^1_{\alpha_1} < 0\}\ , \\ \mathcal{Z} &= \mathcal{Z}_0 \sqcup \mathcal{Z}_1\ .. \end{aligned}$$

The accelerated algorithm is the first return map to $\mathcal{Z}$ of the slow algorithm.

Till now, we have considered $\lambda^* := \Sigma \lambda_\alpha = 1$ as the natural normalization for the length data. Actually, in the sequel, a different normalization seems preferable. As in 4.1, for $\lambda_{\alpha_\varepsilon} > \lambda_{\alpha_{1-\varepsilon}}$, set

$$\begin{cases} \hat{\lambda}_\alpha = \lambda_\alpha \text{ if } \alpha \neq \alpha_\varepsilon , \\ \hat{\lambda}_{\alpha_\varepsilon} = \lambda_{\alpha_\varepsilon} - \lambda_{\alpha_{1-\varepsilon}} . \end{cases}$$

Define then $\hat{\lambda}^* := \sum_\alpha \hat{\lambda}_\alpha = \lambda^* - \lambda_{\alpha_{1-\varepsilon}}$; we will normalize by $\{\hat{\lambda}^* = 1\}$.

6.3 The Absolutely Continuous Invariant Measure

Consider the accelerated algorithm acting on the region $\mathcal{Z}$ of the (λ, τ) space. It is invertible (up to a codimension one subset) and acts by unimodular matrices. Therefore the restriction m_0 of Lebesgue measure to $\mathcal{Z}$ is invariant. The area function $A = \sum_\alpha \lambda_\alpha h_\alpha$ is also invariant, where $h = -\Omega\tau$.

We now use the Teichmüller flow U^t to have the horizontal length $\hat{\lambda}^*$ also invariant. More precisely, let $(\pi_0, \pi_1, \lambda, \tau) \in \mathcal{Z}$, with image $(\bar{\pi}_0, \bar{\pi}_1, \bar{\lambda}, \bar{\tau})$ under the accelerated algorithm. Set

$$t(\lambda) = 2(\log \hat{\lambda}^* - \log \hat{\bar{\lambda}}^*) ,$$

$$\bar{G}(\pi_0, \pi_1, \lambda, \tau) = (\bar{\pi}_0, \bar{\pi}_1, U^{t(\lambda)}(\bar{\lambda}, \bar{\tau})) ,$$

and call $\bar{G}$ the normalized basic step for (the natural extension of) the accelerated algorithm. The measure m_0 is still invariant under $\bar{G}$ because m_0 is invariant under the Teichmüller flow and t is constant along the orbits of the flow. The area function A is still invariant. The length function $\hat{\lambda}^*$ is now also invariant by construction. Define

$$\mathcal{Z}^{(1)} = \mathcal{Z} \cap \{A \leq 1\} ,$$

and denote by m_1 the restriction of m_0 to $\mathcal{Z}^{(1)}$. We now project to $\mathcal{C}(\mathcal{D})$ (cf. 4.3) : we obtain a map

$$G(\pi_0, \pi_1, \lambda) = (\bar{\pi}_0, \bar{\pi}_1, e^{\frac{1}{2}t(\lambda)}\bar{\lambda})$$

and a measure m_2, image of m_1 by the projection, which is invariant under G. As $\hat{\lambda}^*$ is still invariant under G, we can restrict, by homogeneity, the measure m_2 to $\{\hat{\lambda}^* = 1\}$ and get the measure m invariant under G, that we are looking for. We will now check its properties.

6.4 Computation of a Volume

The density of the measure m_2 (w.r.t Lebesgue measure in λ space) is given by the volume of the fiber of the projection sending m_1 onto m_2. Therefore, we have to compute the volumes of

$$\Gamma_\varepsilon \cap \{A \leq 1\}$$

where

$$\Gamma_0 = \{Im\ \xi^0_\alpha > 0\ ,\ \forall\, \alpha \in \mathcal{A}\ ,\ Im\ \xi^1_\alpha < 0\ ,\ \forall\, \alpha \neq \alpha_1\}\ ,$$

$$\Gamma_1 = \{Im\ \xi^0_\alpha > 0\ ,\ \forall\, \alpha \neq \alpha_0\ ,\ Im\ \xi^1_\alpha < 0\ ,\ \forall\, \alpha \in \mathcal{A}\}\ ,$$

and $\lambda_{\alpha_\varepsilon} > \lambda_{\alpha_{1-\varepsilon}}$.

The computation is symmetric and we only consider the case $\varepsilon = 0$. We write the polyhedral cone Γ_0 in τ-space as a union of finitely many disjoint **simplicial** cones Γ up to a codimension 1 subset ; for each Γ, we choose a basis $\tau^{(1)}, \dots, \tau^{(d)}$ of $\mathbf{R}^{\mathcal{A}}$ with volume 1 which generates Γ :

$$\Gamma = \{\sum_{j=1}^{d} t_j\, \tau^{(j)}\ ,\ t_j \geq 0\}\ .$$

We have

$$\mathrm{vol}_d(\Gamma \cap \{\Sigma\, \lambda_\alpha\, h_\alpha \leq 1\}) = (d\,!)^{-1} \prod_1^d (\Sigma\, \lambda_\alpha\, h^{(j)}_\alpha)^{-1}\ ,$$

where $h^{(j)} = -\Omega\tau^{(j)}$. This gives for the density $\mathcal{X}$ of m_2 the formula

$$(*) \qquad \mathcal{X}_{\pi_0,\pi_1}(\lambda) = (d\,!)^{-1} \sum_{\varepsilon} \sum_{\Gamma} \prod_1^d (\Sigma\, \lambda_\alpha\, h^{(j)}_\alpha)^{-1}\ .$$

To estimate further the density, we write, when $\varepsilon = 0$:

$$\hat{\lambda}_{\alpha_0} = \lambda_{\alpha_0} - \lambda_{\alpha_1}\ ,$$
$$\hat{h}_{\alpha_1} = h_{\alpha_0} + h_{\alpha_1}\ ,$$

and $\hat{\lambda}_\alpha = \lambda_\alpha, \hat{h}_\alpha = h_\alpha$ otherwise. We have

$$\sum_\alpha \lambda_\alpha\, h^{(j)}_\alpha = \sum_\alpha \hat{\lambda}_\alpha\, \hat{h}^{(j)}_\alpha$$

and define

$$W_j = \{\alpha \in \mathcal{A},\ \hat{h}^{(j)}_\alpha \neq 0\}\ .$$

6.5 The Key Combinatorial Lemma ([V2], [Z1])

PROPOSITION – *Let X be a subset of $\mathcal{A}$, non empty and distinct from $\mathcal{A}$. Let E_X be the subspace of $\mathbf{R}^{\mathcal{A}}$ generated by the $\tau \in \Gamma_0$ such that $h = -\Omega\tau$ satisfies $\hat{h}_\alpha = 0$ for all $\alpha \in X$. Then the codimension of E_X is $> \#X$.*

COROLLARY – $\#\{j, W_j \cap X = \emptyset\} + \#X < d$.

Proof of corollary– One has $W_j \cap X = \emptyset$ iff $\tau^{(j)} \in E_X$, and the $\tau^{(j)}$ are linearly independent. □

Proof of proposition – As usual, we denote by $\alpha_0, \alpha_1, \alpha'_0, \alpha'_1$ the elements such that $\pi_\varepsilon(\alpha_\varepsilon) = d, \pi_\varepsilon(\alpha'_\varepsilon) = 1$. We write the $\hat{h}_\alpha$ in terms of those $(-1)^\varepsilon \, Im\, \xi^\varepsilon_\alpha, (\varepsilon, \alpha)$ which are nonnegative, *i.e.* with $(\varepsilon, \alpha) \neq (1, \alpha_1)$.

We have

$$\hat{h}_\alpha = Im\ \xi^0_\alpha - Im\ \xi^1_\alpha = Im \xi^0_{\beta_0} - Im\ \xi^1_{\beta_1}$$

for $\alpha \neq \bar{\alpha}_0, \bar{\alpha}_1, \alpha_1$; we have denoted by β_0, β_1 the elements such that $\pi_\varepsilon(\beta_\varepsilon) = \pi_\varepsilon(\alpha) - 1$. The same formula still holds for $\alpha = \bar{\alpha}_1$ and $\alpha = \bar{\alpha}_0 \neq \alpha_1$, with the convention that $Im\ \xi^0_{\beta_0} = 0$ if $\alpha = \bar{\alpha}_0$ and $Im\ \xi^1_{\beta_1} = 0$ if $\alpha = \bar{\alpha}_1$. For $\alpha = \alpha_1$, we have

$$\begin{aligned} \hat{h}_{\alpha_1} &= Im\ \xi^0_{\alpha_1} + Im \xi^1_{\alpha_0} \\ &= Im\ \xi^0_{\beta_0} = Im\ \xi^1_{\beta_1} + Im\ \xi^0_{\gamma_0} + Im\ \xi^1_{\gamma_1} \end{aligned}$$

with $\pi_\varepsilon(\beta_\varepsilon) = \pi_\varepsilon(\alpha_1) - 1, \pi_\varepsilon(\gamma_\varepsilon) = \pi_\varepsilon(\alpha_0) - 1$.

From these formulas, we define subsets $\mathcal{A}(\varepsilon, \alpha) \subset \mathcal{A}$, (with $\mathcal{A}(1, \alpha) \subset \mathcal{A} - \{\alpha_1\}$) such that $\hat{h}_\alpha = 0$ implies $Im \xi^\varepsilon_\beta = 0$ for $\beta \in \mathcal{A}(\varepsilon, \alpha)$: we have

$$\begin{aligned}
&\mathcal{A}(0,\alpha) = \{\beta_0, \alpha\} && \text{if } \alpha \neq \bar{\alpha}_0, \alpha_1\ , \\
&\mathcal{A}(0,\bar{\alpha}_0) = \{\bar{\alpha}_0\} && \text{if } \bar{\alpha}_0 \neq \alpha_1\ , \\
&\mathcal{A}(0,\alpha_1) = \{\gamma_0, \beta_0, \alpha_1\}\ , && \\
&\mathcal{A}(0,\bar{\alpha}_0) = \mathcal{A}(1,\alpha_1) = \{\gamma_0, \bar{\alpha}_0 = \alpha_1\} && \text{if } \bar{\alpha}_0 = \alpha_1\ ; \\
&\mathcal{A}(1,\alpha) = \{\beta_1, \alpha\} && \text{if } \alpha \neq \bar{\alpha}_1, \alpha_1\ , \\
&\mathcal{A}(1,\bar{\alpha}_1) = \{\bar{\alpha}_1\} && \\
&\mathcal{A}(1,\alpha_1) = \begin{cases} \{\gamma_1, \beta_1, \alpha_0\} & \text{if } \alpha_0 \neq \bar{\alpha}_1\ , \\ \{\beta_1, \alpha_0 = \bar{\alpha}_1\} & \text{if } \alpha_0 = \bar{\alpha}_1\ . \end{cases} &&
\end{aligned}$$

CLAIM – One has

$$\bigcup_X \mathcal{A}(0,\alpha) \supset X$$

and equality holds only if $X = \{\alpha, \pi_0(\alpha) < k\}$ for some $k \leq \pi_0(\alpha_1)$ or $k = d$.

Similarly, one has, if $\alpha_1 \notin X$

$$\bigcup_X \mathcal{A}(1,\alpha) \supset X$$

and equality holds only if $X = \{\alpha, \pi_1(\alpha) < k\}$, for some $k \leq d$.

The assertions of the claim are immediate from the definitions of $\mathcal{A}(\varepsilon, \alpha)$. We can now conclude the proof of the proposition. If $\hat{h}_\alpha = 0$ for all $\alpha \in X$, we have $Im\ \xi^\varepsilon_\beta = 0$ for all $\beta \in \bigcup_X \mathcal{A}(\varepsilon, \alpha)$. When either $\bigcup_X \mathcal{A}(0, \alpha)$ or $\bigcup_X \mathcal{A}(1, \alpha)$ is strictly larger than X, we obtain the conclusion of the proposition. Otherwise, by the first half of the claim, we must have $X = \{\alpha, \pi_0(\alpha) < k\}$ for some $k \leq \pi_0(\alpha_1)$ or $k = d$. If $k \leq \pi_0(\alpha_1), \alpha_1 \notin X$ and the second part of the claim would give $X = \{\alpha, \pi_1(\alpha) < k\}$, contradicting admissibility. Finally, in the remaining case $X = \mathcal{A} - \{\alpha_0\}$, one has $h_\alpha = 0$ for all $\alpha \in \mathcal{A}$ (because $\hat{h}_{\alpha_1} = h_{\alpha_1} + h_{\alpha_0}$) and $\tau \equiv 0$. □

6.6 Checking Integrability

From the formula (*) in section 6.4

$$\mathcal{X}_{\pi_0,\pi_1}(\lambda) = \sum_{\varepsilon,\Gamma} \mathcal{X}_\Gamma(\lambda)\ ,$$

$$\mathcal{X}_\Gamma(\lambda) = (d\,!)^{-1} \prod_1^d (\Sigma\ \hat{\lambda}_\alpha\ \hat{h}^{(j)}_\alpha)^{-1}\ ,$$

we deduce the estimate, for each Γ :

$$c^{-1} \leq \mathcal{X}_\Gamma(\lambda) \prod_{j=1}^d (\sum_{W_j} \hat{\lambda}_\alpha) \leq c\,. \tag{6.1}$$

When we restrict to $\{\hat{\lambda}^* = 1\}$, the density up to a constant factor is given by the same formula. Let us decompose the simplex $\Delta := \{\lambda, \hat{\lambda}_\alpha > 0, \hat{\lambda}^* = 1\}$ in the following way : the set of indices is

$$\mathcal{N} = \{\mathbf{n} = (n_\alpha)_{\alpha\in\mathcal{A}} \in \mathbf{N}^{\mathcal{A}}, \min_\alpha n_\alpha = 0\}\,.$$

For each $\mathbf{n} \in \mathcal{N}$, denote by $\Delta(\mathbf{n})$ the set of $(\lambda_\alpha)_{\alpha\in\mathcal{A}} \in \Delta$ such that $\hat{\lambda}_\alpha \geq \frac{1}{2d}$ if $n_\alpha = 0$, and

$$\frac{1}{2d}\ 2^{1-n_\alpha} > \hat{\lambda}_\alpha \geq \frac{1}{2d}\ 2^{-n_\alpha}$$

if $n_\alpha > 0$. We have a partition

$$\Delta = \bigsqcup_{\mathcal{N}} \Delta(\mathbf{n})\,.$$

Clearly, we have, for $\mathbf{n} \in \mathcal{N}$

$$c^{-1} \leq (\text{vol}\ \Delta(\mathbf{n}))\ 2^{\Sigma n_\alpha} \leq c\,. \tag{6.2}$$

On the other hand, for $\lambda \in \Delta(\mathbf{n})$ and Γ as above, one obtains from (6.1) that

$$c^{-1} \leq \mathcal{X}_\Gamma(\lambda)\, 2^{-\sum_{j=1}^{d} \min_{W_j} n_\alpha} \leq c\,. \tag{6.3}$$

With fixed $\mathbf{n}$, let $0 = n^0 < n^1 < \ldots$ be the values taken by the n_α and $V^i \subset \mathcal{A}$ the set of indices with $n_\alpha \geq n^i$. On one side, one has

$$\begin{aligned}\sum_\alpha n_\alpha &= \sum_{i\geq 0} n^i(\#(V^i - V^{i+1})) \\ &= \sum_{i>0} (n^i - n^{i-1})\#V^i\,.\end{aligned}$$

On the other side, let $\tilde{V}^i$ be the set of j such that $W_j \subset V^i$; one has $\min_{W_j} n_\alpha = n^i$ iff $j \in \tilde{V}^i - \tilde{V}^{i+1}$ hence

$$\begin{aligned}\sum_{j=1}^{d} \min_{W_j} n_\alpha &= \sum_{i\geq 0} n^i(\#(\tilde{V}^i - \tilde{V}^{i+1})) \\ &= \sum_{i>0} (n^i - n^{i+1})\#\tilde{V}^i\,.\end{aligned}$$

By the Corollary of 6.5, one has

$$\#\,\tilde{V}^i < \#V^i$$

as long as $0 < \#V^i < d$. This shows that

$$\sum_\alpha n_\alpha - \sum_{j=1}^{d} \min_{W_j} n_\alpha \geq |\mathbf{n}|_\infty := \max_\alpha n_\alpha\,.$$

The last estimate, introduced into (6.2), (6.3), gives

$$(\mathrm{vol}\,\Delta(\mathbf{n}))\, \max_{\Delta(\mathbf{n})} \mathcal{X}_\Gamma \leq c\, 2^{-|\mathbf{n}|_\infty}\,. \tag{6.4}$$

The integrability of $\mathcal{X}_\Gamma$ over Δ now follows from the fact that the number of $\mathbf{n} \in \mathcal{N}$ with $|\mathbf{n}|_\infty = N$ is of order N^{d-2}.

At the same time, we can see that the matrix $Z \in SL(\mathbf{Z}^{\mathcal{A}})$ such that

$$\lambda = Z\hat{\lambda}$$

is such that $\log \|Z\|$ is integrable for the invariant measure m. We use as a norm the supremum of the coefficients. We have, for all $k \in \mathbf{N}$ (when $\varepsilon = 0$; the case $\varepsilon = 1$ is symmetric)

$$\|Z\| > k \Longleftrightarrow \hat{\lambda}_{\alpha_0} > k \sum_{\pi_1\alpha > \pi_1\alpha_0} \hat{\lambda}_\alpha\,,$$

and therefore

$$\|Z\| > (2d)2^{N-1} \Longrightarrow \lambda \in \bigcup_{|\mathbf{n}|_\infty \geq N} \Delta(\mathbf{n}) .$$

This implies that

$$\int_{\|Z\| \geq 2^N} \mathcal{X}_\Gamma \leq c N^{d-2} 2^{-N}$$

for all $N > 0$. Therefore $\|Z\|^\rho$ for $\rho < 1$ and a fortiori $\log \|Z\|$ are integrable for the invariant measure m.

This integrability property puts us in position of applying Oseledets theorem and start studying the ergodic properties of the continuous fraction algorithm. However, we will restrain us to do that here.

References

1. J. Coffrey "Some remarks concerning an example of a minimal, non uniquely ergodic interval exchange map" *Math. Z.* **199** (1988) 577-580.
2. G. Forni "Solutions of the cohomological equation for area-preserving flows on compact surfaces of higher genus" *Annals of Mathematics* **146** (1997) 295-344.
3. G. Forni "Deviation of ergodic averages for area-preserving flows on surfaces of higher genus" *Annals of Mathematics* **155** (2002) 1-103.
4. A. Katok and A.M. Stepin "Approximations in Ergodic Theory" *Russ. Math. Surv.* **22** (1967) 77-102.
5. M. Keane "Interval exchange transformations" *Math. Z.* **141** (1975) 25-31.
6. M. Keane "Non-ergodic interval exchnage transformations" *Isr. J. Math.* **26** (1977) 188-196.
7. S.P. Kerckhoff "Simplicial systems for interval exchange maps and measured foliations" *Ergod. Th. Dynam. Sys.* **5** (1985) 257-271.
8. H.B. Keynes and D. Newton "A "Minimal", Non-Uniquely Ergodic Interval Exchange Transformation" *Math. Z.* **148** (1976) 101-105.
9. R. Krikorian "Déviations de moyennes ergodiques, d'après Forni, Kontsevich, Zorich" *Séminaire Bourbaki* 2003-2004, 56ème année, exposé n^0 927, novembre 2003.
10. M. Kontsevich and A. Zorich "Connected components of the moduli spaces of Abelian differentials with prescribed singularities" *Inv. Math.* **153** (2003) 631-678.
11. H. Masur "Interval exchange transformations and measured foliations" *Annals of Mathematics* **115** (1982) 169-200.
12. S. Marmi, P. Moussa and J-C. Yoccoz "On the cohomological equation for interval exchange maps", *C. R. Math. Acad. Sci. Paris* **336** (2003) 941-948.
13. S. Marmi, P. Moussa and J-C. Yoccoz "The cohomological equation for Roth type interval exchange maps", *J. Amer. Math. Soc.* **18** (2005) 779-882.
14. G. Rauzy "Echanges d'intervalles et transformations induites" *Acta Arit.* (1979) 315-328.
15. M. Rees "An alternative approach to the ergodic theory of measured foliations" *Ergod. th. Dyn. Sys.* **1** (1981) 461-488.

16. W. Veech "Interval exchange transformations" *Journal d'Analyse Mathématique* **33** (1978) 222-272.
17. W. Veech "Gauss measures for transformations on the space of interval exchange maps" *Ann. of Math.* **115** (1982) 201-242.
18. W. Veech "The Teichmuller geodesic flow" *Ann. of Math.* **124** (1986) 441-530.
19. W. Veech "The metric theory of interval exchange transformation I. Generic spectral properties" *Amer. J. of Math.* **106** (1984) 1331–1359
20. W. Veech "The metric theory of interval exchange transformation II. Approximation by primitive interval exchanges" *Amer. J. of Math.* **106** (1984) 1361–1387
21. W. Veech "The metric theory of interval exchange transformation III. The Sah Arnoux Fathi invariant" *Amer. J. of Math.* **106** (1984) 1389–1421
22. A. Zorich "Finite Gauss measure on the space of interval exchange transformations. Lyapunov exponents" *Annales de l'Institut Fourier* Tome 46, fasc. 2 (1996) 325-370.
23. A. Zorich "Deviation for interval exchange transformations" *Ergod. th. Dyn. Sys.* **17** (1997), 1477-1499.
24. A. Zorich "On Hyperplane Sections of Periodic Surfaces" *Amer. Math. Soc. Translations* **179** (1997) 173-189.
25. A. Zorich "How Do the Leaves of a Closed 1-form Wind Around a Surface ?" in Pseudoperiodic Topology, V. Arnold, M. Kontsevich and A. Zorich editors, *Amer. Math. Soc. Translations* **197** (1999) 135-178.

Flat Surfaces

Anton Zorich

IRMAR, Université de Rennes 1, Campus de Beaulieu, 35042 Rennes, France
Anton.Zorich@univ-rennes1.fr

Summary. Various problems of geometry, topology and dynamical systems on surfaces as well as some questions concerning one-dimensional dynamical systems lead to the study of closed surfaces endowed with a flat metric with several cone-type singularities. Such flat surfaces are naturally organized into families which appear to be isomorphic to moduli spaces of holomorphic one-forms.

One can obtain much information about the geometry and dynamics of an individual flat surface by studying both its orbit under the Teichmüller geodesic flow and under the linear group action. In particular, the Teichmüller geodesic flow plays the role of a time acceleration machine (renormalization procedure) which allows to study the asymptotic behavior of interval exchange transformations and of surface foliations.

This survey is an attempt to present some selected ideas, concepts and facts in Teichmüller dynamics in a playful way.

57M50, 32G15 (37D40, 37D50, 30F30)

Key words: Flat surface, billiard in polygon, Teichmüller geodesic flow, moduli space of Abelian differentials, asymptotic cycle, Lyapunov exponent, interval exchange transformation, renormalization, Teichmüller disc, Veech surface

1 Introduction

These notes correspond to lectures given first at Les Houches and later, in an extended version, at ICTP (Trieste). As a result they keep all blemishes of oral presentations. I rush to announce important theorems as facts, and then I deduce from them numerous corollaries (which in reality are used to prove these very keystone theorems). I omit proofs or replace them by conceptual ideas hiding under the carpet all technicalities (which sometimes constitute the main value of the proof). Even in the choice of the subjects I poach the most fascinating issues, ignoring those which are difficult to present no matter how important the latter ones are. These notes also contain some philosophical discussions and hopes which some emotional speakers like me include in their talks and which one, normally, never dares to put into a written text.

I am telling all this to warn the reader that this playful survey of some selected ideas, concepts and facts in this area cannot replace any serious introduction in the subject and should be taken with reservation.

As a much more serious accessible introduction I can recommend a collection of introductory surveys of A. Eskin [E], G. Forni [For2], P. Hubert and T. Schmidt [HuSdt5] and H. Masur [Ma7], organized as a chapter of the Handbook of Dynamical Systems. I also recommend recent surveys of H. Masur and S. Tabachnikov [MaT] and of J. Smillie [S]. The part concerning renormalization and interval exchange transformations is presented in the article of

J.-C. Yoccoz [Y] of the current volume in a much more responsible way than my introductory exposition in Sect. 5.

1.1 Flat Surfaces

There is a common prejudice which makes us implicitly associate a metric of constant positive curvature to a sphere, a metric of constant zero curvature to a torus, and a metric of constant negative curvature to a surface of higher genus. Actually, any surface can be endowed with a flat metric, no matter what the genus of this surface is... with the only reservation that this flat metric will have several singular points. Imagine that our surface is made from plastic. Then we can flatten it from the sides pushing all curvature to some small domains; making these domains smaller and smaller we can finally concentrate all curvature at several points.

Consider the surface of a cube. It gives an example of a perfectly flat sphere with eight conical singularities corresponding to eight vertices of the cube. Note that our metric is nonsingular on edges: taking a small neighborhood of an interior point of an edge and unfolding it we get a domain in a Euclidean plane, see Fig. 1. The illusion of degeneration of the metric on the edges comes from the singularity of the embedding of our flat sphere into the Euclidean space $\mathbb{R}^3$.

However, the vertices of the cube correspond to actual *conical singularities* of the metric. Taking a small neighborhood of a vertex we see that it is isometric to a neighborhood of the vertex of a cone. A flat cone is characterized by the *cone angle*: we can cut the cone along a straight ray with an origin at the vertex of the cone, place the resulting flat pattern in the Euclidean plane and measure the angle between the boundaries, see Fig. 1. Say, any vertex of the cube has cone angle $3\pi/2$ which is easy to see since there are three squares adjacent to any vertex, so a neighborhood of a vertex is glued from three right angles.

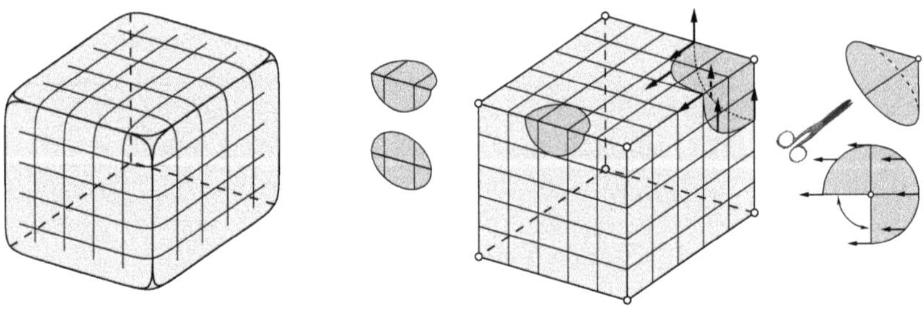

Fig. 1. The surface of the cube represents a flat sphere with eight conical singularities. The metric *does not* have singularities on the edges. After parallel transport around a conical singularity a vector comes back pointing to a direction different from the initial one, so this flat metric has nontrivial *holonomy*

Having a manifold (which is in our case just a surface) endowed with a metric it is quite natural to study *geodesics*, which in a flat metric are locally isometric to straight lines.

General Problem. *Describe the behavior of a generic geodesic on a flat surface. Prove (or disprove) that the geodesic flow is ergodic*[1] *on a typical (in any reasonable sense) flat surface.*

Does any (almost any) flat surface has at least one closed geodesic which does not pass through singular points?

If yes, are there many closed geodesics like that? Namely, find the asymptotics for the number of closed geodesics shorter than L as the bound L goes to infinity.

Believe it or not there has been no (even partial) advance in solving this problem. The problem remains open even in the simplest case, when a surface is a sphere with only three conical singularities; in particular, it is not known, whether any (or even almost any) such flat sphere has *at least one* closed geodesic. Note that in this particular case, when a flat surface is a flat sphere with three conical singularities the problem is a reformulation of the corresponding billiard problem which we shall discuss in Sect. 2.1.

1.2 Very Flat Surfaces

A general flat surface with conical singularities much more resembles a general Riemannian manifold than a flat torus. The reason is that it has nontrivial *holonomy*.

Locally a flat surface is isometric to a Euclidean plane which defines a *parallel transport* along paths on the surface with punctured conical points. A parallel transport along a path homotopic to a trivial path on this punctured surface brings a vector tangent to the surface to itself. However, if the path is not homotopic to a trivial one, the resulting vector turns by some angle. Say, a parallel transport along a small closed path around a conical singularity makes a vector turn exactly by the cone angle, see Fig. 1. (Exercise: perform a parallel transport of a vector around a vertex of a cube.)

Nontrivial linear holonomy forces a generic geodesic to come back and to intersect itself again and again in different directions; geodesics on a flat torus (which has trivial linear holonomy) exhibit radically different behavior. Having chosen a direction to the North, we can transport it to any other point of the torus; the result would not depend on the path. A geodesic on the torus emitted in some direction will forever keep going in this direction. It will either close up producing a regular closed geodesic, or will never intersect itself. In the latter case it will produce a dense irrational winding line on the torus.

[1] In this context *"ergodic"* means that a typical geodesic will visit any region in the *phase space* and, moreover, that in average it will spend a time proportional to the volume of this region; see Appendix A for details.

Fortunately, the class of flat surfaces with trivial linear holonomy is not reduced to flat tori. Since we cannot advance in the General Problem from the previous section, from now on we confine ourselves to the study of these *very flat* surfaces (often called *translation surfaces*): that is, to closed orientable surfaces endowed with a flat metric having a finite number of conical singularities and having trivial linear holonomy.

Triviality of linear holonomy implies, in particular, that all cone angles at conical singularities are integer multiples of 2π. Locally a neighborhood of such a conical point looks like a "*monkey saddle*", see Fig. 2.

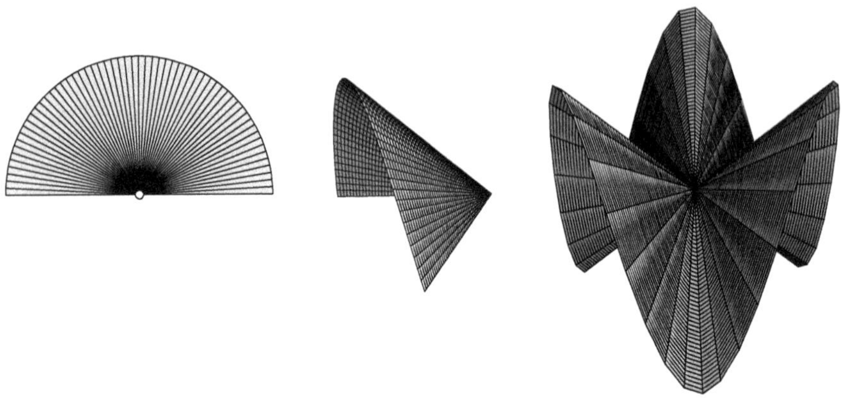

Fig. 2. A neighborhood of a conical point with a cone angle 6π can be glued from six metric half discs

As a first example of a nontrivial very flat surface consider a regular octagon with identified opposite sides. Since identifications of the sides are isometries, we get a well-defined flat metric on the resulting surface. Since in our identifications we used only parallel translations (and no rotations), we, actually, get a very flat (translation) surface. It is easy to see (check it!) that our gluing rules identify all vertices of the octagon producing a single conical singularity. The cone angle at this singularity is equal to the sum of the interior angles of the octagon, that is to 6π.

Figure 3 is an attempt to convince the reader that the resulting surface has genus two. We first identify the vertical sides and the horizontal sides of the octagon obtaining a torus with a hole of the form of a square. To simplify the drawing we slightly cheat: namely, we consider another torus with a hole of the form of a square, but the new square hole is turned by $\pi/4$ with respect to the initial one. Identifying a pair of horizontal sides of the hole by an isometry we get a torus with two holes (corresponding to the remaining pair of sides, which are still not identified). Finally, isometrically identifying the pair of holes we get a surface of genus two.

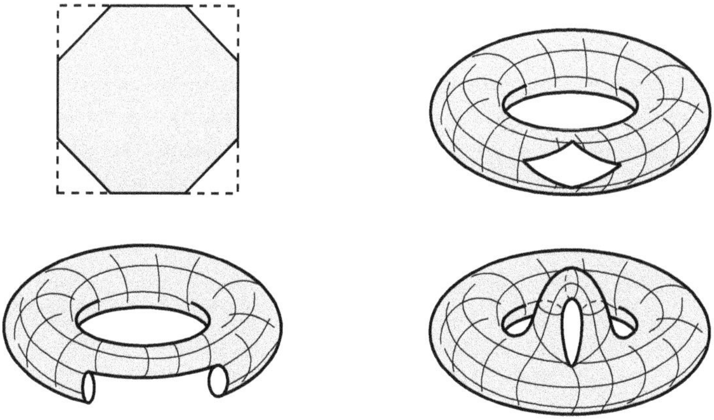

Fig. 3. Gluing a pretzel from a regular octagon

Convention 1. From now on by a *flat surface* we mean a closed oriented surface with a flat metric having a finite number of conical singularities, such that the metric has trivial linear holonomy. Moreover, we always assume that the flat surface is endowed with a distinguished direction; we refer to this direction as the "direction to the North" or as the "*vertical direction*".

The convention above implies, in particular, that if we rotate the octagon from Fig. 3 (which changes the "direction to the North") and glue a flat surface from this rotated octagon, this will give us a different flat surface.

We make three exceptions to Convention 1 in this paper: billiards in general polygons considered at the beginning Sect. 2.1 give rise to flat metrics with nontrivial linear holonomy. In Sect. 3.2 we consider flat tori forgetting the direction to the North.

Finally, in Sect. 8.1 we consider *half-translation surfaces* corresponding to flat metrics with holonomy group $\mathbb{Z}/2\mathbb{Z}$. Such flat metric is a slight generalization of a *very flat* metric: a parallel transport along a loop may change the direction of a vector, that is a vector $\mathbf{v}$ might return as $-\mathbf{v}$ after a parallel transport.

1.3 Synopsis and Reader's Guide

These lectures are an attempt to give some idea of what is known (and what is not known) about flat surfaces, and to show what an amazing and marvellous object a flat surface is: problems from dynamical systems, from solid state physics, from complex analysis, from algebraic geometry, from combinatorics, from number theory, ... (the list can be considerably extended) lead to the study of flat surfaces.

Section 2. Motivations

To give an idea of how flat surfaces appear in different guises we give some motivations in Sect. 2. Namely, we consider billiards in polygons, and, in particular, billiards in rational polygons (Sect. 2.1) and show that the consideration of billiard trajectories is equivalent to the consideration of geodesics on the corresponding flat surface. As another motivation we show in Sect. 2.2 how the electron transport on Fermi-surfaces leads to study of foliation defined by a closed 1-form on a surface. In Sect. 2.3 we show that under some conditions on the closed 1-form such a foliation can be "straightened out" into an appropriate flat metric. Similarly, a Hamiltonian flow defined by the corresponding multivalued Hamiltonian on a surface follows geodesics in an appropriate flat metric.

Section 3. Basic Facts

A reader who is not interested in motivations can proceed directly to Sect. 3 which describes the basic facts concerning flat surfaces. For most of applications it is important to consider not only an individual flat surface, but an entire family of flat surfaces sharing the same topology: genus, number and types of conical singularities. In Sect. 3.1 we discuss deformations of flat metric inside such families. As a model example we consider in Sect. 3.2 the family of flat tori. In Sect. 3.3 we show that a flat structure naturally determines a complex structure on the surface and a holomorphic one-form. Reciprocally, a holomorphic one-form naturally determines a flat structure. The dictionary establishing correspondence between geometric language (in terms of the flat metrics) and complex-analytic language (in terms of holomorphic one-forms) is very important for the entire presentation; it makes Sect. 3.3 more charged than an average one. In Sect. 3.4 we continue establishing correspondence between families of flat surfaces and strata of moduli spaces of holomorphic one-forms. In Sect. 3.5 we describe the action of the linear group $SL(2, \mathbb{R})$ on flat surfaces – another key issue of this theory.

We complete Sect. 3 with an attempt to present the following general principle in the study of flat surfaces. In order to get some information about an individual flat surface it is often very convenient to find (the closure of) the orbit of corresponding element in the *family* of flat surfaces under the action of the group $SL(2, \mathbb{R})$ (or, sometimes, under the action of its diagonal subgroup). In many cases the structure of this orbit gives comprehensive information about the initial flat surface; moreover, this information might be not accessible by a direct approach. These ideas are expressed in Sect. 3.6. This general principle is illustrated in Sect. 3.7 presenting Masur's criterion of unique ergodicity of the directional flow on a flat surface. (A reader not familiar with the ergodic theorem can either skip this last section or read an elementary presentation of ergodicity in Appendix A.)

Section 4. Topological Dynamics of Generic Geodesics

This section is independent from the others; a reader can pass directly to any of the further ones. However, it gives a strong motivation for renormalization discussed in Sect. 5 and in the lectures by J.-C. Yoccoz [Y] in this volume. It can also be used as a formalism for the study of electron transport mentioned in Sect. 2.2.

In Sect. 4.1 we discuss the notion of *asymptotic cycle* generalizing the *rotation number* of an irrational winding line on a torus. It describes how an "irrational winding line" on a surface of higher genus winds around a flat surface in average. In Sect. 4.2 we heuristically describe the further terms of approximation, and we complete with a formulation of the corresponding result in Sect. 4.3.

In fact, this description is equivalent to the description of the deviation of a directional flow on a flat surface from the ergodic mean. Sect. 4.3 involves some background in ergodic theory; it can be either omitted in the first reading, or can be read accompanied by Appendix B presenting the multiplicative ergodic theorem.

Section 5. Renormalization

This section describes the relation between the Teichmüller geodesic flow and *renormalization* for *interval exchange transformations* discussed in the lectures of J.-C. Yoccoz [Y] in this volume. It is slightly more technical than other sections and can be omitted by a reader who is not interested in the proof of the Theorem from Sect. 4.3.

In Sect. 5.1 we show that interval exchange transformations naturally arise as the *first return map* of a directional flow on a flat surface to a transversal segment. In Sect. 5.2 we perform an explicit computation of the *asymptotic cycle* (defined in Sect. 4.1) using interval exchange transformations. In Sect. 5.3 we present a conceptual idea of *renormalization*, a powerful technique of acceleration of motion along trajectories of the directional flow. This idea is illustrated in Sect. 5.4 in the simplest case where we interpret the Euclidean algorithm as a renormalization procedure for rotation of a circle.

We develop these ideas in Sect. 5.5 describing a concrete geometric renormalization procedure (called *Rauzy–Veech induction*) applicable to general flat surfaces (and general interval exchange transformations). We continue in Sect. 5.6 with the elementary formalism of *multiplicative cocycles* (see also Appendix B). Following W. Veech we describe in Sect. 5.7 *zippered rectangles* coordinates in a family of flat surfaces and describe the action of the Teichmüller geodesic flow in a fundamental domain in these coordinates. We show that the first return map of the Teichmüller geodesic flow to the boundary of the fundamental domain corresponds to the Rauzy–Veech induction. In Sect. 5.8 we present a short overview of recent results of G. Forni, M. Kontsevich, A. Avila and M. Viana concerning the spectrum of *Lyapunov exponents* of the corresponding cocycle (completing the proof of the Theorem from

Sect. 4.3). As an application of the technique developed in Sect. 5 we show in Sect. 5.9 that in the simplest case of tori it gives the well-known encoding of a continued fraction by a cutting sequence of a geodesic on the upper half-plane.

Section 6. Closed geodesics

This section is basically independent of other sections; it describes the relation between closed geodesics on individual flat surfaces and "cusps" on the corresponding moduli spaces. It might be useful for those who are interested in the global structure of the moduli spaces.

Following A. Eskin and H. Masur we formalize in Sect. 6.1 the counting problems for closed geodesics and for saddle connections of bounded length on an individual flat surface. In Sect. 6.2 we present the Siegel–Veech Formula and explain a relation between the counting problem and evaluation of the volume of a tubular neighborhood of a "cusp" in the corresponding moduli space. In Sect. 6.3 we describe the structure of a simplest cusp. We describe the structure of general "cusps" (the structure of *principal boundary* of the moduli space) in Sect. 6.4. As an illustration of possible applications we consider in Sect. 6.5 billiards in rectangular polygons.

Section 7 Volume of the Moduli Space

In Sect. 7.1 we consider very special flat surfaces, so called *square-tiled surfaces* which play a role of *integer points* in the moduli space. In Sect. 7.2 we present the technique of A. Eskin and A. Okounkov who have found an asymptotic formula for the number of square-tiled surfaces glued from a bounded number of squares and applied these results to evaluation of volumes of moduli spaces.

As usual, Sect. 7 is independent of others; however, the notion of a square-tiled surface appears later in the discussion of *Veech surfaces* in Sect. 9.5–9.8.

Section 8. Crash Course in Teichmüller Theory

We proceed in Sect. 8 with a very brief overview of some elementary background in Teichmüller theory. Namely, we discuss in Sect. 8.1 the *extremal quasiconformal map* and formulate the *Teichmüller theorem*, which we use in Sect. 8.2 we to define the distance between complex structures (*Teichmüller metric*). We finally explain why the action of the diagonal subgroup in $SL(2;\mathbb{R})$ on the space of flat surfaces should be interpreted as the *Teichmüller geodesic flow*.

Section 9. Main Conjecture and Recent Results

In this last section we discuss one of the central problems in the area – a conjectural structure of *all* orbits of $GL^+(2,\mathbb{R})$. The main hope is that the closure of *any* such orbit is a nice complex subvariety, and that in this sense the moduli spaces of holomorphic 1-forms and the moduli spaces of quadratic differentials resemble homogeneous spaces under an action of a unipotent group. In this section we also present a brief survey of some very recent results related to this conjecture obtained by K. Calta, C. McMullen and others.

We start in Sect. 9.1 with a geometric description of the $GL^+(2,\mathbb{R})$-action in Sect. 9.1 and show why the projections of the orbits (so-called *Teichmüller discs*) to the moduli space $\mathcal{M}_g$ of complex structures should be considered as *complex geodesics*. In Sect. 9.2 we present some results telling that analogous "complex geodesics" in a homogeneous space have a very nice behavior. It is known that the moduli spaces are *not* homogeneous spaces. Nevertheless, in Sect. 9.3 we announce one of the main hopes in this field telling that in the context of the closures of "complex geodesics" the moduli spaces behave as if they were.

We continue with a discussion of two extremal examples of $GL^+(2,\mathbb{R})$-invariant submanifolds. In Sect. 9.4 we describe the "largest" ones: the connected components of the strata. In Sect. 9.5 we consider flat surfaces S (called *Veech surfaces*) with the "smallest" possible orbits: the ones which are closed. Since recently the list of known Veech surfaces was very short. However, K. Calta and C. McMullen have discovered an infinite family of Veech surfaces in genus two and have classified them. Developing these results C. McMullen has proved the main conjecture in genus two. These results of K. Calta and C. McMullen are discussed in Sect. 9.7. Finally, we consider in Sect. 9.8 the classification of Teichmüller discs in $\mathcal{H}(2)$ due to P. Hubert, S. Lelièvre and to C. McMullen.

Section 10. Open Problems

In this section we collect open problems dispersed through the text.

Appendix A. Ergodic Theorem

In Appendix A we suggest a two-pages exposition of some key facts and constructions in ergodic theory.

Appendix B. Multiplicative Ergodic Theorem

Finally, in Appendix B we discuss the Multiplicative Ergodic Theorem which is mentioned in Sect. 4 and used in Sect. 5.

We start with some elementary linear-algebraic motivations in Sect. B.1 which we apply in Sect. B.2 to the simplest case of a "linear" map of a multi-dimensional torus. This examples give us intuition necessary to formulate in Sect. B.3 the *multiplicative ergodic theorem*. Morally, we associate to an ergodic dynamical system a matrix of *mean differential* (or of *mean monodromy* in some cases). We complete this section with a discussion of some basic properties of *Lyapunov exponents* playing a role of logarithms of eigenvalues of the "mean differential" ("mean monodromy") of the dynamical system.

1.4 Acknowledgments

I would like to thank organizers and participants of the workshop *"Frontiers in Number Theory, Physics and Geometry"* held at Les Houches, organizers and participants of the activity on *Algebraic and Topological Dynamics* held at MPI, Bonn, and organizers and participants of the workshop on *Dynamical Systems* held at ICTP, Trieste, for their interest, encouragement, helpful remarks and for fruitful discussions. In particular, I would like to thank A. Avila, C. Boissy, J.-P.Conze, A. Eskin, G. Forni, P. Hubert, M. Kontsevich, F. Ledrappier, S. Lelièvre, H. Masur, C. McMullen, Ya. Pesin, T. Schmidt, M. Viana, Ya. Vorobets and J.-C. Yoccoz.

These notes would be never written without tactful, friendly and persistent pressure and help of B. Julia and P. Vanhove.

I would like to thank MPI für Mathematik at Bonn and IHES at Bures-sur-Yvette for their hospitality while preparation of these notes. I highly appreciate the help of V. Solomatina and of M.-C. Vergne who prepared several most complicated pictures. I am grateful to M. Duchin and to G. Le Floc'h for their kind permission to use the photographs. I would like to thank the The State Hermitage Museum and Succession H. Matisse/VG Bild–Kunst for their kind permission to use "La Dance" of H. Matisse as an illustration in Sect. 6.

2 Eclectic Motivations

In this section we show how flat surfaces appear in different guises: we consider billiards in polygons, and, in particular, billiards in rational polygons. In Sect. 2.1 we show that consideration of billiard trajectories is equivalent to the consideration of geodesics on the corresponding flat surface. As another motivation we show in Sect. 2.2 how the electron transport on Fermi-surfaces leads to the study of foliation defined by a closed 1-form on a surface. In Sect. 2.3 we show, that under some conditions on the closed 1-form such foliation can be "straightened up" in an appropriate flat metric. Similarly, a Hamiltonian flow defined by the corresponding multivalued Hamiltonian on a surface follows geodesics in an appropriate flat metric.

2.1 Billiards in Polygons

Billiards in General Polygons

Consider a *polygonal billiard table* and an ideal billiard ball which reflects from the walls of the table by the "optical" rule: the angle of incidence equals the angle after the reflection. We assume that the mass of our ideal ball is concentrated at one point; there is no friction, no spin.

We mostly consider *regular trajectories*, which do not pass through the corners of the polygon. However, one can also study trajectories emitted from one corner and trapped after several reflections in some other (or the same) corner. Such trajectories are called the *generalized diagonals*.

To simplify the problem let us start our consideration with billiards in *triangles*. A triangular billiard table is defined by angles α, β, γ (proportional rescaling of the triangle does not change the dynamics of the billiard). Since $\alpha + \beta + \gamma = \pi$ the family of triangular billiard tables is described by two real parameters.

It is difficult to believe that the following Problem is open for many decades.

Problem (Billiard in a Polygon).

1. *Describe the behavior of a generic regular billiard trajectory in a generic triangle, in particular, prove (or disprove) that the billiard flow is ergodic*[2]*;*
2. *Does any (almost any) billiard table has at least one regular periodic trajectory? If the answer is affirmative, does this trajectory survive under deformations of the billiard table?*
3. *If a periodic trajectory exists, are there many periodic trajectories like that? Namely, find the asymptotics for the number of periodic trajectories of length shorter than L as the bound L goes to infinity.*

It is easy to find a special closed regular trajectory in an acute triangle: see the left picture at Fig. 4 presenting the *Fagnano trajectory*. This periodic trajectory is known for at least two centuries. However, it is not known whether any (or at least almost any) obtuse triangle has a periodic billiard trajectory.

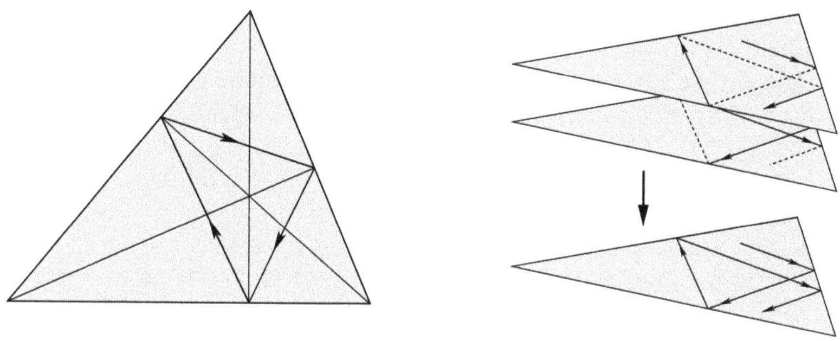

Fig. 4. Fagnano trajectory and Fox–Kershner construction

Obtuse triangles with the angles $\alpha \leq \beta < \gamma$ can be parameterized by a point of a "simplex" Δ defined as $\alpha + \beta < \pi/2$, $\alpha, \beta > 0$. For some obtuse triangles the existence of a regular periodic trajectory is known. Moreover, some

[2]On behalf of the Center for Dynamics and Geometry of Penn State University, A. Katok promised a prize of 10.000 euros for a solution of this problem.

of these periodic trajectories, called *stable periodic trajectories*, survive under small deformations of the triangle, which proves existence of periodic trajectories for some regions in the parameter space Δ (see the works of G. Galperin, A. M. Stepin, Ya. Vorobets and A. Zemliakov [GaStVb1], [GaStVb2], [GaZe], [Vb1]). It remains to prove that such regions cover the entire parameter space Δ. Currently R. Schwartz is in progress of extensive computer search of stable periodic trajectories hoping to cover Δ with corresponding computer-generated regions.

Now, following Fox and Kershner [FxKr], let us see how billiards in polygons lead naturally to geodesics on flat surfaces.

Place two copies of a polygonal billiard table one atop the other. Launch a billiard trajectory on one of the copies and let it jump from one copy to the other after each reflection (see the right picture at Fig. 4). Identifying the boundaries of the two copies of the polygon we get a connected path ρ on the corresponding topological sphere. Projecting this path to any of the two "polygonal hemispheres" we get the initial billiard trajectory.

It remains to note that our topological sphere is endowed with a flat metric (coming from the polygon). Analogously to the flat metric on the surface of a cube which is nonsingular on the edges of the cube (see Sect. 1.1 and Fig. 1), the flat metric on our sphere is nonsingular on the "equator" obtained from the identified boundaries of the two equal "polygonal hemispheres". Moreover, let x be a point where the path ρ crosses the "equator". Unfolding a neighborhood of a point x on the "equator" we see that the corresponding fragment of the path ρ unfolds to a straight segment in the flat metric. In other words, the path ρ is a geodesic in the corresponding flat metric.

The resulting flat metric is *not very flat*: it has nontrivial linear holonomy. The conical singularities of the flat metric correspond to the vertices of the polygon; the cone angle of a singularity is twice the angle at the corresponding vertex of the polygon.

We have proved that every geodesic on our flat sphere projects to a billiard trajectory and every billiard trajectory lifts to a geodesic. This is why General Problem from Sect. 1.1 is so closely related to the Billiard Problem.

Two Beads on a Rod and Billiard in a Triangle

It would be unfair not to mention that billiards in polygons attracted a lot of attention as (what initially seemed to be) a simple model of a Boltzman gas. To give a flavor of this correspondence we consider a system of two elastic beads confined to a rod placed between two walls, see Fig. 5. (Up to the best of my knowledge this construction originates in lectures of Ya. G. Sinai [Sin].)

The beads have different masses m_1 and m_2 they collide between themselves, and also with the walls. Assuming that the size of the beads is negligible we can describe the configuration space of our system using coordinates $0 < x_1 \leq x_2 \leq a$ of the beads, where a is the distance between the walls. Rescaling the coordinates as

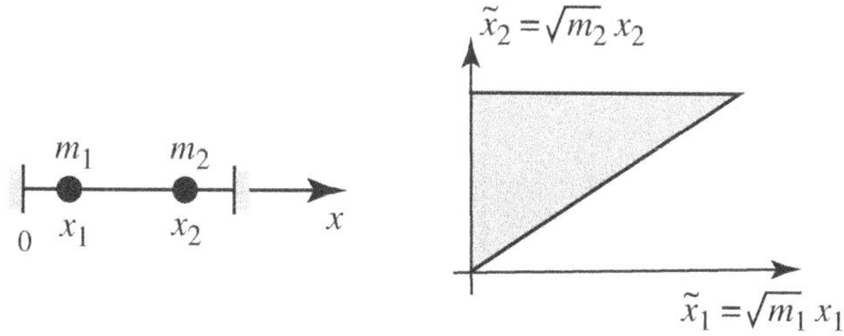

Fig. 5. Gas of two molecules in a one-dimensional chamber

$$\begin{cases} \tilde{x}_1 &= \sqrt{m_1} x_1 \\ \tilde{x}_2 &= \sqrt{m_2} x_2 \end{cases}$$

we see that the configuration space in the new coordinates is given by a right triangle Δ, see Fig. 5. Consider now a trajectory of our dynamical system. We leave to the reader the pleasure to prove the following elementary Lemma:

Lemma. *In coordinates $(\tilde{x}_1, \tilde{x}_2)$ trajectories of the system of two beads on a rod correspond to billiard trajectories in the triangle Δ.*

Billiards in Rational Polygons

We have seen that taking two copies of a polygon we can reduce the study of a billiard in a general polygon to the study of geodesics on the corresponding flat surface. However, the resulting flat surface has *nontrivial* linear holonomy, it is *not* "very flat".

Nevertheless, a more restricted class of billiards, namely, billiards in *rational polygons*, lead to "very flat" (translation) surfaces.

A polygon is called *rational* if all its angles are rational multiples of π. A billiard trajectory emitted in some direction will change direction after the first reflection, then will change direction once more after the second reflection, etc. However, for any given billiard trajectory in a rational billiard the set of possible directions is *finite*, which make billiards in rational polygons so different from general ones.

As a basic example consider a billiard in a rectangle. In this case a generic trajectory at any moment goes in one of four possible directions. Developing the idea with the general polygon we can take *four* (instead of two) copies of our billiard table (one copy for each direction). As soon as our trajectory hits the wall and changes the direction we make it jump to the corresponding copy of the billiard – the one representing the corresponding direction.

By construction each side of every copy of the billiard is identified with exactly one side of another copy of the billiard. Upon these identifications the four copies of the billiard produce a closed surface and the unfolded billiard trajectory produces a connected line on this surface. We suggest to the reader

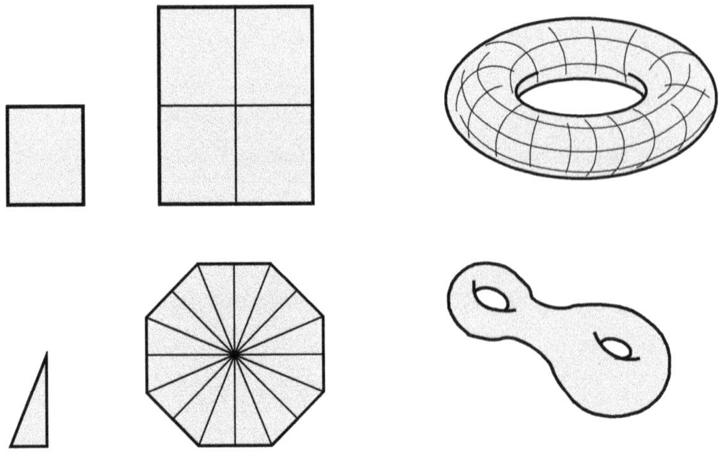

Fig. 6. Billiard in a rectangle corresponds to directional flow on a flat torus. Billiard in a right triangle $(\pi/8, 3\pi/8, \pi/2)$ leads to directional flow on a flat surface obtained from the regular octagon

to check that *the resulting surface is a torus and the unfolded trajectory is a geodesic on this flat torus*, see Fig. 6.

A similar unfolding construction (often called *Katok–Zemliakov construction*) works for a billiard in any rational polygon. Say, for a billiard in a right triangle with angles $(\pi/8, 3\pi/8, \pi/2)$ one has to take 16 copies (corresponding to 16 possible directions of a given billiard trajectory). Appropriate identifications of these 16 copies produce a regular octagon with identified opposite sides (see Fig. 6). We know from Sect. 1.2 and from Fig. 3 that the corresponding flat surface is a "very flat" surface of genus two having a single conical singularity with the cone angle 6π.

Exercise. What is the genus of the surface obtained by Katok–Zemliakov construction from an isosceles triangle $(3\pi/8, 3\pi/8, \pi/4)$? How many conical points does it have? What are the cone angles at these points? Hint: this surface *can not* be glued from a regular octagon.

It is quite common to unfold a rational billiard in two steps. We first unfold the billiard table to a polygon, and then identify the appropriate pairs of sides of the resulting polygon. Note that the polygon obtained in this intermediate step is not canonical.

Show that a generic billiard trajectory in the right triangle with angles $(\pi/2, \pi/5, 3\pi/10)$ has 20 directions. Show that both polygons at Fig. 7 can be obtained by Katok–Zemliakov construction from 20 copies of this triangle. Verify that after identification of parallel sides of these polygons we obtain isometric very flat surfaces (see also [HuSdt5]). What genus, and what conical points do they have? What are the cone angles at these points?

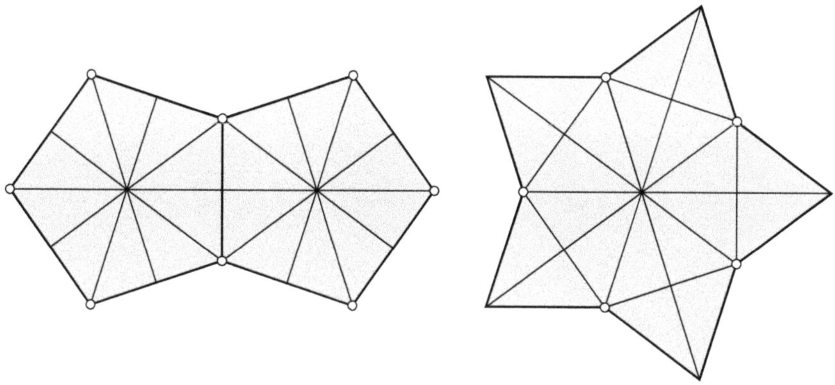

Fig. 7. We can unfold the billiard in the right triangle $(\pi/2, \pi/5, 3\pi/10)$ into different polygons. However, the resulting very flat surfaces are the same (see also [HuSdt5])

Note that in comparison with the initial construction, where we had only two copies of the billiard table we get a more complicated surface. However, what we gain is that in this new construction our flat surface is actually "very flat": it has trivial linear holonomy. It has a lot of consequences; say, due to a Theorem of H. Masur [Ma4] it is possible to find a regular periodic geodesic on *any* "very flat" surface. If the flat surface was constructed from a billiard, the corresponding closed geodesic projects to a regular periodic trajectory of the corresponding billiard which solves part of the Billiard Problem for billiards in rational polygons.

We did not intend to present in this section any comprehensive information about billiards, our goal was just to give a motivation for the study of flat surfaces. A reader interested in billiards can get a good idea on the subject from a very nice book of S. Tabachnikov [T]. Details about billiards in polygons (especially rational polygons) can be found in the surveys of E. Gutkin [Gu1], P. Hubert and T. Schmidt [HuSdt5], H. Masur and S. Tabachnikov [MaT] and J. Smillie [S].

2.2 Electron Transport on Fermi-Surfaces

Consider a periodic surface $\tilde{M}^2$ in $\mathbb{R}^3$ (i.e. a surface invariant under translations by any integer vector in $\mathbb{Z}^3$). Such a surface can be constructed in a fundamental domain of a cubic lattice, see Fig. 8, and then reproduced repeatedly in the lattice. Choose now an affine plane in $\mathbb{R}^3$ and consider an intersection line of the surface by the plane. This intersection line might have some closed components and it may also have some unbounded components. The question is *how does an unbounded component propagate in* $\mathbb{R}^3$?

The study of this subject was suggested by S. P. Novikov about 1980 (see [N]) as a mathematical formulation of the corresponding problem

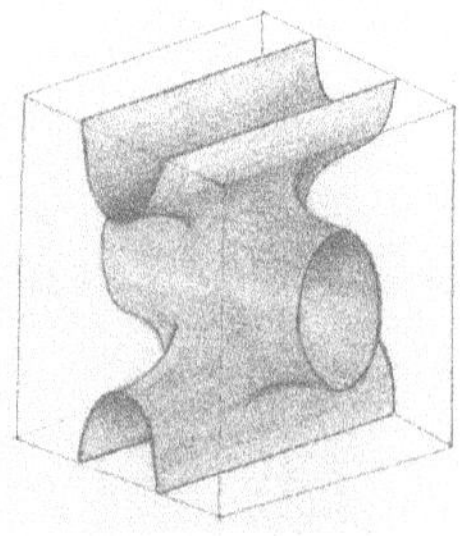

Fig. 8. Riemann surface of genus 3 embedded into a torus $\mathbb{T}^3$

concerning electron transport in metals. A periodic surface represents a *Fermi-surface*, affine plane is a plane orthogonal to a magnetic field, and the intersection line is a trajectory of an electron in the so-called *inverse lattice.*

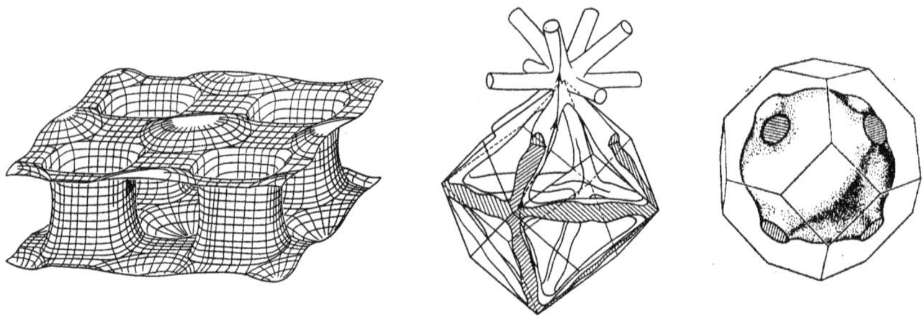

Fig. 9. Fermi surfaces of tin, iron and gold correspond to Riemann surfaces of high genera. (Reproduced from [AzLK] which cites [AGLP] and [WYa] as the source)

It was known since extensive experimental research in the 50s and 60s that Fermi-surfaces may have fairly complicated shape, see Fig. 9; it was also known that open (i.e. unbounded) trajectories exist, see Fig. 10, however, up to the beginning of the 80s there were no general results in this area.

In particular, it was not known whether open trajectories follow (in a large scale) the same direction, whether there might be some scattering (trajectory comes from infinity in one direction and then after some scattering goes to infinity in some other direction, whether the trajectories may even exhibit some chaotic behavior?

Let us see now how this problem is related to flat surfaces.

First note that passing to a quotient $\mathbb{R}^3/\mathbb{Z}^3 = \mathbb{T}^3$ we get a closed orientable surface $M^2 \subset \mathbb{T}^3$ from the initial periodic surface $\tilde{M}^2$. Say, identifying the opposite sides of a unit cube at Fig. 8 we get a closed surface M^2 of genus $g = 3$.

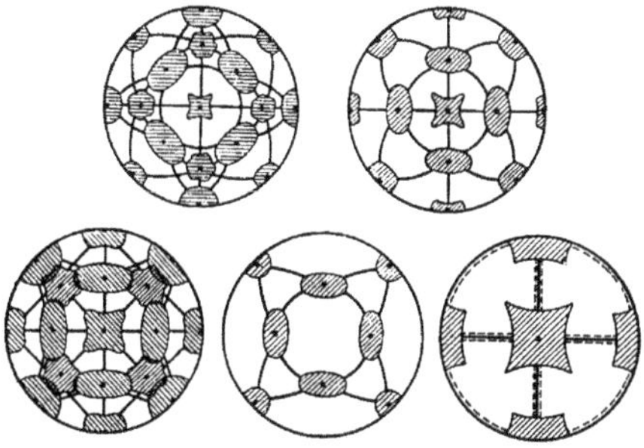

Fig. 10. Stereographic projection of the magnetic field directions (shaded regions and continuous curves) which give rise to open trajectories for some Fermi-surfaces (experimental results in [AzLK])

We are interested in plane sections of the initial periodic surface $\tilde{M}^2$. This plane sections can be viewed as level curves of a linear function $f(x, y, z) = ax + by + cz$ restricted to $\tilde{M}^2$.

Consider now a closed differential 1-form $\tilde{\omega} = a\,dx + b\,dy + c\,dz$ in $\mathbb{R}^3$ and its restriction to $\tilde{M}^2$. A closed 1-form defines a codimension-one foliation on a manifold: locally one can represent a closed one-form as a differential of a function, $\tilde{\omega} = df$; the foliation is defined by the levels of the function f. We prefer to use the 1-form $\tilde{\omega} = a\,dx + b\,dy + c\,dz$ to the linear function $f(x, y, z) = ax + by + cz$ because we cannot push the function $f(x, y, z)$ into a torus $\mathbb{T}^3$ while the 1-form $\omega = a\,dx + b\,dy + c\,dz$ is well-defined in $\mathbb{T}^3$. Moreover, after passing to a quotient over the lattice $\mathbb{R}^3 \to \mathbb{R}^3/\mathbb{Z}^3$ the plane sections of $\tilde{M}^2$ project to the leaves of the foliation defined by restriction of the closed 1-form ω in $\mathbb{T}^3$ to the surface M^2.

Thus, our initial problem can be reformulated as follows.

Problem (Novikov's Problem on Electron Transport). *Consider a foliation defined by a linear closed 1-form $\omega = a\,dx + b\,dy + c\,dz$ on a closed surface $M^2 \subset \mathbb{T}^3$ embedded into a three-dimensional torus. How do the leaves of this foliation get unfolded, when we unfold the torus $\mathbb{T}^3$ to its universal cover $\mathbb{R}^3$?*

The foliation defined by a closed 1-form on a surface is a subject of discussion of the next section. We shall see that under some natural conditions such a foliation can be "straightened up" to a geodesic foliation in an appropriate flat metric.

The way in which a geodesic on a flat surface gets unfolded in the universal Abelian cover is discussed in detail in Sect. 4.

To be honest, we should admit that 1-forms as in the Problem above usually do not satisfy these conditions. However, a surface as in this Problem can be decomposed into several components, which (after some surgery) already satisfy the necessary requirements.

References for Details

There is a lot of progress in this area, basically due to S. Novikov's school, and especially to I. Dynnikov, and in many cases Novikov's Problem on electron transport is solved.

In [Zo1] the author proved that for a given Fermi-surface and for an open dense set of directions of planes any open trajectory is bounded by a pair of parallel lines inside the corresponding plane.

In a series of papers I. Dynnikov applied a different approach: he fixed the direction of the plane and deformed a Fermi-surface inside a family of level surfaces of a periodic function in $\mathbb{R}^3$. He proved that for all but at most one level any open trajectory is also bounded between two lines.

However, I. Dynnikov has constructed a series of highly elaborated examples showing that in some cases an open trajectory can "fill" the plane. In particular, the following question is still open. Consider the set of directions of those hyperplanes which give "nontypical" open trajectories. Is it true that this set has measure zero in the space $\mathbb{RP}^2$ of all possible directions? What can be said about Hausdorff dimension of this set?

For more details we address the reader to papers [D1], [D2] and [NM].

2.3 Flows on Surfaces and Surface Foliations

Consider a closed 1-form on a closed orientable surface. Locally a closed 1-form ω can be represented as the differential of a function $\omega = df$. The level curves of the function f locally define the leaves of the closed 1-form ω. (The fact that the function f is defined only up to a constant does not affect the structure of the level curves.) We get a foliation on a surface.

In this section we present a necessary and sufficient condition which tells when one can find an appropriate flat metric such that the foliation defined by the closed 1-form becomes a foliation of parallel geodesics going in some fixed direction on the surface. This criterion was given in different context by different authors: [Clb], [Kat1], [HbMa]. We present here one more formulation of the criterion. Morally, it says that *the foliation defined by a closed 1-form ω can be "straightened up" in an appropriate flat metric if and only if the form ω does not have closed leaves homologous to zero.* In the remaining part of this section we present a rigorous formulation of this statement.

Note that a closed 1-form ω on a closed surface necessarily has some critical points: the points where the function f serving as a "local antiderivative" $\omega = df$ has critical points.

The first obstruction for "straightening" is the presence of minima and maxima: such critical points should be forbidden. Suppose now that the closed 1-form has only isolated critical points and all of them are "saddles" (i.e. ω does not have minima and maxima). Say, a form defined in local coordinates as df, where $f = x^3 + y^3$ has a saddle point in the origin $(0, 0)$ of our coordinate chart, see Fig. 11.

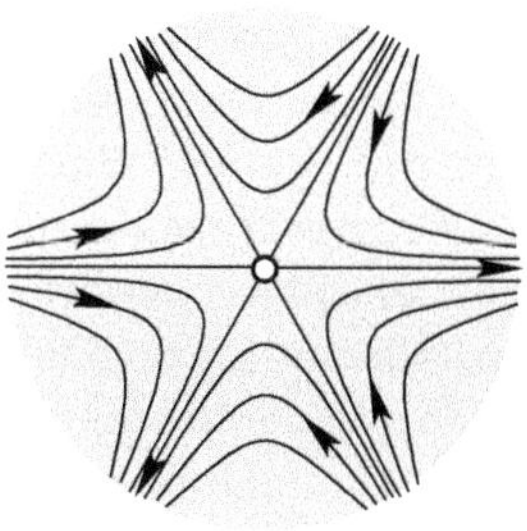

Fig. 11. Horizontal foliation in a neighborhood of a saddle point. Topological (nonmetric) picture

There are several singular leaves of the foliation landing at each saddle; say, a saddle point from Fig. 11 has six prongs (separatrices) representing the critical leaves. Sometimes a critical leaf emitted from one saddle can land to another (or even to the same) saddle point. In this case we say that the foliation has a *saddle connection.*

Note that the foliation defined by a closed 1-form on an oriented surface gets natural orientation (defined by grad(f) and by the orientation of the surface). Now we are ready to present a rigorous formulation of the criterion.

Theorem. *Consider a foliation defined on a closed orientable surface by a closed 1-form ω. Assume that ω does not have neither minima nor maxima but only isolated saddle points. The foliation defined by ω can be represented as a geodesic foliation in an appropriate flat metric if and only if any cycle obtained as a union of closed paths following in the positive direction a sequence of saddle connections is not homologous to zero.*

In particular, if there are no saddle connections at all (provided there are no minima and maxima) it can always be straightened up. (In slightly different terms it was proved in [Kat1] and in [HbMa].)

Note that saddle points of the closed 1-form ω correspond to conical points of the resulting flat metric.

One can consider a closed 1-form ω as a multivalued Hamiltonian and consider corresponding Hamiltonian flow along the leaves of the foliation defined by ω. On the torus $\mathbb{T}^2$ is was studied by V. I. Arnold [Ald2] and by K. Khanin and Ya. G. Sinai [KhSin].

3 Families of Flat Surfaces and Moduli Spaces of Abelian Differentials

In this section we present the generalities on flat surfaces. We start in Sect. 3.1 with an elementary construction of a flat surface from a polygonal pattern.

This construction explicitly shows that any flat surface can be deformed inside an appropriate *family* of flat surfaces. As a model example we consider in Sect. 3.2 the family of flat tori. In Sect. 3.3 we show that a flat structure naturally determines a complex structure on the surface and a holomorphic one-form. Reciprocally, a holomorphic one-form naturally determines a flat structure. The dictionary establishing correspondence between geometric language (in terms of the flat metrics) and complex-analytic language (in terms of holomorphic one-forms) is very important for the entire presentation; it makes Sect. 3.3 more charged than an average one. In Sect. 3.4 we continue with establishing correspondence between families of flat surfaces and strata of moduli spaces of holomorphic one-forms. In Sect. 3.5 we describe the action of the linear group $SL(2,\mathbb{R})$ on flat surfaces – another key issue of this theory.

We complete Sect. 3 with an attempt to present the following general principle in the study of flat surfaces. In order to get some information about an individual flat surface it is often very convenient to find (the closure of) the orbit of the corresponding element in the *family* of flat surfaces under the action of the group $SL(2,\mathbb{R})$ (or, sometimes, under the action of its diagonal subgroup). In many cases the structure of this orbit gives a comprehensive information about the initial flat surface; moreover, this information might be not accessible by a direct approach. These ideas are expressed in Sect. 3.6. This general principle is illustrated in Sect. 3.7 presenting Masur's criterion of unique ergodicity of the directional flow on a flat surface. (A reader not familiar with ergodic theorem can either skip this last section or read an elementary presentation of ergodicity in Appendix A.)

3.1 Families of Flat Surfaces

In this section we present a construction which allows to obtain a large variety of flat surfaces, and, moreover, allows to continuously deform the resulting flat structure. Later on we shall see that this construction is even more general than it may seem at the beginning: it allows to get *almost all* flat surfaces in any *family* of flat surfaces sharing the same geometry (i.e. genus, number and types of conical singularities). The construction is strongly motivated by an analogous construction in the paper of H. Masur [Ma3].

Consider a collection of vectors $\mathbf{v}_1,\ldots,\mathbf{v}_n$ in $\mathbb{R}^2$ and construct from these vectors a broken line in a natural way: a j-th edge of the broken line is represented by the vector $\mathbf{v}_j$. Construct another broken line starting at the same point as the initial one by taking the same vectors but this time in the order $v_{\pi(1)},\ldots,v_{\pi(n)}$, where π is some permutation of n elements.

By construction the two broken lines share the same endpoints; suppose that they bound a polygon as in Fig. 12. Identifying the pairs of sides corresponding to the same vectors v_j, $j=1,\ldots,n$, by parallel translations we obtain a flat surface.

The polygon in our construction depends continuously on the vectors $\mathbf{v}_i$. This means that the topology of the resulting flat surface (its genus g, the

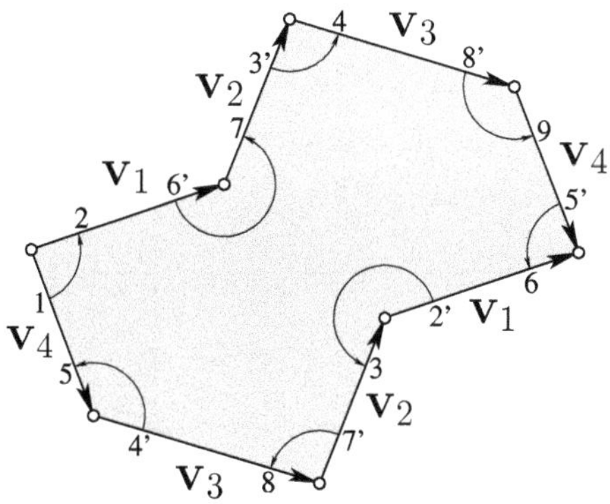

Fig. 12. Identifying corresponding pairs of sides of this polygon by parallel translations we obtain a flat surface

number m and the types of the resulting conical singularities) do not change under small deformations of the vectors $\mathbf{v}_i$. Say, we suggest to the reader to check that the flat surface obtained from the polygon presented in Fig. 12 has genus two and a single conical singularity with cone angle 6π.

3.2 Toy Example: Family of Flat Tori

In the previous section we have seen that a flat structure can be deformed. This allows to consider a flat surface as an element of a *family* of flat surfaces sharing a common geometry (genus, number of conical points). In this section we study the simplest example of such family: we study the family of flat tori. This time we consider the family of flat surfaces *globally*. We shall see that it has a natural structure of a noncompact complex-analytic manifold (to be more honest – *orbifold*). This "baby family" of flat surfaces, actually, exhibits all principal features of any other family of flat surfaces, except that the family of flat tori constitutes a homogeneous space endowed with a nice hyperbolic metric, while general families of flat surfaces *do not* have the structure of a homogeneous space.

To simplify consideration of flat tori as much as possible we make two exceptions from the usual way in which we consider flat surfaces. Temporarily (only in this section) we forget about the choice of the direction to North: in this section two isometric flat tori define the same element of the family of all flat tori. Another exception concerns normalization. Almost everywhere below we consider the area of any flat surface to be normalized to one (which can be achieved by a simple homothety). In this section it would be more convenient for us to apply homothety in the way that the shortest closed geodesic on our flat torus would have length 1. Find the closed geodesic which is next after the shortest one in the length spectrum. Measure the angle ϕ, where $0 \le \phi \le \pi$ between these two geodesics; measure the length r of the second geodesic and

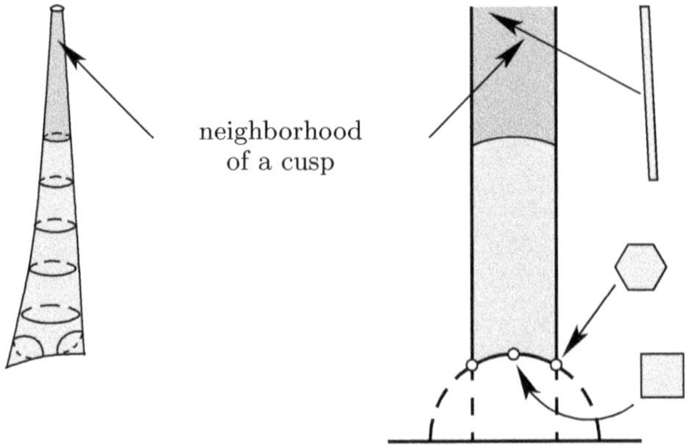

Fig. 13. Space of flat tori

mark a point with polar coordinates (r, ϕ) on the upper half-plane. This point encodes our torus.

Reciprocally, any point of the upper half-plane defines a flat torus in the following way. (Following the tradition we consider the upper half-plane as a complex one.) A point $x + iy$ defines a parallelogram generated by vectors $\mathbf{v}_1 = (1, 0)$ and $\mathbf{v}_2 = (x + iy)$, see Fig. 13. Identifying the opposite sides of the parallelogram we get a flat torus.

To make this correspondence bijective we have to be sure that the vector $\mathbf{v}_1$ represents the shortest closed geodesic. This means that the point $(x + iy)$ representing $\mathbf{v}_2$ cannot be inside the unit disc. The condition that $\mathbf{v}_2$ represents the geodesic which is next after the shortest one in the length spectrum implies that $-1/2 \le x \le 1/2$, see Fig. 13. Having mentioned these two hints we suggest to the reader to prove the following Lemma.

Lemma 1. *The family of flat tori is parametrized by the shadowed fundamental domain from Fig. 13, where the parts of the boundary of the fundamental domain symmetric with respect to the vertical axis* $(0, iy)$ *are identified.*

Note that topologically we obtain a sphere punctured at one point: the resulting surface has a *cusp*. Tori represented by points close to the cusp are "disproportional": they are very narrow and very long. In other words they have an abnormally short geodesic.

Note also that there are two special points on our modular curve: they correspond to points with coordinates $(0 + i)$ and $\pm 1/2 + i\sqrt{3}/2$. The corresponding tori has extra symmetry, they can be represented by a square and by a regular hexagon with identified opposite sides correspondingly. The surface glued from the fundamental domains has "corners" at these two points.

There is an alternative more algebraic approach to our problem. Actually, the fundamental domain constructed above is known as a *modular curve*, it parameterizes the *space of lattices of area one* (which is isomorphic to the space of flat tori). It can be seen as a double quotient

$$SO(2,\mathbb{R}) \backslash SL(2,\mathbb{R}) / SL(2,\mathbb{Z}) = \mathbb{H} / SL(2,\mathbb{Z})$$

3.3 Dictionary of Complex-Analytic Language

We have seen in Sect. 3.1 how to construct a flat surfaces from a polygon as on Fig. 12.

Note that the polygon is embedded into a complex plane $\mathbb{C}$, where the embedding is defined up to a parallel translation. (A rotation of the polygon changes the vertical direction and hence, according to Convention 1 it changes the corresponding flat surface.)

Consider the natural coordinate z in the complex plane. In this coordinate the parallel translations which we use to identify the sides of the polygon are represented as

$$z' = z + const \tag{3.1}$$

Since this correspondence is holomorphic, it means that our flat surface S with punctured conical points inherits the complex structure. It is an exercise in complex analysis to check that the complex structure extends to the punctured points.

Consider now a holomorphic 1-form dz in the initial complex plane. When we pass to the surface S the coordinate z is not globally defined anymore. However, since the changes of local coordinates are defined by the rule (3.1) we see that $dz = dz'$. Thus, the holomorphic 1-form dz on $\mathbb{C}$ defines a holomorphic 1-form ω on S which in local coordinates has the form $\omega = dz$. Another exercise in complex analysis shows that the form ω has *zeroes* exactly at those points of S where the flat structure has conical singularities.

In an appropriate local coordinate w in a neighborhood of zero (different from the initial local coordinate z) a holomorphic 1-form can be represented as $w^d\, dw$, where d is called the *degree* of zero. The form ω has a zero of degree d at a *conical point* with cone angle $2\pi(d+1)$.

Recall the formula for the sum of degrees of zeroes of a holomorphic 1-form on a Riemann surface of genus g:

$$\sum_{j=1}^{m} d_j = 2g - 2$$

This relation can be interpreted as the formula of Gauss–Bonnet for the flat metric.

Vectors $\mathbf{v}_j$ representing the sides of the polygon can be considered as complex numbers. Let $\mathbf{v}_j$ be joining vertices P_j and P_{j+1} of the polygon. Denote by ρ_j the resulting path on S joining the points $P_j, P_{j+1} \in S$. Our interpretation of $\mathbf{v}_j$ as of a complex number implies the following obvious relation:

$$\mathbf{v}_j = \int_{P_j}^{P_{j+1}} dz = \int_{\rho_j} \omega \tag{3.2}$$

Note that the path ρ_j represents a *relative cycle*: an element of the relative homology group $H_1(S, \{P_1, \ldots, P_m\}; \mathbb{Z})$ of the surface S relative to the finite collection of conical points $\{P_1, \dots, P_m\}$. Relation (3.2) means that $\mathbf{v}_j$ represents a *period* of ω: an integral of ω over a relative cycle ρ_j.

Note also that the flat area of the surface S equals the area of the original polygon, which can be measured as an integral of $dx \wedge dy$ over the polygon. Since in the complex coordinate z we have $dx \wedge dy = \frac{i}{2} dz \wedge d\bar{z}$ we get the following formula for the flat area of S:

$$area(S) = \frac{i}{2} \int_S \omega \wedge \bar{\omega} = \frac{i}{2} \sum_{j=1}^{g} (A_j \bar{B}_j - \bar{A}_j B_j) \tag{3.3}$$

Here we also used the Riemann bilinear relation which expresses the integral $\int_S \omega \wedge \bar{\omega}$ in terms of *absolute periods* A_j, B_j of ω, where the absolute periods A_j, B_j are the integrals of ω with respect to some symplectic basis of cycles.

An individual flat surface defines a pair: (complex structure, holomorphic 1-form). A family of flat surfaces (where the flat surfaces are as usual endowed with a choice of the vertical direction) corresponds to a *stratum* $\mathcal{H}(d_1, \dots, d_m)$ in the *moduli space of holomorphic 1-forms*. Points of the stratum are represented by pairs (point in the moduli space of complex structures, holomorphic 1-form in the corresponding complex structure having zeroes of degrees $d_1, \dots, d_m$).

The notion "stratum" has the following origin. The moduli space of pairs (holomorphic 1-form, complex structure) forms a natural vector bundle over the moduli space $\mathcal{M}_g$ of complex structures. A fiber of this vector bundle is a vector space $\mathbb{C}^g$ of holomorphic 1-forms in a given complex structure. We already mentioned that the sum of degrees of zeroes of a holomorphic 1-form on a Riemann surface of genus g equals $2g - 2$. Thus, the total space $\mathcal{H}_g$ of our vector bundle is stratified by subspaces of those forms which have zeroes of degrees exactly $d_1, \dots, d_m$, where $d_1 + \dots + d_m = 2g - 2$. Say, for $g = 2$ we have only two partitions of number 2, so we get two strata $\mathcal{H}(2)$ and $\mathcal{H}(1,1)$. For $g = 3$ we have five partitions, and correspondingly five strata $\mathcal{H}(4), \mathcal{H}(3,1), \mathcal{H}(2,2), \mathcal{H}(2,1,1), \mathcal{H}(1,1,1,1)$.

$$\begin{array}{ccc} \mathcal{H}(d_1, \dots, d_m) & \subset & \mathcal{H}_g \\ & & \downarrow \\ & & \mathcal{M}_g \end{array} \tag{3.4}$$

Every stratum $\mathcal{H}(d_1, \dots, d_m)$ is a complex-analytic orbifold of dimension

$$\dim_{\mathbb{C}} \mathcal{H}(d_1, \dots, d_m) = 2g + m - 1 \tag{3.5}$$

Note, that an individual stratum $\mathcal{H}(d_1, \dots, d_m)$ does not form a fiber bundle over $\mathcal{M}_g$. For example, according to our formula, $\dim_{\mathbb{C}} \mathcal{H}(2g-2) = 2g$, while $\dim_{\mathbb{C}} \mathcal{M}_g = 3g - 3$.

We showed how the geometric structures related to a flat surface define their complex-analytic counterparts. Actually, this correspondence goes in two directions. We suggest to the reader to make the inverse translation: to start with complex-analytic structure and to see how it defines the geometric one. This correspondence can be summarized in the following dictionary.

Table 1. Correspondence of geometric and complex-analytic notions

Geometric language	Complex-analytic language
flat structure (including a choice of the vertical direction)	complex structure + a choice of a holomorphic 1-form ω
conical point with a cone angle $2\pi(d+1)$	zero of degree d of the holomorphic 1-form ω (in local coordinates $\omega = w^d\, dw$)
side $\mathbf{v}_j$ of a polygon	relative period $\int_{P_j}^{P_{j+1}} \omega = \int_{\mathbf{v}_j} \omega$ of the 1-form ω
area of the flat surface S	$\frac{i}{2}\int_S \omega\wedge\bar\omega =$ $= \frac{i}{2}\sum_{j=1}^{g}(A_j\bar B_j - \bar A_j B_j)$
family of flat surfaces sharing the same types $2\pi(d_1+1),\dots,2\pi(d_m+1)$ of cone angles	stratum $\mathcal{H}(d_1,\dots,d_m)$ in the moduli space of Abelian differentials
coordinates in the family: vectors $\mathbf{v}_i$ defining the polygon	coordinates in $\mathcal{H}(d_1,\dots,d_m)$: collection of relative periods of ω, i.e. cohomology class $[\omega]\in H^1(S,\{P_1,\dots,P_m\};\mathbb{C})$

3.4 Volume Element in the Moduli Space of Holomorphic One-Forms

In the previous section we have considered vectors $\mathbf{v}_1,\dots,\mathbf{v}_n$ determining the polygon from which we glue a flat surface S as on Fig. 12. We have identified these vectors $\mathbf{v}_j\in\mathbb{R}^2\sim\mathbb{C}$ with complex numbers and claimed (without proof) that under this identification $\mathbf{v}_1,\dots,\mathbf{v}_n$ provide us with local coordinates in the corresponding family of flat surfaces. We identify every such family with a stratum $\mathcal{H}(d_1,\dots,d_m)$ in the moduli space of holomorphic 1-forms. In complex-analytic language we have locally identified a neighborhood of a "point" (complex structure, holomorphic 1-form ω) in the corresponding stratum with a neighborhood of the cohomology class $[\omega]\in H^1(S,\{P_1,\dots,P_m\};\mathbb{C})$.

Note that the cohomology space $H^1(S,\{P_1,\dots,P_m\};\mathbb{C})$ contains a natural integer lattice $H^1(S,\{P_1,\dots,P_m\};\mathbb{Z}\oplus\sqrt{-1}\,\mathbb{Z})$. Consider a linear volume element $d\nu$ in the vector space $H^1(S,\{P_1,\dots,P_m\};\mathbb{C})$ normalized in such a way that the volume of the fundamental domain in the "cubic" lattice

$$H^1(S,\{P_1,\dots,P_m\};\mathbb{Z}\oplus\sqrt{-1}\,\mathbb{Z})\subset H^1(S,\{P_1,\dots,P_m\};\mathbb{C})$$

is equal to one. In other terms

$$d\nu = \frac{1}{J}\,\frac{1}{(2\sqrt{-1})^n}\,d\mathbf{v}_1 d\bar{\mathbf{v}}_1 \dots d\mathbf{v}_n d\bar{\mathbf{v}}_n,$$

where J is the determinant of a change of the basis $\{\mathbf{v}_1, \dots, \mathbf{v}_n\}$ considered as a basis in the first relative *homology* to some "symplectic" basis in the first relative homology.

Consider now the real hypersurface

$$\mathcal{H}_1(d_1, \dots, d_m) \subset \mathcal{H}(d_1, \dots, d_m)$$

defined by the equation $area(S) = 1$. Taking into consideration formula (3.3) for the function $area(S)$ we see that the hypersurface $\mathcal{H}_1(d_1, \dots, d_m)$ defined as $area(S) = 1$ can be interpreted as a "unit hyperboloid" defined in local coordinates as a level of the indefinite quadratic form (3.3) .

The volume element $d\nu$ can be naturally restricted to a hyperplane defined as a level hypersurface of a function. We denote the corresponding volume element on $\mathcal{H}_1(d_1, \dots, d_m)$ by $d\nu_1$.

Theorem (H. Masur. W. A. Veech). *The total volume*

$$\int_{\mathcal{H}_1(d_1,\dots,d_m)} d\nu_1$$

of every stratum is finite.

The values of these volumes were computed only recently by A. Eskin and A. Okounkov [EOk], twenty years after the Theorem above was proved in [Ma3], [Ve3] and [Ve8]. We discuss this computation in Sect. 7.

3.5 Action of $SL(2,\mathbb{R})$ on the Moduli Space

In this section we discuss a property of flat surfaces which is, probably, the most important in our study: we show that the linear group acts on every family of flat surfaces, and, moreover, acts *ergodically* (see Append. A for discussion of the notion of ergodicity). This enables us to apply tools from dynamical systems and from ergodic theory.

Consider a flat surface S and consider a polygonal pattern obtained by unwrapping it along some geodesic cuts. For example, one can assume that our flat surface S is glued from a polygon $\Pi \subset \mathbb{R}^2$ as on Fig. 12. Consider a linear transformation $g \in GL^+(2,\mathbb{R})$ of the plane $\mathbb{R}^2$. It changes the shape of the polygon. However, the sides of the new polygon $g\Pi$ are again arranged into pairs, where the sides in each pair are parallel and have equal length (different from initial one), see Fig. 14. Thus, identifying the sides in each pair by a parallel translation we obtain a new flat surface gS.

It is easy to check that the surface gS does not depend on the way in which S was unwrapped to a polygonal pattern Π. It is clear that all topological characteristics of the new flat surface gS (like genus, number and types

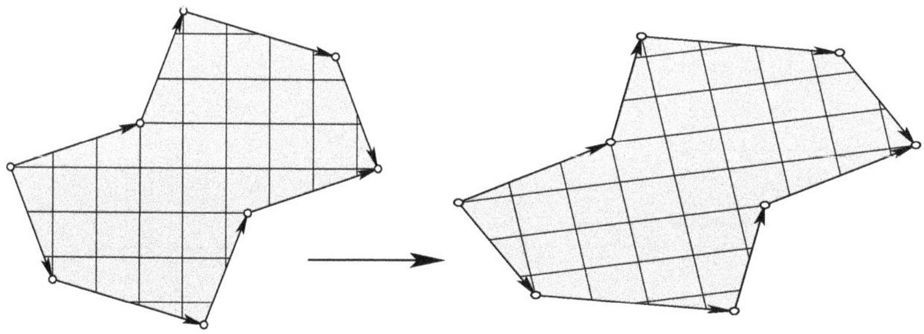

Fig. 14. Action of the linear group on flat surfaces

of conical singularities) are the same as those of the initial flat surface S. Hence, we get a continuous action of the group $GL^+(2,\mathbb{R})$ on each stratum $\mathcal{H}(d_1,\dots,d_m)$.

Considering the subgroup $SL(2,\mathbb{R})$ of area preserving linear transformations we get the action of $SL(2,\mathbb{R})$ on the "unit hyperboloid" $\mathcal{H}_1(d_1,\dots,d_m)$. Considering the diagonal subgroup $\begin{pmatrix} e^t & 0 \\ 0 & e^{-t} \end{pmatrix} \subset SL(2,\mathbb{R})$ we get a continuous action of this one-parameter subgroup on each stratum $\mathcal{H}(d_1,\dots,d_m)$. This action induces a natural flow on the stratum, which is called the *Teichmüller geodesic flow*.

Key Theorem (H. Masur. W. A. Veech). *The action of the groups* $SL(2,\mathbb{R})$ *and* $\begin{pmatrix} e^t & 0 \\ 0 & e^{-t} \end{pmatrix}$ *preserves the measure* $d\nu_1$. *Both actions are ergodic with respect to this measure on each connected component of every stratum* $\mathcal{H}_1(d_1,\dots,d_m)$.

This theorem might seem quite surprising. Consider almost any flat surface S as in Fig. 12. "Almost any flat surface" is understood as "corresponding to a set of parameters $\mathbf{v}_1,\dots,\mathbf{v}_4$ of full measure; here the vectors $\mathbf{v}_i$ define the polygon Π from Fig. 12.

Now start contracting the polygon Π it in the vertical direction and expanding it in the horizontal direction with the same coefficient e^t. The theorem says, in particular, that for an appropriate $t \in \mathbb{R}$ the deformed polygon will produce a flat surface $g_t S$ which would be arbitrary close to the flat surface S_0 obtained from the regular octagon as on Fig. 3 since a trajectory of almost any point under an ergodic flow is everywhere dense (and even "well distributed"). However, it is absolutely clear that acting on our initial polygon Π from Fig. 12 with expansion-contraction we never get close to a regular octagon... Is there a contradiction?..

There is no contradiction since the statement of the theorem concerns flat surfaces and not polygons. In practice this means that we can apply expansion-contraction to the polygon Π, which does not change too much the shape of the polygon, but radically changes the flat structure. Then we can change

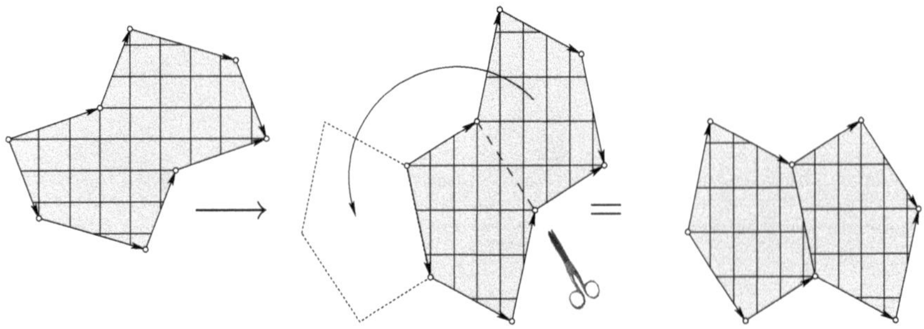

Fig. 15. The first modification of the polygon changes the flat structure while the second one just changes the way in which we unwrap the flat surface

the way in which we unwrap the flat surface $g_t S$ (see Fig. 15). This radically changes the shape of the polygon, but *does not change at all* the flat structure!

3.6 General Philosophy

Now we are ready to describe informally the basic idea of our approach to the study of flat surfaces. Of course it is not universal; however, in many cases it appears to be surprisingly powerful.

Suppose that we need some information about geometry or dynamics of an individual flat surface S. Consider the element S in the corresponding family of flat surfaces $\mathcal{H}(d_1, \dots, d_m)$. Denote by $\mathcal{C}(S) = \overline{GL^+(2,\mathbb{R})\, S} \subset \mathcal{H}(d_1, \dots, d_m)$ the closure of the $GL^+(2,\mathbb{R})$-orbit of S in $\mathcal{H}(d_1, \dots, d_m)$. *In numerous cases knowledge about the structure of $\mathcal{C}(S)$ gives a comprehensive information about geometry and dynamics of the initial flat surface S. Moreover, some delicate numerical characteristics of S can be expressed as averages of simpler characteristics over $\mathcal{C}(S)$.*

The remaining part of this survey is an attempt to show some implementations of this general philosophy. The first two illustrations would be presented in the next section.

We have to confess that we do not tell all the truth in the formulation above. Actually, there is a hope that this philosophy extends much further. A closure of an orbit of an abstract dynamical system might have extremely complicated structure. According to the most optimistic hopes, the closure $\mathcal{C}(S)$ of the $GL^+(2,\mathbb{R})$-orbit of *any* flat surface S is a nice complex-analytic variety. Moreover, according to the most daring conjecture it would be possible to classify all these $GL^+(2,\mathbb{R})$-invariant subvarieties. For genus two the latter statements were recently proved by C. McMullen (see [McM2] and [McM3]) and partly by K. Calta [Clt].

We discuss this hope in more details in Sect. 9, in particular, in Sect. 9.3. We complete this section by a Theorem which supports the hope for some nice and simple description of orbit closures.

Theorem (M. Kontsevich). *Suppose that a closure $\mathcal{C}(S)$ in $\mathcal{H}(d_1, \dots, d_m)$ of a $GL^+(2,\mathbb{R})$-orbit of some flat surface S is a complex-analytic subvariety. Then in cohomological coordinates $H^1(S, \{P_1, \dots, P_m\}; \mathbb{C})$ it is represented by an affine subspace.*

3.7 Implementation of General Philosophy

In this section we present two illustrations showing how the "general philosophy" works in practice.

Consider a directional flow on a flat surface S. It is called *minimal* when the closure of any trajectory gives the entire surface. When a directional flow on a flat torus is minimal, it is necessarily ergodic, in particular, any trajectory in average spends in any subset $U \subset \mathbb{T}^2$ a time proportional to the area (measure) of the subset U. Surprisingly, for surfaces of higher genera a directional flow can be minimal but not ergodic! Sometimes it is possible to find some special direction with the following properties. The flow in this direction is minimal. However, the flat surface S might be decomposed into a disjoint union of several subsets V_i of positive measure in such a way that some trajectories of the directional flow prefer one subset to the others. In other words, the average time spent by a trajectory in the subset V_i is not proportional to the area of V_i anymore. (The original ideas of such examples appear in [Ve1], [Kat1], [Sat] [Kea2]; see also [MaT] and especially [Ma7] for a very accessible presentation of such examples.)

Suppose that we managed to find a direction on the initial surface S_0 such that the flow in this direction is minimal but not ergodic (with respect to the natural Lebesgue measure). Let us apply a rotation to S_0 which would make the corresponding direction vertical. Consider the resulting flat surface S (see Convention 1 in Sect. 1.2). Consider the corresponding "point" $S \in \mathcal{H}(d_1, \dots, d_m)$ and the orbit $\{g_t S\}_{t\in\mathbb{R}}$ of S under the action of the diagonal subgroup $g_t = \begin{pmatrix} e^t & 0 \\ 0 & e^{-t} \end{pmatrix}$.

Recall that the stratum (or, more precisely, the corresponding "unit hyperboloid") $\mathcal{H}_1(d_1, \dots, d_m)$ is never compact, it always contains "cusps": regions where the corresponding flat surfaces have very short saddle connections or very short closed geodesics (see Sect. 3.2).

Theorem (H. Masur). *Consider a flat surface S. If the vertical flow is minimal but not ergodic with respect to the natural Lebesgue measure on the flat surface then the trajectory $g_t S$ of the Teichmüller geodesic flow is divergent, i.e. it eventually leaves any fixed compact subset $K \subset \mathcal{H}_1(d_1, \dots, d_m)$ in the stratum.*

Actually, this theorem has an even stronger form.

A stratum $\mathcal{H}_1(d_1, \ldots, d_m)$ has "cusps" of two different origins. A flat surface may have two distinct zeroes get very close to each other. In this case S has a short saddle connection (or, what is the same, a short relative period). However, the corresponding Riemann surface is far from being degenerate. The cusps of this type correspond to "simple noncompactness": any stratum $\mathcal{H}_1(d_1, \ldots, d_m)$ is adjacent to all "smaller" strata $\mathcal{H}_1(d_1 + d_2, d_3, \ldots, d_m), \ldots$.

Another type of degeneration of a flat surface is the appearance of a short closed geodesic. In this case the underlying Riemann surface is close to a degenerate one; the cusps of this second type correspond to "essential noncompactness".

To formulate a stronger version of the above Theorem consider the natural projection of the stratum $\mathcal{H}(d_1, \ldots, d_m)$ to the moduli space $\mathcal{M}_g$ of complex structures (see (3.4) in Sect. 3.3). Consider the image of the orbit $\{g_t S\}_{t \in \mathbb{R}}$ in $\mathcal{M}_g$ under this natural projection. By the reasons which we explain in Sect. 8 it is natural to call this image a *Teichmüller geodesic*.

Theorem (H. Masur). *Consider a flat surface S. If the vertical flow is minimal but not ergodic with respect to the natural Lebesgue measure on the flat surface then the "Teichmüller geodesic" $g_t S$ is divergent, i.e. it eventually leaves any fixed compact subset $K \subset \mathcal{M}$ in the moduli space of complex structures and never visits it again.*

This statement (in a slightly different formulation) is usually called *Masur's criterion of unique ergodicity* (see Sect. A for discussion of the notion *unique ergodicity*).

As a second illustrations of the "general philosophy" we present a combination of *Veech criterion* and of a Theorem of J. Smillie.

Recall that closed regular geodesics on a flat surface appear in families of parallel closed geodesics. When the flat surface is a flat torus, any such family covers all the torus. However, for surfaces of higher genera such families usually cover a cylinder filled with parallel closed geodesic of equal length. Each boundary of such a cylinder contains a conical point. Usually a geodesic emitted in the same direction from a point outside of the cylinder is dense in the complement to the cylinder or at least in some nontrivial part of the complement. However, in some rare cases, it may happen that the entire surface decomposes into several cylinders filled with parallel closed geodesics going in some fixed direction. This is the case for the vertical or for the horizontal direction on the flat surface glued from a regular octagon, see Fig. 3 (please check). Such direction is called *completely periodic*.

Theorem (J. Smillie; W. A. Veech). *Consider a flat surface S. If its $GL^+(2, \mathbb{R})$-orbit is closed in $\mathcal{H}(d_1, \ldots, d_m)$ then a directional flow in any direction on S is either completely periodic or uniquely ergodic.*

(see Sect. A for the notion of *unique ergodicity*). Note that unique ergodicity implies, in particular, that any orbit which is not a *saddle connection*, (i.e.

which does not hit the singularity both in forward and in backward direction) is everywhere dense. We shall return to this Theorem in Sect. 9.5 where we discuss *Veech surfaces.*

4 How Do Generic Geodesics Wind Around Flat Surfaces

In this section we study geodesics on a flat surface S going in generic directions on S. Such geodesics are dense on S; moreover, it is possible to show that they wind around S in a relatively regular manner. Namely, it is possible to find a cycle $c \in H_1(S;\mathbb{R})$ such that in some sense a long piece of geodesic pretends to wind around S repeatedly following this *asymptotic cycle* c.

In Sect. 4.1 we study the model case of the torus and give a rigorous definition of the asymptotic cycle. Then we study the asymptotic cycles on general flat surfaces. In Sect. 4.2 we study "further terms of approximation". The asymptotic cycle describes the way in which a geodesic winds around the surface in average. In Sect. 4.2 we present an empirical description of the deviation from average. The corresponding rigorous statements are formulated in Sect. 4.3. Some ideas of the proof of this statement are presented in Sect. 5.

4.1 Asymptotic Cycle

Asymptotic Cycle on a Torus

As usual we start from the model case of the torus. We assume that our flat torus is glued from a square in the natural way. Consider an irrational direction on the torus; any geodesic going in this direction is dense in the torus.

Fix a point x_0 on the torus and emit a geodesic in the chosen direction. Wait till it winds for some time around the torus and gets close to the initial point x_0. Join the endpoints of the resulting piece of geodesic by a short path. We get a closed loop on the torus which defines a cycle c_1 in the first homology group $H_1(\mathbb{T}^2;\mathbb{Z})$ of the torus. Now let the initial geodesic wind around the torus for some longer time; wait till it get close enough to the initial point x_0 and join the endpoints of the longer piece of geodesic by a short path. We get a new cycle $c_2 \in H_1(\mathbb{T}^2;\mathbb{Z})$. Considering longer and longer geodesic segments we get a sequence of cycles $c_i \in H_1(\mathbb{T}^2;\mathbb{Z})$.

For example, we can choose a short segment X going through x_0 orthogonal (or just transversal) to the direction of the geodesic. Each time when the geodesic crosses X we join the crossing point with the point x_0 along X obtaining a closed loop. Consecutive return points $x_1, x_2, \dots$ define a sequence of cycles $c_1, c_2, \dots$, see Fig. 16.

For the torus case we can naturally identify the universal covering space $\mathbb{R}^2 \to \mathbb{T}^2$ with the first homology group $H_1(\mathbb{T}^2;\mathbb{R}) \simeq \mathbb{R}^2$. Our irrational

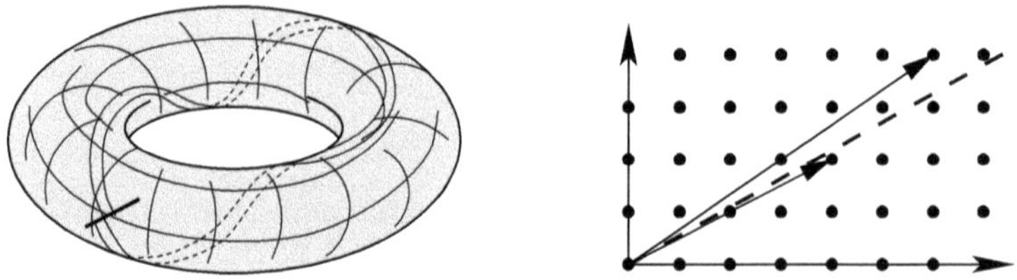

Fig. 16. A sequence of cycles approximating a dense geodesic on a torus

geodesic unfolds to an irrational straight line $\mathcal{V}_1$ in $\mathbb{R}^2$ and the sequence of cycles $c_1, c_2, \dots$ becomes a sequence of integer vectors $\mathbf{v}_1, \mathbf{v}_2, \dots \in \mathbb{Z}^2 \subset \mathbb{R}^2$ approximating $\mathcal{V}_1$, see Fig. 16.

In particular, it is not surprising that there exists the limit

$$\lim_{n\to\infty} \frac{c_n}{\|c_n\|} = c \tag{4.1}$$

Under our identification $H_1(\mathbb{T}^2;\mathbb{R}) \simeq \mathbb{R}^2$ the cycle c represents a unit vector in direction $\mathcal{V}_1$.

Let the area of the torus be normalized to one. Let the interval X, which we use to construct the sequence $c_1, c_2, \dots$, be orthogonal to the direction of the geodesic. Denote by $|X|$ its length. The following limit also exists and is proportional to the previous one:

$$\lim_{n\to\infty} \frac{1}{n}\, c_n = \frac{1}{|X|} \cdot c \tag{4.2}$$

The cycles obtained as limits (4.1) and (4.2) are called *asymptotic cycles*. They show how the corresponding irrational geodesic winds around the torus *in average*. It is easy to see that they do not depend on the starting point x.

The notion "asymptotic cycle" was introduced by S. Schwartzman [Schw].

Asymptotic Cycle on a Surface of Higher Genus

We can apply the same construction to a geodesic on a flat surface S of higher genus. Having a geodesic segment $X \subset S$ and some point $x \in X$ we emit from x a geodesic orthogonal to X. From time to time the geodesic would intersect X. Denote the corresponding points as $x_1, x_2, \dots$. Closing up the corresponding pieces of the geodesic by joining the endpoints x_0, x_j with a path going along X we again get a sequence of cycles $c_1, c_2, \dots$.

Proposition 1. *For any flat surface S of area one and for almost any direction α on it any geodesic going in direction α is dense on S and has an asymptotic cycle which depends only on α.*

In other words, for almost any direction the limit

$$\lim_{n\to\infty} \frac{1}{n}\, c_n = \frac{1}{|X|} \cdot c$$

exists and the corresponding asymptotic cycle c does not depend on the starting point $x_0 \in S$.

This proposition is an elementary corollary from the following theorem of S. Kerckhoff, H. Masur and J. Smillie [KMaS] (which is another Key Theorem in this area).

Theorem (S. Kerckhoff, H. Masur, J. Smillie). *For any flat surface S the directional flow in almost any direction is ergodic.*

In this case the asymptotic cycle has the same dynamical interpretation as for the torus: it shows how a geodesic going in the chosen direction winds around the surface S *in average.*

Remark. Note that the asymptotic cycle $c \in H_1(S, \mathbb{R})$ also has a topological interpretation. Assume for simplicity that c corresponds to the vertical direction. Let ω be the holomorphic 1-form corresponding to the flat structure on S (see Sect. 3.3). Then the closed 1-form $\omega_0 = \mathrm{Re}(\omega)$ defines the vertical foliation and $c = D[\omega_0]$ is Poincaré dual to the cohomology class of ω_0. Choosing other ergodic directions on the flat surface S we get asymptotic cycles in the two-dimensional subspace $\langle D[\omega_0], D[\omega_1]\rangle_{\mathbb{R}} \subset H_1(S, \mathbb{R})$ spanned by homology classes dual to cocycles $\omega_0 = \mathrm{Re}(\omega)$ and $\omega_1 = \mathrm{Im}(\omega)$.

4.2 Deviation from Asymptotic Cycle

We have seen in the previous section that a sequence of cycles $c_1, c_2, \ldots$ approximating long pieces of an "irrational" geodesic on a flat torus $\mathbb{T}^2$ and on a flat surface S of higher genus exhibit similar behavior: their norm grows (approximately) linearly in n and their direction approaches the direction of the asymptotic cycle c. Note, however, that for the torus the cycles c_n live in the two-dimensional space $H_1(\mathbb{T}^2; \mathbb{R}) \simeq \mathbb{R}^2$, while for the surface of higher genus $g \geq 2$ the cycles live in the larger space $H_1(S; \mathbb{R}) \simeq \mathbb{R}^{2g}$. In particular, they have "more room" for deviation from the asymptotic direction.

Namely, observing the right part of Fig. 16 we see that all vectors c_n follow the line $\mathcal{V}_1$ spanned by the asymptotic cycle c rather close: the norm of projection of c_n to the line orthogonal to $\mathcal{V}_1$ is uniformly bounded (with respect to n and to the choice of the starting point x_0).

The situation is different for surfaces of higher genera. Choose a hyperplane $\mathcal{S} \perp c$ in $H_1(S, \mathbb{R})$ as a screen orthogonal (transversal) to the asymptotic cycle c and consider a projection to this screen parallel to c. Projections of cycles c_n would not be uniformly bounded anymore. There is no contradiction since if the norms of these projections grow sublinearly, then the directions of the cycles c_n still tend to direction of the asymptotic cycle c.

Let us observe how the projections are distributed in the screen $\mathcal{S}$. Figure 17 shows results of numerical experiments where we take a projection of

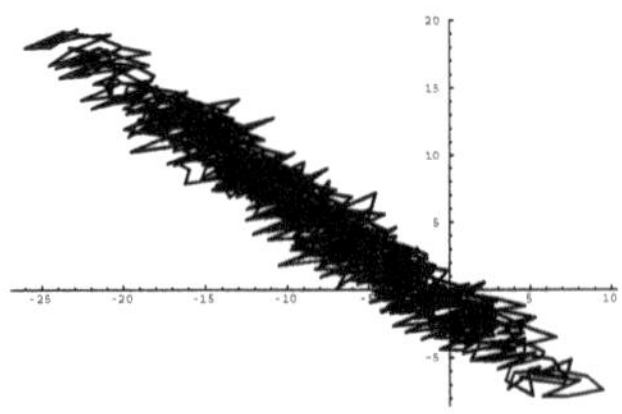

Fig. 17. Projection of a broken line joining the endpoints of $c_1, c_2, ..., c_{100000}$ to a screen orthogonal to the asymptotic cycle. Genus $g = 3$

a broken line joining the endpoints of $c_1, c_2, \ldots, c_{100000}$ and we take a two-dimensional screen orthogonal to c to make the picture more explicit.

We see that the distribution of projections of cycles c_n in the screen $\mathcal{S}$ is anisotropic: the projections accumulate along some line. This means that in the original space $\mathbb{R}^{2g}$ the vectors c_n deviate from the asymptotic direction $\mathcal{V}_1$ not arbitrarily but along some two-dimensional subspace $\mathcal{V}_2 \supset \mathcal{V}_1$, see Fig. 18.

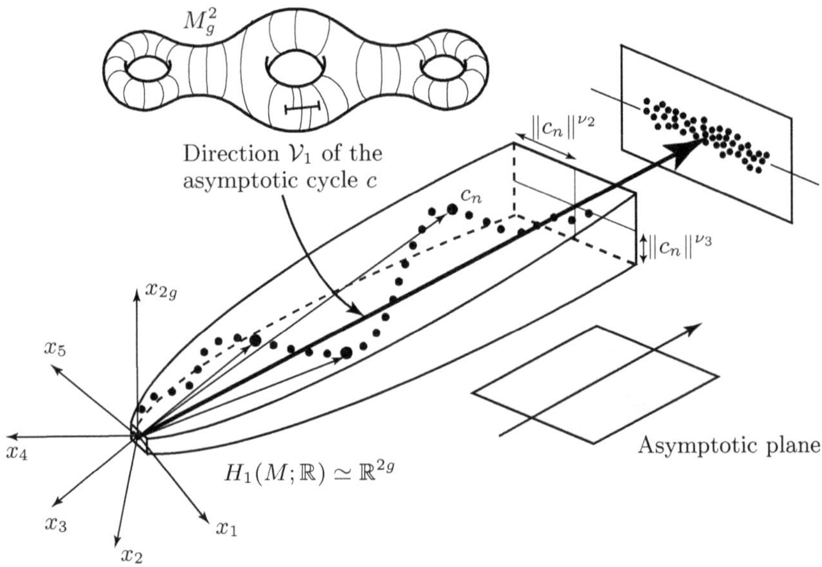

Fig. 18. Deviation from the asymptotic direction

Moreover, measuring the norms of the projections $proj(c_n)$ to the screen $\mathcal{S}$ orthogonal to $\mathcal{L}_1 = \langle c \rangle_{\mathbb{R}}$, we get

$$\limsup_{n\to\infty} \frac{\log \|proj(c_n)\|}{\log n} = \nu_2 < 1$$

In other words the vector c_n is located approximately in the subspace $\mathcal{V}_2$, and the distance from its endpoint to the line $\mathcal{V}_1 \subset \mathcal{V}_2$ is bounded by $const \cdot |c_n\|^{\nu_2}$, see Fig. 18.

Consider now a new screen $\mathcal{S}_2 \perp \mathcal{V}_2$ orthogonal to the plane $\mathcal{V}_2$. Now the screen $\mathcal{S}_2$ has codimension two in $H_1(S,\mathbb{R}) \simeq \mathbb{R}^{2g}$. Considering the projections of c_n to $\mathcal{S}_2$ we eliminate the asymptotic directions $\mathcal{V}_1$ and $\mathcal{V}_2$ and we see how do the vectors c_n deviate from $\mathcal{V}_2$. On the screen $\mathcal{S}_2$ we see the same picture as in Fig. 17: the projections are located along a one-dimensional subspace.

Coming back to the ambient space $H_1(S,\mathbb{R}) \simeq \mathbb{R}^{2g}$, this means that in the first term of approximation all vectors c_n are aligned along the one-dimensional subspace $\mathcal{V}_1$ spanned by the asymptotic cycle. In the second term of approximation, they can deviate from $\mathcal{V}_1$, but the deviation occurs mostly in the two-dimensional subspace $\mathcal{V}_2$, and has order $\|c\|^{\nu_2}$ where $\nu_2 < 1$. In the third term of approximation we see that the vector c_n may deviate from the plane $\mathcal{V}_2$, but the deviation occurs mostly in a three-dimensional space $\mathcal{V}_3$ and has order $\|c\|^{\nu_3}$ where $\nu_3 < \nu_2$.

Going on we get further terms of approximation. However, getting to a subspace $\mathcal{V}_g$ which has half the dimension of the ambient space we shall see that, in a sense, there is no more deviation from $\mathcal{V}_g$: the distance from any c_n to $\mathcal{V}_g$ is uniformly bounded.

Note that the intersection form endows the space $H_1(S,\mathbb{R}) \simeq \mathbb{R}^{2g}$ with a natural symplectic structure. It can be checked that the resulting g-dimensional subspace $\mathcal{V}_g$ is a *Lagrangian* subspace for this symplectic form.

4.3 Asymptotic Flag and "Dynamical Hodge Decomposition"

A rigorous formulation of phenomena described in the previous section is given by the following Theorem proved[3] by the author in [Zo3] and [Zo4].

Following Convention 1 we always consider a flat surface together with a choice of direction which by convention is called the *vertical direction*, or *direction to the North*. Using an appropriate homothety we normalize the area of S to one, so that $S \in \mathcal{H}_1(d_1,\dots,d_m)$.

We chose a point $x_0 \in S$ and a horizontal segment X passing through x_0; by $|X|$ we denote the length of X. We consider a geodesic ray γ emitted from x_0 in the vertical direction. (If x_0 is a saddle point, there are several outgoing vertical geodesic rays; choose any of them.) Each time when γ intersects X we join the point x_n of intersection and the starting point x_0 along X producing a closed path. We denote the homology class of the corresponding loop by c_n.

Let ω be the holomorphic 1-form representing S; let g be genus of S. Choose some Euclidean metric in $H_1(S;\mathbb{R}) \simeq \mathbb{R}^{2g}$ which would allow to measure a distance from a vector to a subspace. Let by convention $\log(0) = -\infty$.

[3]Actually, the theorem was initially proved under certain hypothesis on the Lyapunov exponents of the Teichmüller geodesic flow. These conjectures were proved later by G. Forni and in the most complete form by A. Avila and M. Viana; see the end of this section and especially Sect. 5.8

Theorem. *For almost any flat surface S in any stratum $\mathcal{H}_1(d_1,\dots,d_m)$ there exists a flag of subspaces*

$$\mathcal{V}_1 \subset \mathcal{V}_2 \subset \cdots \subset \mathcal{V}_g \subset H_1(S;\mathbb{R})$$

in the first homology group of the surface with the following properties.

Choose any starting point $x_0 \in X$ in the horizontal segment X. Consider the corresponding sequence $c_1, c_2, \dots$ of cycles.

— The following limit exists

$$|X| \lim_{n\to\infty} \frac{1}{n} c_n = c,$$

where the nonzero asymptotic cycle $c \in H_1(M_g^2;\mathbb{R})$ is Poincaré dual to the cohomology class of $\omega_0 = \mathrm{Re}[\omega]$, and the one-dimensional subspace $\mathcal{V}_1 = \langle c\rangle_{\mathbb{R}}$ is spanned by c.

— For any $j = 1, \dots, g-1$ one has

$$\limsup_{n\to\infty} \frac{\log dist(c_n, \mathcal{V}_j)}{\log n} = \nu_{j+1}$$

and

$$|dist(c_n, \mathcal{V}_g)| \le const,$$

where the constant depends only on S and on the choice of the Euclidean structure in the homology space.

The numbers $2, 1+\nu_2, \dots, 1+\nu_g$ are the top g Lyapunov exponents of the Teichmüller geodesic flow on the corresponding connected component of the stratum $\mathcal{H}(d_1,\dots,d_m)$; in particular, they do not depend on the individual generic flat surface S in the connected component.

A reader who is not familiar with *Lyapunov exponents* can either read about them in Appendix B or just consider the numbers ν_j as some abstract constants which depend only on the connected component $\mathcal{H}^{comp}(d_1,\dots,d_m)$ containing the flat surface S.

It should be stressed, that the theorem above was initially formulated in [Zo4] as a conditional statement: under the conjecture that $\nu_g > 0$ there exist a Lagrangian subspace $\mathcal{V}_g$ such that the cycles are in a bounded distance from $\mathcal{V}_g$; under the further conjecture that all the exponents ν_j, for $j = 1, 2, \dots, g$, are distinct, there is a *complete* Lagrangian flag (i.e. the dimensions of the subspaces $\mathcal{V}_j$, where $j = 1, 2, \dots, g$, rise each time by one). These two conjectures were later proved by G. Forni [For1] and by A. Avila and M. Viana [AvVi]. We discuss their theorems in Sect. 5.8.

Another remark concerns the choice of the horizontal segment X. By convention it is chosen in such way that the trajectories emitted in the vertical direction (in direction to the North) from the endpoints of X hit the conical points before the first return to X. Usually we just place the left endpoint of X at the conical point.

Omitting this condition and considering a continuous family of horizontal subintervals X_t of variable length (say, moving continuously one of the endpoints), the theorem stays valid for a subset of X_t of full measure.

5 Renormalization for Interval Exchange Transformations. Rauzy–Veech Induction

In this section we elaborate a powerful time acceleration machine which allows to study the asymptotic cycles described in Sect. 4. Following the spirit of this survey we put emphasis on geometric ideas and omit proofs. This section can be considered as a geometric counterpart of the article of J.-C. Yoccoz [Y] in the current volume.

I use this opportunity to thank M. Kontsevich for numerous ideas and conjectures which were absolutely crucial for my impact in this theory: without numerous discussions with M. Kontsevich papers [Zo2] and [Zo3], probably, would be never written.

5.1 First Return Maps and Interval Exchange Transformations

Our goal is to study cycles obtained from long pieces of "irrational" geodesic on a flat surface by joining their endpoints along a transversal segment X. To perform this study we elaborate some simple machine which generates the cycles, and then we accelerate this machine to obtain very long cycles in a rather short time.

Consider *all* geodesics emitted from a transverse segment X in the same generic direction and let each of them come back to X for the first time. We get a *first return map* $T : X \to X$ which is interesting by itself and which deserves a separate discussion. Its properties play a crucial role in our study. (See Appendix A for general properties of the first return map.)

As usual let us start with a model case of a flat torus. Take a meridian of the torus as a transversal X and emit from X a directional flow. Every geodesic comes back to X inducing the *first return map* $T : X \to X$, which in this case isometrically rotates the meridian X along itself by an angle which depends on the direction of the flow, see Fig. 19.

Assume that our flat torus $\mathbb{T}^2$ is glued from a unit square. Let us replace now a meridian of the torus by a generic geodesic segment X orthogonal to the direction of the flow. From every point of X we emit a geodesic in direction orthogonal to X and wait till it hits X for the first time. We again obtain a first return map $T : X \to X$, but this time the map T is slightly more complicated.

To study this map it is convenient to unfold the torus into a plane. The map T is presented at Fig. 20. It chops X into three pieces and then shuffles them sending the left subinterval to the right, the right subinterval to the left

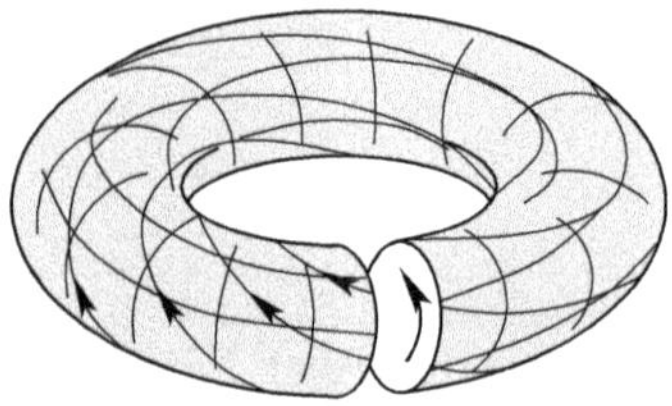

Fig. 19. The first return map of a meridian to itself induced by a directional flow is just a twist

and keeping the middle one in the middle but shifting it a bit. The map T gives an example of an *interval exchange transformation*.

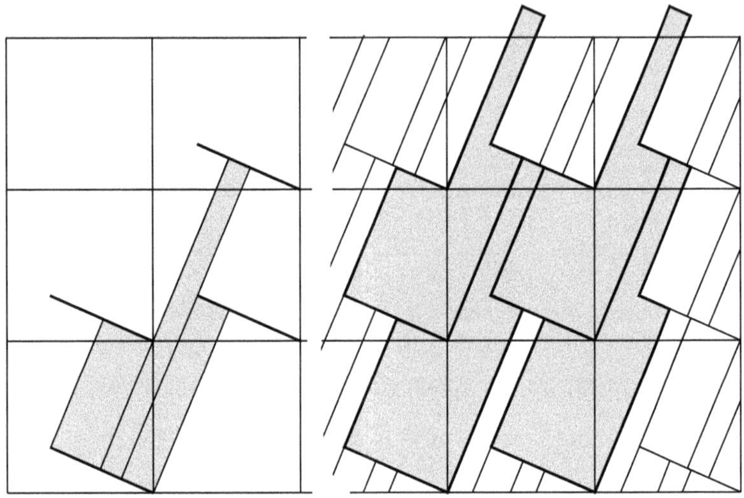

Fig. 20. Directional flow on a torus. The first return map of a segment to itself is an interval exchange transformation of three subintervals

Note that when the direction of the flow is *irrational*, the geodesics emitted from X again cover the entire torus $\mathbb{T}^2$ before coming back to X. The torus is get ripped into three rectangles based on the three subintervals in which T chops X. The corresponding building of three rectangles gives a new fundamental domain representing the torus: one can see at Fig. 20 that it tiles the plane. Initially we glued our flat torus from a square; the building under consideration gives another way to unwrap $\mathbb{T}^2$ into a polygon. We recommend to the reader to check that identifying the two pairs of corresponding vertical sides of the building and gluing the top horizontal sides of the rectangles to the bottom of X as prescribed by the interval exchange transformation T we get the initial torus.

Consider now a flat surface of genus higher than one. Say, consider a flat surface of genus $g = 2$ as on Fig. 21. We suggest to the reader to check that this flat surface has a single conical singularity with a cone angle 6π (see Fig. 2). To study a directional flow choose as before a geodesic segment $X \subset S$ orthogonal to the direction of the flow and consider the first return map $T : X \to X$ induced by the flow; see Fig. 21. We see that X is chopped into a larger number of subintervals (in comparison with the torus case), namely, for our choice of X it is chopped into four subintervals.

Now we observe a new phenomenon: trajectories emitted from some points of X hit the conical point and our directional flow splits at this point. Since in our particular case the cone angle at the conical point is $6\pi = 3 \cdot 2\pi$ there are *three* trajectories in direction $\mathbf{v}$ which hit it. The corresponding points at which X is chopped are marked with bold dots. The remaining discontinuity point of X corresponds to a trajectory which hits the endpoint of X.

Our construction with a segment X transversal to the flow and with trajectories of the flow emitted from X and followed till their first return to X trims a braid from the flow. Conical points play the role of a comb which splits the flow into several locks and then trims them in a different order. Note, however, that if we follow the flow till the second return to X it will pass through the comb twice, and thus will be generically split already into seven locks. (If you are interested in details, think why this second return has *seven* and not *eight* locks and what sort of genericity we need).

Similarly, the interval exchange transformation T of the base interval X can be compared to a shuffling machine. Imagine that X represents a stock of cards. We split the stock into n parts of fixed widths and shuffle the parts in a different order (given by some permutation π of n elements). At the second iteration we again split the new stock in the parts of the same widths as before and shuffle the parts according to the same permutation π, etc. Note, that even if the permutation π is such that $\pi^2 = id$, say, $\pi = (4, 3, 2, 1)$, the second iteration T^2 is not an identical transformation provided the widths $\lambda_1, \dots, \lambda_4$ are not symmetric: for a generic choice of $\lambda_1, \dots, \lambda_4$ the interval exchange transformation T^2 has 6 discontinuities (and hence 7 subintervals under exchange).

Exercise. Consider an interval exchange transformation $T(\lambda, \pi)$ corresponding to the permutation $\pi = (4, 3, 2, 1)$. Choose some generic values of the lengths $\lambda_1, \dots, \lambda_4$ of subintervals and construct T^2 and T^3.

5.2 Evaluation of the Asymptotic Cycle Using an Interval Exchange Transformation

Now we can return to our original problem. We want to study long pieces of leaves of the vertical foliation. Fix a horizontal segment X and emit a vertical trajectory from some point $x \in X$. When the trajectory intersects X for the first time join the corresponding point $T(x)$ to the original point x

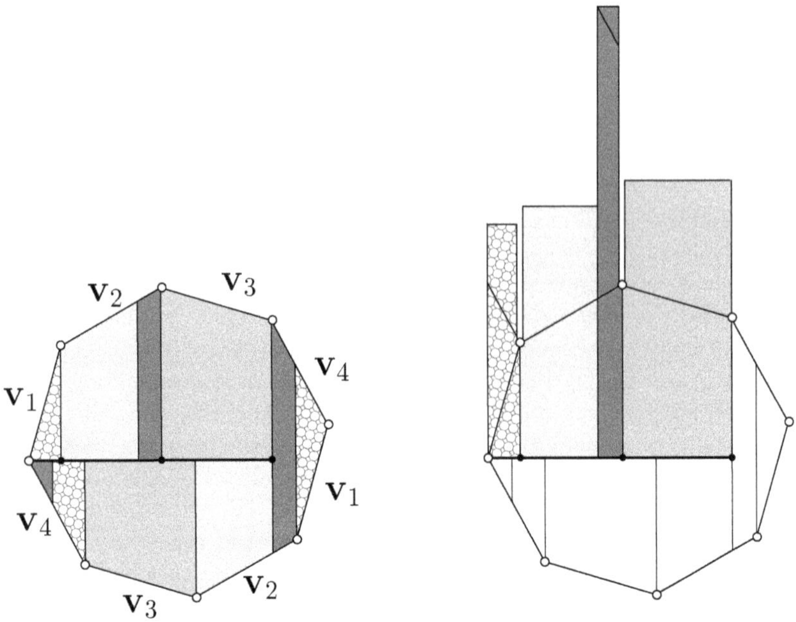

Fig. 21. The first return map $T : X \to X$ of a geodesic segment X defined by a directional flow decomposes the surface into four rectangles "zippered" along singular trajectories

along X to obtain a closed loop. Here $T : X \to X$ denotes the first return map to the transversal X induced by the vertical flow. Denote by $c(x, 1)$ the corresponding cycle in $H_1(S;\mathbb{Z})$. Following the vertical trajectory further on we shall return to X once again. Joining x and the point $T(T(x))$ of the second return to X along X we obtain the second cycle $c(x, 2)$. We want to describe the cycle $c(x, N)$ obtained after a very large number N of returns.

Actually, we prefer to close up a piece of trajectory going from $x \in X$ to the first return point $T(x) \in X$ in a slightly different way. Instead of completing the path joining the endpoints it is more convenient to close this piece of trajectory joining both points x and $T(x)$ to the left endpoint of X along X (see Fig. 22). This modified path defines the same homology cycle $c(x, 1)$ as the closed path for which the points x and $T(x)$ are joined directly.

Consider now the "first return cycle" $c(x, 1)$ as a function $c(x) = c(x, 1)$ of the starting point $x \in X$. Let the interval exchange transformation $T : X \to X$ decompose X into n subintervals $X_1 \sqcup \cdots \sqcup X_n$. It is easy to see that the function $c(x)$ is piecewise constant: looking at Fig. 22 one can immediately verify that if two points x_1 and x_2 are not separated by a discontinuity point (i.e. if they belong to the same subinterval X_j) they determine homologous (and even homotopic) cycles $c(x_1) = c(x_2)$. Each subinterval X_j determines its own cycle $c(X_j)$.

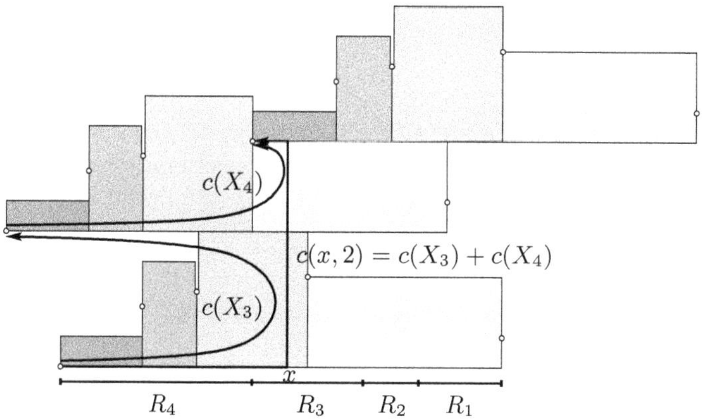

Fig. 22. Decomposition of a long cycle into the sum of basic cycles. (We have unfolded the flat surface along vertical trajectory emitted from the point x)

The vertical trajectory emitted from a point $x \in X$ returns to X at the points $T(x), T^2(x), \ldots, T^N(x)$. It is easy to see that the cycle $c(x,2)$ corresponding to the second return can be represented as a sum $c(x,2) = c(x) + c(T(x))$, see Fig. 22. Similarly the cycle $c(x,N)$ obtained by closing up a long piece of vertical trajectory emitted from $x \in X$ and followed up to N-th return to X can be represented as a sum

$$c(x,N) = c(x) + c(T(x)) + \cdots + c(T^{N-1}(x)) \tag{5.1}$$

According to the fundamental Theorem of S. Kerckhoff, H. Masur and J. Smillie [KMaS], for any flat surface the directional flow in almost any direction is ergodic, and even uniquely ergodic. Hence, the same is true for the corresponding interval exchange transformation. Applying the ergodic theorem (see Appendix A) to the sum (5.1) and taking into consideration that $c(x)$ is a piecewise-constant function we get

$$c(x,N) \sim N \cdot \frac{1}{|X|} \int_X c(x)\, dx = N \cdot \frac{1}{|X|} \Big(\lambda_1 c(X_1) + \cdots + \lambda_n c(X_n) \Big)$$

where $c(X_j)$ denotes the "first return cycle" for the points x in the subinterval X_j, see Fig. 22. This gives an explicit formula for the *asymptotic cycle*

$$c = \lim_{N\to\infty} \frac{c(x,N)}{N} = \frac{1}{|X|} \Big(\lambda_1 c(X_1) + \cdots + \lambda_n c(X_n) \Big) \tag{5.2}$$

Note that the asymptotic cycle does not depend on the starting point $x \in X$.

Exercise. Show that the paths $\mathbf{v}_1, \ldots, \mathbf{v}_4$ (see Fig. 21) represent a basis of cycles of the corresponding flat surface S of genus $g = 2$. Show that the first return cycles $c(X_j)$ determine another basis of cycles and that one can pass from one basis to the other using the following relations (see Fig. 21):

$$c(X_1) = \mathbf{v}_1 - \mathbf{v}_4 \qquad\qquad c(X_2) = \mathbf{v}_1 - \mathbf{v}_3 - \mathbf{v}_4$$
$$c(X_3) = 2\mathbf{v}_1 + \mathbf{v}_2 - \mathbf{v}_4 \qquad\qquad c(X_4) = \mathbf{v}_1 + \mathbf{v}_2 + \mathbf{v}_3 - \mathbf{v}_4$$

Express the asymptotic cycle in the basis $\mathbf{v}_j$ in terms of the lengths λ_j and then in terms of the angle by which the regular octagon is turned with respect to the standard presentation. Locally the vertical foliation goes to the North. And globally?

5.3 Time Acceleration Machine (Renormalization): Conceptual Description

In the previous section we have seen why all trajectories of a typical directional flow wind around the surface following the same asymptotic cycle. We have also found an effective way to evaluate this asymptotic cycle: we have seen that it is sufficient to find an interval exchange transformation $T : X \to X$ induced on any transverse segment X as the first return map of the directional flow, and then to determine the "first return cycles" $c(X_j)$, see Fig. 22. The linear combination of the cycles $c(X_j)$ taken with weights proportional to the lengths $\lambda_j = |X_j|$ of subintervals gives the asymptotic cycle, see (5.2).

Let us proceed now with a more delicate question of deviation of a trajectory of the directional flow from the asymptotic direction. Without loss of generality we may assume that the directional flow under consideration is the vertical flow. We know that a very long cycle $c(x, N)$ corresponding to a large number N of returns of the trajectory to the horizontal segment X stretches in the direction approaching the direction of the asymptotic cycle c. We want to describe how $c(x, N)$ deviates from this direction (see Sect. 4.2 and Sect. 4.3).

We have already seen that as soon as we have evaluated the "first return cycles" $c(X_j)$, complete information about cycles representing long pieces of trajectories of the directional flow is encoded in the corresponding trajectory $x, T(x), \dots, T^{N-1}(x)$ of the interval exchange transformation; see (5.1).

An interval exchange transformations gives an example of a parabolic dynamical system which is neither completely regular (like rotation of a circle) nor completely chaotic (like geodesic flow on a compact manifold of constant negative curvature, which is a typical example of a *hyperbolic* system). In some aspects interval exchange transformations are closer to rotations of a circle: say, as it was proved by A. Katok, an interval exchange transformation is never mixing [Kat2] (though, as it was very recently proved by A. Avila and G. Forni in [AvFor], generically it is weakly mixing). However, the behavior of deviation from the ergodic mean resembles the behavior of a chaotic system.

Our principal tool in the study of interval exchange transformations exploits certain self-similarity of these maps. Choosing a shorter horizontal interval X' we make the vertical flow wind for a long time before the first return to X'. However, the new first return map in a sense would not be more complicated than the initial one: it would be again an interval exchange transformation $T' : X' \to X'$ of the same (or almost the same) number of subintervals.

To check the latter statement let us study the nature of the points of discontinuity of the first return map $T : X \to X$. An interior point $x \in X$ is a point of discontinuity either if the forward vertical trajectory of x hits one of the endpoints of X (as on Fig. 20) or if the forward vertical trajectory of x hits the conical point before coming back to X (see Fig. 21). A conical point having the cone angle $2\pi(d+1)$ has $d+1$ incoming vertical trajectories which land to this conical point. (Say, the flat surface represented on Fig. 21 has a single conical point with the cone angle 6π; hence this conical point has 3 incoming vertical trajectories.) Following them at the backward direction till the first intersection with X we find $d+1$ points of discontinuity on X (see Fig. 21). Thus, all conical points taken together produce $\sum_j (d_j+1)$ points of discontinuity on X.

Generically two more points of discontinuity come from the backward trajectories of the endpoints of X. However, in order to get as small number of discontinuity points as possible we can choose X in such way that either backward or forward trajectory of each of the two endpoints hits some conical point before coming back to X. This eliminates these two additional discontinuity points.

Convention 2. From now on we shall always choose any horizontal subinterval X in such way that the interval exchange transformation $T : X \to X$ induced by the first return of the vertical flow to X has the minimal possible number

$$n = \sum_j (d_j + 1) + 1 = 2g + (\text{number of conical points}) - 1$$

of subintervals under exchange.

In the formula above we used the Gauss–Bonnet formula telling that $\sum_j d_j = 2g - 2$, where g is the genus of the surface.

Following Convention 2 we shall usually place the left endpoint of the horizontal interval X at the conical singularity. This leaves a discrete choice for the position of the right endpoint.

Renormalization

We apply the following strategy in our study of cycles $c(x, N)$. Choose some horizontal segment X satisfying Convention 2. Consider vertical trajectories, which hit conical points. Follow them in backward direction till the first intersection with X. Consider the resulting decomposition $X = X_1 \sqcup \cdots \sqcup X_n$, the corresponding interval exchange transformation $T : X \to X$ and the "first return cycles" $c(X_j)$.

Consider a smaller subinterval $X' \subset X$ satisfying Convention 2. Apply the above procedure to X'; let $X' = X'_1 \sqcup \cdots \sqcup X'_n$ be the corresponding decomposition of X'. We get a new partition of our flat surface into a collection of n rectangles based over subintervals $X'_1 \sqcup \cdots \sqcup X'_n$.

By construction the vertical trajectories of any two points $x_0, x \in X'_k$ follow the same high and narrow rectangle R'_k of the new building up to their first return to X'. This implies that the corresponding new "first return cycles" $c'(x_0) = c'(x)$ are the same and equal to $c'(X'_k)$.

Both vertical trajectories of $x_0, x \in X'_k$ intersect the initial interval X many times before first return to X'. However, since these trajectories stay together, they visit the same intervals X_{j_k} in the same order $j_0, j_1, \ldots, j_l$ (the length $l = l(k)$ of this trajectory depends on the subinterval X'_k).

This means that we can construct an $n \times n$-matrix B_{jk} indicating how many times a vertical trajectory emitted from a point $x \in X'_k$ have visited subinterval X_j before the first return to X'. (By convention the starting point counts, while the first return point does not.) Here $X = X_1 \sqcup \cdots \sqcup X_n$ is the partition of the initial "long" horizontal interval X and $X' = X'_1 \sqcup \cdots \sqcup X'_n$ is the partition of the new "short" subinterval X'.

Having computed this integer matrix B we can represent new "first return cycles" $c'(X'_k)$ in terms of the initial "first return cycles" $c(X_j)$ as

$$c'(X'_k) = B_{1k}c(X_1) + \cdots + B_{nk}c(X_n) \tag{5.3}$$

Moreover, it is easy to see that the lengths $\lambda'_k = |X'_k|$ of subintervals of the new partition are related to the lengths $|X_j|$ of subintervals of the initial partition by a similar relation

$$\lambda_j = B_{j1}\lambda'_1 + \cdots + B_{jn}\lambda'_n. \tag{5.4}$$

Note that to evaluate matrix B we, actually, do not need to use the vertical flow: the matrix B is completely determined by the initial interval exchange transformation $T : X \to X$ and by the position of the subinterval $X' \subset X$.

What we gain with this construction is the following. To consider a cycle $c(x, N)$ representing a long piece of leaf of the vertical foliation we followed the trajectory $x, T(x), \ldots, T^N(x)$ of the initial interval exchange transformation $T : X \to X$ and applied formula (5.1). Passing to a shorter horizontal interval $X' \subset X$ we can follow the trajectory $x, T'(x), \ldots, (T')^{N'}(x)$ of the new interval exchange transformation $T' : X' \to X'$ (provided $x \in X'$). Since the subinterval X' is much shorter than X we cover the initial piece of trajectory of the vertical flow in a smaller number N' of steps. In other words, passing from T to T' we accelerate the time: it is easy to see that the trajectory $x, T'(x), \ldots, (T')^{N'}(x)$ follows the trajectory $x, T(x), \ldots, T^N(x)$ but jumps over many iterations of T at a time.

Of course this approach would be non efficient if the new first return map $T' : X' \to X'$ would be much more complicated than the initial one. But we know that passing from T to T' we stay within a family of interval exchange transformations of the fixed number n of subintervals, and, moreover, that the new "first return cycles" and the lengths of the new subintervals are expressed in terms of the initial ones by means of the $n \times n$-matrix B, which depends only on the choice of $X' \subset X$ and which can be easily computed.

Our strategy can be formalized as follows. In the next two sections we describe a simple explicit algorithm (generalizing Euclidean algorithm) called Rauzy–Veech induction which canonically associates to an interval exchange transformation $T : X \to X$ some specific subinterval $X' \subset X$ and, hence, a new interval exchange transformation $T' : X' \to X'$. This algorithm can be considered as a map from the *space of all interval exchange transformations* of a given number n of subintervals to itself. Applying recursively this algorithm we construct a sequence of subintervals $X = X^{(0)} \supset X^{(1)} \supset X^{(2)} \supset \dots$ and a sequence of matrices $B = B(X^{(0)}), B(X^{(1)}), \dots$ describing transitions form interval exchange transformation $T^{(r)} : X^{(r)} \to X^{(r)}$ to interval exchange transformation $T^{(r+1)} : X^{(r+1)} \to X^{(r+1)}$. Rewriting equations (5.3) and (5.4) in a matrix form we get:

$$\begin{gathered}
\begin{pmatrix} c(X_1^{(r+1)}) \\ \dots \\ c(X_n^{(r+1)}) \end{pmatrix} = \begin{pmatrix} B(X^{(r)}) \end{pmatrix}^T \cdot \begin{pmatrix} c(X_1^{(r)}) \\ \dots \\ c(X_n^{(r)}) \end{pmatrix} \\
\begin{pmatrix} \lambda_1(X^{(r+1)}) \\ \dots \\ \lambda_n(X^{(r+1)}) \end{pmatrix} = \begin{pmatrix} B(X^{(r)}) \end{pmatrix}^{-1} \cdot \begin{pmatrix} \lambda_1(X^{(r)}) \\ \dots \\ \lambda_n(X^{(r)}) \end{pmatrix}
\end{gathered} \tag{5.5}$$

Taking a product $B^{(s)} = B(X^{(0)}) \cdot B(X^{(1)}) \cdot \dots \cdot B(X^{(s-1)})$ we can immediately express the "first return cycles" to a microscopic subinterval $X^{(s)}$ in terms of the initial "first return cycles" to X by a linear expression analogous to (5.5). Note, however, that before coming back to this microscopic subinterval $X^{(s)}$ the vertical flow has to travel for enormously long time. The first return cycle to this very short subinterval $X^{(s)}$ represents the cycle $c(x, N)$ corresponding to very long trajectory $x, T(x), ..., T^N(x)$ of the initial interval exchange transformation with $N \sim \exp(const \cdot s)$. In other words, our renormalization procedure plays a role of a time acceleration machine: instead of following patiently the trajectory $x, T(x), ..., T^N(x)$ of the initial interval exchange transformation for the exponential time $N \sim \exp(const \cdot s)$ we obtain the cycle $c(x, N)$ applying only s steps of renormalization!

One can argue that in this way we can describe only very special parts of vertical trajectories: those which start and end at the same microscopically small subinterval $X^{(s)} \subset X$. This can be overdone by the following technique. Consider an enormously long trajectory $x, T(x), \dots, T^N(x)$ which starts and finishes at some generic points of X. One can choose $s(N)$ in such way that the trajectory would get to $X^{(s)}$ relatively soon (in comparison with its length N); then would return back to $X^{(s)}$ many times; and would reach the last point $T^N(x)$ relatively fast after the last visit to $X^{(s)}$. That means that essentially (up to negligibly short starting part and ending part) one can assume that the entire trajectory starts at $X^{(s)}$ and ends at $X^{(s)}$ (returning to this subinterval many times).

This simple idea can be developed and rigorously arranged (see [Zo4] for details). To avoid overloading of this survey with technicalities we consider only a simplified problem giving a comprehensive description of the first return cycles to $X^{(s)}$. The nature of the asymptotic flag is especially transparent in this case.

5.4 Euclidean Algorithm as a Renormalization Procedure in Genus One

To illustrate the idea of renormalization we start with the "elementary" case, when the Riemann surface is a torus, the foliation is a standard irrational foliation, and the initial transversal X is a meridian. We have seen at Fig. 19 that in this case the first return map $T : X \to X$ is just a rotation of a circle.

Consider rotation of a circle $T : S^1 \to S^1$ by an angle α. Let the length of the circle be normalized to one. Consider trajectory $x, Tx, T^2x, \ldots$ of a point x (see Fig. 23). Denote the length of the arc (x, Tx) by $\lambda = \alpha/(2\pi)$.

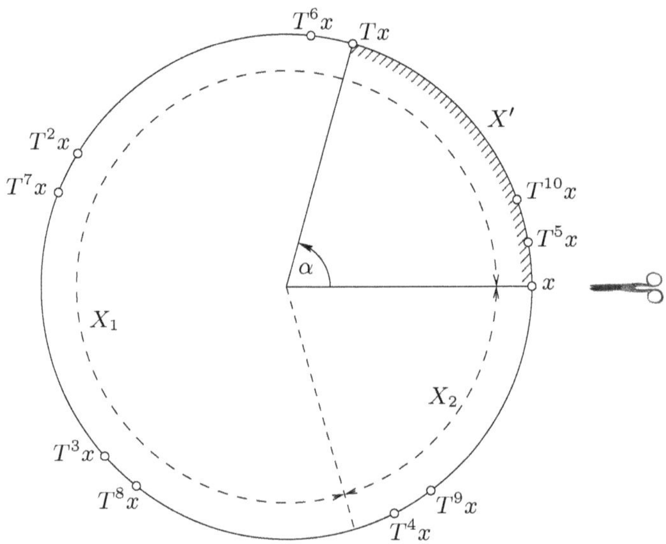

Fig. 23. Renormalization for rotation of a circle leads to Euclidean algorithm and to Gauss measure

Cutting the circle at the point x we get an interval X; the rotation of the circle generates a map of the interval X to itself which we denote by the same symbol $T : X \to X$. Unbend X isometrically to get a horizontal interval of unit length in such way that the counterclockwise orientation of the circle gives the standard positive orientation of the horizontal interval. The map T acts on X as follows: it cuts the unit interval X into two pieces $X_1 \sqcup X_2$ of lengths $|X_1| = 1 - \lambda$ and $|X_2| = \lambda$ and interchanges the pieces preserving the orientation, see Fig. 23. In other words, the map T is an interval exchange transformation of two subintervals. (To avoid confusion we stress that $X_2 = [T^{-1}x, x]$ and *not* $X = [x, Tx]$.)

Suppose now that we are looking at X in the microscope which shows only the subinterval $X' = [x, Tx[$ (corresponding to the sector of angle α at Fig. 23). Consider the trajectory $x, Tx, T^2x, \ldots$ of the point x which is the left extremity of X_1. For the particular rotation represented at Fig. 23 the points Tx, T^2x, T^3x, T^4x are outside of the sector of our vision; the next point of the trajectory which we see in X' is the point T^5x. This is the first return point $T'x = T^5x$ to the subinterval X'. Following the trajectory $T^5x, T^6x, \ldots$ further on we would not see several more points and then we shall see $T^{10}x = T'(T'x)$. This is the second return to the subinterval X'.

Note that the distance between x and $T'x = T^5x$ is the same as the distance between $T'x = T^5x$ and $T'(T'x) = T^{10}x$; it equals $(1 - \{1/\lambda\}) \cdot \lambda$, where $\{\ \}$ denotes the fractional part of a real number. It is easy to see that $T' : X' \to X'$ is again an interval exchange transformation of *two* subintervals $X'_1 \sqcup X'_2$. The lengths of subintervals are $|X'_1| = \{1/\lambda\} \cdot \lambda$ and $(1 - \{1/\lambda\}) \cdot \lambda$. After identification of the endpoints the segment X' becomes a circle and the map T' becomes a rotation of the circle T'. Having started with a rotation T in a *counterclockwise* direction by the angle $\alpha = 2\pi \cdot \lambda$ we get a rotation T' in a *clockwise* direction by the angle $\alpha' = 2\pi \cdot \left\{\frac{1}{\lambda}\right\}$ (please verify).

One should not think that $T' = T^5$ identically. It is true for the points of first subinterval, $T'|_{X'_1} = T^5$. However, for the points of the second subinterval X'_2 we have $T'|_{X'_2} = T^4$. In other words, for the points of the sector α, which are close to the extremity Tx, *four* iterations of T bring them back to the sector. Thus, the matrix $B(X')$ of number of visits to subintervals has the form

$$\begin{pmatrix} B_{11} & B_{12} \\ B_{21} & B_{22} \end{pmatrix} = \begin{pmatrix} 4 & 3 \\ 1 & 1 \end{pmatrix}$$

(please draw $X_1 \sqcup X_2$ and $X'_1 \sqcup X'_2$ and verify). We remind that B_{jk} indicates how many times a vertical trajectory emitted from a point $x \in X'_k$ have visited subinterval X_j before the first return to X', where by convention the starting point counts, while the first return point does not.

Thus we get a renormalization procedure as described in the previous section: confine the map T to a smaller subinterval X'; consider the resulting first return map T'; rescale X' to have unit length. Having started with an interval exchange transformation T of two intervals of lengths $(1-\lambda, \lambda)$, where $\lambda \in (0, 1)$ we get (after rescaling) an interval exchange transformation T' of two intervals of lengths $\{1/\lambda\}, 1 - \{1/\lambda\}$. Or, in terms of rotations, having started with a *counterclockwise* rotation by the angle $\alpha = 2\pi\lambda$ we get a *clockwise* rotation by the angle $\alpha' = 2\pi \cdot \left\{ \frac{1}{\lambda} \right\}$.

One can recognize Euclidean algorithm in our renormalization procedure. Consider the "space of rotations", where rotations are parametrized by the angle $2\pi\lambda$, $\lambda \in [0; 1[$. The map

$$g : \lambda \mapsto \left\{ \frac{1}{\lambda} \right\} \tag{5.6}$$

can be considered as a map from "the space of rotations" to itself, or what is the same, a map from "the space of interval exchange transformations of two subintervals" to itself. The map g is ergodic with respect to the invariant probability measure

$$d\mu = \frac{1}{\log 2} \cdot \frac{d\lambda}{(\lambda + 1)} \tag{5.7}$$

on the parameter space $\lambda \in [0; 1[$ which is called the *Gauss measure.* This map is intimately related with the development of λ into a continued fraction

$$\lambda = \cfrac{1}{n_1 + \cfrac{1}{n_2 + \cfrac{1}{n_3 + \cfrac{1}{\cdots}}}}$$

We shall see another renormalization procedure related to map (5.6) in the next sections, in particular, in Sect. 5.9.

5.5 Rauzy–Veech Induction

In the previous section we have seen an example of a renormalization procedure for interval exchange transformations of two intervals. In this section we consider a similar renormalization procedure which now works for interval exchanges of any number of subintervals. As we have seen in the previous section, we do not need to keep information about the flat surface to describe the renormalization algorithm. Nevertheless, we prefer to keep track of zippered rectangles decomposition of the surface corresponding to the sequence of the horizontal subintervals $X = X^{(0)} \supset X^{(1)} \supset \ldots$ in order to preserve geometric spirit of the algorithm.

Consider a flat surface S; choose a horizontal interval X satisfying Convention 2; consider the corresponding decomposition of the surface into *zippered rectangles* as on Fig. 21. Let $X_1 \sqcup \cdots \sqcup X_n$ be the corresponding decomposition of the horizontal segment X in the base; let $\lambda_j = |X_j|$ denote the widths of subintervals.

Convention 3. We associate to a decomposition of a flat surface into rectangles a permutation π in such way that the top horizontal segments of the rectangles are glued to the bottom side of the interval X in the order $\pi^{-1}(1),\dots,\pi^{-1}(n)$.

Example. In the example presented at Fig. 21 the four rectangles $R_1,\dots,R_4$ appear at the bottom side of X in the order R_3, R_1, R_4, R_2, so we associate to this way of gluing a permutation

$$\begin{pmatrix} 1\,2\,3\,4 \\ 3\,1\,4\,2 \end{pmatrix} = (3,1,4,2)^{-1} = (2,4,1,3) = \pi$$

Exercise. Show that intersection indices $c(X_i) \circ c(X_j)$ of the "first return cycles" (see Sect. 5.2) are given by the following skew-symmetric matrix $\Omega(\pi)$ defined by the permutation π:

$$\Omega_{ij}(\pi) = \begin{cases} 1 & \text{if } i < j \text{ and } \pi^{-1}(i) > \pi^{-1}j \\ -1 & \text{if } i > j \text{ and } \pi^{-1}(i) < \pi^{-1}j \\ 0 & \text{otherwise} \end{cases} \tag{5.8}$$

Evaluate $\Omega(\pi)$ for the permutation in the Example above and compare the result with a direct calculation for the cycles $c(X_j)$, $j = 1,\dots,4$, computed in the Exercise at the end of Sect. 5.2.

Compare now the width λ_n of the rightmost rectangle R_n with the width $\lambda_{\pi^{-1}(n)}$ of the rectangle which is glued to the rightmost position at the bottom of X. As a new subinterval $X' \subset X$ consider the subinterval X', which has the same left extremity as X, but which is shorter than X by $\min(\lambda_n, \lambda_{\pi^{-1}(n)})$.

The situation when $\lambda_n > \lambda_{\pi^{-1}(n)}$ is represented at Fig. 24; the situation when $\lambda_n < \lambda_{\pi^{-1}(n)}$ is represented at Fig. 25.

By construction the first return map $T' : X \to X'$ has the same number n of subintervals in its decomposition. Observing Fig. 24 and 24 one can see that in the first case, when $\lambda_n > \lambda_{\pi^{-1}(n)}$, the new decomposition $X'_1 \sqcup \dots \sqcup X'_n$ is obtained from the original decomposition $X_1 \sqcup \dots \sqcup X_n$ by shortening the last interval by $\lambda_{\pi^{-1}(n)}$ from the right. In the second case, when $\lambda_n < \lambda_{\pi^{-1}(n)}$, the new decomposition $X'_1 \sqcup \dots \sqcup X'_n$ is obtained from the original decomposition $X_1 \sqcup \dots \sqcup X_n$ by eliminating the last subinterval X_n and by partitioning the subinterval $X_{\pi^{-1}(n)}$ into two ones of lengths $\lambda_{\pi^{-1}(n)} - \lambda_n$ and λ_n correspondingly.

The order in which the rectangles of the new building are glued to the bottom of the interval X' changes. The new permutation π' can be described as follows. Consider the initial permutation π as a pair of orderings of a finite set: a "top" ordering $1, 2, \dots, n$ (corresponding to the ordering of the rectangles along the top side of the base interval X) and a "bottom" ordering $\pi^{-1}(1),\dots,\pi^{-1}(m)$ (corresponding to the ordering of the rectangles along the bottom side of the base interval X). In the first case, when $\lambda_n > \lambda_{\pi^{-1}(n)}$, the new permutation π' corresponds to the modification of the *bottom* ordering

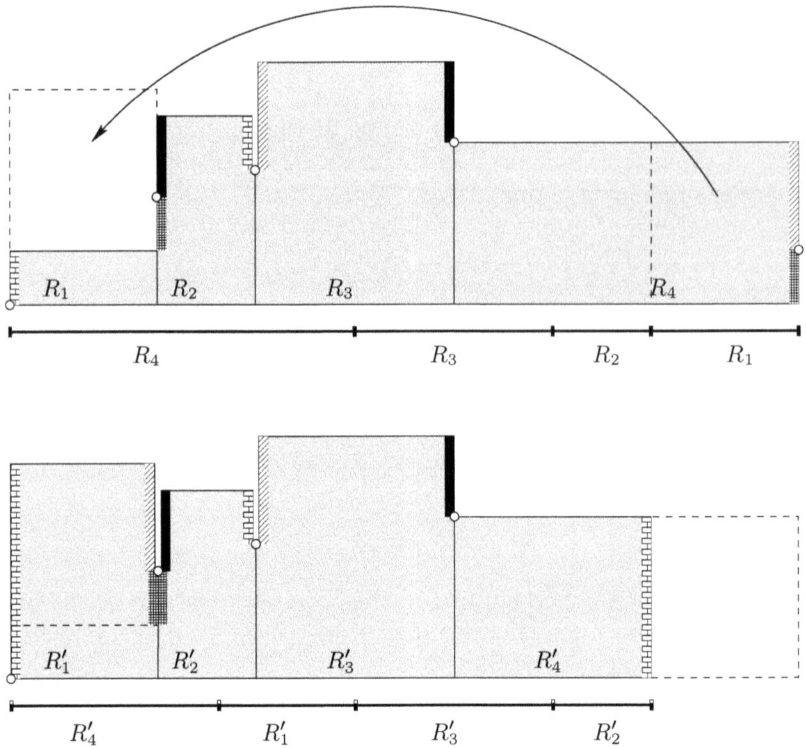

Fig. 24. Type I modification: the rightmost rectangle R_4 on top of X is wider than the rectangle $R_1 = R_{\pi^{-1}(4)}$ glued to the rightmost position at the bottom of X

by cyclically moving one step forward those letters occurring after the image of the last letter in the bottom line, i.e., after the letter n. In the second case, when $\lambda_n < \lambda_{\pi^{-1}(n)}$, the new permutation π' corresponds to the modification of the *top* ordering by cyclically moving one step forward those letters occurring after the image of the last letter in the top line, i.e., after the letter $\pi^{-1}(n)$.

Example. For the initial buildings at both Figures 24 and 25 the permutation π corresponding to the initial interval exchange transformation $T : X \to X$ is the same and equals

$$\pi = \begin{pmatrix} 1\ 2\ 3\ 4 \\ 4\ 3\ 2\ 1 \end{pmatrix}$$

Our modification produces permutation

$$\begin{pmatrix} 1 & 2 & & 3 & & 4 \\ 4 & 3 & \to & 2 & \to & 1 \end{pmatrix} = \begin{pmatrix} 1\ 2\ 3\ 4 \\ 4\ 1\ 3\ 2 \end{pmatrix} = \pi'$$

in the first case (when $\lambda_n > \lambda_{\pi^{-1}(n)}$) and permutation

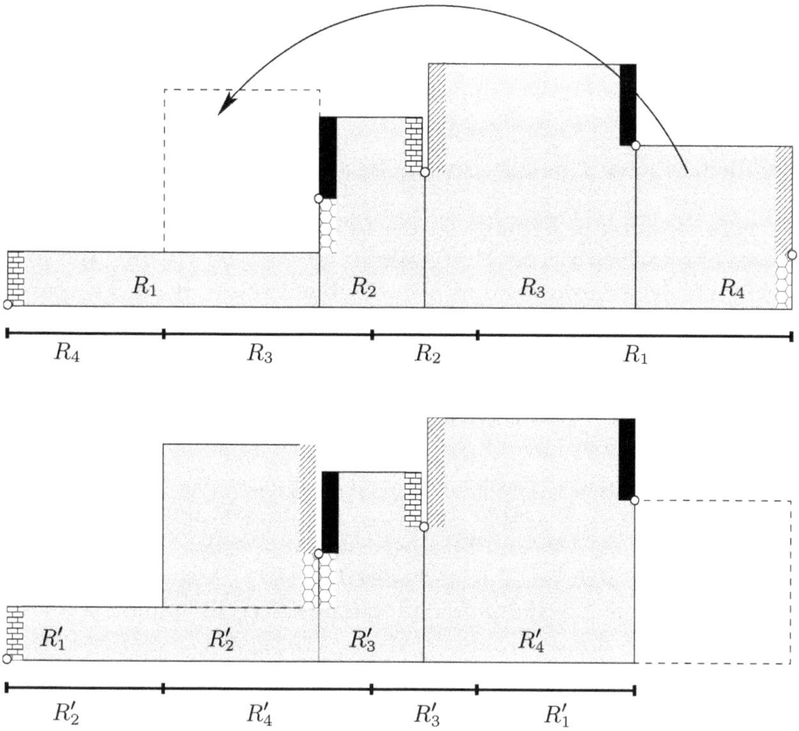

Fig. 25. Type II modification: the rightmost rectangle R_4 on top of X is narrower than the rectangle $R_1 = R_{\pi^{-1}(4)}$ glued to the rightmost position at the bottom of X

$$\begin{pmatrix} 1 & \curvearrowleft 2 \to & 3 \to & 4 \\ 4 & 3 & 2 & 1 \end{pmatrix} = \begin{pmatrix} 1\,4\,2\,3 \\ 4\,3\,2\,1 \end{pmatrix} \sim \begin{pmatrix} 1\,2\,3\,4 \\ 2\,4\,3\,1 \end{pmatrix} = \pi'$$

in the second case (when $\lambda_n < \lambda_{\pi^{-1}(n)}$).

Note that in the second case (when $\lambda_n < \lambda_{\pi^{-1}(n)}$) passing to the new decomposition $X'_1 \sqcup \cdots \sqcup X'_n$ we have to change the initial enumeration of the subintervals though physically all subintervals but one stay unchanged. Another choice would be to assign "names" to subintervals once and forever. Under the first choice the permutations

$$\begin{pmatrix} 1\,4\,2\,3 \\ 4\,3\,2\,1 \end{pmatrix} \text{ and } \begin{pmatrix} 1\,2\,3\,4 \\ 2\,4\,3\,1 \end{pmatrix}$$

coincide; under the latter choice they become different permutations. The article [Y] in the current volume adopts the second convention.

Similarly to the case of interval exchange transformations of two intervals the induction procedure described above can be described entirely in terms of interval exchange transformations $T : X \to X$ and $T' : X' \to X'$. Historically it was proposed by G. Razy [Ra] in these latter form and then was interpreted by W. Veech [Ve3] in terms of zippered rectangles. (Actually, the zippered rectangles decomposition has appeared and was first studied in [Ve3].)

5.6 Multiplicative Cocycle on the Space of Interval Exchanges

The renormalization procedure constructed in Sect. 5.4 gives a map g from the space of rotations of a circle to itself, or, in other terms, from the space of interval exchange transformations of two subintervals to itself. The permutation π corresponding to an interval exchange transformations of two intervals $X_1 \sqcup X_2$ is always equal to $\pi = (2,1)$, so such interval exchange transformation can be parametrized by a single real parameter $\lambda \in (0,1)$, where $\lambda = |X_1|$. Here we assume that the total length $|X_1| + |X_2| = |X|$ of the interval X is normalized as $|X| = 1$.

An interval exchange transformation of n subintervals $X = X_1 \sqcup \cdots \sqcup X_n$ is parametrized by a collection $\lambda_1, \ldots, \lambda_n$ of positive numbers representing the lengths of subintervals and by a permutation $\pi \in \mathfrak{S}_n$. Assuming that the total length of the interval X is normalized as $|X| = 1$ we see that the *space of interval exchange transformations* is parametrized by a finite collection of $(n-1)$-dimensional simplices $\Delta^{n-1} = \{(\lambda_1, \ldots, \lambda_n) \mid \lambda_1 + \cdots + \lambda_n = 1; \lambda_j > 0\}$, where each simplex corresponds to some fixed permutation π.

As a collection of permutations one can consider all permutations obtained from a given one by applying recursively the modifications described at the end of the previous section. Such collection of permutations is called a *Rauzy class* $\mathfrak{R} \subset \mathfrak{S}_n$. Figure 26 illustrates the Rauzy class of the permutation $(4,3,2,1)$, where the arrows indicated modifications of the first and of the second type.

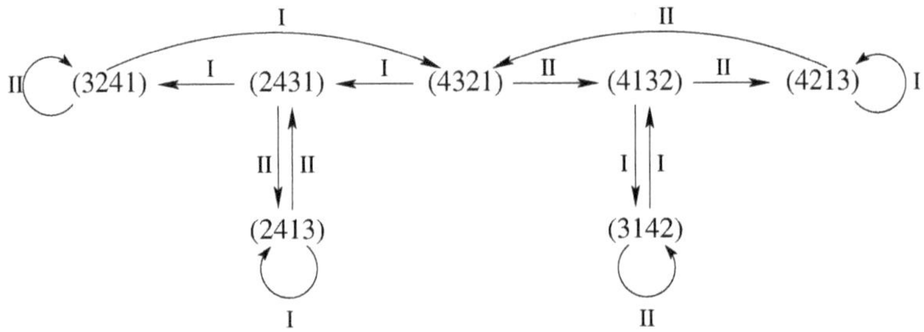

Fig. 26. Rauzy class of permutation $(4,3,2,1)$

The renormalization procedure described in the previous section (combined with rescaling of the resulting interval X' to the unit length) defines a map

$$\mathcal{T} : \Delta^{n-1} \times \mathfrak{R} \to \Delta^{n-1} \times \mathfrak{R}$$

for each space of interval exchange transformations to itself. The following Theorem of W. A. Veech (see [Ve3]) is crucial in this story.

Key Theorem (W. A. Veech). *The map $\mathcal{T}$ is ergodic with respect to absolutely continuous invariant measure.*

Remark. It is easy to see that the Rauzy class presented at Fig. 26 does not depend on the starting permutation. Actually, the same is true for any Rauzy class. Moreover, for almost any flat surface $S \in \mathcal{H}^{comp}(d_1, \dots, d_m)$ in any connected component of any stratum a finite set of all permutations realizable by first return maps to all possible horizontal segments X satisfying Convention 2 does not depend on the surface S. This set is a disjoint union of a finite collection of corresponding Rauzy classes $\mathfrak{R}_1, \dots, \mathfrak{R}_j$, where j is the number of distinct entries $d_{i_1} < d_{i_2} < \dots < d_{i_j}$. The union $\mathfrak{R}_{ex} = \mathfrak{R}_1 \sqcup \dots \sqcup \mathfrak{R}_j$ is called the *extended Rauzy class*; it depends only on the connected component $\mathcal{H}^{comp}(d_1, \dots, d_m)$. In particular, connected components of the strata are characterized by the extended Rauzy classes, where the latter ones can be described in purely combinatorial terms.

With the Theorem above we have almost accomplished our scheme for a renormalization procedure. There is only one trouble with the map $\mathcal{T}$: the measure mentioned in the Theorem is infinite (the total measure of the space of interval exchange transformations is infinite). This technical problem can be fixed by the following trick. We shall modify the renormalization algorithm described in the previous section by making several modifications of the zippered rectangle at a time. At a single step of the new algorithm $\mathcal{G}$ we apply several steps of the previous one $\mathcal{T}$. Namely, we keep going as soon as we apply consecutive transformations $\mathcal{T}$ of the same type I or of the same type II. Conceptually it does not change the renormalization procedure, but now the renormalization develops faster than before. The following example illustrates the correspondence between the renormalization procedures $\mathcal{T}$ and $\mathcal{G}$:

$$
\begin{array}{ccccccccccccc}
(\lambda,\pi) & \xrightarrow{I} & \mathcal{T}(\lambda,\pi) & \xrightarrow{I} & \mathcal{T}^2(\lambda,\pi) & \xrightarrow{I} & \mathcal{T}^3(\lambda,\pi) & \xrightarrow{II} & \mathcal{T}^4(\lambda,\pi) & \xrightarrow{II} & \mathcal{T}^5(\lambda,\pi) & \xrightarrow{I} & \cdots \\
\| & & & & & & \| & & & & \| & & \\
(\lambda,\pi) & \longrightarrow & & & & & \mathcal{G}(\lambda,\pi) & \longrightarrow & & & \mathcal{G}^2(\lambda,\pi) & \rightarrow & \cdots
\end{array}
$$

The accelerated procedure $\mathcal{G}$ was introduced in [Zo2], where the following Theorem was proved.

Theorem. *The map $\mathcal{G}$ is ergodic with respect to absolutely continuous invariant probability measure on each space of zippered rectangles.*

Now, when we have elaborated almost all necessary tools we are ready to give an idea of the proof of the Theorem from Sect. 4.3 concerning asymptotic flag. The last element which is missing is some analysis of the matrices $B(\lambda, \pi)$ in (5.5).

Multiplicative Cocycle

In our interpretation of a renormalization procedure as a map on the space of interval exchange transformations we have to consider matrix B as a matrix-valued function $B(\lambda, \pi)$ on the space $\Delta^{n-1} \times \mathfrak{R}$ of interval exchange transformations. Our goal (as it was outlined in Sect. 5.3) is to describe the properties of the products $B^{(s)} = B(X^{(0)}) \cdot B(X^{(1)}) \cdot \dots \cdot B(X^{(s)})$ of matrices corresponding to successive steps of renormalization.

We shall keep the same notation $B(\lambda, \pi)$ for matrices corresponding to our fast renormalization procedure $\mathcal{G}(\lambda, \pi)$. Consider the product $B^{(s)}(\lambda, \pi)$ of values of $B(\cdot)$ taken along the orbit $(\lambda, \pi), \mathcal{G}(\lambda, \pi), \dots, \mathcal{G}^s(\lambda, \pi)$ of length s of the map $\mathcal{G}$:

$$B^{(s)}(\lambda, \pi) = B(\lambda, \pi) \cdot B(\mathcal{G}(\lambda, \pi)) \cdot \dots \cdot B(\mathcal{G}^s(\lambda, \pi))$$

Such $B^{(s)}(\lambda, \pi)$ is called a *multiplicative cocycle* over the map $\mathcal{G}$:

$$B^{(p+q)}(\lambda, \pi) = B^{(p)}(\lambda, \pi) \cdot B^{(q)}(\mathcal{G}^p(\lambda, \pi))$$

Using ergodicity of $\mathcal{G}$ we can apply multiplicative ergodic theorem (see Appendix B) to describe properties of $B^{(s)}$. Morally, the multiplicative ergodic theorem tells that for large values of s the matrix $B^{(s)}(\lambda, \pi)$ should be considered as a matrix conjugate to the s-th power of some *constant* matrix. (See Appendix B for a rigorous formulation.) The logarithms of eigenvalues of this constant matrix are called *Lyapunov exponents* of the multiplicative cocycle.

Recall that matrix $B^T(\lambda, \pi)$ was defined as a matrix representing the new "first return cycles" in terms of the old ones, see (5.5). Actually, it can be also interpreted as a matrix representing a change of a basis in the first relative cohomology $[\mathrm{Re}\,\omega] = (\lambda_1, \dots, \lambda_n) \in H^1(S, \{\text{conical singularities}\}; \mathbb{R})$. It is easy to check that it respects the (degenerate) symplectic form: the intersection form (5.8). Note that symplectic matrices have certain symmetry of eigenvalues. In particular, it follows from the general theory that the corresponding Lyapunov exponents have the following symmetry:

$$\theta_1 > \theta_2 \geq \theta_3 \geq \dots \geq \theta_g \geq \underbrace{0 = 0 = \dots = 0}_{number\ of\ conical\ points\ -1} \geq -\theta_g \geq \dots \geq -\theta_2 > -\theta_1$$

Note that the first return cycles actually belong to the absolute homology group $H_1(S; \mathbb{Z}) \subset H_1(S; \mathbb{R}) \simeq \mathbb{R}^{2g}$. Passing to this $2g$-dimensional space we get matrices which already preserve a nondegenerate symplectic form. They define the following subcollection

$$\theta_1 > \theta_2 \geq \theta_3 \geq \dots \geq \theta_g \geq -\theta_g \geq \dots \geq -\theta_3 > -\theta_2 > -\theta_1$$

of Lyapunov exponents.

The rest is an elementary linear algebra. We want to describe how do the large powers s of a symplectic matrix with eigenvalues

$$\exp(\theta_1) > \exp(\theta_2) \geq \cdots \geq \exp(\theta_g) \geq \exp(-\theta_g) \geq \cdots \geq \exp(-\theta_2) \geq \exp(-\theta_1)$$

act on a $2g$-dimensional symplectic space.

We know that the Lyapunov exponent $\theta_g \geq 0$ is nonnegative. Assume[4] that it is actually strictly positive: $\theta_g > 0$. Then for half of dimensions our linear map is expanding and for half of dimensions it is contracting. In particular, under the assumption that $\theta_g > 0$ we conclude that the linear map $B^{(s)}(\lambda, \pi)$ projects all homology space to a Lagrangian subspace (spanned by eigenvectors corresponding to positive Lyapunov exponents).

Assuming that the spectrum of Lyapunov exponents is simple, that is assuming that

$$\theta_1 > \theta_2 > \theta_3 > \cdots > \theta_g$$

we get the entire picture of deviation. A generic vector in the homology space stretches along the principal eigenvector (the one corresponding to the eigenvalue θ_1) with a factor $\exp(s\theta_1)$; it expands along the next eigenvector with a factor $\exp(s\theta_2)$, etc, up to the order g; its deviation from the Lagrangian subspace spanned by the first g eigenvectors tends to zero. Hence, the norm l of the image of a generic vector under s-th power of our linear map is of the order $l \sim \exp(s\theta_1)$; its deviation from the subspace $\mathcal{V}_1$, which is spanned by the top eigenvector, is of the order $\exp(s\theta_2) = l^{\frac{\theta_2}{\theta_1}}$; its deviation from the subspace $\mathcal{V}_2$ spanned by the two top eigenvectors is of the order $\exp(s\theta_3) = l^{\frac{\theta_3}{\theta_1}}$, etc; there is no deviation from the Lagrangian subspace spanned by the top g eigenvectors. In particular, the exponent ν_j responsible for deviation from the subspace $\mathcal{V}_{j-1}$ from the Theorem in Sect. 4.3 is obtained by normalization of the Lyapunov exponent θ_j by the leading Lyapunov exponent θ_1:

$$\nu_j = \frac{\theta_j}{\theta_1}. \tag{5.9}$$

This completes the proof of the Theorem in the case when the vertical trajectory starts and ends at the same microscopic horizontal interval (in other words, when the piece of trajectory is "almost closed"). Applying some additional (relatively involved) ergodic machinery one can complete the proof of the Theorem for arbitrary long pieces of vertical trajectories; see [Zo4] for a complete proof.

I would like to stress that the original Theorem proved in [Zo4] is conditional: the statement about Lagrangian subspace was proved modulo conjecture that $\theta_g > 0$; the statement about a complete Lagrangian flag was proved modulo conjectural simplicity of the spectrum of Lyapunov exponents.

Positivity $\theta_g > 0$ was proved by G. Forni in [For1], and simplicity of the spectrum was recently proved by A. Avila and M. Viana [AvVi]; see more details in Sect. 5.8. As it was shown in [Zo2] the proof of the strict inequality $\theta_1 > \theta_2$ immediately follows from results of W. A. Veech.

[4] Actually, this assumption is a highly nontrivial Theorem of G. Forni [For1]; see below, see also Sect. 5.8

Exercise. In analogy with what was done in Sect. 5.4 consider the Rauzy–Veech induction in the torus case applying it to interval exchange transformations of two subintervals. We have seen that in this case the space of interval exchange transformations is just an interval $(0,1)$. Find an explicit formulae for the Rauzy–Veech renormalization map $\mathcal{T}:(0,1)\to(0,1)$ and for the "fast" renormalization map $\mathcal{G}:(0,1)\to(0,1)$. Explain why the invariant measure is infinite for the map $\mathcal{T}$. Find a relation between $\mathcal{G}$ and the Gauss map $g: x\mapsto \left\{\frac{1}{x}\right\}$. Let

$$\frac{p_s}{q_s}=\cfrac{1}{n_1+\cfrac{1}{n_2+\cfrac{1}{\dots+\cfrac{1}{n_s}}}}$$

be the s-th best rational approximation of a real number $x\in(0,1)$. In the torus case the spectrum of Lyapunov exponents reduces to a single pair $\theta_1>-\theta_1$. Show that for almost all $x\in(0,1)$ the Lyapunov exponent θ_1 (called in number theory the *Lévy constant*, see [Lv]) is responsible for the growth rate of the denominator of the continued fraction expansion of x:

$$\lim_{s\to\infty}\frac{\log q_s}{s}=\theta_1=\frac{\pi^2}{12\log 2}$$

5.7 Space of Zippered Rectangles and Teichmüller Geodesic Flow

We have proved the Theorem about an asymptotic Lagrangian flag of subspaces responsible for deviation of the cycles $c(x,N)$ from asymptotic direction. We have also proved that the exponents ν_j responsible for the quantitative description of the deviation are expressed in terms of the Lyapunov exponents of the multiplicative cocycle corresponding to our renormalization procedure: $\nu_j=\theta_j/\theta_1$.

There remains a natural question why should we choose this particular renormalization procedure and not a different one. One more natural question is what is the relation between renormalization procedure and the flow induced by the action of the diagonal subgroup $\begin{pmatrix}\exp(t) & 0\\ 0 & \exp(-t)\end{pmatrix}$ on the space of flat surfaces which we agreed to call the *Teichmüller geodesic flow*; this relation was announced in Sect. 4.3. This section answers to these questions which are, actually, closely related.

In our presentation we follow the fundamental paper [Ve3] of W. A. Veech; the material at the end of the section is based on the paper [Zo2] developing the initial paper [Ve3].

Space of Zippered Rectangles

We have seen that locally a flat surface S can be parametrized by a collection of relative periods of the holomorphic 1-form ω representing the flat surface, i.e. we can choose a small domain containing $[\omega]$ in the relative cohomology group $H^1(S,\{P_1,\dots,P_m\};\mathbb{C})$ as a coordinate chart in the corresponding stratum $\mathcal{H}(d_1,\dots,d_m)$.

Decomposition of a flat surface into zippered rectangles gives another system of local coordinates in the stratum. Namely, choose a horizontal segment X satisfying Convention 2 from Sect. 5.3 and consider the corresponding decomposition of S into zippered rectangles. Let $\lambda_1,\dots,\lambda_n$ be the widths of the rectangles, $h_1,\dots,h_n$ their heights, and $a_1,\dots,a_m$ be the altitudes responsible for the position of singularities (we zip the neighboring rectangles R_j and R_{j+1} from the bottom up to the altitude a_j and then the rectangles split at the singularity, see Figures 21, 24, 25). There is one more parameter describing a decomposition of a flat surface into zippered rectangles: a permutation $\pi \in \mathfrak{S}_n$. This latter parameter is discrete.

The vertical parameters h_j, a_k and are not independent: they satisfy some linear equations and inequalities. Varying the continuous parameters λ, h, a respecting the linear relations between parameters h and a we get a new set of coordinates in the stratum. These coordinates were introduced and studied by W. A. Veech in [Ve3]. In particular, it was proved that for any $(\lambda,\pi) \in \Delta^{n-1}\times\mathfrak{R}$ in the space of interval exchange transformations there is an n-dimensional open cone of solutions (h,a). In other words, having any interval exchange transformation $T: X \to X$ one can always construct a flat surface S and a horizontal segment $X \subset S$ inside it such that the first return of the vertical flow to X gives the initial interval exchange transformations. Moreover, there is a n-dimensional family of such flat surfaces – *suspensions* over the interval exchange transformation $T: X \to X$ (see [Ma3], [Ve3]).

Is there a canonical decomposition of a flat surface into zippered rectangles? A choice of horizontal segment $X \subset S$ completely determines a decomposition of a generic surface S into zippered rectangles, so our question is equivalent to the problem of a canonical choice of a horizontal segment X satisfying Convention 2. The choice which we propose is *almost* canonical; it leaves an arbitrariness of finite order which is the same for almost all S in the stratum. Here is the choice. Let us place the left extremity of the horizontal segment X at one of the conical singularities, and let us choose the length $|X|$ of the segment in such way that X would be the shortest possible interval satisfying Convention 2 and condition $|X| \geq 1$.

In practice the interval X can be constructed as follows: start with a sufficiently long horizontal interval having its left extremity at a conical point and satisfying Convention 2. Apply the "slow" Rauzy–Veech algorithm as long as the resulting subinterval has length at least 1. For almost all flat surfaces after finite number of steps we obtain the desired interval X.

The surface S has finite number of conical singularities; each conical singularity has finite number of horizontal prongs, so we get arbitrariness of finite order. Thus, the resulting *space of zippered rectangles* can be essentially viewed as a (ramified) covering over the corresponding connected component $\mathcal{H}^{comp}(d_1, \ldots, d_m)$ of the stratum. Passing to a codimension one subspace Ω defined by the condition $\lambda \cdot h = 1$ we get a space of zippered rectangles of area one covering the space $\mathcal{H}_1^{comp}(d_1, \ldots, d_m)$ of flat surfaces of area one. Consider a codimension-one subspace $\Upsilon \subset \Omega$ of zippered rectangles which have unit area, and which have the base X of length one, $|X| = \lambda_1 + \cdots + \lambda_n = 1$. The space Υ has a natural structure of a fiber bundle over the space $\Delta^{n-1} \times \mathfrak{R}$ of interval exchange transformations: we associate to a zippered rectangle $(\lambda, h, a, \pi) \in \Upsilon$ the interval exchange transformation (λ, π).

We would like to emphasize an interpretation of Ω as a *fundamental domain* in the space of *all* zippered rectangles of area one. As a fundamental domain Ω is defined by the additional condition on the base: X is the shortest possible interval satisfying Convention 2 such that $|X| \geq 1$. In this interpretation Υ is the boundary of the fundamental domain. Starting with an arbitrary zippered rectangle representation satisfying Convention 2 we can apply several steps of Rauzy–Veech algorithm (see Fig. 24 and Fig. 25), which does not change the surface S. After several iterations we get to the fundamental domain Ω.

We would use the same notation for Ω considered as a fundamental domain and for Ω considered as a quotient, when two boundary components of Ω are identified by the modification of zippered rectangles as on Fig. 24 and Fig. 25.

Teichmüller Geodesic Flow and its First Return Map to a Cross-section

Zippered rectangles coordinates are extremely convenient when working with the Teichmüller geodesic flow, which we identify with the action of the diagonal subgroup $g_t = \begin{pmatrix} \exp(t) & 0 \\ 0 & \exp(-t) \end{pmatrix}$. Namely, g_t expands the horizontal parameters λ by the factor $\exp(t)$ and contracts the vertical parameters h, a by the same factor.

Consider a zippered rectangle $S = (\lambda, h, a, \pi) \in \Upsilon$ with the base X of unit length. Applying g_t to S with t continuously increasing from $t = 0$ we shall eventually make the length of the base of the deformed zippered rectangle $g_t S = (\exp(t)\lambda, \exp(-t)h, \exp(-t)a; \pi)$ too long and thus we shall get outside of the fundamental domain Ω. It is not difficult to determine an exact time t_0 when it will happen. We get to the boundary of the fundamental domain Ω at the time

$$t_0(S) = -\log\left(1 - \min(\lambda_n, \lambda_{\pi^{-1}(n)})\right). \tag{5.10}$$

The time t_0 is chosen in such way that applying to the zippered rectangle $g_{t_0} S$ one step of the Rauzy–Veech induction (see Fig. 24 and Fig. 25) we get a new zippered rectangle with the base X' of unit length. To verify formula (5.10) for t_0 it is sufficient to note that expansion-contraction commutes with Rauzy–Veech induction. Thus, to evaluate t_0 we can *first* apply one step of the Rauzy–Veech induction and *then* apply expansion-contraction for an appropriate time, which would bring us back to Υ, i.e. which would make the length of the base of the new building of zippered rectangles equal to one.

In other words, starting at a point $S \in \Upsilon$ and following the flow for the time $t_0(S)$ we get to the boundary of the fundamental domain in the space of zippered rectangles and we have to instantly jump back to the point of Υ identified with $g_{t_0}S$. One can recognize in this construction the first return map $\mathcal{S} : \Upsilon \to \Upsilon$ defined by the flow g_t on the section Υ: at the time $t_0(S)$ the flow g_t emitted from a point $S \in \Upsilon$ returns back to the codimension-one subspace Υ transversal to the flow.

Morally one should consider the map $\mathcal{S}$ as a map on some subspace of flat surfaces. Note, that $\mathcal{S}$ is not applicable to points of flat surfaces, it associates to a flat surface taken as a whole another flat surface taken as a whole.

We see now that the Rauzy–Veech renormalization procedure $\mathcal{S} : \Upsilon \to \Upsilon$ performed on the level of zippered rectangles is nothing but discrete version of the Teichmüller geodesic flow. Namely $\mathcal{S}$ is the first return map of the Teichmüller geodesic flow to a section Υ. By construction the Rauzy–Veech induction $\mathcal{T} : \Delta^{n-1} \times \mathfrak{R} \to \Delta^{n-1} \times \mathfrak{R}$ on the space of interval exchange transformations is just a projection of $\mathcal{S}$. In other words, the following diagram

$$\begin{array}{ccc} \Upsilon(\mathfrak{R}) & \xrightarrow{\mathcal{S}} & \Upsilon(\mathfrak{R}) \\ \downarrow & & \downarrow \\ \Delta^{n-1} \times \mathfrak{R} & \xrightarrow{\mathcal{T}} & \Delta^{n-1} \times \mathfrak{R} \end{array}$$

is commutative, and the invariant measure on the space $\Delta^{n-1} \times \mathfrak{R}$ of interval exchange transformations is a push forward of the natural invariant measure on the space Υ of zippered rectangles.

Choice of a Section

Now we can return to the questions addressed at the beginning of this section. Ignoring an algorithmic aspect of the choice of renormalization procedure we see that conceptually, it is defined by a section of the Teichmüller geodesic flow. In particular, the "fast" renormalization procedure $\mathcal{G} : \Delta^{n-1} \times \mathfrak{R} \to \Delta^{n-1} \times \mathfrak{R}$ defined in the previous section corresponds to a choice of a subsection $\Upsilon' \subset \Upsilon$. Luckily it has a simple algorithmic representation in terms of modification of the interval exchange transformation $T(\lambda, \pi)$, and, moreover, it has a simple description in terms of coordinates λ, h, a, π in the space of zippered rectangles given by an extra condition for the parameter a_n.

Recall that parameters a_j are responsible for the position of singularities: we zip the neighboring rectangles R_j and R_{j+1} from the bottom up to the altitude a_j, see Fig. 21. In particular, by construction all a_j for $j = 1, \ldots, n-1$ are positive. Parameter a_n is, however, different from the others: the rectangle

R_n is the rightmost rectangle in the collection. If there is a conical singularity located at the right side of this rightmost rectangle (see, for example, the zippered rectangle decomposition of the flat surface on the top part of Fig. 24), then parameter a_m is positive; it indicates as usual at what height is located the singularity. However, the right side of the rightmost rectangle might contain no singularity. This means that the singularity is located on the corresponding vertical trajectory *below* the zero level of the base X. The rectangle which is glued to X from below at the rightmost position is the rectangle $R_{\pi^{-1}(n)}$; the singularity is located on the right side of this rectangle (see, for example, the zippered rectangle decomposition of the flat surface on the bottom part of Fig. 24). In this case we let a_n be negative indicating how low we have to descend along downward vertical trajectory emitted from the right endpoint of X to hit the singularity.

The subsection Υ' is defined by the following extra condition

$$\begin{aligned}\Upsilon' = &\{ (\lambda, h, a, \pi) \in \Upsilon \mid a_n > 0 \text{ when } \lambda_n > \lambda_{\pi^{-1}(n)} \} \sqcup \\ \sqcup &\{ (\lambda, h, a, \pi) \in \Upsilon \mid a_n < 0 \text{ when } \lambda_n < \lambda_{\pi^{-1}(n)} \}\end{aligned}$$

Exercise. Check which zippered rectangles at Figures 21, 24, 25 satisfy the condition $a_n \cdot (\lambda_n - \lambda_{\pi^{-1}(n)}) > 0$ and which do not.

It can be verified (see [Zo2]) that the section Υ' is still a fiber bundle over the space of zippered rectangles and that the corresponding first return map $\mathcal{S}' : \Upsilon' \to \Upsilon'$ projects to the map $\mathcal{G}$:

$$\begin{array}{ccc} \Upsilon'(\mathfrak{R}) & \xrightarrow{\mathcal{S}'} & \Upsilon'(\mathfrak{R}) \\ \downarrow & & \downarrow \\ \Delta^{n-1} \times \mathfrak{R} & \xrightarrow{\mathcal{G}} & \Delta^{n-1} \times \mathfrak{R} \end{array} \tag{5.11}$$

Exercise. Verify that the definition of the renormalization procedure $\mathcal{G}$ as a projection of the first return map of the Teichmüller geodesic flow to Υ' matches the intrinsic definition of $\mathcal{G}$ given in Sect. 5.6.

Different choices of the section also explain why the invariant measure on the space of interval exchange transformations $\Delta^{n-1} \times \mathfrak{R}$ was infinite for the Rauzy–Veech induction $\mathcal{T}$ while is finite for the "fast" renormalization procedure $\mathcal{G}$. As a model case consider a directional flow on a torus and two different sections to this flow. Taking as a section the line Y represented on the left picture of Fig. 27 we get a section of infinite measure though the measure of the torus is finite and the flow is very nice. Taking as a section a finite piece $Y' \subset Y$ as on the right side of Fig. 27 we get a section of finite measure.

Similarly, the component $\mathcal{H}_1^{comp}(d_1, \dots, d_m)$ of the stratum has finite volume and hence the space of zippered rectangles Ω which is a finite covering

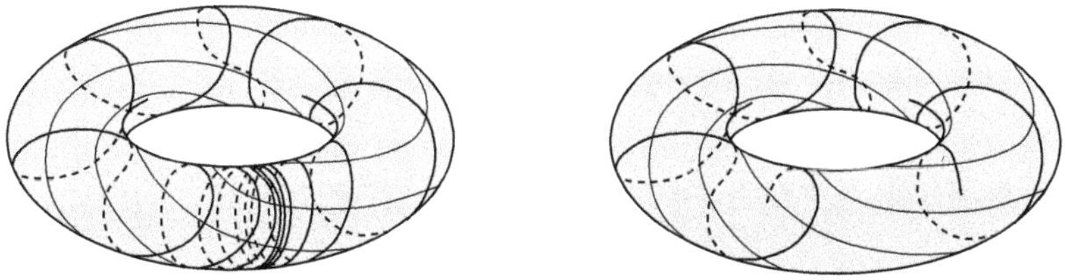

Fig. 27. The section Y on the left picture has infinite measure though the measure of the space is finite. The subsection $Y' \subset Y$ on the right picture has finite measure. In both cases the first return map of the ergodic flow to the section is ergodic, but the mean return time to the left subsection is zero

over $\mathcal{H}_1^{comp}(d_1, \ldots, d_m)$ also has finite volume. However, the initial section Υ has infinite "hyperarea" while section Υ' already has finite "hyperarea".

We complete this section with a several comments concerning Lyapunov exponents. Though these comments are too brief to give a comprehensive proof of the relation between exponents ν_j responsible for the deviation and the Lyapunov exponents of the Teichmüller geodesic flow, they present the key idea, which can be completed by an elementary calculation.

It is clear that the Lyapunov exponents of the Teichmüller geodesic flow g_t on the stratum $\mathcal{H}_1^{comp}(d_1, \ldots, d_m)$ coincide with the Lyapunov exponents on its finite covering Ω. The Lyapunov exponents of the flow g_t differ from the Lyapunov exponents of the first return map $\mathcal{S}'$ to the section Υ' only by a scaling factor representing the "hyperarea" of Υ'. The map $\mathcal{G}$ on the space of zippered rectangles coincides with the restriction of the map $\mathcal{S}'$ on the zippered rectangles restricted to horizontal parameters λ and discrete parameter π. Thus, the Lyapunov exponents of $\mathcal{G}$ form a subcollection of the Lyapunov exponents of $\mathcal{S}'$ corresponding to the subspace of horizontal parameters λ. It remains to note that the Lyapunov exponents of the map $\mathcal{G}$ are related to the Lyapunov exponents of the cocycle $B(\lambda, \pi)$ just by the scaling factor, and that we have already expressed the exponents ν_j responsible for the deviation in terms of the Lyapunov exponents of the multiplicative cocycle B, see (5.9). Matching all the elements of this chain together we get a representation of the exponents ν_j in terms of the Lyapunov exponents of the Teichmüller geodesic flow given in Sect. 4.3.

Zippered rectangles and Lyapunov exponents: more serious reading. More details on Rauzy classes $\mathfrak{R}$, zippered rectangles, Lyapunov exponents of the Teichmüller geodesic flow and their relation might be found in original papers of G. Rauzy [Ra], W. Veech [Ve3], [Ve6] and the author [Zo2], [Zo4].

5.8 Spectrum of Lyapunov Exponents (after M. Kontsevich, G. Forni, A. Avila and M. Viana)

It should be mentioned that the statement that the subspace $\mathcal{V}_g$, such that $|dist(c(x,N), \mathcal{V}_g)| \leq const$ for any N, has dimension *exactly* g was formulated in the original paper [Zo4] as a conditional statement. It was based on the conjecture that the Lyapunov exponent ν_g is strictly positive. This conjecture was later proved by G. Forni in [For1].

Theorem (G. Forni). *For any connected component $\mathcal{H}^{comp}(d_1,\dots,d_m)$ of any stratum of Abelian differentials the first g Lyapunov exponents of the Teichmüller geodesic flow are strictly greater than 1:*

$$1+\nu_g > 1$$

As an indication why this positivity is not something which should be taken for granted we would like to give some precision about related results of G. Forni.

The Lyapunov exponents of the Teichmüller geodesic flow play the role of (logarithms of) eigenvalues of a virtual "average monodromy of the tangent bundle along the flow". Instead of considering the tangent bundle to $\mathcal{H}(d_1,\dots,d_m)$ one can consider another vector bundle intimately related to the tangent bundle. This vector bundle has the space $H^1(S;\mathbb{R})$ as a fiber. Since we know how to identify the lattices $H^1(S;\mathbb{Z}) \subset H^1(S;\mathbb{R})$ and $H^1(S';\mathbb{Z}) \subset H^1(S';\mathbb{R})$ in the fibers over two flat surfaces S and S' which are close to each other in $\mathcal{H}(d_1,\dots,d_m)$, we know how to transport the fiber $H^1(S;\mathbb{R})$ over the "point" S in the base $\mathcal{H}(d_1,\dots,d_m)$ to the fiber $H^1(S';\mathbb{R})$ over the "point" S'. In other words, we have a canonical connection (called Gauss–Manin connection) in the vector bundle. Hence we can again study the "average monodromy of the fiber along the flow". It is not difficult to show that the corresponding Lyapunov exponents are related to the Lyapunov exponents of the Teichmüller flow. Namely, the new collection of Lyapunov exponents has the form:

$$1 \geq \nu_2 \geq \dots \geq \nu_g \geq -\nu_g \geq \dots \geq -\nu_2 \geq -1$$

In particular, the collection of Lyapunov exponents of the Teichmüller geodesic flow can be obtained as follows: take two copies of the collection above; add $+1$ to all the entries in one copy; add -1 to all entries in another copy; take the union of the resulting collections. The theorem of G. Forni tells that for any connected component of any stratum we have $\nu_g > 0$.

Consider now some $SL(2,\mathbb{R})$-invariant subvariety $\mathcal{N} \subset \mathcal{H}(d_1,\dots,d_m)$. Consider the restriction of the vector bundle with the fiber $H^1(S;\mathbb{R})$ to $\mathcal{N}$. We can compute the "average monodromy of the fiber along the Teichmüller flow" restricted to $\mathcal{N}$. It gives a new collection of Lyapunov exponents. Since the "holonomy" preserves the natural symplectic form in the fiber, the collection will be again symmetric:

$$1 \geq \nu'_2 \geq \dots \geq \nu'_g \geq -\nu'_g \geq \dots \geq -\nu'_2 \geq -1$$

G. Forni has showed [For2] that there are examples of invariant subvarieties $\mathcal{N}$ such that all ν'_j, $j=2,\dots,g$, are equal to zero! Moreover, G. Forni explicitly describes the locus where the monodromy does not change the fiber (or it exterior powers) too much, and where one may get multiplicities of Lyapunov exponents.

Another conditional statement in the original paper [Zo4] concerns *strict* inclusions $\mathcal{V}_1 \subset \mathcal{V}_2 \subset \cdots \subset \mathcal{V}_g \subset H_1(S;\mathbb{R})$. It was based on the other conjecture claiming that the Lyapunov exponents have simple spectrum. The first strict inequality $\nu_1 > \nu_2$ is an elementary corollary of general results of Veech; see [Zo3]. The other strict inequalities are much more difficult to prove. Very recently A. Avila and M. Viana [AvVi] have announced a proof of simplicity of the spectrum (5.12) for any connected component of any stratum proving the conjecture which was open for a decade.

Theorem (A. Avila, M. Viana). *For any connected component of any stratum $\mathcal{H}^{comp}(d_1, \dots, d_m)$ of Abelian differentials the first g Lyapunov exponents are distinct:*

$$1 + \nu_1 > 1 + \nu_2 > \cdots > 1 + \nu_g \tag{5.12}$$

Sum of the Lyapunov exponents

Currently there are no methods of calculation of Lyapunov exponents for general dynamical systems. The Teichmüller geodesic flow does not make an exception: there is some knowledge of approximate values of the numbers ν_j obtained by computer simulations for numerous low-dimensional strata, but there is no approach leading to explicit evaluation of these numbers with exception for some very special cases.

Nevertheless, for any connected component of any stratum (and, more generally, for any $GL^+(2;\mathbb{R})$-invariant suborbifold) it is possible to evaluate the *sum* of the Lyapunov exponents $\nu_1 + \dots \nu_g$, where g is the genus. The formula for this sum was discovered by M. Kontsevich in [Kon]; it is given in terms of the following natural structures on the strata $\mathcal{H}(d_1, \dots, d_m)$.

There is a natural action of $\mathbb{C}^*$ on every stratum of the moduli space of holomorphic 1-forms: we can multiply a holomorphic form ω by a complex number. Let us denote by $\mathcal{H}_{(2)}(d_1, \dots, d_m)$ the quotient of $\mathcal{H}(d_1, \dots, d_m)$ over $\mathbb{C}^*$. The space $\mathcal{H}_{(2)}(d_1, \dots, d_m)$ can be viewed as the space of flat surfaces of unit area *without* choice of distinguished direction.

There are two natural holomorphic vector bundles over $\mathcal{H}_{(2)}(d_1, \dots, d_m)$. The first one is the $\mathbb{C}^*$-bundle $\mathcal{H}(d_1, \dots, d_m) \to \mathcal{H}_{(2)}(d_1, \dots, d_m)$. The second one is the $\mathbb{C}^g$-bundle, which fiber is composed of all holomorphic 1-forms in the complex structure corresponding to a flat surface $S \in \mathcal{H}_{(2)}(d_1, \dots, d_m)$. Both bundles have natural curvatures; we denote by γ_1 and γ_2 the corresponding closed curvature 2-forms.

Finally, there is a natural closed codimension two form β on every stratum $\mathcal{H}_{(2)}(d_1, \dots, d_m)$. To construct β consider the natural volume form Ω on $\mathcal{H}(d_1, \dots, d_m)$. Four generators of the Lie algebra $\mathfrak{gl}(2;\mathbb{R})$ define four distinguished vectors in the tangent space $T_S\mathcal{H}(d_1, \dots, d_m)$ at any "point" $S \in \mathcal{H}(d_1, \dots, d_m)$. Plugging these four vectors in the first four arguments of the volume form Ω we get a closed codimension four form on $\mathcal{H}(d_1, \dots, d_m)$. It is easy to check that this form can be pushed forward along the $\mathbb{C}^*$-fibers of the bundle $\mathcal{H}(d_1, \dots, d_m) \to \mathcal{H}_{(2)}(d_1, \dots, d_m)$ resulting in the closed codimension two form on the base of this fiber bundle.

Theorem (M. Kontsevich). *For any connected component of any stratum the sum of the first g Lyapunov exponents can be expressed as*

$$\nu_1 + \dots + \nu_g = \frac{\int \beta \wedge \gamma_2}{\int \beta \wedge \gamma_1},$$

where the integration is performed over the corresponding connected component of $\mathcal{H}_{(2)}(d_1, \dots, d_m)$.

As it was shown by G. Forni, this formula can be generalized for other $GL^+(2;\mathbb{R})$-invariant submanifolds.

The proof is based on two observations. The first one generalizes the fact that dynamics of the geodesic flow on the hyperbolic plane is in some sense equivalent to dynamics of random walk. One can replace Teichmüller geodesics by geodesic broken lines consisting of geodesic segments of unit length. Having a broken line containing n geodesic segments with the endpoint at the point S_n we emit from S_n a new geodesic in a random direction and stop at the distance one from S_n at the new point S_{n+1}. This generalization suggested by M. Kontsevich was formalized and justified by G. Forni.

Consider the vector bundle over the moduli space of holomorphic 1-forms with the fiber $H^1(S;\mathbb{R})$ over the "point" S. We are interested in the sum $\nu_1 + \dots + \nu_g$ of Lyapunov exponents representing mean monodromy of this vector bundle along random walk. It follows from standard arguments concerning Lyapunov exponents that this sum corresponds to the top Lyapunov exponent of the exterior power of order g of the initial vector bundle. In other words, we want to measure the average growth rate of the norm of a g-dimensional subspace in $H^1(S;\mathbb{R})$ when we transport it along trajectories of the random walk using the Gauss–Manin connection.

Fix a Lagrangian subspace L in the fiber $H^1(S_0;\mathbb{R})$ over a "point" S_n. Consider the set of points located at the Teichmüller distance 1 from S_n. Transport L to each point S_{n+1} of this "unit sphere" along the corresponding geodesic segment joining S_n with S_{n+1}; measure the logarithm of the change of the norm of L; take the average over the "unit sphere". The key observation of M. Kontsevich in [Kon] is that for an appropriate choice of the norm this average growth rate is the same for all Lagrangian subspaces L in $H^1(S_0;\mathbb{R})$ and depends only on the point S_n. A calculation based on this observation gives the formula above.

Actually, formula above can be rewritten in a much more explicit form (which is a work in progress). The values of the sum given by this more explicit formula perfectly match numerical simulations. The table below gives the values of the sums of Lyapunov exponents for some low-dimensional strata;

this computation uses the results of A. Eskin and A. Okounkov [EOk] for the volumes of the strata.

Conjectural values of $\nu_1 + \cdots + \nu_g$ for some strata

$\mathcal{H}(2)$	$\mathcal{H}(1,1)$	...	$\mathcal{H}(4,1,1)$	...	$\mathcal{H}(1,1,1,1,1,1)$	$\mathcal{H}(1,1,1,1,1,1,1,1)$
$\frac{4}{3}$	$\frac{3}{2}$	...	$\frac{1137}{550}$	...	$\frac{839}{377}$	$\frac{235\,761}{93\,428}$

In particular, since $\nu_1 = 1$ this information gives the exact value of the only nontrivial Lyapunov exponent ν_2 for the strata in genus two. Some extra arguments show that $\nu_2 = 1/3$ for the stratum $\mathcal{H}(2)$ and for any $GL^+(2;R)$-invariant submanifold in it; $\nu_2 = 1/2$ for the stratum $\mathcal{H}(1,1)$ and for any $GL^+(2;R)$-invariant submanifold in it.

5.9 Encoding a Continued Fraction by a Cutting Sequence of a Geodesic

We have seen that renormalization for a rotation of a circle (or equivalently for an interval exchange transformation of two subintervals) leads to the Euclidean algorithm which can be considered in this guise as a particular case of the fast Rauzy–Veech induction.

The multiplicative cocycle

$$B^{(s)} = \begin{pmatrix} 1 & n_1 \\ 0 & 1 \end{pmatrix} \cdot \begin{pmatrix} 1 & 0 \\ n_2 & 1 \end{pmatrix} \cdots \begin{pmatrix} 1 & n_{2k-1} \\ 0 & 1 \end{pmatrix} \cdot \begin{pmatrix} 1 & 0 \\ n_{2k} & 1 \end{pmatrix} \cdots$$

considered in section 5.6 corresponds to the decomposition of a real number $x \in (0,1)$ into continued fraction, $x = [0; n_1, n_2, \ldots]$,

$$x = \cfrac{1}{n_1 + \cfrac{1}{n_2 + \cfrac{1}{\cdots}}}$$

A flat surface which realizes an interval exchange transformation of two subintervals is a flat torus. The the moduli space of flat tori can be naturally identified with $SL(2,\mathbb{R})/SL(2,\mathbb{Z})$ which in its turn can be naturally identified with the unit tangent bundle to the modular surface $\mathbb{H}^2/SL(2,\mathbb{Z})$ see Sect. 3.2. Moreover, the Teichmüller metric on the space of tori coincides with the hyperbolic metric on $\mathbb{H}^2$, and the Teichmüller geodesic flow on the moduli space of flat tori coincides with the geodesic flow on the modular surface.

Hence, the construction from the previous section suggests that the Euclidean algorithm corresponds to the following geometric procedure. There should be a section Υ in the (covering of) the unit tangent bundle to the modular surface and its subsection $\Upsilon' \subset \Upsilon$ such that the trajectory of the geodesic flow emitted from a point of Υ' returns to Υ' after n_1 intersections with Υ, then after n_2 intersections with Υ, etc. In other words there is a natural way to code a continued fraction by a sequence of intersections (so called "cutting sequence") of the corresponding geodesic with some sections $\Upsilon' \subset \Upsilon$.

Actually, a geometric coding of a continued fraction by a cutting sequence of a geodesics on a surface is known since the works of J. Nielsen and E. Artin in 20s and 30s. The study of the geometric coding was developed in the 80s and 90s by C. Series, R. Adler, L. Flatto and other authors. We refer to the expository paper [Ser] of C. Series for detailed description of the following geometric coding algorithm.

Consider a tiling of the upper half plane with isometric hyperbolic triangles as at Fig. 29. A fundamental domain of the tiling is a triangle with vertices at $0, 1$ and ∞; the corresponding quotient surface is a triple cover over the standard modular surface (see Fig. 47). This triangulation of $\mathbb{H}^2$ by ideal triangles is also known as *Farey tessellation.*

Consider a real number $x \in (0,1)$. Consider any geodesic γ landing to the real axis at x such that γ intersects with the imaginary axis; let iy be the point of intersection. Let us follow the geodesic γ starting from iy in direction of x. Each time when we cross a triangle of our tiling let us note by the symbol L the situation when we have a single vertex on the left and two vertices on the right (see Fig. 28) and by the symbol R the symmetric situation.

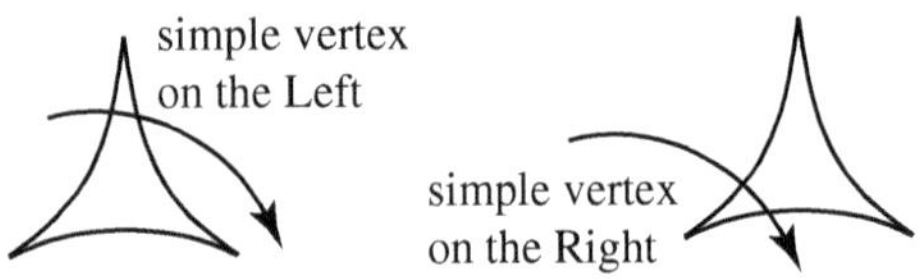

Fig. 28. Coding rule: when we cross a triangle leaving one vertex on the left and two on the right we write symbol L; when there is one vertex on the right and two on the left we write symbol R

Example. Following the geodesic γ presented at Fig. 29 from some iy to $x = (\sqrt{85} - 5)/10) \approx 0.421954$ we get a sequence $R, R, L, L, R, L, L, R, R, L, \dots$ which we abbreviate as $R^2L^2R^1L^2R^2L^1 \dots$.

Theorem (C. Series). *Let $x \in (0,1)$ be irrational. Let γ be a geodesic emitted from some iy and landing at x; let $R^{n_1}L^{n_2}R^{n_3}L^{n_4} \dots$ be the corresponding cutting sequence. Then $x = [0; n_1, n_2, n_3, n_4, \dots]$.*

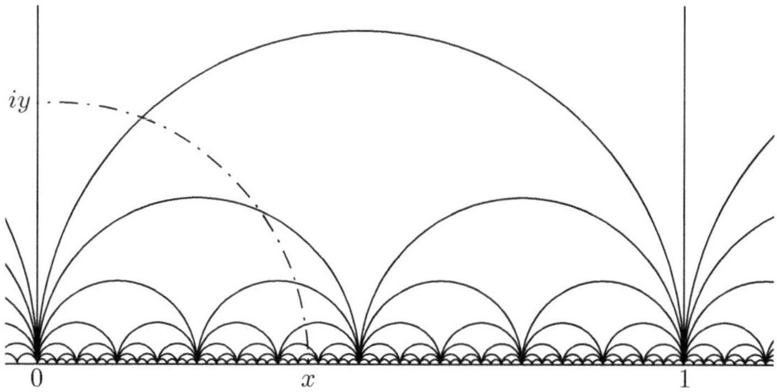

Fig. 29. (After C. Series.) The cutting sequence defined by this geodesic starts with $R, R, L, L, R, L, L, R, R, L, \ldots$ which we abbreviate as $R^2L^2R^1L^2R^2L^1 \ldots$. The real number $x \in (0,1)$ at which lands the geodesic has continued fraction expansion $x = [0; 2, 2, 1, 2, 2, 1, \ldots]$

Geometric symbolic coding: more serious reading. I can strongly recommend a paper of P. Arnoux [Arn] which clearly and rigorously explains the idea of suspension in the spirit of diagram (5.11) handling the particular case of Euclidean algorithm and of geodesic flow on the Poincaré upper half-plane. As a survey on geometric coding we can recommend the survey of C. Series [Ser] (as well as other surveys in this collection on related subjects).

6 Closed Geodesics and Saddle Connections on Flat Surfaces

Emitting a geodesic in an irrational direction on a flat torus we get an irrational winding line; emitting it in a rational direction we get a closed geodesic. Similarly, for a flat surface of higher genus a countable dense set of directions corresponds to closed geodesics.

In this section we study how many closed regular geodesics of bounded length live on a generic flat surface S. We consider also *saddle connections* (i.e. geodesic segments joining pairs of conical singularities) and count them.

We explain a curious phenomenon concerning saddle connections and closed geodesics on flat surfaces: they often appear in pairs, triples, etc of parallel saddle connections (correspondingly closed geodesics) of equal length.

When all saddle connections (closed geodesics) in such *configuration* become short the corresponding flat surface starts to degenerate and gets close to the boundary of the moduli space. Thus, a description of possible configurations of parallel saddle connections (closed geodesics) gives us a description of the multidimensional "cusps" of the strata.

6.1 Counting Closed Geodesics and Saddle Connections

Closed Regular Geodesics Versus Irrational Winding Lines

Consider a flat torus obtained by identifying pairs of opposite sides of a unit square. A geodesic emitted in an irrational direction (one with irrational slope) is an irrational winding line; it is dense in the torus. A geodesic emitted in a rational direction is closed; all parallel geodesic are also closed, so directional flow in a rational direction fills the torus with parallel periodic trajectories. The set of rational directions has measure zero in the set of all possible directions. In this sense directions representing irrational winding lines are typical and directions representing closed geodesic – nontypical.

The situation with flat surfaces of higher genera $g \geq 2$ is similar in many aspects, though more complicated in details. For example , for any flat surfaces S almost all directions are "irrational"; any geodesic emitted in an irrational direction is dense in the surface. Actually, even stronger statement is true:

Theorem (S. Kerckhoff, H. Masur, J. Smillie). *For any flat surface directional flow in almost any direction is uniquely ergodic.*

For the torus the condition that directional flow is *minimal* (that is any trajectory going in this direction is dense in the torus) is equivalent to the condition that the flow is *uniquely ergodic* (the natural Lebesgue measure induced by the flat structure is the only finite measure invariant under directional flow; see Appendix A for details). Surprisingly a directional flow on a surface of higher genus (already for $g = 2$) might be *minimal* but not *uniquely ergodic*! Namely, for some directions which give rise to a minimal directional flow it might be possible to divide the surface into two parts (of nonzero measure) in such way that some trajectories would mostly stay in one part while other trajectories would mostly stay in the other.

Closed geodesics on flat surfaces of higher genera also have some similarities with ones on the torus. Suppose that we have a regular closed geodesic passing through a point $x_0 \in S$. Emitting a geodesic from a nearby point x in the same direction we obtain a parallel closed geodesic of the same length as the initial one. Thus, closed geodesics also appear in families of parallel closed geodesics. However, in the torus case every such family fills the entire torus while a family of parallel regular closed geodesics on a flat surfaces of higher genus fills only part of the surface. Namely, it fills a flat cylinder having a conical singularity on each of its boundaries.

Exercise. Find several periodic directions on the flat surface from Fig. 12. Find corresponding families of parallel closed geodesics. Verify that each of the surfaces from Fig. 44 decomposes under the vertical flow into three cylinders (of different circumference) filled with periodic trajectories. Find these cylinders.

Counting Problem

Take an arbitrary loop on a torus. Imagine that it is made from a stretched elastic cord. Letting it contract we get a closed regular geodesic (may be winding several times along itself). Now repeat the experiment with a more complicated flat surface. If the initial loop was very simple (or if we are extremely lucky) we again obtain a regular closed geodesic. However, in general we obtain a closed broken line of geodesic segments with vertices at a collection of conical points.

Similarly letting contract an elastic cord joining a pair of conical singularities we usually obtain a broken line composed from several geodesic segments joining conical singularities. In this sense torus is very different from a general flat surface.

A geodesic segment joining two conical singularities and having no conical points in its interior is called *saddle connection*. The case when boundaries of a saddle connection coincide is not excluded: a saddle connection might join a conical point to itself.

Convention 4. In this paper we consider only saddle connections and closed regular geodesics. We never consider broken lines formed by several geodesic segments.

Now we are ready to formulate the Counting Problem. Everywhere in this section we normalize the area of flat surfaces to one.

Counting Problem. *Fix a flat surface S. Let $N_{sc}(S,L)$ be the number of saddle connections on S of length at most L. Let $N_{cg}(S,L)$ be the number of maximal cylinders filled with closed regular geodesics of length at most L on S. Find asymptotics of $N_{sc}(S,L)$ and $N_{cg}(S,L)$ as $L \to \infty$.*

It was proved by H. Masur (see [Ma5] and [Ma6]) that for any flat surface S counting functions $N(S,L)$ grow quadratically in L. Namely, there exist constants $0 < const_1(S) < const_2(S) < \infty$ such that

$$const_1(S) \le N(S,L)/L^2 \le const_2(S)$$

for L sufficiently large. Recently Ya. Vorobets has obtained in [Vb2] uniform estimates for the constants $const_1(S)$ and $const_2(S)$ which depend only on the genus of S.

Passing from *all* flat surfaces to *almost all* surfaces in a given connected component of a given stratum one gets a much more precise result; see [EMa].

Theorem (A. Eskin and H. Masur). *For almost all flat surfaces S in a given connected component of a stratum $\mathcal{H}(d_1,\dots,d_m)$ the counting functions $N_{sc}(S,L)$ and $N_{cg}(S,L)$ have exact quadratic asymptotics*

$$\lim_{L\to\infty} \frac{N_{sc}(S,L)}{\pi L^2} = const_{sc} \qquad \lim_{L\to\infty} \frac{N_{cg}(S,L)}{\pi L^2} = const_{cg} \tag{6.1}$$

where Siegel–Veech constants $const_{sc}$ *and* $const_{cg}$ *are the same for almost all flat surfaces in the component* $\mathcal{H}_1^{comp}(d_1, \ldots, d_m)$.

We multiply denominator by π to follow a conventional normalization.

Phenomenon of Higher Multiplicities

Let us discuss now the following problem. Suppose that we have a regular closed geodesic on a flat surface S. Memorize its direction, say, let it be the North-West direction. (Recall that by Convention 1 in Sect. 1.2 we can place a compass at any point of the surface and it will tell us what is the direction to the North.) Consider the maximal cylinder filled with closed regular geodesics parallel to ours. Take a point x outside this cylinder and emit a geodesic from x in the North-West direction. There are two questions.
- *How big is the chance to get a closed geodesic?*
- *How big is the chance to get a closed geodesic of the same length as the initial one?*

Intuitively it is clear that the answer to the first question is: "the chances are low" and to the second one "the chances are even lower". This makes the following Theorem (see [EMaZo]) somehow counterintuitive:

Theorem (A. Eskin, H. Masur, A. Zorich). *For almost all flat surfaces* S *from any stratum different from* $\mathcal{H}_1(2g-2)$ *or* $\mathcal{H}_1(d_1, d_2)$ *the function* $N_{two\_cyl}(S, L)$ *counting the number of families of parallel regular closed geodesics filling two distinct maximal cylinders has exact quadratic asymptotics*

$$\lim_{L\to\infty} \frac{N_{two\_cyl}(S, L)}{\pi L^2} = const_{two\_cyl}$$

where Siegel–Veech constants $const_{two\_cyl} > 0$ *depends only on the connected component of the stratum.*

For almost all flat surface S *in any stratum one cannot find a single pair of parallel regular closed geodesics on* S *of different length.*

There is general formula for the Siegel–Veech constant $const_{two\_cyl}$ and for similar constants which gives explicit numerical answers for all strata in low genera. Recall that the *principal stratum* $\mathcal{H}(1, \ldots, 1)$ is the only stratum of fill dimension in $\mathcal{H}_g$; it is the stratum of holomorphic 1-forms with simple zeros (or, what is the same, of flat surfaces with conical angles 4π at all cone points). Numerical values of the Siegel–Veech constants for the principal stratum are presented in Table 2.

Comparing these values we see, that our intuition was not quite misleading. Morally, in genus $g = 4$ a closed regular geodesic belongs to a one-cylinder family with "probability" 97.1%, to a two-cylinder family with "probability" 2.8% and to a three-cylinder family with "probability" only 0.1% (where "probabilities" are calculated proportionally to Siegel–Veech constants).

Table 2. Siegel–Veech constants $const_{n\_cyl}$ for the principal stratum $\mathcal{H}_1(1,\dots,1)$

	$g=1$	$g=2$	$g=3$	$g=4$
single cylinder	$\frac{1}{2}\cdot\frac{1}{\zeta(2)}\approx 0.304$	$\frac{5}{2}\cdot\frac{1}{\zeta(2)}\approx 1.52$	$\frac{36}{7}\cdot\frac{1}{\zeta(2)}\approx 3.13$	$\frac{3150}{377}\cdot\frac{1}{\zeta(2)}\approx 5.08$
two cylinders	–	–	$\frac{3}{14}\cdot\frac{1}{\zeta(2)}\approx 0.13$	$\frac{90}{377}\cdot\frac{1}{\zeta(2)}\approx 0.145$
three cylinders	–	–	–	$\frac{5}{754}\cdot\frac{1}{\zeta(2)}\approx 0.00403$

In theorem above we discussed closed regular geodesics. A similar phenomenon is true for saddle connections. Recall that the cone angle at a conical point on a flat surface is an integer multiple of 2π. Thus, at a point with a cone angle $2\pi n$ every direction is presented n times. Suppose that we have found a saddle connection of length going from conical point P_1 to conical point P_2. Memorize its direction (say, the North-West direction) and its length l. Then with a "nonzero probability" (understood in the same sense as above) emitting a geodesic from P_1 in one of the remaining $n-1$ North-West directions we make it hit P_2 at the distance l. More rigorously, the Siegel–Veech constant counting configurations of two parallel saddle connections of equal length joining P_1 to P_2 is nonzero.

The explicit formula for any Siegel–Veech constant from [EMaZo] can be morally described as the follows. Up to some combinatorial factor responsible for dimensions, multiplicities of zeroes and possible symmetries any Siegel–Veech constant can be obtained as a limit

$$c(\mathcal{C}) = \lim_{\varepsilon\to 0}\frac{1}{\pi\varepsilon^2}\,\frac{\operatorname{Vol}(\text{“}\varepsilon\text{-neighborhood of the cusp } \mathcal{C}\text{”})}{\operatorname{Vol}\mathcal{H}_1^{comp}(d_1,\dots,d_m)} \tag{6.2}$$

where $\mathcal{C}$ is a particular *configuration* of saddle connections or closed geodesics.

Say, as a configuration $\mathcal{C}$ one can consider a configuration of two maximal cylinders filled with parallel closed regular geodesics of equal lengths. The ε-neighborhood of the corresponding cusp is the subset of those flat surfaces $S\in\mathcal{H}_1^{comp}(d_1,\dots,d_m)$ which have at least one pair of cylinders filled with parallel closed geodesics of length shorter than ε.

As another example one can consider a configuration of three parallel saddle connections of equal lengths on $S\in\mathcal{H}_1(1,1,4,8)$ joining zero P_1 of degree 4 (having cone angle 10π) to zero P_2 of degree 8 (having cone angle 18π) separated by angles $2\pi, 2\pi, 6\pi$ at P_1 and by angles $6\pi, 10\pi, 2\pi$ at P_2. The ε-neighborhood $\mathcal{H}_1^{\varepsilon}(1,1,4,8)\subset\mathcal{H}_1(1,1,4,8)$ of the corresponding cusp is the subset of those flat surfaces in $\mathcal{H}_1(1,1,4,8)$ which have at least one triple of saddle connections as described above of length shorter than ε.

We explain the origin of the key formula (6.2) in the next section. In section 6.4 we give an explanation of appearance of higher multiplicities.

Other counting problems (after Ya. Vorobets)

Having a flat surface S of unit area we have studied above the number of maximal cylinders $N_{cg}(S,L)$ filled with closed regular geodesics of length at most L on S. (In this setting when we get in some direction several parallel maximal cylinders of equal perimeter, we count each of them.) In the paper [Vb2] Ya. Vorobets considered other counting problems.

In particular, among all maximal periodic cylinders of length at most L (as above) he counted the number $N_{cg,\sigma}(S,L)$ of those ones, which have area greater than σ. He also counted the total sum $N_{area}(S,L)$ of areas of all maximal cylinders of perimeter at most L and the number $N_x(S,L)$ of regular periodic geodesics of length at most L passing trough a given point $x \in S$.

Ya. Vorobets has also studied how the maximal cylinders filled with closed geodesics are distributed with respect to their direction and their area. He considered the induced families of probability measures on the circle S^1, on the unit interval $[0,1]$ and on their product $S^1 \times [0;1]$. Given a subset $U \subset S_1$, $V \subset [0;1]$, $W \subset S^1 \times [0;1]$ the corresponding measures $dir_L(U), ar_L(V), pair_L(W)$ tell the proportion of cylinders of bounded perimeter having direction in U, area in V, or the pair (*direction, area*) in W correspondingly.

Using the general approach of A. Eskin and H. Masur, Ya. Vorobets has proved in [Vb2] existence of exact quadratic asymptotics for the counting functions introduced above. He has computed the corresponding Siegel–Veech constants in terms of the Siegel–Veech constant $const_{cg}$ in (6.1) and found the asymptotic distributions of directions and areas of the cylinders:

Theorem (Ya. Vorobets). *For almost any flat surface S of unit area in any connected component of any stratum $\mathcal{H}_1^{comp}(d_1,\dots,d_m)$ and for almost any point x of S one has*

$$\lim_{L\to\infty}\frac{N_{cg,\sigma}(S,L)}{\pi L^2}=c_{cg,\sigma}\qquad \lim_{L\to\infty}\frac{N_{area}(S,L)}{\pi L^2}=c_{area}\qquad \lim_{L\to\infty}\frac{N_x(S,L)}{\pi L^2}=c_x,$$

where $c_{cg,\sigma}=(1-\sigma)^{2g-3+m}\cdot const_{cg}$ *and* $c_{area}=c_x=\dfrac{const_{cg}}{2g-2+m}$.

For almost any flat surface S of unit area one has the following week convergence of measures:

$$dir_L\to\varphi\qquad ar_L\to\rho\qquad pair_L\to\varphi\times\rho,$$

where φ is the uniform probability measure on the circle and ρ is the probability measure on the unit interval $[0;1]$ with the density $(2g-3+m)(1-x)^{2g-4+m}dx$.

Directional flow: more serious reading. Theorem of S. Kerckhoff, H. Masur and J. Smillie is proved in [KMaS]. An example of minimal but not uniquely ergodic interval exchange transformations is constructed by W. Veech in [Ve1] (using different terminology); an independent example (also using different terminology) was constructed at the same time by V. I. Oseledets. For flows such examples are constructed in the paper of A. Katok [Kat1] and developed by E. Sataev in [Sat]. Another example was discovered by M. Keane [Kea2]. For alternative approach to the study of unique ergodicity of interval exchange transformations see the paper of M. Boshernitzan [Ber2]. A very nice construction of minimal but not uniquely ergodic interval exchange transformations (in a language which is very close to the language of this paper) can be found in the survey of H. Masur and S. Tabachnikov [MaT] or in the survey of H. Masur [Ma7].

6.2 Siegel–Veech Formula

We start from a slight formalization of our counting problem. As usual we start with a model case of the flat torus. As usual we assume that our flat torus is glued from a unit square. We count closed regular geodesics on $\mathbb{T}^2$ of a bounded length. To mimic count of saddle connections we mark two points $P_1 \neq P_2$ on $\mathbb{T}^2$ and count geodesic segments of bounded length joining P_1 and P_2.

Our formalization consists in the following construction. Consider an auxiliary Euclidian plane $\mathbb{R}^2$. Having found a regular closed geodesic on $\mathbb{T}^2$ we note its direction α and length l and draw a vector in $\mathbb{R}^2$ in direction α having length l. We apply a similar construction to "saddle connections". The endpoints of corresponding vectors form two discrete subsets in $\mathbb{R}^2$ which we denote by V_{cg} and V_{sc}.

It is easy to see that for the torus case a generic choice of P_1 and P_2 generates a set V_{sc} which is just a shifted square lattice, see Fig. 30. The set V_{cg} is a subset of *primitive* elements of the square lattice, see Fig. 30. Since we count only regular closed geodesics which do not turn many times around themselves we cannot obtain elements of the form (kn_1, kn_2) with $k, n_1, n_2 \in \mathbb{Z}$.

The corresponding counting functions $N_{sc}(\mathbb{T}^2, L)$ and $N_{sc}(\mathbb{T}^2, L)$ correspond to the number of element of V_{sc} and V_{cg} correspondingly which get to a disc of radius L centered in the origin. Both functions have exact quadratic asymptotics. Denoting by $\chi_L(v)$, where $v \in \mathbb{R}^2$ the indicator function of such disc we get

$$N_{sc}(\mathbb{T}^2, L) = \textstyle\sum_{v\in V_{sc}} \chi_L(v) \sim 1 \cdot \pi L^2 \tag{6.3}$$

$$N_{cg}(\mathbb{T}^2, L) = \textstyle\sum_{v\in V_{cg}} \chi_L(v) \sim \frac{1}{\zeta(2)} \cdot \pi L^2$$

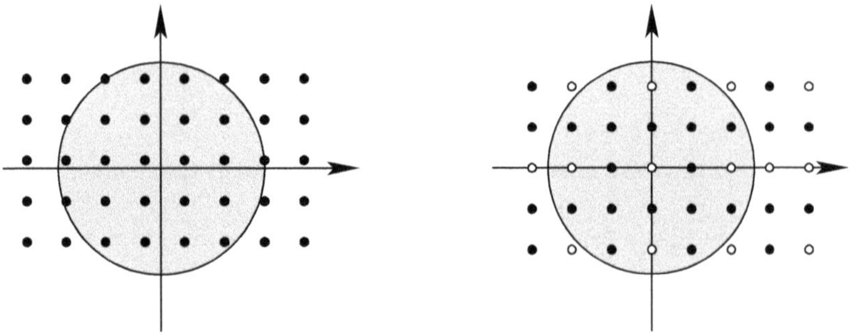

Fig. 30. Sets V_{sc} and V_{cg} for the flat torus

The coefficients in quadratic asymptotics define the corresponding Siegel–Veech constants $const_{sc} = 1$ and $const_{cg} = \dfrac{1}{\zeta(2)} = \dfrac{6}{\pi^2}$. (Note that here we count every geodesic twice: once with one orientation and the other one with the opposite orientation. This explains why in this normalization we obtain the value of $const_{cg}$ twice as much as $const_{cg}$ for genus one in Table 2.)

Consider now a more general flat surface S. Fix the geometric type of *configuration* $\mathcal{C}$ of saddle connections or closed geodesics. By definition all saddle connections (closed geodesics) in $\mathcal{C}$ are parallel and have equal length. Thus, similar to the torus case, every time we see a collection of saddle connections (closed geodesics) of geometric type $\mathcal{C}$ we can associate to such collection a vector in $\mathbb{R}^2$. We again obtain a discrete set $V(S) \subset \mathbb{R}^2$.

Now fix $\mathcal{C}$ and apply this construction to every flat surface S in the stratum $\mathcal{H}_1(d_1, \dots, d_m)$. Consider the following operator $f \mapsto \hat{f}$ generalizing (6.3) from functions with compact support on $\mathbb{R}^2$ to functions on $\mathcal{H}_1(d_1, \dots, d_m)$:

$$\hat{f}(S) = \sum_{v \in V} f(v)$$

Lemma (W. Veech). *The functional*

$$f \mapsto \int_{\mathcal{H}_1^{comp}(d_1,\dots,d_m)} \hat{f}(S) d\nu_1$$

is $SL(2,\mathbb{R})$*-invariant.*

Having proved convergence of the integral above the Lemma follows immediately from invariance of the measure $d\nu_1$ under the action of $SL(2,\mathbb{R})$ and from the fact that $V(gS) = gV(s)$ for any flat surface S and any $g \in SL(2,\mathbb{R})$.

Now note that there very few $SL(2,\mathbb{R})$-invariant functionals on functions with compact support in $\mathbb{R}^2$. Actually, there are two such functionals, and the other ones are linear combinations of these two. These two functionals are the value of $f(0)$ at the origin and the integral $\int_{\mathbb{R}^2} f(x,y)dx\,dy$. It is possible to see that the value $f(0)$ at the origin is irrelevant for the functional from the Lemma above. Hence it is proportional to $\int_{\mathbb{R}^2} f(x,y)dx\,dy$.

Theorem (W. Veech). *For any function $f : \mathbb{R}^2 \to \mathbb{R}$ with compact support one has*

$$\frac{1}{\operatorname{Vol} \mathcal{H}_1^{comp}(d_1, \ldots, d_m)} \int_{\mathcal{H}_1^{comp}(d_1,\ldots,d_m)} \hat{f}(S) d\nu_1 = C \int_{\mathbb{R}^2} f(x,y) dx\, dy \quad (6.4)$$

Here the constant C in (6.4) *does not depend* on f; it depends only on the connected component $\mathcal{H}_1^{comp}(d_1, \ldots, d_m)$ and on the geometric type $\mathcal{C}$ of the chosen configuration.

Note that it is an *exact* equality. In particular, choosing as $f = \chi_L$ the indicator function of a disc of radius L we see that for any given $L \in \mathbb{R}_+$ the *average* number of saddle connections not longer than L on flat surfaces $S \in \mathcal{H}_1^{comp}(d_1, \ldots, d_m)$ is *exactly* $C \cdot \pi L^2$, where C does not depend on L.

It would be convenient to introduce a special notation for such $\hat{f}$. Let

$$\hat{f}_L(S) = \sum_{v \in V} \chi_L(v)$$

The Theorem of Eskin and Masur [EMa] cited above tells that for large values of L one gets $\hat{f}_L(S) \sim c(\mathcal{C})\pi L^2$ for almost all individual flat surfaces $S \in \mathcal{H}_1^{comp}(d_1, \ldots, d_m)$ and that the corresponding constant $c(\mathcal{C})$ coincides with the constant C above.

Formula (6.4) can be applied to $\hat{f}_L$ for any value of L. In particular, instead of considering large L we can choose a very small value $L = \varepsilon$. The corresponding function $\hat{f}_\varepsilon(S)$ counts how many collections of parallel ε-short saddle connections (closed geodesics) of the type $\mathcal{C}$ we can find on the flat surface S.

Usually there are no such saddle connections (closed geodesics), so for most flat surfaces $\hat{f}_\varepsilon(S) = 0$. For some surfaces there is exactly one collection like this. We denote the corresponding subset by $\mathcal{H}_1^{\varepsilon,thick}(\mathcal{C}) \subset \mathcal{H}_1(d_1, \ldots, d_m)$. Finally, for the surfaces from the remaining (very small) subset $\mathcal{H}_1^{\varepsilon,thin}(\mathcal{C})$ one has several collections of short saddle connections (closed geodesics) of the type $\mathcal{C}$. Thus,

$$\hat{f}_\varepsilon(S) = \begin{cases} 0 & \text{for most of the surfaces } S \\ 1 & \text{for } S \in \mathcal{H}_1^{\varepsilon,thick}(\mathcal{C}) \\ > 1 & \text{for } S \in \mathcal{H}_1^{\varepsilon,thin}(\mathcal{C}) \end{cases}$$

and we can rewrite (6.4) for $\hat{f}_\varepsilon$ as

$$
\begin{aligned}
c(\mathcal{C})\cdot\pi\varepsilon^2 &= c(\mathcal{C})\int_{\mathbb{R}^2}\chi_\varepsilon(x,y)dx\,dy = \\
&= \frac{1}{\operatorname{Vol}\mathcal{H}_1^{comp}(d_1,\dots,d_m)}\int_{\mathcal{H}_1^{comp}(d_1,\dots,d_m)}\hat f_\varepsilon(S)\,d\nu_1 = \\
&= \frac{1}{\operatorname{Vol}\mathcal{H}_1^{comp}(d_1,\dots,d_m)}\int_{\mathcal{H}_1^{\varepsilon,thick}(\mathcal{C})} 1\,d\nu_1 + \\
&\quad + \frac{1}{\operatorname{Vol}\mathcal{H}_1^{comp}(d_1,\dots,d_m)}\int_{\mathcal{H}_1^{\varepsilon,thin}(\mathcal{C})}\hat f_\varepsilon(S)\,d\nu_1
\end{aligned}
$$

It can be shown that though $\hat f_\varepsilon(S)$ might be large on $\mathcal{H}_1^{\varepsilon,thin}(\mathcal{C})$ the measure of this subset is so small (it is of the order ε^4 that the last integral above is negligible in comparison with the previous one; namely it is $o(\varepsilon^2)$. (This is a highly nontrivial result of Eskin and Masur [EMa].) Taking into consideration that

$$\int_{\mathcal{H}_1^{\varepsilon,thick}(\mathcal{C})} 1\,d\nu_1 = \operatorname{Vol}\mathcal{H}_1^{\varepsilon,thick}(\mathcal{C}) = \operatorname{Vol}\mathcal{H}_1^{\varepsilon}(\mathcal{C}) + o(\varepsilon^2)$$

we can rewrite the chain of equalities above as

$$c(\mathcal{C})\cdot\pi\varepsilon^2 = \frac{\operatorname{Vol}\mathcal{H}_1^{\varepsilon}(\mathcal{C})}{\operatorname{Vol}\mathcal{H}_1^{comp}(d_1,\dots,d_m)} + o(\varepsilon^2)$$

which is equivalent to (6.2).

Baby Case: Closed Geodesics on the Torus

As an elementary application we can prove that proportion of primitive lattice points among all lattice points is $1/\zeta(2)$. In other words, applying (6.2) we can prove asymptotic formula (6.3) for the number of primitive lattice points in a disc of large radius L. As we have seen at Fig. 30 this number equals to the number $N_{cg}(\mathbb{T}^2,L)$ of families of oriented closed geodesics of length bounded by L on the standard torus $\mathbb{T}^2$.

We want to apply (6.2) to prove the following formula for the corresponding Siegel–Veech constant c_{cg}^+ (where superscript $+$ indicates that we are counting *oriented* geodesics on $\mathbb{T}^2$).

$$c_{cg}^+ = \lim_{L\to\infty}\frac{N_{cg}(\mathbb{T}^2,L)}{\pi L^2} = \frac{1}{\zeta(2)} = \frac{6}{\pi^2}$$

Note that the moduli space $\mathcal{H}_1(0)$ of flat tori is a total space of a unit tangent bundle to the modular surface (see Sect. 3.2, Fig. 13; see also (9.1) in Sect. 9.1 for geometric details). Modular surface can be considered as a space of flat tori of unit area without choice of direction to the North.

Measure on this circle bundle disintegrates to the product measure on the fiber and the hyperbolic measure on the modular curve. In particular, $\operatorname{Vol}(\mathcal{H}_1(0)) = \pi\cdot\pi/3$, where $\pi/3$ is the hyperbolic area of the modular surface.

Similarly, $\mathrm{Vol}(\mathcal{H}_1^\varepsilon(0)) = \pi \cdot \mathrm{Area}(\mathrm{Cusp}(\varepsilon))$, where $\mathrm{Cusp}(\varepsilon)$ is a subset of the modular surface corresponding to those flat tori of unit area which have a geodesic shorter than ε (see Fig. 13).

Showing that $\mathrm{Area}(\mathrm{Cusp}(\varepsilon)) \approx \varepsilon^2$ we apply (6.2) to get

$$c_{cg} = \lim_{\varepsilon\to 0} = \frac{1}{\pi\varepsilon^2}\,\frac{\mathrm{Area}(\mathrm{Cusp}(\varepsilon))}{\mathrm{Area}(\text{Modular surface})} = \frac{1}{\pi\varepsilon^2}\,\frac{\varepsilon^2 + o(\varepsilon^2)}{\pi/3} = \frac{1}{2\zeta(2)}.$$

Note that the Siegel–Veech constant c_{cg} corresponds to counting *nonoriented* closed geodesics on $\mathbb{T}^2$. Thus, finally we obtain the desired value $c_{cg}^+ = 2c_{cg}$.

In the next section we give an idea of how one can compute $\mathrm{Vol}\,\mathcal{H}_1^\varepsilon(\mathcal{C})$ in the simplest case, In Sect. 6.4 we describe the phenomenon of higher multiplicities and discuss the structure of typical cusps of the moduli spaces $\mathcal{H}(d_1,\dots,d_m)$.

6.3 Simplest Cusps of the Moduli Space

In this section we consider the simplest "cusp" $\mathcal{C}$ on a stratum $\mathcal{H}(d_1,\dots,d_m)$ and evaluate $\mathrm{Vol}\,\mathcal{H}_1^\varepsilon(\mathcal{C})$ for this cusp. Namely, we assume that the flat surface has at least two distinct conical points $P_1 \neq P_2$; let $2\pi(d_1+1), 2\pi(d_1+1)$ be corresponding cone angles. As a *configuration* $\mathcal{C}$ we consider a configuration when we have a single saddle connection ρ joining P_1 to P_2 and no other saddle connections on S parallel to ρ. In our calculation we assume that the conical points on every $S \in \mathcal{H}(d_1,\dots,d_m)$ have names; we count only saddle connections joining P_1 to P_2.

Consider some $S \in \mathcal{H}_1^{\varepsilon,thick}(\mathcal{C}) \subset \mathcal{H}(d_1,\dots,d_m)$, that is a flat surface S having a single saddle connection joining P_1 to P_2 which is not longer than ε and having no other short saddle connections or closed geodesics.

We are going to show that there is a canonical way to shrink the saddle connection on $S \in \mathcal{H}_1^{\varepsilon,thick}(\mathcal{C})$ coalescing two conical points into one. We shall see that, morally, this provides us with an (almost) fiber bundle

$$\begin{array}{c} \mathcal{H}_1^{\varepsilon,thick}(d_1,d_2,d_3,\dots,d_m) \\ \Big\downarrow \tilde{D}_\varepsilon^2 \\ \mathcal{H}_1(d_1+d_2,d_3,\dots,d_m) \end{array} \tag{6.5}$$

where $\tilde{D}_\varepsilon^2$ is a ramified cover of order (d_1+d_2+1) over a standard metric disc of radius ε. Moreover, we shall see that the measure on $\mathcal{H}_1^{\varepsilon,thick}(d_1,d_2,d_3,\dots,d_m)$ disintegrates into a product of the standard measure on $\tilde{D}_\varepsilon^2$ and the natural measure on $\mathcal{H}_1(d_1+d_2,d_3,\dots,d_m)$. The latter would imply the following simple answer to our problem:

$$\begin{aligned} &\mathrm{Vol}(\text{"}\varepsilon\text{-neighborhood of the cusp } \mathcal{C}\text{"}) = \\ &\qquad = \mathrm{Vol}\,\mathcal{H}_1^\varepsilon(\mathcal{C}) = \mathrm{Vol}\,\mathcal{H}_1^\varepsilon(d_1,d_2,d_3,\dots,d_m) \approx \\ &\qquad\qquad \approx (d_1+d_2+1)\cdot \pi\varepsilon^2 \cdot \mathrm{Vol}\,\mathcal{H}_1(d_1+d_2,d_3,\dots,d_m) \end{aligned} \tag{6.6}$$

Instead of contracting an isolated short saddle connection to a point we prefer to create it breaking a conical point $P' \in S'$ of degree $d_1 + d_2$ on a surface $S' \in \mathcal{H}_1(d_1 + d_2, d_3, \ldots, d_m)$ into two conical points of degrees d_1 and d_2 joined by a short saddle connections. We shall see that this surgery is invertible, and thus we shall get a coalescing construction. In the remaining part of this section we describe this surgery following [EMaZo].

Breaking up a Conical Point into Two

We assume that the initial surface $S' \in \mathcal{H}_1(d_1 + d_2, d_3, \ldots, d_m)$ does not have any short saddle connections or short closed geodesics.

Consider a metric disc of a small radius ε centered at the point P', i.e. the set of points Q' of the surface S' such that Euclidean distance from Q' to the point P' is less than or equal to ε. We suppose that $\varepsilon > 0$ is chosen small enough, so that the ε-disc does not contain any other conical points of the metric; we assume also, that the disc which we defined in the metric sense is homeomorphic to a topological disc. Then, metrically our disc has a structure of a regular cone with a cone angle $2\pi(d_1 + d_2 + 1)$; here $d_1 + d_2$ is the multiplicity of the zero P'. Now cut the chosen disc (cone) out of the surface. We shall modify the flat metric inside it preserving the metric at the boundary, and then paste the modified disc (cone) back into the surface.

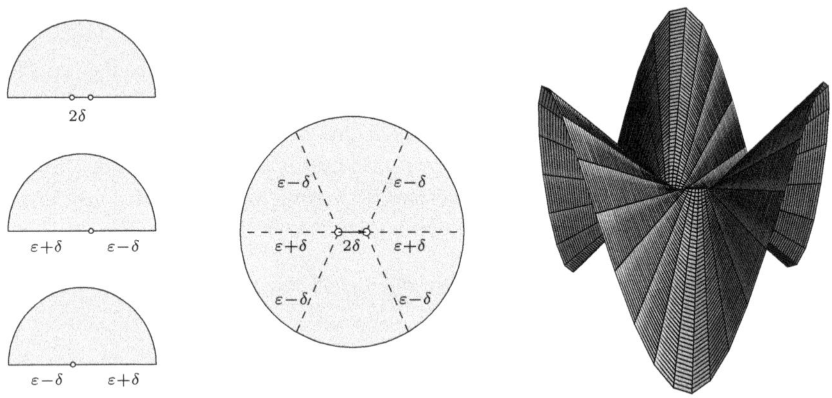

Fig. 31. Breaking up a conical point into two (after [EMaZo])

Our cone can be glued from $2(k_i + 1)$ copies of standard metric half-discs of the radius ε, see the picture at the top of Fig. 31. Choose some small δ, where $0 < \delta < \varepsilon$ and change the way of gluing the half-discs as indicated on the bottom picture of Fig. 31. As patterns we still use the standard metric half-discs, but we move slightly the marked points on their diameters. Now we use two special half-discs; they have two marked points on the diameter at the distance δ from the center of the half disc. Each of the remaining $2k_i$ half-discs has a single marked point at the distance δ from the center of the half-disc. We are alternating the half-discs with the marked point moved to the right and to the left from the center. The picture shows that all the lengths along identifications are matching; gluing the half-discs in this latter way we obtain a topological disc with a flat metric; now the flat metric has two cone-type singularities with the cone angles $2\pi(d_1 + 1)$ and $2\pi(d_2 + 1)$.

Note that a small tubular neighborhood of the boundary of the initial cone is isometric to the corresponding tubular neighborhood of the boundary of the resulting object. Thus we can paste it back into the surface. Pasting it back we can turn it by any angle φ, where $0 \le \varphi < 2\pi(d_1 + d_2 + 1)$.

We described how to break up a zero of multiplicity $d_1 + d_2$ of an Abelian differential into two zeroes of multiplicities d_1, d_2. The construction is local; it is parameterized by the two free real parameters (actually, by one complex parameter): by the small distance 2δ between the newborn zeroes, and by the direction φ of the short geodesic segment joining the two newborn zeroes. In particular, as a parameter space for this construction one can choose a ramified covering of degree $d_1 + d_2 + 1$ over a standard metric disc of radius ε.

6.4 Multiple Isometric Geodesics and Principal Boundary of the Moduli Space

In this section we give an explanation of the phenomenon of higher multiplicities, we consider typical degenerations of flat surfaces and we discuss how can one use configurations of parallel saddle connections or closed geodesics to determine the orbit of a flat surface S.

Multiple Isometric Saddle Connections

Consider a collection of saddle connections and closed geodesics representing a basis of relative homology $H_1(S, \{P_1, \dots, P_m\}; \mathbb{Z})$ on a flat surface S. Recall, that any geodesic on S goes in some constant direction. Recall also that by Convention 1 any flat surface is endowed with a distinguished direction to the North, so we can place a compass at any point of S and determine in which direction goes our geodesic. Thus, every closed geodesic or saddle connection determines a vector $\mathbf{v}_j \in \mathbb{R}^2 \simeq \mathbb{C}$ which goes in the same direction and have the same length as the corresponding geodesic element. Finally recall that collection of planar vectors $\{\mathbf{v}_1, \dots, \mathbf{v}_{2g+m-1}\}$ considered as complex numbers provide us with a local coordinate system in $\mathcal{H}(d_1, \dots, d_m)$. In complex-analytic language these coordinates are the *relative periods* of holomorphic 1-form representing the flat surface S, namely $\mathbf{v}_j = \int_{\rho_j} \omega$, where ρ_j is the corresponding geodesic element (saddle connection or closed geodesic).

We say that two geodesic elements γ_1, γ_2 (saddle connections or closed geodesics) are *homologous* on a flat surface S if they determine the same homology classes in $H_1(S, \{P_1, \dots, P_m\}; \mathbb{Z})$. In other words, γ_1 is homologous to γ_2 if cutting S by γ_1 and γ_2 we break the surface S into two pieces. For example, the saddle connections $\gamma_1, \gamma_2, \gamma_3$ on the right surface at the bottom of Fig. 32 are homologous.

The following elementary observation is very important for the sequel. Since the holomorphic 1-form ω representing S is closed, homologous geodesic elements γ_1, γ_2 define the same period:

$$\int_{\gamma_1} \omega = \mathbf{v} = \int_{\gamma_2} \omega$$

We intensively used the following result of H. Masur and J. Smillie [MaS] in our considerations in the previous section.

Theorem (H. Masur, J. Smillie). *There is a constant M such that for all $\varepsilon, \kappa > 0$ the subset $\mathcal{H}_1^{\varepsilon}(d_1, \dots, d_m)$ of $\mathcal{H}_1(d_1, \dots, d_m)$ consisting of those flat surfaces, which have a saddle connection of length at most ε, has volume at most $M\varepsilon^2$.*

The volume of the set of flat surfaces with a saddle connection of length at most ε and a nonhomologous saddle connection with length at most κ is at most $M\varepsilon^2\kappa^2$.

Morally, this Theorem says that a subset corresponding to one complex parameter with norm at most ε has measure of order ε^2, and a subset corresponding to two complex parameter with norm at most ε has measure of order ε^4. In particular, this theorem implies that $\operatorname{Vol} \mathcal{H}_1^{\varepsilon, thin}(d_1, \dots, d_m)$ is of the order ε^4.

In the previous section we considered the subset of flat surfaces $S \in \mathcal{H}_1^{\varepsilon}(d_1, d_2, d_3, \dots, d_m)$ having a single short saddle connection joining zeroes of degrees d_1 and d_2. We associated to such surface S a new surface $S' \in \mathcal{H}_1^{\varepsilon}(d_1 + d_2, d_3, \dots, d_m)$ in the smaller stratum. Note that, morally, surfaces S and S' have the same periods with a reservation that S has one more period than S': the extra small period represented by our short saddle connection.

Metrically surfaces S and S' are almost the same: having a surface S we know how to contract our short saddle connection to a point; having a surface S' and an abstract short period $\mathbf{v} \in \mathbb{C} \simeq \mathbb{R}^2$ we know how to break the corresponding zero on S' into two zeroes joined by a single short saddle connection realizing period $\mathbf{v}$. (In the latter construction we have some additional discrete freedom: we can break the zeroes in direction $\mathbf{v}$ in $d_1 + d_2 + 1$ different ways.)

Our construction does not work when we have two nonhomologous short geodesic elements on the surface S. But we do not care since according to the Theorem of H, Masur and J. Smillie the subset $\mathcal{H}_1^{\varepsilon, thin}(d_1, \dots, d_m)$ of such surfaces has very small measure (of the order ε^4).

Now consider a slightly more general surgery represented by Fig. 32. We take three distinct flat surfaces, we break a zero on each of them as it was done in the previous section. We do it coherently using the same direction

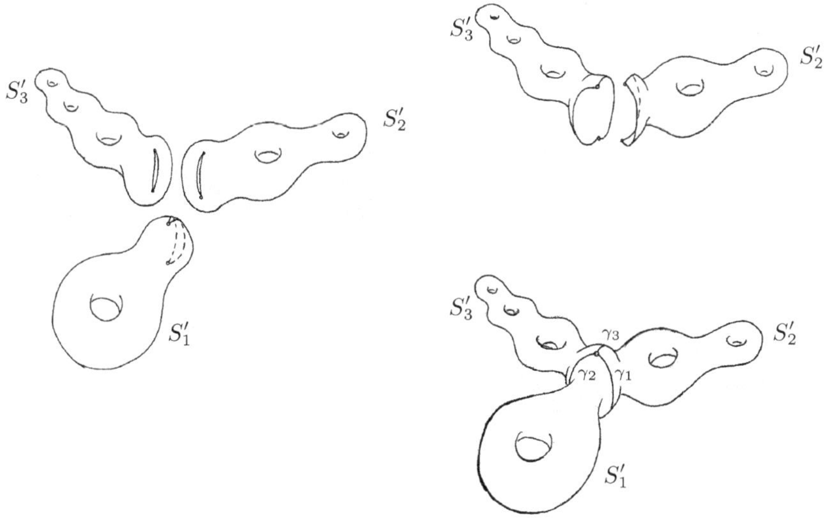

Fig. 32. Multiple homologous saddle connections; topological picture

and the same distance δ on each surface (left part of Fig. 32). Then we slit each surface along the newborn saddle connection and glue the surfaces in a close compound surface as indicated on the right part of Fig. 32.

The resulting surface has three short parallel saddle connections of equal length. By construction they are *homologous*: cutting along any two of them we divide the surface into two parts. Thus, the resulting surface again has only one short period! Note that a complete collection of periods of the compound surface can be obtained as disjoint union

$$\text{periods of } S'_1 \sqcup \text{periods of } S'_2 \sqcup \text{periods of } S'_3 \sqcup \text{newborn short period}$$

Hence, any flat surface S with three short homologous saddle connections and no other short geodesic elements can be viewed as a nonconnected flat surface $S'_1 \sqcup S'_2 \sqcup S'_3$ plus a memory about the short period $\mathbf{v} \in \mathbb{C}$ which we use when we break the zeroes (plus some combinatorial arbitrariness of finite order).

The moduli space of disconnected flat surfaces of the same type as $S'_1 \sqcup S'_2 \sqcup S'_3$ has one dimension less than the original stratum $\mathcal{H}(d_1, \ldots, d_m)$. Our considerations imply the following generalization of formula (6.6) for the volume of "ε-neighborhood" of the corresponding cusp:

$$\mathrm{Vol}(\text{“}\varepsilon\text{-neighborhood of the cusp } \mathcal{C}\text{”}) = \mathrm{Vol}\,\mathcal{H}_1^{\varepsilon}(\mathcal{C}) =$$
$$k \cdot \mathrm{Vol}\,\mathcal{H}_1(\text{stratum of } S'_1) \cdot \mathrm{Vol}\,\mathcal{H}_1(\text{stratum of } S'_2) \cdot \mathrm{Vol}\,\mathcal{H}_1(\text{stratum of } S'_3) \cdot \pi\varepsilon^2$$

where the factor k is an explicit constant depending on dimensions, possible symmetries, and combinatorics of multiplicities of zeroes. In particular, we get a subset of order ε^2.

Now for any flat surface $S_0 \in \mathcal{H}(d_1, \ldots, d_m)$ we can state a counting problem, where we shall count only those saddle connections which appear in configuration $\mathcal{C}$ of triples of homologous saddle connections breaking S_0 into three surfaces of the same topological and geometric types as S'_1, S'_2, S'_3. Applying literally same arguments as in Sect. 6.2 and 6.3 we can show that such number of triples of homologous saddle connections of length at most L has quadratic growth rate and that the corresponding Siegel–Veech constant $c(\mathcal{C})$ can be expressed by the same formula as above:

$$c(\mathcal{C}) = \lim_{\varepsilon \to 0} \frac{1}{\pi \varepsilon^2} \frac{\text{Vol}(\text{“}\varepsilon\text{-neighborhood of the cusp } \mathcal{C}\text{”})}{\text{Vol}\, \mathcal{H}_1^{comp}(d_1, \ldots, d_m)} =$$

$$= (\text{combinatorial factor}) \frac{\prod_{k=1}^{n} \text{Vol}\, \mathcal{H}_1(\text{stratum of } S'_k)}{\text{Vol}\, \mathcal{H}_1^{comp}(d_1, \ldots, d_m)}$$

Principal Boundary of the Strata

The results above give a description of typical degenerations of flat surfaces. A flat surface gets close to the boundary of the moduli space when some geodesic element (or a collection of geodesic elements) get short. Morally, we have described something like “faces” of the boundary, while there are still “edges”, etc, representing degenerations of higher codimension. Flat surfaces which are close to this “principal boundary” of a stratum $\mathcal{H}(d_1, \ldots, d_m)$ have the following structure.

If the short geodesic element is a saddle connection joining two distinct zeroes, then the surface looks like the one at Fig. 32. It can be decomposed to several flat surfaces with slits glued cyclically one to another. The boundaries of the slits form short saddle connections on the compound surface. All these saddle connections join the same pair of points; they have the same length and direction. They represent homologous cycles in the relative homology group $H_1(S, \{P_1, \ldots, P_m\}; \mathbb{Z})$.

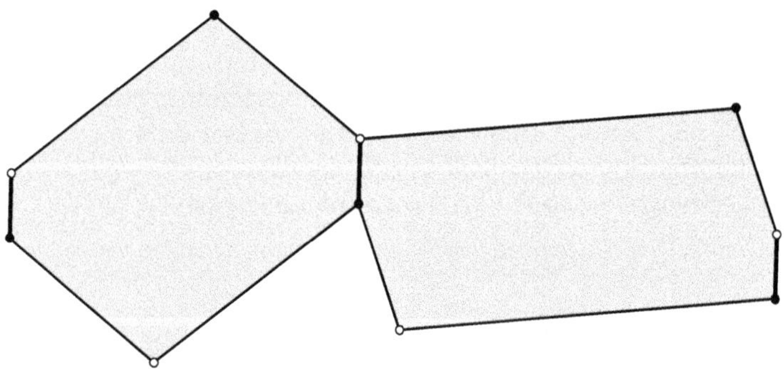

Fig. 33. Flat surface with two short homologous saddle connections. Any small deformation of this surface also has a pair of short homologous saddle connections

Figure 33 represents a flat surface $S_0 \in \mathcal{H}(1,1)$ unfolded to a polygon. The two short bold segments represent two homologous saddle connections. The reader can easily check that on any surface $S \in \mathcal{H}(1,1)$ obtained from S_0 by a small deformation one can find a pair of short parallel saddle connections of equal length. Cutting S by these saddle connections we get a pair of tori with slits.

We did not discuss in the previous section the case when the short geodesic element is a regular closed geodesic (or a saddle connection joining a conical point to itself). Morally, it is similar to the case of saddle connections, but technically it slightly more difficult. A flat surface near the principal boundary of this type is presented on the right of Fig. 34.

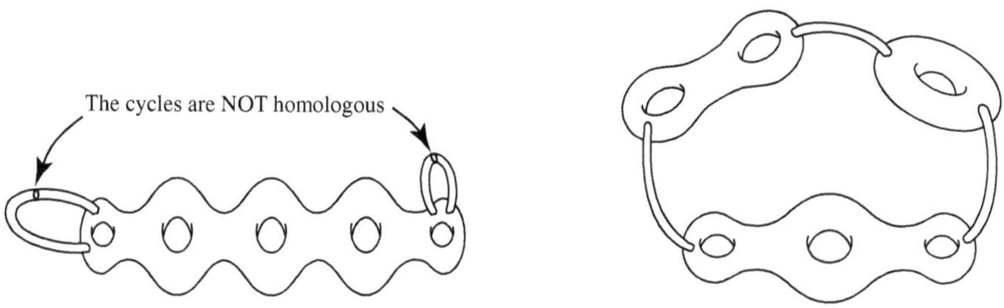

Fig. 34. A typical (on the right) and nontypical (on the left) degenerations of a flat surface. Topological picture

Similar to the case of saddle connections, the surface can be also decomposed to a collection of well-proportional flat surfaces $S'_1, \dots, S'_n$ of lower genera. Each surface S'_k has a pair of holes. Each of these holes is realized by a saddle connection joining a zero to itself. The surfaces are cyclically glued to a "necklace", where two neighboring surfaces might be glued directly or by a narrow cylinder. Since the waist curves of all these cylinders and all saddle connections representing boundaries of surfaces S'_k are homologous, the corresponding closed geodesics on S are parallel and have equal length. A more artistic[5] image of a surface, which is located closed to the boundary of a stratum is represented on Fig. 35.

The surface on the left of Fig. 34 is close to an "edge"of the moduli space in the sense that it represents a "nontypical" degeneration: a degeneration of codimension two. This surface has two nonhomologous closed geodesics shorter than ε. Due to the Theorem of H. Masur and J. Smillie cited above, the subset $\mathcal{H}_1^{\varepsilon,thin}(d_1, \dots, d_m)$ of such surfaces has measure of the order ε^4.

Configurations of Saddle Connections and of Closed Geodesics as Invariants of Orbits

Consider a flat surface S_0 and consider its orbit under the action of $SL(2,\mathbb{R})$. It is very easy to construct this orbit locally for those elements of the group

Fig. 35. Flat surface near the principal boundary[5]

$GL^+(2,\mathbb{R})$ which are close to identity. It is a fairly complicated problem to construct this orbit globally in $\mathcal{H}(d_1,\ldots,d_m)$ and to find its closure. Ergodic theorem of H. Masur and W. Veech (see Sect. 3.5) tells that for almost every surface S_0 the closure of the orbit of S_0 coincides with the embodying connected component of the corresponding stratum. But for some surfaces the closures of the orbits are smaller. Sometimes it is possible to distinguish such surfaces looking at the configurations of parallel closed geodesics and saddle connections. Say, for *Veech surfaces* which will be discussed in Sect. 9.5 the orbit of $SL(2,\mathbb{R})$ is already closed in the stratum, so Veech surfaces are very special. This property has an immediate reflection in behavior of parallel closed geodesics and saddle connections: as soon as we have a saddle connection or a closed geodesic in some direction on a Veech surface, *all* geodesics in this direction are either closed or (finite number of them) produce saddle connections.

Thus, it is useful to study configurations of parallel closed geodesics on a surface (which includes the study of proportions of corresponding maximal cylinders filled with parallel regular closed geodesics) to get information about the closure of corresponding orbit.

One can also use configurations of parallel closed geodesics on a flat surface to determine those connected component of the stratum, to which belongs our surface S_0. Some configurations (say, $g-1$ tori connected in a "necklace" by a chain of cylinders, compare to Fig. 35) are specific for some connected components and never appear for surfaces from other connected components. We return to this discussion in the very end of Sect. 9.4 where we use this idea to distinguish connected components of the strata in the moduli space of quadratic differentials.

[5]H. Matisse: La Danse. The State Hermitage Museum, St. Petersburg. (c) Succession H. Matisse/VG Bild-Kunst, Bonn, 2005

6.5 Application: Billiards in Rectangular Polygons

Consider now a problem of counting *generalized diagonals* of bounded length or a problem of counting closed billiard trajectories of bounded length in a billiard in a rational polygon Π. Apply Katok–Zemliakov construction (see Sect. 2.1) and glue a very flat surface S from the billiard table Π. Every generalized diagonal (trajectory joining two corners of the billiard, possibly, after reflections from the sides) unfolds to a saddle connection and every periodic trajectory unfolds to a closed regular geodesic.

It is very tempting to use the results described above for the counting problem for the billiard. Unfortunately, the technique elaborated above is not applicable to billiards directly. The problem is that flat surfaces coming from billiards form a subspace of large codimension in any stratum of flat surfaces; in particular, this subspace has measure zero. Our "almost all" technique does not see this subspace.

However, the problems are related in some special cases; see [EMaScm] treating billiard in a rectangle with a barrier. As another illustration we consider billiards in "rectangular polygons". These results represent the work [AthEZo] which is in progress. We warn the reader that we are extremely informal in the remaining part of this section.

Rectangular Polygons

Figure 36 suggests several examples of *rectangular polygons*. The "polygons" are allowed to have ramification points at the boundary, with restriction that the angles at ramification points are integer multiples of $\pi/2$. Note that we do not identify the side P_5P_6 with a part of the side P_4P_5 in the right polygon. This polygon should be considered with a cut along the side P_5P_6. The corresponding billiard has a "barrier" along the side P_5P_6.

Fig. 36. Rectangular polygons

Formally speaking, by a *rectangular polygon* Π we call a topological disc endowed with a flat metric, such that the boundary $\partial\Pi$ is presented by a finite broken line of geodesic segments and such that the angle between any two consecutive sides equals $k\pi/2$, where $k \in \mathbb{N}$.

Consider now our problem for a standard rectangular billiard table (the proportions of the sides do not matter). We emit a trajectory from some corner of the table and want it arrive to another corner after several reflections from the sides.

When our trajectory reflects from a side it is convenient to prolong it as a straight line by making a reflection of the rectangle with respect to the corresponding side (see Katok–Zemliakov construction in Sect. 2.1). Unfolding our rectangular table we tile the plane $\mathbb{R}^2$ with a rectangular lattice. Our problem can be reformulated as a problem of counting primitive lattice points (see the right part of Fig. 30).

We are emitting our initial trajectory from some fixed corner of the billiard. It means that in the model with a lattice in the plane we are emitting a straight line from the origin inside one of the four quadrants. Thus, we are counting the asymptotics for the number of primitive lattice points in the intersection of a coordinate quadrant with a disc of large radius L centered at the origin. This gives us 1/4 of the number of all primitive lattice points. Note that in our count we have fixed the initial corner, but we let our trajectory hit any of the remaining three corners. Thus, if we count only those generalized diagonals which are launched from some prescribed corner P_i and arrive to a prescribed corner P_j (different from initial one) we get 1/3 of the previous number. Hence, the number $N_{ij}(L)$ of generalized diagonals joining P_i with P_j is 1/12 of the number $N_{cg}(\mathbb{T}^2, L) \sim (1/\zeta(2)) \cdot \pi L^2$ of primitive lattice points, see (6.3).

In our calculation we assumed that the billiard table has area one. It is clear that asymptotics for our counting function is homogeneous with respect to the area of the table. Adjusting our formula for a rectangular billiard table of the area different from 1 we get the following answer for the number of generalized diagonals of length at most L joining prescribed corner P_i to a prescribed corner P_j different from the first one:

$$N_{ij}(L) \approx \frac{1}{2\pi} \cdot \frac{L^2}{\text{Area of the billiard table}} \tag{6.7}$$

Now, having studied a model case, we announce two examples of results from [AthEZo] concerning rectangular polygons.

Consider a family of rectangular polygons having exactly $k \geq 0$ angles $3\pi/2$ and all other angles $\pi/2$ (see Fig. 37). Consider a generic billiard table in this family (in the measure-theoretical sense). Fix any two corners $P_i \neq P_j$ having angles $\pi/2$. The number $\tilde{N}_{ij}(L)$ of generalized diagonals of length at most L joining P_i to P_j is approximately the same as for a rectangle:

$$\tilde{N}_{ij}(L) \sim \frac{1}{2\pi} \cdot \frac{L^2}{\text{Area of the billiard table}}$$

We have to admit that we are slightly cheating here: the equivalence "$\sim$" which we can currently prove is weaker than "$\approx$" in (6.7); nevertheless, we do not want to go into technical details.

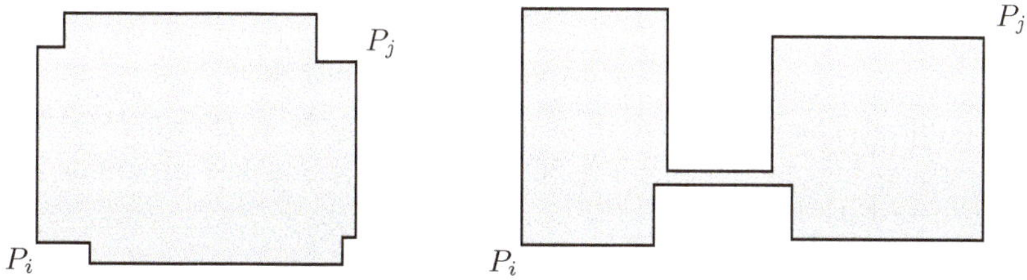

Fig. 37. Family of rectangular polygons of the same geometry and the same area. The shape of these polygons is quite different. Nevertheless for both billiard tables the number of trajectories of length at most L joining the right-angle corner P_i to the right-angle corner P_j is approximately the same as the number of trajectories of length at most L joining two right-angle corners of a rectangle of the same area

Note that the shape of the polygon within the family may vary quite considerably, see Fig. 37, and this does not affect the asymptotic formula. However, the answer changes drastically when we change the family. For rectangular polygons having several angles of the form $n\pi$ the constant in quadratic asymptotics is more complicated. This is why we do not expect any elementary proof of this formula (our proof involves evaluation of corresponding Siegel–Veech constant).

Actually, naive intuition does not help in counting problems of this type. Consider, for example, an L-shaped billiard table as on Fig. 38

The angle at the vertex P_0 is $3\pi/2$ which is *three* times larger than the angle $\pi/2$ at the other five vertices $P_1, \ldots, P_5$. However, the number

©Moon Duchin

Fig. 38. L-shaped billiard table

$$\tilde{N}_{0j}(L) \sim \frac{2}{\pi} \cdot \frac{L^2}{\text{Area of the billiard table}}$$

of generalized diagonals of length at most L joining P_0 to P_j, where $1 \le j \le 5$, is *four* times bigger than the number $\tilde{N}_{ij}(L)$ of generalized diagonals joining two corners with the angles $\pi/2$.

7 Volume of Moduli Space

In Sect. 3.4 we defined a volume element in the stratum $\mathcal{H}(d_1, \ldots, d_m)$. We used linear volume element in cohomological coordinates $H^1(S, \{P_1, \ldots, P_m\}; \mathbb{C})$ normalized in such way that a fundamental domain of the lattice

$$H^1(S, \{P_1, \ldots, P_m\}; \mathbb{Z} \oplus \sqrt{-1}\,\mathbb{Z}) \subset H^1(S, \{P_1, \ldots, P_m\}; \mathbb{C})$$

has unit volume. The unit lattice does not depend on the choice of cohomological coordinates, its vertices play the role of *integer points* in the moduli space $\mathcal{H}(d_1, \ldots, d_m)$. In Sect. 7.1 we suggest a geometric interpretation of flat surfaces representing integer points of the strata.

Using this interpretation we give an idea for counting the volume ("hyperarea") of the hypersurface $\mathcal{H}_1(d_1, \ldots, d_m) \subset \mathcal{H}(d_1, \ldots, d_m)$ of flat surfaces of area one. We apply the strategy which can be illustrated in a model example of evaluation of the area of a unit sphere. We first count the asymptotics for the number $N(R)$ of integer points inside a ball of huge radius R. Clearly $N(R)$ corresponds to the volume of the ball, so if we know the asymptotics for $N(R)$ we know the formula for the volume $\mathrm{Vol}(R)$ of the ball of radius R. The derivative $\frac{d}{dR}\mathrm{Vol}(R)\big|_{R=1}$ gives us the area of the unit sphere.

Similarly, to evaluate the "hyperarea of a unit hyperboloid" $\mathcal{H}_1(d_1, \ldots, d_m)$ it is sufficient to count the asymptotics for the number of integer points inside a "hyperboloid" $\mathcal{H}_R(d_1, \ldots, d_m)$ of huge "radius" R. The role of the "radius" R is played by the positive homogeneous real function $R = area(S)$ defined on $\mathcal{H}(d_1, \ldots, d_m)$.

Note that the volume $\nu\big(\mathcal{H}_{\le R}(d_1, \ldots, d_m)\big)$ of a domain bounded by the "hyperboloid" $\mathcal{H}_R(d_1, \ldots, d_m)$ is a homogeneous function of R of the weight $\dim_{\mathbb{R}} \mathcal{H}(d_1, \ldots, d_m)/2$ while the volume of a ball of radius R is a homogeneous function of R of the weight which equals the dimension of the space. This difference in weights explains the factor 2 in the formula below:

$$\begin{aligned}\mathrm{Vol}\left(\mathcal{H}_1(d_1, \ldots, d_m)\right) &= 2\frac{d}{dR}\,\nu\big(\mathcal{H}_{\le R}(d_1, \ldots, d_m)\big)\Big|_{R=1} = \\ &= \dim_{\mathbb{R}}(\mathcal{H}_1(d_1, \ldots, d_m)) \cdot \nu\big(\mathcal{H}_{\le 1}(d_1, \ldots, d_m)\big) \qquad (7.1)\end{aligned}$$

This approach to computation of the volumes was suggested by A. Eskin and A. Okounkov and by M. Kontsevich and the author. However, the straightforward application of this approach, described in Sect. 7.1, gives the values of the volumes only for several low-dimensional strata. The general solution of the problem was found by A. Eskin and A. Okounkov who used in addition powerful methods of representation theory. We give an idea of their method in Sect. 7.2.

7.1 Square-tiled Surfaces

Let us study the geometric properties of the flat surfaces S represented by "integer points" $S \in H^1(S, \{P_1, \ldots, P_m\}; \mathbb{Z} \oplus \sqrt{-1}\,\mathbb{Z})$ in cohomological coordinates. Let ω be the holomorphic one-form representing such flat surface S. Since $[\omega] \in H^1(S, \{P_1, \ldots, P_m\}; \mathbb{Z} \oplus \sqrt{-1}\,\mathbb{Z})$ all periods of ω (including relative periods) belong to $\mathbb{Z} \oplus \sqrt{-1}\,\mathbb{Z}$. Hence the following map f_ω from the flat surface S to the standard torus $\mathbb{T}^2 = \mathbb{C}/(\mathbb{Z} \oplus \sqrt{-1}\,\mathbb{Z})$ is well-defined:

$$f_\omega : P \mapsto \left(\int_{P_1}^{P} \omega \right) \bmod \mathbb{Z} \oplus \sqrt{-1}\,\mathbb{Z}\,,$$

where P_1 is one of the conical points. It is easy to check that f_ω is a ramified covering, moreover, it has exactly m ramification points, and the ramification points are exactly the zeros $P_1, \ldots, P_m$ of ω. Consider the flat torus $\mathbb{T}^2$ as a unit square with the identified opposite sides. The covering $f_\omega : S \to \mathbb{T}^2$ induces a tiling of the flat surface S by unit squares. Note that all unit squares are endowed with the following additional structure: we know exactly which edge is top, bottom, right, and left; adjacency of the squares respects this structure in a natural way: we glue vertices to vertices and edges to edges, moreover, the right edge of a square is always identified to the left edge of some square and top edge is always identified to the bottom edge of some square. We shall call a flat surface with such tiling a *square-tiled surface*. The problem of counting the volume of $\mathcal{H}_1(d_1, \ldots, d_m)$ is equivalent to the following problem: how many square-tiled surfaces of a given geometric type (determined by number and types of conical singularities) can we construct using at most N unit squares. Say, Fig. 46 gives the list of all square-tiled surfaces of genus $g > 1$ glued from at most 3 squares; they all belong to $\mathcal{H}(2)$.

In terms of the coverings our Problem can be formulated as follows. Consider the ramified coverings $p : S \to \mathbb{T}^2$ over the standard torus $\mathbb{T}^2$. Fix the number m of branching points, and ramification degrees $d_1, \ldots, d_m)$ at these points. Assume that all ramification points $P_1, \ldots, P_m$ project to the same point of the torus $\mathbb{T}^2$. Enumerate ramified coverings of any given ramification type having at most $N \gg 1$ sheets. Here pairs of coverings forming commutative diagrams as below are identified:

$$\begin{array}{ccc} S & \longleftrightarrow & S \\ & \searrow \quad \swarrow & \\ & \mathbb{T}^2 & \end{array} \tag{7.2}$$

Counting of Square-tiled Tori

Let us count the number of square-tiled tori tiled by at most $N \gg 1$ squares. In this case our square-tiled surface has no singularities at all: we have a flat torus tiled with the unit squares in a regular way. Cutting our flat torus by a horizontal waist curve we get a cylinder with a waist curve of length $w \in \mathbb{N}$ and a height $h \in \mathbb{N}$, see Fig. 39. The number of squares in the tiling equals $w \cdot h$. We can reglue the torus from the cylinder with some integer twist t, see Fig. 39. Making an appropriate Dehn twist along the waist curve we can reduce the value of the twist t to one of the values $0, 1, \ldots, w-1$. In other words, fixing the integer perimeter w and height h of a cylinder we get w different square-tiled tori.

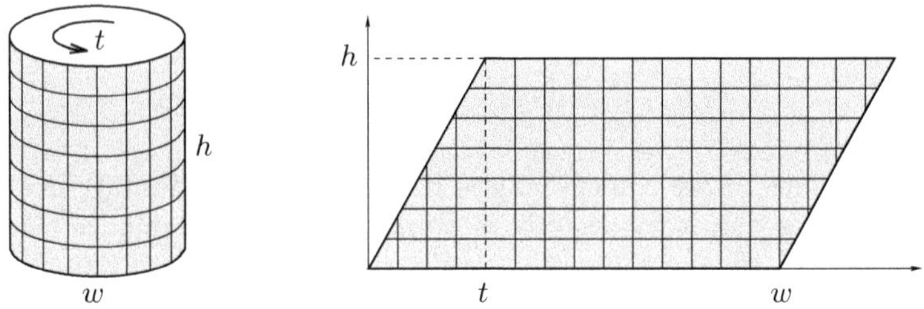

Fig. 39. A square-tiled surface is decomposed into several cylinders. Each cylinder is parametrized by its width (perimeter) w and height h. Gluing the cylinders together we get also a twist parameter t, where $0 \le t < w$, for each cylinder

Thus the number of square tiled tori constructed by using at most N squares is represented as

$$\nu(\mathcal{H}_{\le N}(0)) \sim \sum_{\substack{w,h\in\mathbb{N}\\ w\cdot h\le N}} w = \sum_{\substack{w,h\in\mathbb{N}\\ w\le \frac{N}{h}}} w \approx \sum_{h\in\mathbb{N}} \frac{1}{2}\cdot\left(\frac{N}{h}\right)^2 = \frac{N^2}{2}\cdot\zeta(2) = \frac{N^2}{2}\cdot\frac{\pi^2}{6}$$

Actually, some of the tori presented by the first sum are equivalent by an affine diffeomorphism, so we are counting them twice, or even several times. Say, the tori $w = 2; h = 1; t = 0$ and $w = 1; h = 2; t = 0$ are equivalent. However, this happens relatively rarely, and this correction does not affect the leading term, so we simply neglect it.

Applying the derivative $2\left.\dfrac{d}{dN}\right|_{N=1}$ (see (7.1)) we finally get the following value for the volume of the space of flat tori

$$\mathrm{Vol}(\mathcal{H}_1(0)) = \frac{\pi^2}{3}$$

Decomposition of a Square-Tiled Surface into Cylinders

Let us study the geometry of square-tiles surfaces. Note that all leaves of both horizontal and vertical foliation on every square-tiled surface are closed. In particular the union of all horizontal critical leaves (the ones adjacent to conical points) forms a finite graph Γ. The collection $P_1, \ldots, P_m$ of conical points forms the set of the vertices of this graph; the edges of the graph are formed by horizontal saddle connections. The complement $S - \Gamma$ is a union of flat cylinders.

For example, for the square-tiled surfaces from Fig. 46 we get the following decompositions into horizontal cylinders. We have one surface composed from a single cylinder filled with closed horizontal trajectories; this cylinder has width (perimeter) $w = 3$ and hight $h = 1$. Two other surfaces are composed from two cylinders. The heights of the cylinders are $h_1 = h_2 = 1$, the widths are $w_1 = 1$ and $w_2 = 2$ correspondingly. Observing the two-cylinder surfaces at Fig. 46 we see that they differ by the *twist* parameter t_2 (see Fig. 39) of the wider cylinder: in one case $t_2 = 0$ and in the other case $t_2 = 1$. By construction the width w_i and height h_i of any cylinder are strictly positive integers; the value of the twist t_i is a nonnegative integer bounded by the width of the cylinder: $0 \leq t_i < w_i$.

Separatrix Diagrams

Let us study in more details the graphs Γ of horizontal saddle connections.

We start with an informal explanation. Consider the union of all saddle connections for the horizontal foliation, and add all critical points (zeroes of ω). We obtain a finite oriented graph Γ. Orientation on the edges comes from the canonical orientation of the horizontal foliation. Moreover, graph Γ is drawn on an oriented surface, therefore it carries so called *ribbon structure* (even if we forget about the orientation of edges), i.e. on the star of each vertex P a cyclic order is given, namely the counterclockwise order in which edges are attached to P. The direction of edges attached to P alternates (between directions toward P and from P) as we follow the counterclockwise order.

It is well known that any finite ribbon graph Γ defines canonically (up to an isotopy) an oriented surface $S(\Gamma)$ with boundary. To obtain this surface we replace each edge of Γ by a thin oriented strip (rectangle) and glue these strips together using the cyclic order in each vertex of Γ. In our case surface $S(\Gamma)$ can be realized as a tubular ε-neighborhood (in the sense of transversal measure) of the union of all saddle connections for sufficiently small $\varepsilon > 0$.

The orientation of edges of Γ gives rise to the orientation of the boundary of $S(\Gamma)$. Notice that this orientation is *not* the same as the canonical orientation of the boundary of an oriented surface. Thus, connected components of the boundary of $S(\Gamma)$ are decomposed into two classes: positively and negatively oriented (positively when two orientations of the boundary components coincide and negatively, when they are different). The complement to the tubular ε-neighborhood of Γ is a finite disjoint union of open cylinders foliated by oriented circles. It gives a decomposition of the set of boundary circles $\pi_0(\partial(S(\Gamma)))$ into pairs of components having opposite signs of the orientation.

Now we are ready to give a formal definition:

A *separatrix diagram* is a finite oriented ribbon graph Γ, and a decomposition of the set of boundary components of $S(\Gamma)$ into pairs, such that

- the orientation of edges at any vertex is alternated with respect to the cyclic order of edges at this vertex;
- there is one positively oriented and one negatively oriented boundary component in each pair.

Notice that ribbon graphs which appear as a part of the structure of a separatrix diagram are very special. Any vertex of such a graph has even degree, and the number of boundary components of the associated surface with boundary is even. Notice also, that in general the graph of a separatrix diagram is *not* planar.

Any separatrix diagram $(\Gamma, pairing)$ defines a closed oriented surface together with an embedding of Γ (up to a homeomorphism) into this surface. Namely, we glue to the surface with boundary $S(\Gamma)$ standard oriented cylinders using the given pairing.

In pictures representing diagrams we encode the pairing on the set of boundary components painting corresponding domains in the picture by some colors (textures in the black-and-white text) in such a way that every color appears exactly twice. We will say also that paired components have the *same color*.

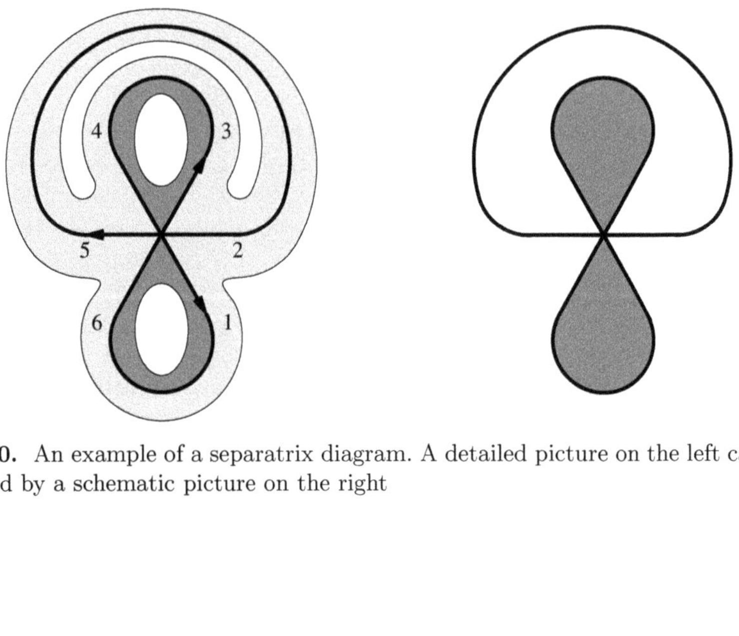

Fig. 40. An example of a separatrix diagram. A detailed picture on the left can be encoded by a schematic picture on the right

Example. The ribbon graph presented at Figure 40 corresponds to the horizontal foliation of an Abelian differential on a surface of genus $g = 2$. The Abelian differential has a single zero of degree 2. The ribbon graph has two pairs of boundary components.

Any separatrix diagram represents an orientable measured foliation with only closed leaves on a compact oriented surface without boundary. We say that a diagram is *realizable* if, moreover, this measured foliation can be chosen as the horizontal foliation of some Abelian differential. Lemma below gives a criterion of realizability of a diagram.

Assign to each saddle connection a real variable standing for its "length". Now any boundary component is also endowed with a "length" obtained as sum of the "lengths" of all those saddle connections which belong to this component. If we want to glue flat cylinders to the boundary components, the lengths of the components in every pair should match each other. Thus for every two boundary components paired together (i.e. having the same color) we get a linear equation: "the length of the positively oriented component equals the length of the negatively oriented one".

Lemma. *A diagram is realizable if and only if the corresponding system of linear equations on "lengths" of saddle connections admits strictly positive solution.*

The proof is obvious.

Example. The diagram presented at Fig. 40 has three saddle connections, all of them are loops. Let p_{16}, p_{52}, p_{34} be their "lengths". There are two pairs of boundary components. The corresponding system of linear equations is as follows:

$$\begin{cases} p_{34} = p_{16} \\ p_{16} + p_{52} = p_{34} + p_{52} \end{cases}$$

Exercise. Check that two separatrix diagrams at Fig. 41 are realizable, and one – not. Check that there are no other realizable separatrix diagrams for the surfaces from the stratum $\mathcal{H}(2)$. Find all realizable separatrix diagrams for the stratum $\mathcal{H}(1,1)$.

Counting of Square-tiled Surfaces in $\mathcal{H}(2)$

To consider one more example we count square-tiled surfaces in the stratum $\mathcal{H}(2)$. We have seen that in this stratum there are only two realizable separatrix diagrams; they are presented on the left and in the center of Fig. 41.

Consider those square tiled surfaces from $\mathcal{H}(2)$ which correspond to the left diagram from Fig. 41. In this case the ribbon graph corresponding to the separatrix diagram has single "top" and single "bottom" boundary component, so our surface is glued from a single cylinder. The waist curve of the

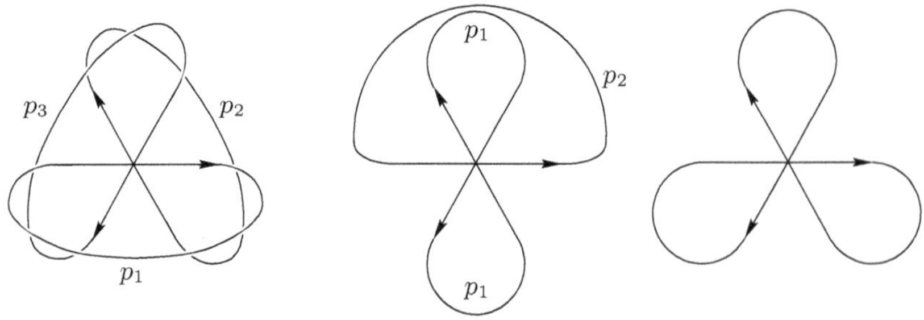

Fig. 41. The separatrix diagrams from left to right a) represent a square-tiled surface glued from one cylinder of width $w = p_1 + p_2 + p_3$; b) represent a square-tiled surface glued from two cylinder of widths $w_1 = p_1$ and $w_2 = p_1 + p_2$; c) do not represent any square-tiled surface

cylinder is of the length $w = p_1 + p_2 + p_3$, where p_1, p_2, p_3 are the lengths of the separatrix loops. As usual, denote the height of the cylinder by h. The twist t of the cylinder has an integer value in the interval $[0, w-1]$. Thus the number of surfaces of this type of the area bounded by N is asymptotically equivalent to the sum

$$\frac{1}{3} \sum_{\substack{p_1,p_2,p_3,h \in \mathbb{N} \\ (p_1+p_2+p_3)h \le N}} (p_1 + p_2 + p_3) \sim \frac{N^4}{24} \cdot \zeta(4) = \frac{N^4}{24} \cdot \frac{\pi^4}{90}$$

(see more detailed computation in [Zo5]). The coefficient $1/3$ compensates the arbitrariness of the choice of enumeration of p_1, p_2, p_3 preserving the cyclic ordering. Similar to the torus case we have neglected a small correction coming from counting equivalent surfaces several times.

Exercise. Check that for $p_1 = p_2 = p_3 = 1$ all possible values of the twist $t = 0, 1, 2$ give equivalent flat surfaces; see also Fig. 46

Consider now a ribbon graph corresponding to the middle diagram from Fig. 41. It has two "top" and two "bottom" boundary components. Thus, topologically, we can glue in a pair of cylinders in two different ways. However, to have a flat structure on the resulting surface we need to have equal lengths of "top" and "bottom" boundary components. These lengths are determined by the lengths of the corresponding separatrix loops. It is easy to check that one of the two possible gluings of cylinders is forbidden: it implies that one of the separatrix loops has zero length, and hence the surface is degenerate.

The other gluing is realizable. In this case there is a pair of separatrix loops of equal lengths p_1, see Fig. 41. The square-tiled surface is glued from two cylinders: one having a waist curve $w_1 = p_1$, and the other one having waist curve $w_2 = p_1 + p_2$. Denote the heights and twists of the corresponding cylinders by h_1, h_2 and t_1, t_2. The twist of the first cylinder has the value in the interval $[0, w_1 - 1]$; the twist of the second cylinder has the value in the interval $[0, w_2 - 1]$. Thus the number of surfaces of 2-cylinder type of the area bounded by N is asymptotically equivalent to the sum

$$\sum_{\substack{p_1,p_2,h_1,h_2\\ p_1h_1+(p_1+p_2)h_2\le N}} p_1(p_1+p_2) = \sum_{\substack{p_1,p_2,h_1,h_2\\ p_1(h_1+h_2)+p_2h_2\le N}} p_1^2 + p_1p_2$$

It is not difficult to represent these two sums in terms of the *multiple zeta values* $\zeta(1,3)$ and $\zeta(2,2)$ (see detailed computation in [Zo5]). Applying the relations $\zeta(1,3) = \zeta(4)/4$ and $\zeta(2,2) = 3\zeta(4)/4$ we get the following asymptotic formula for our sum:

$$\sum_{\substack{p_1,p_2,h_1,h_2\\ p_1(h_1+h_2)+p_2h_2\le N}} p_1^2+p_1p_2 \sim \frac{N^4}{24}(2\cdot\zeta(1,3)+\zeta(2,2)) = \frac{N^4}{24}\cdot\frac{5}{4}\cdot\zeta(4) = \frac{N^4}{24}\cdot\frac{5}{4}\cdot\frac{\pi^4}{90}$$

Joining the impacts of the two diagrams and applying $2\cdot\left.\frac{d}{dN}\right|_{N=1}$ (see (7.1)) we get the following value for the volume of the stratum $\mathcal{H}(2)$:

$$\mathrm{Vol}(\mathcal{H}_1(2)) = \frac{\pi^4}{120}$$

Separatrix diagrams: more serious reading. Technique of separatrix diagrams is extensively used by K. Strebel in [Str] and in some articles like [KonZo].

7.2 Approach of A. Eskin and A. Okounkov

It is time to confess that evaluation of the volumes of the strata by means of naive counting square-tiled surfaces suggested in the previous section is not efficient in general case. The number of realizable separatrix diagrams grows and it is difficult to express the asymptotics of the sums for individual separatrix diagrams in reasonable terms (say, in terms of multiple zeta values). In general case the problem was solved using the following approach suggested by A. Eskin and A. Okounkov in [EOk].

Consider a square-tiled surface $S \in \mathcal{H}(d_1, \dots, d_m)$. Enumerate the squares in some way. For the square number j let $\pi_r(j)$ be the number of its neighbor to the right and let $\pi_u(j)$ be the number of the square atop the square number j. Consider the commutator $\pi' = \pi_r\pi_u\pi_r^{-1}\pi_u^{-1}$ of the resulting permutations. When the total number of squares is big enough, for most of the squares

Geometrically the resulting permutation π' corresponds to the following path: we start from a square number j, then we move one step right, one step up, one step left, one step down, and we arrive to $\pi'(j)$. When the total number of squares is large, then for majority of the squares such path brings us back to the initial square; for these values of index j we get $\pi'(j) = j$. However, we may have more than 4 squares adjacent to a conical singularity $P_k \in S$. For the squares adjacent to a singularity the path right-up-left-down does not bring us back to the initial square. It is easy to check that the commutator $\pi' = \pi_r \pi_u \pi_r^{-1} \pi_u^{-1}$ is represented as a product of m cycles of lengths $(d_1+1), \dots, (d_m+1)$ correspondingly.

We conclude that a square-tiled surface $S \in \mathcal{H}(d_1, \dots, d_m)$ can be defined by a pair of permutations π_r, π_u, such that the commutator $\pi_r \pi_u \pi_r^{-1} \pi_u^{-1}$ decomposes into given number m of cycles of given lengths $(d_1+1), \dots, (d_m+1)$. Choosing another enumeration of the squares of the same square-tiled surface S we get two new permutations $\tilde{\pi}_r, \tilde{\pi}_u$. Clearly the permutations in this new pair are conjugate to the initial ones by means of the same permutation ρ responsible for the change of enumeration of the squares: $\tilde{\pi}_r = \rho \pi_r \rho^{-1}, \tilde{\pi}_u = \rho \pi_u \rho^{-1}$.

We see that the problem of enumeration of square-tiled surfaces can be reformulated as a problem of enumeration of pairs of permutations of at most N elements such that their commutator decomposes into a given number of cycles of given lengths. Here the pairs of permutations are considered up to a simultaneous conjugation. This problem was solved by S. Bloch and A. Okounkov by using methods of representation theory. However, it is not directly applicable to our problem. Describing the square-tiled surfaces in terms of pairs of permutations one has to add an additional explicit constraint that the resulting square-tiled surface is *connected*! Taking a random pair of permutations of very large number $N \gg 1$ of elements realizing some fixed combinatorics of the commutator we usually get a disconnected surface!

The necessary correction is quite nontrivial. It was performed by A. Eskin and A. Okounkov in [EOk]. In the further paper A. Eskin, A. Okounkov and R. Pandharipande [EOkPnd] give the volumes of all individual connected components of the strata; see also very nice and accessible survey [E].

For a given square-tiled surface S denote by $Aut(S)$ its automorphism group. Here we count only those automorphisms which isometrically send each square of the tiling to another square. For most of the square-tiled surfaces $Aut(S)$ is trivial; even for those rare square-tiled surfaces, which have nontrivial inner symmetries the group $Aut(S)$ is obviously finite. We complete this section with the following arithmetic Theorem which confirms two conjectures of M. Kontsevich.

Theorem (A. Eskin, A. Okounkov, R. Pandharipande). *For every connected component of every stratum the generating function*

$$\sum_{N=1}^{\infty} q^N \sum_{\substack{N\text{-square-tiled} \\ \text{surfaces } S}} \frac{1}{|Aut(S)|}$$

is a quasimodular form, i.e. a polynomial in Eisenstein series $G_2(q)$, $G_4(q)$, $G_6(q)$.

Volume $\mathrm{Vol}(\mathcal{H}_1^{comp}(d_1, \ldots, d_m)$ *of every connected component of every stratum is a rational multiple of* π^{2g}*, where* g *is the genus,* $d_1 + \cdots + d_m = 2g - 2$.

8 Crash Course in Teichmüller Theory

In this section we present the Teichmüller theorem about extremal quasiconformal map and define Teichmüller metric. This enables us to explain finally in what sense the "Teichmüller geodesic flow" (which we initially defined in terms of the action of the subgroup of diagonal matrices in $SL(2, \mathbb{R})$ on flat surfaces) is a *geodesic* flow.

8.1 Extremal Quasiconformal Map

Coefficient of Quasiconformality

Consider a closed topological surface of genus g and two complex structures on it. Let S_0 and S_1 be the corresponding Riemann surfaces. When the complex structures are different there are no conformal maps from S_0 to S_1. A smooth map $f : S_0 \to S_1$ (or, being more precise, its derivative Df) sends an infinitesimal circle at $x \in S_0$ to an infinitesimal ellipse, see Fig. 42.

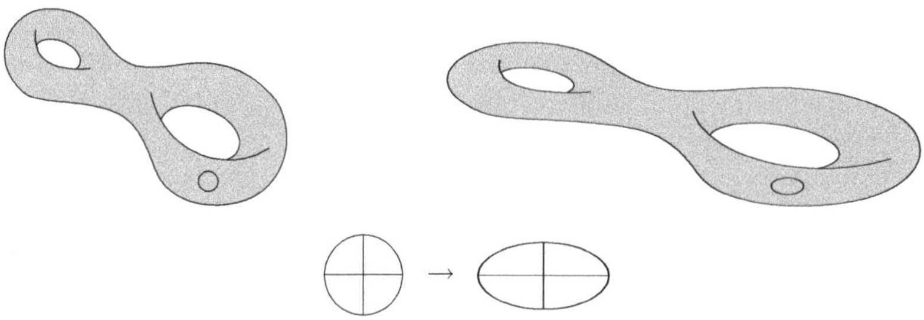

Fig. 42. Quasiconformal map

Coefficient of quasiconformality of f at $x \in S_0$ is the ratio

$$K_x(f) = \frac{a}{b}$$

of demi-axis of this ellipse. *Coefficient of quasiconformality* of f is

$$K(f) = \sup_{x \in S_0} K_x(f)$$

Though S_0 is a compact Riemann surface we use sup and not max since the smooth map f is allowed to have several isolated critical points where $Df = 0$ (and not only $\det(Df) = 0$) and where $K_x(f)$ is not defined.

Half-translation Structure

There is a class of flat metrics which is slightly more general than the *very flat* metrics which we consider in this paper. Namely, we can allow to a flat metric to have the most simple nontrivial linear holonomy which is only possible: we can allow to a tangent vector to change its sign after a parallel transport along some closed loops (see the discussion of linear holonomy in Sect. 1.2).

Surfaces endowed with such flat structures are called *half-translation surfaces*. A holomorphic one-form (also called a holomorphic differential or an Abelian differential) is an analytic object representing a *translation* surface (in our terminology, a *very flat* surface). A holomorphic *quadratic* differential is an analytic object representing a half-translation surface.

In local coordinate w a quadratic differential has the form $q = q(w)(dw)^2$. In other words, the tensor rule for q is

$$q = q(w)(dw)^2 = q\left(w(w')\right) \cdot \left(\frac{dw}{dw'}\right)^2 \cdot (dw')^2 \tag{8.1}$$

under a change of coordinate $w = w(w')$.

One should not confuse $(dw)^2$ with a wedge product $dw \wedge dw$ which equals to zero! It is just a tensor of the type described by the tensor rule (8.1). In particular, any holomorphic one-form defined in local coordinates as $\omega = \omega(w)dw$ canonically defines a quadratic differential $\omega^2 = \omega^2(w)(dw)^2$.

Reciprocally, a holomorphic quadratic differential $q = q(w)(dw)^2$ locally defines a pair of holomorphic one forms $\pm\sqrt{q(w)}\, dw$ in any simply-connected domain where $q(w) \neq 0$. However, for a generic holomorphic quadratic differential neither of these 1-forms is globally defined: trying to extend the local form $\omega_+ = \sqrt{q(w)}\, dw$ along a closed path we may get back with the form $\omega_- = -\sqrt{q(w)}\, dw$.

Recall that there is a bijection between *very flat* (=translation) surfaces and holomorphic 1-forms. There is a similar bijection between half-translation surfaces and holomorphic quadratic differentials, where similar to the "very flat" case a flat surface corresponding to a quadratic differential is polarized: it is endowed with canonical vertical and horizontal directions. (They can be defined locally using the holomorphic one-forms $\omega_\pm = \pm\sqrt{q(w)}\, dw$.) Note, however, that we cannot distinguish anymore between direction to the North and to the South, or between direction to the East and to the West unless the quadratic differential q is a global square of a holomorphic 1-form ω. In particular, the vertical and horizontal foliations are *nonorientable* for generic quadratic differentials.

Teichmüller Theorem

Choose any two complex structures on a topological surface of genus $g \geq 1$; let S_0 and S_1 be the corresponding Riemann surfaces. Developing ideas of Grötzsch Teichmüller has proved a Theorem which we adapt to our language.

Note that *flat structure* used in the formulation of the Theorem below is slightly more general than one considered in Sect. 1.2 and in Convention 1: it corresponds to a half-translation structure and to a holomorphic *quadratic* differential (see above in this section). In particular, speaking about a flat metric compatible with a given complex structure we mean a flat metric corresponding to a quadratic differential holomorphic in the given complex structure.

Theorem (Teichmüller). *For any pair S_0, S_1 of Riemann surfaces of genus $g \geq 1$ there exist an extremal map $f_0 : S_0 \to S_1$ which minimizes the coefficient of quasiconformality $K(f)$.*

For this extremal map f_0 the coefficient of quasiconformality is constant on S_0 (outside of a finite collection of singular points)

$$K_x(f_0) = K(f_0) \quad \forall x \in S_0 - \{P_1, \ldots, P_m\}$$

One can choose a flat metric (half-translation structure) compatible with the complex structure in which foliation along big (correspondingly small) demi-axis of ellipses is the horizontal (correspondingly vertical) foliation in the flat metric.

In flat coordinates the extremal map f_0 is just expansion-contraction with coefficient $\sqrt{K}$.

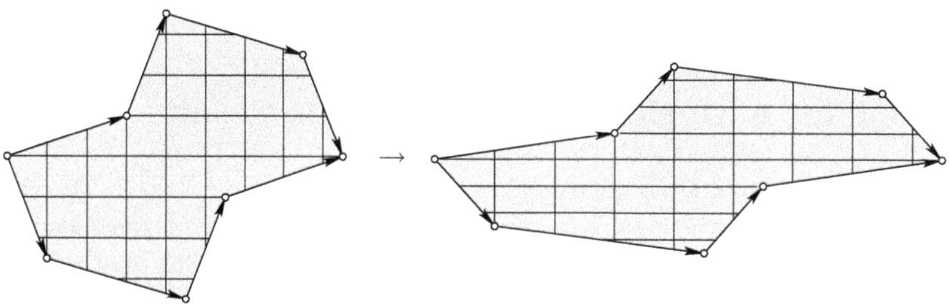

Fig. 43. In flat coordinates the extremal map f_0 is just an expansion-contraction linear map

8.2 Teichmüller Metric and Teichmüller Geodesic Flow

Now everything is ready to define the *Teichmüller metric*. In this metric we measure the distance between two complex structures as

$$dist(S_0, S_1) = \frac{1}{2} \log K(f_0),$$

where $f_0 : S_0 \to S_1$ is the extremal map.

It means that a holomorphic quadratic differential defines a direction of deformation of the complex structure and a geodesic in the Teichmüller metric.

Namely, a holomorphic quadratic differential defines a flat metric. A one-parameter family of maps which in the flat coordinates are defined by diagonal matrices

$$g_t = \begin{pmatrix} e^t & 0 \\ 0 & e^{-t} \end{pmatrix}$$

is a one-parameter family of extremal maps, so it forms a Teichmüller geodesic. According to the definition above we have

$$dist(S_0, g_t S_0) = t.$$

Remark. Note that the Teichmüller metric is not a Riemannian metric but a Finsler metric: it does not correspond to a quadratic form in the tangent space, but just to a norm which depends continuously on the point of the space of complex structures.

It is known that the space of complex structures on a surface of genus $g \geq 2$ has complex dimension $3g - 3$. We have seen, that the space of pairs (complex structure, holomorphic quadratic differential) can be identified with a tangent space to the space of complex structures, in particular, it has complex dimension $6g-6$. Taking into consideration the functorial behavior of the space of pairs (complex structure, holomorphic quadratic differential) one can check, that it should be identified with a total space of a *cotangent* bundle.

9 Hope for a Magic Wand and Recent Results

This section is devoted to one of the most challenging problems in the theory of flat surfaces: to the problem of complete classification of the closures of *all* orbits of $GL^+(2, \mathbb{R})$ on the moduli spaces of Abelian (and quadratic) differentials. This problem was very recently solved for genus two in the works of K. Calta and of C. McMullen; we give a short survey of their results in Sect. 9.7 and 9.8.

9.1 Complex Geodesics

In this section we are following the geometric approach of C. McMullen developed in [McM2] and [McM3].

Fix the genus g of the surfaces. We have seen in the previous section that we can identify the space $\mathcal{Q}$ of pairs (complex structure, holomorphic quadratic differential) with the total space of the (co)tangent bundle to the moduli space $\mathcal{M}$ of complex structures. Space $\mathcal{H}$ of pairs (complex structure, holomorphic quadratic differential) can be identified with a subspace in $\mathcal{Q}$ of those quadratic differentials, which can be represented as global squares of holomorphic 1-forms. This subspace forms a vector subbundle of special directions in the (co)tangent space which we denote by the same symbol $\mathcal{H}$.

The "unit hyperboloid" $\mathcal{H}_1 \subset \mathcal{H}$ of holomorphic 1-forms corresponding to flat surfaces of unit area can be considered as a subbundle of *unit vectors* in $\mathcal{H}$. It is invariant under the Teichmüller geodesic flow – the geodesic flow for the Teichmüller metric.

One can check that an $SL(2;\mathbb{R})$-orbit in $\mathcal{H}_1$ descends to a commutative diagram

$$\begin{array}{ccc} SL(2;\mathbb{R}) & \longrightarrow & \mathcal{H}_1 \\ \downarrow & & \downarrow \\ SL(2;\mathbb{R})/SO(2;\mathbb{R}) \simeq \mathbb{H}^2 & \longrightarrow & \mathcal{M}, \end{array} \tag{9.1}$$

which we interpret as

$$\begin{array}{ccc} \begin{pmatrix} \text{Unit tangent} \\ \text{bundle to} \\ \text{hyperbolic plane} \end{pmatrix} & \longrightarrow & \begin{pmatrix} \text{Unit tangent} \\ \text{subbundle to} \\ \text{moduli space} \end{pmatrix} \\ \downarrow & & \downarrow \\ \begin{pmatrix} \text{Hyperbolic} \\ \text{plane} \end{pmatrix} & \longrightarrow & \begin{pmatrix} \text{Moduli} \\ \text{space} \end{pmatrix} \end{array}$$

Moreover, it can be checked that the map $\mathbb{H}^2 \to \mathcal{M}$ in this diagram is an isometry with respect to the standard hyperbolic metric on $\mathbb{H}^2$ and Teichmüller metric on $\mathcal{M}$. Thus, following C. McMullen it is natural to call the projections of $SL(2;\mathbb{R})$-orbits to $\mathcal{M}$ (which coincide with the images of $\mathbb{H}^2$) *complex geodesics*. Another name for these projections is *Teichmüller discs*.

9.2 Geometric Counterparts of Ratner's Theorem

Though it is proved that the moduli space of complex structures is not a hyperbolic manifold (see [Ma1]) there is a strong hope that with respect to $SL(2,\mathbb{R})$-action on $\mathcal{H}$ and on $\mathcal{Q}$ the moduli space behaves as if it is.

In this section we present several facts about group actions on homogeneous spaces and several related facts about geodesic submanifolds. We warn the reader that our selection is not representative; it opens only a narrow slit to the fascinating world of interactions of group actions, rigidity, hyperbolic geometry, dynamics and number theory.

We start with an informal formulation of part of Ratner's Theorem (see much better exposition adopted to our subject in the survey of A. Eskin [E]).

A discrete subgroup Γ of a Lie group G is called a *lattice* if a homogeneous space G/Γ has finite volume.

Theorem (M. Ratner). *Let G be a connected Lie group and U a connected subgroup generated by unipotent elements. Then, for any lattice $\Gamma \subset G$ and any $x \in G/\Gamma$ the closure of the orbit Ux in G/Γ is an orbit of some closed algebraic subgroup of G.*

We would like to point out why this theorem is so remarkably powerful. Considering a dynamical system, even an ergodic one, it is possible to get a lot of information about a generic (in measure-theoretical sense) trajectory. However, usually there are plenty of trajectories having rather particular behavior. It is sufficient to consider geodesic flow on a surface with cusps to find trajectories with closures producing fairly wild sets. Ratner's theorem proves, that the closure of *any* orbit of a unipotent group acting on a homogeneous space is a nice homogeneous space.

Ratner's theorem has numerous important relations with geometry of homogeneous spaces. As an illustration we have chosen a result of N. Shah [Sh] and a generalization of his result for noncompact hyperbolic manifolds obtained by T. Payne [Pa].

Theorem (N. Shah). *In a compact manifold of constant negative curvature, the closure of a totally geodesic, complete (immersed) submanifold of dimension at least* 2 *is a totally geodesic immersed submanifold.*

Theorem (T. Payne). *Let* $f : M_1 \to M_2$ *be a totally geodesic immersion between locally symmetric spaces of noncompact type, with* M_2 *of finite volume. Then the closure of the image of* f *is an immersed submanifold. Moreover, when* M_1 *and* M_2 *have the same rank, the closure of the image is a totally geodesic submanifold.*

9.3 Main Hope

Main Conjecture and its Possible Applications

If only the moduli space of complex structures $\mathcal{M}$ would be a homogeneous space we would immediately apply the Theorem above to diagram (9.1) and would solve considerable part of our problems. But it is not. Nevertheless, there is a strong hope for an analogous Theorem.

Problem. *Classify the closures of* $GL^+(2,\mathbb{R})$*-orbits in* $\mathcal{H}_g$ *and in* $\mathcal{Q}_g$. *Classify the orbit closures of the unipotent subgroup* $\begin{pmatrix} 1 & t \\ 0 & 1 \end{pmatrix}_{t\in\mathbb{R}}$ *on* $\mathcal{H}_g$ *and on* $\mathcal{Q}_g$.

The following Conjecture is one of the key conjectures in this area.

Conjecture. *The closure* $\mathcal{C}(S)$ *of a* $GL^+(2,\mathbb{R})$*-orbit of any flat surface* $S \in \mathcal{H}$ *(or* $S \in \mathcal{Q}$*) is a complex-algebraic suborbifold.*

Remark. We do not discuss here the problems related with possible *compactifications* of the moduli spaces $\mathcal{H}_g$ and $\mathcal{Q}_g$. A complex-analytic description of a compactification of $\mathcal{Q}_g$ can be found in the papers of J. Fay [Fay] and H. Masur [Ma2].

Recall that according to Theorem of M. Kontsevich (see Sect. 3.6) any $GL^+(2,\mathbb{R})$-invariant complex suborbifold in $\mathcal{H}$ is represented by an affine subspace in cohomological coordinates. Thus, if the Conjecture above is true, the structure of orbit closures of the action of $GL^+(2,\mathbb{R})$ on $\mathcal{H}$ and on $\mathcal{Q}$ (and of $SL(2,\mathbb{R})$ on $\mathcal{H}_1$ and on $\mathcal{Q}_1$) would be as simple as in the case of homogeneous spaces.

We have not discussed the aspects of Ratner's Theorem concerning the measures; it states more than we cited above. Actually, not only orbit closures have a nice form, but also invariant ergodic measures; namely, all such measures are just the natural measures supported on orbits of closed subgroups. Trying to make a parallel with Ratner's Theorem one should extend the Conjecture above to invariant measures.

In the most optimistic hopes the study of an individual flat surface $S \in \mathcal{H}(d_1,\dots,d_m)$ would look as follows. (Frankly speaking, here we slightly exaggerate in our scenario, but after all we are describing the dreams.) To describe all geometric properties of a flat surface S we first find the orbit closure $\mathcal{C}(S) = \overline{GL^+(2,\mathbb{R})\,S} \subset \mathcal{H}(d_1,\dots,d_m)$ (our optimistic hope assumes that there is an efficient way to do this). Then we consult a (conjectural) classification list and find $\mathcal{C}(S)$ in some magic table which gives all information about $\mathcal{C}(S)$ (like volume, description of cusps, Siegel–Veech constants, Lyapunov exponents, adjacency to other invariant subspaces, etc). Using this information we get answers to all possible questions which one can ask about the initial flat surface S.

Billiards in rational polygons give an example of possible implementation of this optimistic scenario. Fixing angles of the polygon which defines a billiard table we can change the lengths of its sides. We get a family $\mathcal{B}$ of polygons which induces a family $\tilde{\mathcal{B}}$ of flat surfaces obtained by Katok–Zemlyakov construction (see Sect. 2.1). This family $\tilde{\mathcal{B}}$ belongs to some fixed stratum $\tilde{\mathcal{B}} \subset \mathcal{H}(d_1,\dots,d_m)$. However, it has a nontrivial codimension in the stratum, so $\tilde{\mathcal{B}}$ has measure zero and one cannot use ergodic theorem naively to get any information about billiards in corresponding polygons. Having a version of ergodic theorem which treats *all* orbits (like in Ratner's Theorem) presumably it would be possible to get a powerful tool for the study of rational billiards.

Exercise. Consider the family $\mathcal{B}$ of billiard tables as on Fig. 37. Determine the stratum $\mathcal{H}(d_1,\dots,d_m)$ to which belong the corresponding flat surfaces and compute the codimension of the resulting family $\tilde{\mathcal{B}} \subset \mathcal{H}(d_1,\dots,d_m)$.

Content of Remaining Sections

The Conjecture above is trivial for genus one, since in this case the "Teichmüller space of Riemann surfaces of genus one" coincides with an upper half-plane, and the entire space coincides with a single Teichmüller disc (image of $\mathbb{H}^2$ in diagram (9.1)).

Very recently C. McMullen proved the Conjecture in genus two [McM2], [McM3], and this is a highly nontrivial result. We give a short report of revolutionary results of K. Calta [Clt] and of C. McMullen [McM2]–[McM6] in Sect. 9.7 below. However, their techniques, do not allow any straightforward generalizations to higher genera: Riemann surfaces of genus two are rather special, in particular, every such surface is hyperelliptic.

In the next two sections 9.4 and 9.5 we try to give an idea of what is known about invariant submanifolds in higher genera (which is an easy task since, unfortunately, little is known).

Having an invariant submanifold $\mathcal{K} \subset \mathcal{H}_g$ (or $\mathcal{K} \subset \mathcal{Q}_g$) in genus g one can construct a new invariant submanifold $\tilde{\mathcal{K}} \subset \mathcal{H}_{\tilde{g}}$ (correspondingly $\tilde{\mathcal{K}} \subset \mathcal{Q}_{\tilde{g}}$) in higher genus $\tilde{g} > g$ replacing every $S \in \mathcal{K}$ by an appropriate ramified covering $\tilde{S}$ over S of some fixed combinatorial type. We do not want to specify here what does a "fixed combinatorial type" mean; what we claim is that having an invariant manifold $\mathcal{K}$ there is some procedure which allows to construct a whole bunch of new invariant submanifolds $\tilde{\mathcal{K}}$ for higher genera $\tilde{g} > g$.

In some cases all quadratic differentials in the invariant submanifold $\tilde{\mathcal{K}}$ obtained by a ramified covering construction from some $\mathcal{K} \subset \mathcal{Q}_g$ might become global squares of Abelian differentials. Hence, using special ramified coverings one can construct $GL^+(2,\mathbb{R})$-invariant submanifolds $\tilde{\mathcal{K}} \subset \mathcal{H}_{\tilde{g}}$ from invariant submanifolds $\mathcal{K} \subset \mathcal{Q}_g$.

What is really interesting to understand is what invariant manifolds form the "roots" of such constructions. Such invariant manifolds are often called the *primitive* ones.

In the following two sections 9.4 and 9.5 we consider the two extremal classes of primitive invariant submanifolds: the largest ones and the smallest ones. Namely, in Sect. 9.4 we present a classification of connected components of the strata $\mathcal{H}(d_1, \dots, d_m)$. It follows from ergodicity of $SL(2,\mathbb{R})$-action on $\mathcal{H}_1(d_1, \dots, d_m)$ that the orbit closure of almost any flat surface in $\mathcal{H}(d_1, \dots, d_m)$ coincides with the embodying connected component of $\mathcal{H}(d_1, \dots, d_m)$.

In section 9.5 we consider the smallest possible $GL^+(2,\mathbb{R})$-invariant submanifolds: those which correspond to closed orbits. Teichmüller discs obtained as projections of such orbits to the moduli space $\mathcal{M}$ of complex structures form the "closed complex geodesics" — Riemann surfaces with cusps.

9.4 Classification of Connected Components of the Strata

In order to formulate the classification theorem for connected components of the strata $\mathcal{H}(d_1, \dots, d_m)$ we need to describe the classifying invariants. There are two of them: *spin structure* and *hyperellipticity*. Both notions are applicable only to part of the strata: flat surfaces from the strata $\mathcal{H}(2d_1, \dots, 2d_m)$ have *even* or *odd spin structure*. The strata $\mathcal{H}(2g-2)$ and $\mathcal{H}(g-1, g-1)$ have special *hyperelliptic connected component*.

Spin Structure

Consider a flat surface S from a stratum $\mathcal{H}(2d_1,\dots,2d_m)$. Let $\rho : S^1 \to S$ be a smooth closed path on S; here S^1 is a standard circle. Note that at any point of the surfaces S we know where is the "direction to the North". Hence, at any point $\rho(t) = x \in S$ we can apply a compass and measure the direction of the tangent vector $\dot{x}$. Moving along our path $\rho(t)$ we make the tangent vector turn in the compass. Thus we get a map $G(\rho) : S^1 \to S^1$ from the parameter circle to the circumference of the compass. This map is called the *Gauss map*. We define the *index* $\operatorname{ind}(\rho)$ of the path ρ as a degree of the corresponding Gauss map (or, in other words as the algebraic number of turns of the tangent vector around the compass) taken modulo 2.

$$\operatorname{ind}(\rho) = \deg G(\rho) \mod 2$$

It is easy to see that $\operatorname{ind}(\rho)$ does not depend on parameterization. Moreover, it does not change under small deformations of the path. Deforming the path more drastically we may change its position with respect to conical singularities of the flat metric. Say, the initial path might go on the left of P_k and its deformation might pass on the right of P_k. This deformation changes the $\deg G(\rho)$. However, if the cone angle at P_k is of the type $2\pi(2d_k+1)$, then $\deg G(\rho) \mod 2$ does not change! This observation explains why $\operatorname{ind}(\rho)$ is well-defined for a free homotopy class $[\rho]$ when $S \in \mathcal{H}(2d_1,\dots,2d_m)$ (and hence, when all cone angles are odd multiples of 2π).

Consider a collection of closed smooth paths $a_1, b_1, \dots, a_g, b_g$ representing a symplectic basis of homology $H_1(S, \mathbb{Z}/2\mathbb{Z})$. We define the *parity of the spin-structure* of a flat surface $S \in \mathcal{H}(2d_1,\dots,2d_m)$ as

$$\phi(S) = \sum_{i=1}^{g} (\operatorname{ind}(a_i)+1)(\operatorname{ind}(b_i)+1) \mod 2 \tag{9.2}$$

Lemma. *The value $\phi(S)$ does not depend on the choice of symplectic basis of cycles $\{a_i, b_i\}$. It does not change under continuous deformations of S in $\mathcal{H}(2d_1,\dots,2d_m)$.*

Lemma above shows that the parity of the spin structure is an invariant of connected components of strata of those Abelian differentials, which have zeroes of even degrees.

Exercise. Consider two flat surfaces presented at Fig. 44. They are obtained by a surgery which attaches a handle to a flat surface obtained from a regular octagon. Note, however, that the handles are attached in two different ways (see the identifications of vertical sides). Check that both surfaces belong to the same stratum $\mathcal{H}(4)$.

Consider a symplectic basis of cycles of the initial surface (corresponding to the regular octagon) realized by paths which do not pass through the

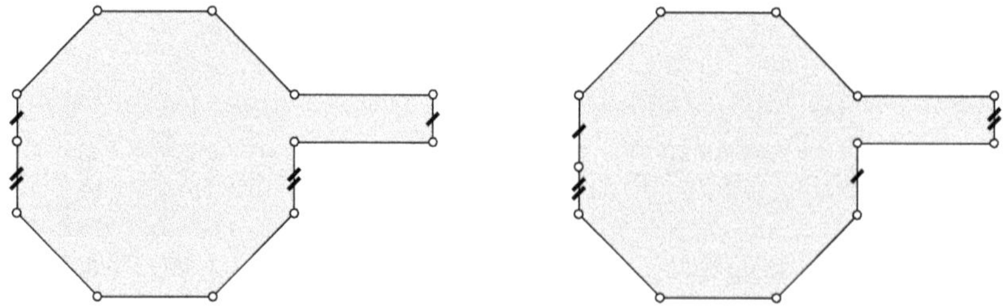

Fig. 44. Attaching a handle to the flat surface $S_0 \in \mathcal{H}(2)$ in two different ways we get two flat surfaces in $\mathcal{H}(4)$ with different parities of spin structure. Hence the resulting flat surfaces live in different connected components of $\mathcal{H}(4)$

conical singularity. Show that a symplectic basis of cycles for each of two new surfaces can be obtained by completion of the initial basis with a pair of cycles a, b representing the attached handle, where the cycle a is the waist curve of the handle. Calculate $\operatorname{ind}(a)$ and $\operatorname{ind}(b)$ for each of two surfaces. Check that $ind(b)$ are different, and thus our two flat surfaces have different parities of spin structure and hence belong to different connected components of $\mathcal{H}(4)$.

Spin structure: more serious reading. We have hidden under the carpet geometry of the "spin structure" defining the "parity-of-the-spin-structure". The reader can find details in [KonZo] and in original papers of M.Atiyah [At], J. Milnor [Mil], D.Mumford [Mum] and D. Johnson [J]. Recent paper of C. McMullen [McM4] contains further applications of spin structures to flat surfaces.

Hyperellipticity

A flat surface S may have a symmetry; one specific family of such flat surfaces, which are "more symmetric than others" is of a special interest for us. Recall that there is a one-to-one correspondence between flat surfaces and pairs (Riemann surface M , holomorphic 1-form ω), see Sect. 3.3. When the corresponding Riemann surface is *hyperelliptic* the *hyperelliptic involution* $\tau : M \to M$ acts on any holomorphic 1-form ω as $\tau^*\omega = -\omega$.

We say that a flat surface S is a *hyperelliptic flat surface* if there is an isometry $\tau : S \to S$ such that τ is an involution, $\tau \circ \tau = \operatorname{id}$, and the quotient surface S/τ is a topological sphere. In flat coordinates differential of such involution obviously satisfies $D\tau = -\operatorname{Id}$.

Exercise. Check that the flat surface S from Fig. 12 is hyperelliptic, and that the central symmetry of the polygon induces the hyperelliptic involution of S.

In a general stratum $\mathcal{H}(d_1, \dots, d_m)$ hyperelliptic surfaces form a small subspace of nontrivial codimension. However, there are two special strata, namely $\mathcal{H}(2g-2)$ and $\mathcal{H}(g-1, g-1)$, for which hyperelliptic surfaces form entire *hyperelliptic connected components* $\mathcal{H}^{hyp}(2g-2)$ and $\mathcal{H}^{hyp}(g-1, g-1)$ correspondingly.

Note that in the stratum $\mathcal{H}(g-1,g-1)$ there are hyperelliptic flat surfaces of two different types. A hyperelliptic involution $\tau S \to S$ may fix the conical points or might interchange them. It is not difficult to show that for surfaces from the *connected component* $\mathcal{H}^{hyp}(g-1,g-1)$ the hyperelliptic involution *interchanges* the conical singularities.

The remaining family of those hyperelliptic flat surfaces in $\mathcal{H}(g-1,g-1)$, for which the hyperelliptic involution keeps the saddle points fixed, forms a subspace of nontrivial codimension in the complement $\mathcal{H}(g-1,g-1)-\mathcal{H}^{hyp}(g-1,g-1)$. Thus, the hyperelliptic connected component $\mathcal{H}^{hyp}(g-1,g-1)$ does not coincide with the space of all hyperelliptic flat surfaces.

Classification Theorem for Abelian Differentials

Now, having introduced the classifying invariants we can present the classification of connected components of strata of Abelian differentials.

Theorem (M. Kontsevich and A. Zorich). *All connected components of any stratum of Abelian differentials on a curve of genus $g \geq 4$ are described by the following list:*

The stratum $\mathcal{H}(2g-2)$ has three connected components: the hyperelliptic one, $\mathcal{H}^{hyp}(2g-2)$, and two nonhyperelliptic components: $\mathcal{H}^{even}(2g-2)$ and $\mathcal{H}^{odd}(2g-2)$ corresponding to even and odd spin structures.

The stratum $\mathcal{H}(2d,2d)$, $d \geq 2$ has three connected components: the hyperelliptic one, $\mathcal{H}^{hyp}(2d,2d)$, and two nonhyperelliptic components: $\mathcal{H}^{even}(2d,2d)$ and $\mathcal{H}^{odd}(2d,2d)$.

All the other strata of the form $\mathcal{H}(2d_1,\dots,2d_m)$ have two connected components: $\mathcal{H}^{even}(2d_1,\dots,2d_m)$ and $\mathcal{H}^{odd}(2d_1,\dots,2d_n)$, corresponding to even and odd spin structures.

The stratum $\mathcal{H}(2d-1,2d-1)$, $d \geq 2$, has two connected components; one of them: $\mathcal{H}^{hyp}(2d-1,2d-1)$ is hyperelliptic; the other $\mathcal{H}^{nonhyp}(2d-1,2d-1)$ is not.

All the other strata of Abelian differentials on the curves of genera $g \geq 4$ are nonempty and connected.

In the case of small genera $1 \leq g \leq 3$ some components are missing in comparison with the general case.

Theorem. *The moduli space of Abelian differentials on a curve of genus $g = 2$ contains two strata: $\mathcal{H}(1,1)$ and $\mathcal{H}(2)$. Each of them is connected and coincides with its hyperelliptic component.*

Each of the strata $\mathcal{H}(2,2)$, $\mathcal{H}(4)$ of the moduli space of Abelian differentials on a curve of genus $g = 3$ has two connected components: the hyperelliptic one, and one having odd spin structure. The other strata are connected for genus $g = 3$.

Since there is a one-to-one correspondence between connected components of the strata and *extended Rauzy classes* (see Sect. 5.6 and paper [Y] in this collection) the Classification Theorem above classifies also the extended Rauzy classes.

Classification Theorem for Quadratic Differentials

Note that for any partition $d_1 + \cdots + d_m = 2g - 2$ of a positive even integer $2g - 2$ the stratum $\mathcal{H}(d_1, \ldots, d_m)$ of Abelian differentials is nonempty. For meromorphic quadratic differentials with at most simple poles there are four *empty* strata! Namely,

Theorem (H. Masur and J. Smillie). *Consider a partition of the number $4g - 4$, where $g \geq 0$ into integers $d_1 + \cdots + d_m$ with all $d_j \in \mathbb{N} \cup \{-1\}$. The corresponding stratum $\mathcal{Q}(d_1, \ldots, d_m)$ is non-empty with the following four exceptions:*

$$\mathcal{Q}(\emptyset), \mathcal{Q}(1,-1) \ (\textit{in genus } g = 1) \quad \textit{and} \quad \mathcal{Q}(4), \mathcal{Q}(1,3) \ (\textit{in genus } g = 2)$$

Classification of connected components of the strata of meromorphic quadratic differentials with at most simple poles was recently obtained by E. Lanneau [Lan].

Theorem (E. Lanneau). *Four exceptional strata $\mathcal{Q}(-1,9)$, $\mathcal{Q}(-1,3,6)$, $\mathcal{Q}(-1,3,3,3)$ and $\mathcal{Q}(12)$ of meromorphic quadratic differentials contain exactly two connected components; none of them hyperelliptic.*

Three series of strata

$$\begin{array}{ll}
\mathcal{Q}(2(g-k)-3, 2(g-k)-3, 2k+1, 2k+1) & k \geq -1,\ g \geq 1,\ g-k \geq 2 \\
\mathcal{Q}(2(g-k)-3, 2(g-k)-3, 4k+2) & k \geq 0,\ g \geq 1 \ \textit{and}\ g-k \geq 1 \\
\mathcal{Q}(4(g-k)-6, 4k+2) & k \geq 0,\ g \geq 2 \ \textit{and}\ g-k \geq 2
\end{array}$$

contain hyperelliptic connected components. The strata from these series in genera $g \geq 3$ and the strata $\mathcal{Q}(-1,-1,3,3)$, $\mathcal{Q}(-1,-1,6)$ in genus $g = 2$ contain exactly two connected components; one of them – hyperelliptic, the other one – not.

The remaining strata from these series, namely, $\mathcal{Q}(1,1,1,1)$, $\mathcal{Q}(1,1,2)$, $\mathcal{Q}(2,2)$ in genus $g = 2$ and $\mathcal{Q}(1,1,-1,-1)$, $\mathcal{Q}(-1,-1,2)$ in genus $g = 1$ coincide with their hyperelliptic connected component.

All other strata of meromorphic quadratic differentials with at most simple poles are connected.

Recall that having a meromorphic quadratic differential with at most simple poles one can associate to it a surface with a flat metric which is slightly more general that our usual *very* flat metric (see Sect. 8.1 for a discussion of half-translation structures).

It is easy to verify whether a half-translation surface belongs to a hyperelliptic component or not. However, currently there is no simple and efficient way to distinguish half-translation surfaces from the four exceptional components.

Problem. *Find an invariant of the half-translation structure which would be easy to evaluate and which would distinguish half-translation surfaces from different connected components of the four exceptional strata* $\mathcal{Q}(-1,9)$, $\mathcal{Q}(-1,3,6)$, $\mathcal{Q}(-1,3,3,3)$ *and* $\mathcal{Q}(12)$.

Currently there are two ways to determine to which of the two connected components of an exceptional stratum belongs a flat surface S.

The first approach suggests to find a "generalized permutation" for an analog of the first return map of the vertical flow to a horizontal segment and then to find it in one of the two extended Rauzy classes (see Sect. 5.6) corresponding to two connected components. Note, however, that already for the stratum $\mathcal{Q}(-1,9)$ the corresponding Rauzy classes contain 97 544, and 12 978 generalized permutations; the Rauzy classes for the components of $\mathcal{Q}(12)$ contain already 894 117 and 150 457 elements.

In the second approach one studies configurations of saddle connections (see Sect. 6.4) on the surface S and tries to find a configuration which is forbidden for one of the two connected components of the corresponding stratum.

For example, for surfaces from one of the two connected components of $\mathcal{Q}(-1,9)$ as soon as we have a saddle connection joining the simple pole with the zero we necessarily have a closed geodesic going in the same direction. Thus, if we manage to find on a surface $S \in \mathcal{Q}(-1,9)$ a saddle connection which is not accompanied by a parallel closed geodesic, S belong to the other connected component of $\mathcal{Q}(-1,9)$.

9.5 Veech Surfaces

For almost every flat surface S in any stratum $\mathcal{H}_1(d_1,\dots,d_m)$ the orbit $SL(2,\mathbb{R})\cdot S$ is dense in the stratum and for any $g_1 \neq g_2 \in SL(2,\mathbb{R})$ we have $g_1 S \neq g_2 S$. However, some flat surfaces have extra symmetries. When a flat surface S_0 has an affine automorphism, i.e. when for some $g_0 \in SL(2,\mathbb{R})$ we get $g_0 S = S$ the orbit of S_0 is smaller than usual.

The *stabilizer* $Stab(S) \in SL(2;\mathbb{R})$, that is a subgroup of those $g \in SL(2,\mathbb{R})$ for which $gS = S$, is called the *Veech group* of the flat surface S and is denoted $SL(S)$. In representation of the flat surface S in terms of a pair (Riemann surface X, holomorphic 1-form ω on it) the Veech group is denoted as $SL(X,\omega)$ following the notation of C. McMullen [McM2].

Some exceptional flat surfaces S possess very large group of symmetry and their orbits are very small. The flat surfaces having the largest possible symmetry group are called *Veech surfaces.* More precisely, a flat surface is called a *Veech surface* if its Veech group $SL(S)$ is a lattice in $SL(2,\mathbb{R})$ (that is the quotient $SL(2,\mathbb{R})/SL(S)$ has finite volume).

Theorem (J. Smillie). *An* $SL(2,\mathbb{R})$*-orbit of a flat surface* S *is closed in* $\mathcal{H}_1(d_1,\dots,d_m)$ *if and only if* S *is a Veech surface.*

Forgetting polarization (direction to the "North") of a flat surface we get a *Teichmüller disc* of S (see (9.1) and the comments below it)

$$SO(2,\mathbb{R})\backslash SL(2,\mathbb{R})/SL(S) = \mathbb{H}/SL(S)$$

A flat surface S is a Veech surface if its Teichmüller disc $\mathbb{H}^2/SL(S)$ has finite volume. However, even for a Veech surface the Teichmüller disc is never compact: it necessarily contains at least one cusp. The Teichmüller discs of Veech surfaces can be considered as *closed complex geodesics* (see discussion at the end of Sect. 9.1).

Consider an elementary example. As a flat surface take a flat torus obtained from a unit square. Fig. 45 shows why the element $g_+ = \begin{pmatrix} 1 & 1 \\ 0 & 1 \end{pmatrix} \in SL(2,\mathbb{Z})$ belongs to a stabilizer of S.

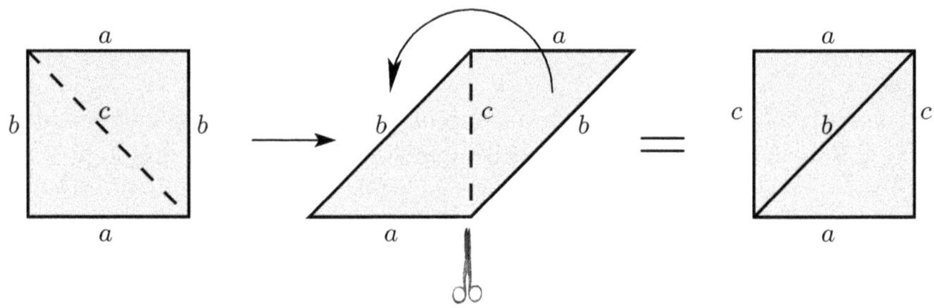

Fig. 45. This linear transformation belongs to the Veech group of $\mathbb{T}^2$

Similarly the element $g_- = \begin{pmatrix} 1 & 0 \\ 1 & 1 \end{pmatrix} \in SL(2,\mathbb{R})$ also belongs to the Veech group $SL(\mathbb{T}^2)$ of $\mathbb{T}^2$. Since the group $SL(2,\mathbb{Z})$ is generated by g_+ and g_- we conclude that $SL(2,\mathbb{Z}) \subset SL(\mathbb{T}^2)$. It is easy to check that, actually, $SL(2,\mathbb{Z}) = SL(\mathbb{T}^2)$. As the Teichmüller disc of $\mathbb{T}^2$ we get the modular curve $\mathbb{H}^2/SL(2,\mathbb{Z})$ (see Fig. 13) which, actually, coincides with the moduli space of complex structures on the torus.

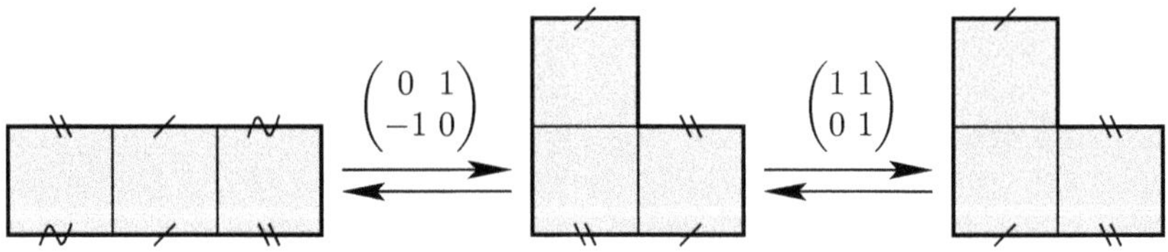

Fig. 46. There are three 3-square-tiled surfaces in $\mathcal{H}(2)$. Our picture shows that they all belong to the same $SL(2;\mathbb{Z})$-orbit

Consider a slightly more complicated example.

Exercise. Verify that square-tiled surfaces presented at Fig. 46 belong to the stratum $\mathcal{H}(2)$. Show that there are no other 3-square-tiled surfaces. Verify that the linear transformations indicated at Fig. 46 act as it is described on the Figure; check that the surfaces belong to the same $SL(2,\mathbb{R})$-orbit. Find Veech groups of these three surfaces. Show that these flat surfaces are Veech surfaces. Verify that the corresponding Teichmüller disc is a triple cover over the modular curve (see Fig. 47).

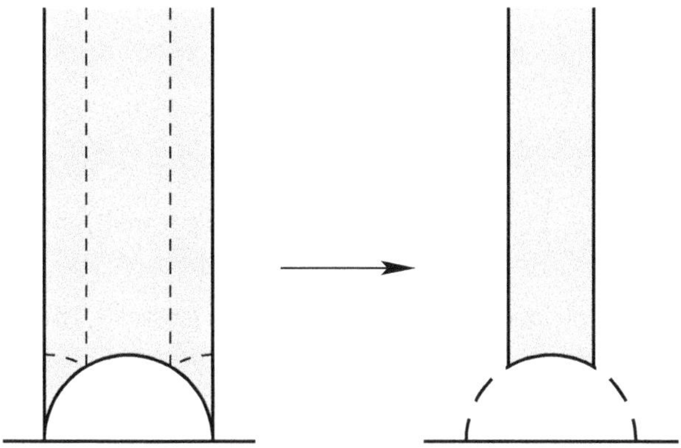

Fig. 47. Teichmüller discs of a 3-square-tiled surface is a triple cover over the modular curve

Primitive Veech Surfaces

It is not difficult to generalize the Exercise above and to show that any *square-tiled surface* (see Sect. 7.1) is necessarily a Veech surface.

A square-tiled surface is a ramified covering over a flat torus, such that all ramification points project to the same point on the flat torus, which is a Veech surface. One can generalize this observation. Having a Veech surface S one can construct a ramified covering $\tilde{S} \to S$ such that all ramification points on $\tilde{S}$ project to conical singularities on S. One can check that any such $\tilde{S}$ is a Veech surface. Thus, having a Veech surface we can construct a whole bunch of Veech surfaces in higher genera.

Veech surfaces which cannot be obtained from simpler Veech surfaces by the covering construction are called *primitive*. For a long time (and till recent revolution in genus two, see Sect. 9.7, the list of known primitive Veech surfaces was very short. Very recently C. McMullen has found infinitely many Veech surfaces in genera 3 and 4 as well, see [McM7]. All other known primitive Veech surfaces of genus $g > 2$ can be obtained by Katok–Zemlyakov construction (see Sect. 2.1) from triangular billiards of the first three types in the list below:

$$\left(\frac{\pi}{n}, \frac{n-1}{2n}\pi, \frac{n-1}{2n}\pi\right), \quad \text{for } n \geq 6 \text{ (discovered by W. Veech)}$$

$$\left(\frac{\pi}{n}, \frac{\pi}{n}, \frac{n-2}{n}\pi\right), \quad \text{for } n \geq 7 \text{ (discovered by W. Veech)}$$

$$\left(\frac{\pi}{n}, \frac{\pi}{2n}, \frac{2n-3}{2n}\pi\right), \quad \text{for } n \geq 4 \text{ (discovered by Ya. Vorobets)}$$

$$\left(\frac{\pi}{3}, \frac{\pi}{4}, \frac{5\pi}{12}\right) \quad \text{(discovered by W. Veech))}$$

$$\left(\frac{\pi}{3}, \frac{\pi}{5}, \frac{7\pi}{15}\right) \quad \text{(discovered by Ya. Vorobets)}$$

$$\left(\frac{2\pi}{9}, \frac{3\pi}{9}, \frac{4\pi}{9}\right) \quad \text{(discovered by R. Kenyon and J. Smillie)}$$

$$\left(\frac{\pi}{3}, \frac{\pi}{12}, \frac{7\pi}{12}\right) \quad \text{(discovered by W. P. Hooper)}$$

— The flat surface corresponding to the isosceles triangle with the angles $\pi/n, (n-1)\pi/(2n), (n-1)\pi/(2n)$ belongs to the hyperelliptic component $\mathcal{H}^{hyp}(2g-2)$ when $n = 2g$ and to the hyperelliptic component $\mathcal{H}^{hyp}(g-1, g-1)$ when $n = 2g+1$. The surface can be unwrapped to the regular $2n$-gon with opposite sides identified by parallel translations, see Fig. 21.

— The flat surface corresponding to the isosceles triangle with the angles $\pi/n, \pi/n, (n-2)\pi/(2n)$ belongs to the hyperelliptic component $\mathcal{H}^{hyp}(2g-2)$ when $n = 2g+1$ and to the hyperelliptic component $\mathcal{H}^{hyp}(g-1, g-1)$ when $n = 2g+2$. The surface can be unwrapped to a pair of regular n-gons glued by one side. Each side of one polygon is identified by a parallel translation with the corresponding side of the other polygon, see Fig. 7.

— The flat surface corresponding to the obtuse triangle with the angles $\pi/n, \pi/(2n), (2n-3)\pi/(2n)$ belongs to one of two nonhyperelliptic components of the stratum $\mathcal{H}(2g-2)$ where $n = g+1$.

— The flat surface corresponding to the acute triangle $\pi/3, \pi/4, 5\pi/12$ belongs to the nonhyperelliptic component $\mathcal{H}^{odd}(4)$; here $g = 3$.

— The flat surface corresponding to the acute triangle $\pi/3, \pi/5, 7\pi/15$ belongs to the nonhyperelliptic component $\mathcal{H}^{even}(6)$; here $g = 4$.

— The flat surface corresponding to the acute triangle $2\pi/9, 3\pi/9, 4\pi/9$ belongs to the stratum $\mathcal{H}(3, 1)$; here $g = 3$.

— The flat surface corresponding to the obtuse triangle $\pi/3, \pi/12, 7\pi/12$ belongs to the stratum $\mathcal{H}(6)$; here $g = 4$ (the information that this is a Veech surface is taken from [McM7]).

The details on unwrapping of these surfaces and on cylinder decompositions of some of them can be found in the paper of Ya. Vorobets [Vb1].

It is proved that unwrapping triangular billiards in other acute, rectangular or isosceles triangles does not give new Veech surfaces in genera $g > 2$ (see [KenS], [Pu], [Vb1] and further references in these papers). For obtuse triangles the question is open.

We discuss genus $g = 2$ separately in the next section: very recently K. Calta [Clt] and C. McMullen [McM2] have found a countable family of primitive Veech surfaces in the stratum $\mathcal{H}(2)$ and proved that the list is complete. However, even in genus $g = 2$ the situation with the stratum $\mathcal{H}(1,1)$ is drastically different: using the results of M. Moeller [Mo1]–[Mo3] very recently C. McMullen has proved [McM6] the following result.

Theorem (C. McMullen). *The only primitive Veech surface in the stratum $\mathcal{H}(1,1)$ is the surface represented by the regular decagon with identified opposite sides.*

Thus, it is not clear, what one should expect as a solution of the following general problem.

Problem. *Find all primitive Veech surfaces.*

An algebro-geometric approach to Veech surfaces suggested by M. Möller in [Mo1] and [Mo2] might help to shed some light on this Problem.

Veech surfaces: more serious reading. I recommend the survey paper [HuSdt5] of P. Hubert and T. Schmidt as an introduction to Veech surfaces. A canonical reference for square-tiled surfaces (also called *arithmetic Veech surfaces*) is the paper of E. Gutkin and C. Judge [GuJg]. More information about Veech surfaces can be found in the pioneering paper of W. Veech [Ve7] and in the paper of Ya. Vorobets [Vb1]. For the most recent results concerning Veech groups and geometry of the Teichmüller discs see the original papers of P. Hubert and T. Schmidt [HuSdt1], [HuSdt2], [HuSdt3], [HuSdt4], of C. McMullen [McM1], [McM7] and of P. Hubert and S. Lelièvre [HuLe1], [HuLe2].

9.6 Kernel Foliation

In this section we describe some natural holomorphic foliation on the moduli space of Abelian differentials. In higher genera little is known about this foliation (though it seems to be worth of study). We use this foliation in the next section to describe $GL(2,\mathbb{R})$-invariant submanifolds of "intermediate type" discovered by K. Calta and by C. McMullen in genus two.

We have seen that any stratum $\mathcal{H}(d_1,\dots,d_m)$ can be locally parameterized by a collection of basic *relative periods* of the holomorphic one-form ω, or, in other words, that a neighborhood $\mathcal{U}([\omega]) \subset H^1(S;\{P_1,\dots,P_m\};\mathbb{C})$ gives a local chart in $\mathcal{H}(d_1,\dots,d_m)$.

Let $S \in \mathcal{H}(1,1)$. Let closed paths a_1, a_2, b_1, b_2 represent a basis of cycles in $H_1(S;\mathbb{Z})$. Any path c joining conical singularities P_1 and P_2 represents a *relative cycle* in $H_1(S,\{P_1,P_2\};\mathbb{Z})$. Let $A_1, A_2, B_1, B_2, C \in \mathbb{C}$ be the *periods* of ω: the integrals of ω over a_1, a_2, b_1, b_2, c correspondingly.

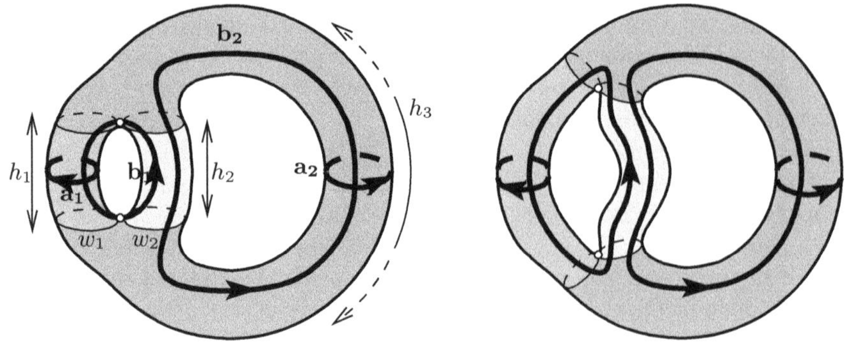

Fig. 48. A deformation of a flat surface inside the kernel foliation keeps the absolute periods unchanged

Example. The collection of cycles a_i, b_i, $i = 1, 2$, on the surfaces from Fig. 48 represent a basis of cycles in $H_1(S;\mathbb{Z})$. All horizontal geodesics on these surfaces are closed; each surface can be decomposed into three cylinders filled with horizontal geodesics. Let $w_1, w_2, w_3 = w_1 + w_2$ be the widths (perimeters) of these cylinders; h_1, h_2, h_3 be their heights; t_1, t_2, t_3 their twists, see Fig. 39. It is easy to check that

$$A_1 = \int_{a_1} \omega = -w_1 \qquad B_1 = \int_{b_1} \omega = (t_1 - t_2) + \sqrt{-1}(h_2 - h_1)$$

$$A_2 = \int_{a_2} \omega = -(w_1 + w_2) \qquad B_2 = \int_{b_2} \omega = (t_2 + t_3) - \sqrt{-1}(h_2 + h_3)$$

The *kernel foliation* in $\mathcal{H}(1,1)$ is the foliation defined in local coordinates by equations

$$\begin{cases} A_1 &= const_{11} \\ A_2 &= const_{21} \end{cases} \qquad \begin{cases} B_1 = const_{12} \\ B_2 = const_{22} \end{cases}$$

In other words, this is a foliation which is obtained by fixing *all* absolute periods and changing the relative period $C = \int_{P_1}^{P_2} \omega$. Similarly, the *kernel foliation* in arbitrary stratum $\mathcal{H}(d_1, \dots, d_m)$ is a foliation which in cohomological coordinates is represented by parallel complex $(m-1)$-dimensional affine subspaces obtained by changing all relative periods while fixing the absolute ones.

Passing to a finite cover over $\mathcal{H}(d_1, \dots, d_m)$ we can assume that all zeroes $P_1, \dots, P_m$ are *named* (i.e. having two zeroes P_j, P_k of same degrees, we know exactly which of the two is P_j and which is P_k). Now we can fix an arbitrary subcollection of zeroes and define a kernel "subfoliation" along relative periods corresponding to chosen subcollection.

Recall that the area of a flat surface is expressed in terms of the absolute periods (see Riemann bilinear relation in Table 1 in Sect. 3.3). Thus, moving along leaves of kernel foliation we do not change the area of the surface. In particular, we can consider the kernel foliation as a foliation of the "unit hyperboloid" $\mathcal{H}_1(d_1, \ldots, d_m)$.

Exercises on Kernel Foliation

Exercise. To deform the flat surface on the left of Fig. 48 along the kernel foliation we have to keep all A_1, A_2, B_1, B_2 unchanged. Hence, we cannot change the widths (perimeters) of the cylinders, since they are expressed in terms of A_1 and A_2. Increasing the height of the second cylinder by ε we have to *increase* the height of the first cylinder by the same amount ε to keep $B_1 = (t_1 - t_2) + \sqrt{-1}\,((h_2 + \varepsilon) - (h_1 + \varepsilon))$ unchanged; we also have to *decrease* the height h_3 of the third cylinder by ε to preserve the value of B_2. Similarly, *increasing* the twist t_1 by δ we have to *increase* the twist t_2 by the same amount δ and to *decrease* t_3 by δ.

Exercise. It is convenient to consider the kernel foliation in the total moduli space $\mathcal{H}_g$ of all holomorphic 1-forms without subdivision of $\mathcal{H}_g$ into strata $\mathcal{H}(d_1, \ldots, d_m)$, where $\sum_j d_j = 2g - 2$. In particular, to deform a surface $S \in \mathcal{H}(2) \subset \mathcal{H}_2$ along the kernel foliation we have to break the double zero into two simple zeroes preserving the absolute periods. The corresponding surgery is presented at Fig. 31.

The leaves of the kernel foliation are naturally endowed with a flat structure, which has conical singularities at the points of intersection of the leaf with the smaller strata and with degenerate strata.

Assuming that the zeroes P_1, P_2 of a surface $S \in \mathcal{H}(1,1)$ are *named* show that the intersection of the kernel foliation with the stratum $\mathcal{H}(2)$ corresponds to a conical point with the cone angle 6π, while the intersections with the two strata of degenerate flat surfaces (determine which ones) are just the regular points of the flat structure.

In the exercise below we use a polygonal representation of a flat surface (compare to Fig. 8 in the paper [Clt] of K. Calta).

Exercise. Consider a regular decagon. Imagine that there are springs inside its sides so that we can shrink or expand the sides keeping them straight segments. Imagine that we hammer a nail in the center of each side. Though the centers of the sides are now fixed our decagon is still flexible: we can pull a vertex and the whole frame will follow, see Fig. 49b. We assume that under any such deformation each nail stays exactly in the middle of the corresponding side.

Prove that the deformed polygon is again centrally symmetric with the same center of symmetry. Prove that the opposite sides of the deformed polygon are parallel and have equal length. Prove that the resulting flat surface lives in the same leaf of the kernel foliation. Show that the "nails" (the centers of the sides) and the center of symmetry are the Weierstrass points of the corresponding Riemann surface (fixed points of the hyperelliptic involution).

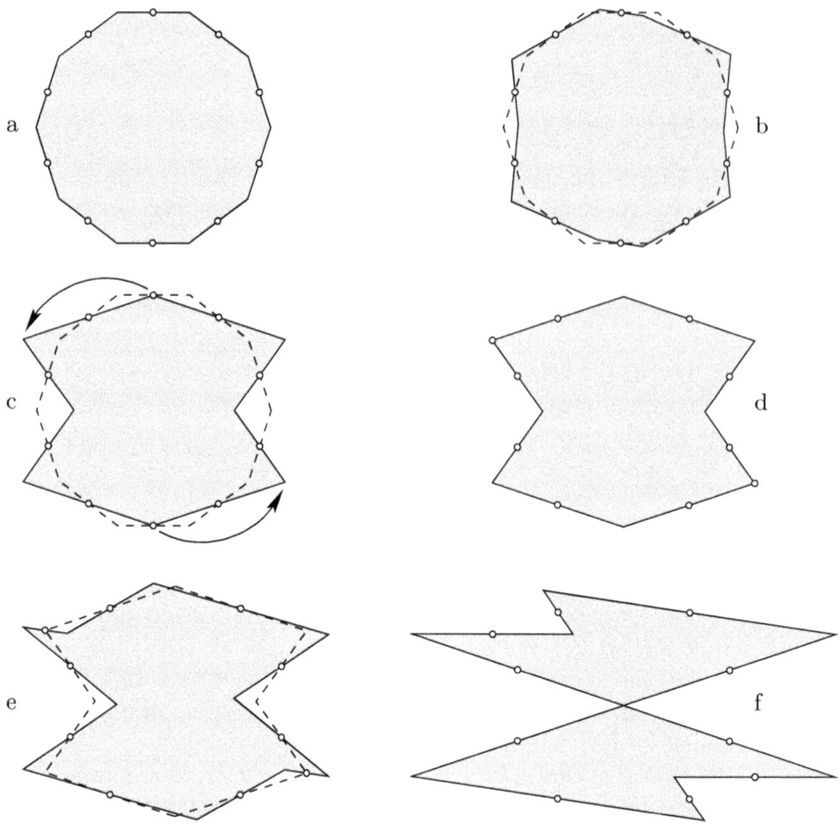

Fig. 49. This cartoon movie represents a path living inside a leaf of the kernel foliation. When at the stage c it lands to a surface from the stratum $\mathcal{H}(2)$ we remove the nails from one pair of vertices and hammer them in the other symmetric pair of vertices (stage d). Then we continue the trip inside the kernel foliation

Move a vertex towards the center of the side which this vertex bounds. The side becomes short (see the upper side on Fig. 49b) and finally contracts to a point (Fig. 49c). What flat surface do we get?

Show that the deformation presented at Fig. 49c brings us to a surface S_0 in the stratum $\mathcal{H}(2)$. Remove the pair of clues, which is hammered at the vertices of the resulting octagon. Hammer them to another pair of symmetric vertices (see Fig. 49d). We can declare that we have a new centrally-symmetric decagon with a pair of sides of zero lengths. Stretching this pair of sides and making them have positive length (see Fig. 49e) we continue our trip inside the kernel foliation. Show that for a given small value $\mathbf{v} \in \mathbb{C}$ of the saddle connection joining two conical singularities, there are exactly three different surfaces obtained as a small deformation of the surface $S_0 \in \mathcal{H}(2)$ as on

Fig. 49c along the leaf of the kernel foliation and having a saddle connection **v**. (Use previous Exercise and Fig. 31.)

Following our path in the kernel foliation we get to the surface S_1 as on Fig. 49f. Is this surface singular? To what stratum (singular stratum) it belongs?

Compact Leaves of the Kernel Foliation

Let us show that the leaf $\mathcal{K}$ of the kernel foliation passing through a square-tiled surface $S(\omega_0)$ of genus two is a compact square-tiled surface. To simplify notations suppose that the absolute periods of the holomorphic 1-form ω_0, representing the flat surface $S(\omega_0)$, generate the entire integer lattice $\mathbb{Z} + \sqrt{-1}\,\mathbb{Z}$.

Consider the "relative period" map $p : \mathcal{K} \to \mathbb{T}^2$ from the corresponding leaf $\mathcal{K}$ of the kernel foliation containing $S(\omega_0)$ to the torus. The map p associates to a flat surface $S(\omega) \in \mathcal{K}$ the relative period C taken modulo integers,

$$\mathcal{K} \ni S(\omega) \overset{p}{\mapsto} C = \int_{P_1}^{P_2} \omega \mod \mathbb{Z} + \sqrt{-1}\,\mathbb{Z} \quad \in \quad \mathbb{C}/(\mathbb{Z} + \sqrt{-1}\,\mathbb{Z}) = \mathbb{T}^2$$

Since the flat surface $S(\omega)$ belongs to the same leaf $\mathcal{K}$ as the square-tiled surface $S(\omega_0) \in \mathcal{K}$, the absolute periods of ω are the same as the ones of ω_0, and hence the integral above taken modulo integers does not depend on the path on $S(\omega)$ joining P_1 and P_2. It is easy to check that the map p is a finite ramified covering over the torus $\mathbb{T}^2$, and thus the leaf $\mathcal{K}$ is a square-tiled surface.

Those flat surfaces $S(\omega) \in \mathcal{K}$, which have integer relative period $C \in \mathbb{Z} + \sqrt{-1}\,\mathbb{Z}$, have *all* periods in $\mathbb{Z} + \sqrt{-1}\,\mathbb{Z}$. Hence, these flat surfaces are square-tiled. Since $S(\omega)$ and $S(\omega_0)$ have the same area, the number N of squares tiling $S(\omega)$ and $S(\omega_0)$ is the same. Thus, $\mathcal{K}$ has a structure of a square-tiled surface such that the vertices of the tiling are represented by N-square-tiled surfaces $S(\omega) \in \mathcal{K}$.

To discuss the geometry of $\mathcal{K}$ we need to agree about enumeration of zeroes P_1, P_2 of a surface $S \in \mathcal{H}(1,1)$. We choose the convention where the zeroes are *named*. That is, given two zeroes of order 1 we know which of them is P_1 and which of them is P_2. Under this convention the square-tiled surface $\mathcal{K}$ is a translation surface; it is represented by a *holomorphic one-form*. (Accepting the other convention we would obtain the quotient of $\mathcal{K}$ over the natural involution exchanging the names of the zeroes. In this latter case the zeroes of $S \in \mathcal{H}(1,1)$ are not distinguishable; the leaf of the kernel foliation gives a flat surface represented by a *quadratic differential*.)

The lattice points of the square-tiled surface $\mathcal{K}$ are represented by N-square-tiled surfaces $S \in \mathcal{K}$ of several types. We have the lattice points represented by N-square tiled surfaces from $\mathcal{H}(1,1)$. These points are the regular points of the flat metric on $\mathcal{K}$.

There are points of intersection of $\mathcal{K}$ with $\mathcal{H}(2)$. Such point $S(\omega) \in \mathcal{K} \cap \mathcal{H}(2)$ is always represented by a square-tiled surface, and hence gives a vertex of the tiling of $\mathcal{K}$. We have seen (see Fig. 31) that given a surface $S(\omega) \in \mathcal{H}(2)$ and a small complex period C one can construct three different flat surfaces $S(\omega_1), S(\omega_2), S(\omega_3) \in \mathcal{H}(1,1)$ with the same absolute periods as ω and with the relative period C. Thus, the points $S(\omega) \in \mathcal{K} \cap \mathcal{H}(2)$ correspond to conical points of $\mathcal{K}$ with the cone angles $3 \cdot 2\pi$ when the zeroes are named (and with the cone angle 3π, when they are not named). The total number of such points was computed in the paper of A. Eskin, H. Masur and M. Schmol [EMaScm]; it equals

$$\text{number of conical points on } \mathcal{K} = \frac{3}{8}(N-2)N^2 \prod_{p\,|\,N}\left(1-\frac{1}{p^2}\right) \tag{9.3}$$

There remain vertices of the tiling of $\mathcal{K}$ represented by degenerate square-tiled surfaces $S(\omega)$. It is not difficult to show that these points are regular for the flat metric on $\mathcal{K}(\omega)$ when the zeroes are named. (They correspond to conical singularities with the cone angle π, when the zeroes are not named). The degenerate N-square-tiled surfaces $S(\omega)$ are of two types. It might be an N-square-tiled torus with two points of the tiling identified. It might be a pair of square-tiled tori with a vertex of the tiling on one torus identified with a vertex of the tiling on the other torus. Here the total number of squares used to tile these two tori is N. The total number of the vertices of the tiling of $\mathcal{K}$ of this type is computed in the paper [Schl1]; it equals

$$\text{number of special points on } \mathcal{K} = \frac{1}{24}(5N+6)N^2 \prod_{p\,|\,N}\left(1-\frac{1}{p^2}\right)$$

Summarizing we conclude that the translation surface $\mathcal{K}$ lives in the stratum $\mathcal{H}(\underbrace{2,\dots,2}_{k})$, where the number k of conical points is given by formula (9.3).

We complete this section with an interpretation of a compact leaf $\mathcal{K}$ as a space of torus coverings; this interpretation was introduced by A. Eskin, H. Masur and M. Schmol in [EMaScm] and developed by M. Schmoll in [Schl1], [Schl2].

We have seen that a nondegenerate flat surface $S(\omega_0) \in \mathcal{H}(1,1)$ representing a vertex of the square tiling of $\mathcal{K}$ is an N-square-tiled surface. Hence, $S(\omega)$ is a ramified covering over the standard torus $\mathbb{T}^2$ of degree N having two simple ramification points, which project to the same point of the torus. A non vertex point $S(\omega) \in \mathcal{K}$ is also a ramified covering over the standard torus $\mathbb{T}^2$. To see this consider once more the period map, but this time applied to $S(\omega)$:

$$S(\omega) \ni P \overset{proj}{\mapsto} \int_{P_1}^{P} \omega \mod \mathbb{Z} + \sqrt{-1}\,\mathbb{Z} \quad \in \quad \mathbb{C}/(\mathbb{Z} + \sqrt{-1}\,\mathbb{Z}) = \mathbb{T}^2.$$

Since all absolute periods of ω live in $\mathbb{Z} + \sqrt{-1}\,\mathbb{Z}$ the integral taken modulo integers does not depend on the path joining the marked point (conical singularity) P_1 with a point P of the flat surface $S(\omega)$. The map *proj* is a ramified covering.

The degree of the covering can be computed as the ratio of areas of $S(\omega)$ and of the torus $\mathbb{T}^2$, which gives N. The covering has precisely two simple ramification points, which are the conical points P_1, P_2 of $S(\omega)$. This time they project to two different points of the torus.

Recall that by convention we assume that the absolute periods of ω generate the entire lattice $\mathbb{Z} + \sqrt{-1}\,\mathbb{Z}$. They corresponds to primitive covers: the ones which do not quotient through a larger torus.

Proposition (M. Schmoll). *Consider primitive branched covers over the standard torus $\mathbb{T}^2$. Fix the degree N of the cover. Let the cover have exactly two simple branch points.*

The space of such covers is connected; its natural compactification coincides with the corresponding leaf $\mathcal{K}$ of the kernel foliation. The Veech group of the square-tiled surface $\mathcal{K}$ coincides with $SL(2,\mathbb{Z})$.

Connectedness of the space of covers is not quite obvious (actually, it was proved earlier in other terms by W. Fulton [Ful]). An observation above shows, that the leaf $\mathcal{K}$ coincides with a connected component of the space of covers. Thus, connectedness of the space of covers implies that this space coincides with $\mathcal{K}$. The group $SL(2,\mathbb{Z})$ acts naturally on the space of covers; in particular it maps the space of covers to itself. This implies that $SL(2,\mathbb{Z})$ belongs to the Veech group of the square-tiled surface $\mathcal{K}$. It is easy to show, that it actually coincides with $SL(2,\mathbb{Z})$.

Corollary (M. Schmoll). *Consider square-tiled surfaces $S(\omega)$ of genus two such that the absolute periods of ω span the entire integer lattice $\mathbb{Z} + \sqrt{-1}\,\mathbb{Z}$. For any given $N > 3$ all such N-square tiled surfaces belong to the same compact connected leaf $\mathcal{K}(N)$ of the kernel foliation.*

For more information on kernel foliation of square-tiled surfaces in genus two see the papers of A. Eskin, H. Masur and M. Schmol [EMaScm] and of M. Schmoll [Schl1], [Schl2]. In particular, the latter papers propose a beautiful formula for Siegel–Veech constants of any flat surface $S \in \mathcal{K}(N)$ in terms of geometry of the cylinder decomposition of the square-tiled surface $\mathcal{K}(N)$.

9.7 Revolution in Genus Two (after K. Calta and C. McMullen)

In this section we give an informal survey of recent revolutionary results in genus $g = 2$ due to K. Calta [Clt] and to C. McMullen [McM2].

Using different methods they found a countable collection of primitive Veech surfaces in the stratum $\mathcal{H}(2)$, proved that this collection describes all Veech surfaces, and gave efficient algorithms which recognize and classify Veech surfaces in $\mathcal{H}(2)$.

This result is in a sharp contrast with the Theorem of C. McMullen [McM6] cited above, which tells that in the other stratum $\mathcal{H}(1,1)$ in genus $g = 2$ there is *only one* primitive Veech surface.

This discovery of an infinite family of primitive Veech surfaces in the stratum $\mathcal{H}(2)$ is also in a sharp contrast with our poor knowledge of primitive Veech surfaces in higher genera: as we have seen in the previous section, primitive Veech surfaces in higher genera $g \geq 3$ are currently known only in some special strata (mostly hyperelliptic), and even in these special strata we know only finite number of primitive Veech surfaces (basically, only one).

Another remarkable result is a discovery by K. Calta and by C. McMullen of nontrivial examples of invariant submanifolds of intermediate dimension: larger than closed orbits and smaller than the entire stratum.

One more revolutionary result in genus two is a Classification Theorem due to C. McMullen [McM3] which proves that a closure of *any* $GL^+(2,\mathbb{R})$-orbit is a nice complex-analytic variety which is either an entire stratum, or which has one of the types mentioned above.

Algebro-geometric Approach To avoid overloading of this survey I had to sacrifice beautiful algebro-geometric part of this story developed by C. McMullen; the reader is addressed to original papers [McM2]–[McM6] and to a short overview presented in [HuSdt5].

Periods of Veech Surfaces in Genus $g = 2$

If S is a Veech surface then the flat surface gS is also a Veech surface for any $g \in GL^+(2,\mathbb{R})$. Thus, speaking about a finite or about a countable collection of Veech surfaces we, actually, choose some family of representatives $\{S_k\}$ of the orbits $GL^+(2,\mathbb{R}) \cdot S$ of Veech surfaces.

The question, which elements of our collection $\{S_k\}$ belong to the same $GL^+(2,\mathbb{R})$-orbit and which ones belong to different orbits is a matter of a separate nontrivial study. A solution was found by C. McMullen in [McM4]; it is briefly presented in the next Sect. 9.8. In this section we present effective algorithm due to K. Calta and to C. McMullen which enables to determine whether a given flat surface in $\mathcal{H}(2)$ is a Veech surface.

Following K. Calta we say that a flat surface S can be *rescaled* to a flat surface S' if S and S' belong to the same $GL^+(2,\mathbb{R})$-orbit. We say that a flat surface S is *quadratic* if for any homology cycle $c \in H_1(S;\mathbb{Z})$ we have

$$\int_c \omega = (p + q\sqrt{d}) + i(r + s\sqrt{d}), \quad \text{where } d \in \mathbb{N}, \quad p,q,r,s \in \mathbb{Q}$$

In other words, we say that a flat surface S defined by a holomorphic 1-form ω is *quadratic* if all periods of ω live in $\mathbb{Q}(\sqrt{d}) + i\mathbb{Q}(\sqrt{d})$.

We can considerably restrict the area of our search using the following Lemma of W. Thurston.

Lemma (W. Thurston). *Any Veech surface in genus $g = 2$ (no matter primitive or not, in the stratum $\mathcal{H}(2)$ or $\mathcal{H}(1,1)$) can be rescaled to a quadratic surface.*

Using this Lemma K. Calta suggest the following algorithm deciding whether a given flat surface $S \in \mathcal{H}(2)$ is a Veech surface or not.

Algorithm of Calta

Recall, that if S is a Veech surface (of arbitrary genus), then by Veech alternative (see Sect. 3.7) a directional flow in any direction is either minimal or completely periodic, that is a presence of a closed geodesic going in some direction implies that all geodesics going in this direction are periodic. Moreover, it was proved by Veech that as soon as there a saddle connection going in some direction, this direction is also completely periodic. In both cases the surface decomposes into a finite collection of cylinders; each boundary component of each cylinder contains a conical singularity (see Sect. 7.1).

The algorithm works as follows. Having a flat surface $S \in \mathcal{H}(2)$ it is easy to find *some* closed geodesic on S (which is allowed to be a closed geodesic saddle connection). Since "rescaling" the surface S (i.e. applying a linear transformation from $GL^+(2,\mathbb{R})$) we replace a Veech surface by a Veech surface, we can turn S in such way that the direction of the closed geodesic will become horizontal. We denote the resulting surface by the same symbol S.

Since S lives in $\mathcal{H}(2)$ it has a single conical point P with the cone angle 6π. In particular, there are exactly three geodesics leaving the conical point in the positive horizontal direction (to the East). If at least one of these three horizontal geodesics does not come back to P the surface S is *not* a Veech surface. Otherwise our test continues.

As we have seen in Sect. 7.1 there are two possible ways in which three horizontal geodesics emitted from P to the East can return to P. Either all three geodesics return at the angle 3π, or one of them returns at the angle 3π and two others return at the angle π, see Fig. 41. In both cases all horizontal geodesics are closed. In the first case the surface decomposes into a single cylinder; in the second case the surface is glued from two cylinders, see Sect. 7.1.

If the surface is decomposed into a single cylinder, it is sufficient to compare the lengths p_1, p_2, p_3 of three horizontal saddle connections, see Fig. 41. The flat surface S is a Veech surface if and only if p_1, p_2, p_3 are commensurable. Moreover, if p_1, p_2, p_3 are commensurable, we can rescale S to a square-tiled surface. It can be done in several elementary steps. First we rescale S in the horizontal direction making p_1, p_2, p_3 rational and then integer. Then we rescale S in the vertical direction making the height h of the cylinder integer. Finally, we apply an appropriate parabolic linear transformation $\begin{pmatrix} 1 & s \\ 0 & 1 \end{pmatrix}$. It does not change neither p_1, p_2, p_3 nor h, but when $s \in \mathbb{R}$ varies the twist t

also varies continuously (see Fig. 39) taking all values in $\mathbb{R}$; in particular, we can achieve $t = 0$. We get a square-tiled surface.

Consider the second case, when the surface decomposes into two cylinders. Let $w_1, w_2, h_1, h_2, t_1, t_2$ be the widths (perimeters), heights and twists of this cylinders correspondingly (compare to Sect. 7.1). In a complete analogy with the one-cylinder case we can rescale the surface horizontally, then vertically, and finally apply an appropriate parabolic linear transformation in order to make the width (perimeter) w_1 and height h_1 of the first cylinder equal to one, $h_1 = w_1 = 1$, and the twist $t_1 = 0$ equal to zero. Applying an appropriate Dehn twist to the second cylinder we can assure $0 \leq t_2 < w_2$.

If after our rescaling all parameters w_2, h_2, t_2 characterizing the second cylinder do not get to the same quadratic field $\mathbb{Q}(\sqrt{d})$ for some $d \in \mathbb{N}$, the surface S is not a Veech surface.

If w_2, h_2, t_2 are rational (i.e. if d is a complete square), the surface S can be rescaled to a square-tiled surface.

The remaining case is treated by one of the key Theorems in the paper of K. Calta [Clt].

Theorem (K. Calta). *Let all parameters $w_j, h_j, t_j,\ j = 1,2$ of a two-cylinder decomposition of a flat surface $S \in \mathcal{H}(2)$ belong to the same quadratic field $\mathbb{Q}(\sqrt{d})$ with $d \in \mathbb{N}$ not a complete square. Then S is a Veech surface if and only if the parameters satisfy the following system of equations:*

$$\begin{cases} w_1\bar{h}_1 & = -w_2\bar{h}_2, \\ \bar{w}_1 t_1 + \bar{w}_2 t_2 & = w_1\bar{t}_1 + w_2\bar{t}_2, \end{cases} \tag{9.4}$$

(where the bar denotes conjugation $\overline{p + q\sqrt{d}} = p - q\sqrt{d}$ in $\mathbb{Q}(\sqrt{d})$ with $p, q \in \mathbb{Q}$).

Actually, we kept the system of equation above as it is written in the original paper [Clt]. In this form it can be adopted to a more general normalization of parameters: it is sufficient to rescale surface S to bring all $w_j, h_j, t_j,\ j = 1,2$ to a quadratic field.

Remark. Similar *necessary* conditions for Veech surfaces in $\mathcal{H}(2)$ were obtained by D. Panov independently of K. Calta and of C. McMullen.

Since by Lemma of Thurston any Veech surface in $\mathcal{H}(2)$ can be rescaled to a quadratic surface, taking a collection of all quadratic surfaces decomposed into two horizontal cylinders satisfying the condition above, we get representatives of the $GL^+(2,\mathbb{R})$-orbits of *all* flat surfaces in $\mathcal{H}(2)$.

Exercise. Show that the Katok–Zemlyakov construction applied to an L-shaped billiard as on Fig. 50 (see also Fig. 38) generates a surface $S \in \mathcal{H}(2)$. Show that this surface is decomposed into two cylinders filled by closed horizontal geodesics and that these cylinders have parameters $w_1 = 2, h_1 =$

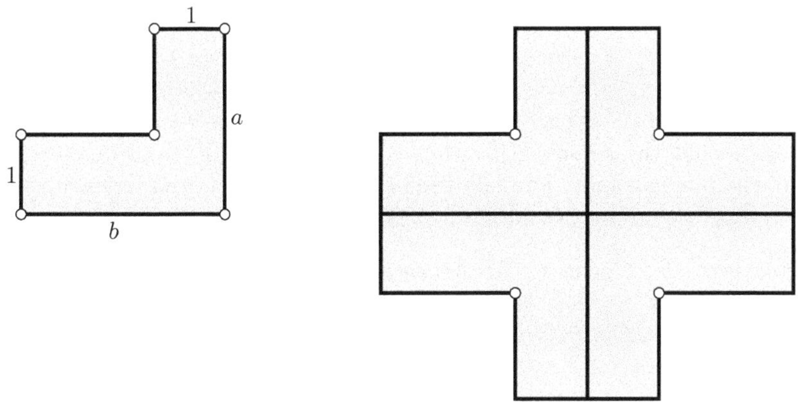

Fig. 50. L-shaped billiard table $P(a, b)$ and its unfolding into a flat surface $S \in \mathcal{H}(2)$ (after C. McMullen [McM2])

$2a - 2, t_1 = 0$ for the first cylinder and $w_2 = 2b, h_2 = 2, t_2 = 0$ for the second cylinder. Using the condition above prove the following Theorem of C. McMullen [McM2]:

Theorem (C. McMullen). *The L-shaped billiard table $P(a, b)$ as on Fig. 50 generates a Veech surface if and only if a and b are rational or*

$$a = x + z\sqrt{d} \quad \text{and} \quad b = y + z\sqrt{d}$$

for some $x, y, z \in \mathbb{Q}$ with $x + y = 1$ and $d \geq 0$ in $\mathbb{Z}$.

Kernel Foliation in Genus 2

The following elementary observation explains our interest to kernel foliation in the content of our study of $GL^+(2, \mathbb{R})$-invariant subvarieties. Let $\mathcal{N} \subset \mathcal{H}(d_1, \dots, d_n)$ be a $GL^+(2, \mathbb{R})$-invariant submanifold. The "germ" of the kernel foliation at $\mathcal{N}$ is equivariant with respect to the action of $GL^+(2, \mathbb{R})$.

In other words this statement can be described as follows. Denote by $t(\delta)$ a translation by δ along kernel foliation defined in a neighborhood of $S \in \mathcal{H}(1, 1)$. Here $\delta \in \mathbb{C}$ is a small parameter. Let $g \in GL^+(2, \mathbb{R})$ be close to identity. Then

$$g \circ t(\delta) \cdot S = t(g\delta) \circ gS$$

where g acts on a complex number δ as on a vector in $\mathbb{R}^2$. Moving along the kernel foliation, and then applying an element g of the group is the same as applying first the same element of the group and then moving along an appropriate translation along the kernel foliation. A similar construction works in a general stratum.

We get the following very tempting picture. Suppose, that we have a closed $GL^+(2, \mathbb{R})$-orbit $\mathcal{N}$ in $\mathcal{H}(1, 1)$ or in $\mathcal{H}(2)$. For example, suppose that $\mathcal{N}$ is an orbit of a Veech surface. Consider a union of leaves of the kernel foliation passing through $\mathcal{N}$. Due to the remark above this is a $GL^+(2,\mathbb{R})$-invariant subset!

The weakness of this optimistic picture is that there is no *a priori* reason to hope that the resulting invariant subset in $\mathcal{H}(1,1)$ would be closed. And here the magic comes. The following statement was proved independently by K. Calta [Clt] and by C. McMullen [McM2].

Theorem (K. Calta; C. McMullen). *For any Veech surface $S_0 \in \mathcal{H}(2)$ the union of leaves of the kernel foliation passing through the $GL^+(2,\mathbb{R})$-orbit of S_0 is a closed $GL^+(2,\mathbb{R})$-invariant complex orbifold $\mathcal{N}$ of complex dimension* 3.

In particular, the complex dimension of the resulting orbifold is the sum of the complex dimension of the $GL^+(2,\mathbb{R})$-orbit of S_0 and of the complex dimension of the kernel foliation $3 = 2 + 1$.

K. Calta has found the following beautiful geometric characterization of flat surfaces living in an invariant subvariety $\mathcal{N}$ as above. Surfaces in any such $\mathcal{N}$ are *completely periodic*: as soon as there is a single closed trajectory in some direction, *all* geodesics going in this direction are closed. This condition is necessary and sufficient condition for a surface to live in an invariant subvariety $\mathcal{N}$ as above. In particular, this shows that not only Veech surfaces have this property.

An algorithm analogous to the algorithm determining Veech surfaces in $\mathcal{H}(2)$ (see above) allows to K. Calta to determine whether a given surface $S \in \mathcal{H}(1,1)$ is completely periodic or not (and hence, whether it belongs to an invariant subvariety $\mathcal{N}$ as above or not). As before one starts with finding *some* closed geodesic or *some* saddle connection. If the surface is completely periodic in the corresponding direction, then all other geodesics going in this direction are periodic and the surface decomposes into cylinders. After an appropriate rotation this periodic direction becomes horizontal. Without loss of generality, we may assume that S decomposes into three cylinders. By w_i, h_i and t_i with $1 \le i \le 3$, we denote the widths, heights and twists. After renumbering, we may assume that $w_3 = w_1 + w_2$. Define $s_1 = h_1 + h_3$, $s_2 = h_2 + h_3$, $\tau_1 = t_1 + t_3$, $\tau_2 = t_2 + t_3$. If the surface is completely periodic its absolute periods can be rescaled to get to $\mathbb{Q}(\sqrt{d}) + i\mathbb{Q}(\sqrt{d})$ (compare to the algorithm for Veech surfaces). Leaving the elementary case when $d \in \mathbb{N}$ is a complete square, the following characteristic equations obtained by K. Calta (analogous to equations (9.4) above) tell whether our flat surface is completely periodic or not:

$$\begin{aligned} w_1 \bar{s}_1 &= -w_2 \bar{s}_2, \\ \bar{w}_1 \tau_1 + \bar{w}_2 \tau_2 &= w_1 \bar{\tau}_1 + w_2 \bar{\tau}_2, \qquad 0 \le \tau_i < w_i + w_3. \end{aligned}$$

Note that we consider any invariant subvariety $\mathcal{N}$ as above as a subvariety in $\mathcal{H}_2 = \mathcal{H}(1,1) \sqcup \mathcal{H}(2)$. Clearly, the intersection $\mathcal{N} \cap \mathcal{H}(1,1)$ and $\mathcal{N} \cap \mathcal{H}(2)$

results in close $GL^+(2,\mathbb{R})$-invariant subvarieties in the corresponding strata. Note that $dim_{\mathbb{C}}\mathcal{N} = 3$. Since $\mathcal{N}$ is a union of leaves of the kernel foliation this implies that $dim_{\mathbb{C}}\mathcal{N} \cap \mathcal{H}(2) = 2$. This means that *any* surface $S \in \mathcal{N} \cap \mathcal{H}(2)$ is a Veech surface!

Classification Theorem of McMullen

We complete this section with a description of the wonderful result of C. McMullen [McM3] realizing a dream of a complete classification of closures of $GL^+(2,\mathbb{R})$-orbits in genus $g = 2$. The classification is astonishingly simple. We slightly reformulate the original Theorem using the notions of kernel foliation and of completely periodic surface (and, hence, using implicitly results of K. Calta [Clt]).

Theorem (C. McMullen).

- *If a surface $S \in \mathcal{H}(2)$ is a Veech surface, its $GL^+(2,\mathbb{R})$-orbit is a closed complex 2-dimensional subvariety;*
- *Closure of $GL^+(2,\mathbb{R})$-orbit of any surface $S \in \mathcal{H}(2)$ which is not a Veech surface is the entire stratum $\mathcal{H}(2)$;*
- *If a surface $S \in \mathcal{H}(1,1)$ is a Veech surface, its $GL^+(2,\mathbb{R})$-orbit is a closed complex 2-dimensional subvariety;*
- *If a surface $S \in \mathcal{H}(1,1)$ is not a Veech surface but is a completely periodic surface, then the closure of its $GL^+(2,\mathbb{R})$-orbit is a closed complex 3-dimensional subvariety $\mathcal{N}$ foliated by leaves of the kernel foliation as described above;*
- *If a surface $S \in \mathcal{H}(1,1)$ is not completely periodic, then the closure of its $GL^+(2,\mathbb{R})$-orbit is the entire stratum $\mathcal{H}(1,1)$.*

Actually, the Theorem above is even stronger: connected components of these invariant submanifolds are basically also classified. We have seen that the $GL^+(2,\mathbb{R})$-orbit of any Veech surfaces in $\mathcal{H}(2)$ has a representative with all periods in a quadratic field. The *discriminant* $D = b^2 - 4c > 0$ is a positive integer: a discriminant of the corresponding quadratic equation $x^2+bx+c=0$ with integer coefficients. The discriminant is an invariant of an $GL^+(2,\mathbb{R})$-orbit. Since for any integer b the number $b^2 \mod 4$ can be either 0 or 1, the discriminant $D \mod 4 = 0, 1$. The values $D = 1, 4$ are not realizable, so the possible values of D are $5, 8, 9, 12, 13, \ldots$.

We postpone the description of results of P. Hubert and S. Lelièvre [HuLe1] and of C. McMullen [McM4] on classification of the orbits of Veech surfaces in $\mathcal{H}(2)$ to the next section. Here we state the following result of C. McMullen [McM3]. By $\mathcal{N}(D)$ denote the 3-dimensional invariant submanifold obtained as a union of leaves of the kernel foliation passing through the orbits of all Veech surfaces corresponding to the given discriminant D.

Theorem (C. McMullen). *The invariant subvariety $\mathcal{N}(D)$ is nonempty and connected for any $D = 0, 1 \mod 4$, $D \in \mathbb{N}$, $D \geq 5$.*

Ergodic Measures

Actually, the Classification Theorem of C. McMullen is even stronger: it also classifies the invariant measures. Consider the "unit hyperboloids" $\mathcal{H}_1(2)$ and $\mathcal{H}_1(1,1)$: the subvarieties of real codimension one representing flat surfaces of area 1. The group $SL(2,\mathbb{R})$ acts on $\mathcal{H}_1(2)$ and $\mathcal{H}_1(1,1)$ preserving the measure induced on these "unit hyperboloids", see Sect. 3.4. Let us discuss what other $SL(2,\mathbb{R})$-invariant measures do we know.

When we have an $SL(2,\mathbb{R})$-invariant subvariety, we can get an invariant measure concentrated on this subvariety. For example, as we know an $SL(2,\mathbb{R})$-orbit of a Veech surface S is closed; it is isomorphic to the quotient $SL(2,\mathbb{R})/\Gamma(S)$, where, by definition of a Veech surface, this quotient has finite volume. Thus, Haar measure on $SL(2,\mathbb{R})$ induces a finite invariant measure on the $SL(2,\mathbb{R})$-orbit of a Veech surface S.

Consider now a "unit hyperboloid" $\mathcal{N}_1 \subset \mathcal{N}$ in the manifold $\mathcal{N}$ obtained as a union of leafs of the kernel foliation passing through $SL(2,\mathbb{R})$-orbit of a Veech surface S. Note that by Riemann bilinear relations the area of a flat surface can be expressed in terms of absolute periods. Thus, moving along the kernel foliation we do not change the area of the surface. We have seen that every leaf of the kernel foliation is flat. Consider the corresponding Euclidean volume element in each leaf. The group $SL(2,\mathbb{R})$ maps leaves of the kernel foliation to leaves and respects this volume element. Thus we get an invariant measure on $\mathcal{N}_1 \subset \mathcal{N}$; near an $SL(2,\mathbb{R})$-orbit of a Veech surface it disintegrates to a product measure.

We have associated to any connected invariant subvariety of each of four types as above a natural $SL(2,\mathbb{R})$-measure supported on it. One more result of C. McMullen in [McM3] tells that there are no other ergodic measures.

Other Properties

The invariant subvarieties have numerous wonderful geometric properties. In particular, their projections to the moduli space $\mathcal{M}$ of complex structures on a surface of genus two are also nice subvarieties. C. McMullen has showed that the $GL^+(2,\mathbb{R})$-orbit of a Veech surface projects to an isometrically immersed algebraic curve and $\mathcal{N}(D)$ projects to a complex surface. Such surfaces (of *complex* dimension two) are called *Hilbert modular surfaces.*

One more surprising phenomenon proved by C. McMullen in [McM3] concerns Veech groups of flat surfaces in genus $g = 2$.

Theorem (C. McMullen). *If the Veech group $\Gamma(S)$ of a flat surface $S \in \mathcal{H}(2)$ contains a hyperbolic element, the flat surface S is a Veech surface; in particular, its Veech group $\Gamma(S)$ is a lattice in $SL(2,\mathbb{R})$.*

If the Veech group $\Gamma(S)$ of a flat surface $S \in \mathcal{H}(1,1)$ contains a hyperbolic element, and the flat surface S is not a Veech surface, then S is completely periodic. In this case the Veech group $\Gamma(S)$ is infinitely generated.

The Veech group of a flat surface S contains a hyperbolic element if and only if S admits an affine pseudoanosov diffeomorphism.

We complete this section with the following natural problem for higher genera $g \geq 3$.

Problem. *Let $\mathcal{K}$ be a $GL^+(2,\mathbb{R})$-invariant subvariety in some stratum of holomorphic one-forms $\mathcal{H}(d_1,\dots,d_m) \subset \mathcal{H}_g$. Consider the union $\mathcal{U}$ of leaves of the kernel foliation passing through $\mathcal{K}$. Is $\mathcal{U}$ a closed subvariety in $\mathcal{H}(1,\dots,1)$? Similar question for other strata.*

9.8 Classification of Teichmüller Discs of Veech Surfaces in $\mathcal{H}(2)$

It is easy to check that any *square-tiled surface* (see Sect. 7.1) is a Veech surface, and thus an $SL(2,\mathbb{R})$-orbit of any square-tiled surface is closed. Such orbit contains other square-tiled surfaces. Since the $SL(2,\mathbb{R})$-action does not change the area of a surface these other square-tiled surfaces are tiled with the same number of squares, see Fig. 46 in Sect. 9.5 and the Exercise related to this Figure.

For a fixed integer n the number of n-square-tiled surfaces is finite. It would be interesting to know (and this is certainly a part of general Problem from Sect. 9.3) how the square-tiled surfaces are arranged into orbits of $SL(2,\mathbb{R})$. Say, we have seen in the previous section that there are exactly three 3-square-tiled surfaces in $\mathcal{H}(2)$ (see Fig. 46) and that they belong to the same orbit. For $n = 4$ there are already nine 4-square-tiled surfaces in $\mathcal{H}(2)$ and they still belong to the same $SL(2,\mathbb{R})$-orbit. The corresponding Teichmüller disc is a 9-fold cover over the modular curve. For $n = 5$ there are twenty seven 5-square-tiled surfaces in $\mathcal{H}(2)$ and they split into two different orbits of $SL(2,\mathbb{R})$.

Generalizing a result of P. Hubert and S. Lelièvre [HuLe1] obtained for prime number n C. McMullen has recently proved the following conjecture of P. Hubert and S. Lelièvre.

Theorem (C. McMullen). *All n-square-tiled surfaces in $\mathcal{H}(2)$, which cannot be tiled with $p \times q$-rectangles with p or q greater than 1, get to the same $SL(2,\mathbb{R})$-orbit when $n \geq 4$ is even and get to exactly two distinct orbits when $n \geq 5$ is odd.*

Actually, C. McMullen has classified in [McM4] the orbits of all Veech surfaces in $\mathcal{H}(2)$. As we have seen in the previous section Veech surfaces in $\mathcal{H}(2)$ are characterized by an integer parameter, called the *discriminant* D. For n-square-tiled surfaces the discriminant equals $D = n^2$.

Veech surfaces which cannot be rescaled to a square-tiled surface are called *nonarithmetic* Veech surfaces. Any nonarithmetic Veech surface in $\mathcal{H}(2)$ can be rescaled to a flat surface having all periods in a quadratic field (see the Lemma of W. Thurston in the previous section). The discriminant corresponding to a nonarithmetic Veech surface is the discriminant of this quadratic field.

Of course a $GL^+(2,\mathbb{R})$-orbit of a nonarithmetic Veech surface might have different representatives S, such that all periods of S belong a quadratic field. Nevertheless, for Veech surfaces in genus $g = 2$ the discriminant is well-defined: it is an invariant of a $GL^+(2,\mathbb{R})$-orbit. The discriminant is a positive integer $D = 0, 1 \mod 4$, $D \geq 5$.

C. McMullen has proved the following classification Theorem [McM4]:

Theorem (C. McMullen). *For $D = 1 \mod 8$, $D > 9$, all Veech surfaces in $\mathcal{H}(2)$ corresponding to discriminant D get to exactly two distinct $GL^+(2,\mathbb{R})$-orbits. For other values $D = 0, 1 \mod 4$, $D \geq 5$, they belong to the same $GL^+(2,\mathbb{R})$-orbit.*

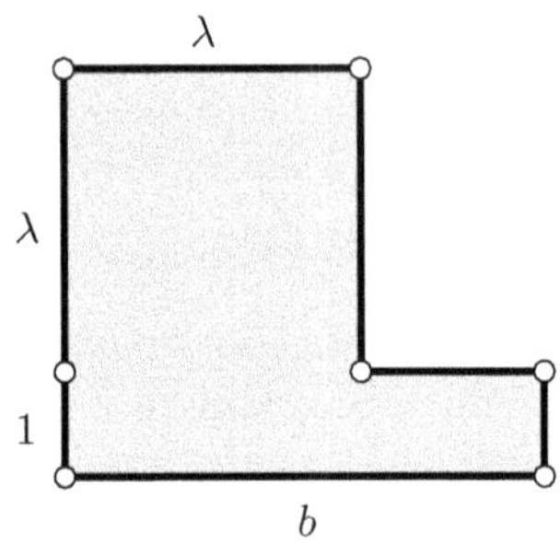

Fig. 51. The L-shaped billiard table $L(b, e)$ generating a "canonical" Veech surface

Moreover, C. McMullen proposed the following canonical representative for any such $GL^+(2,\mathbb{R})$-orbit, see [McM4]. Consider an L-shaped billiard $L(b, e)$ as on Fig. 51, where

$$\begin{cases} b, e \in \mathbb{Z}; \quad \lambda = (e + \sqrt{e^2 + 4b})/2 \\ e = -1, 0 \text{ or } 1 \\ e + 1 < b \\ \text{if } e = 1 \text{ then } b \text{ is even} \end{cases} \tag{9.5}$$

The billiard table $L(b, e)$ generates a flat surface $S(b, e)$ in $\mathcal{H}(2)$, see Fig. 50.

Theorem (C. McMullen). *The flat surface $S(b, e)$ generated by the L-shaped billiard table $L(b, e)$ with parameters b, e satisfying (9.5) is a Veech surface. Any closed $GL^+(2,\mathbb{R})$-orbit in $\mathcal{H}(2)$ is represented by one of such $S(b, e)$ and this representation is unique. The discriminant D of $S(b, e)$ equals $D = e^2 + 4b$.*

Exercise. Using linear transformations and scissors rescale the flat surface obtained from the "double pentagon" (see Fig. 7) to the flat surface obtained from a "golden cross" $P(a, b)$, with $a = b = \dfrac{1 + \sqrt{5}}{2}$; see Fig. 50 for the definition of $P(a, b)$. Using linear transformations and scissors rescale any of these flat surfaces to a surface obtained from the billiard table $L(1, -1)$, see Fig. 51, proving that these Veech surfaces have discriminant $D = 5$. (For solutions see Fig. 4 in [McM2]; see also and [McM5]).

Exercise (T. Schmidt). Using linear transformations and scissors rescale the flat surface obtained from the regular octagon to a surface obtained from the billiard table $L(2, 0)$, see Fig. 51, proving that these Veech surfaces have discriminant $D = 8$. (See [McM5]).

Stratum $\mathcal{H}(1,1)$

We have discussed in details Veech surfaces in $\mathcal{H}(2)$. We did not discussed the Veech surface for the stratum $\mathcal{H}(1,1)$ because for square-tiled surfaces in $\mathcal{H}(1,1)$ (also called *arithmetic Veech surfaces*) the classification of the $GL^+(2,\mathbb{R})$-orbits is knot known yet...

The classification of *nonarithmetic* Veech surfaces in $\mathcal{H}(1,1)$ is, however, known (since very recently), and is quite surprising. Using the results of M. Moeller [Mo1]–[Mo3] C. McMullen has proved in [McM6] the following Theorem.

Theorem (C. McMullen). *Up to a rescaling the only primitive nonarithmetic Veech surface in* $\mathcal{H}(1,1)$ *is the one obtained from the regular decagon by identification of opposite sides. In other words, any primitive nonarithmetic Veech surface in* $\mathcal{H}(1,1)$ *belongs to the* $GL^+(2,\mathbb{R})$*-orbit of the surface obtained from the regular decagon.*

For higher genera nothing is known neither about the number of $SL(2,\mathbb{R})$-orbits of n-square-tiled surfaces, nor about their geometry.

Problem. *Classify orbits of square-tiled surfaces in any stratum, in particular in* $\mathcal{H}(1,1)$.

Square-tiled Surfaces: more serious reading. An elementary introduction can be found in [Zo5]. Paper [HuLe1] of P. Hubert and S.Lelièvre and [McM4] of C. McMullen classify orbits of square-tiled surfaces in $\mathcal{H}(2)$. See also the paper of G. Schmithüsen [Schn] for an algorithm of evaluation of the Veech group of a square-tiled surface and for examples of square-tiled surfaces having $SL(2,\mathbb{R})$ as a Veech group. Another such example due to M. Möller is presented in the survey [HuSdt5] of P. Hubert and T. Schmidt.

10 Open Problems

Flat surfaces with nontrivial holonomy and billiards in general polygons

Problem 1 (Geodesics on general flat surfaces; Sect. 1.1).

Describe behavior of geodesics on general flat surfaces with nontrivial holonomy. Prove (or disprove) that geodesic flow is ergodic on a typical (in any reasonable sense) flat surface.

Does any (almost any) flat surface has at least one closed geodesic which does not pass through singular points?

If yes, are there many regular closed geodesics? Namely, find the asymptotics for the number of closed geodesics of bounded length as a function of the bound.

Problem 2 (Billiards in general polygons; Sect. 2.1).

Describe the behavior of a generic regular billiard trajectory in a generic triangle, in particular, prove (or disprove) that the billiard flow is ergodic.

Does any (almost any) billiard table has at least one regular periodic trajectory? If the answer is affirmative, does this trajectory survive under deformations of the billiard table?

If a periodic trajectory exists, are there many periodic trajectories like that? Namely, find the asymptotics for the number of periodic trajectories of bounded length as a function of the bound.

More problems on billiards can be found in the survey of E. Gutkin [Gu2].

Problem 3 (Renormalization of billiards in polygons; Sect. 2.1 and Sect. 5).

Is there a natural dynamical system (renormalization procedure) acting on the space of billiards in polygons?

Classification of orbit closures in $\mathcal{H}_g$ and $\mathcal{Q}_g$

Problem 4 (Orbit closures for moduli spaces; Sect. 9.3).

Is it true that the closures of $GL^+(2,\mathbb{R})$-orbits in $\mathcal{H}_g$ and $\mathcal{Q}_g$ are always complex-analytic (complex-algebraic?) orbifolds? Classify these closures. Classify ergodic measures for the action of $SL(2,\mathbb{R})$ on "unit hyperboloids".

Suppose that these orbit closures are described by an explicit list. Find natural intrinsic invariants of a flat surface S which would allow to determine the closure of the orbit of S in the list.

To be honest, even having obtained a conjectural classification above, one would need to develop a serious further machinery to get full variety of interesting applications. The situation with the problem below is quite different: a reasonable solution of this problem would immediately give a burst of applications since such a machinery already exists. For the experts interested in ergodic aspects, counting problems, etc, the measure-theoretic analogue of Ratner's Theorem discussed below is the biggest open problem in the area.

*Problem 5 (**A. Eskin**: "Ratner's theorem" for moduli spaces; Sect. 9.3).*

Does the unipotent subgroup of $SL(2,\mathbb{R})$ act nicely on $\mathcal{H}_g$ and $\mathcal{Q}_g$ or not? Are the orbit closures always nice (for example, real-analytic) orbifolds or one can get complicated closures (say, Kantor sets)?

If the action is "nice", classify the closures of orbits of the unipotent subgroup $\begin{pmatrix} 1 & t \\ 0 & 1 \end{pmatrix}_{t\in\mathbb{R}}$ in $\mathcal{H}_g$ and $\mathcal{Q}_g$. Classify ergodic measures for the action of the unipotent subgroup on "unit hyperboloids".

Solve the problem for genus $g = 2$.

The problem above is solved in the particular case when the unipotent flow acts on a $SL(2,\mathbb{R})$-invariant submanifold in $\mathcal{H}_g$ obtained by a ramified covering construction from a Veech surface; see the papers of A. Eskin, H. Masur and M. Schmoll [EMaScm] and of A. Eskin, J. Marklof, D. Witte Morris [EMkWt].

The next problem concerns possibility of construction of $GL^+(2,\mathbb{R})$-invariant submanifolds in higher genera using kernel foliation.

It follows from the results of K. Calta and C. McMullen that for *any* Veech surface $S_0 \in \mathcal{H}(2)$ the union of complex one-dimensional leaves of the kernel foliation passing through the complex two-dimensional $GL^+(2,\mathbb{R})$-orbit $\mathcal{O}(S_0)$ of S_0 is a closed complex orbifold $\mathcal{N}$ of complex dimension 3, see Sect. 9.7. By construction it is $GL^+(2,\mathbb{R})$-invariant.

Problem 6 (Kernel foliation; Sect. 9.7).

Let $\mathcal{O} \subset \mathcal{H}(d_1,\dots,d_m) \subset \mathcal{H}_g$ be a $GL(2,\mathbb{R})$-invariant submanifold (suborbifold). Let $\mathcal{H}(d'_1,\dots,d'_n) \subset \mathcal{H}_g$ be a bigger stratum adjacent to the first one, $d_j = d'_{k_1} + \dots + d'_{k_j}$, $i = 1,\dots,m$.

Consider the closure of the union of leaves of the kernel foliation in $\mathcal{H}(d'_1,\dots,d'_n)$ (or in $\mathcal{H}_g$) passing through $\mathcal{O}$. We get a closed $GL^+(2,\mathbb{R})$-invariant subset $\mathcal{N} \subset \mathcal{H}(d'_1,\dots,d'_n)$ (correspondingly $\mathcal{N} \subset \mathcal{H}_g$).

Is $\mathcal{N}$ a complex-analytic (complex-algebraic) orbifold? When $\mathcal{N}$ does not coincide with the entire connected component of the stratum $\mathcal{H}(d'_1,\dots,d'_n)$ (correspondingly $\mathcal{H}_g$)? When $\dim_{\mathbb{C}} \mathcal{N} = \dim_{\mathbb{C}} \mathcal{O} + (n-m)$ (correspondingly $\dim_{\mathbb{C}} \mathcal{N} = \dim_{\mathbb{C}} \mathcal{O} + (2g-2-m)$? Here $(n-m)$ and $(2g-2-m)$ is the complex dimension of leaves of the kernel foliation in $\mathcal{H}(d'_1,\dots,d'_n)$ and in $\mathcal{H}_g$ correspondingly.

Particular cases of classification problem

Problem 7 (Exceptional strata of quadratic differentials; Sect. 9.4).

Find an invariant which would be easy to evaluate and which would distinguish half-translation surfaces from different connected components of the four exceptional strata $\mathcal{Q}(-1,9)$, $\mathcal{Q}(-1,3,6)$, $\mathcal{Q}(-1,3,3,3)$ and $\mathcal{Q}(12)$.

Problem 8 (Veech surfaces; Sect. 9.5).

Classify primitive Veech surfaces.

Problem 9 (Orbits of square-tiled surfaces; Sect. 9.8).

Classify orbits of square-tiled surfaces in any stratum. Describe their Teichmüller discs.

Same problem for the particular case $\mathcal{H}(1,1)$.

Geometry of individual flat surfaces

Problem 10 (Quadratic asymptotics for any surface; Sect. 6.1).

Is it true that *any* very flat surface has exact quadratic asymptotics for the number of saddle connections and for the number of regular closed geodesics?

Problem 11 (Error term for counting functions; Sect. 6.1).

What can be said about the error term in quadratic asymptotics for counting functions $N(S,L) \sim c \cdot L^2$ on a generic flat surface S? In particular, is it true that the limit

$$\limsup_{L\to\infty} \frac{\log |N(S,L) - c\cdot L^2|}{\log L} \overset{?}{<} 2$$

is strictly less than two? Is it the same for almost all flat surfaces in a given connected component of a stratum?

One of the key properties used by C. McMullen for the classification of the closures of orbits of $GL(2,\mathbb{R})$ in $\mathcal{H}(1,1)$ was the knowledge that on *any* flat surface in this stratum one can find a pair of homologous saddle connections. Cutting the surface along these saddle connections one decomposes the surface into two tori and applies machinery of Ratner theorem.

*Problem 12 (**A. Eskin; C. McMullen**: Decomposition of surfaces; Sect. 6.4).*

Given a connected component of the stratum $\mathcal{H}(d_1,\dots,d_m)$ of Abelian differentials (or of quadratic differentials $\mathcal{Q}(d_1,\dots,d_m)$) find those configurations of homologous saddle connections (or homologous closed geodesics), which are presented at *any* very flat surface S in the stratum.

For quadratic differentials the notion of "homologous" saddle connections (closed geodesics) should be understood in terms of homology with local coefficients, see [MaZo].

Topological, geometric, and dynamical properties of the strata

*Problem 13 (**M. Kontsevich**: Topology of strata; Sect. 3).*

Is it true that strata $\mathcal{H}(d_1,\dots,d_m)$ and $Q(q_1,\dots,q_n)$ are $K(\pi,1)$-spaces (i.e. their universal covers are contractible)?

Problem 14 (Compactification of moduli spaces; Sect. 6.3 and 6.4).

Describe natural compactifications of the moduli spaces of Abelian differentials $\mathcal{H}_g$ and of the moduli spaces of meromorphic quadratic differentials with at most simple poles $\mathcal{Q}_g$. Describe natural compactifications of corresponding strata $\mathcal{H}(d_1,\dots,d_m)$ and $\mathcal{Q}(q_1,\dots,q_n)$.

Problem 15 (Dynamical Hodge decomposition; Sect. 4.3, 5.7 and Appendix B).

Study properties of distributions of Lagrangian subspaces in $H^1(S;\mathbb{R})$ defined by the Teichmüller geodesic flow, in particular, their continuity. Is there any topological or geometric way to define them?

Problem 16 (Lyapunov exponents; Sect. 5.7 and Appendix B).

Study *individual* Lyapunov exponents of the Teichmüller geodesic flow
– for all known $SL(2;\mathbb{R})$-invariant subvarieties;
– for strata in large genera.

Are they related to characteristic numbers of some natural bundles over appropriate compactifications of the strata?

Some other open problems can be found in [HuMSdtZ].

A Ergodic Theorem

We closely follow the presentation in Chapter 1 of [CFSin]. However, for the sake of brevity we do not consider flows; for the flows the theory is absolutely parallel.

Ergodic Theorem

Consider a manifold M^n with a measure μ. We shall assume that the measure comes from a volume form on M^n, and that the total volume (total measure) of M^n is finite. We shall consider only measurable subsets of M^n.

Let $T : M^n \to M^n$ be a smooth map. We do not assume that T is a bijection unless it is explicitly specified. We say that T *preserves measure* μ is for any subset $U \subset M^n$ measure $\mu(T^{-1}U)$ of the preimage coincides with measure $\mu(U)$ of the set. For example the double cover $T : S^1 \to S^1$ of the circle $S^1 = \mathbb{R}/\mathbb{Z}$ over itself defined as $T : x \mapsto 2x \mod 1$ preserves the Lebesgue measure on S^1. In this section we consider only measure preserving maps.

We say that some property is valid *for almost all* points of M^n if it is valid for a subset $U \subset M^n$ of complete measure $\mu(U) = \mu(M^n)$.

A subset $U \subset M^n$ is *invariant* under the map T if the preimage $T^{-1}U$ coincides with U. Thus, a notion of an invariant subset is well-defined even when T is not a one-to-one map. The measure-preserving map $T : M^n \to M^n$ is *ergodic* with respect to μ if any invariant subset has measure 0 or 1. The measure-preserving map $T : M^n \to M^n$ is *uniquely ergodic* with respect to μ if there is no other invariant probability measure.

Note that if T has a fixed point or, more generally, a periodic orbit (that is $T^k(x_0) = x_0$ for some $x_0 \in M^n$ and some $k \in \mathbb{N}$), one can consider an invariant probability measure concentrated at the points of the orbit. Thus, such map T cannot be uniquely ergodic with respect to any Lebesgue equivalent measure.

Now we are ready to formulate the keystone theorem. Consider an integrable function f on M^n and some point x. Consider an orbit of T of length n starting at x. Let us evaluate the values of f at the points of the orbit, and let us calculate the "mean value" $\frac{1}{n}\Big(f(x)+f(Tx)+\cdots+f(T^{n-1}x)\Big)$ with respect to the discrete time k of our dynamical system $\ldots, T^{k-1}x, T^k x, T^{k+1}x, \ldots$.

Ergodic Theorem. *Let $T : M^n \to M^n$ preserve a finite measure μ on M^n. Then for any integrable function f on M^n and for almost all point $x \in M^n$ there exists the time mean: there exist the following limit:*

$$\lim_{n\to\infty} \frac{1}{n}\sum_{k=0}^{n-1} f(T^k x) = \bar{f}(x).$$

The function $\bar{f}(x)$ is integrable and invariant under the map $\bar{f}(Tx) = \bar{f}(x)$. In particular, if T is ergodic, $\bar{f}$ is constant almost everywhere. Moreover,

$$\int_{M^n} \bar{f}\, d\mu = \int_{M^n} f\, d\mu$$

First Return Map

The following theorem allows to construct numerous *induced* dynamical systems which are closely related to the initial one.

Theorem (Poincaré Recurrence Theorem). *For any subset $U \subset M^n$ of positive measure and for almost any starting point $x \in U$ the trajectory $x, Tx, \ldots$ eventually returns to U, i.e. there is some $n \geq 1$ such that $T^n x \in U$.*

The minimal $n = n(x) \in \mathbb{N}$ as above is called the *first return time*. According to Poincaré Recurrence Theorem integer-valued function $n(x)$ is defined almost everywhere in U. Consider the *first return map* $T|_U : U \to U$ defined as $T|_U : x \mapsto T^{n(x)}x$, where $x \in U$. In other words, the map $T|_U$ maps a point $x \in U$ to the point where trajectory $Tx, T^2x, \ldots$ first meets U.

Lemma. *For any subset $U \subset M^n$ of positive measure the first return map $T|_U : U \to U$ preserves measure μ restricted to U. If $T : M^n \to M^n$ is ergodic than $T|_U : U \to U$ is also ergodic.*

The first return time induced by an ergodic map T has the following geometric property.

Kac Lemma. For an ergodic diffeomorphism $T : M^n \to M^n$ and for any subset $U \in M^n$ of positive measure the mean value of the first return time equals to the volume of entire space:

$$\int_U n(x) d\mu = \mu(M^n)$$

Ergodic Theory: more serious reading. There are numerous nice books on ergodic theory. I can recommend a classical textbook of I. P. Cornfeld, S. V. Fomin and Ya. G. Sinai [CFSin] and a recent survey of B. Hasselblatt and A. Katok [HaKat].

B Multiplicative Ergodic Theorem

In this section we discuss multiplicative ergodic theorem and the notion of *Lyapunov exponents* and then we present some basic facts concerning Lyapunov exponents. As an alternative elementary introduction to this subject we can recommend beautiful lectures of D. Ruelle [Ru]. A comprehensive information representing the state-of-the-art in this subject can be found in the very recent survey [BP2].

B.1 A Crash Course of Linear Algebra

Consider a linear transformation $A : \mathbb{R}^n \to \mathbb{R}^n$ represented by a matrix $A \in SL(n, \mathbb{R})$. Assume that A has n distinct eigenvalues $e^{\lambda_1}, \dots, e^{\lambda_n}$; let $\mathbf{v}_1, \dots, \mathbf{v}_n$ be the corresponding eigenvectors. Note that since $\det A = 1$ we get $\lambda_1 + \dots + \lambda_n = 0$; in particular, $\lambda_1 > 0$ and $\lambda_n < 0$.

Consider now a linear transformation $A^N : \mathbb{R}^n \to \mathbb{R}^n$, where N is a very big positive integer. For almost all vectors $\mathbf{v} \in \mathbb{R}^n$ the linear transformation A^N acts roughly as follows: it takes the projection $\mathbf{v}_{proj}$ of $\mathbf{v}$ to the line $\mathcal{V}_1 = Vec(\mathbf{v}_1)$ spanned by the top eigenvector $\mathbf{v}_1$ and then expands it with a factor $e^{N\lambda_1}$. So, roughly, A^N smashes the whole space to the straight line $\mathcal{V}_1$ and then stretches this straight line with an enormous coefficient of expansion $e^{N\lambda_1}$. (Speaking about projection to $\mathcal{V}_1 = Vec(\mathbf{v}_1)$ we mean a projection along the hyperplane spanned by the remaining eigenvectors $\mathbf{v}_2, \dots, \mathbf{v}_n$.)

To be more precise, we have to note that the image of $\mathbf{v}$ would not have exactly the direction of $\mathbf{v}_1$. A better approximation of $A^N(\mathbf{v})$ would give us a vector in a two-dimensional subspace $\mathcal{V}_2 = Vec(\mathbf{v}_1, \mathbf{v}_2)$ spanned by the two top eigenvectors $\mathbf{v}_1, \mathbf{v}_2$. When $\lambda_2 > 0$ the direction of $A^N(\mathbf{v})$ would be very close to the direction of $\mathcal{V}_1$ though the endpoint of $A^N(\mathbf{v})$ might be at a distance of order $e^{N\lambda_2}$ from $\mathcal{V}_1$, which is very large. However, this distance is small in comparison with the length of $A^N(\mathbf{v})$ which is of order $e^{N\lambda_1} \gg e^{N\lambda_2}$.

Further terms of approximation give us subspaces $\mathcal{V}_j$ spanned by the top j eigenvectors, $\mathcal{V}_j = Vec(\mathbf{v}_1, \dots, \mathbf{v}_j)$. Note that starting with some $k \le n$ the eigenvalues e^{λ_k} become strictly smaller than one. This means that the image $A^N(\mathbf{v})$ of *any* fixed vector $\mathbf{v}$ gets exponentially close to the subspace $\mathcal{V}_{k-1}$.

Going into details we have to admit that vectors $\mathbf{v}$ from a subspace of measure zero in the set of directions expose different behavior. Namely, vectors $\mathbf{v}$ from the hyperplane $\mathcal{L}_2$ spanned by $\mathbf{v}_2, \dots, \mathbf{v}_n$ have smaller coefficient of distortion than the generic ones. From this point of view vectors from linear subspaces $\mathcal{L}_j = Vec(\mathbf{v}_j, \mathbf{v}_{j+1}, \dots, \mathbf{n})$ expose more and more exotic behavior; in particular, all vectors from the subspace $\mathcal{L}_k = Vec(\mathbf{v}_k, \mathbf{v}_{j+1}, \dots, \mathbf{v}_n)$ get exponentially contracted.

B.2 Multiplicative Ergodic Theorem for a Linear Map on the Torus

Consider a linear transformation $A : \mathbb{R}^n \to \mathbb{R}^n$ this time represented by an *integer* matrix $A \in SL(n, \mathbb{Z})$ as above. Consider the induced map $F : \mathbb{T}^n \to \mathbb{T}^n$ of the torus $\mathbb{T}^n = \mathbb{R}^n/\mathbb{Z}^n$. This map preserves the natural linear measure on the torus. Consider the N-th iterate F^N of the map F, where N is a *very large* number.

Note, that the differential of F in the natural coordinates on the torus is represented by the constant matrix $D_{x_0}F = A$ for any $x_0 \in \mathbb{T}^n$. Note also that the differential of the N-th iterate of F is represented as a product of N differentials of F along the trajectory $x_0, F(x_0), F(F(x_0)) \dots, F^{N-1}(x_0)$ of x_0:

$$D_x(F^N) = D_{F^{N-1}(x_0)}F \circ \dots \circ D_{F(x_0)}F \circ D_{x_0}F \tag{B.1}$$

Hence, in "linear" coordinates we get $D_x(F^N) = A^N$. Thus, the results of the previous section are literarily applicable to the local analysis of the map F^N, where now vector $\mathbf{v}$ should be interpreted as a tangent vector to the torus $\mathbb{T}^n$. In particular, these results have the following interpretation. If we consider the trajectory $x_0, F(x_0), \dots, F^N(x_0)$ of the initial point and the trajectory $x, F(x), \dots,\ F^N(x)$ of a point x obtained from x_0 by a very small deformation in direction $\mathbf{v}$, then for *most* of the vectors $\mathbf{v}$ trajectories would deviate exponentially fast one from the other; while for *some special* vectors they would approach each other exponentially fast.

Namely, we get a distribution of linear subspaces $\mathcal{L}_k$ in the tangent space to the torus such that deforming the starting point of trajectory in any direction in $\mathcal{L}_k$ we get two exponentially converging trajectories. The subspace $\mathcal{L}_k = Vec(\mathbf{v}_k, \mathbf{v}_{k+1}, \dots, \mathbf{v}_n)$ is spanned by the eigenvectors of the matrix A having eigenvalues, which are smaller than one. This distribution is integrable; it defines a so-called *stable* foliation.

There is also a complementary *unstable* foliation corresponding to the distribution of subspaces $\mathcal{V}_{k-1} = Vec(\mathbf{v}_1, \dots, \mathbf{v}_{k-1})$ spanned by the eigenvectors of the matrix A having eigenvalues, which are greater than one. Passing from the map F to the map F^{-1} the stable and unstable foliations change the roles: stable foliation of F becomes unstable for F^{-1} and vice versa.

When matrix A has an eigenvalue (or several eigenvalues) equal to ± 1, we get also a *neutral* foliation corresponding to the distribution spanned by the corresponding eigenvectors.

Exercise. Evaluate the limit

$$\lim_{N\to\infty} \frac{\log \|D_x(F^N)(\mathbf{v}\|}{N} \tag{B.2}$$

for a tangent vector $\mathbf{v} \in T_{x_0}\mathbb{T}^n$ having a generic direction. What are the possible values of this limit for *any* tangent vector $\mathbf{v} \in T_{x_0}\mathbb{T}^n$? Show that vectors leading to different values of this limit are organized into a flag of subspaces

$\mathcal{L}_1 \supset \mathcal{L}_2 \supset \cdots \supset \mathcal{L}_n$, where we assume that all eigenvalues of the matrix A are positive and distinct. How would this flag change if some eigenvalues would have multiplicities? Would we have a flag of subspaces defined by different values of the limit above for the most general matrix $A \in SL(n,\mathbb{Z})$ (which may have Jordan blocks, complex eigenvalues, multiplicities, ...)?

B.3 Multiplicative Ergodic Theorem

Consider now a smooth measure-preserving map $F : M^n \to M^n$ on a manifold M^n. We consider the case when that the total measure of M^n is finite, and when the map F is *ergodic* with respect to this measure.

Consider some generic point x_0. Let us study, whether we have convergence of the limit (B.2) for tangent vectors $\mathbf{v} \in T_{x_0}M^n$ in this more general situation. We can always trivialize the tangent bundle to M^n on an open subset of full measure. Using this trivialization we can reduce our problem to the study of product of matrices (B.1). This study is now much more difficult than in the previous case since the matrices $D_{F^k(x_0)}F$ are not constant anymore. The following *multiplicative ergodic theorem* formulated for general mappings of general manifolds mimics the simplest situation with a linear map on the torus.

Theorem (Oseledets). *Let a smooth map $F : M^n \to M^n$ be ergodic with respect to a finite measure. Then, there exists a collection of numbers*

$$\lambda_1 > \lambda_2 > \cdots > \lambda_k,$$

such that for almost any point $x \in M$ there is an equivariant filtration

$$\mathbb{R}^n \simeq T_xM^n = \mathcal{L}_1 \supset \mathcal{L}_2 \supset \cdots \supset \mathcal{L}_k \supset \mathcal{L}_{k+1} = \{0\}$$

in the fiber T_xM^n of the tangent bundle at x with the following property. For every $\mathbf{v} \in \mathcal{L}_j - \mathcal{L}_{j+1}$, $j = 1,\ldots,k$, one has

$$\lim_{N\to+\infty} \frac{1}{N} \log \|(DF^N)_x(\mathbf{v})\| = \lambda_j$$

The multiplicative ergodic theorem was proved by V. I. Oseledets [O2]; a similar statement for products of random matrices was proved earlier by H. Furstenber [Fur].

Multiplicative ergodic theorem has several natural generalizations. The theorem essentially describes the behavior of products (B.1) of matrices along trajectories of the map F. Actually, matrices D_xF are not distinguished by any special property. One can consider any matrix-valued function $A : M^n \to GL(m,\mathbb{R})$ and study the products of matrices

$$A(F^{N-1}(x)) \cdot \cdots \cdot A(F(x)) \cdot A(x)$$

along trajectories $x, F(x), \dots, F^{N-1}(x)$ of F. A statement completely analogous to the above Theorem is valid in this more general case provided the matrix-valued function $A(x)$ satisfy some very moderate requirements. Namely, we do not assume that $A(x)$ is continuous or even bounded. The only requirement is that

$$\int_{M^n} \log_+ \|A(x)\| \mu < +\infty,$$

where $\log_+(y) = \max(\log(y), 0)$. When this condition is satisfied one says that $A(x)$ defines an *integrable cocycle*. The numbers $\lambda_1 > \cdots > \lambda_k$ are called the *Lyapunov exponents* of the corresponding cocycle.

Exercise. Formulate a "continuous-time" version of multiplicative ergodic theorem when instead of a map $F : M^n \to M^n$ we have a flow F_t which is ergodic with respect to a finite measure on M^n. Show that under the natural normalization the corresponding Lyapunov exponents coincide with the Lyapunov exponents of the map F_1 obtained as an action of the flow at the time $t = 1$.

Consider a vector bundle over M^n; suppose that the vector bundle is endowed with a flat connection. Formulate a version of multiplicative ergodic Theorem for the natural action of the flow on such vector bundle.

Note that in the latter case the Lyapunov exponents are responsible for the "mean holonomy" of the fiber along the flow. Namely, we take a fiber of the vector bundle and transport it along a very long piece of trajectory of the flow. When the trajectory comes close to the starting point we identify the fibers using the flat connection and we study the resulting linear transformation of the fiber.

Note that the choice of a norm in the fibers V_x is in a sense irrelevant. Consider two norms $\| \ \|$ and $\| \ \|'$ and let

$$c(x) = \min_{\|\mathbf{v}\|=1} \|\mathbf{v}\|' \qquad C(x) = \max_{\|\mathbf{v}\|=1} \|\mathbf{v}\|'.$$

If

$$\int_M \max\left(|\log(c(x)|, |\log(C(x)|\right) \mu < +\infty,$$

then neither the filtration $\mathcal{L}_k(x)$ nor the Lyapunov exponents λ_k do not change when we replace the norm $\| \ \|$ by the norm $\| \ \|'$. In particular, when M is a compact manifold all nonsingular norms are equivalent.

In general, even for smooth maps $F : M \to M$ (flows F_t) the subspaces defined by the terms $\mathcal{L}_k(x) \subset V_x$ of the filtration do not change continuously with respect to a deformation of the base point x. However, these subspaces behave nicely for maps (flows) which have strong hyperbolic behavior (see [Po] for a short introduction; a recent quite accessible textbook [BP1] and a survey [BP2] describing the contemporary status of *Pesin theory*).

Currently there are no general methods of computation of Lyapunov exponents other than numerically. There are some particular situations, say, when the vector bundle has a one-dimensional equivariant subspace, or when F_t is a homogeneous flow on a homogeneous space; in these rather special cases the corresponding Lyapunov exponents can be computed explicitly. However, in general it is extremely difficult to obtain any nontrivial information (positivity, simplicity of spectrum) about Lyapunov exponents.

References

[AGLP] N. E. Alekseevskii, Yu. P. Gaidukov, I. M. Lifshits, V. G. Peschanskii: JETPh, **39** (1960)

[Ald1] V. I. Arnold: Small denominators and problems of stability of motion in classical and celestial mechanics. Russian Math. Surveys **18** no. 6, 85–191 (1963)

[Ald2] V. I. Arnold: Topological and ergodic properties of closed 1-forms with rationally independent periods. Functional Anal. Appl., **25**, no. 2, 81–90 (1991)

[Arn] P. Arnoux: Le codage du flot géodésique sur la surface modulaire. L'Enseignement Mathématique, **40**, 29–48 (1994)

[At] M. Atiyah: Riemann surfaces and spin structures. Ann. scient. ÉNS 4e série, **4**, 47–62 (1971)

[AthEZo] J. Athreya, A. Eskin, A. Zorich: Rectangular billiards and volumes of spaces of quadratic differentials. In preparation.

[AvFor] A. Avila, G. Forni: Weak mixing for interval exchange transformations and translation flows. To appear in Annals of Math. Eprint, `arXiv.math.DS/0406326`, 21 pp (2004)

[AvVi] A. Avila, M. Viana: Simplicity of Lyapunov spectra: proof of the Zorich–Kontsevich conjecture. Eprint `math.DS/0508508`, 36 pp (2005)

[AzLK] M. Ya. Azbel, I. M. Lifshits, M. I. Kaganov: Elektronnaia Teoriya Metallov. "Nauka", 1971; English translation: Electron Theory of Metals. Consultants Bureau (1973)

[BP1] L. Barreira and Ya. Pesin: Lyapunov Exponents and Smooth Ergodic Theory. Univ. Lect. Series 23, Amer. Math. Soc. (2002)

[BP2] L. Barreira and Ya. Pesin: Smooth ergodic theory and nonuniformly hyperbolic dynamics. In: B. Hasselblatt and A. Katok (ed) Handbook of Dynamical Systems, Vol. 1B. Elsevier Science B.V., Amsterdam, 57–264 (2006)

[Ber1] L. Bers: Quasiconformal mappings and Teichmüller's theorem. In collection: R. Nevanlinna et al. (ed) Analytic Functions. Princeton Math. Ser. 24, Princeton University Press, NJ, 89–119 (1960)

[Ber2] L. Bers: Finite dimensional Teichmüller spaces and generalizations. Bull. Amer. Math. Soc. **5**, 131–172 (1981)

[Ber2] M. Boshernitzan: A condition for minimal interval exchange maps to be uniquely ergodic. Duke Math. J. **52**, no. 3, 723–752 (1985)

[Clb] E. Calabi: An intrinsic characterization of harmonic 1-forms, Global Analysis. In: D. C. Spencer and S. Iyanaga (ed) Papers in Honor of K. Kodaira. 101–117 (1969)

[Clt] K. Calta: Veech surfaces and complete periodicity in genus two. J. Amer. Math. Soc., **17** 871-908 (2004)

[CFSin] I. P. Corndeld, S. V. Fomin, Ya. G.Sinai: Ergodic Theory. Springer Verlag New York (1982)

[D1] I. A. Dynnikov: Surfaces in 3-Torus: Geometry of Plane Sections. Progress in Math., Vol. 168, 162–177 Birkhäuser Verlag. Basel (1998)

[D2] I. A. Dynnikov: Geometry of stability zones in S.P.Novikov's problem on semi-classical motion of an electron. Russian Math. Surveys, **54**, no. 1, 21–60 (1999)

[E] A. Eskin: Counting problems in moduli space. In: B. Hasselblatt and A. Katok (ed) Handbook of Dynamical Systems, Vol. 1B. Elsevier Science B.V., Amsterdam, 581–595 (2006)

[EMa] A. Eskin, H. Masur: Asymptotic formulas on flat surfaces. Ergodic Theory and Dynamical Systems, **21:2**, 443–478 (2001)

[EMaScm] A. Eskin, H. Masur, M. Schmoll; Billiards in rectangles with barriers, Duke Math. J., **118** (3), 427–463 (2003)

[EMkWt] A. Eskin, J. Marklof, D. Witte Morris; Unipotent flows on the space of branched covers of Veech surfaces, submitted to Ergod. Theory and Dyn. Syst., **26** (1), 129–162 (2006)

[EMaZo] A. Eskin, H. Masur, A. Zorich: Moduli spaces of Abelian differentials: the principal boundary, counting problems, and the Siegel–Veech constants. Publications de l'IHES, **97:1**, pp. 61–179 (2003)

[EOk] A. Eskin, A. Okounkov: Asymptotics of number of branched coverings of a torus and volumes of moduli spaces of holomorphic differentials. Inventiones Mathematicae, **145:1**, 59–104 (2001)

[EOkPnd] A. Eskin, A. Okounkov, R. Pandharipande: The theta characteristic of a branched covering, `math.AG/0312186`, 22 pp (2003)

[Fay] J. Fay: Theta functions on Riemann surfaces. Lecture Notes in Mathematics, **352**. Springer-Verlag (1973)

[For1] G. Forni: Deviation of ergodic averages for area-preserving flows on surfaces of higher genus. Annals of Math., **155**, no. 1, 1–103 (2002)

[For2] G. Forni: On the Lyapunov exponents of the Kontsevich–Zorich cocycle. In: B. Hasselblatt and A. Katok (ed) Handbook of Dynamical Systems, Vol. 1B. Elsevier Science B.V., Amsterdam, 549–580 (2006)

[Ful] W. Fulton: Hurwitz Schemes and Moduli of Curves. Annals of Math., **90**, 542–575 (1969)

[FxKr] R. H. Fox, R. B. Kershner: Geodesics on a rational polyhedron. Duke Math. J., **2**, 147–150 (1936)

[Fur] H. Furstenberg: Non-commuting random products. Trans. Amer. Math. Soc., **108**, 377–428 (1963)

[GaStVb1] G. Galperin, Ya. B. Vorobets, A. M. Stepin: Periodic billiard trajectories in polygons. Russian Math. Surveys **46**, no. 5, 204–205 (1991)

[GaStVb2] G. Galperin, Ya. B. Vorobets, A. M. Stepin: Periodic billiard trajectories in polygons: generation mechanisms. Russian Math. Surveys **47**, no. 3, 5–80 (1992)

[GaZe] G. Galperin and A. N. Zemliakov: Mathematical billiards. Billiard problems and related problems in mathematics and mechanics. Library "Kvant", **77**. "Nauka", Moscow (in Russian) 288 pp. (1990)

[Gu1] E. Gutkin: Billiards in polygons: surfey of recent results. J. Stat. Phys., **83**, 7–26 (1996)

[Gu2] E. Gutkin: Billiards dynamics: a surfey with the emphasis on open problems. Regular and Chaotic Dynamics, **8**, no.1, 1–13 (2003)

[GuJg] E. Gutkin, C. Judge: Affine mappings of translation surfaces: geometry and arithmetic. Duke Math. J., **103**, no. 2, 191–213 (2000)

[HaKat] B. Hasselblatt and A. B. Katok: Principal structures, In: B. Hasselblatt and A. Katok (ed) Handbook of Dynamical Systems, Vol. 1A, Elsevier Science B.V., 1–203, (2002)

[HbMa] J. Hubbard and H. Masur: Quadratic differentials and measured foliations. Acta Math., **142**, 221–274 (1979)

[HuLe1] P. Hubert, S. Lelièvre: Prime arithmetic Teichmüller discs in $\mathcal{H}(2)$. Israel Journal of Math., **151**, 281–321 (2006)

[HuLe2] P. Hubert, S. Lelièvre: Noncongruence subgroups in $\mathcal{H}(2)$, Internat. Math. Research Notes **2005:1**, 47–64 (2005)

[HuMSdtZ] P. Hubert, H. Masur, T. A. Schmidt, A. Zorich: Problems on billiards, flat surfaces and translation surfaces. In: B. Farb (ed) Problems on Mapping Class Groups and Related Topics, Proc. Symp. Pure Math., Amer. Math. Soc. (2006)

[HuSdt1] P. Hubert, T. A. Schmidt: Veech groups and polygonal coverings. J. Geom. and Phys. **35**, 75–91 (2000)

[HuSdt2] P. Hubert, T. A. Schmidt: Invariants of translation surfaces. Ann. Inst. Fourier (Grenoble) **51**, no. 2, 461–495 (2001)

[HuSdt3] P. Hubert, T. A. Schmidt: Infinitely generated Veech groups. Duke Math. J. **123**, 49–69 (2004)

[HuSdt4] P. Hubert, T. A. Schmidt: Geometry of infinitely generated Veech groups. Conform. Geom. Dyn. **10**, 1–20 (2006)

[HuSdt5] P. Hubert and T. Schmidt: An Introduction to Veech Surfaces. In: B. Hasselblatt and A. Katok (ed) Handbook of Dynamical Systems, Vol. 1B. Elsevier Science B.V., Amsterdam, 501–526 (2006)

[J] D. Johnson: Spin structures and quadratic forms on surfaces. J. London Math. Soc., (2) **22**, 365–373 (1980)

[Kat1] A. Katok: Invariant measures of flows on oriented surfaces. Soviet Math. Dokl., **14**, 1104–1108 (1973)

[Kat2] A. Katok: Interval exchange transformations and some special flows are not mixing, Israel Journal of Mathematics **35**, no. 4, 301–310 (1980)

[KatZe] A. Katok, A. Zemlyakov: Topological transitivity of billiards in polygons. Math. Notes, **18**, 760–764 (1975)

[Kea1] M. Keane: Interval exchange transformations. Math. Z., **141**, 25–31 (1975)

[Kea2] M. Keane: Non-ergodic interval exchange transformations. Israel J. Math. **26**, no. 2, 188–196 (1977)

[KenS] R. Kenyon and J. Smillie: Billiards in rational-angled triangles, Comment. Mathem. Helv. **75**, 65–108 (2000)

[Ker1] S. P. Kerckhoff: The Asymptotic geometry of Teichmüller space. Topology, **19**, 23–41 (1980)

[Ker2] S. P. Kerckhoff: Simplicial systems for interval exchange maps and measured foliations. Ergod. Th. & Dynam. Sys. **5**, 257–271 (1985)

[KMaS] S. Kerckhoff, H. Masur, and J. Smillie: Ergodicity of billiard flows and quadratic differentials. Annals of Math., **124**, 293–311 (1986)

[KhSin] K. M. Khanin, Ya. G. Sinai: Mixing of some classes of special flows over rotations of the circle. Functional Anal. Appl. **26**, no. 3, 155–169 (1992)

[Kon] M. Kontsevich: Lyapunov exponents and Hodge theory. “The mathematical beauty of physics” (Saclay, 1996), (in Honor of C. Itzykson) 318–332, Adv. Ser. Math. Phys., 24. World Sci. Publishing, River Edge, NJ (1997)

[KonZo] M. Kontsevich, A. Zorich: Connected components of the moduli spaces of Abelian differentials. Invent. Math., **153:3**, 631–678 (2003)

[Lan] E. Lanneau: Connected components of the moduli spaces of quadratic differentials. Eprint, `arXiv.math.GT/0506136`, 41 pp (2005)

[Lv] P. Lévy: Sur le développement en fraction continue d’un nombre choisi au hasard. Compositio Mathematica, **3**, 286–303 (1936)

[MmMsY] S. Marmi, P. Moussa, J.-C. Yoccoz: The cohomological equation for Roth–type interval exchange maps. J. Am. Math. Soc. **18**, No.4, 823–872 (2005)

[Ma1] H. Masur: On a class of geodesics in Teichmüller space. Ann. of Math., **102**, 205–221 (1975)

[Ma2] H. Masur: Extension of the Weil-Petersson metric to the boundary of Teichmüller space. Duke Math. Jour., **43**, no. 3, 623–635 (1976)

[Ma3] H. Masur: Interval exchange transformations and measured foliations. Ann. of Math., **115**, 169-200 (1982)

[Ma4] H. Masur: Closed Trajectories for Quadratic Differentials with an Application to Billiards. Duke Math. Jour. **53**, 307–314 (1986)

[Ma5] H. Masur: Lower bounds for the number of saddle connections and closed trajectories of a quadratic differential. In: Holomorphic Functions and Moduli, Vol. I (Berkeley, CA, 1986), 215–228. Math. Sci. Res. Inst. Publ., **10** Springer, New York – Berlin (1988)

[Ma6] H.Masur: The growth rate of trajectories of a quadratic differential. Ergodic Theory and Dynamical Systems, **10**, no. 1, 151–176 (1990)

[Ma7] H. Masur: Ergodic Theory of Translation Surfaces. In: B. Hasselblatt and A. Katok (ed) Handbook of Dynamical Systems, Vol. 1B Elsevier Science B.V., Amsterdam, 527–547 (2006)

[MaS] H. Masur, J. Smillie: Hausdorff dimension of sets of nonergodic foliations. Ann. of Math. **134**, 455–543 (1991)

[MaT] H. Masur and S. Tabachnikov: Rational Billiards and Flat Structures. In: B. Hasselblatt and A. Katok (ed) Handbook of Dynamical Systems, Vol. 1A, Elsevier Science B.V., 1015–1089 (2002)

[MaZo] H. Masur and A. Zorich: Multiple Saddle Connections on Flat Surfaces and Principal Boundary of the Moduli Spaces of Quadratic Differentials, Preprint `math.GT/0402197` 73pp (2004)

[McM1] C. McMullen: Teichmüller geodesics of infinite complexity, Acta Math. **191**, 191–223 (2003)

[McM2] C. McMullen: Billiards and Teichmüller curves on Hilbert modular surfaces. J. Amer. Math. Soc., **16**, no. 4, 857–885 (2003)

[McM3] C. McMullen: Dynamics of $SL_2(\mathbb{R})$ over moduli space in genus two. Annals of Math. (to appear)

[McM4] C. McMullen: Teichmüller curves in genus two: Discriminant and spin. Math. Ann., **333**, no.1, 87–130 (2005)

[McM5] C. McMullen: Teichmüller curves in genus two: The decagon and beyond. J. Reine Angew. Math., **582**, 173–199 (2005)

[McM6] C. McMullen: Teichmüller curves in genus two: Torsion divisors and ratios of sines. Invent. Math. (to appear).

[McM7] C. McMullen: Prym varieties and Teichmüller curves. Duke Math. J., to appear.

[Mil] J. Milnor: Remarks concerning spin manifolds. In: Differential and Combinatorial Topology (in Honor of Marston Morse), Princeton (1965)

[Mo1] M. Möller: Teichmüller curves, Galois actions and GT-relations. Math. Nachrichten **278:9**, 1061–1077 (2005)

[Mo2] M. Möller: Variations of Hodge structures of a Teichmüller curve. Journal of the AMS, **19**, 327–344 (2006)

[Mo3] M. Möller: Periodic points on Veech surfaces and the Mordell–Weil group over a Teichmüller curve. Invent. Math. (to appear). Eprint `math.AG/0410262`, 13 pages (2004)

[Mum] D. Mumford: Theta-characteristics of an algebraic curve. Ann. scient. Éc. Norm. Sup. 4e série, **2**, 181–191 (1971)

[NgRd] A. Nogueira, D. Rudolph: Topological weak-mixing of interval exchange maps. Ergodic Theory Dynam. Systems **17**, no. 5, 1183–1209 (1997)

[N] S. P. Novikov: The Hamiltonian formalism and a multi-valued analogue of Morse theory. Russian Math. Surveys, **37:5**, 1–56 (1982)

[NM] S. P. Novikov, A. Ya. Maltsev: Topological phenomena in normal metals. Letters to JETPh, **63**, No. 10, 809–813 (1996)

[O1] V. I. Oseledets (Oseledec): The spectrum of ergodic automorphisms. Dokl. Akad. Nauk SSSR (Soviet Math. Dokl.) **168**, 1009–1011 (1966)

[O2] V. I. Oseledetc: A Multiplicative Ergodic Theorem. Ljapunov characteristic numbers for dynamical systems. Trans. Moscow Math. Soc. **19**, 197–231 (1968)

[Pa] T. Payne: Closures of totally geodesic immersions into locally symmetric spaces of noncompact type. Proc. Amer. Math. Soc. **127**, no. 3, 829–833 (1999)

[Po] M. Pollicott: Lectures on ergodic theory and Pesin theory on compact manifolds. London Mathematical Society Lecture Note Series, **180**. Cambridge University Press, Cambridge (1993)

[Pu] J.-Ch. Puchta: On triangular billiards, Comment. Mathem. Helv. **76**, 501–505 (2001)

[Ra] G. Rauzy: Echanges d'intervalles et transformations induites. Acta Arith. **34**, 315–328 (1979)

[Ru] D. Ruelle: Chaotic Evolution and Strange Attractors. Cambridge University Press (1989)

[Sat] E. Sataev: The number of invariant measures for flows on orientable surfaces. Izv. Akad. Nauk SSSR Ser. Mat. **39**, no. 4, 860–878 (1975)

[Schn] G. Schmithüsen: An algorithm for finding the Veech group of an origami. Experimental Mathematics **13:4**, 459–472 (2004)

[Schl1] M. Schmoll: Spaces of elliptic differentials. In: "Algebraic and Topological Dynamics", S. Kolyada, Yu. I. Manin and T. Ward (Editors), Contemporary Math., Vol. 385, Amer. Math. Soc., Providence, RI, 303–320 (2005)

[Schl2] M. Schmoll: Moduli spaces of branched covers of Veech surfaces I, II. Eprint, `arXiv.math.GT/0602396`, 40 pp, (2006)

[Schw] S. Schwartzman: Asymptotic cycles. Annals of Math., **66**, 270–284 (1957)

[Ser] C. Series: Geometric methods of symbolic coding. In: T. Bedford, M. Keane, C. Series (ed) Ergodic Theory, Symbolic Dynamics and Hyperbolic Spaces. Oxford University Press, Oxford, 125–151 (1991)

[Sin] Ya. G. Sinai: Dynamical systems with elastic reflections, Russ. Math. Surveys, **68**, 137–189 (1970)

[S] J. Smillie: Dynamics of billiard flow in rational polygons. In: Ya. G. Sinai (ed) Dynamical Systems. Ecyclopedia of Math. Sciences. Vol. 100. Math. Physics 1. Springer Verlag, 360–382 (2000)

[Sh] N. Shah: Closures of totally geodesic immersions in manifolds of constant negative curvature. In: Group theory from a geometrical viewpoint (Trieste, 1990), 718–732, World Sci. Publishing, River Edge, NJ (1991)

[Str] K. Strebel: Quadratic Differentials. Springer-Verlag (1984)

[T] S. Tabachnikov: Billiards. Panoramas and Synthèses. SMF (1995)

[Ve1] W. A. Veech: Strict ergodicity in zero dimensional dynamical systems and the Kronecker-Weyl theorem mod 2. Trans. Amer. Math. Soc. **140**, 1–33 (1969)

[Ve2] W. Veech: Interval exchange transformations. Journ. Anal. Math., **33** 222–278 (1978)

[Ve3] W. A. Veech: Gauss measures for transformations on the space of interval exchange maps. Annals of Math., **115**, 201–242 (1982)

[Ve4] W. A. Veech: The metric theory of interval exchange transformations I. Generic spectral properties. Amer. Journal of Math., **106**, 1331–1359 (1984)

[Ve5] W. A. Veech: The metric theory of interval exchange transformations II. Approximation by primitive interval exchanges. Amer. Journal of Math., **106**, 1361–1387 (1984)

[Ve6] W. A. Veech: Teichmüller geodesic flow. Annals of Math., **124**, 441–530 (1986)

[Ve7] W. A. Veech: Teichmüller curves in modular space, Eisenstein series, and an application to triangular billiards, Inv. Math. **97**, 553–583 (1989)

[Ve8] W. A. Veech: Flat surfaces. Amer. Journal of Math., **115**, 589–689 (1993)

[Ve9] W. A. Veech: Geometric realization of hyperelliptic curves. Chaos, Dynamics and Fractals. Plenum (1995)

[Vb1] Ya. Vorobets: Planar structures and billiards in rational polygons: the Veech alternative. Russian Math. Surveys, **51:5**, 779–817 (1996)

[Vb2] Ya. Vorobets: Periodic geodesics on generic translation surfaces. In: "Algebraic and Topological Dynamics", S. Kolyada, Yu. I. Manin and T. Ward (Editors), Contemporary Math., Vol. 385, Amer. Math. Soc., Providence, RI, 205–258 (2005)

[WYa] S. Wakon, J. Yamashita: J.Phys. Soc. Japan, **21**, 1712 (1966)

[Y] J.-C. Yoccoz: Continuous fraction algorithms for interval exchange maps: an introduction. "Frontiers in Number Theory, Physics and Geometry Vol. 1", P. Cartier; B. Julia; P. Moussa; P. Vanhove (Editors), Springer Verlag, 403–437 (2006)

[Zo1] A. Zorich: The S. P. Novikov problem on the semiclassical motion of an electron in homogeneous Magnetic Field. Russian Math. Surveys, **39:5**, 287–288 (1984)

[Zo2] A. Zorich: Finite Gauss measure on the space of interval exchange transformations. Lyapunov exponents. Annales de l'Institut Fourier, **46:2**, 325–370 (1996)

[Zo3] A. Zorich: Deviation for interval exchange transformations. Ergodic Theory and Dynamical Systems, **17**, 1477–1499 (1997)

[Zo4] A. Zorich: How do the leaves of a closed 1-form wind around a surface. In the collection: "Pseudoperiodic Topology", *AMS Translations*, Ser. 2, vol. 197, AMS, Providence, RI, 135–178 (1999)

[Zo5] A. Zorich: Square tiled surfaces and Teichmüller volumes of the moduli spaces of Abelian differentials. In collection "Rigidity in Dynamics and Geometry", M. Burger, A. Iozzi (Editors), Springer Verlag, 459–471 (2002)

Brjuno Numbers and Dynamical Systems

Guido Gentile

Dipartimento di Matematica, Università di Roma Tre, I-00146 Roma
gentile@mat.uniroma3.it

1 Introduction

We shall consider some elementary analytic dynamical systems widely studied in literature, which are perturbations of integrable (linear) ones. A classical problem is to study the conditions for the perturbed system to be analytically conjugated to the linear one, that is the conditions for the existence of an analytic change of coordinates which maps the perturbed system to the linear one. We shall see that such conditions can be naturally expressed in terms of the Diophantine properties of a suitable parameter.

1.1 Brjuno Function and Brjuno Numbers

For $\omega \in \mathbb{R} \setminus \mathbb{Q}$ define the *Brjuno function* as the solution of the functional equation

$$\begin{cases} B(\omega+1) = B(\omega), \\ B(\omega) = -\log\omega + \omega B(1/\omega), & \text{if } \omega \in (0,1), \end{cases} \tag{1.1}$$

and call ω a *Brjuno number* if $B(\omega) < \infty$. We shall be interested essentially in numbers $\omega \in \mathcal{B} = \{\omega \in \mathbb{R} \setminus \mathbb{Q} \cap (0,1) \mid B(\omega) < \infty\}$.

Note that, if we denote with $\{p_n/q_n\}_{n=0}^{\infty}$ the sequence of convergents of ω and define

$$B_1(\omega) = \sum_{n=0}^{\infty} \frac{\log q_{n+1}}{q_n}, \tag{1.2}$$

there exists a constant C such that $|B(\omega) - B_1(\omega)| < C$ for all $\omega \in \mathcal{B}$ [20].

1.2 Some Dynamical Systems

Siegel's problem. Consider the holomorphic diffeomorphism [19]

$$z' = f(z), \qquad f(z) = \lambda z + O(z^2), \tag{1.3}$$

with $\lambda = e^{2\pi i\omega}$. We shall write $f \in G_\lambda$, where G is the group of germs of holomorphic diffeomorphisms of $(\mathbb{C}, 0)$ and $G_\lambda = \{f \in G \mid f'(0) = \lambda\}$.

Define also $R_\lambda(z) = \lambda z$. We say that $f \in G_\lambda$ is *linearizable* if there exists $h \equiv h_f \in G_1$ such that $f \circ h = h \circ R_\lambda$: in such a case the diffeomorphism (1.3) is *conjugated* to its linear part R_λ: if we set $\alpha' = \lambda\alpha$ then $z' = h(\alpha') = h \circ R_\lambda(\alpha) = f \circ h(\alpha) = f(h(\alpha))$. The problem of finding conditions under which f is linearizable is usually referred to as *Siegel's problem.* The following result holds.

Theorem 1. *If $\omega \in \mathcal{B}$ and $\lambda = e^{2\pi i\omega}$ then $f \in G_\lambda$ is linearizable.*

This follows from a stronger result of Yoccoz. If $\mathbb{D}$ denotes the unit disk in $\mathbb{C}$ call S_λ the topological space of germs of holomorphic diffeomorphisms $f : \mathbb{D} \to \mathbb{C}$ such that $f(0) = 0$, $f'(0) = \lambda$ and f is univalent on $\mathbb{D}$, and, for $\lambda = e^{2\pi i\omega}$, set $r(\omega) = \inf_{f \in S_\lambda} r(f, \omega)$, if $r(f, \omega)$ is the radius of convergence of h_f for $f \in G_\lambda$: then Theorem 1 is implied from the following one [20].

Theorem 2. *There exists a constant C such that $|\log r(\omega) + B(\omega)| < C$ for all $\omega \in \mathcal{B}$. Moreover if $f(z)$ is the* quadratic polynomial $P_\lambda(z) = \lambda z(1 - z/2)$, *for all $\eta > 0$ there exists a constant C_η such that $-B(\omega) - C < \log r(P_\lambda, \omega) < -(1-\eta)B(\omega) + C_\eta$ for a universal constant C and for all $\omega \in \mathcal{B}$.*

Yoccoz also proved that if P_λ is linearizable then every germ $f \in G_\lambda$ is also linearizable [20].

Standard map, semistandard map and generalizations. Consider the dynamical system

$$T_\varepsilon : \begin{cases} x' = x + y + \varepsilon f(x), \\ y' = y + \varepsilon f(x), \end{cases} \tag{1.4}$$

where f is a zero-average analytic function and ε is a small parameter. If f is real for real x, then we shall study T_ε as a map from the cylinder to itself, $T_\varepsilon : \mathbb{T} \times \mathbb{R} \to \mathbb{T} \times \mathbb{R}$, with $\mathbb{T} = \mathbb{R}/2\pi\mathbb{Z}$, otherwise we shall consider T_ε as a map $T_\varepsilon : \mathbb{C}/2\pi\mathbb{Z} \times \mathbb{C} \to \mathbb{C}/2\pi\mathbb{Z} \times \mathbb{C}$. As particular cases we consider the *standard map* and the *semistandard map* [9, 16], obtained by choosing $f(x) = s(x) \equiv \sin x$ and $f(x) = e(x) \equiv \exp(ix)/2i$, respectively. In the case of the semistandard map the parameter ε can be eliminated, but we prefer to keep it in order to make easier the comparison with the standard map.

We look for solutions of the form

$$\begin{cases} x = \alpha + u(\alpha, \varepsilon), \\ y = 2\pi\omega + v(\alpha, \varepsilon), \end{cases} \tag{1.5}$$

such that in the variable α the dynamics is a trivial rotation $\alpha' = \alpha + 2\pi\omega$, with rotation number ω. We see immediately that we can express $v(\alpha, \varepsilon)$ in terms of $u(\alpha, \varepsilon)$ as $v(\alpha, \varepsilon) = u(\alpha, \varepsilon) - u(\alpha - 2\pi\omega, \varepsilon)$, while $u(\alpha, \varepsilon)$ solves the functional equation

$$u(\alpha + 2\pi\omega, \varepsilon) + u(\alpha - 2\pi\omega, \varepsilon) - 2u(\alpha, \varepsilon) = \varepsilon f\left(\alpha + u(\alpha, \varepsilon)\right). \tag{1.6}$$

We say that T_ε admits an *invariant curve* with rotation number ω if there exists an analytic function $u(\alpha, \varepsilon)$ such that (1.5) holds with $\alpha \to \alpha + 2\pi\omega$; the function $u(\alpha, \varepsilon)$ is called the *conjugating function*. The following result holds.

Theorem 3. *If $\omega \in \mathcal{B}$ and ε is small enough then T_ε admits an analytic invariant curve with rotation number ω.*

A more quantitative statement is the following. Given $\omega \in \mathcal{B}$ there exists $r(\omega, f) \in \mathbb{R}^+$ such that there exists a solution of (1.4) of the form (1.5) with u (and hence v) analytic in ε for $|\varepsilon| < r(\omega, f)$: $r(\omega, f)$ is the radius of convergence of the conjugating function for fixed ω and f (see also (2.2) below).

For the semistandard and the standard maps the result above can be strengthened. In fact in such cases, if we set $\rho_0(\omega) = r(e, \omega)$ and $\rho(\omega) = r(s, \omega)$, one can prove the following results.

Theorem 4. *There exist a universal constant C_0 such that $|\log \rho_0(\omega) + 2B(\omega)| < C_0$ for all $\omega \in \mathcal{B}$.*

Theorem 5. *There exist a universal constant C such that $|\log \rho(\omega) + 2B(\omega)| < C$ for all $\omega \in \mathcal{B}$.*

The bounds of Theorem 4 were proved by Davie [10], who also proved that one has $\rho(\omega) \le \rho_0(\omega)$, which implies the upper bound in Theorem 5. In an unpublished paper [11] Davie also showed, by using renormalization group methods, that for all $\eta > 0$ there exists a constant C_η such that one has

$\log \rho(\omega) \geq -(1-\eta)B(\omega) - C_\eta$ for $\omega \in \mathcal{B}$. This is not enough to obtain the lower bound in Theorem 5, which was proved by Berretti and Gentile [4].

Here we want to describe a technique which is suitable for proving the lower bounds considered so far. In particular we shall see that it is very easy to obtain the lower bounds for Siegel's problem and the semistandard map, while the case of the standard map turns out to be more difficult. Such a difficulty is mainly due to the optimality of the bound: if we confined ourselves to show that $\rho(\omega)$ (or even $r(\omega, f)$) is strictly positive for $\omega \in \mathcal{B}$, then the proof would be much easier.

2 Tree Formalism

We use a technique which shows strong analogies with the renormalization group method in quantum field theory. Such a technique has been introduced by Eliasson [12] and Gallavotti [13] to prove existence of KAM tori for quasi-integrable Hamiltonian systems, and then developed and extended in other papers (we refer to the bibliography for some reviews [15, 14, 6]).

Let us consider first the dynamical system (1.4). The conjugating function $u(\alpha, \varepsilon)$ admits a formal expansion – the *Lindstedt series* – of the form

$$u(\alpha,\varepsilon) = \sum_{\nu\in\mathbb{Z}} u_\nu(\varepsilon)\, e^{i\nu\alpha} = \sum_{k\geq 1} u^{(k)}(\alpha)\, \varepsilon^k = \sum_{k\geq 1}\sum_{\nu\in\mathbb{Z}} u_\nu^{(k)}\, e^{i\nu\alpha}\, \varepsilon^k . \tag{2.1}$$

For the standard map, at order k in ε, the Fourier expansion in α contains only frequencies $|\nu| \leq k$, while for the semistandard map one has $\nu = k$. The *radius of convergence* of the Lindstedt series is naturally defined as

$$r(\omega, f) = \inf_{\alpha\in\mathbb{T}} \left(\limsup_{k\to\infty} \left| u^{(k)}(\alpha) \right|^{1/k} \right)^{-1} . \tag{2.2}$$

By inserting (2.1) into the functional equation (1.6), and equating the Taylor-Fourier coefficients we obtain a recursion relation for the coefficients of the Lindstedt series: one has $u_\nu^{(1)} = g(\omega\nu) f_\nu$ and, for $k \geq 2$,

$$u_\nu^{(k)} = g(\omega\nu) \sum_{m\geq 1} \frac{1}{m!} \sum_{\substack{k_1+\ldots+k_m=k-1 \\ \nu_0+\nu_1+\ldots+\nu_m=\nu}} f_{\nu_0}\, (i\nu_0)^m \prod_{j=1}^m u_{\nu_j}^{(k_j)}, \tag{2.3}$$

with $f_{\nu_0} = -i\nu_0\delta_{|\nu_0|,1}/2$ for the standard map and $f_{\nu_0} = -i\nu_0\delta_{\nu_0,1}/2$ for the semistandard map, and

$$g(\omega\nu) = \frac{1}{\gamma(\omega\nu)}, \qquad \gamma(\omega\nu) = 2\left[\cos(2\pi\omega\nu) - 1\right]. \tag{2.4}$$

The denominators $\gamma(\omega\nu)$ can become arbitrarily small for ω irrational: this is a manifestation of the famous *small divisors* problem.

By iterating the recursion relation (2.3), which corresponds to apply to each $u_{\nu_j}^{(k_j)}$ the same decomposition used for $u_\nu^{(k)}$ itself, we see that at the end the coefficients $u_\nu^{(k)}$ can be written as sums of contributions which are represented in terms of tree graphs (or simply *trees*); see for instance Figure 1.

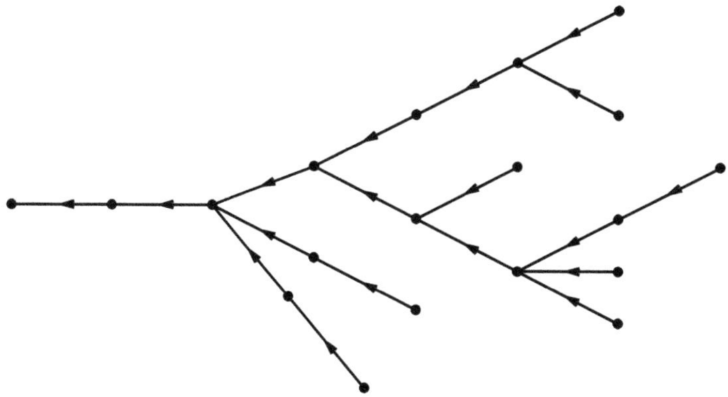

Fig. 1. A tree of order 18. The labels are not shown. All the arrows point toward the point to the extreme left (root). All the other points are called nodes

A tree ϑ is constructed in the following way. Consider a family of $k-1$ *lines* arranged to connect a set of k points called *nodes* (so that no loops arise), and add an extra line connecting one of the nodes to another point called the *root.* Each line carries an arrow pointing toward the root: this induces a partial ordering relation on the nodes (and lines), with the lowest nodes drawn to the right and the root to the left. Given two nodes u_1 and u_2, we write $u_2 \preceq u_1$ if u_1 is along the path of lines connecting u_2 to the root r of the tree (they can coincide: we say that $u_2 \prec u_1$ if they do not).

If a line connects a node u to the right to a point u' to the left we say that the line comes out from u and enters u'. For each node u there are only one line coming out from u and $m_u \geq 0$ entering ones; as there is a one-to-one correspondence between nodes and lines, we can associate to each node u a line ℓ_u coming out from it. By construction there is only one line entering the root: we shall call it the *root line*. Note that each line ℓ_u can be considered the root line of the subtree consisting of the nodes satisfying $w \preceq u$ and of the lines connecting them plus the line connecting u to u', which will be the root of such subtree. The *order* of the tree is defined as the number k of nodes of the tree.

To each node $u \in \vartheta$ we associate a *mode label* $\nu_u \in \mathbb{Z} \setminus \{0\}$, which is a Fourier label of the function $f(x)$; define the *momentum* flowing through the line ℓ_u as $\nu_{\ell_u} = \sum_{w \preceq u} \nu_w$.

Let us denote by $\Theta_{k,\nu}^0$ the set of all trees of order k with momentum ν flowing through the root line, and by $V(\vartheta)$ and $\Lambda(\vartheta)$, respectively, the set of nodes and the set of lines of the tree ϑ. To each node $u \in V(\vartheta)$ we associate a *node factor* $F_v = (i\nu_u)^{m_u+1}/(m_u!2)$, while to each line $\ell \in \Lambda(\vartheta)$ we associate a *propagator* $G_\ell = g(\omega\nu_\ell)$, where $g(\omega\nu)$ is defined in (2.4).

Given a tree $\vartheta \in \Theta^0_{k,\nu}$ define a map $\mathrm{Val} : \Theta^0_{k,\nu} \to \mathbb{R}$ as

$$\mathrm{Val}(\vartheta) = \Big(\prod_{u \in V(\vartheta)} F_v \Big) \Big(\prod_{\ell \in \Lambda(\vartheta)} G_\ell \Big); \tag{2.5}$$

if $\Theta^0_{k,\nu} = \emptyset$ we interpret $\mathrm{Val}(\vartheta) = 0$. We call $\mathrm{Val}(\vartheta)$ the *value* of the tree ϑ. Then the following result holds, establishing a link between the conjugating function and the trees.

Lemma 1. *One can take $u_0^{(k)} = 0$ for all $k \geq 1$. Then $\nu_\ell \neq 0$ $\forall \ell \in \Lambda(\vartheta)$ and one has $u_\nu^{(k)} = \sum_{\vartheta \in \Theta^0_{k,\nu}} \mathrm{Val}(\vartheta)$ for all $k \geq 1$ and for all $\nu \neq 0$.*

The (easy) proof can be carried out by induction. We say that two trees are equivalent if they can be transformed into each other by continuously deforming the lines without crossing. The sum over $\Theta^0_{k,\nu}$ is meant as a sum over all the trees which are not equivalent; that this is the correct way to count the trees follows from the fact that it keeps trace of the combinatorial factors naturally arising from the Taylor expansion (2.1) when we iterate the graphical construction generated by (2.3). The number of elements of $\Theta^0_{k,\nu}$ is bounded by 2^{2k} in the case of the semistandard map and by $2^k 2^{2k}$ in the case of the standard map. So we are left with the problem of proving that the series (2.1) converges.

In the case of Siegel's problem we consider the formal expansion

$$h(\alpha) = \alpha + \sum_{k=2}^{\infty} h^{(k)} \alpha^k \equiv \sum_{k=1}^{\infty} h^{(k)} \alpha^k, \tag{2.6}$$

so that, by writing $f(z) = \lambda z + \sum_{k=2}^{\infty} f_k z^k$ and inserting (2.6) into (1.3), we obtain $h^{(1)} = 1$ and, for $k \geq 2$,

$$h^{(k)} = g(\omega k) \sum_{m=2}^{k} f_m \sum_{k_1 + \ldots + k_m = k} \prod_{j=1}^{m} h^{(k_j)}, \tag{2.7}$$

with

$$g(\omega k) = \frac{1}{\lambda^k - \lambda} = \frac{1}{e^{2\pi i \omega k} - e^{2\pi i \omega}}. \tag{2.8}$$

Therefore also $h^{(k)}$ admits a graphic representation. The trees are defined as before, but with different labels. First of all for each node u one can have either $m_u = 0$ or $m_u \geq 2$, where m_u is the number of lines entering u; one sets $\nu_u = 1$ when $m_u = 0$ and $\nu_u = 0$ when $m_u \geq 2$. The order k of a tree ϑ is defined as $k = \sum_{u \in V(\vartheta)} \nu_u$, while the momentum is defined as before; hence

$k = \nu_{\ell_0}$, if ℓ_0 is the root line. To each node $u \in V(\vartheta)$ we associate a node factor f_{m_u}, and to each line ℓ we associate a propagator G_ℓ, with $G_\ell = g(\omega\nu_\ell)$ if $\nu_\ell \geq 2$ and $G_\ell = 1$ if $\nu_\ell = 1$.

By defining $\mathrm{Val}(\vartheta)$ as in (2.5), with the new definition of the node factors and of the propagators, the following result holds.

Lemma 2. *One has formally* $h^{(k)} = \sum_{\vartheta \in \Theta^0_{k,k}} \mathrm{Val}(\vartheta, \omega)$ *for all* $k \geq 1$.

Again we have to face the problem of proving the convergence of the formal expansion. As the case of Siegel's problem can be essentially dealt with as for the semistandard map (as far as the lower bound on the radius of convergence is concerned) we shall concentrate henceforth on the system (1.4).

3 Renormalization Group and Multiscale Decomposition

To control the product of the propagators in (2.5) one needs a multiscale analysis which can be pursued as follows. We say that a line ℓ has scale n if $\|\omega\nu_\ell\| = \min_{p \in \mathbb{Z}} |\omega\nu_\ell - p|$ is approximately equal to $1/q_{n+1}$. A more formal statement can be obtained by introducing a C^∞ partition of unity. Let $\chi(x)$ a C^∞ non-increasing compact-support function defined on $\mathbb{R}^+$, such that

$$\chi(x) = \begin{cases} 1 & \text{for } x \leq 1, \\ 0 & \text{for } x \geq 2, \end{cases} \tag{3.1}$$

and define $\chi_0(x) = 1 - \chi(96 q_1 x)$ and $\chi_n(x) = \chi(96 q_n x) - \chi(96 q_{n+1} x)$ for each $n \in \mathbb{N}$; then for each line ℓ set

$$G_\ell = g(\omega\nu_\ell) = \sum_{n=0}^{\infty} G_\ell^{(n)}, \qquad G_\ell^{(n)} = \chi_n(\|\omega\nu_\ell\|)\, g(\omega\nu_\ell), \tag{3.2}$$

and call $G_\ell^{(n)}$ the *propagator on scale n*. [To deal with the semistandard map one could simply to introduce a sharp partition of unity (through step functions); however when discussing the standard map one has to develop to second order the propagators, and this explains why we need a smooth function in (3.2).]

Given a tree ϑ, we can associate to each line ℓ of ϑ a scale label n_ℓ, using the multiscale decomposition (3.2) and singling out the summand with $n = n_\ell$. We shall call n_ℓ the *scale label* of the line ℓ, and we shall say that the line ℓ is on scale n_ℓ.

Note that if a line ℓ has momentum ν_ℓ and scale n_ℓ, then

$$\frac{1}{96 q_{n_\ell+1}} \leq \|\omega\nu_\ell\| \leq \frac{1}{48 q_{n_\ell}}, \tag{3.3}$$

provided that one has $\chi_{n_\ell}(\|\omega\nu_\ell\|) \neq 0$. Note also that given a line ℓ at most only two summands in (3.2) are really non-vanishing.

Therefore $u_\nu^{(k)}$ can be rewritten as

$$\begin{aligned} u_\nu^{(k)} &= \sum_{\vartheta \in \Theta_{k,\nu}} \mathrm{Val}(\vartheta), \\ \mathrm{Val}(\vartheta) &= \Big(\prod_{u \in V(\vartheta)} F_v \Big) \Big(\prod_{\ell \in \Lambda(\vartheta)} G_\ell^{(n_\ell)} \Big), \end{aligned} \tag{3.4}$$

where $\Theta_{k,\nu}$ is the set of all trees of order k and with $\nu_{\ell_0} = \nu$ carrying also scale labels (in addition to the node labels); in the case of the semistandard map the number of elements in $\Theta_{k,\nu}$ is bounded by $2^k 2^{2k}$, while in the case of the standard map the number of elements in $\Theta_{k,k}$ is bounded by $2^k 2^k 2^{2k}$.

Given a tree ϑ, a *cluster* T of ϑ on scale n is a maximal connected set of lines of lines on scale $\leq n$ with at least one line on scale n; see Figure 2. We shall say that such lines are internal to T, and write $\ell \in \Lambda(T)$. A node u is called *internal* to T, and we write $u \in V(T)$, if at least one of the lines entering or coming out from it is in T. Each cluster has an arbitrary number $m_T \geq 0$ of entering lines but only one or zero line comes out from it; we shall call *external* to T the lines entering or coming out from T (and which are all on scale $> n$). We shall denote with n_T the scale of the cluster T, and with k_T the number of nodes in T.

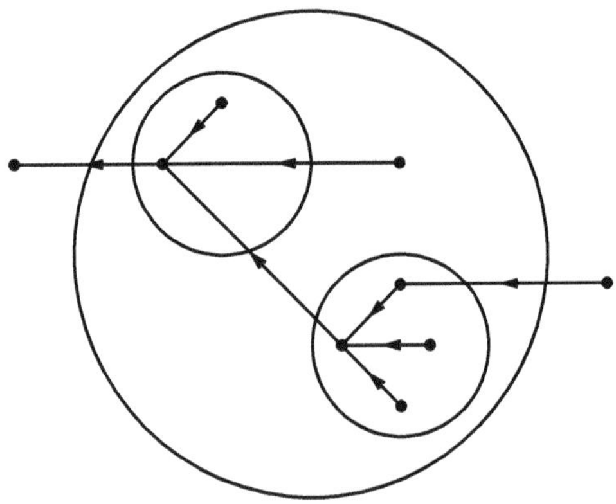

Fig. 2. A tree with its clusters. One should imagine that the tree itself is a cluster including all the other clusters. All the clusters T with $m_T = 1$ are resonances if also the conditions (2) and (3) of the definition of resonance are satisfied

Note that there is an inclusion relation between clusters: the innermost ones are those with highest scale, while the outermost ones are on the lowest scale. The aim of introducing the clusters is to characterize the lines of the trees on the basis of the sizes of the corresponding propagators: the lines which are contained inside the outermost clusters correspond to the propagators with the smallest small divisors, and so on.

If we confine ourselves to the semistandard map the notion of cluster is sufficient to prove the lower bound of Theorem 4: this will be discussed in Section 4. On the contrary in the case of the standard map additional problems arise, due to the fact that for each node $u \in V(\vartheta)$ one has $\nu_u = \pm 1$, whereas $\nu_u = 1$ for the semistandard map. To single out the cases which can give problems, we have to introduce the notion of resonance; see Figure 2.

A cluster T will be called a *resonance* with *resonance-scale* n if the following three properties are verified: (1) $m_T = 1$, (2) $\sum_{u \in V(T)} \nu_u = 0$ and (3) $k_T < q_n$, where n is minimum between the scales of the external lines of T.

The condition (1) means that T has only one entering line, which, by the condition (2) must have the same momentum of the exiting one (so that the scales of the two lines can differ at most by one unit). Of course resonances can not arise in the case of the semistandard map.

Then, because of the presence of resonances, in the case of the standard map there can be some trees in which there can be accumulation of small divisors equal to each other, but, when the values of all trees in $\Theta_{k,\nu}$ are summed together, if we group the trees into suitable classes then some remarkable cancellation mechanisms intervene between them, and the overall contribution still admits a bound of the same kind of that of the semistandard map: we shall be more precise in Section 5.

With the notion of resonance given above we are able to prove the lower bound in Theorem 5 with 4 instead of 2 in front of the Brjuno function; see Section 5. In order to obtain 2 a more careful analysis is needed: in particular the cancellation mechanisms have to be extended to a larger class of trees; we shall give some ideas about the proof at the end of Section 5.

4 Lower Bound for the Semistandard Map

Let $\{q_n\}_{n=0}^{\infty}$ be the denominators of the convergents of ω. Then one has [18]

$$\frac{1}{2q_{n+1}} < \|\omega q_n\| < \frac{1}{q_n}, \tag{4.1a}$$

$$\|\omega \nu\| > \|\omega q_n\| \qquad \forall |\nu| < q_{n+1},\ |\nu| \neq q_n. \tag{4.1b}$$

Let us denote by $N_n(\vartheta)$ the number of lines $\ell \in \Lambda(\vartheta)$ with scale $n_\ell = n$. Then we want to prove that, in the case of the semistandard map one has

$$N_n(\vartheta) \leq \frac{k}{q_n} + \frac{8k}{q_{n+1}}, \tag{4.2}$$

which immediately implies the lower bound of Theorem 4. In fact (4.2), inserted into (3.4), gives

$$
\begin{aligned}
|\mathrm{Val}(\vartheta)| &\le \left(\frac{1}{2}\right)^k \prod_{n=0}^{\infty} \left(c q_{n+1}^2\right)^{N_n(\vartheta)} \\
&\le \left(\frac{c}{2}\right)^k \exp\left[2k \sum_{n=0}^{\infty} \left(\frac{\log q_{n+1}}{q_n} + \frac{8 \log q_{n+1}}{q_{n+1}}\right)\right],
\end{aligned}
\tag{4.3}
$$

where, in the first line, $1/2$ is a bound on the node factor F_v, while cq_{n+1}^2 is a bound on the propagator of a line on scale n, with c a constant. Then $|u_k^{(k)}| \le 2^{3k}\left(ce^{16D}/2\right)^k e^{2B_1(\omega)k}$, where D is a universal constant bounding $\sum_{n=1}^{\infty} q_n^{-1} \log q_n$ for any irrational ω with convergents $\{p_n/q_n\}$, so that, by using the definition (2.2), the lower bound $\log \rho_0(\omega) + 2B_1(\omega) > -C$ follows for some constant C.

So it remains to check the bound (4.2). One can prove inductively on the order k the following result.

Lemma 3. *One has*

$$
\begin{cases}
N_n(\vartheta) = 0, & \text{if } k < q_n, \\
N_n(\vartheta) \le 2k/q_n - 1, & \text{if } k \ge q_n,
\end{cases}
\tag{4.4}
$$

for all $n \ge 0$ and for all $\vartheta \in \Theta_{k,k}$.

The first bound in (4.4) is immediately implied by the property (4.1). Roughly speaking the idea behind the proof of the second bound is the following. First of all note that the propagators G_ℓ are large for $\|\omega\nu_\ell\|$ small. However, even if the quantities $\|\omega\nu_\ell\|$ can become very small for ν_ℓ large enough, they cannot "accumulate" too much. In fact once a line ℓ_1 on scale n (i.e. with momentum ν_{ℓ_1} such that $\|\omega\nu_{\ell_1}\|$ is of order $1/q_{n+1}$) has been obtained, in order to have again a line ℓ_2 on the same scale along the path connecting ℓ_1 to the root, one needs many nodes between the two lines (i.e. many nodes preceding ℓ_2 and following ℓ_1), as each node contributes a mode label 1 to the momentum ν_{ℓ_2} and to have $\|\omega(\nu_{\ell_2} - \nu_{\ell_1})\| \le \|\omega\nu_{\ell_1}\| + \|\omega\nu_{\ell_2}\| = O(1/q_{n+1})$ requires $\nu_{\ell_2} - \nu_{\ell_1}$ to be large enough, by (4.1); but $\nu_{\ell_2} - \nu_{\ell_1}$ is exactly the number of nodes between ℓ_1 and ℓ_2. Therefore the number of lines on a fixed scale n can not grow indefinitely.

Unfortunately the bound (4.4) is not enough to obtain (4.2) because of the factor 2 instead of 1. This is not only a technical problem: it is not difficult to provide explicit examples of trees for which the bound (4.4) with 1 instead of 2 is false. In fact what we can prove is that Lemma 3 can be improved into the following one.

Lemma 4. *If $q_{n+1} > 4q_n$ one has*

$$
\begin{cases}
N_n(\vartheta) = 0, & \text{if } k < q_n, \\
N_n(\vartheta) \le k/q_n, & \text{if } q_n \le k < q_{n+1}/4, \\
N_n(\vartheta) \le k/q_n + 8k/q_{n+1} - 1, & \text{if } k \ge q_{n+1}/4,
\end{cases}
\tag{4.5}
$$

for all $n \ge 0$ and for all $\vartheta \in \Theta_{k,k}$.

Again the proof is by induction; the details can be found in literature [6]. A basic result in order to deduce Lemma 4 is the following one [10].

Lemma 5. *Given* $\nu \in \mathbb{Z}$ *such that* $\|\omega\nu\| \le 1/4q_n$*, then (1) either* $\nu = 0$ *or* $|\nu| \ge q_n$*, and (2) either* $|\nu| \ge q_{n+1}/4$ *or* $\nu \in q_n\mathbb{Z}$*.*

The proof is elementary, but the result plays an essential role in proving Lemma 4; in particular it allows to distinguish between the two latter bounds of (4.5) by treating in a different way the cases $k < q_{n+1}/4$ and $k \ge q_{n+1}/4$.

5 Lower Bound for the Standard Map

In the case of the standard map, once a line ℓ_1 on a large scale n has appeared, another line ℓ_2 on the same scale can be easily obtained once more along the path connecting ℓ_1 to the root, without going much further along the tree. For instance it is enough to have between ℓ_1 and ℓ_2 two nodes u_1 and u_2 with $\nu_{u_1} = -\nu_{u_2} = 1$, in order to have $\nu_{\ell_2} = \nu_{u_1} + \nu_{u_2} + \nu_{\ell_1} = \nu_{\ell_1}$, so that the line connecting u_1 to u_2 forms a resonance T with external lines ℓ_1 and ℓ_2 (as all the conditions of the definition are satisfied) and the line ℓ_2 coming out from T can have the same scale n; see Figure 3. So we can prove a bound like the one of Lemma 3, but with $N_n(\vartheta)$ replaced by $N_n^*(\vartheta)$, if the latter denotes the number of lines $\ell \in \Lambda(\vartheta)$ with scale $n_\ell = n$ which do not come out from any resonance (we shall call them *non-resonant lines*).

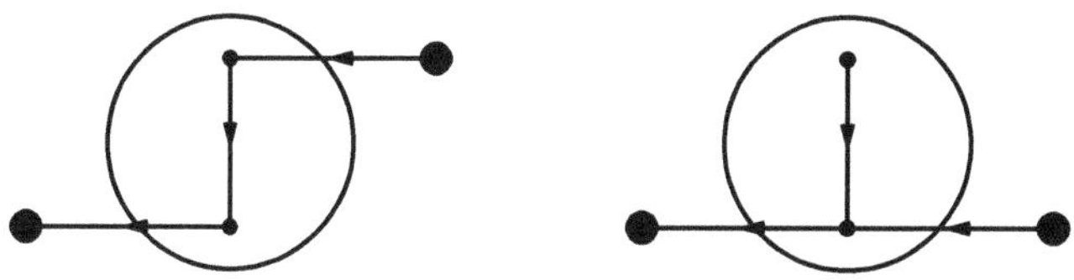

Fig. 3. Trees obtained by shifting the entering line of a resonance T with two nodes u_1 and u_2 with opposite mode labels. The black balls represent the remaining parts of the tree and the labels are not shown

The key remark to deal with the resonances is that, when summing the tree values over all possible trees as in (3.4), all terms containing the same resonance cancel almost exactly: more precisely, for any resonance T, the values of the trees containing that resonance cancel to order 2 in $\|\omega\nu_{\ell_1}\|$, if ℓ_1 is the line entering T. [To exploit the cancellations we have to derive twice the propagators: here the smoothness of the compact support functions comes in.] The criterion to single out the trees between which the cancellation operates is the following: given a tree ϑ with a resonance T, consider the class $\mathcal{F}_T(\vartheta)$ of all trees obtained by detaching the line entering T and re-attaching it to all the nodes inside T, and for each of such trees consider also the tree obtained by reverting the sign of the mode labels of the nodes contained in T (i.e. by replacing each $\nu_u = 1$ with $\nu_u = -1$ and vice versa); see for an example Figure 3.

But a second order cancellation produces a factor proportional to $\|\omega\nu_{\ell_1}\|^2$ which is exactly of the same size of the propagator of the line ℓ_2 (recall that $\nu_{\ell_2} = \nu_{\ell_1}$). In other words, given a line ℓ_1 on a very large scale n, it is possible to create another line ℓ_2 with the same scale adding only a few nodes (for instance 2 in the example above), but in such a way a resonance arises, and the cancellation mechanism described above produces a gain factor which compensates the propagator of the newly added line ℓ_2: this means that also in the case of the standard map there cannot be any accumulation of small divisors.

The conclusion is that the propagators corresponding to the non-resonant lines can be controlled as in the case of the semistandard map (through a minor extension of Lemma 3), while the propagators corresponding to the lines coming out from some resonance (the *resonant* lines) are in fact compensated by the gain factors produced by the cancellation mechanism. This is enough to prove the convergence of the perturbative series (hence the existence of the corresponding invariant curve), but produces again the extra factor 2 for the lower bound of the radius of convergence.

The first thing one can think about is to try to extend also Lemma 4 to cover the case of the non-resonant lines for the standard map. But this does not work: with the previous definition of resonance the bounds (4.5) fail to be satisfied. Then one can try to generalize the definition of resonance by enlarging the class of graphs to use in order to exploit the cancellations.

The condition (2) will be certainly retained, as it is needed for the cancellation mechanism to work. On the contrary we eliminate the condition $m_T = 1$: we allow any number of entering lines (see Figure 4). Note that such conditions are suggested from the study of the conjugating function for complex values of the rotation number ω; see Section 6.

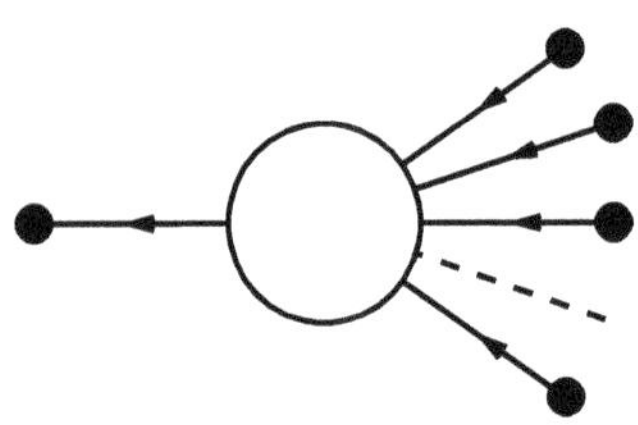

Fig. 4. A resonance, according to the new definition, can have an arbitrary number of entering lines. The black balls represent the remaining parts of the tree and the structure of the cluster is not shown

The definition of resonance which is obtained at the end is rather involved, and we give it here only for completeness. Given a cluster T denote with n_T^i the minimum of the scales of the lines entering T (if any) and with n_T^o the scale of the line exiting T (if any). A cluster T of ϑ, with one exiting line and at least one entering line, will be called a *resonance* with *resonance-scale* $n = \min\{n_T^i, n_T^o\}$, if (1) $\sum_{u \in V(T)} \nu_u = 0$, (2) all the lines entering V are on the same scale except at most one which can be on a higher scale, (3) $n_T^i \le n_T^o$ if $m_T \ge 2$, and $|n_T^i - n_T^o| \le 1$ for $m_T = 1$, (4) $k_T < q_n$, (5) $m_T = 1$ if $q_{n+1} \le 4q_n$, (6) if $q_{n+1} > 4q_n$ and $m_T \ge 2$, denoting by k_0 the sum of the orders of the subtrees of order $< q_{n+1}/4$ entering T, either there is only one subtree of order $k_1 \ge q_{n+1}/4$ entering T and $k_0 < q_{n+1}/8$, or there is no such subtree and $k_0 < q_{n+1}/4$.

Of course we shall not give here a proof how to extend Lemma 4 to the number of non-resonant lines (with the new definition of resonance), for which we refer to the literature [4]. An important remark is however that if we try to prove a bound like (4.5) for $N_n^*(\vartheta)$ we have to restrict the definition of non-resonant lines, in order to eliminate some cases which we are not able to control (of course all such cases are not present in the case of the semistandard map because of the condition (1) in the definition of resonance). What is left out at the end is exactly the class of graphs verifying all the conditions (1)÷(6) listed above.

6 Extensions, Conclusions and Open Problems

The analysis of the previous sections extends to more general functions f in (1.4), and analyticity of the conjugating function can be proved for any analytic f, at least as far as we are not looking for optimal bounds. This means that the proof of Theorem 3 can be easily obtained for any analytic function f in (1.4), but no analogous of Theorems 4 and 5 is known for functions f different from those of the semistandard and the standard maps.

Some light can be shed on the problem by studying the conjugating function for $\omega \in \mathbb{C}$. Suppose for instance that, in the case of the standard map, a bound like that of Theorem 5 still holds for complex rotation numbers. One can take ω of the form $\omega = p/q + i\eta$, with $\eta \in \mathbb{R}$ tending to zero (more generally one can consider rotation numbers tending to a rational value along any path of the complex plane non-tangentially to the real axis): since the Brjuno function $B(p/q + i\eta)$ diverges as $q^{-1} \log|\eta|^{-1}$ [17] one can expect that $\rho(p/q + i\eta)$ goes to zero as $|\eta|^{2/q}$. In fact this can be proved, and one can also prove that, by rescaling $\varepsilon \to (2\pi\eta)^{2/q}\varepsilon$ and letting η go to zero, then the conjugating function admits a limit function which can be easily expressed in terms of elliptic functions [2]. An analogous result has been also proved for the semistandard map and Siegel's problem [7].

The advantage to study the conjugating function for complex rotation number is that it is by far easier. Then one can ask how the radius of convergence $r(\omega, f)$ behaves for rotation numbers of the form $\omega = p/q + i\eta$ when more general functions f are taken into account. What is found [3, 5] is that the scaling properties of the radius of convergence strongly depend on the perturbation: one finds the that $r(p/q + i\eta, f)$ goes to zero as $|\eta|^{2/q(f)}$ for an integer $q(f)$ which can be obtained from the solution of a suitable Diophantine problem. Generically one has $q(f) = 1$: in particular this means that the standard map is not generic in this respect. Another consequence is that a bound like those of Theorems 4 and 5 can not hold for any analytic function in general, not even by replacing the factor 2 with another integer or real number. It is an open problem to see if, given a function f in (1.4), there exists a suitable function, generalizing the Brjuno function, in terms of which one can bound the radius of convergence.

Coming back to the the standard map, another interesting problem is to see what happens by keeping real the parameter ε. In other words one can study until which real value of $\varepsilon_c(\omega)$ the invariant curve with rotation number ω exists for the standard map; of course one has $\varepsilon_c(\omega) \geq \rho(\omega)$ and numerical evidence [1] shows that one can have $\varepsilon_c(\omega) > \rho(\omega)$. Note that a problem of this can kind does not arise for the semistandard map and Siegel's problem, where the analyticity domains (in ε and in α, respectively) are just disks.

The value $\varepsilon_c(\omega)$ is called the *critical function*: the problem of determining $\varepsilon_c(\omega)$ is much difficult, and it can not be studied with the techniques described so far (which work well for ε inside the analyticity domain) There is some numerical (and even some partial theoretical) results on the subject [8], which seem to suggest that if a bound analogous to that of Theorem 5 holds, with the factor 2 replaced by some other number β, then β is likely to be 1. But the situation is not so clear and the problem is still open. Note also that this is a problem for which it is very difficult to obtain results from numerical investigations, as Brjuno numbers (which are not necessarily Diophantine) are very hard to deal with from a numerical point of view.

References

[1] A. Berretti, C. Falcolini, G. Gentile: *The shape of analyticity domains of Lindstedt series: the standard map*, Phys. Rev. E **64** (2001), no. 1, 015202(R).

[2] A. Berretti, G. Gentile: *Scaling properties for the radius of convergence of Lindstedt series: the standard map*, J. Math. Pures Appl. (9) **78** (1999), no. 2, 159–176.

[3] A. Berretti, G. Gentile: *Scaling properties for the radius of convergence of Lindstedt series: generalized standard maps*, J. Math. Pures Appl. (9) **79** (2000), no. 7, 691–713.

[4] A. Berretti, G. Gentile: *Bryuno function and the standard map*, Comm. Math. Phys. **220** (2001), no. 3, 623–656.

[5] A. Berretti, G. Gentile: *Non-universal behaviour of scaling properties for generalized semistandard and standard maps*, Nonlinearity **14** (2001), no. 5, 1029–1039.

[6] A. Berretti, G. Gentile: *Renormalization group and field theoretic techniques for the analysis of the Lindstedt series*, Regul. Chaotic Dyn. **6** (2001), no. 4, 389–420.

[7] A. Berretti, S. Marmi, D. Sauzin: *Limit at resonances of linearizations of some complex analytic dynamical systems*, Ergodic Theory Dynam. Systems **20** (2000), no. 4, 963–990.

[8] T. Carletti, J. Laskar: *Scaling law in the standard map critical function. Interpolating Hamiltonian and frequency map analysis*, Nonlinearity **13** (2000), no. 6, 2033–2061.

[9] B.V. Chirikov: *A universal instability of many dimensional oscillator systems*, Phys. Rep. **52** (1979), no. 5, 264–379.

[10] A.M. Davie: *The critical function for the semistandard map*, Nonlinearity **7** (1994), no. 1, 219–229.

[11] A.M. Davie: *Renormalization for area preserving maps*, unpublished.

[12] L.H. Eliasson: *Absolutely convergent series expansions for quasi-periodic motions*, Math. Phys. Electron. J. **2** (1996), paper 4, 1–33 (electronic).

[13] G. Gallavotti: *Twistless KAM tori*, Comm. Math. Phys. **164** (1994), no. 1, 145–156.

[14] G. Gentile: *Diagrammatic techniques in perturbation theory, and applications.* In A. Degasperis and G. Gaeta (eds), *Proceedings of "Symmetry and Perturbation Theory II"* (Rome, 16–22 December 1998), 59–78, World Scientific (1999).

[15] G. Gentile, V. Mastropietro: *Methods for the analysis of the Lindstedt series for KAM tori and renormalizability in classical mechanics. A review with some applications*, Rev. Math. Phys. **8** (1996), no. 3, 393–444.

[16] J.M. Greene: *A method for determining a stochastic transition*, J. Math. Phys. **20** (1979), 1183–1201.

[17] S. Marmi, P. Moussa, J.-Ch. Yoccoz: *Complex Brjuno functions*, J. Amer. Math. Soc. **14** (2001), no. 4, 783–841.

[18] W.M. Schmidt: *Diophantine approximation*, Lecture Notes in Mathematics **785**, Springer, Berlin, 1980.

[19] C.L. Siegel: *Iterations of analytic functions*, Ann. of Math. **43** (1943), no. 4, 607–612.

[20] J.-C. Yoccoz: *Théorème de Siegel, Nombres de Bruno et Polinômes Quadratiques*, Astérisque **231** (1995), 3–88.

Some Properties of Real and Complex Brjuno Functions

Stefano Marmi[1], Pierre Moussa[2], and Jean-Christophe Yoccoz[3]

[1] Scuola Normale Superiore, Piazza dei Cavalieri 7, I-56126 Pisa, Italy
marmi@sns.it
[2] Service de Physique Théorique, CEA/Saclay, F-91191 Gif-sur-Yvette, France
moussa@spht.saclay.cea.fr
[3] Collège de France, 3 Rue d'Ulm, F-75005 Paris, France

1 Introduction

Let $\alpha \in \mathbb{R} \setminus \mathbb{Q}$ and let $(p_n/q_n)_{n\geq 0}$ be the sequence of the convergents of its continued fraction expansion. A *Brjuno number* is an irrational number α such that $\sum_{n=0}^{\infty} q_n^{-1} \log q_{n+1} < +\infty$.

The importance of Brjuno numbers comes from the study of analytic small divisors problems in dimension one. In the case of germs of holomorphic diffeomorphisms of one complex variable with an indifferent fixed point, extending a previous result of Siegel [29], Brjuno proved [7] that all germs with linear part $\lambda = e^{2\pi i \alpha}$ are linearizable if α is a Brjuno number. Conversely the third author proved that this condition is also necessary [33]. Similar results hold for the local conjugacy of analytic diffeomorphisms of the circle [17, 34, 35] and for some area–preserving maps [22, 10], including the standard family [11, 3, 4].

The set of Brjuno numbers is invariant under the action of the modular group $\mathrm{PGL}\,(2,\mathbb{Z})$ and it can be characterized as the set where the *Brjuno function* $B : \mathbb{R} \setminus \mathbb{Q} \to \mathbb{R} \cup \{+\infty\}$ is finite. This arithmetical function is $\mathbb{Z}$–periodic and satisfies a remarkable functional equation which allows B to be interpreted as a *cocycle* under the action of the modular group. The Brjuno function gives the size (modulus L^∞ functions) of the domain of stability around an indifferent fixed point [2, 33] and it conjecturally plays the same role in many other small divisor problems [26, 27, 23].

In two previous papers [24, 25] we studied the regularity properties of the Brjuno function and we constructed its complex analytic extension to the upper half–plane. Both the real and the complex analysis systematically exploit its relationship with continued fractions and the cocycle relation.

In Section 2 we recall some elementary properties of the continued fraction expansion of a real number and discuss the relationship between continued fractions and a certain monoid $\mathcal{M}$ of the full modular group $\mathrm{GL}\,(2,\mathbb{Z})$. Note that the same monoid arises also in the investigations of Lewis and Zagier [20, 21] concerning period functions and the Selberg zeta function for the Laplace–Beltrami operator on the modular surface. We then describe various automorphic actions of $\mathcal{M}$.

In Section 3 we introduce various types of diophantine conditions including Brjuno numbers. Then we introduce the Brjuno function B and some variants B_σ, $\sigma > 0$, which lead to different kinds of diophantine conditions and establish the functional equations all these Brjuno functions satisfy. This functional equation has the form $(1-T)B(x) = -\log x$ for the Brjuno function and $(1-T)B_\sigma(x) = x^{-1/\sigma}$ for the variants, where T is, roughly speaking, the operator $Tf(x) = xf(x^{-1})$ acting on periodic functions. Then we describe more precisely the relationship of B with the dynamics of quadratic polynomials as established in the works [33, 2] and various conjectures [22, 26, 27, 8].

In Section 4 we explain how the Brjuno functions B and B_σ can be regarded as cocycles under the action of the modular group. First we briefly recall some notions from the cohomology of groups and their applications to ergodic theory of dynamical systems. Then we describe the application to the Brjuno functions.

In Section 5 we give explicit formulas for the complex Brjuno functions $\mathcal{B}$ and $\mathcal{B}_\sigma$ associated to B and B_σ and we state the main results of [25] on the complexification of the Brjuno function B. A sketch of the proof of these formulas is given in Section 6: the key remark is that the action of the operator T can be extended to hyperfunctions. In this section we also introduce a complex analogue of the continued fraction expansion of a real number. The main feature of the complex continued fraction is that it reduces to the real continued fraction when the number is real and it stops after a finite number of iterations when the number is rational or complex. In the latter case the absolute value of the imaginary part of the iterates grows at least exponentially with the number of iterations and when it reaches the value 1/2 the iteration stops. The complex continued fraction can be used to study the behaviour of the complex Brjuno function $\mathcal{B}(z)$ when z is close to $[0,1]$. This is interesting in itself and it was crucial in [25], when applied to the complex Brjuno function $\mathcal{B}$, in order to prove our main results. Our study allows us to prove that the restriction of T to the Hardy space (of function vanishing at infinity) $H^p(\overline{\mathbb{C}} \setminus [0,1]) \cap \mathcal{O}^1(\overline{\mathbb{C}} \setminus [0,1])$, $1 \le p \le +\infty$, is continuous with spectral radius bounded above by $\frac{\sqrt{5}-1}{2} < 1$. This is sufficient, for example, to prove that $\mathcal{B}$ belongs to all Hardy spaces with $p < +\infty$ whereas $\mathcal{B}_\sigma$, $\sigma > 1$, belongs to all spaces with $p < \sigma$.

In an Appendix we recall some elementary properties of the dilogarithm, which arises naturally when considering the complexification of B.

Acknowledgements. The first and third authors are grateful to the organizers of the Les Houches school on Number Theory and Physics for their invitation and support.

2 Continued Fractions, the Modular Group, the Monoid $\mathcal{M}$ and Its Action

2.1 The Gauss Map and Continued Fractions

2.1.1 The continued fractions arise constructing the symbolic dynamics of the Gauss map (as well as for the linear flow on the two–dimensional torus or the geodesic flow on the modular surface). Here we will consider the iteration of the Gauss map

$$A : (0,1) \mapsto [0,1] , \tag{2.1}$$

defined by

$$A(x) = \frac{1}{x} - \left[\frac{1}{x} \right] , \tag{2.2}$$

where as as usua $[x]$ denotes the integer part of x. Let

$$G = \frac{\sqrt{5}+1}{2}\ , \ g = G^{-1} = \frac{\sqrt{5}-1}{2}\ .$$

To each $x \in \mathbb{R} \setminus \mathbb{Q}$ we associate a continued fraction expansion by iterating A as follows. Let

$$\begin{aligned} x_0 &= x - [x]\ , \\ a_0 &= \quad [x]\ , \end{aligned} \tag{2.3}$$

then $x = a_0 + x_0$. We now define inductively for all $n \geq 0$

$$\begin{aligned} x_{n+1} &= A(x_n)\ , \\ a_{n+1} &= \left[\tfrac{1}{x_n}\right] \geq 1\ , \end{aligned} \tag{2.4}$$

thus

$$x_n^{-1} = a_{n+1} + x_{n+1}\ . \tag{2.5}$$

Therefore we have

$$x = a_0 + x_0 = a_0 + \frac{1}{a_1 + x_1} = \ldots = a_0 + \cfrac{1}{a_1 + \cfrac{1}{a_2 + \ddots + \cfrac{1}{a_n + x_n}}}\ , \tag{2.6}$$

and we will write

$$x = [a_0, a_1, \ldots, a_n, \ldots]\ . \tag{2.7}$$

The nth-convergent is defined by

$$\frac{p_n}{q_n} = [a_0, a_1, \ldots, a_n] = a_0 + \cfrac{1}{a_1 + \cfrac{1}{a_2 + \ddots + \cfrac{1}{a_n}}}\ . \tag{2.8}$$

The numerators p_n and denominators q_n are recursively determined by

$$p_{-1} = q_{-2} = 1\ , \quad p_{-2} = q_{-1} = 0\ , \tag{2.9}$$

and for all $n \geq 0$

$$\begin{aligned} p_n &= a_n p_{n-1} + p_{n-2}\ , \\ q_n &= a_n q_{n-1} + q_{n-2}\ . \end{aligned} \tag{2.10}$$

Moreover

$$x = \frac{p_n + p_{n-1}x_n}{q_n + q_{n-1}x_n}\ , \tag{2.11}$$

$$x_n = -\frac{q_n x - p_n}{q_{n-1}x - p_{n-1}}\ , \tag{2.12}$$

$$q_n p_{n-1} - p_n q_{n-1} = (-1)^n\ . \tag{2.13}$$

Let

$$\beta_n = \Pi_{i=0}^{n} x_i = (-1)^n (q_n x - p_n) \quad \text{for } n \geq 0, \quad \text{and } \beta_{-1} = 1\ . \tag{2.14}$$

From the definitions given one easily proves by induction the following [24]

Proposition 1. *For all* $x \in \mathbb{R} \setminus \mathbb{Q}$ *and for all* $n \geq 1$ *one has*

(i) $|q_n x - p_n| = \dfrac{1}{q_{n+1} + q_n x_{n+1}}$, *so that* $\frac{1}{2} < \beta_n q_{n+1} < 1$;

(ii) $\beta_n \leq g^n$ *and* $q_n \geq \frac{1}{2} G^{n-1}$.

2.2 Algebraic Properties: the Monoid $\mathcal{M}$ and Its Relations with Continued Fractions, Farey Intervals and the Modular Group

2.2.1 Notations:

- $G = \mathrm{GL}(2,\mathbb{Z}) = \{\begin{pmatrix} a & b \\ c & d \end{pmatrix}, \ a,b,c,d \in \mathbb{Z}, \ \varepsilon_g := ad - bc = \pm 1\}$;
- H is the subgroup of order 8 of matrices of the form $\begin{pmatrix} \varepsilon & 0 \\ 0 & \varepsilon' \end{pmatrix}$ or $\begin{pmatrix} 0 & \varepsilon \\ \varepsilon' & 0 \end{pmatrix}$, where $\varepsilon, \varepsilon' \in \{-1, +1\}$;
- $\mathcal{M}$ is the monoid with unit $\begin{pmatrix} 1 & 0 \\ 0 & 1 \end{pmatrix}$ made of matrices $g = \begin{pmatrix} a & b \\ c & d \end{pmatrix} \in G$ such that, if $g \neq \mathrm{id}$, we have $d \geq b \geq a \geq 0$ and $d \geq c \geq a$.
- Z is the subgroup of matrices of the form $\begin{pmatrix} 1 & n \\ 0 & 1 \end{pmatrix}$, $n \in \mathbb{Z}$.

2.2.2 Let $g(m) = \begin{pmatrix} 0 & 1 \\ 1 & m \end{pmatrix}$, where $m \geq 1$. Clearly $g(m) \in \mathcal{M}$. Moreover, $\mathcal{M}$ is the *free monoid* generated by the elements $g(m)$, $m \geq 1$: each element g of $\mathcal{M}$ can be written as

$$g = g(m_1) \cdots g(m_r) \ , \quad r \geq 0 \ , \ m_i \geq 1 \ ,$$

and this decomposition is unique (see , Proposition A1.2 in [25] and also [20]).

2.2.3 One has

$$G = Z \cdot \mathcal{M} \cdot H \ ,$$

i.e. the application

$$Z \times \mathcal{M} \times H \to G \ , \quad (z, m, h) \mapsto g = z \cdot m \cdot h$$

is a *bijection.*

2.2.4 The subset $Z \cdot \mathcal{M}$ of G is made of matrices $g = \begin{pmatrix} a & b \\ c & d \end{pmatrix}$ such that $d \geq c \geq 0$ with the following additional restrictions: $a = 1$ if $c = 0$, and, $b = a + 1$ if $d = c = 1$.

2.2.5 Let us consider the usual action of G on $\overline{\mathbb{C}} = \mathbb{C} \cup \{\infty\}$ by homographical transformations: $\begin{pmatrix} a & b \\ c & d \end{pmatrix} \cdot z = \dfrac{az + b}{cz + d}$. The following facts are easy to check:

(i) Equation (2.5) can also be written $x_n = g(a_{n+1}) x_{n+1}$, therefore we have $x_0 = g(a_1) g(a_2) \cdots g(a_n) x_n$.

(ii) The application $g \mapsto g \cdot 1 = \frac{a+b}{c+d}$ is a *bijection* of $Z\mathcal{M}$ over $\mathbb{Q}$ which maps $\mathcal{M}$ onto $\mathbb{Q} \cap (0, 1]$.

(iii) The application $g \mapsto g \cdot 0 = b/d$ maps $Z\mathcal{M}$ onto $\mathbb{Q}$ and each rational number has exactly *two inverse images*. The two elements which map 0 on 1 are $\begin{pmatrix} 1 & 1 \\ 0 & 1 \end{pmatrix}$ and $\begin{pmatrix} 0 & 1 \\ 1 & 1 \end{pmatrix}$.

(iv) The application $g \mapsto g \cdot [0, +\infty]$ is a *bijection* of $Z\mathcal{M}$ on the set of Farey intervals (the convention we adopt here implies that $[n, +\infty]$ is a Farey interval, but $[-\infty, n]$ is not). For the definition and properties of the Farey partition of $[0, 1]$ we refer the reader to [15].

3 Diophantine Conditions, Brjuno Numbers and the Brjuno Function.

3.1 Diophantine Conditions

3.1.1 Let $\gamma > 0$ and $\tau \geq 0$ be two real numbers. A number $x \in \mathbb{R} \setminus \mathbb{Q}$ is **diophantine** of exponent τ and constant γ if and only if for all $p, q \in \mathbb{Z}$, $q > 0$, one has $\left| x - \frac{p}{q} \right| \geq \gamma \, q^{-2-\tau}$.

We denote CD (γ, τ) the set of all irrationals x such that $\left| x - \frac{p}{q} \right| \geq \gamma q^{-2-\tau}$ for all $p, q \in \mathbb{Z}$, $q > 0$. CD (τ) will denote the union $\cup_{\gamma > 0}$CD (γ, τ) and CD $= \cup_{\tau \geq 0}$CD (τ). One has

$$\begin{aligned} \mathrm{CD}\,(\tau) &= \{x \in \mathbb{R} \setminus \mathbb{Q} \mid \; q_{n+1} = \mathrm{O}(q_n^{1+\tau})\} = \quad \{x \in \mathbb{R} \setminus \mathbb{Q} \mid \; a_{n+1} = \mathrm{O}(q_n^{\tau})\} \\ &= \{x \in \mathbb{R} \setminus \mathbb{Q} \mid \; x_n^{-1} = \mathrm{O}(\beta_{n-1}^{-\tau})\} = \{x \in \mathbb{R} \setminus \mathbb{Q} \mid \; \beta_n^{-1} = \mathrm{O}(\beta_{n-1}^{-1-\tau})\} \end{aligned}$$

The complement in $\mathbb{R} \setminus \mathbb{Q}$ of CD is called the set of Liouville numbers. The set of Liouville numbers has zero Lebesgue measure, zero Hausdorff dimension but it is a dense G_δ–set

The sets CD (τ) and CD are both PGL $(2, \mathbb{Z})$–invariant. Moreover if $\tau > 0$ then CD (τ) has full Lebesgue measure. The same holds for $\cap_{\tau > 0}$CD (τ) (Roth numbers).

We will see in the next section that some Brjuno functions can be used to characterize diophantine numbers.

3.1.2 It is easy to show that if x is an algebraic number of degree $n \geq 2$, i.e. $x \in \mathbb{R} \setminus \mathbb{Q}$ is a zero of a monic polynomial with coefficients in $\mathbb{Q}$ and degree n, then $x \in$ CD $(n - 2)$ (Liouville's theorem). Thue improved this result in 1909 showing that $x \in$ CD $(\tau - 1 + n/2)$ for all $\tau > 0$ (see [32], Chapter V, for a very nice discussion of the proof in the cubic case). Actually one can prove that if x is algebraic then $x \in$ CD (τ) for all $\tau > 0$ regardless of the degree, but this is difficult (Roth's theorem).

Using the fact that the continued fraction of $e = \sum_{n=0}^{\infty} \frac{1}{n!}$ is

$$[2, 1, 2, 1, 1, 4, 1, 1, 6, 1, 1, 8, 1, 1, 10, \ldots]$$

one obtains that $e \in \cap_{\tau>0}\mathrm{CD}\,(\tau)$. A proof of the continued fraction expansion of e, which is due to L. Euler, can be found in [18], Chapter V.

The set CD (0) is also called the set of numbers of *constant type* since $x \in \mathrm{CD}\,(0)$ if and only if the sequence of its partial fractions is bounded. It has Hausdorff dimension 1 and zero Lebesgue measure.

3.2 Brjuno Numbers and the Real Brjuno Function

Let $\alpha \in \mathbb{R} \setminus \mathbb{Q}$ and let $(p_n/q_n)_{n\geq 0}$ be the sequence of the convergents of its continued fraction expansion. A *Brjuno number* is an irrational number α such that $\sum_{n=0}^{\infty} q_n^{-1} \log q_{n+1} < +\infty$. All diophantine numbers are Brjuno numbers but also "many" Liouville numbers are Brjuno numbers: for example the number $\sum_{n\geq 1} 10^{-n!}$ is a Brjuno number.

The set of Brjuno numbers is invariant under the action of the modular group $\mathrm{PGL}\,(2,\mathbb{Z})$ and it can be characterized as the set where the *Brjuno function* $B : \mathbb{R} \setminus \mathbb{Q} \to \mathbb{R} \cup \{+\infty\}$ is finite. This arithmetical function is $\mathbb{Z}$–periodic and satisfies the remarkable functional equation

$$B(\alpha) = -\log \alpha + \alpha B\left(\frac{1}{\alpha}\right) \quad , \quad \alpha \in (0,1)\,, \tag{3.1}$$

which allows B to be interpreted as a *cocycle* under the action of the modular group (see the next section). Moreover, the quadratic irrationals have an eventually periodic continued fraction, and one can compute explicitely the Brjuno function for these numbers, which form a countable but dense set of irrationals.

In terms of the continued fraction expansion of α the Brjuno function is defined as follows:

$$B(\alpha) = \sum_{j=0}^{+\infty} \beta_{j-1}(\alpha) \log \alpha_j^{-1}\,, \tag{3.2}$$

where the two sequences $(\beta_j)_{j\geq -1}$ and $(\alpha_j)_{j\geq 0}$ are obtained iterating the Gauss map from $\alpha_0 = \{\alpha\}$, as in (2.12) and (2.14).

In order to study the regularity properties of B in [24] we introduced the linear operator

$$Tf(x) = xf\left(\frac{1}{x}\right) \quad , \quad x \in (0,1) \tag{3.3}$$

acting in the space of $\mathbb{Z}$–periodic measurable functions and we studied the equation

$$(1-T)B_f = f\,, \tag{3.4}$$

so that

$$\begin{aligned} B_f(x+1) &= B_f(x) & \forall x \in \mathbb{R}\,, \\ B_f(x) &= f(x) + xB_f(1/x) & \forall x \in (0,1)\,. \end{aligned} \tag{3.5}$$

The choice $f(x) = -\log\{x\}$ (where $\{\cdot\}$ denotes fractional part) leads to the Brjuno function B. For other choices of the singular behaviour of f at 0 the condition $B_f < +\infty$ leads to different diophantine conditions. For example let $\sigma > 0$ and consider the function

$$B_\sigma(\alpha) = \sum_{j=0}^{+\infty} \beta_{j-1}(\alpha)\alpha_j^{-1/\sigma} . \tag{3.6}$$

The same argument as for the Brjuno function (3.2) shows that B_σ is the solution of the functional equation (3.5) with $f(x) = x^{-1/\sigma}$. Moreover if $B_\sigma(x) < +\infty$ then $x \in \mathrm{CD}\,(\sigma)$. Viceversa, if $x \in \mathrm{CD}\,(\tau)$ then $B_\sigma(x) < +\infty$ for all $\sigma > \tau$.

Some sort of singular behaviour for f at 0 is needed in order to characterize some set of diophantine numbers. Indeed if f is Hölder continuous then B_f is also Hölder continuous and this fact could help to explain the numerical results of Buric, Percival and Vivaldi [6].

Acting on $L^p([0,1])$ the operator T has spectral radius bounded above by $\frac{\sqrt{5}-1}{2}$ (thus $(1-T)$ is invertible). A suitable adaptation of this argument has led us to conclude that the Brjuno function belongs to $\mathrm{BMO}\,(\mathbb{T}^1)$ (bounded mean oscillation, see [13, 14] for its definition and more information). The proof of the first statement is very simple: replacing the Haar measure dx with the invariant measure $\frac{dx}{(1+x)\log 2}$ for the Gauss map one obtains the same L^p spaces since the density is bounded below and above. But now

$$\begin{aligned} \|T^m f\|_{L^p}^p = & \int |(T^m f)(x)|^p \frac{dx}{(1+x)\log 2} &= \int \beta_{m-1}^p |(f\circ A^m)(x)|^p \frac{dx}{(1+x)\log 2} \\ \le g^{p(m-1)} & \int |f(x)|^p \frac{dx}{(1+x)\log 2} &= \qquad g^{p(m-1)} \|f\|_{L^p}^p , \end{aligned}$$

where the inequality is obtained applying Proposition 1 (ii) and the invariance of the measure by A allows one to replace $f \circ A^m$ with f.

By Fefferman's duality theorem BMO is the dual of the Hardy space H^1 thus one can add an L^∞ function to B so that the harmonic conjugate of the sum will also be L^∞ (actually, it is proved in [25] that the harmonic conjugate of B is bounded, see below). This suggests to look for an holomorphic function $\mathcal{B}$ defined on the upper half plane which is $\mathbb{Z}$–periodic and whose trace on $\mathbb{R}$ has for imaginary part the Brjuno function B. The function $\mathcal{B}$ will be called the *complex Brjuno function*.

3.3 The Brjuno Function and Dynamics.

The importance of Brjuno numbers comes from the study of one–dimensional analytic small divisors problems. In the case of germs of holomorphic diffeomorphisms of one complex variable with an indifferent fixed point, extending a previous result of Siegel [29], Brjuno proved [7] that all germs with linear part $\lambda = e^{2\pi i\alpha}$ are linearizable if α is a Brjuno number. Conversely the third author proved that this condition is also necessary [33]. Similar results hold for the local conjugacy of analytic diffeomorphisms of the circle [17, 34, 35] and for some area–preserving maps [22, 10] including the standard family [11, 3, 4].

Another motivation for the introduction of the complex Brjuno function comes from results concerning the linearization of the quadratic polynomial $P_\lambda(z) = \lambda(z - z^2)$ ([33], Chapter II). One has the following results:

(i) there exists a bounded holomorphic function $U : \mathbb{D} \to \mathbb{C}$ such that $|U(\lambda)|$ is equal to the radius of convergence of the normalized linearization of P_λ;
(ii) for all $\lambda_0 \in \mathbb{S}^1$, $|U(\lambda)|$ has a non–tangential limit in λ_0 (which is still equal to the radius of convergence of the normalized linearization of P_{λ_0});
(iii) if $\lambda = e^{2\pi i\alpha}$, $\alpha \in \mathbb{R} \setminus \mathbb{Q}$, P_λ is linearizable if and only if α is a Brjuno number. Moreover there exists a universal constant $C_1 > 0$ such that for all Brjuno numbers α one has

$$B(\alpha) - C_1 \le -\log|U(\lambda)| \le B(\alpha) + C_1 \ .$$

(The upper bound was proved in [33] together with a weaker lower bound: this version is due to [2]).

In [24] the authors proposed the following conjecture (see also [22]): the function defined on the set of Brjuno numbers by $\alpha \mapsto B(\alpha) + \log|U(e^{2\pi i\alpha})|$ extends to a 1/2–Hölder continuous function as α varies in $\mathbb{R}$. For a recent numerical study of this conjecture see [8]. (An analogue of this conjecture for the so–called critical function of the standard map was stated in [26]: see [5] for some numerical evidence). If this were true then the function $-i\mathcal{B}(z) + \log U(e^{2\pi iz})$ would also extend to a Hölder continuous function on $\overline{\mathbb{H}}$.

4 The Real Brjuno Function as a Cocycle

In this Section we show how to interpret the real Brjuno function as a cocycle under the action of $\mathrm{PGL}(2,\mathbb{Z})$ on $\mathbb{R}\setminus\mathbb{Q}$. To this purpose we first introduce some elementary notions borrowed from group cohomology (a subject which has had its most important applications in number theory). The standard reference is [30] but [31] has a nice short introduction to it (p. 104 and p. 222). For a beautiful introduction to the cohomology of $\mathrm{SL}(2,\mathbb{Z})$ and its applications to the theory of periods of modular forms see [36].

4.1 Group Cohomology

Let G be a group, L be an abelian group (later on G will be the modular group $\mathrm{PGL}(2,\mathbb{Z})$ and L the real projective line $\mathbb{P}^1(\mathbb{R})$). To say that L is a G–set means that G acts on L, i.e. there is a homomorphism $G \to \mathrm{Hom}\,(L,L)$ or, equivalently, one has a map $G \times L \to L$, $(g,l) \mapsto g\cdot l$, such that $e\cdot l = l$, $g_1\cdot(g_2\cdot l) = (g_1 g_2)\cdot l$ for all $g_1, g_2 \in G$, $l \in L$ and where e is the neutral element of G. If in addition one has that $g \cdot (l_1 + l_2) = g \cdot l_1 + g \cdot l_2$ we say that L is a G–module. This is equivalent to giving L the structure of a $\mathbb{Z}[G]$–module.

Let $r \in \mathbb{N}$. An element $\varphi \in C^r(G, L) = \mathrm{Map}(G^r, L)$ is called an *r–cochain.* There is a sequence

$$\ldots \to 0 \to 0 \to C^0(G,L) \xrightarrow{d} C^1(G,L) \xrightarrow{d} C^2(G,L) \xrightarrow{d} C^3(G,L) \xrightarrow{d} \ldots$$

where $C^0(G, L) = L$ and the *coboundary operator* d is defined as follows: let $\varphi_i \in C^i(G, L)$, then

$$\begin{aligned}
(d\varphi_0)(g) &= g \cdot \varphi_0 - \varphi_0 \\
(d\varphi_1)(g_1, g_2) &= g_1 \cdot \varphi_1(g_2) - \varphi_1(g_1 g_2) + \varphi_1(g_1) \\
(d\varphi_2)(g_1, g_2, g_3) &= g_1 \cdot \varphi_2(g_2, g_3) - \varphi_2(g_1 g_2, g_3) + \varphi_2(g_1, g_2 g_3) - \varphi_2(g_1, g_2) \\
(d\varphi_3)(g_1, g_2, g_3, g_4) &= g_1 \cdot \varphi_3(g_2, g_3, g_4) - \varphi_3(g_1 g_2, g_3, g_4) + \varphi_3(g_1, g_2 g_3, g_4) \\
&\quad - \varphi_3(g_1, g_2, g_3 g_4) + \varphi_3(g_1, g_2, g_3) \\
\ldots &= \ldots
\end{aligned}$$

It is now easy to guess how does d act on an arbitrary r–cochain. One has $d \circ d = 0$, i.e. $\mathrm{Im}\, d \subset \mathrm{Ker}\, d$.

Definition 1. *An* r–cocycle *is an element of* $Z^r(G, L) = Ker d|_{C^r(G.L)}$. *An* r–coboundary *is an element of* $B^r(G, L) = Im\, d|_{C^{r-1}(G.L)}$. *The* r–th cohomology group $H^r(G, L) = Z^r(G, L)/B^r(G, L)$.

Suppose that one is given a group G, a *set* X on which G acts (i.e. a group homomorphism $G \to \mathrm{End}\,(X)$), a commutative ring A (with multiplicative group A^*) and an A–module M. Let $M^X = \mathrm{Map}\,(X, M)$. (We will be interested later in the case $X = \mathbb{R} \setminus \mathbb{Q}$ and $M = \mathbb{C}$).

Definition 2. *A function* $\chi : G \times X \to A^*$ *is an* automorphic factor *if the application* $G \times M^X \to M^X$ *given by* $(g, \varphi) \mapsto g \cdot \varphi$ *where* $(g \cdot \varphi)(x) = \chi(g^{-1}, x)\varphi(g^{-1} \cdot x)$ *defines a left action of* G *on* M^X, *i.e.* $\chi(g_0 g_1, x) = \chi(g_0, g_1 \cdot x)\chi(g_1, x)$.

The datum of an automorphic factor gives M^X the structure of a G–module and the previous considerations apply. In particular a 1–*cocycle* is a map $c : G \to M^X$ such that $g_0 \cdot c(g_1) - c(g_0 g_1) + c(g_0) = 0$. If we let $\check{c}(g) = c(g^{-1})$, being a 1–cocycle means that

$$\check{c}(g_0 g_1, x) = \chi(g_1, x)\check{c}(g_0, g_1 \cdot x) + \check{c}(g_1, x)\ \forall x \in X\,.$$

The *coboundary* of $\varphi \in M^X$ is $(d\varphi)(g) = g \cdot \varphi - \varphi$.

4.2 Dynamics and Cohomology

The standard situation in dynamical systems (see, e.g. [17]) is the following: the phase space X is a compact metric space and the time evolution is provided by a homeomoerphism $f \in \mathrm{Homeo}(X)$. The dynamical system generated by f gives X the structure of a $\mathbb{Z}$–set and gives to the space of *observables* $\mathcal{C}(X,\mathbb{R})$ the structure of a $\mathbb{Z}$–module (with the trivial choice of automorphic factor $\chi \equiv 1$, but also other choices are conceivable). Thus a 1–coboundary is $d\varphi = \varphi \circ f - \varphi$ where $\varphi \in \mathcal{C}(X,\mathbb{R})$. Since for a $\mathbb{Z}$–action one can always make the identification between observables f and 1–cocycles c (the correspondence being $c(1) \longleftrightarrow f$) the first cohomology groups can be identified with the quotient $\mathcal{C}(X,\mathbb{R})/d\mathcal{C}(X,\mathbb{R})$. This has relevant applications in ergodic theory since, for example, a system is uniquely ergodic if and only if the uniform closure $\overline{d\mathcal{C}(X,\mathbb{R})}$ has codimension one in $\mathcal{C}(X,\mathbb{R})$ (i.e. it is as large as possible since a constant function cannot be a coboundary).

4.3 Action of $\mathbf{PGL(2,\mathbb{Z})}$ on $\mathbb{R}\setminus\mathbb{Q}$

Let us consider $G = \mathrm{PGL}(2,\mathbb{Z})$ and $X = \mathbb{R}\setminus\mathbb{Q}$, the action being given by the homographies. The transformations $T(x) = x+1$ and $S(x) = x^{-1}$ generate $\mathrm{PGL}(2,\mathbb{Z})$. One has the following more precise result:

Proposition 2. *Let $g \in PGL(2,\mathbb{Z})$ and let $x_0 \in \mathbb{R}\setminus\mathbb{Q}$. There exist $r \geq 0$ and elements $g_1,\dots,g_r \in \{S,T,T^{-1}\}$ such that*

(i) $g = g_r \cdots g_1$;
(ii)let $x_i = g_i x_{i-1}$ for $1 \leq i \leq r$, then $x_{i-1} > 0$ if $g_i = S$.

Moreover one can require that $g_i g_{i-1} \neq 1$ for $0 < i \leq r$, and in this case $r, g_1,\dots,g_r$ are uniquely determined.

The previous proposition has the following important consequence for the modular interpretation of the Brjuno function functional equation: it is enough to prescribe an automorphic factor for the $\mathrm{PGL}(2,\mathbb{Z})$–action on functions on $\mathbb{R}\setminus\mathbb{Q}$ giving its values in correspondence of the inversion S just at points belonging to the interval $(0,1)$. More precisely one has

Corollary 1. *Let A be an abelian ring, $t : \mathbb{R}\setminus\mathbb{Q} \to A$, $s : (0,1)\cap(\mathbb{R}\setminus\mathbb{Q}) \to A^*$, where A^* denotes the group of invertible elements of A. There exists a unique automorphic factor χ such that*

$$\begin{aligned}\chi(T,x) &= t(x) \text{ for all} x \in \mathbb{R}\setminus\mathbb{Q} = X\ ,\\ \chi(S,x) &= s(x) \text{ for all} x \in X \cap (0,1)\ .\end{aligned}$$

The same property holds for cocycles: they just need to be prescribed, in correspondence of the inversion, on the interval $(0,1)$.

Corollary 2. *Let A be an abelian ring, χ an automorphic factor, M a A–module, M^X with the structure of $\mathbb{Z}^{[G]}$–module defined by χ. Let*

$$\begin{aligned}\check{c}_T :&\quad X \to M\\ \check{c}_S :&\ X\cap(0,1) \to M\end{aligned}$$

denote two maps. There exists a unique cocycle $\check{c} : G \times M \to M$ *such that*

$$\begin{aligned} \check{c}(T;x) &= \quad \check{c}_T(x) \text{ for all } x \in X \\ \check{c}(S;x) &= \check{c}_S(x) \text{ for all } x \in X \cap (0,1) \,. \end{aligned}$$

Note that one must have

$$\check{c}(T^{-1};x) = -\chi(T^{-1},x)\check{c}_T(x-1) \text{ for all } x \in X \,, \tag{4.1}$$

$$\check{c}(S;x) = -\chi(S,x)\check{c}_S(x^{-1}) \text{ for all } x \in X \,,\; x > 1 \,. \tag{4.2}$$

Moreover, if $g = g_r \dots g_1$ and x_0 are given as in Proposition 2 then

$$\check{c}(g;x_0) = \sum_{i=1}^{r} (\check{c}(g_i, x_{i-1})\chi(g_{i-1} \dots g_1, x_0)) \tag{4.3}$$

4.4 The Real Brjuno Function as a Cocycle

We now apply the results of 4.3 to the functional equation of the Brjuno function (and, more generally, to equations (3.4) where T is replaced by $(T_\nu f)(x) = x^\nu f(x^{-1})$ as in [24]. Let $A = \mathbb{R}$, $t(x) = 1$ and $s(x) = \varepsilon x^\nu$ with $\varepsilon \in \{-1,+1\}$, $\nu \in \mathbb{R}$ and apply Corollary 1. Then

$$\begin{aligned} \chi(T^n,x) &= \quad 1 \,, \quad \text{for all } n \in \mathbb{Z} \,,\; x \in X \\ \chi(S,x) \ &= \varepsilon x^\nu \,, \quad \text{for all } x \in X \,,\; x > 0 \,. \end{aligned}$$

If $x_0 \in (0,1)$, one has seen that $U = T^{-1}STST^{-1}S$, thus

$$\chi(U,x_0) = \varepsilon x_0^\nu \varepsilon \left(\frac{1-x_0}{x_0}\right)^\nu \varepsilon \left(\frac{1}{1-x_0}\right)^\nu = \varepsilon$$

From $U = T^n U T^n$ it follows that

$$\chi(U,x) = \varepsilon$$

for all $x \in X$ and from $S = USU$ follows that

$$\chi(S,x) = \varepsilon\varepsilon|x|^\nu\varepsilon$$

for $x < 0$, or

$$\chi(S,x) = \varepsilon|x|^\nu$$

for all $x \in X$. One concludes that one must have

$$\chi(g,x) = \begin{cases} |cx+d|^\nu & \text{if } \varepsilon = +1 \\ \det(g)|cx+d|^\nu & \text{if } \varepsilon = -1 \end{cases} \tag{4.4}$$

for all $g = \begin{pmatrix} a & b \\ c & d \end{pmatrix} \in \mathrm{PGL}(2,\mathbb{Z})$.

Consider now the functional equations

$$\begin{aligned} B_f^{(\nu)}(x) &= x^\nu B_f^{(\nu)}(1/x) + f(x) \quad , \quad x \in (0,1) \cap \mathbb{R} \setminus \mathbb{Q} \\ B_f^{(\nu)}(x) &= \qquad B_f^{(\nu)}(x+1) \quad , \quad x \in \mathbb{R} \setminus \mathbb{Q} \,, \end{aligned} \tag{4.5}$$

where $f \ : \ (0,1) \cap \mathbb{R} \setminus \mathbb{Q} \to \mathbb{C}$ is given. By Corollary 2, there exists exactly one 1–cocycle $\check{c}_f$ such that

$$\begin{aligned} \check{c}_f(T,x) &= \quad 0 \quad \forall x \in \mathbb{R} \setminus \mathbb{Q} \\ \check{c}_f(S,x) &= f(x) \quad \forall x \in \mathbb{R} \setminus \mathbb{Q} \cap (0,1) \,. \end{aligned} \tag{4.6}$$

The 1–cocycle is a 1–coboundary if and only if the functional equations have a solution $B_f^{(\nu)}$, in which case we have $c_f = -d^0(B_f^{(\nu)})$. These considerations also apply and may become fruitful in case we restrict $\mathbb{C}^{\mathbb{R}\setminus\mathbb{Q}}$ to one of its $\mathbb{C}^{[G]}$–submodules: measurable functions, L^p spaces, BMO, etc..

4.5 The Cocycles Associated to the Brjuno Functions

We can use (4.2) and (4.3) to compute the cocycle associated to $B_f^{(\nu)}$ on the whole real line. From (4.2) and (4.6) we get

$$\check{c}_f(S,x) = -x^\nu f(x^{-1}) \quad \forall x \in \mathbb{R} \setminus \mathbb{Q},\, x > 1 \,.$$

To obtain the values of $\check{c}_f(S,x)$ for $x < 0$ we apply (4.3) to $\check{c}_f(S,U(x))$ where $U(x) = -x$. If $0 < x < 1$, since $U = T^{-1}STST^{-1}S$ and $\chi(U,x) = 1$ we obtain

$$\check{c}_f(S,-x) = \check{c}_f(S,x) + \chi(S,x)\check{c}_f(U,S(x)) - \check{c}_f(U,x) \,. \tag{4.7}$$

On the other hand iterating several times the cocycle condition we get for $0 < x < 1$

$$\begin{aligned} \check{c}_f(U,x) = \chi(S,x)\ &\chi(S,(T^{-1}S)(x))\check{c}_f(S,(TST^{-1}S)(x)) \\ &+\chi(S,x)\check{c}_f(S,(T^{-1}S)(x)) + \check{c}_f(S,x) \end{aligned}$$

thus if $0 < x < 1$ we obtain

$$\check{c}_f(U,x) = f(x) - f(1-x) + \begin{cases} -(1-x)^\nu f\left(\frac{x}{1-x}\right) & \text{if } 0 < x < 1/2 \\ \quad +x^\nu f\left(\frac{1-x}{x}\right) & \text{if } 1/2 < x < 1 \end{cases} \tag{4.8}$$

If $n < x < n+1$ then from the identity $U = T^{-n}UT^n$ one gets

$$\check{c}_f(S,U(x)) = \check{c}_f(S,x) + \chi(S,x)\check{c}_f(U,S(x)) - \check{c}_f(U,x-n) \tag{4.9}$$

$$\check{c}_f(U,x) = \check{c}_f(T^{-n}UT^n,x) = \check{c}_f(U,x-n) \tag{4.10}$$

thus for all $x > 0$ we get

$$\check{c}_f(U,x) = f(x_0) - f(1-x_0) + \begin{cases} -(1-x_0)^\nu f\left(\frac{x_0}{1-x_0}\right) & \text{if } 0 < x_0 < 1/2 \\ +x_0^\nu f\left(\frac{1-x_0}{x_0}\right) & \text{if } 1/2 < x_0 < 1 \end{cases} \tag{4.11}$$

where $x_0 = \{x\}$ is the fractional part of x. Plugging (4.11) into (4.7) we find that if $0 < x < 1$, if we denote $x_1 = \{x^{-1}\}$

$$\begin{aligned} \check{c}_f(S, -x) &= f(1-x) \\ &+x^\nu \left[f(x_1) - f(1-x_1) + \begin{cases} -(1-x_1)^\nu f\left(\frac{x_1}{1-x_1}\right) & \text{if } 0 < x_1 < 1/2 \\ x_1^\nu f\left(\frac{1-x_1}{x_1}\right) & \text{if } 1/2 < x_1 < 1 \end{cases} \right] \\ &+ \begin{cases} -(1-x)^\nu f\left(\frac{x}{1-x}\right) & \text{if } 0 < x < 1/2 \\ +x^\nu f\left(\frac{1-x}{x}\right) & \text{if } 1/2 < x < 1 \end{cases} \end{aligned} \tag{4.12}$$

Finally if $x > 1$ we get

$$\begin{aligned} \check{c}_f(S,-x) &= -x^\nu f(x^{-1}) + f(x^{-1}) - f(1-x^{-1}) \\ &+ \begin{cases} -(1-x^{-1})^\nu f\left(\frac{x^{-1}}{1-x^{-1}}\right) & \text{if } 0 < x^{-1} < 1/2 \\ x^{-\nu} f\left(\frac{1-x^{-1}}{x^{-1}}\right) & \text{if } 1/2 < x^{-1} < 1 \end{cases} \\ &- f(x_0) + f(1-x_0) + \begin{cases} -(1-x_0)^\nu f\left(\frac{x_0}{1-x_0}\right) & \text{if } 0 < x_0 < 1/2 \\ +x_0^\nu f\left(\frac{1-x_0}{x_0}\right) & \text{if } 1/2 < x_0 < 1 \end{cases} \end{aligned} \tag{4.13}$$

where as usual $x_0 = \{x\}$.

5 Complexification. Statement of the Main Theorem

The main result of [25] can be summarized as follows:

Theorem 1. *(i) The complex Brjuno function is given by the series*

$$\begin{aligned} \mathcal{B}(z) = \quad & -\tfrac{1}{\pi} \textstyle\sum_{p/q\in\mathbb{Q}} \Big\{ (p'-q'z) \left[Li_2\left(\frac{p'-q'z}{qz-p}\right) - Li_2\left(-\frac{q'}{q}\right) \right] \\ & +(p''-q''z) \left[Li_2\left(\frac{p''-q''z}{qz-p}\right) - Li_2\left(-\frac{q''}{q}\right) \right] + \tfrac{1}{q} \log \tfrac{q+q''}{q+q'} \Big\} , \end{aligned} \tag{5.1}$$

where $\left[\frac{p'}{q'}, \frac{p''}{q''}\right]$ is the Farey interval such that $\frac{p}{q} = \frac{p'+p''}{q'+q''}$ (with the convention $p' = p-1$, $q' = 1$, $p'' = 1$, $q'' = 0$ if $q = 1$) and $Li_2(z)$ is the dilogarithm of z (see Appendix 1).

(ii) The real part of $\mathcal{B}$ is bounded on the upper half plane and its trace (i.e. non–tangential limit) on $\mathbb{R}$ is continuous at all irrational points and has a decreasing jump of π/q at each rational point $p/q \in \mathbb{Q}$.

(iii) As one approaches the boundary the imaginary part of $\mathcal{B}$ behaves as follows:

– *if α is a Brjuno number then $\Im m\, \mathcal{B}(\alpha+w)$ converges to $B(\alpha)$ as $w \to 0$ in any domain with a finite order of tangency to the real axis;*
– *if α is diophantine one can allow domains with infinite order of tangency.*

If instead of considering the Brjuno function associated to the cocycle determined by the choice $f(x) = -\log x$ in (3.5) we choose $f(x) = x^{-1/\sigma}$ with $\sigma > 1$ this leads to the real Brjuno function B_σ given in (3.6) whereas the corresponding complex Brjuno function $\mathcal{B}_\sigma(z)$ is obtained by a formula analogue of (5.1) where the dilogarithm is replaced by the hypergeometric function

$$\varphi_\sigma(z) = \frac{1}{z}\frac{\sigma}{\sigma-1}F\left(1, 1-\frac{1}{\sigma}, 2-\frac{1}{\sigma}, z^{-1}\right) . \tag{5.2}$$

More precisely one has

$$\begin{aligned}\mathcal{B}_\sigma(z) = -\tfrac{1}{\pi}\textstyle\sum_{p/q\in\mathbb{Q}} \Big\{ &(p'-q'z)\left[\varphi_\sigma\left(\tfrac{qz-p}{p'-q'z}\right) - \varphi_\sigma\left(-\tfrac{q}{q'}\right)\right] \\ &+(p''-q''z)\left[\varphi_\sigma\left(\tfrac{qz-p}{p''-q''z}\right) - \varphi_\sigma\left(-\tfrac{q}{q''}\right)\right] \\ &+\left[-\tfrac{1}{q'}\varphi_\sigma'\left(-\tfrac{q}{q'}\right) + \tfrac{1}{q''}\varphi_\sigma'\left(-\tfrac{q}{q''}\right)\right]\Big\} .\end{aligned} \tag{5.3}$$

The results of Section 6.2 show that $\mathcal{B}$ belongs to all Hardy spaces H^p with $1 \le p < +\infty$ whereas $\mathcal{B}_\sigma$, $\sigma > 1$, belongs to those with $1 \le p < \sigma$.

6 Some Ideas from the Proofs.

6.1 Hyperfunctions and Operator T

6.1.1 We follow here [16], Chapter 9 and [9], Chapitre I. Let K be a non empty compact subset of $\mathbb{R}$. A *hyperfunction with support in K* is a linear functional u on the space $\mathcal{O}(K)$ of functions analytic in a neighborhood of K such that for all neighborhood V of K there is a constant $C_V > 0$ such that

$$|u(\varphi)| \le C_V \sup_V |\varphi| , \quad \forall \varphi \in \mathcal{O}(V) .$$

We denote by $A'(K)$ the space of hyperfunctions with support in K. It is a Fréchet space: a seminorm is associated to each neighborhood V of K.

Let $\mathcal{O}^1(\overline{\mathbb{C}} \setminus K)$ denote the $\mathbb{C}$–vector space of functions holomorphic on $\overline{\mathbb{C}} \setminus K$ and vanishing at $z = \infty$. One has the canonical isomorphism

$$A'(K) \simeq \frac{\mathcal{O}(\mathbb{C}\setminus K)}{\mathcal{O}(\mathbb{C})} \simeq \mathcal{O}^1(\overline{\mathbb{C}} \setminus K) . \tag{6.1}$$

To each $u \in A'(K)$ corresponds $\varphi \in \mathcal{O}^1(\overline{\mathbb{C}} \setminus K)$ given by

$$\varphi(z) = u(c_z) \ , \ \forall z \in \mathbb{C} \setminus K \ , \tag{6.2}$$

where $c_z(x) = \frac{1}{\pi}\frac{1}{x-z}$.

Note that to $u(x) = -\chi_{(0,1)}(x)\log x$ corresponds $\varphi_u(z) = -\frac{1}{\pi}\int_0^1 \frac{\log x}{x-z}dx = -\frac{1}{\pi}\mathrm{Li}_2(z^{-1})$ whereas to $u_\sigma(x) = \chi_{(0,1)}(x)x^{-1/\sigma}$ corresponds $-\frac{1}{\pi}\int_0^1 \frac{x^{-1/\sigma}}{x-z}dx = -\frac{1}{\pi}\varphi_\sigma(z)$ where φ_σ is given by the hypergeometric function (5.2).

Conversely to each $\varphi \in \mathcal{O}^1(\overline{\mathbb{C}} \setminus K)$ corresponds the hyperfunction

$$u(\psi) = \frac{i}{2\pi}\int_\gamma \varphi(z)\psi(z)dz \ , \ \forall \psi \in A \tag{6.3}$$

where γ is any piecewise $\mathcal{C}^1$ path winding around K in the positive direction. We will also use the notation

$$u(x) = \frac{1}{2i}[\varphi(x+i0) - \varphi(x-i0)] \tag{6.4}$$

for short.

6.1.2 Let $\mathbb{T}^1 = \mathbb{R}/\mathbb{Z} \subset \mathbb{C}/\mathbb{Z}$. A *hyperfunction on* $\mathbb{T}$ is a linear funtional U on the space $\mathcal{O}(\mathbb{T}^1)$ of functions analytic in a complex neighborhood of $\mathbb{T}^1$ such that for all neighborhood V of $\mathbb{T}$ there exists $C_V > 0$ such that

$$|U(\Phi)| \le C_V \sup_V |\Phi| \ , \quad \forall \Phi \in \mathcal{O}(V) \ .$$

We will denote $A'(\mathbb{T}^1)$. the Fréchet space of hyperfunctions with support in $\mathbb{T}$. For $U \in A'(\mathbb{T})$, let $\hat{U}(n) := U(e_{-n})$ with $e_n(z) = e^{2\pi i n z}$. The doubly infinite sequence $(\hat{U}(n))_{n\in\mathbb{Z}}$ satisfies

$$|\hat{U}(n)| < C_\varepsilon e^{2\pi|n|\varepsilon} \ . \tag{6.5}$$

for all $\varepsilon > 0$ and for all $n \in \mathbb{Z}$ with a suitably chosen $C_\varepsilon > 0$. Conversely any such sequence is the Fourier expansion of a unique hyperfunction with support in $\mathbb{T}$.

Let $\mathcal{O}_\Sigma$ denote the complex vector space of holomorphic functions Φ : $\mathbb{C} \setminus \mathbb{R} \to \mathbb{C}$, 1–periodic, bounded at $\pm i\infty$ and such that the limit $\Phi(\pm i\infty) := \lim_{\Im m\, z \to \pm\infty} \Phi(z)$ exist and verify $\Phi(+i\infty) = -\Phi(-i\infty)$.

The spaces $A'(\mathbb{T}^1)$ and $\mathcal{O}_\Sigma$ are canonically isomorphic. To each $U \in A'(\mathbb{T}^1)$ corresponds $\Phi \in \mathcal{O}_\Sigma$ given by

$$\Phi(z) = U(C_z) \ , \ \forall z \in \mathbb{C} \setminus K \ , \tag{6.6}$$

where $C_z(x) = \cot \pi(x-z)$. Conversely to each $\Phi \in \mathcal{O}_\Sigma$ corresponds the hyperfunction

$$U(\Psi) = \frac{i}{2}\int_\Gamma \Phi(z)\Psi(z)dz \ , \ \forall \Psi \in A(\mathbb{T}^1) \tag{6.7}$$

where Γ is any piecewise $\mathcal{C}^1$ path winding around a closed interval $I \subset \mathbb{R}$ of length 1 in the positive direction. We will also use the notation

$$U(x) = \frac{1}{2i}[\Phi(x+i0) - \Phi(x-i0)] \tag{6.8}$$

for short.

Euler's formula

$$\frac{1}{\pi}\sum_{n\in\mathbb{Z}} \frac{1}{z-n} = \cot \pi z \tag{6.9}$$

shows that the following diagram commutes:

$$\begin{array}{ccc} A'([0,1]) & \longrightarrow & \mathcal{O}^1(\overline{\mathbb{C}} \setminus [0,1]) \\ \Big\downarrow {\scriptstyle \sum_{\mathbb{Z}}} & & \Big\downarrow {\scriptstyle \sum_{\mathbb{Z}}} \\ A'(\mathbb{T}^1) & \longrightarrow & \mathcal{O}_{\Sigma} \end{array}$$

the horizontal lines are the above mentioned isomorphisms and the sum over integer translates, denoted $\sum_{\mathbb{Z}}$, is defined as follows. If $\varphi \in \mathcal{O}^1(\overline{\mathbb{C}} \setminus [0,1])$ then one can decompose in a unique way

$$\varphi(z) = a_0 \log \frac{z}{z-1} + \varphi_0(z) , \tag{6.10}$$

where $a_0 \in \mathbb{C}$, $\varphi_0 \in \mathcal{O}^2(\overline{\mathbb{C}} \setminus J)$ (i.e. φ_0 has a zero of order at least two at infinity) and we consider the main branch of the logarithm in $\mathbb{C} \setminus \mathbb{R}^-$. We have

$$\sum_{n=-N}^{N} \log \frac{z-n}{z-n-1} = \log \frac{z+N}{z-N-1}$$

and this leads to the definition

$$\sum_{\mathbb{Z}} \varphi(z) := \sum_{\mathbb{Z}} \varphi_0(z) + \begin{cases} -a_0 \pi i & \text{if } \Im m\, z > 0 \\ +a_0 \pi i & \text{if } \Im m\, z < 0 \end{cases} . \tag{6.11}$$

6.1.3 The results of 6.1.1 and 6.1.2 become fruitful in answering to the natural question how to extend the operator T to complex analytic functions. Indeed this is achieved as follows: the operator T extends to the space $A'([0,1])$ of hyperfunctions u with support contained in $[0,1]$ (see [25], section 1.4 for a proof of this fact). Using the canonical isomorphism (6.1), on $\mathcal{O}^1(\overline{\mathbb{C}} \setminus [0,1])$ the formula for T reads

$$(T\varphi)(z) = -z \sum_{m=1}^{\infty} \left[\varphi\left(\frac{1}{z} - m\right) - \varphi(-m) \right] + \sum_{m=1}^{\infty} \varphi'(-m) . \tag{6.12}$$

Formally we have

$$(1-T)^{-1}\varphi(z) = \sum_{r\geq 0} (T^r \varphi)(z) = \sum_{g\in\mathcal{M}} (L_g \varphi)(z) , \tag{6.13}$$

where the monoid

$$\mathcal{M} = \left\{ g = \begin{pmatrix} a & b \\ c & d \end{pmatrix} \in \mathrm{GL}\,(2,\mathbb{Z})\,,\; d \geq b \geq a \geq 0\,,\;\; d \geq c \geq a \right\} \cup \left\{ \begin{pmatrix} 1 & 0 \\ 0 & 1 \end{pmatrix} \right\},$$

acts on $\mathcal{O}^1(\overline{\mathbb{C}} \setminus [0,1])$ according to

$$(L_g\varphi)(z) = (a-cz)\left[\varphi\left(\frac{dz-b}{a-cz}\right) - \varphi\left(-\frac{d}{c}\right)\right] - \det(g)c^{-1}\varphi'\left(-\frac{d}{c}\right)\,. \quad (6.14)$$

The series (6.13) actually converges in $\mathcal{O}^1(\overline{\mathbb{C}} \setminus [0,1])$ to a function $\sum_{\mathcal{M}} \varphi$. To recover a holomorphic periodic function on $\mathbb{H}$ one sums over integer translates:

$$\mathcal{B}_\varphi(z) = \sum_{n\in\mathbb{Z}} \left(\sum_{\mathcal{M}} \varphi \right)(z-n)\,. \quad (6.15)$$

As we have already mentioned, to construct the complex Brjuno function $\mathcal{B}$ one has to take $\varphi_0(z) = -\frac{1}{\pi}\mathrm{Li}_2\left(\frac{1}{z}\right)$ whereas for the functions $\mathcal{B}_\sigma(z)$ one must take $-\frac{1}{\pi}\varphi_\sigma(z)$ as in (5.2).

The proof of formulas (5.1) and (5.3) is the immediate from (6.15): it is enough to use property (iv) of 2.2.5 which relates the matrices in $Z\mathcal{M}$ to rational numbers.

To summarize, in order to construct the complex analytic extension of any of the functions B_f (defined by (3.5)) our strategy is the following:

(i) take the restriction of the periodic function f to the interval $[0,1]$;
(ii) consider its associated hyperfunction u_f and its holomorphic representative $\varphi \in \mathcal{O}^1(\overline{\mathbb{C}} \setminus [0,1])$.

Then the series (6.15) converges to the complex extension $\mathcal{B}_f$ of the function B_f. The main difficulty (unless f belongs to some L^p space, see [25], Section 4.3) would be to recover B_f as non–tangential limit of the imaginary part of $\mathcal{B}_f$ as $\Im m\, z \to 0$. But this is always the case for the functions described in Section 5. , namely the complex Brjuno function $\mathcal{B}$ and its generalizations $\mathcal{B}_\sigma$. This is explained in the next Section where we discuss the action of T on Hardy spaces.

6.2 H^p Estimates

In order to control the action of T on spaces of holomorphic functions we will make use of the following important remark. The open set $\overline{\mathbb{C}} \setminus [0,1]$ is an hyperbolic Riemann surface which is naturally equipped with a Poincaré metric. By the Lemma of Schwarz–Pick (see [1]), given two hyperbolic Riemann surfaces M, N and an analytic map $f : M \to N$ either its differential df contracts the hyperbolic metric or f is a surjective local isometry. In what follows we will denote d_{hyper} the Poincaré metric on the Riemann surface under consideration. Given $\rho > 0$ we denote

$$V_\rho(D_\infty) = \{z \in \overline{\mathbb{C}} \setminus [0,1]\,,\, d_{hyper}(z, D_\infty) < \rho\}\,. \tag{6.16}$$

the ρ–neighborhood of D_∞ in $\overline{\mathbb{C}} \setminus [0,1]$.

It is then easy to check that for all $\rho \geq 0$ and for all $m \geq 1$ if $z \in V_\rho(D_\infty)$ then $\frac{1}{z} - m \in V_\rho(D_\infty)$.

On Hardy spaces one can prove [25] results which are completely analogous to those obtained for the real Brjuno operator in [24]. The Hardy spaces setting is especially useful in order to have guaranteed the existence of almost everywhere non–tangential limits (see, e.g. [12, 13]). For example an easy proof gives

$$\sup_{V_\rho(D_\infty)} |T^r\varphi(z)| \leq c'_\rho \left(\frac{\sqrt{5}-1}{2}\right)^r \sup_{V_\rho(D_\infty)} |\varphi(z)|\,, \tag{6.17}$$

where $\rho \geq 0$ and for all $\varphi \in \mathcal{O}^1(\overline{\mathbb{C}} \setminus [0,1])$.

We may also consider the Hardy space $H^p(D_\infty)$, $1 \leq p < +\infty$ of analytic functions $\varphi : D_\infty \to \mathbb{C}$ such that the subharmonic function $|\varphi|^p$ has a harmonic majorant. It is an immediate consequence of the Riemann mapping theorem that this space is isomorphic to $H^p(\mathbb{D})$. Indeed if h maps D_∞ conformally onto $\mathbb{D}$ one can use the norm

$$\|\varphi\|_{H^p(D_\infty)} = \|\varphi \circ h^{-1}\|_{H^p(\mathbb{D})} = \left(\int_{\partial D_\infty} |\varphi(z)|^p |h'(z)||dz|\right)^{1/p}\,. \tag{6.18}$$

Note that since ∂D_∞ is a rectifiable Jordan curve h extends to a homeomorphism of ∂D_∞ onto $\mathbb{T}^1$ which is conformal almost everywhere. Again one finds that T is a bounded linear operator on $H^p(D_\infty)$ with spectral radius $\leq \frac{\sqrt{5}-1}{2}$.

6.3 Complex Continued Fractions

The introduction of a complex analogue of Gauss' algorithm of continued fraction expansion of a real number is essential for the study of the boundary behaviour of $\sum_{\mathcal{M}}(L_g\varphi)(z)$ and the construction of the complex Brjuno function. Here we recall from [25] a complex version of the continued fraction which has been used there to study the complex Brjuno function $\mathcal{B}$.

6.3.1 We consider the following domains:

$$\begin{aligned}
D_0 &= \left\{z \in \mathbb{C}\,, |z+1| \leq 1\,,\, \Re e\, z \geq \tfrac{\sqrt{3}}{2} - 1\right\}\,,\\
D_1 &= \left\{z \in \mathbb{C}\,, |z| \geq 1\,,\, \left|z - \tfrac{1}{\sqrt{3}}\right| \leq \tfrac{1}{\sqrt{3}}\right\}\,,\\
D &= \{z \in \mathbb{C}\,, |z| \leq 1\,,\, |z-i| \geq 1\,,\, |z+i| \geq 1\,,\, \Re e\, z > 0\}\,,\\
H_0 &= \{z \in \mathbb{C}\,, |z-i| \leq 1\,,\, |z+1| \geq 1\,,\, \Im m\, z \leq 1/2\}\,,\\
H_0' &= \{z \in \mathbb{C}\,,\, \overline{z} \in H_0\}\\
\Delta &= H_0 \cup H_0' \cup D = \{z \in \mathbb{C}\,, |z| \leq 1\,,\, |z+1| \geq 1\,,\, |\Im m\, z| \leq 1/2\}\,,\\
D_\infty &= \overline{\mathbb{C}} \setminus (D_0 \cup \Delta \cup D_1)\\
&= \{|\Im m\, z| > 1/2\} \cup \{\Re e\, z < \tfrac{\sqrt{3}}{2} - 1\} \cup \{\Re e\, z > \tfrac{\sqrt{3}}{2}, |z - \tfrac{\sqrt{3}}{3}| > \tfrac{\sqrt{3}}{3}\}.
\end{aligned}$$

A fundamental property is the following

(i) if $z \notin D \cup D_1$ (in particular if $z \in D_\infty$) then $1/z - m \in D_\infty$ for all $m \geq 1$;
(ii) if $z \in D_1$, then $1/z - 1 \in D_0$ and $1/z - m \in D_\infty$ for all $m \geq 2$.

Observe that

$$SD = \cup_{m \geq 1} (\Delta + m) \ ,$$

where the domains have disjoint interior. Thus, for $z \in D$, we define

$$A(z) = \frac{1}{z} - m = (g(m))^{-1} \cdot z \ , \tag{6.19}$$

(we recall that $g(m) = \begin{pmatrix} 0 & 1 \\ 1 & m \end{pmatrix}$, $m \geq 1$) where $m \geq 1$ is the unique integer such that

$$A(z) \in \Delta \ , \quad |A(z)| < 1 \ .$$

Iterating from $z_0 \in D$, we define

$$z_{i+1} = A(z_i) = A^{i+1}(z_0) \tag{6.20}$$

as long as $z_i = A^i(z) \in D$. The iteration process stops when one of the two following conditions is verified:

(i) $z_l = 0$ for some $l \geq 0$; this happens if and only if $z_0 \in \mathbb{Q}$,
(ii) $z_l \notin (D \cup \{0\})$, for some $l \geq 0$; this happens if and only if $z_0 \notin \mathbb{R}$.

For all $0 \leq i < l$, we will denote m_{i+1} the integer such that

$$z_{i+1} = \frac{1}{z_i} - m_{i+1} \ , \ m_{i+1} \geq 1 \ . \tag{6.21}$$

6.3.2 Let

$$\begin{pmatrix} p_{i-1} & p_i \\ q_{i-1} & q_i \end{pmatrix} = \begin{pmatrix} 0 & 1 \\ 1 & m_1 \end{pmatrix} \cdots \begin{pmatrix} 0 & 1 \\ 1 & m_i \end{pmatrix} \in \mathcal{M} \ , \ 0 \leq i \leq l \ .$$

Then one has the same recurrence relations as for the real continued fraction

$$\begin{aligned} p_{i+1} &= m_{i+1} p_i + p_{i-1} \ , \\ q_{i+1} &= m_{i+1} q_i + q_{i-1} \ , \end{aligned} \tag{6.22}$$

with initial data $p_{-1} = q_0 = 1$ and $p_0 = q_{-1} = 0$. Moreover

$$z_0 = \frac{p_{i-1} z_i + p_i}{q_{i-1} z_i + q_i} \ , \quad z_i = \frac{p_i - q_i z_0}{q_{i-1} z_0 - p_{i-1}} \ , \tag{6.23}$$

and if one poses

$$\beta_i(z_0) = \prod_{j=0}^{i} z_j = (-1)^i (q_i z_0 - p_i) \ , \tag{6.24}$$

then

$$\beta_i(z_0) = \frac{z_i}{q_i + q_{i-1}z_i} = \frac{1}{q_{i+1} + q_i z_{i+1}} . \tag{6.25}$$

Finally one has

$$\begin{aligned}(-1)^i \Im m\, z_0 &= |\beta_{i-1}(z_0)|^2 \Im m\, z_i = |q_i + q_{i-1}z_i|^{-2} \Im m\, z_i ,\\ \tfrac{dz_i}{dz_0} &= (-1)^i(\beta_{i-1}(z_0))^{-2} = (-1)^i(q_i + q_{i-1}z_i)^2 .\end{aligned}$$

Observe that, as $|z_{i+1} + 1| \geq 1$ and $\Re e\, z_{i+1} \geq \frac{\sqrt{3}}{2} - 1$ for $i < l$, we have from (6.26)

$$|\beta_i(z_0)| \leq q_{i+1}^{-1}[\cos\pi/12]^{-1} = \frac{2\sqrt{2}}{1+\sqrt{3}} q_{i+1}^{-1} \tag{6.27}$$

and, as $q_i \leq q_{i+1}$, $|z_{i+1}| \leq 1$,

$$|\beta_i(z_0)| \geq \frac{1}{2} q_{i+1}^{-1} . \tag{6.28}$$

A1. Appendix 1: Some Properties of the Dilogarithm

The classical dilogarithmic series (see [19, 28] for more information) is defined by

$$\mathrm{Li}_2(z) = \sum_{n=1}^{+\infty} \frac{z^n}{n^2} \tag{A1.1}$$

and it is convergent for $|z| \leq 1$. Since $-\log(1-z) = \sum_{n=1}^{+\infty} \frac{z^n}{n}$, the analytic continuation of the dilogarithm to $\mathbb{C} \setminus [1, +\infty)$ is given by

$$\mathrm{Li}_2(z) = -\int_0^z \frac{\log(1-t)}{t} dt = \int_0^z \left(\int_0^t \frac{d\zeta}{1-\zeta}\right) \frac{dt}{t} . \tag{A1.2}$$

From

$$\mathrm{Li}_2\left(\frac{1}{z}\right) = -\int_0^1 \frac{\log t}{z-t} dt , \tag{A1.3}$$

it follows that $\mathrm{Li}_2\left(\frac{1}{z}\right)$ is the Cauchy–Hilbert transform of the real function

$$\varphi_0(t) = \begin{cases} -\log t & \text{if } t \in [0,1] \\ 0 & \text{elsewhere} \end{cases} \tag{A1.4}$$

Note also that

$$\mathrm{ImLi}_2(t \pm i0) = \pm\pi \log t , \tag{A1.5}$$

where $t \in [1, +\infty)$. Moreover

$$|\mathrm{Li}_2(z)| = \mathcal{O}(\log^2 |z|) \text{ as} |z| \to +\infty . \tag{A1.6}$$

References

1. L.V. Ahlfors "Conformal Invariants: Topics in Geometric Function Theory" McGraw–Hill (1973)
2. X. Buff and A. Cheritat, "On the size of quadratic Siegel disks. Part I", preprint Math.DS/0305080
3. A. Berretti and G. Gentile, "Scaling properties for the radius of convergence of the Lindstedt series: the standard map", *J. Math. Pures Appl.* **78** (1999), 159–176
4. A. Berretti and G. Gentile, "Bryuno function and the standard map", *Comm. Math. Phys.* **220** (2001), 623–656
5. A. Berretti and G. Gentile, "Scaling of the critical function for the standard map: some numerical results", preprint mp-arc /03–212 (mathematical physics archive)
6. N. Buric, I. Percival and F. Vivaldi "Critical Function and Modular Smoothing" *Nonlinearity* **3** (1990), 21–37.
7. A. D. Brjuno "Analytical form of differential equations" *Trans. Moscow Math. Soc.* **25** (1971), 131–288; **26** (1972), 199–239.
8. T. Carletti "The 1/2–complex Bruno function and the Yoccoz function. A numerical study of the Marmi-Moussa-Yoccoz conjecture", preprint Math.-DS/0306009
9. A. Cerezo, J. Chazarain and A. Piriou "Introduction aux hyperfonctions" in *Lect. Notes in Math.* **449** (1975), 1–53
10. A.M. Davie "The critical function for the semistandard map" *Nonlinearity* **7** (1994), 219–229.
11. A. M. Davie "Renormalisation for analytic area–preserving maps" University of Edinburgh preprint, (1995).
12. P. L. Duren "Theory of H^p spaces" Academic Press, New York, (1970).
13. J. B. Garnett "Bounded Analytic Functions" Academic Press, New York, (1981).
14. J. Garcia–Cuerva and J.L. Rubio de Francia "Weighted Norm Inequalities and Related Topics" North Holland Mathematical Studies **116**, Amsterdam, (1985).
15. G.H. Hardy and E.M. Wright "An introduction to the theory of numbers" Fifth Edition, Oxford Science Publications (1990).
16. L. Hörmander "The Analysis of Linear Partial Differential Operators I" Grundlehren der mathematischen Wissenschaften **256**, Springer–Verlag, Berlin, Heidelberg, New York, Tokyo (1983).
17. A. Katok and B. Hasselblatt "Introduction to the modern theory of dynamical systems" Encyclopedia of Mathematics and its Applications **54**, Cambridge University Press, (1995).
18. S. Lang "Introduction to Diophantine Approximation" Addison–Wesley (1966)
19. L. Lewin "Polylogarithms and Associated Functions" Elsevier North–Holland, New York, (1981).
20. J. Lewis and D. Zagier "Period functions and the Selberg zeta function for the modular group", in "The Mathematical Beauty of Physics", 83–97, *Adv. Series in Math. Phys.* **24**, World Sci. Publ., River Edge, NJ, (1997)
21. J. Lewis and D. Zagier "Period functions for Maass wave forms" *Ann. Math.* **153** (2001), 191–258
22. S. Marmi "Critical Functions for Complex Analytic Maps" *J. Phys. A: Math. Gen.* **23** (1990), 3447–3474.

23. S. Marmi "An introduction to small divisor problems" Quaderni del Dottorato di Ricerca in Matematica, Pisa (2000) (also available at mp-arc and front.math.ucdavis.edu: math.DS/0009232)
24. S. Marmi, P. Moussa and J.-C. Yoccoz "The Brjuno functions and their regularity properties" *Commun. Math. Phys.* **186** (1997), 265–293.
25. S. Marmi, P. Moussa and J.-C. Yoccoz "Complex Brjuno functions" *Journal of the A.M.S.* **14** (2001) 783–841
26. S. Marmi and J. Stark "On the standard map critical function" *Nonlinearity* **5** (1992) 743–761
27. S. Marmi and J.-C. Yoccoz "Some open problems related to small divisors" in "Dynamical Systems and Small Divisors" *Lecture Notes in Mathematics* **1784** (2002) 175–191
28. J. Oesterlé "Polylogarithmes" *Séminaire Bourbaki n. 762 Astérisque* **216** (1993), 49–67.
29. C.L. Siegel "Iteration of analytic functions" Annals of Mathematics **43** (1942), 807-812.
30. J.-P. Serre "Cohomologie Galoisienne" Lecture Notes in Mathematics, **5**, Springer–Verlag (1973).
31. I.R. Shafarevich "Basic Notions of Algebra" Springer–Verlag (1997)
32. J. Silverman and J. Tate "Rational Points on Elliptic Curves" Undergraduate Texts in Mathematics, Springer–Verlag (1992)
33. J.-C. Yoccoz "Théorème de Siegel, nombres de Bruno et polynômes quadratiques" *Astérisque* **231** (1995), 3–88.
34. J.-C. Yoccoz "An introduction to small divisors problem", in "From number theory to physics", M. Waldschmidt, P. Moussa, J.-M. Luck and C. Itzykson (editors) Springer–Verlag (1992) pp. 659–679.
35. J.-C. Yoccoz "Analytic linearisation of circle diffeomorphisms" in "Dynamical Systems and Small Divisors" *Lecture Notes in Mathematics* **1784** (2002) 125–173
36. D. Zagier "Quelques conséquences surprenantes de la cohomologie de $SL_2(\mathbb{Z})$" in "Leçons de mathématiques d'aujourd'hui" Cassini, Paris (2000), 99–123

Part IV

Appendices

A

List of Participants

List of Participants of the school *Frontiers in Number Theory, Physics and Geometry* held in Les Houches, March 9 - 21 2003 The affiliations of the participants are the one at the time of the school or the one known at the time the book is in press.

Abou Zeid Mohab	(Theory Group, The Blackett Laboratory, Imperial College London, UK)
Berman David	(Department of Applied Mathematics, Cambridge University, UK)
Bern Zvi	(UCLA, USA)
Beukers Frits	(University of Utrecht, The Netherlands)
Bogomolny Eugene	(LPTMS, Orsay, France)
Bohigas Oriol	(LPTMS, Orsay, France)
Bondal Alexei	(Steklov Math. Institute, Moscow, Russia & Université Paris 6, France)
Bonechi Francesco	(INFN, sezione di Firenze, Italy)
Brasselet Jean-Paul	(IML - CNRS, France)
Braun Volker	(Laboratoire de Physique théorique ENS, Paris, France)
Candelas Philip	(Oxford University, UK)
Cantini Luigi	(Scuola Normale Superiore Pisa, Italy)
Cartier Pierre	(Institut Mathématique de Jussieu, CNRS, France)
Connes Alain	(Collège de France, IHES, France)
Conrey Brian	(American Institute of Mathematics, USA)
Conway John	(Princeton University, USA)
Cristadoro Giampaolo	(Dipartimento di Scienze, Univeristà dell'Insubria, sede di Como, Italy)
Cvitanovic Predrag	(Georgia Tech. University, USA)
DeWitt Bryce	(University of Texas, USA)
DeWitt-Morette Cécile	(University of Texas, USA & CA Les Houches)
Dijkgraaf Robbert	(Amsterdam University, The Netherlands)
Di Vecchia Paolo	(Nordita, Danemark)
Elbau Peter	(ETH Zurich, Switzerland)
Frenkel Edward	(University of California, Berkeley, USA)
Fucito Francesco	(INFN sez. Roma 2, Italy)

Gangl Herbert	(Max-Planck-Institut für Mathematik, Bonn, Germany)
Gentile Guido	(Università di Roma III, Italy)
Grange Pascal	(CPHT, École polytechnique, Palaiseau, France)
Gutkin Boris	(SPhT, CEA-Saclay, France)
Harrison Jonathan	(University of Ulm, Germany)
Henry Pierre	(Queen Mary College, University of London, UK)
Julia Bernard	(Laboratoire de Physique théorique ENS - CNRS, Paris, France)
Kaste Peter	(CPHT, École Polytechnique, Palaiseau, France)
Kreimer Dirk	(CNRS-IHES, France)
Kremnizer Kobi	(Tel-Aviv University, Israel)
Lagarias Jeffrey	(AT&T Labs-Research, USA)
Leboeuf Patricio	(LPTMS, Université de Paris XI, Orsay, France)
Marcolli Matilde	(Max Planck Institute for Mathematics, Germany)
Marklof Jens	(School of Mathematics, University of Bristol, UK)
Marmi Stefano	(Scuola Normale Superiore, Pisa, Italy)
Mastrolia Pierpaolo	(Università di Bologna, Italy & Universitaet Karlsruhe, Germany)
Mckay John	(Concordia University, USA)
Moore Gregory	(Rutgers University, USA)
Moussa Pierre	(CEA-Service de Physique Théorique de Saclay, France)
Nahm Werner	(DIAS, Dublin, Ireland & Bonn University, Germany)
Nikeghbali Ashkan	(Laboratoire de probabilité et modèles aléatoires Paris 6, France)
Pakis Stathis	(Queen Mary College, University of London, UK)
Pal Ambrus	(Centre de Recherche Mathématique, Montreal, Quebec)
Paugam Frederic	(IRMAR-Université Rennes 1, France)
Paulot Louis	(Laboratoire de Physique théorique ENS, Paris, France)
Pioline Boris	(LPTHE, Paris, France)
Pollicott Mark	(Manchester University, UK)
Ramachandran Niranjan	(University of Maryland, College Park, USA & MPIM, Bonn, Germany)

Roggenkamp Daniel	(Physikalisches Institut der Universitaet Bonn, Germany)
Schafer-Nameki Sakura	(DAMTP, University of Cambridge, UK)
Scheidegger Emanuel	(Institut fuer theoretische Physik der TU Wien, Austrria)
Soulé Christophe	(CNRS and IHES, France)
Then Holger	(Abteilung Theoretische Physik, Universitaet Ulm, Germany)
Todorov Ivan	(Bulgarian Academy of Sciences, Bulgaria)
Vanhove Pierre	(CEA-Service de Physique Théorique de Saclay, France)
Vasserot Eric	(Université de Cergy-Pontoise, France)
Vershik Anatoly	(St. Petersburg Department of Steklov Mathematical institute, Russia)
Voiculescu Dan-Virgil	(Dept. Math. UC Berkeley, USA)
Voros André	(CEA-Service de Physique Théorique de Saclay, France)
Waldschmidt Michel	(Institut de Mathématiques, Paris 6, France)
Weinzierl Stefan	(Dipartimento di Fisica, Universita di Parma, Italy)
Wendland Katrin	(Mathematics Institute, University of Warwick, UK)
Yao Yi-Jun	(CMAT, École Polytechnique, Palaiseau, France)
Yoccoz Jean-Christophe	(Collège de France, Paris, France)
Zabrodin Anton	(ITEP, Moscow, Russia)
Zabzine Maxim	(INFN section of Florence, Italy)
Zagier Don	(MPIM Bonn, Germany)
Zorich Anton	(Université de Rennes, France & Moscow Independent University, Russia)

Index

GPSR Compliance
The European Union's (EU) General Product Safety Regulation (GPSR) is a set of rules that requires consumer products to be safe and our obligations to ensure this.

If you have any concerns about our products, you can contact us on

ProductSafety@springernature.com

In case Publisher is established outside the EU, the EU authorized representative is:

Springer Nature Customer Service Center GmbH
Europaplatz 3
69115 Heidelberg, Germany

www.ingramcontent.com/pod-product-compliance
Ingram Content Group UK Ltd.
Pitfield, Milton Keynes, MK11 3LW, UK
UKHW020712071026
15590UKWH00010B/482

* 9 7 8 3 6 4 2 0 6 2 1 9 3 *